Experimental and Theoretical Approaches to Actinide Chemistry

Experimental and Theoretical Approaches to Actinide Chemistry

Edited by John K. Gibson and Wibe A. de Jong

Lawrence Berkeley National Laboratory
California, United States

This edition first published 2018

Registered Office(s)
John Wiley & Sons, Inc., 111 River Street, Hoboken, NJ 07030, USA
John Wiley & Sons Ltd, The Atrium, Southern Gate, Chichester, West Sussex, PO19 8SQ, UK

Editorial Office
9600 Garsington Road, Oxford, OX4 2DQ, UK

For details of our global editorial offices, customer services, and more information about Wiley products visit us at www.wiley.com.

Wiley also publishes its books in a variety of electronic formats and by print-on-demand. Some content that appears in standard print versions of this book may not be available in other formats.

Library of Congress Cataloging-in-Publication Data

Names: Gibson, John K. (John Knight), editor. | Jong, Wibe A. de, 1969– editor.
Title: Experimental and theoretical approaches to actinide chemistry / edited by John K. Gibson (Lawrence Berkeley National Laboratory, CA, USA), Wibe A. de Jong (Lawrence Berkeley National Laboratory, CA, USA).
Description: Hoboken, NJ : Wiley, 2018. | Includes bibliographical references and index. |
Identifiers: LCCN 2017041633 (print) | LCCN 2017052767 (ebook) | ISBN 9781119115533 (pdf) | ISBN 9781119115540 (epub) | ISBN 9781119115526 (cloth : alk. paper)
Subjects: LCSH: Actinium compounds. | Actinide elements. | Heavy elements.
Classification: LCC QD181.A2 (ebook) | LCC QD181.A2 E95 2018 (print) | DDC 546/.421–dc23
LC record available at https://lccn.loc.gov/2017041633

Cover Design: Wiley
Cover Images: Background Image ©Arco Images GmbH/Alamy Stock Photo; Left Image Courtesy of W. A. de Jong (LBNL); Middle Image Courtesy of R. Abergel(LBNL); Right Image Courtesy of Prof. T. Albrecht-Schmitt (Florida State University)

Set in 10/12pt Warnock by SPi Global, Pondicherry, India
Printed and bound in Malaysia by Vivar Printing Sdn Bhd

10 9 8 7 6 5 4 3 2 1

Contents

List of Contributors

Rebecca J. Abergel
Chemical Sciences Division
Lawrence Berkeley National Laboratory
California
United States

Peter Agbo
Chemical Sciences Division
Lawrence Berkeley National Laboratory
California
United States

Thomas E. Albrecht-Schmitt
Florida State University
Tallahassee, United States

Rami J. Batrice
Department of Chemistry
Georgetown University, Washington, DC
United States

Marjorie Bertolus
CEA, DEN
Centre de Cadarache, France

N. Bryan
National Nuclear Laboratory
United Kingdom

Aurora E. Clark
Department of Chemistry
Washington State University
United States

M.A. Denecke
The University of Manchester
United Kingdom

Trevor W. Hayton
Department of Chemistry and
Biochemistry
University of California Santa Barbara
United States

Michael C. Heaven
Department of Chemistry
Emory University, Atlanta
United States

S. Kalmykov
Lomonosov Moscow State University
Department of Chemistry – Radiochemistry
Russia

Nikolas Kaltsoyannis
School of Chemistry
The University of Manchester
United Kingdom

Karah E. Knope
Department of Chemistry
Georgetown University, Washington, DC
United States

Rudy Konings
European Commission
Joint Research Centre
Karlsruhe
Germany

Gregory R. Lumpkin
Nuclear Fuel Cycle Research
Australian Nuclear Science and
Technology Organisation
New South Wales, Australia

Matthew L. Marsh
Florida State University
Tallahassee
United States

K. Morris
The University of Manchester
United Kingdom

Kirk A. Peterson
Department of Chemistry
Washington State University
United States

F. Quinto
Karlsruhe Institute of Technology
Institute for Nuclear Waste Disposal
Germany

Julian A. Rees
Chemical Sciences Division
Lawrence Berkeley National Laboratory
California
United States

Jenifer C. Shafer
Department of Chemistry
Colorado School of Mines
United States

Jennifer N. Wacker
Department of Chemistry
Georgetown University
Washington, DC
United States

Ping Yang
Theoretical Division
Los Alamos National Laboratory
United States

Preface

This book delivers a contemporary overview of experimental and computational techniques to probe the chemistry and physics of actinides needed for predicting the fate and controlling the behavior of actinide-containing materials in the ecosystem, as well as for developing new advanced applications in energy, medicine, and forensics.

The actinides are located at the bottom of the periodic table, and comprise the 15 elements actinium (Ac; $Z = 89$) through lawrencium (Lr; $Z = 103$). Placement of the actinides in the periodic table as a homologous series to the lanthanides was proposed in 1944 by Glenn Seaborg when only a few members of the series had yet been discovered. Seaborg accurately hypothesized that the actinides are characterized by filling of the 5f orbital shell, in analogy with filling of the lanthanide 4f orbital shell. The extensive hexavalent chemistry of U had already been established in 1944, this in stark contrast to the dominant trivalent oxidation state of the lanthanides. The substantially more complex, and interesting, chemistry of the 5f actinides continued to develop with stable oxidation states from II to VII having been characterized.

Some of the early actinides find their use primarily in commercial nuclear reactors, and secondarily in nuclear weapons. Worldwide there is substantial interest in the development of new nuclear reactors with advanced fuel cycles, utilizing a wider range of actinides, with the appropriate safety and nonproliferation constraints to meet the energy needs. In addition, there are ongoing critical issues with respect to the environmental cleanup at production sites of nuclear materials, and storage and reprocessing sites for spent reactor fuels. Understanding the chemistry of the actinides is key to the development of next-generation nuclear reactors, novel actinide reprocessing, and recycling technologies, and to safeguard society against radiological contamination. Research on actinides ranges from the fundamental aspects of actinide bonding to the chemistry and stability of actinide complexes in harsh conditions. The challenges of studying actinides are profoundly complex, as they are hazardous and difficult due to radioactivity, they generally reside in harsh environments, and only small quantities of most actinides are available for research. In the last decade, computational chemistry and materials science has provided another avenue to probe the complex actinide chemistry.

This book is divided into nine chapters and attempts to cover the diversity of actinide chemistry. The chapters focus on the multidisciplinary and multimodal nature of actinide chemistry, the interplay between multiple experiments and theory, as well as between basic and applied actinide chemistry. In covering the central aspects of actinide chemistry, the chapters in this book hopefully will provide a much-needed reference work for both researchers entering the field of actinide chemistry and those desirous of introducing new techniques into their current actinide research.

Beginning with the most elementary of species, in Chapter 1 Heaven and Peterson provide an overview of the gas-phase chemistry of small actinide-containing molecules. In Chapter 2, Batrice, Wacker, and Knope focus on the speciation and chemistry of actinide species in solution environments. Marsh and Albrecht-Schmitt discuss the world of inorganic actinide materials, covering a large number of known actinide structures, in Chapter 3. In Chapter 4, Hayton and Kaltsoyannis provide an experimental and computational overview of organometallic actinide complexes with novel oxidation states and ligand types. The remaining five chapters are focused on the various natural and engineering environments where actinides are being encountered. The chemistry behind solvation extraction used in separation processes are discussed in Chapter 5 by Clark, Yang, and Shafer. Konings and Bertolus provide a detailed overview of the behavior and properties of the complex nuclear reactor fuel materials in Chapter 6. Lumpkin discusses nuclear waste forms and the effects of radiation damage in Chapter 7. In Chapter 8, Denecke, Bryan, Kalmykov, Morris, and Quinto present an overview of actinide sources and their behavior and mobility in natural environments. Finally, in Chapter 9, Agbo, Rees, and Abergel gives an overview of actinides in biological systems.

We would like to thank all authors of the chapters for their hard work, outstanding contributions, and patience as we brought all the chapters together. Your work is greatly appreciated. We would like to thank Teresa Eaton for her help in copyediting the manuscript.

1

Probing Actinide Bonds in the Gas Phase: Theory and Spectroscopy

Michael C. Heaven[1] and Kirk A. Peterson[2]

[1] *Department of Chemistry, Emory University, Atlanta, Georgia, United States*
[2] *Department of Chemistry, Washington State University, Pullman, Washington, United States*

1.1 Introduction

Theoretical studies of actinide compounds have two primary goals. The first is to understand the chemical bonding within these materials and their physical properties. A sub-focus within this task is the elucidation of the roles played by the 5*f* electrons. The second goal is to understand the reactivities of actinide compounds. The long-term objective is to develop reliable computational methods for exploring and predicting actinide chemistry. This is highly desirable owing to the practical difficulties in handling the radioactive and short-lived elements within this family.

Actinides pose a severe challenge for computational quantum chemistry models due to the large numbers of electrons and the presence of strong relativistic effects [1–4]. Although small molecules (di- and tri- atomics that contain just one metal atom) can be explored using rigorous theoretical models, the computational cost of this approach is currently too high for most problems of practical interest (e.g., actinide ions interacting with chelating ligands in solution). Consequently, approximate methods are applied. Ab initio calculations can be accelerated by using a single effective core potential to represent the deeply bound electrons of the metal atom [5, 6]. The relativistic effects are folded into this core potential, and the number of electrons explicitly considered by the calculations is greatly reduced. Semi-empirical density functional theory (DFT) methods offer even better computational efficiency. It is, of course, essential that these approximate methods be tested against both accurate experimental data and the results of rigorous "benchmark" calculations.

There are clear advantages for using data obtained from gas-phase measurements for the comparisons with theory. The ideal situation is to evaluate predictions for the bare molecule against experimental data that are untainted by interactions with solvent molecules or a host lattice. Gas-phase spectroscopy can provide accurate determinations of rotational constants, dipole moments, vibrational frequencies, electronic excitation energies, ionization energies, and electron affinities [7–11]. Information concerning the geometric structure and the electronic state symmetries can be derived from the

Experimental and Theoretical Approaches to Actinide Chemistry, First Edition.
Edited by John K. Gibson and Wibe A. de Jong.

rotational energy level patterns, which can only be observed for the unperturbed molecule. The reactivities of actinide-containing species can also be investigated in the gas phase, under conditions that facilitate theoretical comparisons [8, 12–15]. The majority of this work relies on mass spectrometry for the selection of the reactants and identification of the products. In addition to revealing reaction pathways, mass spectrometric experiments provide critical thermodynamic data, such as bond dissociation and ionization energies.

Over the past 30 years, there has been steady progress in the development of relativistic quantum chemistry methods, combined with a dramatic increase in the speed and capacity of computing platforms. On the experimental side, the application of laser-based spectroscopy, guided ion beam, and ion-trap mass spectrometry has significantly advanced our ability to explore the structure, bonding, and reactivity of actinide species. In this chapter, we present an overview of the theoretical and experimental techniques that are currently being used to gain a deeper understanding of actinides through the studies of small molecules in the gas phase. While some background material is presented, the primary focus is on the techniques that are currently being applied and developed. To limit the scope, the experimental section is strictly devoted to gas-phase measurements. There is a large body of excellent spectroscopic work that has been carried out for actinide species trapped in cryogenic rare-gas matrices. The data from these measurements are also of great value for tests of theoretical predictions, as the rare-gas solid is usually a minimally perturbing host. For a review of the matrix work, see Reference [8]

1.2 Techniques for Obtaining Actinide-Containing Molecules in the Gas Phase

The earliest spectroscopic studies of actinide-containing molecules in the gas phase were carried out using compounds that possessed appreciable vapor pressures at room temperature. Hence, the hexafluorides UF_6, NpF_6 and PuF_6 were studied by conventional spectroscopic means [8], with suitable precautions for handling radionuclides. The tetrahalides have lower room temperature vapor pressures, but workable number densities have been obtained using moderate heating of the samples [8]. Studies of thorium oxide (ThO) emission spectra were carried out using ThI_4 as the source of the metal [16]. This reagent was subjected to a 2.45 GHz microwave discharge that was sustained in approximately 0.1 Torr of Ne buffer gas. The oxide was readily formed by the reaction of the discharge-generated Th atoms with the walls of the quartz tube that contained the ThI_4/Ne mixture. Molecular ions can also be generated at workable number densities using discharges with volatile compounds. An example is provided by the recent study of ThF^+ reported by Gresh et al. [17]. ThF_4 was used as the starting reagent, and the vapor pressure was increased by heating the quartz discharge tube to 1193 K. The tube was filled with approximately 5 Torr of He, and the mixture was excited by an AC discharge operated at a frequency of 10 kHz.

More commonly, actinide species are refractory solids that require somewhat extreme conditions for vaporization. Tube furnaces [18, 19], Knudsen ovens [20, 21], discharge sputtering [14], discharges [16, 17, 22], and laser ablation techniques [8, 23–25] have been successfully applied. High-temperature vaporization is exemplified by studies of the electronic spectrum of uranium oxide (UO), carried out using a Knudsen oven to vaporize U-metal samples that had been pre-oxidized by exposure to air [21]. The crucible was heated to temperatures in the range of 2400–2600 K. This was high enough that thermally excited electronic transitions could be observed using conventional emission spectroscopy. The UO vapor pressure generated by the Knudsen oven was sufficient for the recording of absorption and laser-induced fluorescence (LIF) spectra. Resistively heated tube furnaces have also been used for studies of gas-phase UO. The advantage of this approach is that it provides a longer optical path length, and thereby yields spectra with greater signal-to-noise ratios. Extensive high-resolution electronic spectra for UO were recorded by Kaledin et al. [18] using a tube furnace that was heated to 2400 K. More recently, Holt et al. [19] have used a tube furnace to record microwave absorption spectra for UO in the 500–650 GHz range. Transitions between the rotational levels of multiple low-lying electronic states (including the ground state) were observed.

Many of the early mass spectrometric studies of gas-phase actinide molecules were carried out using thermal vaporization in combination with electron impact ionization techniques [26–28]. Surface ionization of uranyl acetate has been used to generate UO^+ and UO_2^+ [29]. In the thorium ion experiments of Armentrout et al. [14, 30], thorium powder was mounted in a cathode held at −2.5 kV. This electrode produced a discharge in a flow of Ar, and the resulting Ar^+ ions generated Th^+ by sputtering. Molecular ions, such as $ThCH_4^+$, were formed in a downstream flow tube. These techniques are suitable for measurements that rely on charged particle detection, but the number densities are usually too small for conventional spectroscopic observations (e.g., absorption or fluorescence detection).

For spectroscopic techniques that are compatible with pulsed generation of the target species, laser vaporization has proved to be a particularly versatile method. The products include cations, anions, and molecules in exotic oxidation states. Duncan [23] has published a very informative review of this technique. Typically, a pulsed laser (Nd/YAG or excimer with 5–10 ns pulse duration) is focused onto the surface of a target to produce a vapor plume. The products are cooled and transported using a flow of an inert carrier gas (e.g., He or Ar). If the plume does not contain the species of interest, the carrier gas can be seeded with a reagent that will produce the desired product. For example, gas-phase UF has been produced by vaporizing U metal into a flowing mixture of He and SF_6 [31].

Charged species are produced by the vaporization process, even in the absence of an externally applied electric field. Cations are formed by thermal processes and photoionization. Some of the liberated electrons attach to neutral molecules, providing a viable flux of negative ions under suitable conditions. External fields can be applied if the initial ion yield is not high enough for the detection technique.

A particularly valuable advantage of the laser vaporization technique is that it can be readily combined with a supersonic jet expansion nozzle. Figure 1.1 (reproduced from Reference [23]) shows this combination for a pulsed solenoid gas valve (labeled "General Valve"). The valve supplies a high-pressure pulse of carrier gas, timed to coincide with the laser vaporization pulse. The entrained plume passes through a reaction channel before expansion into a vacuum chamber. On expansion, the gas temperature and density drops rapidly. The internal degrees of freedom for entrained molecules can be relaxed to population distribution temperatures in the range of 5–20 K. Condensation

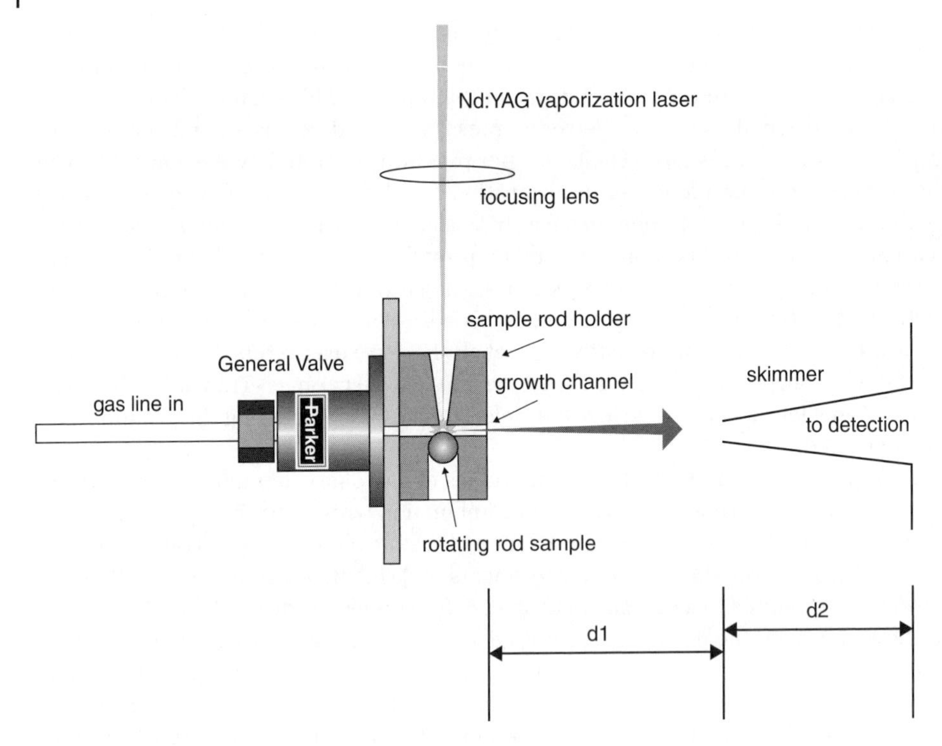

Figure 1.1 The general schematic design of a "standard" laser vaporization cluster source (Reproduced with permission from Duncan, M.A. (2012) Invited review article: Laser vaporization cluster sources. *Review of Scientific Instruments*, **83**, 041101/1–041101/19.)

is prevented by the concomitant drop in gas density. The cooling greatly facilitates spectroscopic studies by dramatically reducing the number of thermally populated internal energy states. In this context, it is useful to note that the rapid expansion often leads to non-equilibrium population distributions. Collisional relaxation rates are sensitive to the energy spacings between levels, and they are most favorable for the smallest energy intervals (provided that there are no symmetry restrictions involved). Transfer between the closely spaced rotational levels is facile. Consequently, low rotational temperatures are easily attained. The more widely spaced vibrational levels require a larger number of collisions for effective relaxation. As the number of collisions is limited by the expansion kinetics, the final population distribution will often show a vibrational temperature that is substantially higher than the rotational temperature. Lastly, the pulsed vaporization process can populate metastable electronically excited states, which can be quite resistive to cooling in the expansion. A study of the jet-cooled Be_2 dimer illustrates this hierarchy [32]. The rotational temperature was 7 K, the vibrational temperature was near 160 K, and an electronically excited triplet state with an energy of 7406 cm^{-1} was substantially populated.

The gas dynamics can also be used to exert some control over the chemistry occurring in the expansion. One way this can be accomplished is by varying the length of the "growth channel" as shown in Figure 1.1. Lengthening this channel increases the time for which the density of reactive species is high enough for chemistry to occur. As a simple example, clusters of metal atoms are often seen following vaporization of a metal target. The mean size of the clusters increases with the length of the growth channel. There are several other factors that influence the chemistry. These include the concentrations of reactants added to the carrier gas, the laser power, wavelength, and pulse duration. Duncan [23] provides a helpful discussion of the effects of these parameters.

Electrospray ionization (ESI) provides a versatile means of producing both cations and anions in the gas phase [33]. ESI is a soft ionization method that can produce complex ions that would not survive the harsh conditions of thermal vaporization or laser ablation. Multiply charged ions can be made, along with partially solvated species that yield insights concerning solution phase speciation. For this technique, actinide salts are first dissolved in a polar solvent (e.g., water, methanol, or acetonitrile). The solution is then forced through a metal capillary tube that is held at a high voltage (on the order of a few kilovolts). Ionization and droplet fragmentation occurs as the liquid exits the capillary tube. The products are de-solvated and then sampled into ion traps and/or mass spectrometers. Gibson and coworkers [10, 13, 34–41] have made extensive use of ESI to study reactions of both bare and solvated actinide ions. Gibson's laboratory has the rare capability of handling samples that are radiologically challenging (e.g., Pa, Pu, Am, and Cm). Spectroscopic studies of cations produced by ESI have been performed by means of action spectroscopy, which permits the application of charged particle detection [24, 42–48]. Anions can also be produced by ESI and characterized by photoelectron spectroscopy. Wang [33] has recently reviewed the application of these techniques to actinide-containing anions.

1.3 Techniques for Spectroscopic Characterization of Gas-Phase Actinide Compounds

1.3.1 Conventional Absorption and Emission Spectroscopy

The volatile hexafluorides UF_6, NpF_6, and PuF_6 have been examined using IR and optical absorption spectroscopy [8, 49–55]. Data for these molecules, recorded under room temperature conditions, are complicated by the thermal population of many low-energy ro-vibronic states. For example, it has been estimated that 99.6% of UF_6 molecules populate vibrationally excited levels at 300 K [53, 56]. This problem has been partially mitigated by working with low-temperature cells, but the temperature dependence of the vapor pressure imposes a practical lower limit of about 160 K. Lower temperatures have been achieved for UF_6 using isentropic gas expansion techniques [57, 58]. Due to the large flow rates needed for this technique, it does not appear to have been applied to NpF_6 or PuF_6. Electronic spectra for the hexafluorides have been examined at room temperature, but the level of congestion yields data that are difficult to assign.

Electronic emission spectra for ThO and UO have been recorded using standard monochromators for dispersion. Thermal excitation of UO [21] and discharge excitation of ThO [16, 59–67] yielded emission intensities that were sufficient for the recording of spectra at the Doppler limit of resolution. Rotationally resolved electronic spectra have been obtained over the range from the near-UV to the near-IR. In these studies, the data were recorded by means of photographic plates or high-gain photomultiplier tubes. Owing to the high-temperature source conditions, many levels of the oxides were significantly populated. In the case of UO, Kaledin et al. [21] reported seeing approximately 500 vibronic bands within the 400–900 nm range. Despite this complexity, several assignable bands were identified. This was possible due to the relatively large rotational constants of the monoxide. Kaledin et al. [21] were also able to record absorption spectra for UO using a continuum light source (Xenon lamp) and a monochromator.

Pure rotational spectroscopy, conducted in the microwave and sub-millimeter wave regions, can provide some of the most accurate molecular constants. At these low transition energies, both the Doppler and lifetime broadenings of the spectral lines are smaller by orders of magnitude, as compared to the visible region. This is conveniently exemplified by UO. For the experiments of Kaledin et al. [21], the Doppler width near 500 nm was 1.3 GHz, and the natural linewidth was approximately 1 MHz. By contrast, the sub-millimeter spectrum of UO, recorded by Holt et al. [19] at a temperature of 2200 K, had a Doppler linewidth of 1.3 MHz, and the natural linewidth was below the effects of pressure broadening. The sub-millimeter wave data for UO were recorded using a frequency modulation technique, and over 280 lines were observed within the 510–652 GHz range. Prior to this study, Cooke and coworkers [68–70] were the first to obtain a microwave spectrum for an actinide-containing molecule (ThO). For these measurements, they used laser ablation and jet expansion cooling to obtain cold gas-phase samples of ThO. A cavity-enhanced absorption technique was used to achieve high sensitivities. Pulsed excitation was used, and the spectra were recovered via Fourier analysis of the time-dependent response. Due to the non-equilibrium nature of jet cooling, they were able to record the pure rotational lines ($J = 0 \rightarrow 1$) for $^{232}Th^{16}O$ vibrational levels ranging from $v = 0$ to 11 (transitions of the $Th^{17}O$ and $Th^{18}O$ isotopes were also examined for lower vibrational levels).

1.3.2 Photoelectron Spectroscopy

Gas-phase photoelectron spectra have been recorded for a small number of Th and U compounds, including UO, UO_2 [71], UF_6 [72], ThX_4, and UX_4 (X = F, Cl, and Br) [73–76]. The vapors were generated by heating solid samples, and the molecules were ionized by means of monochromatic VUV light (e.g., the He II line at 40.8 eV). Ionization energies and spectra for the molecular ions were derived by measuring the kinetic energy distributions of the photoelectrons. This was accomplished using a hemispherical electrostatic analyzer, which is capable of resolving down to about 40 cm^{-1}. This is sufficient to distinguish well-separated vibrational levels and electronically excited states. However, the actinide ions often possess multiple low-lying electronic states, due to the presence of partially filled *d*- or *f*- orbitals on the metal. For molecular ions such as UO^+ and UO_2^+, the low-energy vibrational states are so dense that their photoelectron spectra yield broad, unresolved features [71]. A summary of the gas-phase photoelectron studies of simple actinide compounds is given in Reference [8]. Higher-resolution data for molecular ions have since been obtained using the laser-based methods described in the following sections.

1.3.3 Velocity Modulation and Frequency Comb Spectroscopy

Spectroscopic studies of molecular ions are challenging due to the low densities at which ions can be generated. The concentrations are limited by the reactive nature of the ions, and their Coulombic repulsion. Consequently, ions are nearly always accompanied by much higher concentrations of neutral molecules. In most instances, the spectra for the neutrals will mask the much weaker signals from the ions. One solution to this problem is to use velocity modulation of the ions [77–79]. This technique requires a tunable laser light source that has a linewidth that is comparable to or smaller than the Doppler linewidth of the transition of interest. Ions are generated in a discharge tube using an alternating current with a frequency in the kilohertz range. During each cycle, the ions are accelerated back and forth along the tube. The laser is propagated along the tube axis, and the transmitted intensity is monitored. If the laser is tuned to the edge of an absorption line, the transmitted intensity will be modulated as the ions are Doppler-tuned in and out of resonance. By observing the signal at the modulation frequency (or multiples thereof), the detection can be strongly biased in favor of the ions (transient neutral species may still be detected due to concentration modulation)[79].

Due to the narrow linewidth required for the velocity modulation technique, scanning for new spectral features is a time-consuming process. A dramatic improvement in the data acquisition rate has recently been achieved by combining velocity modulation with frequency comb spectroscopy (for a detailed review of the frequency comb technique, see Reference [80]). Briefly, a femtosecond pulsed laser with a stabilized optical cavity is used to generate a "comb" of laser frequencies. The frequency spacing of the comb is determined by the cavity round trip time, and the comb contains around 10^5 individual frequencies. This radiation is transmitted through the sample, and dispersed using the combination of a virtually imaged phased array (VIPA) disperser (high resolution) and a diffraction grating (low resolution). By this means, every element of the frequency comb is resolved, and their transmitted intensities are recorded. For the detection of ions, velocity modulation and a multipass optical cavity are employed. Figure 1.2 shows the apparatus used by Gresh et al. [17] to record high-resolution spectra for the ThF^+ ion. Bands that were tentatively assigned to the ThO^+ ion were also present in these spectra. Cossel et al. [22] found that data acquisition with the frequency comb system was about 30 times faster than with conventional velocity modulation spectroscopy. To date, ThF^+ is the only actinide-containing species examined using frequency comb spectroscopy. However, it is evident that a wide range of actinide species (ions and neutrals) could be studied using this powerful tool.

1.3.4 LIF Spectroscopy

LIF spectroscopy of electronic transitions is a versatile method that can achieve high resolution (including methods for overcoming Doppler broadening) [79]. In its simplest form, LIF is performed by tuning a laser through absorption bands and detecting the resulting fluorescence without active selection of the detected wavelengths (usually the optical filter used to block scattered laser light, and the characteristics of the photodetector determine the part of the emission spectrum that is most effectively detected). The result is essentially the absorption spectrum, convoluted with the fluorescence quantum yield for the excited state. Pulsed laser excitation works well, and is readily combined with pulsed laser ablation sources. To illustrate the technique, Figure 1.3 shows the main components of the apparatus at Emory University that is used to study the electronic spectra of actinide-containing molecules. The first vacuum chamber of this instrument houses a laser ablation/jet cooling nozzle source. The beam from a pulsed dye laser crosses the expanding jet a few centimeters downstream from the

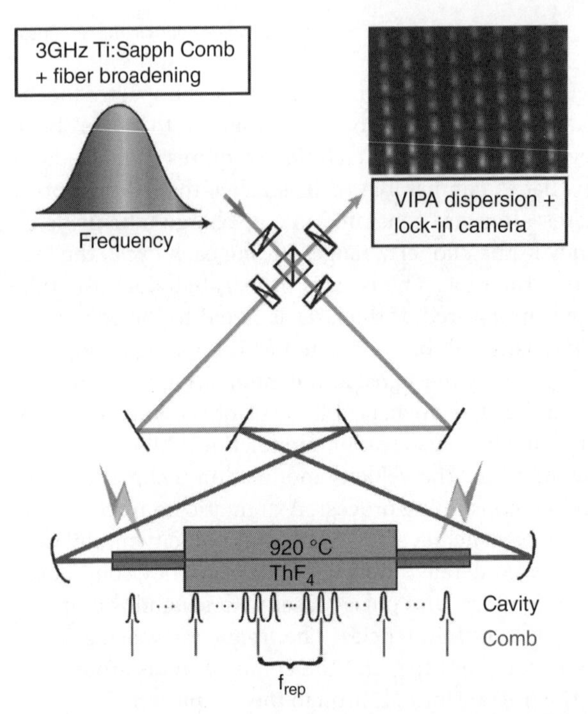

Figure 1.2 The frequency comb/velocity modulation spectrometer. A 3 GHz Ti:Sapphire comb is broadened by a nonlinear fiber and coupled into a bow-tie ring cavity, propagating in either direction through a discharge of ThF_4. Four liquid crystal variable retarders and a polarizing beam splitter control the direction of light propagation through the discharge. After exiting the cavity, the laser light is sent to a cross-dispersive VIPA imaging system with single comb mode resolution and imaged onto a lock-in camera. Approximately 1,500 comb modes are resolved in a single image. (Reproduced with permission from Gresh, D.N., Cossel, K.C., Zhou, Y., Ye, J., and Cornell, E.A. (2016) Broadband velocity modulation spectroscopy of ThF^+ for use in a measurement of the electron electric dipole moment. *Journal of Molecular Spectroscopy*, **319**, 1–9.)

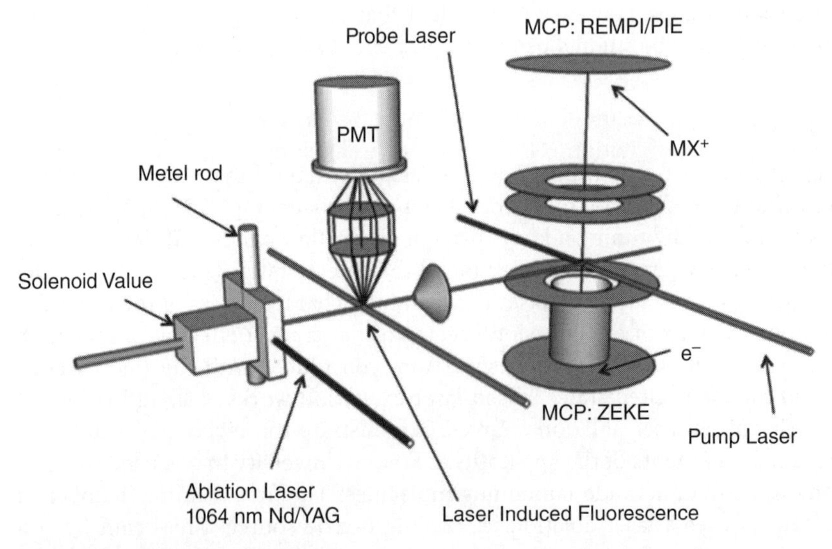

Figure 1.3 Schematic diagram of an apparatus used to record LIF, resonantly enhanced multiphoton ionization (REMPI), PIE, and PFI-ZEKE spectra for gas-phase actinide molecules.

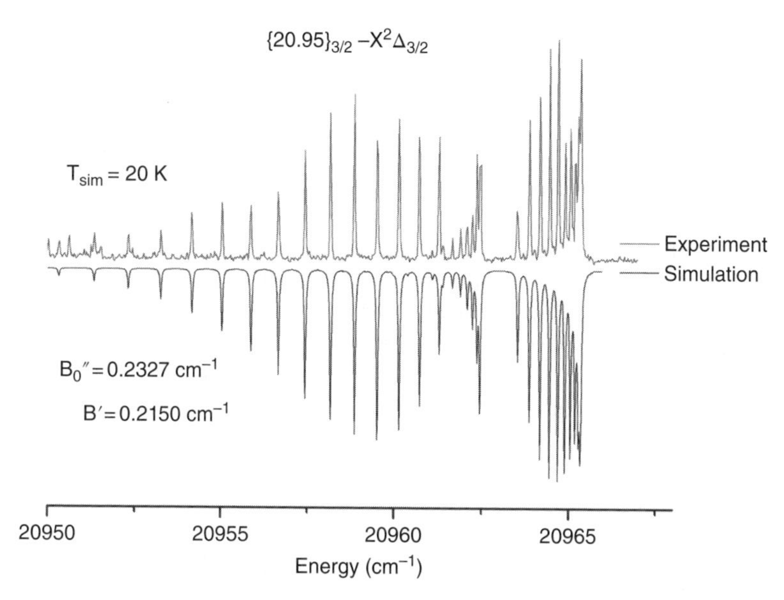

Figure 1.4 Rotationally resolved laser-induced fluorescence (LIF) spectrum of the ThF {20.95}3/2-$X^2\Delta_{3/2}$ band. The downward-going trace is a computer simulation. (*See insert for color representation of the figure.*)

nozzle orifice. The resulting fluorescence is detected along an axis that is perpendicular to the gas flow direction and the probe laser beam. Figure 1.4 shows representative data for ThF. Here, it can be seen that the rotational temperature is low (around 20 K), and the rotational structure is resolved. Note, however, that small splittings associated with breaking of the electronic degeneracy (Ω-doubling) and the hyperfine structure resulting from the nuclear spin of the F atom are not resolved. An advantage of pulsed laser excitation is that the fluorescence can be time-resolved, thereby providing information concerning the lifetimes and transition moments of the excited states.

The resolution of the spectrometer shown in Figure 1.3 is ultimately limited by two constraints. The first is the use of incoherent pulsed laser excitation, which cannot provide a linewidth that is less than that defined by the Fourier transform of the temporal pulse shape. The second limitation is from Doppler broadening. Although the internal temperature within the supersonic flow is low, the molecules are accelerated by the carrier gas as the expansion evolves. If a wide angular spread of the jet is probed by the laser, the supersonic flow velocity (on the order of 1760 ms^{-1} for He) results in appreciable line broadening. This effect can be mitigated by either skimming the core of the expansion into a second vacuum chamber prior to laser excitation, or by masking the fluorescence detection system such that only molecules with low velocities along the laser axis can be observed (a simple slit mask is sufficient). The problem with the laser linewidth can be overcome by switching to a tunable continuous wave (CW) laser. Despite the typically low duty cycle of pulsed laser ablation sources, LIF detection has a high enough sensitivity that the use of a CW probe laser is viable.

The combination of a skimmed molecular beam and a narrow linewidth CW probe laser has been applied to the study of actinide-containing molecules by Steimle and coworkers. To date, this system has been used to examine electronic transitions of ThO

[81], ThS [82], UO [83], and UF [84]. The resolution achieved (around 40 MHz) has been close to the limit imposed by the natural linewidth ($\Delta\nu = \frac{1}{2\pi\tau}$, where τ is the excited state lifetime). This high resolution has permitted measurements of line splitting induced by externally applied electric and magnetic fields. These data yield accurate values for the electric and magnetic dipole moments of both the upper and lower electronic states.

LIF spectra of jet-cooled molecules yield extensive information on electronically excited states, but the data for the ground state may be limited to the properties of the zero-point level. A more extended view of the electronic ground state can be obtained by spectral resolution of the LIF (referred to as dispersed fluorescence in the following text). In addition, the dispersed fluorescence spectra may also reveal the presence of low-energy electronically excited states. Such measurements are particularly effective when the excitation laser is tuned to a single vibronic level of an electronically excited state. The range of ground state vibrational levels observed is dictated by the square of the overlap integral between the upper and lower state vibrational wavefunctions (the Franck–Condon factors). When other electronic states are present, transitions to those that satisfy the usual selection rules are observed. As an example, Figure 1.5 shows a dispersed fluorescence spectrum for ThS, obtained by exciting an upper electronic state with $\Omega = 1$ (where Ω is the unsigned projection of the electronic angular momentum along the bond axis) [85]. The bands on the left-hand side of this spectrum reveal the vibrational structure of the $X^1\Sigma^+$ ground state, while the bands at higher energy are for the $^3\Delta_1$ and $^3\Delta_2$ excited states (the third component of the triplet, $^3\Delta_3$, was not observed as the selection rule is $\Delta\Omega = 0,\pm1$). The range of levels observed in the dispersed fluorescence spectrum can be manipulated, to some extent, through the choice of the initially excited level.

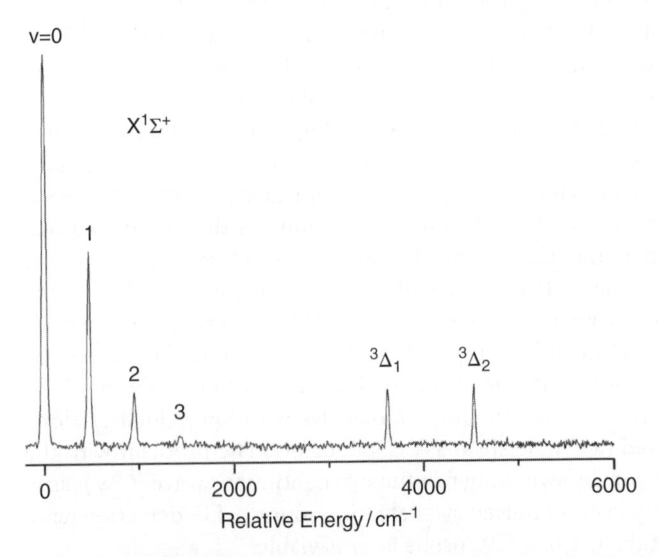

Figure 1.5 Dispersed fluorescence spectrum for ThS recorded using laser excitation of the [22.13]1-$X^1\Sigma^+$ band. (Reproduced with permission from Bartlett, J.H., Antonov, I.O., and Heaven, M.C. (2013) Spectroscopic and Theoretical Investigations of ThS and ThS^+. *Journal of Physical Chemistry A*, **117**, 12042–12048.).

The combination of laser excitation with dispersed fluorescence detection can be used as a two-dimensional method in order to disentangle complex and congested spectra [79]. There are typically two sources of complexity. In laser ablation/jet cooling experiments, the control over the species created is somewhat limited. Consequently, the spectra may include features from several different molecules. The problem is compounded when thermal vaporization or discharge methods are used to obtain gas-phase species. Many states are populated under these conditions, often leading to severe spectral congestion. The dispersed fluorescence technique described earlier, carried out with a fixed excitation wavelength, shows lower state energy levels that are related through their connection to a common upper level. Provided that a sufficiently narrow linewidth is used in excitation, the emission spectrum for a single molecular component of a mixture can be obtained. Once the pattern of low-lying vibronic states has been established, the spectral scanning procedure can be permuted (fix the monochromator, scan the dye laser) or carried out as a coordinated sweep of both the laser and monochromator wavelengths. Spectra recorded with wavelength-selected fluorescence detection (monochromator fixed) show transitions to a specific upper state. This is almost the same information that is obtained using dispersed fluorescence, but restricted to lower levels that are significantly populated. However, a much higher resolution can be achieved via control of the laser linewidth. The excitation spectrum of a specific molecular species, sought under conditions where other species with overlapping spectra are present, can be obtained by a coordinated scan of both the laser and monochromator wavelengths. The tuning rates for both instruments are calculated to keep the energy difference between the absorbed and emitted photons equal to a lower state interval.

An LIF study of the ThF^+ ion illustrates the value of the two-dimensional approach [86]. In these experiments, ThF was generated by laser ablation of Th in the presence of SF_6. Although ThF^+ ions were surely generated by this means, they were not present at a concentration that was sufficient for LIF detection. Additional ThF^+ ions were generated in the downstream region of the jet expansion by means of pulsed laser ionization. The challenge for this experiment was that both ThF and ThF^+ have many overlapping bands in the visible range. ThF dominated the ordinary LIF spectrum, as only a small fraction of the initial ThF was ionized. A further complication was the presence of many intense ThO bands. Metal oxides are nearly always present in laser ablation experiments, and they are very difficult to suppress. Even when high-purity oxygen-free gasses are used, the oxides are routinely observed. The spectrum of ThF^+ was separated from the interfering bands of ThF and ThO by means of the differences in their ground state vibrational intervals ($\Delta G_{1/2}(ThF^+) = 605$, $\Delta G_{1/2}(ThF) = 653$, $\Delta G_{1/2}(ThO) = 891\ cm^{-1}$). The clean ThF^+ LIF spectrum shown in Figure 1.6 was obtained with the monochromator set to observe fluorescence that was $605\ cm^{-1}$ red-shifted relative to the excitation energy [86].

So far, we have described the two-dimensional technique as the sequential recording of one-dimensional cuts (fixed laser or fixed monochromator wavelength slices). This strategy is successful, but the rate of data acquisition is slow. In recent years, the technique has been improved by using monochromators equipped with CCD cameras to record the emission spectrum for each point in the laser excitation spectrum (a method that is conveniently combined with pulsed laser excitation). This allows for a more detailed spectral map to be assembled within a reasonable acquisition time. Recent

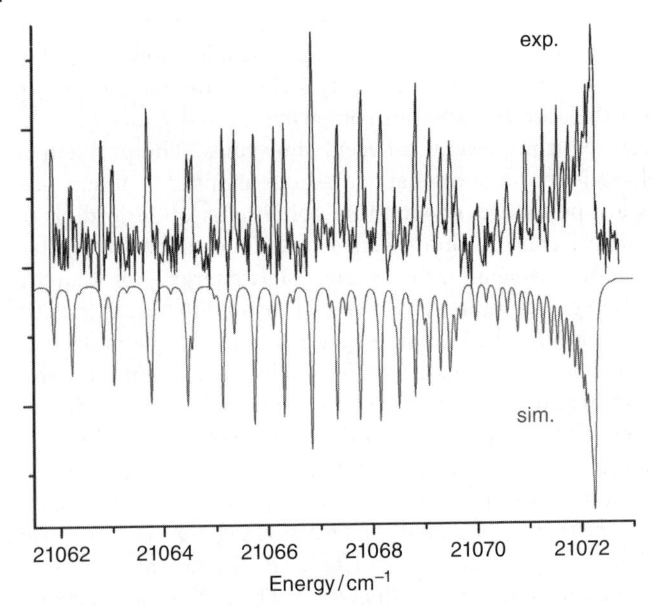

Figure 1.6 Rotationally resolved LIF spectrum of the ThF^+ [21.1]0^+-$X^3\Delta_1$ band. The downward-going trace is a computational simulation of the band. (*See insert for color representation of the figure.*)

work by the Steimle group on the first LIF detection of Th_2 illustrates the advantages of this enhancement [87]. The dimer was produced by pulsed laser vaporization of Th metal. As usual, the LIF spectrum, without wavelength selection of the emission, contained many intense bands of ThO. Figure 1.7 shows the two-dimensional spectrum, where a 2000 cm^{-1} segment of the emission spectrum was recorded for every excitation wavelength sampled. Identification of the Th_2 bands in the two-dimensional map is straightforward. Owing to the low vibrational frequency of Th_2, every band of this molecule produces a characteristic column of features. Recording the fluorescence spectrum using the CCD array also facilitates the characterization of relative transition probabilities, as the relative intensities are immune to drifts in the source conditions [88].

1.3.5 Two-Photon Excitation Techniques

The LIF techniques described earlier rely on one-photon excitation processes. With the high intensities provided by lasers and related coherent light sources, there are many two-photon excitation processes that can be exploited. Two-dimensional techniques can be applied if independently tunable light sources are used.

A recent example of a two-photon excitation measurement is the microwave-optical double resonance (MODR) study of ThS reported by Steimle et al. [89]. The MODR technique is a well-established method for recording microwave spectra with a sensitivity that is determined by the electronic transition (electronic transition probabilities are typically orders of magnitude larger than rotational transition probabilities). Figure 1.8 shows the essential components of the MODR experiment. The molecules of interest are entrained in a molecular beam, and a CW laser is used to excite a single rotational line of an electronic transition. The laser beam is divided into two, so that the

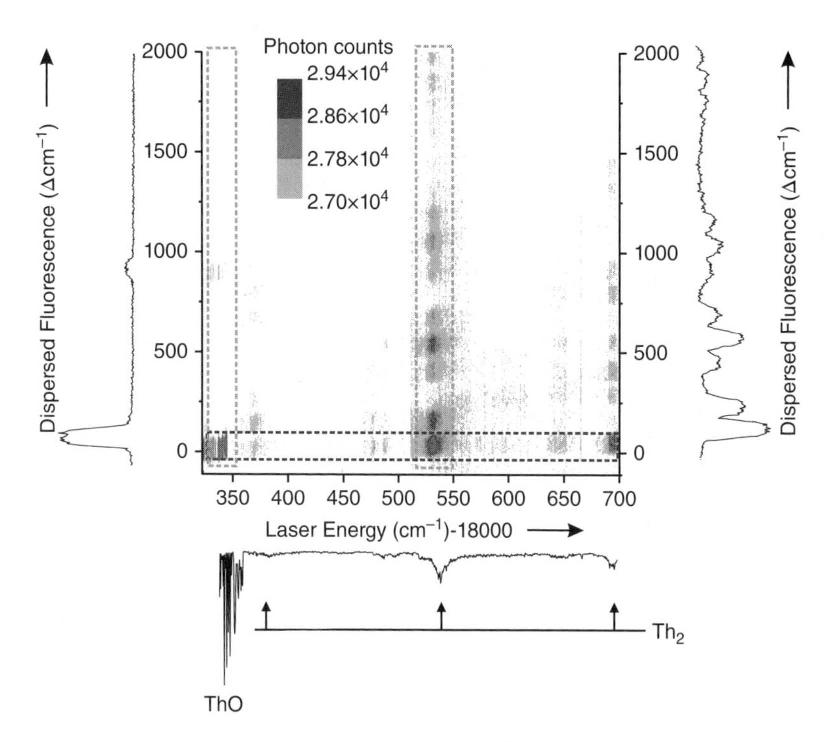

Figure 1.7 2D spectrum over the 18325 – 18700 cm^{-1} laser excitation range of ablated thorium in the presence of a SF6/Ar mixture. Not corrected for laser power variation or the spectral response of the CCD. At the bottom is the on-resonance detected laser excitation spectrum obtained from the vertical integration of the intensities of the horizontal slice marked by the dashed blue rectangle. At the left and right are the dispersed fluorescence spectra resulting from excitation of the (0,0) $F^1\Sigma(+) - X^1\Sigma^+$ band of ThO at 18340 cm^{-1} and Th_2 band at 18530 cm^{-1}, respectively, obtained by horizontal integration of the intensities of the left and right horizontal slices marked in red. (Reproduced with permission from Steimle, T., Kokkin, D.L., Muscarella, S., and Ma, T. (2015) Detection of thorium dimer via two-dimensional fluorescence spectroscopy. *Journal of Physical Chemistry A*, **119**, 9281–9285.) (*See insert for color representation of the figure.*)

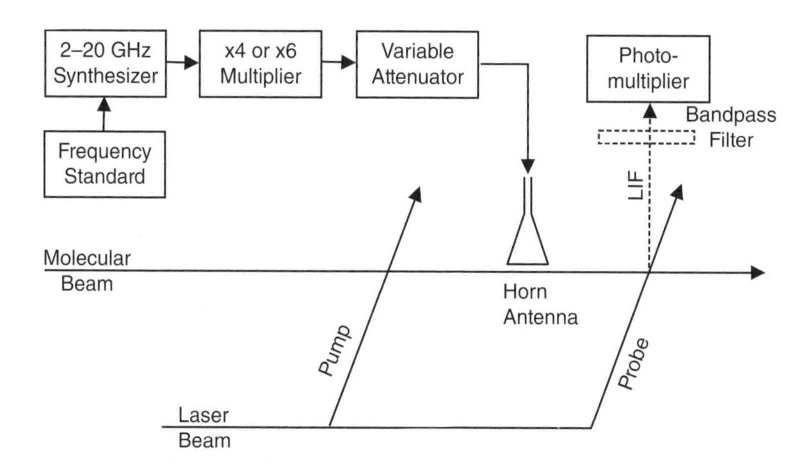

Figure 1.8 A block diagram of the MODR spectrometer. (Reproduced with permission from Steimle, T.C., Zhang, R., and Heaven, M.C. (2015) The pure rotational spectrum of thorium monosulfide, ThS. *Chemical Physics Letters*, **639**, 304–306.)

molecular beam can be excited at two different locations. Most of the power is in the laser beam that is first to intersect the molecular beam. This is used to deplete the population of the rotational state (J) from which excitation occurs. Some distance downstream, the weaker component of the laser beam is used to excite the same rotational line, and the intensity of the fluorescence from this second excitation is monitored. Between the two laser beams, the molecular beam is excited by microwave radiation. When this is resonant with the $J-1 \rightarrow J$ transition, the population transfer increases the LIF signal. By this means, Steimle et al. [89] were able to record pure rotational transitions for ThS ranging from $J=7$ to $J=14$, with a linewidth of 25 kHz. Previous attempts to record the ThS microwave spectrum by using the one-photon excitation techniques that yielded spectra for ThO were unsuccessful. The dipole moment of ThS [82] is slightly larger than that of ThO [81], so it seems that the problem was the production of ThS by laser ablation. We have found that it is easier to make actinide oxides and fluorides with laser ablation, as compared to other chalcogenides or the heavier halogens. The high sensitivity of the MODR method compensates for the low chemical yield, permitting microwave studies of species that are more challenging for gas-phase production.

Resonantly enhanced two-photon ionization techniques have proved to be particularly useful for studies of actinide-containing molecules [7–9]. These methods have the advantage that they can be combined with the mass-resolved detection of ions. In many cases, this mass filter can extract the spectrum for a single species from a complex chemical environment. It can also be used to separate spectra of isotopes, where small isotopic shifts result in congested LIF spectra. Note also that methods that utilize charged particle detection can be far more sensitive than those based on fluorescence detection [79].

The apparatus shown in Figure 1.3 is a typical setup for the recording of mass-resolved, resonantly enhanced multiphoton ionization (REMPI) spectra. The supersonic jet from a laser ablation source is skimmed into the second differentially pumped vacuum chamber (right-hand side of Figure 1.3), forming a well-defined molecular beam. The second chamber houses the components of a Wiley-McLaren time-of-flight mass spectrometer (TOF-MS). Two overlapping pulsed laser beams traverse the molecular beam near the center of the TOF-MS. The lasers have pulse durations of approximately 10 ns, and their synchronization is controlled by a precision pulse delay generator. The ions produced by photoionization are deflected into the flight tube by voltages applied to the repeller and draw-out grids. They are detected by microchannel plates. With a flight path of 40 cm, this system achieves near-unit-mass resolution for masses near 250 amu.

With sequential excitation, there are two options for recording REMPI spectra. In the first scheme, the first laser excites the molecule via a resonant transition. The second laser operates at a fixed photon energy that is high enough to ionize the excited molecules, but not sufficient to ionize from the ground state. The wavelength of the first laser is scanned, and the ion of interest is monitored using the mass spectrometer. It is, of course, straightforward to record the spectra associated with every observed mass peak independently, in a single spectral sweep. The data obtained are equivalent to the absorption spectra for the neutral molecules.

Conversely, the first laser can be fixed on a resonant transition and the second laser scanned in order to locate the ionization threshold (the plot of ion current versus photon energy is often referred to as a photoionization efficiency (PIE) curve).

This measurement can provide a reasonably accurate ionization energy, with uncertainties on the order of ±30 cm^{-1} (±4 meV). The choice of the first laser resonance can be used to ensure that the signal is exclusively associated with ground state molecules (often a particular rotational level of the zero-point level). Tuning the second laser above the ionization limit can result in a very structured spectrum. The structure is due to the presence of strong auto-ionizing resonances that are embedded in the ionization continuum. The resonances are members of Rydberg series that converge to the rovibronic states of the molecular ion. The high-*n* Rydberg states that exist just beneath each threshold for producing the ion in a particular v, J state are unusually long-lived. This property can be exploited in order to obtain spectroscopic data for the molecular ion, using a technique known as pulsed-field ionization zero electron kinetic energy (PFI-ZEKE) spectroscopy [90–92]. In this experiment, a pulsed laser is used to excite the molecule to a small group of high-*n* Rydberg states under field-free conditions. For final ion states that lie above the ionization threshold (e.g., MX^+ $v = 1$), the laser pulse will also cause direct photoionization and rapid auto-ionization. Electrons that are produced by these fast processes are allowed to dissipate during a time interval of a few microseconds. A weak pulsed electric field is then applied. Molecules that are in the high-*n* Rydberg states are ionized by this pulse, and the electrons are accelerated to a microchannel plate detector. Note that PFI-ZEKE spectroscopy can be carried out using single-photon excitation, but the resolution is substantially improved if resonant two-color excitation is employed. The resolution of the two-photon technique, which is determined by the magnitude of the pulsed field, can be as low as 0.5 cm^{-1}. These measurements provide accurate ionization energies and well-resolved spectra for the molecular ions. Rotational resolution has been achieved for diatomic ions.

A valuable facet of the PFI-ZEKE technique is that there are few selection rules governing the ion final states can be observed. For example, Goncharov et al. [93] were able to characterize ten low-lying electronic states of UO^+, with Ω values ranging from 0.5 to 5.5. The ability to recover complete maps of low-lying states has proved to be critical for obtaining a deeper understanding of actinide bonding motifs. Such studies of UO^+ and UF^+ [31] show that the U 5*f* orbitals do not participate significantly in the bonding of these ions.

1.3.6 Anion Photodetachment Spectroscopy

Just as PFI-ZEKE spectroscopy can provide information on the low-lying states of a molecular cation, anion photodetachment spectroscopy can reveal the low-energy states of a neutral molecule [33, 94–96]. An advantage of working with anions is that the species of interest can be mass-selected prior to laser excitation. The study of small metal clusters illustrates problems that can be circumvented by working with anions. Previously we have attempted to record spectra for Th_n and U_n with $n = 2$ or 3 using the mass-resolved REMPI method, in the hope of probing actinide metal–metal bonds. The target ions were readily observed, but their visible range excitation spectra did not exhibit a resolvable structure. Larger metal clusters were present in the laser ablation plume, and it is most probable that these clusters suffered facile multiphoton fragmentation processes. This caused the effectively featureless spectra of the larger clusters to contribute to the signal from the M_2^+ and M_3^+ ions. This problem can be averted when mass selection can be applied prior to laser excitation.

Anion photodetachment measurements use a fixed wavelength laser for detachment. The ejected electrons have kinetic energies (eKE) that are determined by the expression

$$eKE = h\upsilon - EA - E_{Int} \tag{1.1}$$

where hυ is the photon energy, EA is the electron affinity of the neutral molecule, and E_{Int} is the internal energy of the neutral molecule (this model assumes that the anions are cooled to a low internal temperature prior to photodetachment). Hence, the electron kinetic energy distribution encodes the internal energy states of the molecule. A variety of methods have been used to measure the electron kinetic energy distributions. This proves to be a technically challenging problem that is usually the step that limits the resolution. Prior to the development of electron imaging methods, magnetic bottle spectrometers were state of the art, achieving resolutions on the order of 5–10 meV (40–80 cm^{-1}). A considerable improvement in resolution, down to a few cm^{-1}, has been realized by using velocity-map imaging techniques to characterize the detached electrons [94, 95]. With this method, an electrostatic lens is used to project the photodetached electrons along the axis of a flight tube, and image them on a position-sensitive detector (MCPs combined with a phosphor screen and a digital camera). To facilitate the description of this method, Figure 1.9 shows the velocity-map imaging spectrometer used by Wang and coworkers [95].

The components on the left-hand side of this diagram (up to the valve labeled GV2) are used for generation of the anions by laser ablation, followed by time-of-flight mass selection of the target anion. The components for velocity-map imaging (VMI) are housed in the final chamber on the right-hand side of the figure. To understand the images that this instrument produces, consider a photodetachment process that ejects electrons without a preferred direction (isotropic angular distribution). The electrons associated with a specific final state of the molecule will have a fixed velocity, and their

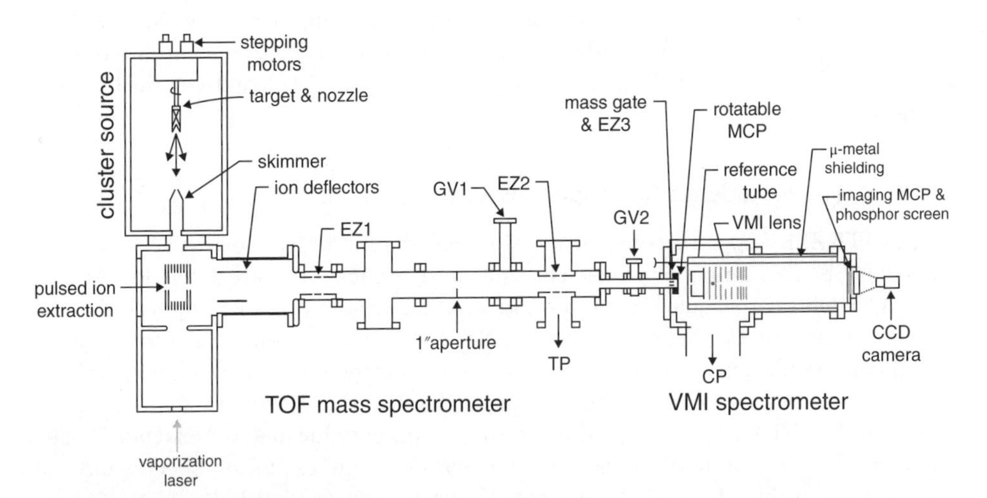

Figure 1.9 Overall schematic view of the high-resolution photoelectron spectroscopy apparatus for size-selected clusters using velocity-map imaging (VMI). (Reproduced with permission from Leon, I., Yang, Z., Liu, H.-T., and Wang, L.-S. (2014) The design and construction of a high-resolution velocity-map imaging apparatus for photoelectron spectroscopy studies of size-selected clusters. *Review of Scientific Instruments*, **85**, 083106/1–083106/12.)

radial distance from the point of detachment will increase linearly with time (an expanding sphere of electrons). This group of electrons is projected along the flight tube at a velocity that is determined by the biasing of the electrostatic (VMI) lens. When the electrons reach the detector, they produce a disk image owing to the reduction of the three-dimensional spatial distribution to a two-dimensional image. The diameter of this disc is determined by the electron kinetic energy and the time needed to travel to the detector. If multiple final states of the molecule are produced, the image consists of multiple concentric discs. The use of linearly polarized laser light can produce spatial velocity distributions that are cylindrically symmetric about the polarization vector, but otherwise anisotropic. Several computational methods have been developed to permit the reconstruction of the three-dimensional velocity distributions from the two-dimensional images [95]. When present, the anisotropy of the distribution can provide insights concerning the symmetry of the orbital from which the electron was removed.

The kinematics leading to image formation are such that the energy resolution is highest for slow-moving electrons [94]. Hence, the best resolution is achieved by tuning the detachment laser to a photon energy that is not much above the threshold for the channel that is being examined (referred to as slow-electron VMI (SEVI)). This is nicely illustrated by Leon et al. [95], who used photodetachment of Bi^- to demonstrate how the resolution improves (down to $1.2\,cm^{-1}$) as the kinetic energy imparted to the electron is reduced from 165.5 to $5.2\,cm^{-1}$.

Factors other than the image resolution limit the energy resolution in studies of molecular anions. In general, anions are rather fragile species. They are easily destroyed in laser ablation sources if the carrier gas pressure is too high. In many cases, the low pressures imposed by the anion instability result in poor cooling. The resolution of the SEVI technique is then compromised by the large number of internal energy states of the anion that are populated. Electrospray ionization has also been used to generate anions for SEVI measurements. This is a versatile source, but it usually produces anions under conditions that are at or above ambient temperatures. The strategy to solve this source temperature problem has been to accumulate the anions in ion traps and subject them to buffer gas cooling by low-pressure He. The trap electrodes can be cooled to temperatures as low as 4.4 K. After a period of accumulation and cooling, the anion packet is injected into the SEVI spectrometer (see Figure 1 of Reference [95]). Wang and coworkers have examined UOs [11, 97, 98], fluorides [99–101], and chlorides [102, 103] using the SEVI technique. As an example, Figure 1.10 shows the data for electron detachment from UF_5^- [101]. The structure present in this spectrum corresponds to the vibrational levels of UF_5. This figure illustrates the advantages of ion-trap cooling and the use of near-threshold photon energies.

1.3.7 Action Spectroscopy

In the present context, action spectroscopy refers to schemes where photoexcitation is detected by observing a molecular fragmentation event. Through mass spectrometric detection of the fragments, action spectroscopy is a powerful tool for probing molecular ions that can only be produced at low number densities. For uranyl-containing species, a particularly successful series of gas-phase action spectroscopy experiments has been carried out using the free electron laser (FELIX) that is currently located at Radbound University in the Netherlands [10, 44–48, 104, 105]. This facility is capable of

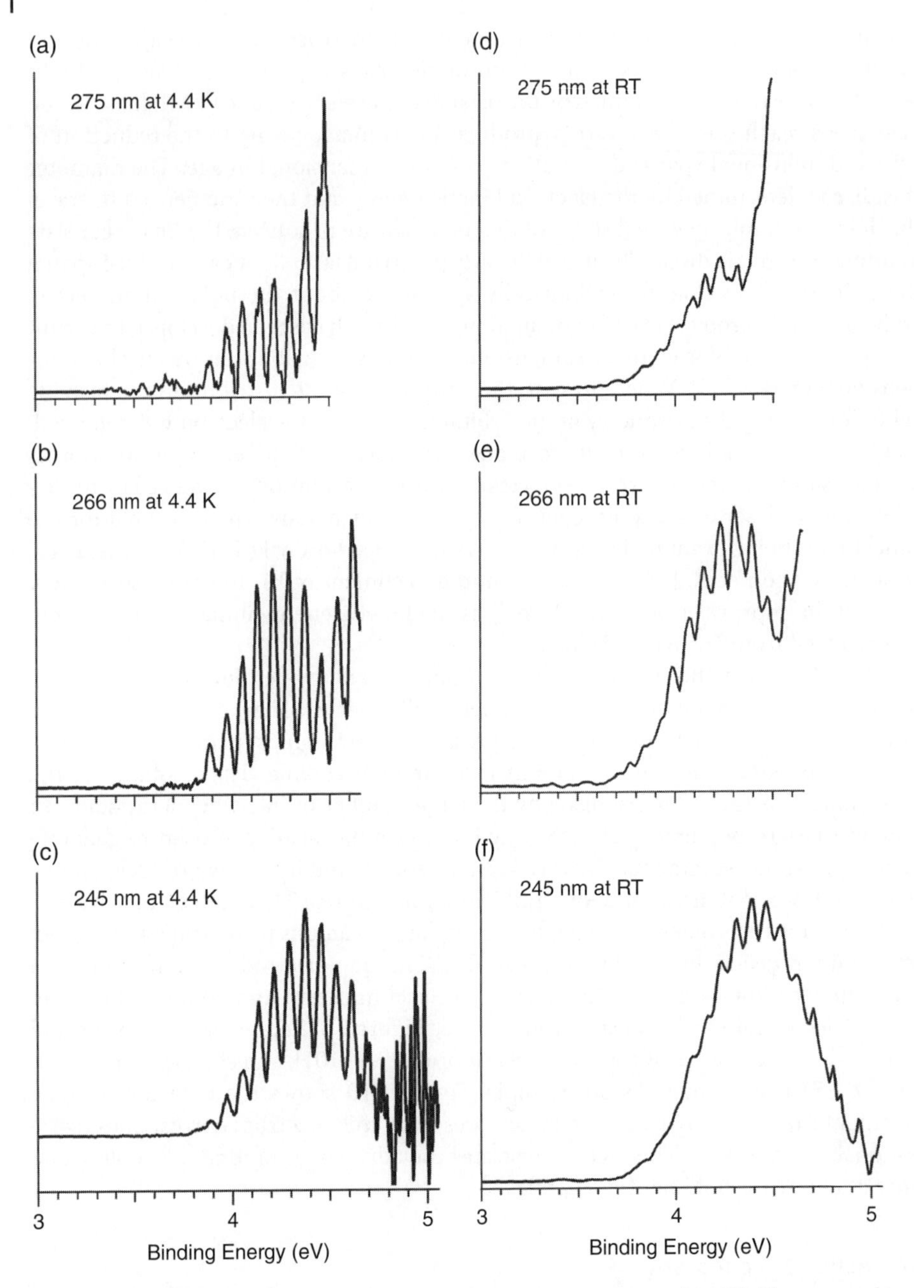

Figure 1.10 Photoelectron spectra of UF_5^- at an ion-trap temperature of 4.4 K (left), compared to those taken at room temperature (right) at (a) and (d) 275 nm (4.508 eV), (b) and (e) 266 nm (4.661 eV), and (c) and (f) 245 nm (5.061 eV). (Reproduced with permission from Dau, P.D., Liu, H.-T., Huang, D.-L., and Wang, L.-S. (2012) Note: Photoelectron spectroscopy of cold UF_5^-. *Journal of Chemical Physics*, **137**, 116101/1–116101/2.)

producing intense tunable radiation throughout the infrared region of the spectrum. For this experiment, various $(UO_2^{2+})L_n$ complexes (where L is a ligand or solvent molecule) have been generated in the gas phase by means of electrospray ionization. Uranyl nitrate solutions have proved to be suitable sources for UO_2^{2+}. The initial studies involved complexes of the uranyl ion with acetone and acetonitrile [48]. The complexes were accumulated in a hexapole trap and then injected into a Fourier transform ion cyclotron resonance mass spectrometer (FT-ICR-MS). A tailored frequency sweep of the ICR-MS was used to eject all ions except the species of interest. Once isolated, the ions were irradiated by intense IR pulses from the free electron laser. If the IR was resonant with a vibrational transition, a high-order multiphoton absorption process was triggered. The energy imparted by this event was sufficient to cause fragmentation of the parent ion. For example, the $(UO_2^{2+})(acetone)_4$ complex fragmented to $(UO_2^{2+})(acetone)_3$ + acetone when vibrational transitions in the range of 900–1800 cm^{-1} were excited. The FT-ICR-MS was used to characterize the fragmentation, and the relative IR absorption strengths were determined from the dissociation yields. Although fragmentation is induced by multiphoton processes, it has been shown that the absorption of the first photon controls the fragmentation. Hence, the data obtained are equivalent to conventional IR absorption spectra. The spectra reported to date have a resolution on the order of 15–20 cm^{-1}. This is sufficient to discern trends in the ion-ligand interactions. For the acetone and acetonitrile complexes, Groenewold et al. [48] clearly observed the effects of electron donation from the ligand on the uranyl asymmetric stretch. Over the past decade, complexes of uranyl with acetone [47, 48], acetonitrile [47, 48], water [47], ammonia [47] simple alcohols [46, 47], nitrate [45, 104], and the chelating ligand TMOGA [10] have been investigated.

The FELIX facility has yielded excellent spectroscopic data, but the need to carry out the measurements using this unique instrument is a significant practical limitation. Action spectroscopy using commercial laser sources at shorter wavelengths (below 3.5 μm) has been a productive endeavor for many years. The possibility of working at longer wavelengths in a typical university laboratory has been realized through improvements in the design of optical parametric oscillators (OPOs) and the availability of new materials for difference frequency down-conversion. The intensities from these systems are sufficient for robust excitation of single-photon transitions, and low-order multiphoton transitions may also be observed. However, the table-top laser systems are not powerful enough to drive the high-order multiphoton processes that are exploited in the FELIX experiments.

Duncan and coworkers have used IR action spectroscopy to study uranium oxide and uranium carbonyl cations [24, 42, 43, 106]. A pulsed laser ablation source was used to generate these species. Mass selection of the target ion and identification of the photofragments was accomplished using a reflectron time-of-flight mass spectrometer. The reflectron consists of two flight tubes and an electrostatic mirror. The latter reflects the ions back along a direction that is 5° away from the incident path. Duncan and coworkers installed a set of deflection electrodes in the first flight tube. These electrodes are used to mass-select the ions that are allowed into the mirror. Within the mirror, at the turning region of the ion trajectories, the ions are excited by the IR laser beam. The second limb of the reflectron then provides a time-of-flight mass analysis of the products.

Ricks et al. [42, 43] used this technique to examine the IR spectra of $U^+(CO)_n$ and $UO_2^+(CO)_m$ complexes with values of $n \leq 10$ and $m \leq 7$. They excited the CO stretch modes of all complexes and the asymmetric stretch mode of the UO_2^+ moiety. The resulting fragmentation patterns showed that the $U^+(CO)_8$ and $UO_2^+(CO)_5$ complexes were particularly stable, indicative of the species with the highest coordination numbers. These results were in accordance with the predictions of DFT calculations. Ricks et al. [24] also examined the UO_4^+ and UO_6^+ ions. These are exotic species in that U is not expected to exhibit an oxidation state greater than six. However, a stable ion such as UO_2^+ can readily bind additional O_2 molecules under the low-temperature conditions of a free-jet expansion. Spectroscopic measurements for these oxide ions were complicated by the fact that IR excitation of the bare ions did not produce a measureable degree of fragmentation. Consequently, action spectra were obtained using a rare-gas atom tagging technique [24]. The ablation source was operated under conditions that generated the $UO_4^+Ar_2$ and $UO_6^+Ar_2$ van der Waals complexes, and the IR transitions were detected by observing the loss of one Ar atom. Interestingly, the complexes that started with just one Ar atom did not dissociate. The central assumption of the tagging method is that the van der Waals interactions will not significantly perturb the properties of the molecule (similar to the assumption made for cryogenic rare-gas matrix isolation experiments). Ricks et al. [24] were able to show that UO_4^+ was a charge transfer complex of the form $(UO_2^{2+})(O_2^-)$, while the second O_2 unit of UO_6^+ is weakly bound to the UO_4^+ core.

Pillai, Duncan, and coworkers [106] have also used their reflectron mass spectrometer to observe direct photodissociation of complexes consisting of benzene molecules bound to U^+, UO^+, and UO_2^+. Mass-selected clusters were photofragmented by the third harmonic from an Nd/YAG laser (355 nm). For $U^+(C_6H_6)$, they observed the fragments U^+, $U^+(C_4H_2)$, and $U^+(C_2H_2)$, showing that ligand decomposition was an important decay channel. The other clusters investigated, $U^+(C_6H_6)_n$, n = 2,3, $UO^+(C_6H_6)$, and $UO_2^+(C_6H_6)$, all exhibited simple intact loss of one C_6H_6 unit. Pillai et al. [106] noted that these trends were consistent with other studies of the relative gas-phase reactivities of U^+, UO^+, and UO_2^+.

1.3.8 Bond Energies and Reactivities from Mass Spectrometry

The spectroscopic techniques described in the preceding sections provide accurate information concerning the bonding mechanisms, electronic structure, equilibrium geometries, and vibrational modes of actinide-containing molecules. Basic thermodynamic information can be derived from these data via the application of the theorems of statistical mechanics. Another thermodynamic property, the ionization energy (IE), can be measured accurately using spectroscopic methods. However, bond energies and relative reactivities are not readily extracted from spectroscopic measurements alone.

Some of the earliest studies of actinides in the gas phase were carried out using classical mass spectrometry techniques. Measurements of ionization energies were made using thermal vaporization of the actinide compound, followed by electron impact ionization [107–110]. This method gave useful estimates of the energies, but, when high temperatures were needed for vaporization, the values could be compromised by the presence of thermally excited molecules that lowered the appearance potentials. Other thermodynamic properties were deduced from analyses of the equilibrium distributions of various species in the vapor phase, generated over a range of temperatures. Electron impact ionization mass spectrometry was used to characterize relative concentrations [13–15, 26–28, 30, 110–114]. Second and third law models were then used to extract the enthalpies and bond energies. Much of this work was recently reviewed for actinide oxides by Kovacs et al. [115].

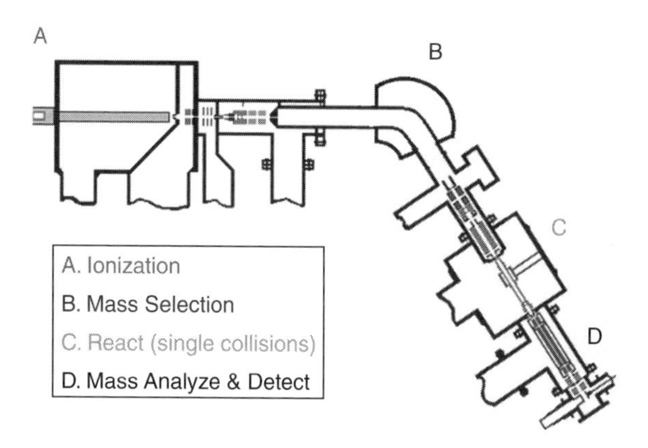

Figure 1.11 Guided ion-beam apparatus for studies of dissociative and reactive collisions with controlled collision energies. See text for details.

Bond dissociation energies for ionic species ($AB^+ \rightarrow A + B^+$) can be determined with good accuracy using collision-induced dissociation (CID) techniques. Armentrout and coworkers [25, 29, 116] have applied guided ion-beam CID to a wide range of metal-containing ions, including Th and U species. An example of the instrumentation used for CID measurements is shown in Figure 1.11. The two differentially pumped vacuum chambers shown on the left-hand side of this schematic are used for ion generation and ion-beam formation. Armentrout and coworkers have used laser ablation and a variety of discharge methods to generate gas-phase ions. In the next chambers, marked as region B in the figure, the ions are accelerated and mass-resolved using a magnetic sector mass spectrometer. After mass selection, the ion velocities are adjusted to the desired value, and they are injected into a collision cell (region C in Figure 1.11). The pressure of the inert collision gas (e.g., Ar or Xe) is chosen to yield single-collision events. Finally, the fragments are characterized using a quadrupole mass spectrometer (region D). One of the first applications of this technique to actinides was the study of UO^+ and $UO_2{}^+$ bond dissociation energies by Armentrout and Beauchamp [116]. The error ranges were on the order of 0.4 eV for bond energies of 8.0 and 7.7 eV. Note that the bond dissociation energy for a neutral molecule (D_0(A-B)) can be derived from a value for $D_0(A - B^+)$ by the IEs for AB and B, using the thermodynamic cycle $D_0(A - B) = D_0(A - B^+) + IE(AB) - IE(B)$.

The apparatus shown in Figure 1.11 has also been used in extensive studies of reactive collisions. Drawing on a recent example, Figure 1.12 shows data for the collisions between Th^+ and CD_4. Here, it can be seen that the lowest energy channel is the ejection of D_2 to form Th^+CD_2. Several other channels open at collision energies close to 2 eV. In this same study, Cox et al. [14] used $Th^+CH_4 + Xe$ CID to measure the $Th^+CH_4 \rightarrow Th^+ + CH_4$ dissociation energy.

Other MS techniques have been used to investigate actinide species that include transuranic (Np-Es) and radiologically hazardous (Pa) materials. Gibson, Marçalo, and coworkers have developed the specialized facilities needed for these experiments [8, 12, 117]. Initially, Gibson and coworkers used laser ablation as the means to generate gas-phase ions. The ablation products were allowed to react with a reagent at low pressures (near single-collision conditions), and the products were analyzed by means of a TOF-MS. This technique was termed laser ablation with prompt reaction and detection

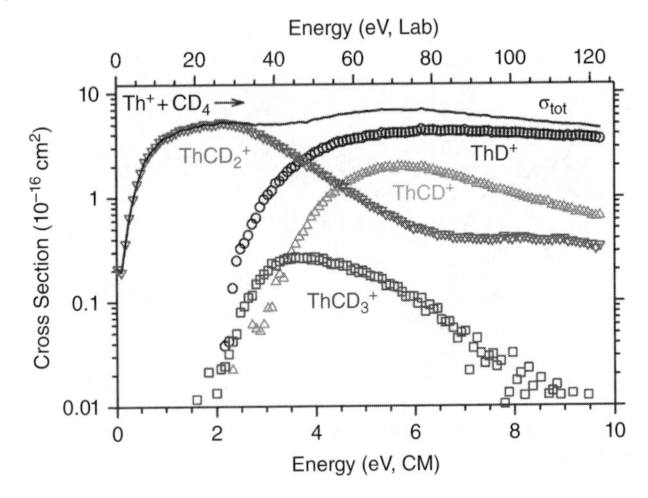

Figure 1.12 Cross sections for the reaction between Th^+ and CD_4 as a function of energy in the center of mass (lower *x* axis) and lab (upper *x* axis) frames of reference. (Reproduced with permission from Cox, R.M., Armentrout, P.B., and de Jong, W.A. (2015) Activation of CH_4 by Th^+ as studied by guided ion-beam mass spectrometry and quantum chemistry. *Inorganic Chemistry*, **54**, 3584–3599.) (*See insert for color representation of the figure.*)

(LAPRD) [118,119]. Bond dissociation and ionization energies have been obtained by a bracketing approach. As charge exchange processes of the kind $A^+ + B \rightarrow A + B^+$ do not have barriers, this method works well for determination of the first IE. Bond dissociation energies are more difficult to determine by bracketing studies. Consider a reaction of the form $MO^+ + OR \rightarrow MO_2^+ + R$. If the reaction is observed, it is evident that the O-MO^+ bond energy exceeds that of O-R. However, as there may be a barrier to the reaction, a negative result does not necessarily imply that the O-R bond energy exceeds that of O-MO^+. Reliable results can be obtained by combining the bracketing data with the predictions of electronic structure calculations.

Prior to the LAPRD studies of actinides, Fourier transform ion cyclotron resonance (FT-ICR) MS had been used to examine the reactions of Th^+ and U^+ [8]. Marçalo, Gibson, and coworkers [117] used similar instruments to examine the IEs, bond dissociation energies, and reaction pathways of many radiologically active actinides. In these experiments, ions are exposed to charge exchange or reaction partners within the ion trap. MS/MS double resonance techniques can then be used to observe the time-dependent loss of parent ions and the growth of charged product species. In favorable cases, reaction rate constants and branching ratios can be determined. Quadrupole ion-trap (QIT) MS provides another versatile technique for studying gas-phase actinide ion chemistry. Gibson and coworkers have had considerable success using a commercial QIT-MS with ESI [10, 13, 34–41, 120–123]. This system has been configured for work with radiologically active species, but it has also provided valuable new results for both Th and U compounds. As for the ICR system, the QIT can be programmed to trap selected ions, excite them, permit the addition of collision partners, and mass-analyze the charged reaction products. A notable example of the work with this system is provided by a recent study of the oxo exchange reactions between AnO_2^+ and H_2O for An = Pa and U [39]. $Pa^{16}O_2^+$ and $U^{16}O_2^+$ ions were generated by ESI of methanol/ethanol

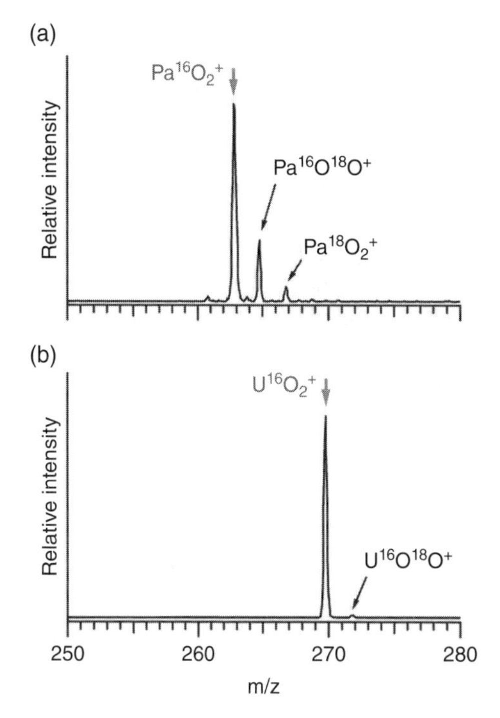

Figure 1.13 Mass spectra acquired after 0.5 s exposure to a constant $H^{18}O$ pressure for (a) $Pa^{16}O^{2+}$ and (b) $U^{16}O^{2+}$. Exchange of the first and second O is apparent for PaO^{2+}, whereas only very minor exchange of one O, is observed for UO^{2+}.

solutions. These ions were reacted with $H_2{}^{18}O$ in the QIT, and the results are shown in Figure 1.13. Here, it is immediately apparent that the oxo exchange with PaO_2^+ is much faster than the exchange with UO_2^+. The intermediate for this exchange is the bis-hydroxo species $AnO(OH)_2^+$, and it was suggested that it is the relative stability of this ion that controls the reaction rate. To test the stability of the $PaO_3H_2^+$ intermediate, this ion was trapped and exposed to a low pressure of acetonitrile. It had been established that ligated water could be replaced by the more strongly basic acetonitrile ligand, but that the bond strength was not sufficient for replacement of a hydroxyl group. This experiment confirmed that most of the $PaO_3H_2^+$ in the trap was of the bis-hydroxo form. The same measurements for $UO_3H_2^+$ showed facile replacement of the H_2O ligand. Further probes of the ion stabilities were carried out using CID. The interpretation of the MS/MS results were confirmed by electronic structure calculations that also provided structural information for the intermediates [124].

1.4 Considerations for Characterizing Actinide-Containing Molecules in the Gas Phase by Ab Initio Methods

In addition to the usual considerations of the electron correlation method and basis set, quantum chemical calculations on actinide-containing molecules present a number of additional difficulties that are somewhat reminiscent of the transition metals. The actinides can exist in a number of oxidation states and have nearly degenerate 5f, 6d, 7s, and 7p orbitals. This leads to a high density of electronic states and often strong multideterminantal character even in their electronic ground states. This is particularly true for free radical actinide molecules, which can involve open 5f, 6d, and 7s shells. In addition, since the 6s and 6p orbitals have very similar radial extents as the 5f, these electrons must also be treated as valence along with the 5f, 6d, and 7s, leading to a large number of electrons to be treated even in frozen-core calculations, for example, already 14 for just the U atom. These same characteristics that make the electron correlation problem difficult also put strong demands on the underlying 1-particle basis set since a given valence set needs to represent orbitals from the 6s out to the 7p with occupied angular momenta ranging from s to f. Of course, for these heavy elements, the treatment of relativistic effects, both spin-orbit and particularly scalar, is mandatory even for a qualitative description.

1.4.1 Electron Correlation Methods

Among the ab initio methods commonly used for actinide species, DFT has played a major role due to its relatively modest computational cost and good accuracy. In regards to spectroscopy, it has been most widely used to provide quality estimates of structures and harmonic vibrational frequencies. Most studies have used the B3LYP hybrid exchange-correlation functional [125, 126], although generalized gradient approximation (GGA) functionals such as PW91 [127] and PBE [128] have also been utilized in the literature. Besides its relatively low computational cost, another advantage of DFT methods is the general availability of analytical gradients and even analytical hessians, which greatly facilitates the efficient calculation of minima, transition states, and harmonic frequencies. The main drawback of DFT methods, however, particularly in regard to predicting spectroscopic properties, is the lack of a route for systematic improvement of the results. Thus, it is critical to attempt an assessment of the accuracy of the calculated quantities of interest by carrying out analogous calculations on very similar species or carrying out high-accuracy benchmark calculations.

Currently the most accurate spectroscopic properties obtainable from ab initio methods are afforded, at least in principle, by wavefunction-based methods since they can be systematically improved toward the exact solution of a particular relativistic Hamiltonian. As in molecules not involving heavy elements, the coupled cluster singles and doubles with perturbative triples method, CCSD(T) [129], is still one of the main methods of choice for high accuracy. Its computational cost is much higher than DFT (approximately N^7 vs. N^4 where N is a measure of the system size), but it can still be applied to actinide molecules containing up to about 10 non-hydrogen atoms using modern computational resources. Of course, CCSD(T) can only be utilized for molecules where the wavefunction is well described by the Hartree–Fock determinant. This can quickly become problematic as the number of unpaired electrons, particularly in the open 5f shell, increases, which can make lower oxidation states particularly difficult. The successive removal of F from UF_6 provides an excellent example. The electron configuration on U for the formal oxidation state varies from $5f^07s^0$ in UF_6 to $5f^37s^0$ in UF_3 to $5f^37s^2$ in UF [99]. None of these species has nominally more than three electrons distributed in the 5f shell, but this still leads to significant multideterminantal character for the smaller systems UF_3, UF_2, and UF due to the near-degeneracy of the available spatial components of the 5f orbitals. Curiously, in these cases, this does not lead to large T_1 diagnostics for the previous three molecules (all are ≤ 0.03), but it does strongly present itself in terms of symmetry-breaking problems and instabilities in the calculation of numerical derivatives for harmonic frequencies due to the many close-lying states. In part, this is a typical example related to the expression among computational chemists of "smaller is harder" that applies as well to calculations on undercoordinated transition metal compounds, for example, transition metal diatomics. It is worth noting here as well that since CCSD(T) is based on a single HF determinant, it is generally only applicable to electronic ground states or to excited states of differing spin multiplicities (in the absence of spin-orbit coupling).

For the situation where the wavefunction is not dominated by a single determinant, there are several multireference methods that have found widespread use for actinide spectroscopy. Probably the most ubiquitous in the actinide community is the complete active space self-consistent field (CASSCF) method [130] followed by second-order perturbation theory, that is, CASPT2 [131, 132]. In the CASSCF step, the orbitals must first be partitioned by the user into inactive, active, and external spaces. For the standard "full valence active space," the active space consists of all the molecular orbitals that can be constructed from the valence atomic orbitals of the atoms. The inactive space is defined as the lower-lying core orbitals, and the external space consists of the remaining unoccupied orbitals lying at higher energy above the active space. For example, for the N_2 molecule, the inactive space would correspond to the two doubly occupied MOs consisting of the two 1s core orbitals, and the full valence active space would consist of the eight MOs constructed from the 2s and 2p orbitals. All possible determinants of the desired space and spin symmetry are then constructed with the valence electrons (10 in the N_2 example) using the active space orbitals; that is, a full configuration interaction (FCI) calculation is carried out just within the active space orbitals, and the MO coefficients are also simultaneously optimized (generally including the inactive orbitals, which are constrained to be doubly occupied in all determinants).

A full valence CASSCF can be a fairly black-box procedure, particularly for molecules consisting of just main group elements, since the choice of active space orbitals can often be done in an automated, unambiguous manner. Unfortunately for most actinide molecules, the total number of valence molecular orbitals can become prohibitively large, particularly if both the 5f and 6d shells are deemed important (which is often the case). In particular, the number of determinants involved in the CAS has a factorial dependence on the number of active electrons and active space size, which can quickly become prohibitively expensive. Hence, some truncation of the active space is often necessary, which relies on a fair amount of chemical intuition and testing to ensure that important near-degeneracy effects are recovered in a balanced manner. As a rule of thumb, all orbitals should be included that are involved in a bond-breaking process or low-lying electronic states of interest, particularly including those that can interact strongly with the ground state. For example in the previous work by Gagliardi and co-workers [133] on oxides of Pa, for PaO it was possible to distribute nine electrons into 16 active orbitals, which consisted of the 7s, 6d, and 5f orbitals of Pa and the 2p orbitals of O. Not surprisingly, extending this same strategy to PaO_2 was found to be too demanding, and the active space was truncated by omitting the two lowest energy π_g bonding and corresponding antibonding orbitals, as well as one additional σ_g^* antibonding orbital. This yielded a nine electrons in 14 orbitals (9/14) CASSCF. Small molecules involving multiple actinide atoms are even more challenging as evidenced by the work of Roos et al. on actinide dimers [134]. On the basis of several trial studies, an active space for U_2 was reduced from the full valence 12 electrons in 26 orbitals space

(7s, 6d 5f) to 6 electrons in 20 orbitals. This was deemed sufficient near equilibrium but lacked sufficient flexibility for full potential curves. More complicated electron/orbital partitioning schemes can also be carried out that restrict the allowed excitations between subsets of orbitals. These methods are referred to as restricted active space [135] (RAS) or generalized active space [136] (GAS) approaches and can lead to a much smaller number of determinants (or CSFs) in the expansion of the wavefunction.

Just a CASSCF calculation alone, however, does not yield quantitative results since it mostly recovers just near-degeneracy effects, that is, non-dynamical electron correlation. Additional electron correlation, in particular dynamical correlation, can be recovered in a subsequent multiconfigurational second-order perturbation theory calculation with a reference function given by the CASSCF wavefunction. This method has been shown to yield reliable accuracy for actinide systems, although it can suffer from the "intruder state" problem which can cause serious convergence issues. These can generally be mitigated by a redefinition of the active space (adding more orbitals) or through appropriate level shifts. A computationally more expensive alternative to CASPT2 that can also use CASSCF as a reference function is the multireference configuration interaction (MRCI) method [137]. Unlike CASPT2, it is not size extensive, but this can be accounted for in an approximate manner via a multireference Davidson correction (MRCI + Q) [138]. Table 1.1 demonstrates the intricacies of assessing the accuracy of these methods for the $^6L_{11/2} \leftarrow {}^4I_{9/2}$ atomic excitation energy of the U^+ atom using a CASSCF of five electrons in 13 orbitals (5f, 6d, 7s) [139]. Spin-orbit coupling was included via the state interacting approach as discussed below. At the CASSCF level of theory, the wrong ground state is obtained by over 4000 cm^{-1}, while CASPT2 yields the correct ground state although it overestimates the experimental excitation energy (289 cm^{-1}) by about 850 cm^{-1}. Both MRCI and MRCI + Q yield the wrong ground state, although the effect of the + Q correction is very large, increasing the excitation energy relative to MRCI by nearly 1100 cm^{-1} and leaving the MRCI + Q result 971 cm^{-1} below the experimental value. Upon including the 5s, 5p, and 5d electrons into the correlation space at the CASPT2 level, however, the excitation energy is further increased compared to the valence correlated result by nearly 1100 cm^{-1}. This yields a CASPT2 excitation energy much larger than experiment (by 1921 cm^{-1}), but if this core-valence correlation correction is applied to the MRCI + Q result, excellent agreement with experiment is achieved. Unfortunately, this level of calculation becomes much too computationally demanding even for a study of the UF diatomic.

Table 1.1 Errors with respect to experiment (cm^{-1}) at the extrapolated complete basis set limit for the $^6L_{11/2} \leftarrow {}^4I_{9/2}$ transition of U^+. The 60-electron PP of Dolg and Cao [140] was used throughout.

Method	Valence correlation	Valence +5s5p5d[a]
CASSCF	−4573	—
CASPT2	+849	+1921
MRCI	−2058	−989[a]
MRCI + Q	−971	+115[a]

a Core correlation effect taken from the CASPT2 calculations.

As an alternative to the CASSCF/CASPT2 and MRCI methods, multireference coupled cluster theory has also been used in accurate calculations involving actinides. The most successful of these is the Fock space implementation, that is, FS-CCSD [141]. Its advantage is that the full electronic spectrum can be computed in a single calculation with a computational cost similar to standard CCSD. One of its main practical limitations, however, is that the desired open-shell state(s) must be within two electrons of a closed-shell reference determinant. Starting from this correlated reference state, denoted sector (0,0), an orbital active space or model (*P*) space is chosen in which to either add *n* electrons, sector (0,*n*), or remove *m* electrons, sector (*m*,0), and the correlation energy is solved at each step. The electronic spectrum of a molecule containing one or two electrons added to the unoccupied *P* space involves calculating sectors (0,1) and (0,2), respectively, while the spectrum for the molecule with one to two fewer electrons than the reference involves removing these electrons from the occupied space, that is, sectors (1,0) and (2,0). The electronic spectrum of the reference molecule itself can also be obtained by calculating sector (1,1). By its construction, FS-CCSD takes into account the multireference character of a state but is limited to just single and double excitations. In some cases, convergence issues are experienced in standard FS-CCSD due to intruder states, and this led to the development of the intermediate Hamiltonian (IH-) FS-CCSD method [142], where the *P* space is itself divided into two subspaces: a primary space and a buffer space. For example, Infante et al. [143] used the IH-FS-CCSD method to study the electronic states of UO_2 and UO_2^+. The UO_2^{2+} molecule was used as the closed-shell reference state, while UO_2^+ and UO_2 involved solving for sectors (0,1) and (0,2), respectively. The main model space consisted of the 7*s*, two of the five 6*d*, and six of the seven 5*f* orbitals of U.

1.4.2 Relativistic Effects

As noted earlier, the incorporation of effects due to relativity is essential in any calculation containing actinide atoms. There are roughly two strategies in use in regards to actinide calculations. In one case, a spin-free, scalar relativistic Hamiltonian is first employed to generate orbitals and perhaps also a basis of correlated electronic states, with spin-orbit coupling (SOC) then being introduced after the fact. In the second option, scalar and spin-orbit relativistic effects are included together from the beginning in either full four-component (positron and electron states) or two-component (only electron states) treatments. Both have their advantages and disadvantages in terms of their ease of interpretation, computational cost, and overall attainable accuracy. Scalar relativistic Hamiltonians in common use for actinide chemistry include the all-electron Douglas–Kroll–Hess [144, 145] (DKH) and exact two-component (X2C) Hamiltonians [146], as well as relativistic effective core potentials (RECPs) or pseudopotentials (PPs) [5]. Mostly commonly, the second-order DKH Hamiltonian (DKH2) is used, but recently it has been shown that DKH3 has a non-negligible effect even on energy differences and geometries compared to DKH2 if high accuracy is desired. The X2C approach is essentially analogous to infinite-order DKH, but is not as readily available as DKH. The PP approach has been very successful in actinide applications, generally yielding results in very good agreement with comparable all-electron treatments. The most accurate PPs for actinides subsume only 60 or 68 electrons in the effective potential, leaving at least the 5d, 6s, 6p, 6d, 5f, and 7s electrons explicitly in the valence space.

The first strategy of the above approaches to SOC has seen the widest application use in actinide chemistry, and in most cases this employs a state interacting approach [147, 148] where spin-orbit coupling is introduced via degenerate perturbation theory. In the first stage of these calculations, a set of electronic states is computed in the absence of SOC using a scalar-only relativistic Hamiltonian ($\hat{H}_{el}$). The final spin-orbit eigenstates are then obtained by explicitly constructing and diagonalizing $\hat{H}_{el} + \hat{H}_{SO}$ in a basis of these electronic states, which are eigenfunctions of $\hat{H}_{el}$. The results, of course, can be very sensitive to the number of basis states included as well as the level of theory used to compute them. Since the accuracy of the off-diagonal SO matrix elements is generally less sensitive to the level of theory than the spin-free electronic state separations themselves, this method easily allows for combining different sources for diagonal and off-diagonal elements of $\hat{H}_{el} + \hat{H}_{SO}$. One advantage of this approach is that the final SO eigenstates can be easily interpreted in terms of the contributions from electronic states of well-defined spin multiplicities (the basis states). In typical applications to actinide molecules, a set of scalar-only electronic states of various spin multiplicities are first computed using the CASPT2 method, and the SO matrix elements between them are obtained at the CASSCF level of theory with two-electron contributions to the SOC terms approximately obtained via the atomic mean-field integral (AMFI) [149] approximation. For example, in recent calculations of the low-lying electronic states of the UFO molecule to assist in the interpretation of the photoelectron spectrum of UFO^-, Roy et al. [100] utilized the all-electron second-order Douglas–Kroll–Hess (DKH2) scalar relativistic Hamiltonian with the CASPT2 method to compute up to 40 doublet and 40 quartet states with SO matrix elements determined by CASSCF. In this same work, a relativistic pseudopotential (PP) was also employed for both the zeroth-order CASPT2 states and CASSCF SO matrix elements, with both methods (DKH2 and PP) yielding a very consistent set of UFO electronic state separations and UFO^- vertical detachment energies.

A different strategy is presented by the SO-MRCI method, whereby the orbitals are determined in a scalar-only HF, CASSCF, or MCSCF (multiconfiguration self-consistent field) calculation, after which a single and double excitation SO-MRCISD calculation is carried out in conjunction with a PP-based SO operator employing double group symmetry [150]. This approach is very effective at recovering important SO effects, even for nominally closed-shell ground states where the state interacting approach can be less effective, but the resulting CI expansions grow very rapidly with system size. For instance, for the UO^+ molecule, Pitzer and coworkers [151] correlated 13 electrons in an MCSCF active space that included 10 orbitals, resulting in a SO-MRCI in a triple-zeta basis set with 33 million spin eigenfunctions. Extending the correlation treatment to also include the 3σ and 4σ orbitals formed from the U6pσ and O2pσ atomic orbitals increased the size of the resulting CI to a prohibitive 96 million CSFs. It should be noted that as in the state interacting approach, the resulting accuracy can be limited when the orbitals are strongly affected by SOC.

In terms of relativistic calculations with scalar relativity and SOC treated on the same footing, full four-component (4-c) calculations utilizing the Dirac–Coulomb (DC) or Dirac–Coulomb–Breit (DCB) Hamiltonians [152] in MRCI or CCSD(T) calculations [153, 154] are the gold standard, but this can be prohibitively expensive for many molecular calculations. Alternatively, the small and large components of the 4-c wavefunction can be decoupled, with the small component discarded, leading to so-called two-component (2-c) Hamiltonians. For all-electron wavefunction-based methods, the two-component version of the X2C Hamiltonian [146] has been used extensively for calculations involving actinides, while the zeroth-order regular approximation [155] (ZORA) has been used for 2-c DFT. In the 2c-X2C approach the two-electron spin-same-orbit and spin-other-orbit corrections to SOC are obtained by either AMFIs or the molecular mean-field approach [156]. It is also possible to accurately use some PPs and their associated one-electron SO parameters, for example, the newer Stuttgart–Cologne small-core PPs [140, 157], in variational 2-c calculations, and these have also seen considerable use.

1.4.3 Basis Sets

Even with sophisticated treatments of electron correlation and relativity, the overall accuracy can still be strongly dependent on the quality of the one-particle basis set used to describe the molecular orbitals or spinors. This is just as true for PP-based calculations as it is for all-electron cases. Particularly for the accurate prediction of relative energies, such as bond dissociation enthalpies, being able to approach or extrapolate to the complete basis set (CBS) limit in a reliable, systematic manner is very valuable to minimize basis set incompleteness errors. The design and optimization of basis sets for actinide elements are challenging in themselves, as addressed in a recent review [158]. They must be able to describe electron correlation effects for electrons in the semi-core 6s and 6p orbitals up through the valence 7s in conjunction with either an all-electron or PP relativistic Hamiltonian. They must also be flexible enough to describe a large number of oxidation states and low-lying electronic states of the actinide atom. In order to provide angular correlation of the 5f electrons, already g-type functions are necessary and thus should be included even in a polarized valence double-zeta (VDZ) basis set. Relative energetics approaching chemical accuracy (1–3 kcal/mol) must also include expanding the correlation space to at least the 5d electrons, which requires further extensions of standard valence basis sets. With the exception of the Slater-type basis sets found in the ADF program [159] and used most often in relativistic DFT calculations, basis sets for actinides are constructed with Gaussian-type functions.

For use in all-electron, correlated calculations on actinides, there are three families of basis sets now in common use, the correlation consistent basis sets of Dyall [160] for use in fully relativistic four-component calculations (dyall.vnz), the atomic natural orbital (ANO) basis sets of Roos et al. [161] (ANO-RCC) for use with the DKH2 scalar relativistic Hamiltonian, and the correlation consistent basis sets of Peterson and coworkers [162, 163] constructed with the DKH3 Hamiltonian (cc-pVnZ-DK3). Each of these choices include basis sets ranging from small valence DZ up to quadruple-zeta (QZ) quality, with the Dyall and Peterson sets including extensions for 5s5p5d correlation. Both of the latter family of sets have the advantage of being amenable to standard CBS extrapolation techniques for the HF and correlation energies [164]. In regards to actinide DZ basis sets, the SARC sets of Pantazis and Neese [165] have been particularly designed for use in DFT calculations using the DKH2 and ZORA scalar relativistic Hamiltonians.

Gaussian basis sets constructed for use with relativistic PPs or ECPs include the correlation consistent cc-pVDZ and cc-pVTZ sets of Pitzer and coworkers [166] for use with the 68-electron PPs of Christiansen et al. [167], as well as the DZ quality basis sets originally accompanying the Stuttgart quasirelativistic 60-electron PPs of Küchle et al. [168] (the latter combination is also referred to as SDD in the literature) and the QZ set also subsequently developed for the latter PP [169]. In addition to the QZ-quality basis sets accompanying the new energy-consistent multiconfigurational Dirac–Hartree–Fock (MCDHF)-adjusted (60-electron) PPs of Dolg and coworkers [140, 157] (Ac–U), Peterson [162] has recently reported correlation consistent basis sets from DZ through QZ (cc-pVnZ-PP), together with core correlation sets (cc-pwCVnZ-PP), for use with these latter PPs for Th through U. Since these sets are very similar in construction to the all-electron DKH3 correlation consistent sets noted above, these two families of basis sets have been used in CCSD(T) calculations to demonstrate the high attainable accuracy of these PPs [162, 170].

The most common quantum chemistry programs used for actinide calculations (typical methods used given in parentheses) include MOLCAS [171] (CASSCF and SO-CASPT2), MOLPRO [172] (single-reference coupled cluster, CASSCF, SO-CASPT2, SO-MRCI), DIRAC [173] (two- and four-component relativistic HF, DFT, MRCI, and coupled cluster), and GAUSSIAN [174] (DFT).

1.5 Computational Strategies for Accurate Thermodynamics of Gas-Phase Actinide Molecules

Given the challenges of experimentally determining gas phase bond enthalpies and heats of formation of actinide-containing molecules, computational quantum chemistry can play a very important role in accurately determining the thermodynamic stability of actinide species, particularly those whose short lifetime or radioactivity makes laboratory studies difficult. As is now well known in light main group chemistry, the reliable prediction of thermochemical properties to chemical accuracy (1 kcal/mol) by ab initio methods requires a systematic treatment of both electron correlation and basis set incompleteness errors [175]. Typically, this implies using at least two correlation consistent basis sets of increasing size, generally at least triple- and quadruple-zeta, with extrapolation to the CBS limit using high levels of electron correlation such as CCSD(T). Corrections for outer-core electron correlation and relativistic effects are then included together with accurate harmonic frequencies for zero-point vibrations and thermal corrections. This collection of calculations is the minimal basis for most composite thermochemistry calculations in common use for lighter elements, such as FPA [176], Wn [177], HEAT [178], ccCA [179], and FPD [175,180]. This is not to say that accurate results cannot be obtained by simply choosing one particular method and basis set, although by its very nature this implies some fortuitous, and perhaps significant, cancellation of errors that can lead to unexpected results.

Recently one of the present authors has extended the composite FPD thermochemistry methodology to molecules containing actinide elements [162]. Versions based on both relativistic PPs and the all-electron DKH3 Hamiltonian were investigated, with the all-electron approach yielding the highest average accuracy. Utilizing geometries optimized at the frozen-core DKH3-CCSD(T)/cc-pVTZ-DK3 level of theory, the following procedure was adopted:

- The Hartree–Fock energy was computed with cc-pwCVnZ-DK3 [162] and aug-cc-pcWVnZ-DK [181–183] basis sets on the actinide and lighter elements, respectively, with $n = \text{T}$ and Q. These energies were extrapolated to the HF CBS limit by [184] $E_n = E_{\text{CBS}} + A(n+1)e^{-6.57\sqrt{n}}$ ($n = 3$ for TZ and $n = 4$ for QZ).

- The CCSD(T) correlation energy with valence (6s6p5f6d7s) and outer-core (5s5p5d) electrons of the actinide atom together with the valence and outer-core electrons of the lighter elements were calculated with cc-pwCV*n*Z-DK3 and aug-cc-pwCV*n*Z-DK [185] basis sets with $n = \text{T}$ and Q. The correlation energy was extrapolated to the CBS limit via [186] $E_n = E_{\text{CBS}} + \frac{B}{(n+\frac{1}{2})^4}$. In practice, this step was obtained by carrying out separate valence and core correlation calculations in order to assess the impact of the core correlation. Ideally, a CBS limit for valence correlation would be combined with a smaller basis set core correlation calculation for the core-valence correlation effect, but it was found that the effect of 5s5p5d correlation was very slowly convergent with basis set and required extrapolation.
- Contributions due to SOC were calculated using completely uncontracted cc-pVDZ-DK3 and aug-cc-pVDZ-DK basis sets with the X2C Hamiltonian [146], which includes atomic-mean-field two-electron spin-same-orbit corrections. Two-component, frozen-core coupled cluster calculations were then carried out, either CCSD(T) or FS-CCSD on the basis of how well the wavefunction was dominated by a single determinant. The contribution of the Gaunt term of the Breit interaction was evaluated in all-electron X2C or 4-c calculations at the Dirac–Hartree–Fock level. Atomic SOC corrections for light elements were generally obtained by *J*-averaging of the experimental term energies.
- The leading quantum electrodynamic (QED) contributions from the Lamb shift, that is, the self-energy and vacuum polarization, were obtained via a model potential approach proposed by Pyykkö and Zhao [187]. This local potential was added to the one-electron DKH3 Hamiltonian in frozen-core CCSD(T)/cc-pwCVDZ-DK3/aug-cc-pVDZ-DK calculations and compared to the analogous calculation without this modification.
- Harmonic vibrational frequencies calculated at the CCSD(T)/cc-pVDZ-DK3/aug-cc-pVDZ-DK level of theory were utilized for both the zero-point vibrational corrections and thermal corrections, the latter being obtained from standard ideal gas partition functions.

Table 1.2 shows representative results [162] using the above scheme for the atomization energy of ThO_2 and the first two bond dissociation enthalpies (BDEs) of UF_6 (all relative to 298 K). As explicitly shown in Reference [162], the CBS limit of Table 1.2 for the ThO_2 atomization energy was larger than the QZ result by nearly 4 kcal/mol, while the UF_6 BDEs were only larger by a few tenths of a kcal/mol. In all three cases, the effect of correlating the 5s5p5d electrons of Th and U is non-negligible, affecting the energetics by 2.0 to 2.7 kcal/mol. It is worth mentioning that this is not all due to the 5d electrons; 5s5p correlation accounted for up to 1 kcal/mol of these values. Obviously, the effects due to SOC are essential to include, particularly for the ThO_2 atomization energy, due to the large SO of Th atom, and the bond enthalpy of UF_5, since both UF_5 and UF_4 are open-shell species. The calculated effects due to QED were small but certainly not negligible for the atomization energy of ThO_2. In all cases, the final enthalpies are well within the experimental uncertainties.

Table 1.2 Summary of composite FPD thermochemistry contributions and comparison to experiment (in kcal/mol).

Process	CBS[TQ]	ΔCV[a]	ΔSO[b]	ΔQED	ΔZPE	ΔH(0)	ΔH(298)	ΔH(expt)
$ThO_2 \rightarrow Th + 2O$	374.5	2.7	−8.4	+0.8	−2.4	366.8	369.0	367.8 ± 3.2[c]
$UF_6 \rightarrow UF_5 + F$	80.7	−2.1	−3.0	−0.4	−1.8	73.4	73.9	71.0 ± 2.9[d]
								75.3 ± 4[c]
$UF_5 \rightarrow UF_4 + F$	104.8	−2.0	−6.4	−0.4	−1.6	94.3	95.7	98.0 ± 3.6[d]
								92.5 ± 5.3[c]

a CCSD(T)/CBS[wTQ] with 5s5p5d correlation on Th and U as well as 1s correlation of O and F.
b Includes small contributions due to the Gaunt interaction.
c From the experimental heats of formation; see Reference [162] for details.
d Reference [188].

The FPD values discussed above are based on the CCSD(T) method, which is, of course, not an exact treatment of electron correlation. Going beyond CCSD(T) involves carrying out CCSDT [189, 190] and CCSDTQ [191] or CCSDT(Q) [192] calculations, which are nearly prohibitively expensive for actinide-containing molecules given the large intrinsic basis set sizes and number of valence electrons to correlate. It is feasible, however, for select diatomic molecules, such as ThC^+ and ThO. In the latter case, a CCSDTQ/cc-pVDZ-DK3(Th)/aug-cc-pVDZ-DK(O) calculation involved about 4.3 billion configurations [193]. The resulting impacts on the dissociation energies of ThC^+ and ThO, which were calculated in support of guided ion-beam experiments [193], were largest for the CCSDT − CCSD(T) correction, −1.2 kcal/mol for both molecules, while the effects of quadruple excitations, CCSDTQ − CCSDT, were just +0.3 and +0.06 kcal/mol for ThC^+ and ThO, respectively. Given the apparently fast convergence of the coupled cluster sequence for these molecules, the final dissociation energies were estimated to be accurate to well under 1 kcal/mol.

A complication in accurately determining the BDEs of actinide molecules, although this occurs more generally in the calculation of heats of formation, is that the atoms are often much more difficult to treat with single determinant-based methods like CCSD(T) than their molecules. For instance, the ground electronic state of U has a valence electron configuration of $5f^3 6d^1 7s^2$, which is not very amenable to HF-based calculations. One route around this is to revert to an isodesmic reaction approach using molecules with well-known heats of formation, such as UF_6 [194]. This was recently used to good effect by Bross and Peterson [170] to determine the heats of formation for several U-containing molecules using a similar CCSD(T) FPD scheme as outlined above. As is well known from the ab initio thermochemistry of lighter molecules, the use of isodesmic reactions can also be accurately used to avoid carrying out high-level composite treatments such as FPD. Thanthiriwatte et al. [195] used isodesmic reactions to determine the BDEs of ThF_4, as well as its cations and anions. Using the frozen-core CCSD(T) method with the older 60 electron PP of Küchle et al. [168] on Th with the segmented contracted basis set of Cao and Dolg [169] together with the aug-cc-pVTZ set [181, 182] on F, their resulting dissociation energies were within the experimental error bars (around 3 kcal/mol) in most cases. They also carried out B3LYP calculations, which agreed well with CCSD(T) in these cases.

As mentioned earlier, DFT has been applied extensively in molecular actinide chemistry, and this includes the calculation of BDEs. Perhaps not surprisingly, its performance is mixed, partly because this method's accuracy is just as dependent on the accurate treatment of relativistic effects, particularly SOC, as wavefunction-based approaches, but also due to sensitivity in the choice of functional. For example, Peralta et al. [196] used the hybrid B3LYP and PBEh [197, 198] functionals together with the DKH3 Hamiltonian to calculate the BDEs of the series UF_n and UCl_n with $n = 1–6$. With their best treatment of SOC, which was a self-consistent treatment with approximate screened-nuclear two-electron SO contributions (SNSO), the agreement with experiment ranged from under 1 to 11 kcal/mol (ignoring the very poor agreement for UF and UCl, which was presumably due to the difficulties with atomic U as previously discussed). Their SNSO contributions seem to strongly underestimate the SOC effects on the BDEs in comparison to the X2C coupled cluster results of Reference [162]. Use of the latter values to correct the scalar results of Peralta et al. [196] improves the agreement with experiment for UF_6 (SOC correction of −3.2 kcal/mol compared to −0.3 kcal/mol) while moving the UF_5 result (SOC correction of −6.7 kcal/mol compared to −0.9 kcal/mol) strongly away from what was originally very good agreement. The choice of functional also had an impact in some cases, with differences of 9 and 10 kcal/mol for UF_4 and UCl_2, respectively, between B3LYP and PBEh. Armentrout and coworkers [14, 30, 193] have also used various DFT functionals in comparison to CCSD(T) using relativistic PPs in support of guided ion-beam mass spectrometry experiments. In calculations on $Th(CH_n)^+$ species ($n = 1–4$)[14], the average error with respect to experiment for all three functionals used (B3LYP, B3PW91 [125, 127], and BHandHLYP [174]) was about 8 kcal/mol compared to about 6.5 kcal/mol for their frozen-core CCSD(T) calculations with the cc-pVQZ-PP basis set [162] on Th (and cc-pVTZ on the light atoms). Individual variations among the different functionals were much larger, however, with BHandHLYP exhibiting the largest differences ranging up to nearly 24 kcal/mol for $ThCH^+$.

Of course, all of the above strategies can be used for other energy differences, such as ionization potentials and electron affinities. Table 1.3 shows FPD results for the IE of ThO [193], which is very accurately known from experiment, 152.2603(6) kcal/mol [199]. The agreement of the FPD composite approach with experiment is nearly perfect in this case. It is worth recognizing that the effect of SOC on the IE, +0.74 kcal/mol, is rather large considering that the ionization involves a removal of a 7s electron from the

Table 1.3 Summary of composite FPD calculations of the ionization energy of ThO (in kcal/mol).

CBS[TQ][a]	ΔCV[b]	ΔSO[c]	ΔQED[d]	ΔT[e]	ΔQ[f]	ΔZPE[g]	IE(0)	IE(expt)
150.92	+0.18	+0.74	−0.31	−0.02	+0.33	+0.08	152.29	152.2603

a CCSD(T)/CBS[TQ] within the frozen-core approximation (6s6p6d7s of Th, 2s2p of O correlated).
b CCSD(T)/CBS[wTQ] correlation effect of correlating the 5s5p5d electrons of Th and 1s of O.
c X2C, 2c-CCSD(T) with uncontracted cc-pVDZ-DK3/aug-cc-pVDZ basis sets.
d DKH3-CCSD(T)/cc-pwCVDZ-DK3/aug-cc-pwCVDZ-DK with a QED model potential.
e CCSDT − CCSD(T) in cc-pVTZ-DK3/aug-cc-pVTZ-DK basis sets using DKH3.
f CCSDTQ − CCSDT in cc-pVDZ-DK3/aug-cc-pVDZ-DK basis sets using DKH3.
g Using DKH3-CCSD(T)/cc-pVDZ-DK3/aug-cc-pVDZ-DK harmonic frequencies.

closed-shell ThO molecule. This is consistent, however, with the uranium isodesmic reaction enthalpies of Bross and Peterson [170], where SOC contributions were on the order of 1 kcal/mol even though all species were formally closed-shell singlets.

As in isodesmic reactions, the calculation of ionization potentials or electron affinities can often benefit from cancellation of errors since the two species involved often have very similar electronic structures. As an example, Gagliardi and coworkers [200] applied both the SO-CASPT2 and X2C-CCSD(T) methods with basis sets of just valence TZ and DZ quality, respectively, to obtain ionization potentials for both AnO and AnO_2 species (An = Th, U, Np, Pu, Am, Cm). Their results compared very well with the experimental values determined by Marçalo and Gibson [201] by FTICR/MS. In addition, Gagliardi and coworkers [202] also have reviewed the performance of several wavefunction-based methods with DFT on ionization potentials and bond energies of actinide monoxides and dioxides, albeit with ANO-RCC basis sets of just TZ quality. Surprisingly, their results demonstrated very similar accuracy between several DFT functionals with either CASPT2 or CCSD(T); however, the experimental uncertainties used for comparison were often relatively large, that is, 3–17 kcal/mol.

1.6 Ab Initio Molecular Spectroscopy of Gas-Phase Actinide Compounds

In principle, the accurate ab initio calculation of spectroscopic properties could follow the same procedure as described above for ab initio thermochemistry, except that now many configurations need to be sampled on the potential energy surface (PES) rather than just at the global minimum and dissociation asymptote(s). The need for many energy evaluations, however, can turn a computationally expensive thermochemistry calculation into something intractable for an accurate spectroscopy study. This is particularly true for situations where multireference methods are required, which is nearly always the case for excited electronic states. Particularly in these cases, sampling configurations away from the PES minimum necessarily involves lowering the symmetry, which can greatly increase the computational cost. Of course, in cases where single determinant-based methods are appropriate, analytical derivative techniques can sometimes be used to obtain geometries and harmonic frequencies. This strategy has been used most extensively at the DFT level of theory for ground state actinide chemistry.

1.6.1 Pure Rotational and Ro-Vibrational Spectroscopy

As an example of using the FPD composite scheme for spectroscopic properties, Table 1.4 shows the CCSD(T) spectroscopic constants for the $^1\Sigma^+$ ground electronic state of ThO. The values were obtained by fitting seven energies distributed about the equilibrium geometry to polynomials in simple displacement coordinates, and the resulting derivatives were subjected to a Dunham analysis. The frozen-core (FC) calculations involved cc-pVnZ-DK3 (Th) and aug-cc-pVnZ-DK (O) basis sets (n = D–Q, denoted VnZ below) with the DKH3 scalar relativistic Hamiltonian. The core correlation calculations (CV) involved additional correlation of the 5s5p5d electrons of Th and the 1s electrons of O using cc-pwCVnZ-DK3 and aug-cc-pwCVnZ-DK basis sets (n = D-Q, denoted wCnZ

Table 1.4 Calculated CCSD(T) spectroscopic constants of $X^1\Sigma^+$ ThO compared to experiment.

	R_e (Å)	ω_e (cm^{-1})	$\omega_e x_e$ (cm^{-1})
Valence correlation			
VDZ	1.8671	859.37	2.16
VTZ	1.8524	883.58	2.30
VQZ	1.8488	887.03	2.25
CBS[TQ][a]	1.8471	888.80	2.22
ΔCV[b]			
wCVDZ	+0.0001	+3.21	-0.01
wCVTZ	−0.0037	+6.01	+0.03
wCVQZ	−0.0053	+7.12	+0.01
CBS[wTQ]	−0.0062	+7.77	+0.00
ΔSOC	−0.0001	+3.75	+0.04
Final values	1.8399	901.2	2.26
Experiment	1.840186[c]	895.77[d]	2.39[d]

a The FC CBS limits using the wCVnZ basis sets were 1.8462 Å, 889.70 cm^{-1}, and 2.22 cm^{-1} for R_e, ω_e, and $\omega_e x_e$, respectively. These values were used in the final composite results.
b The wCVnZ basis sets were used in both frozen-core and core correlation calculations of the near-equilibrium potential energy function.
c Reference [70].
d Reference [16].

below). CBS extrapolations were carried out as outlined in the previous section. The SOC calculations utilized the two-component X2C Hamiltonian in relativistic CCSD(T) calculations with uncontracted cc-pVDZ-DK3 and aug-cc-pVDZ basis sets with spin-same-orbit two-electron contributions using the AMFI approximation.

As seen in Table 1.4, the final composite spectroscopic constants are in excellent agreement with experiment, with the bond length within 0.0003 Å and the harmonic frequency within ~5 cm^{-1}. Improvement of these final results would presumably require electron correlation beyond CCSD(T). The convergence with basis set is smooth and relatively rapid, with the VTZ results being with 0.005 Å and about 5 cm^{-1} from the FC CBS limits of R_e and ω_e. Correlation of the 5s5p5d electrons of Th is not an insignificant effect, although the bond length and harmonic frequency are lowered and raised, respectively, by just 0.006 and 7.8 cm^{-1}. As expected for this nominally closed-shell molecule, the effect of SOC is relatively minor.

Among actinide diatomic molecules, ThO is an anomaly in that its wavefunction is dominated by a single determinant. A more typical example is the UO molecule, which involves two close-lying electronic states, one correlating to U^{2+} in a $5f^3 7s$ configuration and another with U^{2+} in a $5f^2 7s^2$ configuration [18]. High-level spin-orbit MRCI calculations have been carried out by Pitzer and coworkers [151] on UO and UO^+ using TZ-quality basis sets and small-core relativistic PPs. These calculations aptly demonstrate how complicated such "simple" molecules can be. The ground state of UO was calculated to be an $\Omega = 4$ state (correlating to the $5f^3 7s$ configuration) consistent with experiment with a bond length agreeing to within about 0.01 Å, which is to be expected with a basis set of this size. Earlier, Hirao and coworkers [203] also computed the spectroscopic properties of UO and UO^+ but at the DKH2-CASPT2 level of theory with an ANO-RCC-QZP basis set [161]. They also determined the ground state to be $\Omega = 4$ with 71% attributable to the ^5I state. Their calculated bond length and harmonic frequency were in excellent agreement with experiment, which might be partially attributed to the larger basis set used.

Beyond the examples of diatomic molecules, very few anharmonic force fields have been calculated for actinide polyatomic molecules, although anharmonic vibrational anharmonicity constants at the CCSD(T) level have been previously reported for both UO_2^{2+} and ThO_2 [204]. In regard to calculating harmonic frequencies of actinide molecules, Odoh and Schreckenbach [205] have demonstrated the importance of using small-core PPs compared to large-core ones (the latter with 5d in the core) in DFT calculations of UO_3 and UF_6. In particular, the small-core Stuttgart PPs yielded B3LYP equilibrium geometries and harmonic frequencies in good agreement with full four-component B3LYP calculations. In regard to the performance of different functionals, at least in the case of UF_6, B3LYP and PBE0 calculations yielded harmonic frequencies in better agreement with experiment compared to PBE, BPBE [125, 128], or BLYP [125, 126]. In one of the earliest applications of CASPT2 to actinide molecules, Gagliardi and Roos [206] found that in the case of several U(V) and U(VI) triatomics, CASPT2 gave much better agreement with experiment compared to B3LYP when VTZ basis sets were used; that is, the largest relative error for CASPT2 was just 4% in a basis with g-type functions, while it reached 10% for B3LYP.

While not strictly gas phase, there has been a particularly fruitful interplay between theory and experiment in the area of matrix isolation infrared spectroscopy [8]. Recent calculations of structures and harmonic frequencies have mostly been carried out using the combination of B3LYP with the Stuttgart 60-electron PP of Küchle et al. [168] and the VQZ basis set from Cao and Dolg [169]. One very interesting example is the linear CUO molecule, which was first observed [207] in a Ne matrix with an assignment assisted by calculated harmonic frequencies and isotope shifts at the scalar relativistic DFT level using the PW91 functional with a TZ Slater-type-orbital (STO) basis set. The ground state was assigned at that point as $^1\Sigma^+$, on the basis of the close correspondence of the DFT frequencies from that electronic state with experiment. The story became more interesting, however, when experiments in an Ar matrix were carried out, and the vibrational spectrum was very different as compared to that in neon [208]. Gas-phase DFT calculations [208] indicated that a low-lying $^3\Phi$ state had vibrational frequencies very similar to those observed for CUO in an Ar matrix. Hence, it was proposed that the more polarizable Ar directly interacted with the CUO molecule, inverting the order of these two states and resulting in a $^3\Phi$ ground state. This seemed confirmed by further experiments in a matrix consisting of 1% argon in neon [209]. More extensive ab initio calculations by Roos et al. [210] at the DKH2-CASPT2/ANO-RCC-QZP level of theory, however, seemed to put this interpretation strongly in doubt. In particular, their work placed the $^3\Phi$ state below the $^1\Sigma^+$, and the inclusion of SOC, which had not been done yet to that point, further stabilized the $\Omega = 2$ component of the $^3\Phi$ compared to the $\Omega = 0$ arising from the $^1\Sigma^+$. Thus, the $^3\Phi_2$ state was predicted to be the ground state of CUO with an equilibrium excitation energy to the $^1\Sigma^+$ of over 12 kcal/mol, which could not explain the experimental observations in the neon matrix.

The following year, however, Infante and Visscher [211] used both relativistic (scalar and SO) ZORA, as well as full four-component (4-c) Dirac-Coulomb-CCSD(T) and FS-CCSD, to investigate the two lowest electronic states of UCO. Their scalar-only ZORA results were consistent with the original calculations of Andrews et al., yielding a $^1\Sigma^+$ ground state, but the inclusion of SOC reversed the order of states, which contradicts the interpretation of the experiments. However, at the 4-c coupled cluster level, either CCSD(T) or FS-CCSD, the $^1\Sigma_0^+$ state was computed to lie lower than the $^3\Phi_2$ by about 13.9 kcal/mol. This agreed well with experiment in that several heavy noble gas atoms were required to reverse the ordering of states. More recently, Yang et al. [212] carried out PP-based SO-MRCI calculations on CUO and obtained results consistent with the coupled cluster values, albeit with a smaller excitation energy of just 3.9 kcal/mol.

1.6.2 Electronic Spectroscopy

One particular area in the gas phase spectroscopy of actinide molecules that ab initio electronic structure methods can very effectively contribute to is in the interpretation and prediction of their complicated electronic spectra. In recent years, the workhorse in this area has been the CASPT2 method using the state interacting approach for SOC effects. This method and general strategy has been effective for interpreting/predicting both absorption and emission spectra, as well as negative ion photodetachment spectra. The strength of the CASPT2 method, as mentioned earlier, is that it is amenable to any spin state. This differs from the IH-FS-CCSD method, which can handle at most two electrons outside a closed-shell reference state. There have been a few studies that have directly compared CASPT2 to multireference CCSD, but these have been complicated by the use of different basis sets, relativistic Hamiltonians, or numbers of correlated electrons. In their study on UO_2 and UO_2^+, Infante et al. [143] carried out a detailed comparison of CASPT2 and IH-FS-CCSD for the related atomic ions U^{5+} and U^{4+} using identical basis sets and 4-c relativistic Hamiltonians. A total of 52 electrons were correlated for U^{4+}; that is, a [Kr] core was used. A summary of some of their error statistics is given in Table 1.5 for the U^{4+} ion for electronic transitions ranging from 4,000 to 100,000 cm^{-1}.

As discussed in Reference [143], one disadvantage of the IH-FS-CCSD approach for U^{4+} is the use of orbitals from the highly charged U^{6+} system. The majority of the orbital relaxation is recovered from the single excitation amplitudes, but this is incomplete

Table 1.5 Mean unsigned errors (in cm^{-1}) with respect to experiment for the U^{4+} ion from Reference [143] using the universal basis set of Malli et al. [213].

Transition type	Experimental energy range (cm^{-1})	CASPT2 (DC[a])	CASPT2 (DCB[b])	IH-FSCCSD (DC[a])	IH-FSCCSD (DCB[b])
$5f^2$	<44,000	814	825	514	357
5f6d	<68,000	3,467	6,024	2,824	2,110
5f7s	<103,000	657	3,610	3,548	2,680

a 4-c Dirac–Coulomb Hamiltonian
b 4-c Dirac–Coulomb–Breit Hamiltonian

without higher excitation operators or a more expanded active space. The effects of this are observed in Table 5, where the errors in the excitation energies for IH-FS-CCSD involving the more diffuse 6d and 7s orbitals are much larger than those involving the 5f alone. Upon comparing CASPT2 (with an active space involving the 5f, 6d, and 7s orbitals) to IH-FS-CCSD with the DC Hamiltonian, a very similar average accuracy is seen for the $5f^2$ transitions, but larger and smaller errors are found with CASPT2 for 5f6d and 5f7s transitions, respectively. Interestingly, the average errors of CASPT2 increase strongly in the cases of the 5f6d and 5f7s transitions (transition energies typically overestimated) when the more complete DCB Hamiltonian is employed, while this Hamiltonian improves the agreement of IH-FS-CCSD on average for these same transitions (transition energies generally underestimated). Hence, some significant cancellation of errors seem to occur in the CASPT2 case, similar to the example of U^+ given above in Table 1.1. It is not clear, however, how this comparison would change if only the valence electrons were correlated. These challenges for an atomic system, however, do help underscore the difficulties involved in reliably computing accurate electronic spectra of molecular actinide species.

1.7 Summary and Outlook

The interplay between gas-phase experimental studies and computational models has been steadily advancing our understanding of actinide bonding and reactivity. On the experimental side, a variety of laser spectroscopic and ion-beam/ion-trap mass spectrometric techniques are now yielding a wealth of incisive new data. In parallel, relativistic quantum chemistry methods are providing critical insights that cannot be obtained from the experimental data alone. For example, the permanent electric dipole of a molecule encodes information regarding the ground state wavefunction, but the nuanced physical significance can only be recovered through comparisons with theory. Consider ThS, where the experimental values for the dipole moment and bond length are 4.58(10) D and 2.3436(7) Å, respectively [82]. A high-level theoretical calculation (multireference configuration interaction) predicts that the ground state is an admixture of 56% $Th^{2+}(7s^2)S^{2-}(3s^23p^6)$ with 17% $Th^+(7s^26d)S^-(3s^23p^5)$ [85]. Clearly, the dipole moment will be sensitive to the relative weights of these configurations, and the predicted dipole moment (4.2 D) is in reasonable agreement with the experimental data. This correlation, along with the reliable prediction of both the bond length and vibrational frequency, confirms the validity of the calculations. It is then of interest to ask how well does a single-reference DFT calculation perform for this multi-configurational ground state? For ThS, the answer is that the B3LYP scheme does remarkably well [82], giving results of comparable quality to MRCI. Further study of ThS and related molecules can be used to explore the underlying reasons for the success, and the situations that might lead to failure of the less rigorous models. Turning to an example where the 5*f* electrons are of importance, experimental and theoretical studies of UF and UF^+ show that the $5f^3$ configuration on the metal ion retains its atomic character in the molecule/molecular ion, with minimal participation in the covalent bonding [31, 139]. Significant challenges still exist for theory, however, when multi-determinantal wavefunctions are present, particularly when SOC is included. Previously reliable results have been obtained using CASPT2, but new developments at both ends of the accuracy scale need further development, particularly for molecules containing transuranium elements. These future developments will need accurate gas-phase spectroscopic data in order to assess their ranges of applicability.

Thermodynamic properties derived from mass spectrometric methods are validating new composite theoretical models for bond dissociation energies, ionization energies, and heats of formation [170, 214]. Studies of gas-phase reactions indicate that actinide $f \rightarrow d$ electron promotion energies are important in determining reactivities [8]. The application of action spectroscopy techniques, in combination with theoretical modeling, is expanding our knowledge of ligand binding, coordination numbers, and the local environments of solvated actinide ions [10, 24, 42–44, 104, 105]. Trends are beginning to emerge from these studies, and the information needed for the development of truly predictive models is being assembled. For the immediate future, we anticipate that rigorous studies of isolated molecules will accelerate progress in our understanding of actinide chemistry, permitting a dramatic increase in the size and complexity of actinide systems that can be meaningfully explored by the methods of relativistic quantum chemistry.

Acknowledgments

Michael Heaven gratefully acknowledges support from the US Department of Energy (grant DE-FG02-01ER15153-12) for the actinide research carried out at Emory University. Kirk Peterson also gratefully acknowledges support from the US Department of Energy under grant DE-FG02-12ER16329.

References

1 Pyykkö, P. (2012) The physics behind chemistry and the periodic table. *Chemical Reviews (Washington, DC, U. S.)*, **112**, 371–384.
2 Pyykkö, P. (2012) Relativistic effects in chemistry: More common than you thought. *Annual Reviews of Physical Chemistry*, **63**, 45–64.
3 Schwerdtfeger, P. (2014) Relativity and chemical bonding. *Chemical Bond*, **1**, 383–404.
4 Pepper, M. and Bursten, B.E. (1991) The electronic structure of actinide-containing molecules: A challenge to applied quantum chemistry. *Chemical Reviews*, **91**, 719–741.
5 Dolg, M. and Cao, X. (2012) Relativistic pseudopotentials: Their development and scope of applications. *Chemical Reviews (Washington, DC, U. S.)*, **112**, 403–480.
6 Cao, X. and Dolg, M. (2011) Pseudopotentials and model potentials. *WIREs Comput. Mol. Sci.*, **1**, 200–210.
7 Heaven, M.C., Barker, B.J., and Antonov, I.O. (2014) Spectroscopy and structure of the simplest Actinide bonds. *Journal of Physical Chemistry A*, **118**, 10867–10881.
8 Heaven, M.C., Gibson, J.K., and Marçalo, J. (2010) Molecular spectroscopy and reactions of the actinides in the gas phase and cryogenic matrices, in *The Chemistry of the Actinide and Transactinide Elements*, L.R. Morss, N.M. Edelstein, and J. Fuger, Editors. Springer, Dordrecht, the Netherlands. pp. 4079–4156.
9 Heaven, M.C. (2006) Probing actinide electronic structure using fluorescence and multiphoton ionization spectroscopy. *Physical Chemistry Chemical Physics*, **8**, 4497–4509.

10 Gibson, J.K., Hu, H.-S., van Stipdonk, M.J., Berden, G., Oomens, J., and Li, J. (2015) Infrared multiphoton dissociation spectroscopy of a gas-phase complex of uranyl and 3-oxa-glutaramide: An extreme red-shift of the $[O=U=O]^{2+}$ asymmetric stretch. *Journal of Physical Chemistry A*, **119**, 3366–3374.
11 Li, W.-L., Su, J., Jian, T., Lopez, G.V., Hu, H.-S., Cao, G.-J., Li, J., and Wang, L.-S. (2014) Strong electron correlation in UO_2^-. A photoelectron spectroscopy and relativistic quantum chemistry study. *Journal of Chemical Physics*, **140**, 094306/1–094306/9.
12 Gibson, J.K., Haire, R.G., Marçalo, J., Santos, M., Leal, J.P., Pires de Matos, A., Tyagi, R., Mrozik, M.K., Pitzer, R.M., and Bursten, B.E. (2007) FTICR/MS studies of gas-phase actinide ion reactions: Fundamental chemical and physical properties of atomic and molecular actinide ions and neutrals. *European Physical Journal D*, **45**, 133–138.
13 Dau, P.D., Armentrout, P.B., Michelini, M.C., and Gibson, J.K. (2016) Activation of carbon dioxide by a terminal uranium-nitrogen bond in the gas-phase: A demonstration of the principle of microscopic reversibility. *Physical Chemistry Chemical Physics*, **18**, 7334–7340.
14 Cox, R.M., Armentrout, P.B., and de Jong, W.A. (2015) Activation of CH_4 by Th^+ as studied by guided ion beam mass spectrometry and quantum chemistry. *Inorganic Chemistry*, **54**, 3584–3599.
15 Armentrout, P.B. (2013) The power of accurate energetics (or thermochemistry: What is it good for?). *Journal of the American Society for Mass Spectrometry*, **24**, 173–185.
16 Edvinsson, G., Selin, L.E., and Aslund, N. (1965) Band spectrum of ThO. *Arkiv foer Fysik*, **30**, 283–319.
17 Gresh, D.N., Cossel, K.C., Zhou, Y., Ye, J., and Cornell, E.A. (2016) Broadband velocity modulation spectroscopy of ThF^+ for use in a measurement of the electron electric dipole moment. *Journal of Molecular Spectroscopy*, **319**, 1–9.
18 Kaledin, L.A., McCord, J.E., and Heaven, M.C. (1994) Laser spectroscopy of UO: Characterization and assignment of states in the 0- to 3-eV Range, with a comparison to the electronic structure of ThO. *Journal of Molecular Spectroscopy*, **164**, 27–65.
19 Holt, J., Neese, F.N., De Lucia, F.C., Medvedev, I., and Heaven, M.C. (2014) The submillimeter spectrum OF UO, in *69th International Symposium on Molecular Spectroscopy*. University of Illinois.
20 Kaledin, L.A., Kulikov, A.N., Kobylyanskii, A.I., Shenyavskaya, E.A., and Gurvich, L.V. (1987) Reciprocal arrangements of the groups of low-lying states of uranium monoxide molecules. *Zhurnal Fizicheskoi Khimii*, **61**, 1374–1376.
21 Kaledin, L.A., Shenyavskaya, E.A., and Gurvich, L.V. (1986) Electronic spectra of uranium monoxide molecules. *Zhurnal Fizicheskoi Khimii*, **60**, 1049–1050.
22 Cossel, K.C., Gresh, D.N., Sinclair, L.C., Coffey, T., Skripnikov, L.V., Petrov, A.N., Mosyagin, N.S., Titov, A.V., Field, R.W., Meyer, E.R., Cornell, E.A., and Ye, J. (2012) Broadband velocity modulation spectroscopy of HfF^+: Towards a measurement of the electron electric dipole moment. *Chemical Physics Letters*, **546**, 1–11.
23 Duncan, M.A. (2012) Invited review article: Laser vaporization cluster sources. *Review of Scientific Instruments*, **83**, 041101/1–041101/19.
24 Ricks, A.M., Gagliardi, L., and Duncan, M.A. (2011) Uranium oxo and superoxo cations revealed using infrared spectroscopy in the gas phase. *Journal of Physical Chemistry Letters*, **2**, 1662–1666.
25 Loh, S.K., Hales, D.A., Li, L., and Armentrout, P.B. (1989) Collision-induced dissociation of iron cluster ions (Fe_n^+) ($n=2-10$) with xenon: Ionic and neutral iron binding energies. *Journal of Chemical Physics*, **90**, 5466–5485.

26 Lau, K.H. and Hildenbrand, D.L. (1982) Thermochemical properties of the gaseous lower valent fluorides of uranium. *Journal of Chemical Physics*, **76**, 2646–2652.
27 Lau, K.H. and Hildenbrand, D.L. (1984) Thermochemical studies of the gaseous uranium chlorides. *Journal of Chemical Physics*, **80**, 1312–1317.
28 Lau, K.H. and Hildenbrand, D.L. (1990) High-temperature equilibrium studies of the gaseous thorium chlorides. *Journal of Chemical Physics*, **92**, 6124–6130.
29 Armentrout, P.B. and Beauchamp, J.L. (1980) Reactions of U^+ and UO^+ ions with diatomic oxygen, carbon monoxide, carbon dioxide, carbon oxide sulfide, carbon disulfide and water-d2. *Chemical Physics*, **50**, 27–36.
30 Cox, R.M., Armentrout, P.B., and de Jong, W.A. (2016) Reactions of $Th^+ + H_2$, D_2, and HD studied by guided ion beam tandem mass spectrometry and quantum chemical calculations. *Journal of Physical Chemistry B*, **120**, 1601–1614.
31 Antonov, I.O. and Heaven, M.C. (2013) Spectroscopic and theoretical investigations of UF and UF^+. *Journal of Physical Chemistry A*, **117**, 9684–9694.
32 Merritt, J.M., Kaledin, A.L., Bondybey, V.E., and Heaven, M.C. (2008) The ionization energy of Be_2, and spectroscopic characterization of the $(1)^3\Sigma^+{}_u$, $(2)^3\Pi_g$, and $(3)^3\Pi_g$ states. *Physical Chemistry Chemical Physics*, **10**, 4006–4013.
33 Wang, L.-S. (2015) Perspective: Electrospray photoelectron spectroscopy: From multiply-charged anions to ultracold anions. *Journal of Chemical Physics*, **143**, 040901/1–040901/14.
34 Van Stipdonk, M.J., O'Malley, C., Plaviak, A., Martin, D., Pestok, J., Mihm, P.A., Hanley, C.G., Corcovilos, T.A., Gibson, J.K., and Bythell, B.J. (2016) Dissociation of gas-phase, doubly-charged uranyl-acetone complexes by collisional activation and infrared photodissociation. *International Journal of Mass Spectrometry*, **396**, 22–34.
35 Maurice, R., Renault, E., Gong, Y., Rutkowski, P.X., and Gibson, J.K. (2015) Synthesis and structures of plutonyl nitrate complexes: Is plutonium heptavalent in $PuO_3(NO_3)^{2-}$? *Inorganic Chemistry*, **54**, 2367–2373.
36 Lucena, A.F., Lourenco, C., Michelini, M.C., Rutkowski, P.X., Carretas, J.M., Zorz, N., Berthon, L., Dias, A., Conceicao Oliveira, M., Gibson, J.K., and Marçalo, J. (2015) Synthesis and hydrolysis of gas-phase lanthanide and actinide oxide nitrate complexes: A correspondence to trivalent metal ion redox potentials and ionization energies. *Physical Chemistry Chemical Physics*, **17**, 9942–9950.
37 Lucena, A.F., Carretas, J.M., Marçalo, J., Michelini, M.C., Gong, Y., and Gibson, J.K. (2015) Gas-phase reactions of molecular oxygen with uranyl(V) anionic complexes-synthesis and characterization of new superoxides of uranyl(VI). *Journal of Physical Chemistry A*, **119**, 3628–3635.
38 Gong, Y., de Jong, W.A., and Gibson, J.K. (2015) Gas phase uranyl activation: Formation of a uranium nitrosyl complex from uranyl azide. *Journal of the American Chemical Society*, **137**, 5911–5915.
39 Dau, P.D., Wilson, R.E., and Gibson, J.K. (2015) Elucidating protactinium hydrolysis: The relative stabilities of $PaO_2(H_2O)^+$ and $PaO(OH)_2{}^+$. *Inorganic Chemistry*, **54**, 7474–7480.
40 Dau, P.D. and Gibson, J.K. (2015) Halide abstraction from halogenated acetate ligands by actinyls: A competition between bond breaking and bond making. *Journal of Physical Chemistry A*, **119**, 3218–3224.
41 Dau, P.D., Carretas, J.M., Marçalo, J., Lukens, W.W., and Gibson, J.K. (2015) Oxidation of actinyl(V) complexes by the addition of nitrogen dioxide is revealed via the replacement of acetate by nitrite. *Inorganic Chemistry*, **54**, 8755–8760.

42 Ricks, A.M., Reed, Z.E., and Duncan, M.A. (2011) Infrared spectroscopy of mass-selected metal carbonyl cations. *Journal of Molecular Spectroscopy*, **266**, 63–74.

43 Ricks, A.M., Gagliardi, L., and Duncan, M.A. (2010) Infrared spectroscopy of extreme coordination: The carbonyls of U^+ and UO_2^+. *Journal of the American Chemical Society*, **132**, 15905–15907.

44 Groenewold, G.S., van Stipdonk, M.J., Oomens, J., de Jong, W.A., Gresham, G.L., and McIlwain, M.E. (2010) Vibrational spectra of discrete UO_2^{2+} halide complexes in the gas phase. *International Journal of Mass Spectrometry*, **297**, 67–75.

45 Groenewold, G.S., Oomens, J., de Jong, W.A., Gresham, G.L., McIlwain, M.E., and van Stipdonk, M.J. (2008) Vibrational spectroscopy of anionic nitrate complexes of UO_2^{2+} and Eu^{3+} isolated in the gas phase. *Physical Chemistry Chemical Physics*, **10**, 1192–1202.

46 Groenewold Gary, S., van Stipdonk Michael, J., de Jong Wibe, A., Oomens, J., Gresham Garold, L., McIlwain Michael, E., Gao, D., Siboulet, B., Visscher, L., Kullman, M., and Polfer, N. (2008) Infrared spectroscopy of dioxouranium(V) complexes with solvent molecules: Effect of reduction. *ChemPhysChem*, **9**, 1278–1285.

47 Groenewold Gary, S., Gianotto Anita, K., McIlwain Michael, E., Stipdonk Michael, J.V., Kullman, M., Moore David, T., Polfer, N., Oomens, J., Infante, I., Visscher, L., Siboulet, B., and Jong Wibe, A.D. (2008) Infrared spectroscopy of discrete uranyl anion complexes. *Journal of Physical Chemistry B*, **112**, 508–521.

48 Groenewold, G.S., Gianotto, A.K., Cossel, K.C., van Stipdonk, M.J., Moore, D.T., Polfer, N., Oomens, J., de Jong, W.A., and Visscher, L. (2006) Vibrational spectroscopy of mass-selected $[UO_2(\text{ligand})_n]^{2+}$ complexes in the gas phase: Comparison with theory. *Journal of the American Chemical Society*, **128**, 4802–4813.

49 Gasner, E.L. and Frlec, B. (1968) Raman spectrum of neptunium hexafluoride. *Journal of Chemical Physics*, **49**, 5135–5137.

50 Beitz, J.V., Williams, C.W., and Carnall, W.T. (1982) Fluorescence studies of neptunium and plutonium hexafluoride vapors. *Journal of Chemical Physics*, **76**, 2756–2757.

51 Person, W.B., Kim, K.C., Campbell, G.M., and Dewey, H.J. (1986) Absolute intensities of infrared-active fundamentals and combination bands of gaseous plutonium hexafluoride and neptunium hexafluoride. *Journal of Chemical Physics*, **85**, 5524–5528.

52 Kim, K.C. and Mulford, R.N. (1989) The combination bands ($\nu_1 + \nu_3$, $\nu_2 + \nu_3$) and overtone band ($3\nu_3$) of neptunium hexafluoride. *Chemical Physics Letters*, **159**, 327–330.

53 Kim, K.C. and Mulford, R.N. (1990) Vibrational properties of actinide (uranium, neptunium, plutonium, americium) hexafluoride molecules. *Journal of Molecular Structure THEOCHEM*, **66**, 293–299.

54 Mulford, R.N., Dewey, H.J., and Barefield, J.E., II, (1991) Fluorescence and absorption spectroscopy of the near-infrared vibronic transitions in matrix-isolated neptunium fluoride (NpF_6). *Journal of Chemical Physics*, **94**, 4790–4796.

55 Mulford, R.N. and Kim, K.C. (1996) Measurement and analysis of the Fourier transform spectra of the ν_3 fundamental and $\nu_1 + \nu_3$ combination of NpF_6. *Journal of Molecular Spectroscopy*, **176**, 369–374.

56 Aldridge, J.P., Brock, E.G., Filip, H., Flicker, H., Fox, K., Galbraith, H.W., Holland, R.F., Kim, K.C., Krohn, B.J., and et al. (1985) Measurement and analysis of the

infrared-active stretching fundamental (ν_3) of uranium hexafluoride. *Journal of Chemical Physics*, **83**, 34–48.
57 McDowell, R.S., Asprey, L.B., and Paine, R.T. (1974) Vibrational spectrum and force field of uranium hexafluoride. *Journal of Chemical Physics*, **61**, 3571–3580.
58 McDowell, R.S., Reisfeld, M.J., Nereson, N.G., Krohn, B.J., and Patterson, C.W. (1985) The $\nu_1 + \nu_3$ combination band of uranium-238 hexafluoride. *Journal of Molecular Spectroscopy*, **113**, 243–249.
59 Edvinsson, G., Bornstedt, A.V., and Nylen, P. (1968) Rotational analysis for a perturbed $^1\Pi$ state in thorium oxide. *Arkiv foer Fysik*, **38**, 193–218.
60 Edvinsson, G. and Jonsson, J. (1991) *On the K-X and M-X Band Systems of Thorium Monoxide (ThO)*. Dep. Phys.,Univ. Stockholm, Stockholm, Sweden. 30 pp.
61 Edvinsson, G. and Lagerqvist, A. (1984) Rotational analysis of yellow and near infrared bands in thorium(II) oxide. *Physica Scripta*, **30**, 309–320.
62 Edvinsson, G. and Lagerqvist, A. (1985) Rotational analysis of two red band systems in the thorium oxide (ThO) spectrum. *Physica Scripta*, **32**, 602–610.
63 Edvinsson, G. and Lagerqvist, A. (1985) A low-lying $\Omega = 2$ state in the thorium monoxide (ThO) molecule. *Journal of Molecular Spectroscopy*, **113**, 93–104.
64 Edvinsson, G. and Lagerqvist, A. (1987) Rotational analysis of some violet and green bands in the thorium oxide (ThO) spectrum. *Journal of Molecular Spectroscopy*, **122**, 428–439.
65 Edvinsson, G. and Lagerqvist, A. (1988) Two band systems of thorium oxide (ThO) in the near ultraviolet. *Journal of Molecular Spectroscopy*, **128**, 117–125.
66 Edvinsson, G. and Lagerqvist, A. (1990) Two new band systems in thorium oxide (ThO). *Physica Scripta*, **41**, 316–320.
67 Edvinsson, G. and Selin, L.E. (1964) Band spectrum of thorium oxide. *Physics Letters*, **9**, 238–239.
68 Dewberry, C.T., Etchison, K.C., and Cooke, S.A. (2007) The pure rotational spectrum of the actinide-containing compound thorium monoxide. *Physical Chemistry Chemical Physics*, **9**, 4895–4897.
69 Dewberry, C.T., Etchison, K.C., Grubbs, G.S., Powoski, R.A., Serafin, M.M., Peebles, S.A., and Cooke, S.A. (2007) Oxygen-17 hyperfine structures in the pure rotational spectra of SrO, SnO, BaO, HfO and ThO. *Physical Chemistry Chemical Physics*, **9**, 5897–5901.
70 Long, B.E., Novick, S.E., and Cooke, S.A. (2014) Measurement of the $J = 1\text{-}0$ pure rotational transition in excited vibrational states of $X^1\Sigma$ thorium (II) oxide, ThO. *Journal of Molecular Spectroscopy*, **302**, 1–2.
71 Allen, G.C., Baerends, E.J., Vernooijs, P., Dyke, J.M., Ellis, A.M., Feher, M., and Morris, A. (1988) High-temperature photoelectron spectroscopy: A study of uranium and uranium oxides (UO and UO_2). *Journal of Chemical Physics*, **89**, 5363–5372.
72 Maartensson, N., Malmquist, P.A., Svensson, S., and Johansson, B. (1984) The electron spectrum of uranium hexafluoride recorded in the gas phase. *Journal of Chemical Physics*, **80**, 5458–5464.
73 Beeching, L.J., Dyke, J.M., Morris, A., and Ogden, J.S. (2001) Study of the electronic structure of the actinide tetrabromides $ThBr_4$ and UBr_4 using ultraviolet photoelectron spectroscopy and density functional calculations. *Journal of Chemical Physics*, **114**, 9832–9839.

74 Boerrigter, P.M., Snijders, J.G., and Dyke, J.M. (1988) A reassignment of the gas-phase photoelectron spectra of the actinide tetrahalides uranium tetrafluoride, uranium tetrachloride, thorium tetrafluoride and thorium tetrachloride by relativistic Hartree-Fock-Slater calculations. *Journal of Electron Spectroscopy and Related Phenomena*, **46**, 43–53.

75 Dyke, J.M., Fayad, N.K., Morris, A., Trickle, I.R., and Allen, G.C. (1980) A study of the electronic structure of the actinide tetrahalides uranium tetrafluoride, thorium tetrafluoride, uranium tetrachloride, and thorium tetrachloride using vacuum ultraviolet photoelectron spectroscopy and SCF-Xα scattered wave calculations. *Journal of Chemical Physics*, **72**, 3822–3827.

76 Gagliardi, L., Skylaris, C.-K., Willetts, A., Dyke, J.M., and Barone, V. (2000) A density functional study of thorium tetrahalides. *Physical Chemistry Chemical Physics*, **2**, 3111–3114.

77 Sinclair, L.C., Cossel, K.C., Coffey, T., Ye, J., and Cornell, E.A. (2011) Frequency comb velocity-modulation spectroscopy. *Physical Review Letters*, **107**, 093002/1–093002/4.

78 Stephenson Serena, K. and Saykally Richard, J. (2005) Velocity modulation spectroscopy of ions. *Chemical Reviews*, **105**, 3220–3234.

79 Demtröder, W. (2008) *Laser Spectroscopy: Vol. 2: Experimental Techniques*. 4 ed. Vol. 2. Berlin: Springer-Verlag.

80 Stowe, M.C., Thorpe, M.J., Pe'er, A., Ye, J., Stalnaker, J.E., Gerginov, V., and Diddams, S.A. (2008) Direct frequency comb spectroscopy. *Advances in Atomic, Molecular, and Optical Physics*, **55**, 1–60.

81 Wang, F., Le, A., Steimle, T.C., and Heaven, M.C. (2011) Communication: The permanent electric dipole moment of thorium monoxide, ThO. *Journal of Chemical Physics*, **134**, 031102/1–031102/3.

82 Le, A., Heaven, M.C., and Steimle, T.C. (2014) The permanent electric dipole moment of thorium sulfide, ThS. *Journal of Chemical Physics*, **140**, 024307/1–024307/5.

83 Heaven, M.C., Goncharov, V., Steimle, T.C., Ma, T., and Linton, C. (2006) The permanent electric dipole moments and magnetic g factors of uranium monoxide. *Journal of Chemical Physics*, **125**, 204314/1–204314/11.

84 Linton, C., Adam, A.G., and Steimle, T.C. (2014) Stark and Zeeman effect in the [18.6]3.5 – X(1)4.5 transition of uranium monofluoride, UF. *Journal of Chemical Physics*, **140**, 214305/1–214305/7.

85 Bartlett, J.H., Antonov, I.O., and Heaven, M.C. (2013) Spectroscopic and Theoretical Investigations of ThS and ThS^+. *Journal of Physical Chemistry A*, **117**, 12042–12048.

86 Barker, B.J., Antonov, I.O., Heaven, M.C., and Peterson, K.A. (2012) Spectroscopic investigations of ThF and ThF^+. *Journal of Chemical Physics*, **136**, 104305/1–104305/9.

87 Steimle, T., Kokkin, D.L., Muscarella, S., and Ma, T. (2015) Detection of thorium dimer via two-dimensional fluorescence spectroscopy. *Journal of Physical Chemistry A*, **119**, 9281–9285.

88 Kokkin, D.L., Steimle, T.C., and DeMille, D. (2014) Branching ratios and radiative lifetimes of the U, L, and I states of thorium oxide. *Physical Review A Atomic, Molecular, and Optical Physics*, **90**, 062503/1–062503/10.

89 Steimle, T.C., Zhang, R., and Heaven, M.C. (2015) The pure rotational spectrum of thorium monosulfide, ThS. *Chemical Physics Letters*, **639**, 304–306.

90 Schlag, E.W. (1998) *ZEKE Spectroscopy*. 256 pp.

91 Linton, C., Simard, B., Loock, H.P., Wallin, S., Rothschopf, G.K., Gunion, R.F., Morse, M.D., and Armentrout, P.B. (1999) Rydberg and pulsed field ionization-zero electron kinetic energy spectra of YO. *Journal of Chemical Physics*, **111**, 5017–5026.

92 Yang, D.-S. (2011) High-resolution electron spectroscopy of gas-phase metal-aromatic complexes. *Journal of Physical Chemistry Letters*, **2**, 25–33.

93 Goncharov, V., Kaledin, L.A., and Heaven, M.C. (2006) Probing the electronic structure of UO^+ with high-resolution photoelectron spectroscopy. *Journal of Chemical Physics*, **125**, 133202/1–133202/8.

94 Neumark, D.M. (2008) Slow electron velocity-map imaging of negative ions: Applications to spectroscopy and dynamics. *Journal of Physical Chemistry A*, **112**, 13287–13301.

95 Leon, I., Yang, Z., Liu, H.-T., and Wang, L.-S. (2014) The design and construction of a high-resolution velocity-map imaging apparatus for photoelectron spectroscopy studies of size-selected clusters. *Review of Scientific Instruments*, **85**, 083106/1–083106/12.

96 Kim, J.B., Weichman, M.L., and Neumark, D.M. (2015) Low-lying states of FeO and FeO^- by slow photoelectron spectroscopy. *Molecular Physics*, Ahead of Print.

97 Su, J., Li, W.-L., Lopez, G.V., Jian, T., Cao, G.-J., Li, W.-L., Schwarz, W.H.E., Wang, L.-S., and Li, J. (2016) Probing the electronic structure and chemical bonding of mono-uranium oxides with different oxidation states: UO_x^- and UO_x (x = 3–5). *Journal of Physical Chemistry A*, **120**, 1084–1096.

98 Czekner, J., Lopez, G.V., and Wang, L.-S. (2014) High resolution photoelectron imaging of UO^- and UO_2^- and the low-lying electronic states and vibrational frequencies of UO and UO_2. *Journal of Chemical Physics*, **141**, 244302/1–244302/8.

99 Li, W.-L., Hu, H.-S., Jian, T., Lopez, G.V., Su, J., Li, J., and Wang, L.-S. (2013) Probing the electronic structures of low oxidation-state uranium fluoride molecules UF_x^- (x = 2–4). *Journal of Chemical Physics*, **139**, 244303/1–244303/8.

100 Roy, S.K., Jian, T., Lopez, G.V., Li, W.-L., Su, J., Bross, D.H., Peterson, K.A., Wang, L.-S., and Li, J. (2016) A combined photoelectron spectroscopy and relativistic ab initio studies of the electronic structures of UFO^- and UFO. *Journal of Chemical Physics*, **144**, 084309/1–084309/11.

101 Dau, P.D., Liu, H.-T., Huang, D.-L., and Wang, L.-S. (2012) Note: Photoelectron spectroscopy of cold UF_5^-. *Journal of Chemical Physics*, **137**, 116101/1–116101/2.

102 Su, J., Dau, P.D., Liu, H.-T., Huang, D.-L., Wei, F., Schwarz, W.H.E., Li, J., and Wang, L.-S. (2015) Photoelectron spectroscopy and theoretical studies of gaseous uranium hexachlorides in different oxidation states: UCl_6^{q-} (q = 0–2). *Journal of Chemical Physics*, **142**, 134308/1–134308/13.

103 Su, J., Dau, P.D., Xu, C.-F., Huang, D.-L., Liu, H.-T., Wei, F., Wang, L.-S., and Li, J. (2013) A joint photoelectron spectroscopy and theoretical study on the electronic structure of UCl_5^- and UCl_5. *Chemistry – Asian Journal*, **8**, 2489–2496.

104 Groenewold, G.S., van Stipdonk, M.J., Oomens, J., de Jong, W.A., and McIlwain, M.E. (2011) The gas-phase bis-uranyl nitrate complex $[(UO_2)_2(NO_3)_5]^-$: Infrared spectrum and structure. *International Journal of Mass Spectrometry*, **308**, 175–180.

105 Groenewold, G.S., de Jong, W.A., Oomens, J., and van Stipdonk, M.J. (2010) Variable denticity in carboxylate binding to the uranyl coordination complexes. *Journal of the American Society for Mass Spectrometry*, **21**, 719–727.

106 Pillai, E.D., Molek, K.S., and Duncan, M.A. (2005) Growth and photodissociation of $U^+(C_6H_6)_n$ (n = 1–3) and $UO_m^+(C_6H_6)$ (m = 1,2) complexes. *Chemical Physics Letters*, **405**, 247–251.

107 Rauh, E.G. and Ackermann, R.J. (1975) First ionization potentials of neptunium and neptunium monoxide. *Journal of Chemical Physics*, **62**, 1584.

108 Rauh, E.G. and Ackermann, R.J. (1974) First ionization potentials of some refractory oxide vapors. *Journal of Chemical Physics*, **60**, 1396–1400.

109 Ackermann, R.J. and Rauh, E.G. (1973) Preparation and characterization of the metastable monoxides of thorium and uranium. *Journal of Inorganic and Nuclear Chemistry*, **35**, 3787–3794.

110 Hildenbrand, D.L. (1977) Thermochemistry of gaseous uranium pentafluoride and uranium tetrafluoride. *Journal of Chemical Physics*, **66**, 4788–4794.

111 Hildenbrand, D.L. (1988) Equilibrium measurements as a source of entropies and molecular constant information. *Pure and Applied Chemistry*, **60**, 303–307.

112 Hildenbrand, D.L. and Lau, K.H. (1991) Redetermination of the thermochemistry of gaseous uranium fluorides (UF_5, UF_2, and UF). *Journal of Chemical Physics*, **94**, 1420–1425.

113 Kleinschmidt, P.D. and Hildenbrand, D.L. (1979) Thermodynamics of the dimerization of gaseous uranium(V) fluoride. *Journal of Chemical Physics*, **71**, 196–201.

114 Lau, K.H. and Hildenbrand, D.L. (1987) Thermochemistry of the gaseous uranium bromides UBr through UBr_5. *Journal of Chemical Physics*, **86**, 2949–2954.

115 Kovacs, A., Konings, R.J.M., Gibson, J.K., Infante, I., and Gagliardi, L. (2015) Quantum chemical calculations and experimental investigations of molecular actinide oxides. *Chemical Reviews (Washington, DC, U. S.)*, **115**, 1725–1759.

116 Armentrout, P.B. and Beauchamp, J.L. (1980) Collision-induced dissociation of UO^+ and UO_2^+ ions. *Chemical Physics*, **50**, 21–25.

117 Gibson, J.K. and Marçalo, J. (2006) New developments in gas-phase actinide ion chemistry. *Coordination Chemistry Reviews*, **250**, 776–783.

118 Gibson, J.K. (1997) Gas-phase f-element organometallic chemistry: Reactions of cyclic hydrocarbons with Th^+, U^+, ThO^+, UO^+, and lanthanide ions, Ln^+. *Organometallics*, **16**, 4214–4222.

119 Gibson, J.K. (2002) Gas-phase chemistry of actinide ions: Probing the distinctive character of the 5f elements. *International Journal of Mass Spectrometry*, **214**, 1–21.

120 Van Stipdonk, M.J., Michelini, M.d.C., Plaviak, A., Martin, D., and Gibson, J.K. (2014) Formation of bare UO_2^{2+} and NUO^+ by fragmentation of gas-phase uranyl-acetonitrile complexes. *Journal of Physical Chemistry A*, **118**, 7838–7846.

121 Lucena, A.F., Odoh, S.O., Zhao, J., Marçalo, J., Schreckenbach, G., and Gibson, J.K. (2014) Oxo-exchange of gas-phase uranyl, neptunyl, and plutonyl with water and methanol. *Inorganic Chemistry*, **53**, 2163–2170.

122 Gong, Y. and Gibson, J.K. (2014) Crown ether complexes of uranyl, neptunyl, and plutonyl: Hydration differentiates inclusion versus outer coordination. *Inorganic Chemistry*, **53**, 5839–5844.

123 Gong, Y., Hu, H.-S., Rao, L., Li, J., and Gibson, J.K. (2013) Experimental and theoretical studies on the fragmentation of gas-phase uranyl-, neptunyl-, and plutonyl-diglycolamide complexes. *Journal of Physical Chemistry A*, **117**, 10544–10550.

124 Vasiliu, M., Peterson, K.A., Gibson, J.K., and Dixon, D.A. (2015) Reliable potential energy surfaces for the reactions of H_2O with ThO_2, PaO_2^+, UO_2^{2+}, and UO_2^+. *Journal of Physical Chemistry A*, **119**, 11422–11431.

125 Becke, A.D. (1993) Density-functional thermochemistry. III. The role of exact exchange. *Journal of Chemical Physics*, **98**, 5648.

126 Lee, C., Yang, W., and Parr, R.G. (1988) Development of the Colle-Salvetti correlation-energy formula into a functional of the electron density. *Physical Review B*, **37**, 785.

127 Perdew, J.P. and Wang, Y. (1992) Accurate and simple analytic representation of the electron gas correlation energy. *Physical Review B*, **45**, 13244–13249.

128 Perdew, J.P., Burke, K., and Ernzerhof, M. (1996) Generalized gradient approximation made simple. *Physical Review Letters*, **77**, 3865–3868.

129 Raghavachari, K., Trucks, G.W., Pople, J.A., and Head-Gordon, M. (1989) A fifth-order perturbation comparison of electron correlation theories. *Chemical Physics Letters*, **157**, 479–483.

130 Roos, B., Taylor, P., and Siegbahn, P.E.M. (1980) A complete active space SCF method (CASSCF) using a density matrix formulated super-CI approach. *Chemical Physics*, **48**, 157–173.

131 Andersson, K., Malmqvist, P.-A., and Roos, B.O. (1992) Second-order perturbation theory with a complete active space self-consistent field reference function. *Journal of Chemical Physics*, **96**, 1218.

132 Andersson, K., Malmqvist, P.A., Roos, B.O., Sadlej, A.J., and Wolinski, K. (1990) Second-order perturbation-theory with a CASSCF reference function. *Journal of Physical Chemistry*, **94**, 5483–5488.

133 Kovacs, A., Infante, I., and Gagliardi, L. (2013) Theoretic study of the electronic spectra of neutral and cationic PaO and PaO_2. *Structural Chemistry*, **24**, 917–925.

134 Roos, B.O., Malmqvist, P.-A., and Gagliardi, L. (2006) Exploring the actinide-actinide bond: Theoretical studies of the chemical bond in Ac_2, Th_2, Pa_2, and U_2. *Journal of the American Chemical Society*, **128**, 17000–17006.

135 Olsen, J., Roos, B.O., Jørgensen, P., and Jensen, H.J.A. (1988) Determinant based configuration interaction algorithms for complete and restricted configuration interaction spaces. *Journal of Chemical Physics*, **89**, 2185.

136 Fleig, T., Olsen, J., and Marian, C.M. (2001) The generalized active space concept for the relativistic treatment of electron correlation. I. Kramers-restricted two-component configuration interaction. *Journal of Chemical Physics*, **114**, 4775–4790.

137 Werner, H.-J. and Knowles, P.J. (1988) An efficient internally contracted multiconfiguration-reference configuration interaction method. *Journal of Chemical Physics*, **89**, 5803–5814.

138 Langhoff, S.R. and Davidson, E.R. (1974) Configuration interaction calculations on the nitrogen molecule. *International Journal of Quantum Chemistry*, **8**, 61–72.

139 Bross, D.H. and Peterson, K.A. (2015) Theoretical spectroscopy study of the low-lying electronic states of UX and UX^+, X = F and Cl. *Journal of Chemical Physics*, **143**, 184313.

140 Dolg, M. and Cao, X. (2009) Accurate relativistic small-core pseudopotentials for actinides. Energy adjustment for uranium and first applications to uranium hydride†. *Journal of Physical Chemistry A*, **113**, 12573–12581.

141 Visscher, L., Eliav, E., and Kaldor, U. (2001) Formulation and implementation of the relativistic Fock-space coupled cluster method for molecules. *Journal of Chemical Physics*, **115**, 9720–9726.
142 Landau, A., Eliav, E., and Kaldor, U. (1999) Intermediate Hamiltonian Fock-space coupled-cluster method. *Chemical Physics Letters*, **313**, 399–403.
143 Infante, I., Eliav, E., Vilkas, M.J., Ishikawa, Y., Kaldor, U., and Visscher, L. (2007) A Fock space coupled cluster study on the electronic structure of the UO_2, UO_2^+, U^{4+}, and U^{5+} species. *Journal of Chemical Physics*, **127**, 124308.
144 Jansen, G. and Hess, B.A. (1989) Revision of the Douglas-Kroll transformation. *Physical Review A*, **39**, 6016.
145 Wolf, A., Reiher, M., and Hess, B.A. (2002) The generalized Douglas-Kroll transformation. *Journal of Chemical Physics*, **117**, 9215–9226.
146 Iliaš, M. and Saue, T. (2007) An infinite-order two-component relativistic Hamiltonian by a simple one-step transformation. *Journal of Chemical Physics*, **126**, 064102.
147 Malmqvist, P.A., Roos, B.O., and Schimmelpfennig, B. (2002) The restricted active space (RAS) state interaction approach with spin-orbit coupling. *Chemical Physics Letters*, **357**, 230–240.
148 Hess, B.A., Marian, C.M., and Peyerimhoff, S.D. (1995) Ab initio calculation of spin-oribit effects in molecules including electron correlation, in *Modern Electronic Structure Theory, Part I*, D.R. Yarkony, Editor. World Scientific, Singapore. pp. 152–278.
149 Hess, B.A., Marian, C.M., Wahlgren, U., and Gropen, O. (1996) A mean-field spin-orbit method applicable to correlated wavefunctions. *Chemical Physics Letters*, **251**, 365–371.
150 Yabushita, S., Zhang, Z., and Pitzer, R.M. (1999) Spin-orbit configuration interaction using the graphical unitary group approach and relativistic core potential and spin-orbit operators. *Journal of Physical Chemistry A*, **103**, 5791–5800.
151 Tyagi, R., Zhang, Z., and Pitzer, R.M. (2014) Electronic spectrum of the UO and UO^+ molecules. *Journal of Physical Chemistry A*, **118**, 11758–11767.
152 Saue, T. (2011) Relativistic Hamiltonians for chemistry: A primer. *ChemPhysChem*, **12**, 3077–3094.
153 Knecht, S., Jensen, H.J.A., and Fleig, T. (2010) Large-scale parallel configuration interaction. II. Two- and four-component double-group general active space implementation with application to BiH. *Journal of Chemical Physics*, **132**, 014108.
154 Visscher, L., Lee, T.J., and Dyall, K.G. (1996) Formulation and implementation of a relativistic unrestricted coupled-cluster method including noniterative connected triples. *Journal of Chemical Physics*, **105**, 8769–8776.
155 van Lenthe, E., Baerends, E.J., and Snijders, J.G. (1993) Relativistic regular two-component Hamiltonians. *Journal of Chemical Physics*, **99**, 4597.
156 Sikkema, J., Visscher, L., Saue, T., and Iliaš, M. (2009) The molecular mean-field approach for correlated relativistic calculations. *Journal of Chemical Physics*, **131**, 124116–.
157 Weigand, A., Cao, X., Hangele, T., and Dolg, M. (2014) Relativistic small-core pseudopotentials for actinium, thorium, and protactinium. *Journal of Physical Chemistry A*, **118**, 2519–2530.

158 Peterson, K.A. and Dyall, K.G. (2015) Gaussian Basis sets for lanthanide and actinide elements: Strategies for their development and use, in *Computational Methods in Lanthanide and Actinide Chemistry*, M. Dolg, Editor. John Wiley & Sons, Chichester, West Sussex, UK. pp. 195–216.

159 Velde, G., Bickelhaupt, F., Baerends, E.J., Guerra, C., van Gisbergen, S., Snijders, J., and Ziegler, T. (2001) Chemistry with ADF. *Journal of Computational Chemistry*, **22**, 931–967.

160 Dyall, K.G. (2007) Relativistic double-zeta, triple-zeta, and quadruple-zeta basis sets for the actinides Ac–Lr. *Theoretical Chemistry Accounts*, **117**, 491–500.

161 Roos, B.O., Lindh, R., Malmqvist, P.A., Veryazov, V., and Widmark, P.O. (2005) New relativistic ANO basis sets for actinide atoms. *Chemical Physics Letters*, **409**, 295–299.

162 Peterson, K.A. (2015) Correlation consistent basis sets for actinides. I. The Th and U atoms. *Journal of Chemical Physics*, **142**, 074105.

163 Feng, R. and Peterson, K.A. (2017) Correlation consistent basis sets for actinides. II. The atoms Ac and Np-Lr. *Journal of Chemical Physics*, **147**, 084108.

164 Feller, D., Peterson, K.A., and Hill, J.G. (2011) On the effectiveness of CCSD(T) complete basis set extrapolations for atomization energies. *Journal of Chemical Physics*, **135**, 044102.

165 Pantazis, D.A. and Neese, F. (2011) All-electron scalar relativistic basis sets for the actinides. *Journal of Chemical Theory and Computation*, **7**, 677–684.

166 Blaudeau, J.-P., Brozell, S.R., Matsika, S., Zhang, Z., and Pitzer, R.M. (2000) Atomic orbital basis sets for use with effective core potentials. *International Journal of Quantum Chemistry*, **77**, 516–520.

167 Christiansen, P.A., Lee, Y.S., and Pitzer, K.S. (1979) Improved ab initio effective core potentials for molecular calculations. *Journal of Chemical Physics*, **71**, 4445–4450.

168 Küchle, W., Dolg, M., Stoll, H., and Preuss, H. (1994) Energy-adjusted pseudopotentials for the actinides – parameter sets and test calculations for thorium and thorium monoxide. *Journal of Chemical Physics*, **100**, 7535–7542.

169 Cao, X. and Dolg, M. (2004) Segmented contraction scheme for small-core actinide pseudopotential basis sets. *Journal of Molecular Structure: THEOCHEM*, **673**, 203–209.

170 Bross, D.H. and Peterson, K.A. (2014) Composite thermochemistry of gas phase U(VI)-containing molecules. *Journal of Chemical Physics*, **141**, 244308.

171 Lindh, R., Molcas 8.0: F. Aquilante, J. Autschbach, R. K. Carlson, L. F. Chibotaru, M. G. Delcey, L. De Vico, I. Fdez. Galvan, N. Ferré, L. M. Frutos, L. Gagliardi, M. Garavelli, A. Giussani, C. E. Hoyer, G. Li Manni, H. Lischka, D. Ma, P Å Malmqvist, T. Müller, A. Nenov, M. Olivucci, T. B. Pedersen, D. Peng, F. Plasser, B. Pritchard, M. Reiher, I. Rivalta, I. Schapiro, J. Segarra-Marti, M. Stenrup, D. G Truhlar, L. Ungur, A. Valentini, S. Vancoillie, V. Veryazov, V. P. Vysotskiy, O. Weingart, F. Zapata, R. Lindh (2016). MOLCAS 8. *Journal of Computational Chemistry*, **37**, 506–541.

172 Werner, H.-J., MOLPRO, version 2015.1, a package of ab initio programs, H.-J. Werner, P. J. Knowles, G. Knizia, F. R. Manby, M. Schütz, P. Celani, W. Györffy, D. Kats, T. Korona, R. Lindh, A. Mitrushenkov, G. Rauhut, K. R. Shamasundar, T. B. Adler, R. D. Amos, A. Bernhardsson, A. Berning, D. L. Cooper, M. J. O. Deegan, A. J. Dobbyn, F. Eckert, E. Goll, C. Hampel, A. Hesselmann, G. Hetzer, T. Hrenar, G. Jansen, C. Köppl, Y. Liu, A. W. Lloyd, R. A. Mata, A. J. May, S. J. McNicholas, W. Meyer, M. E. Mura, A. Nicklass, D. P. O'Neill, P. Palmieri, D. Peng, K. Pflüger, R. Pitzer,

M. Reiher, T. Shiozaki, H. Stoll, A. J. Stone, R. Tarroni, T. Thorsteinsson, and M. Wang, see http://www.molpro.net.

173 Visscher, L., Jensen, H.J.A., Bast, R., and Saue, T., DIRAC, a relativistic ab initio electronic structure program, Release DIRAC15 (2015), written by R. Bast, T. Saue, L. Visscher, and H. J. Aa. Jensen, with contributions from V. Bakken, K. G. Dyall, S. Dubillard, U. Ekstroem, E. Eliav, T. Enevoldsen, E. Fasshauer, T. Fleig, O. Fossgaard, A. S. P. Gomes, T. Helgaker, J. Henriksson, M. Iliaš, Ch. R. Jacob, S. Knecht, S. Komorovsky, O. Kullie, J. K. Laerdahl, C. V. Larsen, Y. S. Lee, H. S. Nataraj, M. K. Nayak, P. Norman, G. Olejniczak, J. Olsen, Y. C. Park, J. K. Pedersen, M. Pernpointner, R. Di Remigio, K. Ruud, P. Salek, B. Schimmelpfennig, J. Sikkema, A. J. Thorvaldsen, J. Thyssen, J. van Stralen, S. Villaume, O. Visser, T. Winther, and S. Yamamoto (see http://www.diracprogram.org).

174 Fox, D.J., Gaussian 09, Revision E.01, M. J. Frisch, G. W. Trucks, H. B. Schlegel, G. E. Scuseria, M. A. Robb, J. R. Cheeseman, G. Scalmani, V. Barone, B. Mennucci, G. A. Petersson, H. Nakatsuji, M. Caricato, X. Li, H. P. Hratchian, A. F. Izmaylov, J. Bloino, G. Zheng, J. L. Sonnenberg, M. Hada, M. Ehara, K. Toyota, R. Fukuda, J. Hasegawa, M. Ishida, T. Nakajima, Y. Honda, O. Kitao, H. Nakai, T. Vreven, J. A. Montgomery, Jr., J. E. Peralta, F. Ogliaro, M. Bearpark, J. J. Heyd, E. Brothers, K. N. Kudin, V. N. Staroverov, R. Kobayashi, J. Normand, K. Raghavachari, A. Rendell, J. C. Burant, S. S. Iyengar, J. Tomasi, M. Cossi, N. Rega, J. M. Millam, M. Klene, J. E. Knox, J. B. Cross, V. Bakken, C. Adamo, J. Jaramillo, R. Gomperts, R. E. Stratmann, O. Yazyev, A. J. Austin, R. Cammi, C. Pomelli, J. W. Ochterski, R. L. Martin, K. Morokuma, V. G. Zakrzewski, G. A. Voth, P. Salvador, J. J. Dannenberg, S. Dapprich, A. D. Daniels, Ö. Farkas, J. B. Foresman, J. V. Ortiz, J. Cioslowski, and D. J. Fox, Gaussian, Inc., Wallingford CT, 2009.

175 Peterson, K.A., Feller, D., and Dixon, D.A. (2012) Chemical accuracy in ab initio thermochemistry and spectroscopy: Current strategies and future challenges. *Theoretical Chemistry Accounts*, **131**, 1079.

176 Schuurman, M.S., Muir, S.R., Allen, W.D., and Schaefer, H.F., III, (2004) Towamochemistry: Focal point analysis of the heat of formation of NCO and [H,N,C,O] isomers. *Journal of Chemical Physics*, **120**, 11586.

177 Karton, A., Rabinovich, E., Martin, J.M.L., and Ruscic, B. (2006) W4 theory for computational thermochemistry: In pursuit of confident sub-kJ/mol predictions. *Journal of Chemical Physics*, **125**, 144108.

178 Harding, M.E., Vazquez, J., Ruscic, B., Wilson, A.K., Gauss, J., and Stanton, J.F. (2008) High-accuracy extrapolated ab initio thermochemistry. III. Additional improvements and overview. *Journal of Chemical Physics*, **128**, 114111.

179 DeYonker, N.J., Grimes, T., Yokel, S., Dinescu, A., Mintz, B., Cundari, T.R., and Wilson, A.K. (2006) The correlation-consistent composite approach: Application to the G3/99 test set. *Journal of Chemical Physics*, **125**, 104111.

180 Feller, D., Peterson, K.A., and Dixon, D.A. (2008) A survey of factors contributing to accurate theoretical predictions of atomization energies and molecular structures. *Journal of Chemical Physics*, **129**, 204105.

181 Kendall, R.A., Dunning Jr, T.H., and Harrison, R.J. (1992) Electron affinities of the first-row atoms revisited. Systematic basis sets and wave functions. *Journal of Chemical Physics*, **96**, 6796.

182 Dunning Jr, T.H. (1989) Gaussian basis sets for use in correlated molecular calculations. I. The atoms boron through neon and hydrogen. *Journal of Chemical Physics*, **90**, 1007.

183 de Jong, W.A., Harrison, R.J., and Dixon, D.A. (2001) Parallel Douglas-Kroll energy and gradients in NWChem: Estimating scalar relativistic effects using Douglas-Kroll contracted basis sets. *Journal of Chemical Physics*, **114**, 48–53.

184 Karton, A. and Martin, J.M.L. (2006) Comment on: "Estimating the Hartree-Fock limit from finite basis set calculations" [Jensen F (2005) *Theor Chem Acc* 113, 267]. *Theoretical Chemistry Accounts*, **115**, 330–333.

185 Peterson, K.A. and Dunning, T.H., Jr. (2002) Accurate correlation consistent basis sets for molecular core-valence correlation effects. The second row atoms Al–Ar, and the first row atoms B–Ne revisited. *Journal of Chemical Physics*, **117**, 10548–10560.

186 Martin, J.M.L. (1996) Ab initio total atomization energies of small molecules – Towards the basis set limit. *Chemical Physics Letters*, **259**, 669–678.

187 Pyykkö, P. and Zhao, L.-B. (2003) Search for effective local model potentials for simulation of quantum electrodynamic effects in relativistic calculations. *Journal of Physics B: Atomic, Molecular and Optical Physics*, **36**, 1469.

188 Hildenbrand, D.L. and Lau, K.H. (1991) Redetermination of the thermochemistry of gaseous UF_5, UF_2, and UF. *Journal of Chemical Physics*, **94**, 1420–1425.

189 Noga, J. and Bartlett, R.J. (1987) The full CCSDT model for molecular electronic structure. *Journal of Chemical Physics*, **86**, 7041–7050.

190 Watts, J.D. and Bartlett, R.J. (1990) The coupled-cluster single, double, and triple excitation model for open-shell single reference functions. *Journal of Chemical Physics*, **93**, 6104–6105.

191 Kucharski, S.A. and Bartlett, R.J. (1992) The coupled-cluster single, double, triple, and quadruple excitation method. *Journal of Chemical Physics*, **97**, 4282–4288.

192 Kucharski, S.A. and Bartlett, R.J. (1998) Noniterative energy corrections through fifth-order to the coupled cluster singles and doubles method. *Journal of Chemical Physics*, **108**, 5243.

193 Cox, R.M., Citir, M., Armentrout, P.B., Battey, S.R., and Peterson, K.A. (2016) Bond energies of ThO^+ and ThC^+: A guided ion beam and quantum chemical investigation of the reactions of thorium cation with O_2 and CO. *Journal of Chemical Physics*, accepted.

194 Guillaumont, R., Fanghånel, T., Neck, V., Fuger, J., Palmer, D.A., Grenthe, I., and Rand, M.H. (2003) *Chemical Thermodynamics 5: Update on the Chemical Thermodynamics of Uranium, Neptunium, Plutonium, Americium and Technetium*. Amsterdam: Elsevier.

195 Thanthiriwatte, K.S., Wang, X., Andrews, L., Dixon, D.A., Metzger, J., Vent-Schmidt, T., and Riedel, S. (2014) Properties of ThF_x from infrared spectra in solid argon and neon with supporting electronic structure and thermochemical calculations. *Journal of Physical Chemistry A*, **118**, 2107–2119.

196 Peralta, J.E., Batista, E.R., Scuseria, G.E., and Martin, R.L. (2005) All-electron hybrid density functional calculations on UF_n and UCl_n (n = 1-6). *Journal of Chemical Theory and Computation*, **1**, 612–616.

197 Adamo, C. and Barone, V. (1999) Toward reliable density functional methods without adjustable parameters: The PBE0 model. *Journal of Chemical Physics*, **110**, 6158–6170.

198 Ernzerhof, M. and Scuseria, G.E. (1999) Assessment of the Perdew–Burke–Ernzerhof exchange-correlation functional. *Journal of Chemical Physics*, **110**, 5029–5036.

199 Goncharov, V. and Heaven, M.C. (2006) Spectroscopy of the ground and low-lying excited states of ThO^+. *Journal of Chemical Physics*, **124**, 064312.

200 Infante, I., Kovacs, A., La Macchia, G., Shahi, A.R.M., Gibson, J.K., and Gagliardi, L. (2010) Ionization energies for the actinide mono- and dioxides series, from Th to Cm: Theory versus experiment. *Journal of Physical Chemistry A*, **114**, 6007–6015.

201 Marçalo, J. and Gibson, J.K. (2009) Gas-phase energetics of actinide oxides: An assessment of neutral and cationic monoxides and dioxides from thorium to curium. *Journal of Physical Chemistry A*, **113**, 12599–12606.

202 Averkiev, B., Mantina, M., Valero, R., Infante, I., Kovacs, A., Truhlar, D., and Gagliardi, L. (2011) How accurate are electronic structure methods for actinoid chemistry? *Theoretical Chemistry Accounts*, **129**, 657–666.

203 Paulovič, J., Gagliardi, L., Dyke, J.M., and Hirao, K. (2005) A theoretical study of the gas-phase chemi-ionization reaction between uranium and oxygen atoms. *Journal of Chemical Physics*, **122**, 144317.

204 Jackson, V.E., Craciun, R., Dixon, D.A., Peterson, K.A., and de Jong, W.A. (2008) Prediction of vibrational frequencies of UO_2^{2+} at the CCSD(T) level. *Journal of Physical Chemistry A*, **112**, 4095–4099.

205 Odoh, S.O. and Schreckenbach, G. (2010) Performance of relativistic effective core potentials in DFT calculations on Actinide compounds. *Journal of Physical Chemistry A*, **114**, 1957–1963.

206 Gagliardi, L. and Roos, B.O. (2000) Uranium triatomic compounds XUY (X,Y = C,N,O): A combined multiconfigurational second-order perturbation and density functional study. *Chemical Physics Letters*, **331**, 229–234.

207 Zhou, M.F., Andrews, L., Li, J., and Bursten, B.E. (1999) Reaction of laser-ablated uranium atoms with CO: Infrared spectra of the CUO, CUO^-, OUCCO, (eta(2)-C-2) UO_2,-and $U(CO)_x$ (x = 1-6) molecules in solid neon. *Journal of the American Chemical Society*, **121**, 9712–9721.

208 Andrews, L., Liang, B.Y., Li, J., and Bursten, B.E. (2000) Ground-state reversal by matrix interaction: Electronic states and vibrational frequencies of CUO in solid argon and neon. *Angewandte Chemie-International Edition*, **39**, 4565.

209 Li, J., Bursten, B.E., Liang, B.Y., and Andrews, L. (2002) Noble gas-actinide compounds: Complexation of the CUO molecule by Ar, Kr, and Xe atoms in noble gas matrices. *Science*, **295**, 2242–2245.

210 Roos, B.O., Widmark, P.O., and Gagliardi, L. (2003) The ground state and electronic spectrum of CUO: A mystery. *Faraday Discussions*, **124**, 57–62.

211 Infante, I. and Visscher, L. (2004) The importance of spin-orbit coupling and electron correlation in the rationalization of the ground state of the CUO molecule. *Journal of Chemical Physics*, **121**, 5783–5788.

212 Yang, T., Tyagi, R., Zhang, Z., and Pitzer, R.M. (2009) Configuration interaction studies on the electronic states of the CUO molecule. *Molecular Physics*, **107**, 1193–1195.

213 Malli, G.L., da Silva, A.B.F., and Ishikawa, Y. (1993) Universal Gaussian basis set for accurate ab initio relativistic Dirac-Fock calculations. *Physical Review A*, **47**, 143–146.

214 Thanthiriwatte, K.S., Vasiliu, M., Battey, S.R., Lu, Q., Peterson, K.A., Andrews, L., and Dixon, D.A. (2015) Gas phase properties of MX_2 and MX_4 (X = F, Cl) for M = Group 4, Group 14, Cerium, and Thorium. *Journal of Physical Chemistry A*, **119**, 5790–5803.

2

Speciation of Actinide Complexes, Clusters, and Nanostructures in Solution

Rami J. Batrice, Jennifer N. Wacker, and Karah E. Knope

Department of Chemistry, Georgetown University, Washington, DC

2.1 Introduction

An understanding of metal ion speciation underpins the development of models that reliably and predictably describe the chemical behavior of a metal in solution. This is especially critical for actinides as the rationalization and control of their behavior under nuclear fuel reprocessing, waste management, and geo- or biochemical conditions necessitates fundamental knowledge of their oxidation state, coordination environment, nuclearity, and potential reactivity. Since the advent of the chemical studies of the actinide elements, researchers have sought to elucidate these aspects of actinide chemistry that are well known to directly influence speciation. Unfortunately, in contrast to d-block metals, research to these ends is often complicated due to several factors such as the range of accessible oxidation states for the actinides and hence the complex chemical behavior the frontier *5f* orbital electrons afford, and in aqueous systems the hydrolysis/condensation chemistry that has historically complicated fundamental investigations into their solution behavior. Nonetheless, the importance of speciation as governed by variables such as solution conditions, inner- and outer-sphere complexation, and redox properties of the actinide ion is plainly apparent and has directed investigations aimed at answering enduring questions relevant to both fundamental and applied aspects of actinide chemistry. Indeed, investigations of the aqueous nature and speciation of actinide complexes, clusters, and nanostructures remain an active field of study in modern chemical research, and as such a number of experimental and theoretical methods for determining speciation have been developed.

In this chapter, selected experimental techniques that have been applied toward understanding the solution-state speciation of actinides in aqueous systems are presented. Analytical methods such as potentiometry, Raman, UV-vis, nuclear magnetic resonance (NMR), or fluorescence spectroscopies provide indirect information about a metal ion's correlations and coordination complexes in solution. Further, small-angle X-ray scattering, a technique that has found much use as a means of interrogating the speciation and dynamics of actinide clusters, is described. Finally, two more recently developed techniques, extended X-ray absorption spectroscopy and high-energy X-ray

Experimental and Theoretical Approaches to Actinide Chemistry, First Edition.
Edited by John K. Gibson and Wibe A. de Jong.

scattering, are highlighted and shown to provide direct metrical information which informs of solution speciation, including metal–ligand coordination modes, distances, and for the latter, coordination numbers and longer-range ordering. For each technique, we present recent examples from the literature that illustrate the utility and limitations of the technique as a means of probing metal ion speciation. In addition, as recent advances in theory and quantum chemical studies have played an important role in understanding actinide speciation, reactivity, solubility, and stability, we have provided examples that display the synergy between experimental and theoretical approaches to actinide speciation in aqueous solution. A summary of the experimental techniques discussed in this chapter is given in Table 2.1.

2.2 Potentiometry

Although not as widely used in recent years, potentiometry remains a reliable means to indirectly probe actinide speciation in aqueous solution. In a given solution of one or more known metals, a potentiometric titration may be performed by adjusting the pH of the solution using either an acidic or basic medium; at each given pH (corresponding to a potential which exists throughout the solution), a particular species is identified to exist; however, considerable variability is often observed even upon repetition of similar titration studies. Nevertheless, this technique may be applied in both a qualitative or quantitative method. In the former approach, the conversion of the material of interest is monitored in a similar manner to a classical redox titration by determining the equivalence point of the transformation. This alone provides useful information regarding the stability of the analyte; however, the identity of the dissolved species may not be determined in the qualitative method. Rather, using this method in conjunction with other analyses (i.e., NMR, optical spectroscopy, SAXS) provides a quantitative avenue of determining both the identity of the inorganic species in solution as well as the pH regime in which it is stable. Using the data obtained from these methods, speciation diagrams may be generated which provide extensive information about the various species accessible as a function of pH and metal ion concentration.

2.2.1 Potentiometric Titrations to Reveal Speciation

In environmental actinide chemistry, the actinide ions have been identified to exist in particular zones of water deposits. Figure 2.1 shows the accessible oxidation states of the actinide metals in various environmental conditions, but more broadly, illustrates the potential complexity in determining actinide speciation in environmental systems. In order to understand the binding of microbes in the suboxic and anaerobic zones, Lütke and coworkers performed potentiometric titrations of the uranyl ion (UO_2^{2+}) in the presence of bacteria, such as those in the *Paenibacillus* genus. (1) Samples were prepared using micromolar concentrations of the uranyl cation in sodium perchlorate solutions containing bacterial samples. By initially titrating from acidic conditions ($pH = \sim 3$) using aqueous sodium hydroxide, the equivalence point of the uranyl transformation was determined (Figure 2.2).

Performing similar titrations over a broad range of pH values in the presence of these bacteria informs of the stability of the various hybrid materials. Given that the

Table 2.1 Summary of experimental methods for the analysis of aqueous actinide speciation.

Technique	Information obtained	Detection limits	Advantages	Disadvantages
Potentiometry (pH)	Identity of dissolved species as a function of pH and metal ion concentration; stability constants; speciation diagrams may be generated when used in combination with other analyses.	User's discretion	Simple experimental setup; inexpensive.	Indirect probe of speciation; requires frequent calibration; limited precision; sensitive to changes in ionic strength; interpretation based on "best-fit" models.
Ultraviolet-visible-near infrared absorption spectroscopy (UV-Vis-NIR)	Identification of characteristic absorption bands of actinide ions based upon the metal and its oxidation state; identify charge-transfer bands; characterize electronic transitions of metal ion.	$\geq 10^{-5}$ M	Determination of metal oxidation state; monitor conditions and chemical environment favoring species formation; characterize complex identity and coordination.	Indirect probe of speciation; limited structural information obtained; must be used with other analytical methods for speciation.
Fluorescence spectroscopy	Identify characteristic radiative emission of actinides; lifetime measurements can give information on product dispersity in solution.	10^{-5}–10^{-12} M	Low concentration; high sensitivity; high specificity; rapid diagnostic ability; rapid-pulse laser may be used for time-resolved measurements.	Most reliable for uranium, americium, and curium; majority of An ions lack radiative decay pathway; requires moderately expensive equipment.
Nuclear Magnetic Resonance (NMR)	Ligand-based structural information that may be used to infer speciation; couplings between nuclei; rates of exchange and equilibria.	10^{-1}–10^{-4} M	Common instrument in most research labs; high-resolution information on ligand structure and dynamics; ligand, pH, and temperature effects easily measured; 100+ isotopes which may be detected.	Indirect probe of speciation; relatively large sample size; nuclei in similar environments can result in spectral crowding; isotopic labeling experiments may be required; paramagnetic metal may induce line broadening and large change in chemical shift.
Raman spectroscopy	Inelastic scattering provides structural information; force constants of bonds; presence of specific functional groups; molecular symmetry.	$\geq 10^{-4}$ M	Common instrument in research labs; most molecules have Raman-active vibrations; small structural changes detectable; may be used with all phases of matter.	Indirect probe of speciation; in the absence of actinyl, difficult to directly detect metal ion; potential photochemical damage; structural data informative, but not necessarily unique.

(*Continued*)

Table 2.1 (Continued)

Technique	Information obtained	Detection limits	Advantages	Disadvantages
X-ray Absorption Near Edge Structure (XANES)	Average metal oxidation state; metal–ligand coordination geometry for selected metals.	~0.5–1 mM (in metal)	Can be run as frozen solution or solid samples; element specific; direct probe of oxidation state.	Can only be used for oxidation state identification in actinides; difficult to deconvolute scattering in heterogeneous samples; synchrotron-based technique.
Extended X-ray Absorption Fine Structure (EXAFS)	Single-ion probe for detailed local coordination environment surrounding metal center (<6 Å); coordination number (10%), oxidation state, correlation distances, and coordination symmetry.	~10 atom %	Direct probe of speciation; element specific; can probe mononuclear and lower-order polynuclear species; metal of interest never "silent"; can be run as powder, solution, or frozen solution.	No long-range information gained; synchrotron-based technique; measured distances have limited resolution (0.1–0.2 Å); requires experienced operator to acquire and process data.
Small-Angle X-ray Scattering (SAXS)	Overall size, shape, and morphology of aggregates ranging from 0.5 to 100 nm; used to monitor dispersity and structural transitions in solution.	1–50 mM	Can be used to analyze amorphous materials; quantitative information on structure obtained; small sample sizes.	Data can be overmodeled and therefore misleading; requires use of numerous structural models; expensive instrumentation.
High-Energy X-ray Scattering (HEXS)	Local and mid-range structural organization up to 8 Å; coordination number; coordination modes; direct probe of mononuclear and larger-order oligomers.	~0.2–0.5 *m*	Direct probe of speciation; discrimination of particular metal–ligand bonds; minimal radiation damage; small sample sizes.	Significant statistical and systematic errors for higher-angle data; total scattering method; synchrotron-based technique; requires experienced operator to acquire and process data.

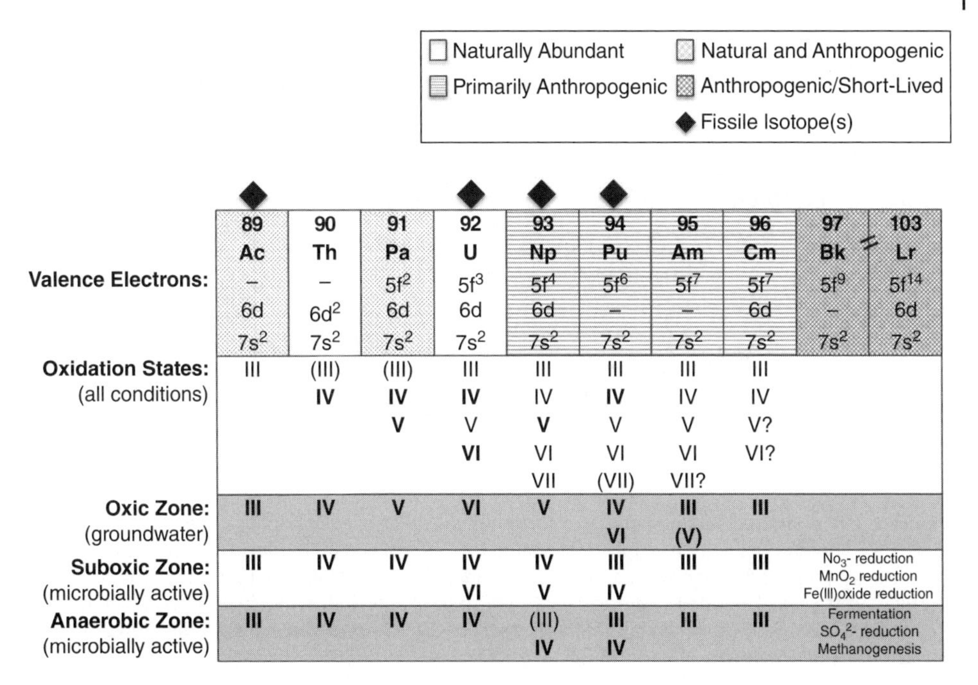

Figure 2.1 Summary of actinide metals and their origin, electron configurations, and common valence states in groundwater. (–) unstable; (?) claimed but unsubstantiated; (bold) most prevalent. Image adapted with permission from Maher K, Bargar JR, Brown GE. Environmental speciation of actinides. *Inorganic Chemistry.* 2013;52(7):3510–3532. Copyright 2013 American Chemical Society.

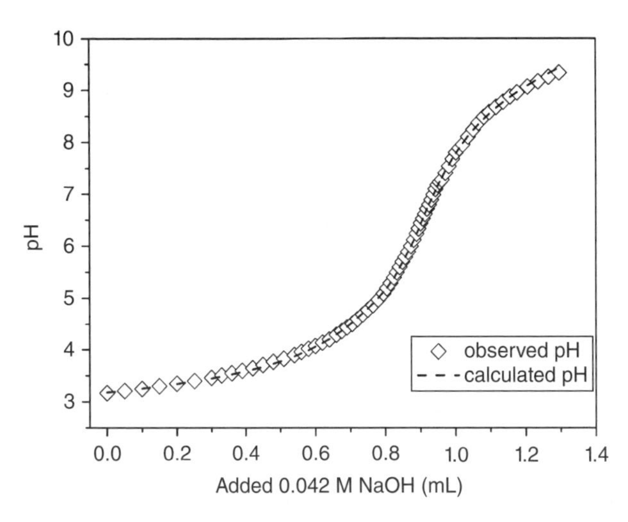

Figure 2.2 Titration of sodium perchlorate solutions of uranyl nitrate and *Paenibacillus* sp. using 0.042 M NaOH and calculated fit. Image adapted with permission from Lütke L, Moll H, Bachvarova V, Selenska-Pobell S, Bernhard G. The U(VI) speciation influenced by a novel Paenibacillus isolate from Mont Terri Opalinus clay. *Dalton Transactions.* 2013;42(19):6979–6988. Copyright 2013 Royal Society of Chemistry.

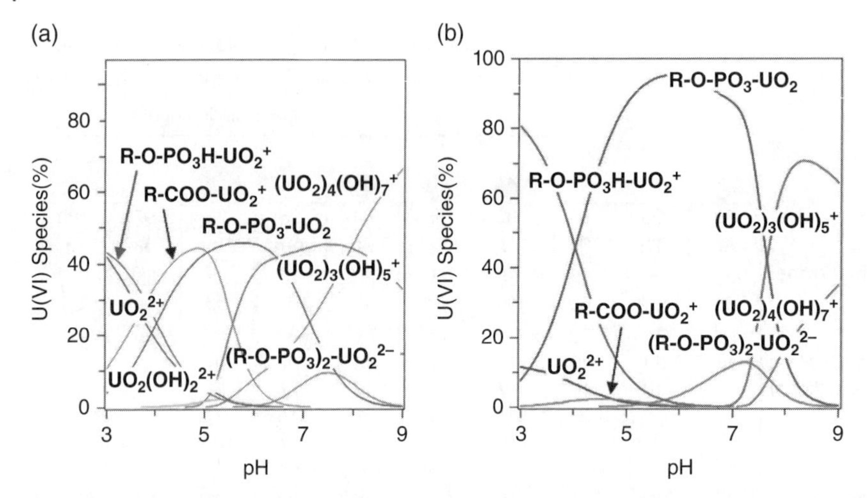

Figure 2.3 Speciation diagram of uranyl cation in the presence of *Paenibacillus sp.* cells at various pH levels. (a) Laboratory conditions applied for analysis by (TRLFS) and (b) model of environmental conditions. System is free of carbon dioxide. Image adapted with permission from Lütke L, Moll H, Bachvarova V, Selenska-Pobell S, Bernhard G. The U(VI) speciation influenced by a novel Paenibacillus isolate from Mont Terri Opalinus clay. *Dalton Transactions.* 2013;42(19):6979–6988. Copyright 2013 Royal Society of Chemistry.

functional groups present in the microbes are known, identification of the species in solution using time-resolved laser-induced fluorescence spectroscopy (TRLFS) was achieved, and the corresponding speciation diagrams generated (Figure 2.3).

Inspection of the speciation diagrams above reveals dissimilar behaviors between the laboratory and simulated environmental conditions. The principal difference between the conditions of these two experiments is the lower concentration of reagents in the environmental model in Figure 2.3b; however, altering of this single variable shows the prevalence of uranyl–phosphate interactions up to approximately neutral pH. This is clearly divergent from the former sample in Figure 2.3a, where the uranyl phosphates are favored only in a narrow pH range at an approximate value of 6. Outside of these described pH ranges, uranyl hydroxides of varying nuclearity are clearly preferred, especially upon moving to higher alkalinity.

Despite the prevalence of uranium in environmental deposits, several of the neighboring actinide metals similarly draw considerable attention owing to their inherent radioactivity and potentially adverse environmental effects. In addition, anthropogenic releases of radionuclides have prompted other researchers to investigate the influence of pH and therefore utilize potentiometric titrations as a means of describing the environmental stability of aqueous actinide compounds. Choppin investigated the identity of actinide ions in simulated natural waters and determined the species present for various metals (2, 3). One excellent example in this study is that of americium ion in oxic water, studied from mildly acidic conditions up to basic pH values (Figure 2.4).

The study performed examined the Am(III) concentration as a function of pH. Under low carbonate concentration and mildly acidic to mildly basic pH, a mixture of bare americium ions and americium-hydroxides are the sole species seen in solution. At pH values of 8 or greater, americium-carbonates begin to emerge, and their concentrations

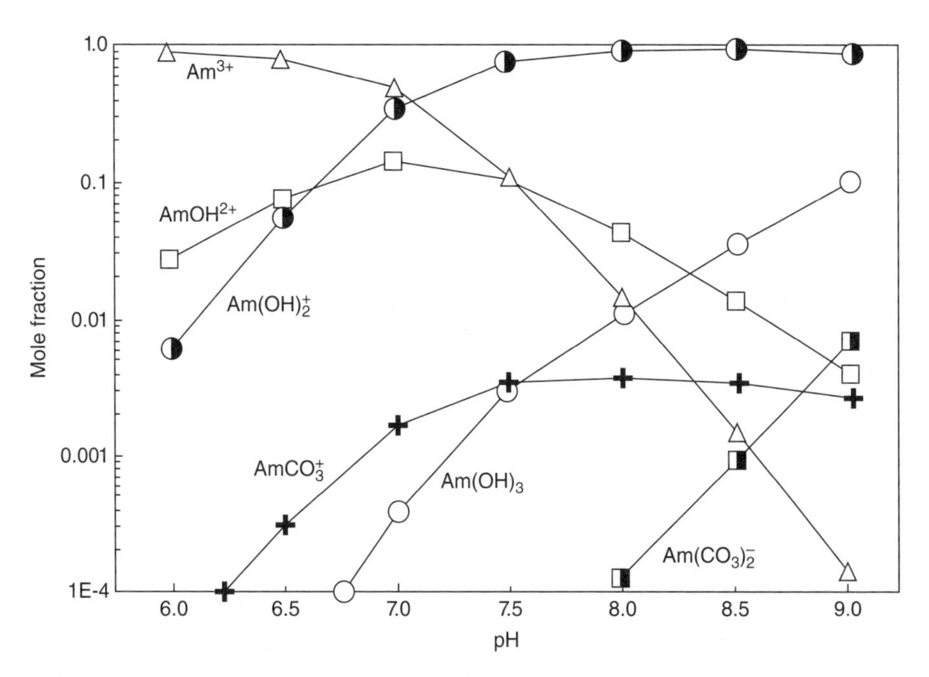

Figure 2.4 Speciation diagram of trivalent americium as a function of pH in seawater. Image adapted with permission from Choppin GR. Actinide speciation in aquatic systems. *Marine Chemistry.* 2006;99(1–4):83–92. Copyright 2006 Elsevier.

increase with increasing solution alkalinity, despite the low concentration of carbonate. In spite of the formation of the carbonato compound, americium-hydroxides remain the dominant species in solution.

Although potentiometry is somewhat less prevalent in modern inorganic research, it remains a powerful tool in the characterization of aqueous actinide species, especially in modeling environmental systems. When used in combination with other analytical techniques such as NMR, optical spectroscopy, or X-ray absorption methods, considerable detail regarding the speciation of actinides in aqueous systems can be determined. One of the most useful, broadly applicable measurements using potentiometric titrations is that of the hydrolysis constants of the actinide series. These values have been summarized in numerous publications, and the resources cited within this publication provide detailed descriptions of the conditions used to elucidate the abovementioned thermodynamic parameters (4, 5).

2.2.2 Overview of Potentiometry in Aqueous Actinide Chemistry

Although potentiometry is an older analytical method, it remains an invaluable technique when properly performed. Potentiometric titrations can be easily performed with materials and equipment often found in the average research laboratory. While a purely potentiometric approach is convenient and inexpensive, reaction solutions are incredibly sensitive to ionic strength, require frequent calibration, and are inundated by generally low precision. The drawbacks described can be ameliorated to a certain extent by also performing spectrophotometric titrations, which eliminate the need for calibration and provide significantly higher precision. However, indicator purity and pH perturbation arising from the indicator used may still adversely affect the quality of measurements. When performed carefully, though, the data extracted from these experiments can provide valuable insight into the pH dependence of actinide speciation in aqueous media.

2.3 Optical Spectroscopy

Of the analytical techniques used to identify and characterize ions of the actinide metals, optical spectroscopy is perhaps one of the earliest and most familiar methods used. The presence of *5f* orbitals, which contrast significantly with either the *4f* or *d*-block valence orbitals, is a unique feature of the actinides. The *5f* electrons are more diffuse when compared to the lanthanide valence electrons, yet significantly more contracted than their outermost *d*-shells. This imparts interesting qualities in that the electronic spectra remain relatively unchanged despite the chemical environment of the metal center, but may still contribute to some degree in structure and bonding. This gives rise to fairly consistent electronic transition energies characteristic of specific actinide ions, while still facilitating the ability to form elaborate structures not often seen using *d*-block or lanthanide metals. Detailed descriptions of the electronic structure of actinide ions have been well studied, and several sources are available that provide the optical spectra for bare actinide ions or simple salts (6). This section will incorporate these data in conjunction with recent studies that have expounded upon early research on actinide optical properties.

2.3.1 UV-vis-NIR Spectroscopy in Actinide Speciation

The absorption bands of actinide ions are often present in well-defined energy regimes with characteristic features based upon the metal and its oxidation state. While the ligand environment and metal symmetry are of significant consequence in transition metal chemistry due to orbital splitting, the valence *f*-electrons of the actinides are less influenced by these factors. This has allowed several works to identify the oxidation state of particular actinide ions in isolated products; however, UV-vis-NIR spectroscopy has also proved useful in monitoring the transformation of actinide mono- and polynuclear systems to elucidate the conditions and chemical environment favoring product formation (7–12).

Using UV-vis-NIR spectroscopy, the analysis and interpretation of spectra from less common actinide metals is easily achieved. Thorium and uranium are by far the most widely studied of the actinide series, largely thanks to their relative abundance as compared with their *5f* neighbors (13), yet despite this, the entire actinide block has been studied using absorption spectroscopy (14, 15). Having a fundamental understanding of the electronic structure and optical properties of these metals and their ions has in turn opened several possibilities to study them using optical absorption spectroscopy. In the investigation of neptunium nanoclusters, Takao and coworkers extracted informative data about the formation of polynuclear species based on the dynamics of the absorption spectra of the reaction solution (Figure 2.5).

The image shown in Figure 2.5a depicts the general structure of the tetravalent neptunium nanocluster within this study, wherein "R" represents the different aliphatic

(a)

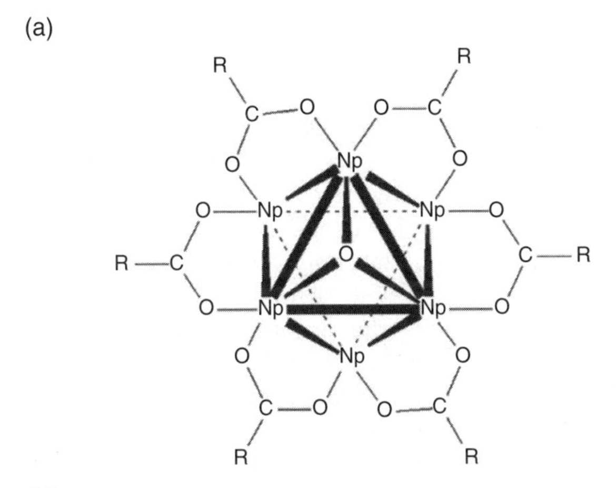

(b)

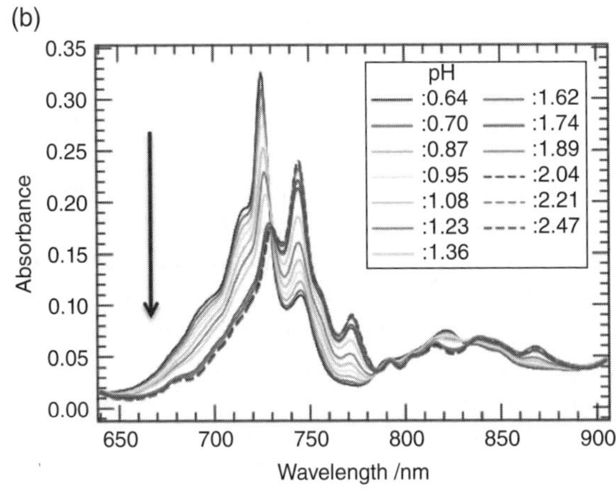

(c)

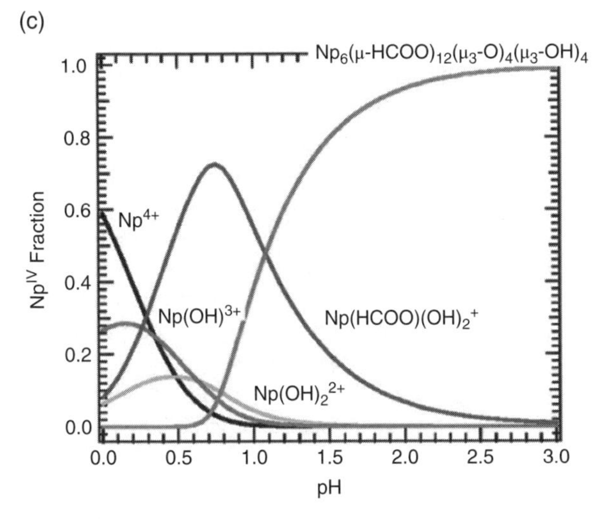

Figure 2.5 (a) Structural model of neptunium(IV) hexanuclear nanocluster. R = H, CH_3, or $CHR'NH_2$; R′ = H, CH_3, or CH_2SH. (b) UV-vis-NIR absorption spectra of the reaction mixture forming the hexanuclear cluster at various pH values. (c) Speciation diagram as a function of pH calculated from experimental stability constants derived from UV-vis-NIR data. Image adapted with permission from Takao K, Takao S, Scheinost AC, Bernhard G, Hennig C. Formation of soluble hexanuclear neptunium(IV) nanoclusters in aqueous solution: Growth termination of actinide(IV) hydrous oxides by carboxylates. *Inorganic Chemistry*. 2012;51(3):1336–1344. Copyright 2012 American Chemical Society.

functional groups used. The chemical structure revealed a body-centered octahedron with each vertex being occupied by a neptunium center, a single oxygen atom occupying the center of the octahedron and bridging the metal centers, and carboxylates bridging adjacent neptunium ions. The Np^{4+} stock solutions were combined with the corresponding carboxylic acids in highly acidic media (pH = 0) and the absorption spectra acquired. The solutions were titrated slowly using dilute ammonia solution to increase the pH, and the UV-vis-NIR spectrum was collected at each interval, providing the stacked spectra shown in Figure 2.5b. The first spectrum acquired at pH 0 shows the characteristic Np^{4+} bands at 730 and 820 nm, but it is evident that with each gradual increase in pH, the intensity of the neptunium signal diminishes, and the dominant absorption at 730 nm undergoes a bathochromic shift of about 20 nm. This type of red shift is well known to occur in the presence of strong cation–cation interactions, and supports the proposed aggregation of neptunium centers with increasing pH (16–18). The spectra reach parity at approximately pH 2, indicating that the cluster formation is essentially complete beyond this point. From determination of the isobestic points, it was evident that key intermediates form before the final hexanuclear cluster. To determine these species, calculations using principal component analysis (PCA) was performed (19). It was found that between the presumably bare neptunium ion and the hexanuclear cluster, three other intermediates form, transitioning from $Np(OH)^{3+}$ to $Np(OH)_2{}^{2+}$ and then $Np(HCOO)(OH)_2{}^{+}$ with increasing pH. Using the data described earlier, the speciation diagram in Figure 2.5c was generated to provide the pH ranges through which the product and intermediates are favored in the equilibrium reaction.

Similar studies were performed using the uranyl cation and identified the conditions that lead to aggregation and resulted in formation of a trimeric species. An early study by Meinrath sought to understand the mechanism of hexavalent uranium hydrolysis and formation of an insoluble $UO_3{\cdot}2H_2O$ using UV-vis spectroscopy (20, 21). In studying the oligomerization of the uranyl cation and absorption properties of the materials obtained, the following UV-vis spectra and speciation diagrams were prepared (Figure 2.6).

The study found that oligomerization to the dimer or trimer resulted in large increases in the molar absorptivity with each increase in nuclearity. In addition, the characteristic bathochromic shift is present, but to a much lower extent (Figure 2.6a). Using similar experimental techniques as those in the formation of the neptunium cluster (*vide supra*), the corresponding speciation diagram in Figure 2.6b was produced. Later, the work of Tsushima et al. repeated this study with the addition of EXAFS studies and DFT calculations to more closely investigate the mechanism of oligomerization to the uranyl trimer (22). The DFT calculations were performed on the B3LYP level, and interestingly, the intermediates proposed were in very good agreement with the previously reported structures by Meinrath. The calculations showed only a minor deviation regarding the geometry of the uranyl dimer ($(UO_2)_2(OH)_2{}^{2+}$), suggesting that the $O=U=O$ angles were slightly bent away from the ideal linear geometry. The proposed structure of the trimer as determined by the quantum chemical calculations showed agreement with Meinrath's 1997 proposed structure ($(UO_2)_3O(OH)^{3+}$), wherein an oxo ion is trapped in the center of the octahedron to create the uranyl and bridging hydroxides.

As illustrated in the examples above, UV-vis-NIR spectroscopy is a convenient and versatile method for the speciation of actinide ions. The characteristic absorption bands of the actinide ions provide a suitable spectroscopic handle to examine the nature of

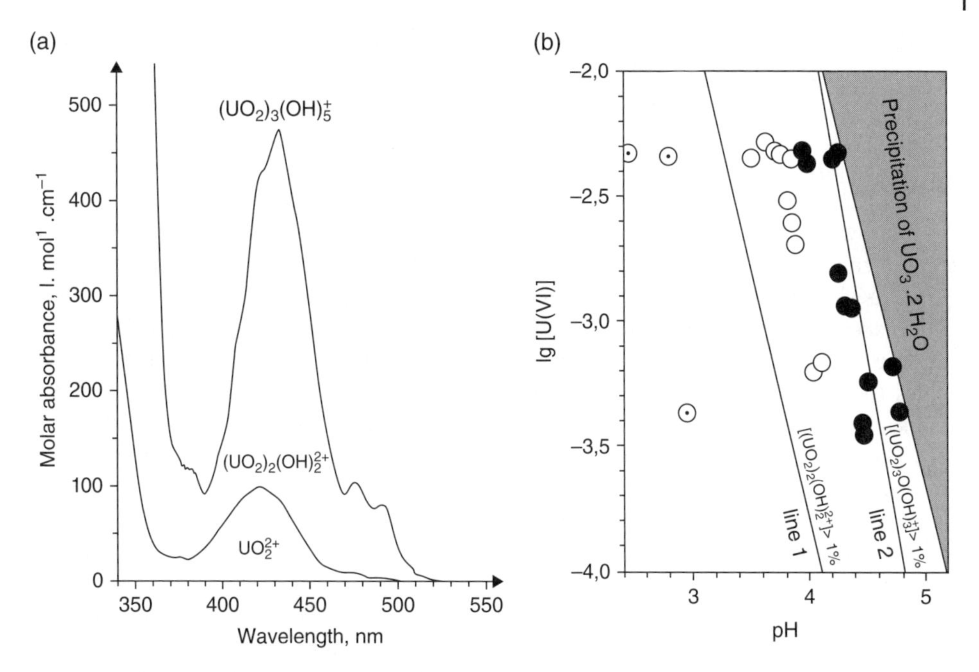

Figure 2.6 (a) UV-vis absorbance spectra of uranyl monomer, dimer, and trimer. (b) Speciation diagram as a function of pH and logarithm of uranyl concentration. Image adapted with permission from Meinrath G. Uranium(VI) speciation by spectroscopy. *Journal of Radioanalytical and Nuclear Chemistry.* 1997;224(1):119–126. Copyright 1997 Springer.

these metals in aqueous solution. While the examples provided here specifically show dynamic features of UV-vis spectra which point to metal ion aggregation into clusters, more basic absorption studies have been used extensively to confirm incorporation of the target metal ion into polynuclear clusters and networks, and to identify actinide metal oxidation states (8–10, 12, 16, 23–29). In recent years, as computing power has increased, the level of quantum chemical calculations has improved sufficiently to the point where stronger predictive power for the chemical pathways and optical properties of actinide clusters may be determined. This use of theory to augment experimental data presents as a valuable tool for future studies of aqueous actinide chemistry and reveals reactive properties that would otherwise be difficult to determine.

2.3.2 Fluorescence Spectroscopy

Just as UV-vis-NIR spectroscopy is used to examine distinctive absorption properties in the actinides, fluorescence spectroscopy can similarly be employed to monitor characteristic electronic transitions as a function of the coordination environment about the metal center, and thereby serves as a versatile tool for examining aqueous actinide speciation. Fluorescence studies are often preferred to absorption spectroscopy due to some intrinsic properties of this analysis. Fluorescence measurements require significantly lower, submicromolar concentrations of analyte in solution, showing a clear advantage over other spectroscopic methods that require significantly higher concentrations to overcome instrumental detection limits. Furthermore, the radiative emission spectra of actinides in solution can yield information on metal coordination environments and ligand exchange dynamics that is difficult to observe in absorption measurements. The sensitivity and short timescale of fluorescence spectroscopy disposes emission measurements to be well suited for the detection of actinide species and chemical processes.

The fluorescence properties of uranium(VI), americium(III), and curium (III) are well known and have been used for a number of speciation studies with the identification of the coordination environment about the metal center being perhaps one of the more useful aspects of fluorescence spectroscopy (30–32). Time-resolved laser-induced fluorescence spectroscopy (TRLFS), in particular, has been used in this regard to examine the speciation of actinides across a number of platforms ranging from small molecules to larger biomolecules; (33) most studies focus on U(VI) and Cm(III), yet the emissive properties of Am are established as are – more recently – the fluorescence of U(IV) and Pa(IV). In this section, we will highlight two examples intended to give an overview of the technique and the information that can be gleaned from fluorescence spectroscopic studies. For more detailed descriptions and further examples, readers are directed to recent reviews of laser-induced spectroscopy as applied to actinide speciation (30, 31).

Key to the safe disposal of nuclear waste is an understanding of how the actinides behave in the environment and under the influence of inorganic and organic molecules available in natural systems. Toward this end, many studies have focused on the coordination chemistry of actinides with those ions found in natural groundwater, including, for example, OH^-, CO_3^{2-}, Cl^-, SO_4^{2-}, and PO_4^{3-}. Both complexation and thermodynamic descriptions of the interactions of the actinides with these ligands have been developed at low, ambient, and high temperatures. Curium, having a high quantum yield, is often used as the trivalent actinide of choice as it permits studies down to submicromolar concentrations (31). To highlight one example, the speciation of curium–sulfate complexes was recently explored as a function of temperature using a custom-built cell for two sets of samples: (i) at sulfate concentrations ranging from 0.006 *m* to 0.365 *m* at constant ionic strength and (ii) at varying ionic strength (1.0–4.0 m) and constant sulfate concentration (34). As shown in Figure 2.7, the emission intensity of the $^6D'_{7/2}$ to $^8S'_{7/2}$ transition, used to track speciation, was found to decrease in all samples by nearly 50% with increasing temperature. Such changes in emission are consistent with previous studies examining Cm emission (and hence coordination) as a function of temperature; however, in those systems without a complexing ligand, the decrease in emission intensity is much more pronounced, with only 10% of the initial emission intensity remaining at 200 °C (34).

The temperature-dependent spectra collected for the four samples at various sulfate concentrations are shown Figure 2.7. Bathochromic shifts in the emission wavelength are observed with both increasing sulfate concentrations and/or increasing temperature. This shift can be attributed to the successive formation of $[Cm(SO_4)_n]^{3-2n}$ (n = 1, 2, 3) complexes as described by the following equilibrium:

$$Cm^{3+} + nSO_4^{2-} \rightleftarrows Cm(SO_4)_n^{\;3-2n};\ n = 1,2,3$$

Furthermore, deconvolution of the emission spectra in this system via PCA can be achieved using single-component spectra of the $[Cm(SO_4)_n]^{3-2n}$ species (Figure 2.8a),

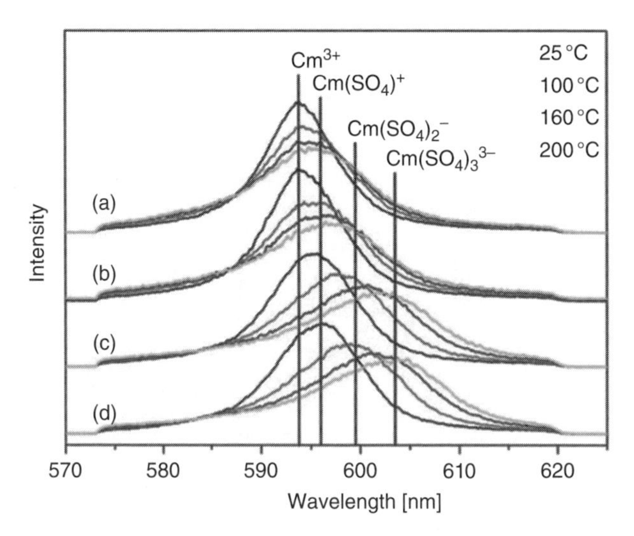

Figure 2.7 Normalized emission spectra of Cm(III) at 25 °C, 100 °C, 160 °C, and 200 °C at ionic strength 2.0 *m* ($NaClO_4$) and sulfate concentrations of (a) 5.00×10^{-3} *m*, (b) 3.10×10^{-2} *m*, (c) 1.44×10^{-1} *m*, and (d) 3.65×10^{-1} *m*. Reused with permission from Skerencak A, Panak PJ, and Fanghänel T. Complexation and thermodynamics of Cm (iii) at high temperatures: the formation of $[Cm(SO_4)_n]^{3-2n}$ (n = 1, 2, 3) complexes at T = 25 °C to 200 °C. *Dalton Transactions*. 2013;42(2):542–549. Copyright © 2012, Royal Society of Chemistry.

accounting for relative fluorescence intensity factors of the various species (34), and yields the speciation profiles shown in Figure 2.8b. These results suggest that that at room temperature and sulfate concentrations $\leq 0.15\,m$, the speciation is mainly governed by the Cm^{3+} aquo ion. As expected, $[Cm(SO_4)_n]^{3-2n}$ forms with increasing sulfate concentration. At higher temperatures and high sulfate concentrations, the Cm^{3+} aquo ion and lower-order $Cm(SO_4)^+$ complexes do not contribute significantly to the speciation.

Like curium, the emissive properties of hexavalent uranium are well known, and as such, many literature examples of using fluorescence spectroscopy to determine U(VI) speciation exist (4). The hydrolysis chemistry of uranium is well established (35), and several time-resolved laser-induced fluorescence spectroscopy (TRLFS) studies have examined the formation of mononuclear and polynuclear uranyl hydroxo complexes as a function of pH and metal concentration, owing to the importance of these species to chemical models that describe the behavior of uranium in aqueous solution (36, 37). Fluorescence spectroscopy has likewise been utilized to examine complexation of U(VI) by both inorganic and organic ligands, the latter including both geochemically and biochemically relevant model compounds. Here, we will focus on one such example, wherein this analytical tool was applied toward understanding the interaction of the uranyl cation with peptidoglycan (PG), the main component of the outer membrane of gram-positive bacteria (33). Barkleit et al. examined the complexation of U(VI) by PG over a range of pH values. In this work, potentiometry was used to determine both the dissociation constants of PG and the stability constants of the uranyl–PG complexes, while TRLFS was used to detect the complexes at trace metal concentrations, which is especially relevant for environmental systems.

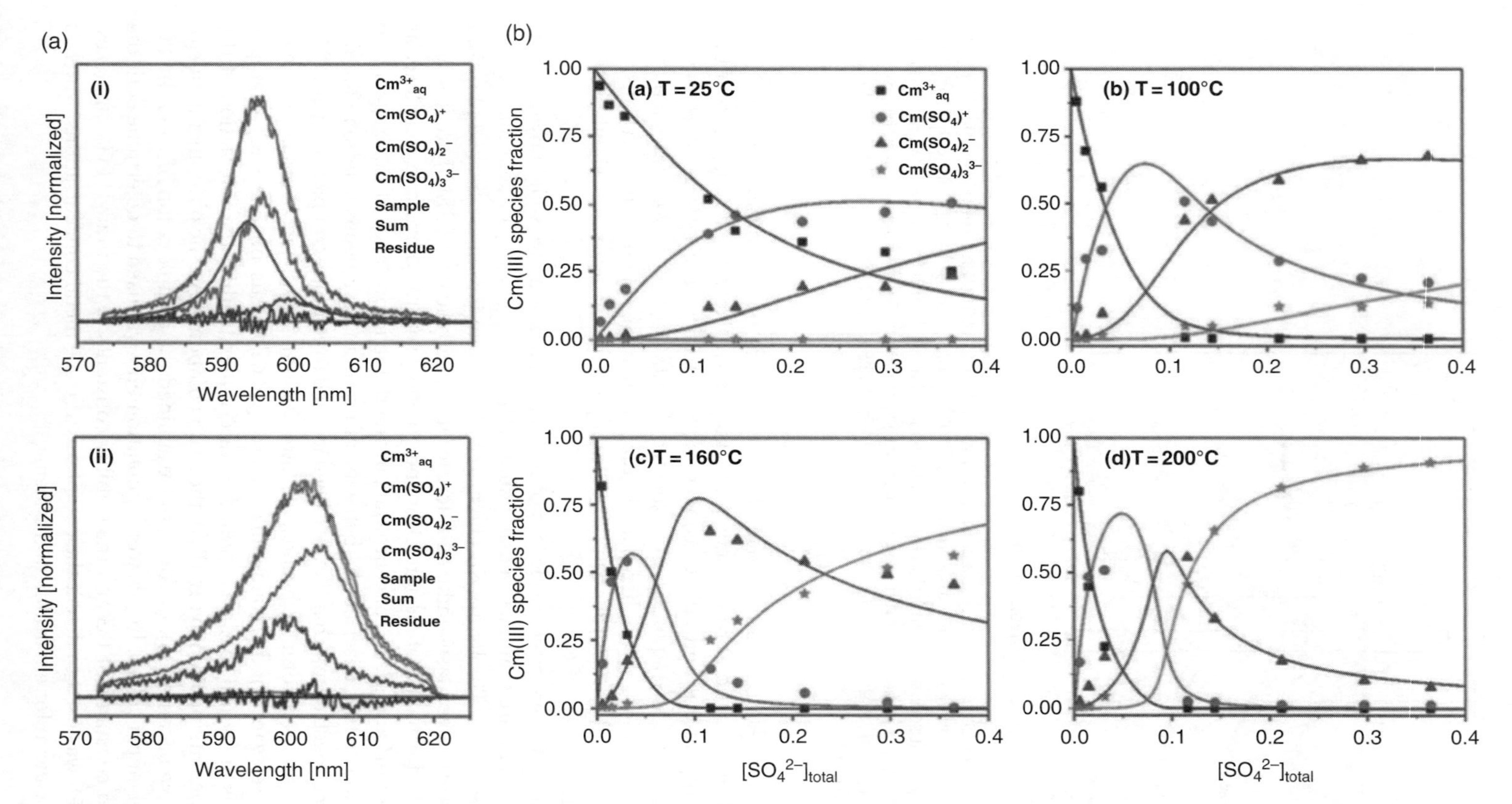

Figure 2.8 (a) Deconvoluted spectra of Cm(III) solutions at sulfate concentrations of 3.10×10^{-2} *m*, ionic strength 2.0 *m* ($NaClO_4$) at (i) T = 25 °C and (ii) T = 200 °C. Single spectra of $[Cm(SO_4)_n]^{3-2n}$ (n = 0, 1, 2, 3) species used for the deconvolution of the of the sample are shown. (b) Experimental (data points) and calculated (lines) distribution of the $[Cm(SO_4)_n]^{3-2n}$ species as a function of sulfate concentration at ionic strength 2.0 *m*. Figures reused with permission from Skerencak A, Panak PJ, Fanghänel T. Complexation and thermodynamics of Cm(iii) at high temperatures: the formation of $[Cm(SO_4)_n]^{3-2n}$ (n = 1, 2, 3) complexes at T = 25 °C to 200 °C. *Dalton Transactions*. 2013;42(2):542–549. Copyright © 2012, Royal Society of Chemistry.

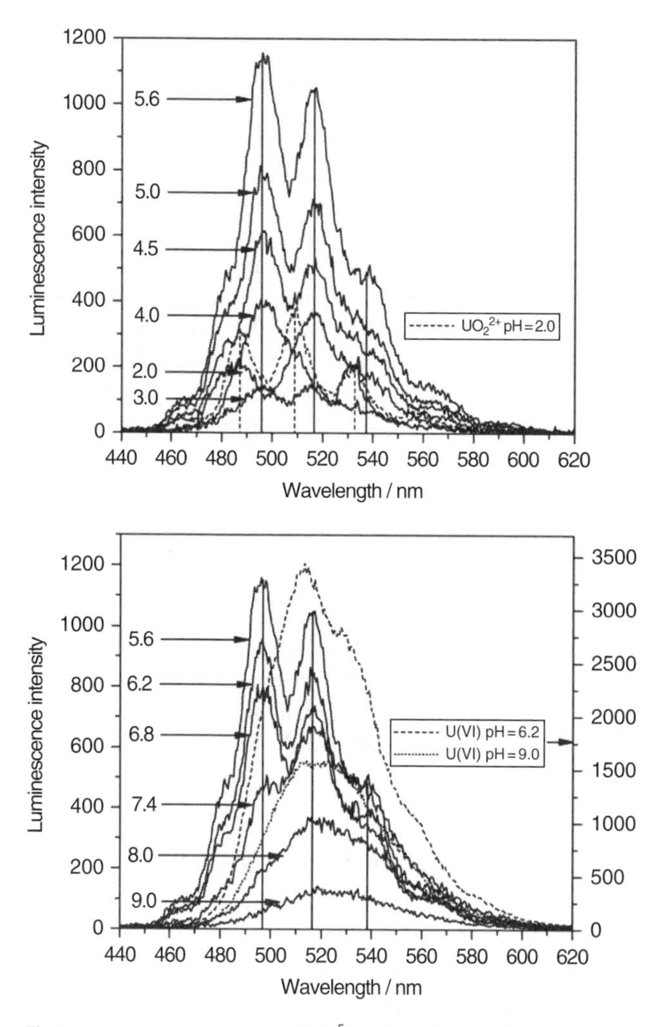

Figure 2.9 Emission spectra of 10^{-5}M U(VI) with 0.1 g/L PG as a function of pH. Image reused with permission from Barkleit A, Moll H, Bernhard G. Complexation of uranium(VI) with peptidoglycan. *Dalton Transactions*. 2009(27):5379–5385. Copyright 2009 Royal Society of Chemistry.

TRLFS spectra were collected for a solution containing 10^{-5}M UO_2^{2+} at a fixed PG concentration (0.1 g/L) as a function of pH (2.0–9.0). These data are presented in Figure 2.9. At a very low pH, no discernible change in the fluorescence spectrum of the free UO_2^{2+} was observed, suggesting that at a low pH, complexation is negligible. With increasing pH, over 2–3, both a decrease in the emission intensity and a red shift are observed. Over pH 3–5.6, the intensity increases, and above pH 5.6, the intensity again decreases. For comparison, the spectra of U(VI) solutions without PG at pH 2.0 and 6.2, wherein the main species are $UO_2^{2+}{}_{(aq)}$, $(UO_2)_3(OH)_5^+$, and $(UO_2)_3(OH)_7^-$, respectively, are shown. Differences between the spectra of UO_2^{2+} with and without PG are easily discernible from these spectra. Changes in the intensity and wavelength of this emission are attributed by the authors to the formation of three species, different from the uranyl hydroxides that form in the absence of PG up to a pH of 7. Further, time-resolved measurements were used to obtain lifetimes from which information on the number of species in solution was drawn.

Although it is not discussed in detail in this section, it is worth noting that information about the coordination environment can be obtained from fluorescence lifetimes (τ), particularly in aqueous systems (3). Water molecules in the first coordination sphere of the metal ion provide non-radiative decay pathways and hence affect the observed lifetimes, which can be related to the number of inner coordination sphere water molecules (34). In addition, an empirical equation determined by Kimura et al. correlates the number of coordinated water molecules (n) with the observable decay rate (k_{obs}) at 25 °C according to the equation $n(H_2O) = 0.65k_{obs} - 0.88$ (38). As a result, the lifetime generally increases as coordinating ligands displace water from the inner coordination sphere. However, at higher temperatures, an increase in lifetimes is observed, even with decreasing n (34).

The fluorescence behavior of the actinides is a valuable feature that can be exploited for various applications in actinide aqueous chemistry. Despite these advantages, though, some shortcomings persist when employing this analysis. One of the more unfortunate of these is the ability to perform fluorescence studies only on a limited number of actinide ions. Moreover, depending on the ligand environment and level of hydration, it is possible that any radiative decay which could be detected in fluorescence studies is quenched by the solvent or ligands (6, 11, 39, 40). In spite of these challenges, advances in our understanding of the actinides have been accomplished, only further illustrating the effective use of fluorescence spectroscopy in the speciation of aqueous actinide species (41–47).

2.3.3 Overview of Optical Spectroscopy in Aqueous Actinide Speciation

The electronic behavior of the actinide ions is well established and remains a reliable method to study actinide incorporation and speciation. In the realm of absorption and emission spectroscopy, each has its respective advantages and limitations. Both techniques are nondestructive and designed to probe the excited state of species, whereas most other analytical methods probe the electronic ground state of species. In addition, the UV and visible light detectors used in both spectroscopic methods are highly sensitive, providing high spectral resolution and fast detection times. In conjunction with the vast advances in quantum chemical calculations, the ability to address and characterize the excited-state properties is more convenient than ever, even in highly complex systems. When considering the limitations of optical spectroscopy, it is generally not possible to obtain an actinide-based optical spectrum for hypervalent actinides. However, notable exceptions to this are the uranyl and neptunyl cations thanks to the characteristic "yl" bonding of the *trans*-dioxo functionality. This necessitates particular consideration when utilizing optical methods for actinide speciation. Moreover, emission spectroscopy is not suited for several of the actinide ions due to the lack of radiative decay pathways for several of these metals. Nevertheless, with proper experimental design, optical spectroscopy can be used very effectively to determine the metal oxidation state, coordination, and speciation of aqueous solutions.

2.4 NMR Spectroscopy

Of the range of analytical methods used to characterize solution-state species, nuclear magnetic resonance (NMR) spectroscopy is arguably one of the most ubiquitous methods used. While NMR is well established as the exemplar for the characterization of organic molecules, this technique is less straightforward when applied to inorganic and metal-organic species, and even less so for actinide complexes, clusters, and extended networks. Nevertheless, various NMR techniques have been developed and utilized to study the formation, identity, and dynamic behavior of actinide-bearing compounds. The ability to probe the structure and dynamics of actinide clusters in the solid state has been demonstrated in recent literature using magic-angle spinning (MAS) NMR spectroscopy (48–50), and in the solution state, complex mechanisms and equilibrium processes can be studied using a variety of NMR methods (51–54). However, the identification and characterization of soluble actinide complexes, clusters, and reactive intermediates has recently drawn considerable attention. These advances arise in part due to the versatility of this analytical method and the ability to detect various isotopes, as well as the high sensitivity of NMR in detecting changes in ligand and coordination environment. While this provides a convenient spectroscopic handle to indirectly probe actinide speciation in solution, the inherent paramagnetic electron configuration of several of the actinide ions, such as U(IV), Np(VI), and Pu(VI), can often result in contact shifts which deviate considerably from expected values (55); early investigations of this phenomenon suggest that higher covalency, and thus, electron contribution into empty *5f* orbitals from coordinated ligands, results in more distinct contact shifts (56). This feature presents additional challenges, especially when investigating organoactinide complexes, yet the use of 1D and 2D NMR spectroscopic techniques still proves highly valuable in aqueous actinide chemistry, and the use of DFT calculations in conjunction with such NMR studies in select examples further improves our understanding of the operative chemical processes. Presented in this section are examples in which NMR can be used to indirectly probe (1) chemical equilibria, (2) the self-assembly of actinide complexes in solution, and (3) product formation and/or evolution.

2.4.1 Probing Chemical Equilibria by NMR

The sensitivity of NMR spectroscopy makes this an ideal experimental technique to observe changes in the reaction media that correspond to shifts in the equilibrium reaction, and several NMR-active nuclei have been exploited to probe such changes. The use of ^{13}C-NMR in actinide cluster chemistry has largely been used to examine pressure-dependent exchange rates of the carbonate ion; however, research by Allen and coworkers studied the effects of increased solution acidity on aggregation and condensation of a uranyl carbonate anion. In the study, it was hypothesized that a tris(carbonato) uranyl tetraanion was activated in the presence of acidic media to generate a trinuclear uranyl species, $(UO_2)_3(CO_3)_6^{6-}$, with concomitant formation of carbon dioxide and water according to the equilibrium presented in Scheme 2.1 (57).

While the reaction depicted in Scheme 2.1 lends itself to a variety of spectroscopic probes to monitor the equilibrium process, carbon NMR using isotopically enriched $Na_2{}^{13}CO_3$ as carbonate source was chosen as a convenient analytical technique to

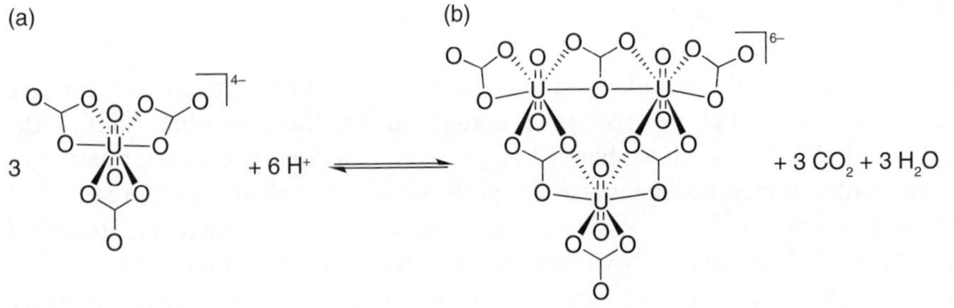

Scheme 2.1 Proposed reaction of tris(carbonato)uranyl tetraanion to form a trinuclear, $(UO_2)_3(CO_3)_6^{6-}$.

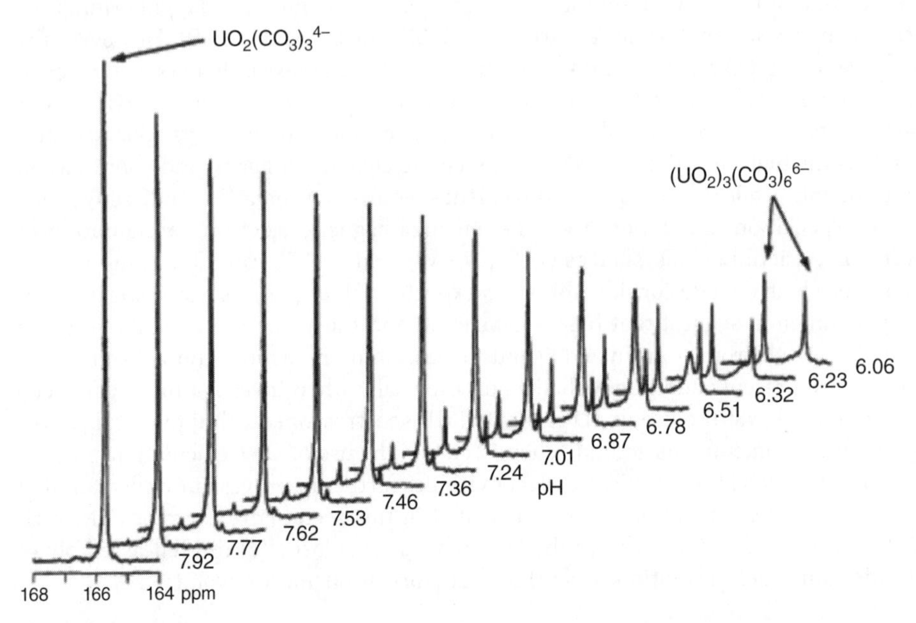

Figure 2.10 ^{13}C-NMR spectra of tris(carbonato)uranyl tetraanion as a function of pH. Increases in the acidity of the solution result in the presence of two well-resolved signals corresponding to bridging and terminal carbonate ligands. The numbers below each spectrum denote pH of solution. Image adapted with permission from Allen PG, Bucher JJ, Clark DL, Edelstein NM, Ekberg SA, Gohdes JW, et al. Multinuclear NMR, Raman, EXAFS, and X-ray diffraction studies of uranyl carbonate complexes in near-neutral aqueous solution. X-ray structure of $[C(NH_2)_3]_6[(UO_2)_3(CO_3)_6]$.cntdot.6.5$H_2O$. *Inorganic Chemistry.* 1995;34(19):4797–4807. Copyright 1995 American Chemical Society.

monitor the condensation of the cluster. In comparing structures (a) and (b) in Scheme 2.1, it is evident that the former complex possesses one spectroscopically unique carbon site, whereas the trimer has two sets of unique carbon atoms arising from bridging and terminal carbonate ions. In order to support the proposed mechanism, samples were titrated with perchloric acid, flame-sealed in NMR tubes, stored at ambient temperature for 48 to 72 h to allow equilibration, and then analyzed by ^{13}C-NMR (Figure 2.10).

Beginning with a slightly basic solution (pH = 7.92) at 0 °C, a single carbon signal is present arising from the three equivalent carbonate moieties bound along the equatorial plane of the uranyl center; as the pH decreases, protolytic displacement of carbonate occurs, resulting in condensation of the uranyl units to create the trimeric species

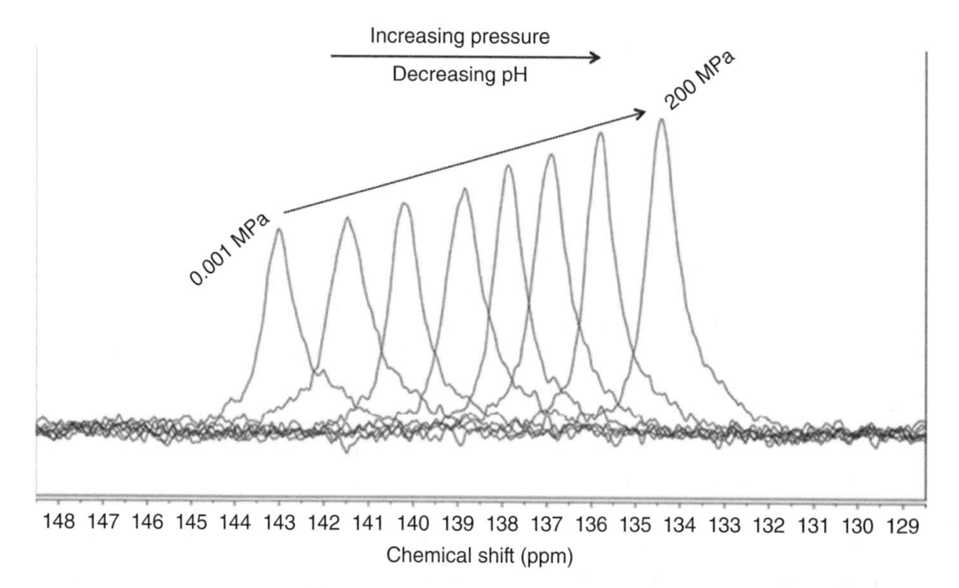

Figure 2.11 Variable pressure ^{13}C-NMR spectra showing the inter-conversion of CN^-/HCN as a function of pressure. The applied pressure ranged from 1 kPa up to 200 MPa. Image adapted with permission from Pilgrim CD, Zavarin M, and Casey WH. Pressure dependence of carbonate exchange with $[NpO_2(CO_3)_3]^{4-}$ in aqueous solutions. *Inorganic Chemistry*. 2017;56(1):661–666. Copyright 2017 American Chemical Society.

shown in Scheme 2.1. The formation of this trimer is evident on the basis of the disappearance of the single carbon signal at 164.9 ppm and the emergence of the two inequivalent resonances of $(UO_2)_3(CO_3)_6{}^{6-}$ at 164.9 and 166.2 ppm. The authors performed variable temperature studies and found that heating the sample to room temperature resulted in line broadening of the high-field signal, suggesting that the resonance at 164.9 ppm arises from the terminally bound carbonate as this ligand would be more labile than the bridging units. More recent studies on aqueous actinide chemistry incorporating ^{13}C-NMR characterization have also probed carbonate binding to a monomeric actinide center (58, 59). This process has also been investigated for a neptunium(VI) complex (7). In the latter study, the electronic configuration of the neptunium ion is $5f^1 6d^0$, presenting a challenge in NMR studies due to the paramagnetism of the central ion. Rather than directly measuring the carbon nucleus of the carbonate ion, potassium cyanide was added to the reaction as a pH probe; by monitoring the pH of the solution, the amount of free versus neptunium-bound carbonate as a function of reaction pressure could be correlated, thus describing the equilibrium of the reaction (Figure 2.11).

According to the proposed reaction, two interrelated equilibrium processes are operative which yield the above spectrum and explain the observed transition. First, the initial equilibrium is the binding of carbonate, wherein free carbonate likely exists in aqueous solution to some extent as bicarbonate. This can in turn react with the neptunyl species with the accompanying liberation of an acidic proton according to Equation 2.1:

$$[\mathrm{Np}](\mathrm{CO_3})_{n-1(aq)} + \mathrm{HCO_3}^-{}_{(aq)} \rightleftharpoons [\mathrm{Np}](CO_3)_{n(aq)} + \mathrm{H}^+{}_{(aq)} \tag{2.1}$$

When this intermediate is formed in the presence of cyanide ion, a second equilibrium occurs forming hydrogen cyanide (Equation 2.2):

$$H^{+}_{(aq)} + CN^{-}_{(aq)} \rightleftharpoons HCN_{(g)} \tag{2.2}$$

If these two equilibrium equations are combined, it is apparent that at a lower pH (higher proton concentration) a shift in the ^{13}C-NMR spectrum of CN^- from the secondary equilibrium process ensues, conducive to a reduction in the concentration of free carbonate through formation of neptunium–carbonate interactions. This proposed reaction is further supported when looking at the experimental pH values of the reaction as determined by the carbon shift in CN^- at 9.77, 9.73, and 9.72, wherein the ratio of free to bound carbonate was 12.3, 9.3, and 7.7, respectively. Analysis of these data in combination with the proposed Le Chatelier equilibrium reveal that formation of the neptunium–carbonate compound is favored by increasing the pressure of the reaction.

Beyond carbon-13, fluorine-19 is another NMR-active nucleus that has gained significant attention in actinide chemistry, perhaps in part due to the strong An-F bonds which form robust actinide-fluorides across a variety of conditions. Under aqueous conditions and in the presence of chelating ligands, actinide-fluorides are often found as mononuclear species. However, under particular conditions, aggregation to small polynuclear compounds (such as dimers or trimers) can be seen, and such equilibrium can be easily monitored by ^{19}F-NMR. Palladino and coworkers showed the capacity of ^{19}F-NMR to elucidate the various binding modes in a ternary system comprised of uranyl ion, ethylenediamine-*N,N′*-diacetate (EDDA), and fluoride (51). A combination of these reagents in aqueous media formed crystals containing $UO_2(EDDA)F^-$ after slow evaporation. This material was used in subsequent equilibrium studies using potentiometric titrations in combination with NMR. A series of ^{19}F-NMR spectra were acquired over a range of basic pH ranges and revealed a decrease in the concentration of $UO_2(EDDA)F^-$ and an increase in the concentration of free fluoride ion (Figure 2.12).

From the ^{19}F-NMR spectra above in addition to ^{1}H-NMR experiments, the researchers deduced the equilibrium reaction shown in Equation 2.3:

$$UO_2(EDDA)F^- + H_2O \rightleftharpoons UO_2(OH)(EDDA)^- + H^+ + F^- \tag{2.3}$$

Interestingly, reduction of the pH to acidic conditions resulted in hydrolysis and condensation of the monomer to form isomers of a uranyl dimer. Upon dissolving a sample of $UO_2(EDDA)F^-$ and titrating to a pH of 5.38, a marked transformation of the uranyl starting material occurs according to ^{19}F-NMR (Figure 2.13).

By a comparison with other spectra of similar studies, the authors identified the uranyl precursor in addition to three uranyl fluorides, but most interesting were the two distinct signals separated by 620 Hz centered at 159 ppm. When considering the large separation and the similar intensity of these two signals, in addition to crystallographic data collected in this study, the resonances were determined to be isomers of $(UO_2)_2(\mu_2\text{-}OH)_2(EDDA)_2F_2^{2-}$ (Figure 2.14).

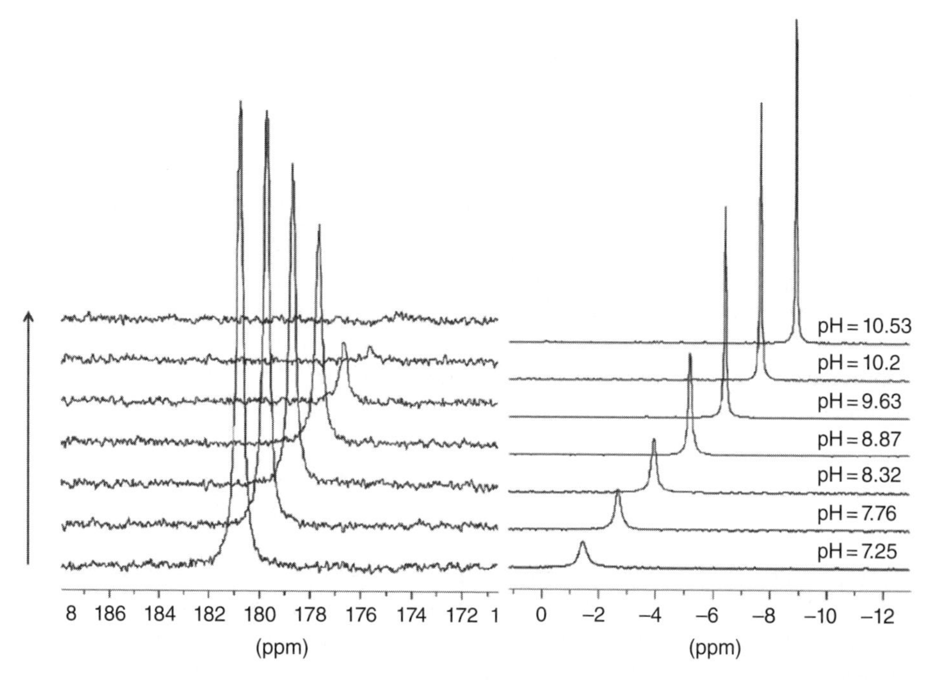

Figure 2.12 pH dependence of the ^{19}F-NMR spectrum of $UO_2(EDDA)F^-$. With increasing pH, free fluoride ion concentration increases (−2 ppm) while $UO_2(EDDA)F^-$ concentration decreases (181 ppm). Image adapted with permission from Palladino G, Szabó Z, Fischer A, Grenthe I. Structure, equilibrium, and ligand exchange dynamics in the binary and ternary dioxouranium(vi)-ethylenediamine-N,N[prime or minute]-diacetic acid-fluoride system: A potentiometric, NMR, and X-ray crystallographic study. *Dalton Transactions.* 2006(43):5176–5183. Copyright 2006 Royal Society of Chemistry.

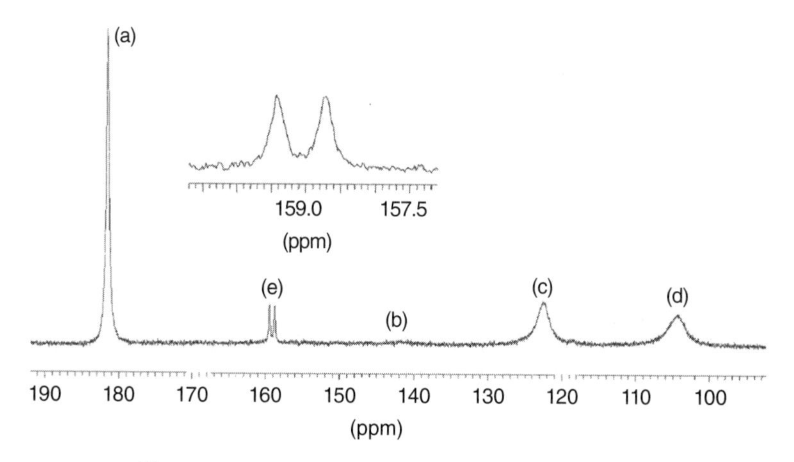

Figure 2.13 ^{19}F-NMR spectrum of $UO_2(EDDA)F^-$ titrated to pH of 5.38. (a) $UO_2(EDDA)F^-$; (b) low-intensity broad peak of $UO_2F_{2(aq)}$; (c) $UO_2F_3^-$; (d) $UO_2F_4^{2-}$; (e) isomers of $(UO_2)_2(\mu_2\text{-}OH)_2(HEDDA)_2F_2^{2-}$. Image adapted with permission from Palladino G, Szabó Z, Fischer A, Grenthe I. Structure, equilibrium and ligand exchange dynamics in the binary and ternary dioxouranium(vi)-ethylenediamine-N,N[prime or minute]-diacetic acid-fluoride system: A potentiometric, NMR and X-ray crystallographic study. *Dalton Transactions.* 2006(43):5176–5183. Copyright 2006 Royal Society of Chemistry.

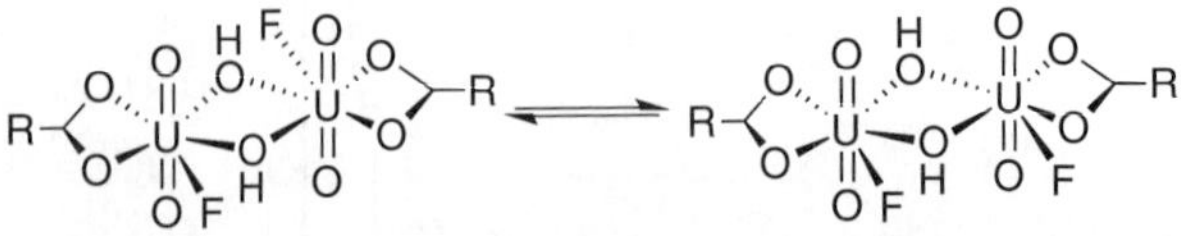

Figure 2.14 Possible isomers of $(UO_2)_2(\mu_2\text{-}OH)_2(EDDA)_2F_2^{2-}$. EDDA is coordinated only through the carboxylate in this model. $R = CH_2NH_2^+CH_2CH_2NH_2^+CH_2COO$. Image adapted with permission from Palladino G, Szabó Z, Fischer A, Grenthe I. Structure, equilibrium, and ligand exchange dynamics in the binary and ternary dioxouranium(vi)-ethylenediamine-N,N[prime or minute]-diacetic acid-fluoride system: A potentiometric, NMR and X-ray crystallographic study. *Dalton Transactions.* 2006(43): 5176–5183. Copyright 2006 Royal Society of Chemistry.

From the experimental ^{19}F-NMR data, the equilibrium constant for the formation of these dimers from the initial $UO_2(EDDA)F^-$ was determined according to Equation 2.4:

$$2UO_2(EDDA)F^- + 2H_2O \rightleftharpoons (UO_2)_2(\mu_2-OH)_2(EDDA)_2F_2^{2-} + 2OH^- \qquad \boxed{\log K = -2.7 \pm 0.1} \tag{2.4}$$

This report illustrates the diverse applications of ^{19}F-NMR spectroscopy in actinide chemistry. The use of this isotope proved advantageous in identifying structural nuances that would otherwise be difficult to detect, as well as in determining the equilibrium processes operative in the formation of the uranyl dimers. While the presence of actinide-fluoride clusters and extended networks are unknown to the best of our knowledge, these techniques still provide utility in various aspects of aqueous actinide chemistry such as the investigation of monomeric species.

2.4.2 Monitoring Product Formation/Evolution by NMR Spectroscopy

In the study of aqueous actinide chemistry, investigations of complex formation and equilibrium processes are deceptively similar and often overlap with one another. While a discussion of equilibrium studies using NMR was discussed earlier, a recent representative example of species evolution monitored by NMR spectroscopy is presented. The dynamic coordination of glycolic acid was monitored using ^1H-NMR in a study of the uranyl cation. In the investigation of a binary system, Szabó and Grenthe monitored the proton signals from coordinated glycolate at various pH levels and used the obtained spectra to infer the inner coordination sphere about the uranyl center (Figure 2.15) (60).

Under acidic conditions (pH = 3.0), a combination of free ligand and either compound ***i***, ***ii***, or ***iii*** (Figure 2.15a) is present in solution; due to the similar chemical environment of these three complexes and the lack of an internal standard to integrate these signals, the authors were unable to determine which of these compounds produced the resonance at 3.9 ppm. The solution was titrated to pH 3.5 using aqueous sodium hydroxide, which resulted in the emergence of an additional resonance at 6.4 ppm which was attributed to compounds ***iv*** or ***v***. The downfield peak continues to grow as the pH increases up to a value of 4.5, wherein a third signal, attributed to the symmetric complex ***vi***, appears at 6.4 ppm. Increasing the pH further to slightly basic conditions (pH 7.2) ultimately results in the exclusive formation of ***vi*** and free ligand. Additional examples of product formation will be presented that similarly use NMR to probe this dynamic process (*vide infra*).

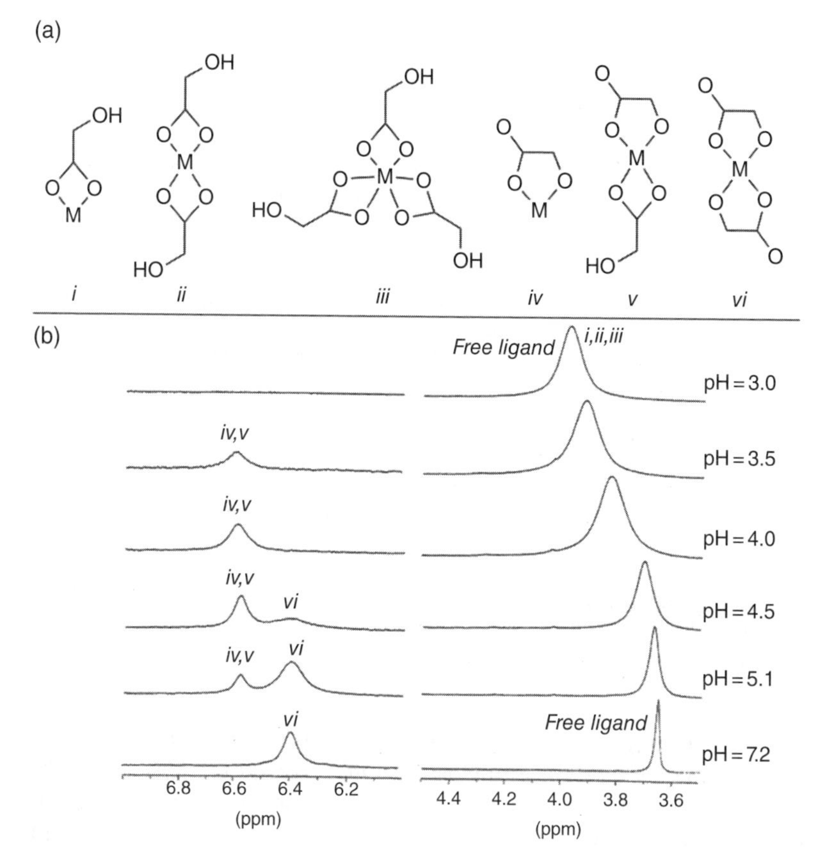

Figure 2.15 (a) Various coordination environments of the uranyl center (M) in the presence of glycolic acid in water. (b) pH-dependent ^{1}H-NMR spectra for a binary uranyl-glycolate system. Images adapted with permission from Szabó Z, Grenthe I. Potentiometric and multinuclear NMR study of the binary and ternary uranium(VI) – L – fluoride systems, where L Is α-hydroxycarboxylate or glycine. *Inorganic Chemistry.* 2000;39(22):5036–5043. Copyright 2000 American Chemical Society.

2.4.3 Monitoring Actinide Self-Assembly by NMR Spectroscopy

One of the main challenges in understanding the chemistry of actinides in aqueous solution that persists today is the characterization of hydrolysis and condensation products, as well as self-assembly processes in solution. Numerous experimental techniques have been applied to observe such transformations, yet NMR still serves as a convenient, in-house analytical tool to discern solution dynamics. In addition to protium and carbon-13, oxygen-17 is an extremely useful nucleus for the study of aqueous actinide chemistry, especially in the detection of hydrolysis and/or condensation products and, as illustrated by the example that follows, cluster self-assembly. In a recent study by Hu et al., the electrolytic formation of a ligand-stabilized thorium hexanuclear cluster was investigated by ^{17}O-NMR methods (61). The use of heteronuclear NMR in this study showed that, contrary to prevailing theories, metal cluster aggregation is strongly influenced by ligand-directing effects. It was found that bulk electrolysis on an acidic solution of $Th(ClO_4)_4$ followed by sequential addition of $H_2{}^{17}O$ and glycine yielded a discrete

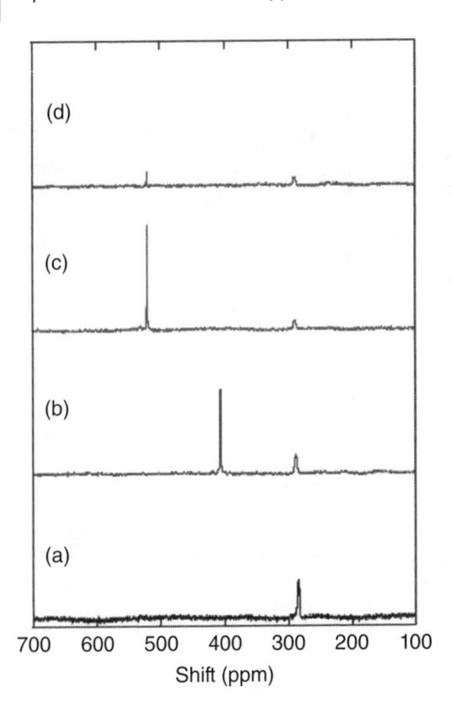

Figure 2.16 ^{17}O-NMR spectra of the reaction described in the text. (a) Initial thorium perchlorate solution in aqueous perchloric acid. Peak at 282 ppm is assigned to naturally abundant oxygen-17 in the perchlorate ion. (b) Reaction mixture after bulk electrolysis at 74 mA. Signal at 400 ppm corresponds to bridging-oxo and -hydroxo species in an intermediate thorium dimer. (c) Spectrum of the thorium dimer after addition of $H_2{}^{17}O$, then glycine. Peak at 512 ppm, as well as small satellite peak at 522 ppm, is attributed to bridging-oxo and -hydroxo units in the hexameric unit. (d) Spectrum of the thorium dimer after addition of glycine followed by $H_2{}^{17}O$. Similar ^{17}O-resonances are seen in spectrum (c). Image adapted with permission from Hu Y-J, Knope KE, Skanthakumar S, Soderholm L. Understanding the ligand-directed assembly of a hexanuclear Th(IV) molecular cluster in aqueous solution. *European Journal of Inorganic Chemistry.* 2013;2013(24):4159–4163. Copyright 2013 Wiley.

oxo/hydroxo-bridged thorium hexanuclear cluster. Without the addition of glycine, the hexanuclear cluster remained absent. Immediately noteworthy is the absence of the thorium hexanuclear cluster when no glycine is added. To support the proposed ligand-directed assembly, ^{17}O-NMR experiments were performed, varying the order of reagent addition (Figure 2.16). By observing the ^{17}O-NMR spectra in the figure, it is evident that the addition of glycine is the determinant step in the formation of the thorium hexamer.

The lower degree of ^{17}O incorporation seen in Figure 2.16d is telling of this process in that once the hexanuclear cluster is formed after addition of glycine, minimal oxygen exchange occurs with the surrounding solution. Further, the researchers performed another ^{17}O-NMR experiment wherein the spectrum was acquired after addition of $H_2{}^{17}O$ without the addition of glycine; in this instance, no peaks were seen beyond 400 ppm, providing further support to the notion that that the thorium hexanuclear cluster quantitatively assembles from dimeric species in the presence of the carboxylate donor ligand, and thus further insight into the formation and speciation of the actinide metal center was gained.

2.4.4 Following Cluster Stability in Solution by NMR Spectroscopy

Of the modern applications of NMR spectroscopy in inorganic chemistry, one of the most basic is the elucidation of the static structure of species in solution. By probing the NMR-active nuclei present in a given compound, a considerable amount of information can be inferred regarding the structural composition of actinide compound and clusters. It is well established that in aqueous actinide systems in particular, slight variations in the pH of the reaction media can result in significant variations in the products obtained. This was aptly illustrated in a study by Dembowski and coworkers which reported the preparation of a uranyl peroxide nanocluster containing pyrophosphate bridging units of the formula $Na_{44}K_6[(UO_2)_{24}(O_2)_{24}(P_2O_7)_{12}][IO_3]_2\cdot 140H_2O$ (**U$_{24}$**); while X-ray and neutron diffraction studies were used to ascertain the solid-state structure of this cluster, solution-state ^{31}P-NMR was used as a suitable technique to determine the stability and dynamic behavior of this compound (Figure 2.17) (62).

The nanocluster shown in Figure 2.17a was prepared first by the synthesis of studtite, followed by addition of sodium pyrophosphate, titration with aqueous tetraethylammonium hydroxide, acidification to pH 7.10 with iodic acid, and evaporation to yield pure **U$_{24}$**. In this structure, cation–anion interactions between potassium and uranyl oxygen atoms within the cage structure give rise to two sets of inequivalent pyrophosphates. The ^{31}P-NMR spectrum of this compound dissolved in D_2O revealed two sets of resonances in a 1:2 ratio at 4.28 and 3.45 ppm, respectively (Figure 2.17b); the presence of the two signals is consistent with the inequivalent set of pyrophosphate groups in the solid-state structure, which is held together largely through the aforementioned cation–anion interactions. However, a more impressive finding in this data is the retention of asymmetry in the presence of solvent water. By knowing the characteristic spectrum of the cluster, the stability of **U$_{24}$** as a function of pH could be easily determined (Figure 2.17c). The in situ experiment was prepared in an analogous manner to the synthesis described above up to the point of titration with triethylammonium hydroxide. At the onset of the NMR study, the solution was strongly alkaline (pH = 11), and the corresponding ^{31}P-NMR spectrum showed no sign of aggregation to the target nanocluster. Upon acidification with iodic acid, **U$_{24}$** begins forming at pH 10 according to the emergence of the ^{31}P signals at 4.28 and 3.45 ppm. As the acidity of the solution continued to increase through the addition of HIO_3, a gradual rise in the product concentration is seen up to pH 6; however, at a pH of 5, the two corresponding NMR signals disappear entirely, indicating rapid decomposition of the nanocluster.

The use of ^{31}P-NMR spectroscopy was useful in this study for the characterization of a single product with discrete chemical moieties, yet when multiple chemical sites or conformers exist, the interpretation of data can become more complex. Nevertheless, this spectroscopic technique can be utilized to detect slight variations in the coordination environment of a single compound, and even to distinguish between similarly constructed conformers. Such techniques were used to this effect in the analysis of previously reported **U$_{22}$** and **U$_{28}$** clusters (Figure 2.18) (63).

The initial study by Qiu et al. presented several pieces of data which described various aspects of these clusters, but a later study from this group investigated the solution-state dynamics of the clusters using ^{31}P-NMR and other multidimensional NMR experiments (64). It was established in this work that the phosphite bridges (which bridge adjacent uranyl sites) possess seven possible orientations of binding, five of which were identified in the crystal structure of **U$_{22}$** and **U$_{28}$** (Figure 2.19). After identification of the phosphite conformers present in the prepared clusters according to ^{31}P-NMR, variable temperature experiments were performed, and the ligand fluctuation was monitored as a function of temperature (Figure 2.20). These proposed chemical shifts and their assignments to the possible conformers were deduced according to their predicted chemical environments; however, two-dimensional NMR studies also confirmed the cluster nuclearity in solution as well as experimentally supporting the ^{31}P peak assignments.

The first of the 2D-NMR experiments performed to determine speciation was diffusion-ordered spectroscopy (DOSY). In this NMR experiment, the translational

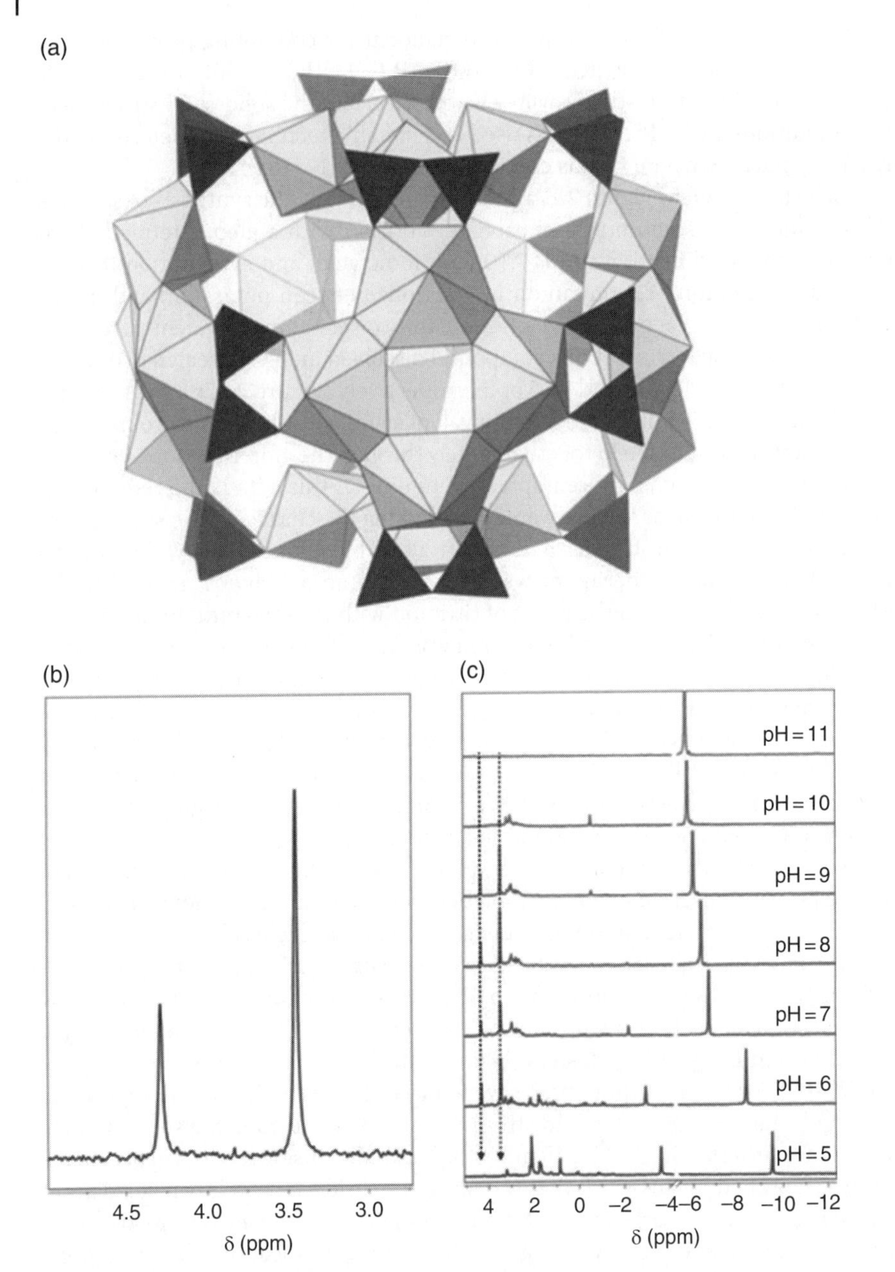

Figure 2.17 (a) Polyhedral representation of $\mathbf{U_{24}}$ nanocluster. White polyhedra represent uranyl centers, and black tetrahedra are pyrophosphate linkers. Na, K, O, I, and H atoms are not shown, for clarity. As can be seen in the image above, two unique environments of pyrophosphate exist; four which are arranged around the equatorial girdle of the cluster, and eight which cap the top and bottom of the structure. (b) ^{31}P-NMR spectrum of pure $\mathbf{U_{24}}$. (c) ^{31}P-NMR study of $\mathbf{U_{24}}$ formation as a function of pH. Image adapted with permission from Dembowski M, Olds TA, Pellegrini KL, Hoffmann C, Wang X, Hickam S, et al. Solution ^{31}P NMR Study of the acid-catalyzed formation of a highly charged $\{U_{24}Pp_{12}\}$ nanocluster, $[(UO_2)_{24}(O_2)_{24}(P_2O_7)_{12}]^{48-}$, and its structural characterization in the solid state using single-crystal neutron diffraction. *Journal of the American Chemical Society.* 2016;138(27): 8547–8553. Copyright 2016 American Chemical Society.

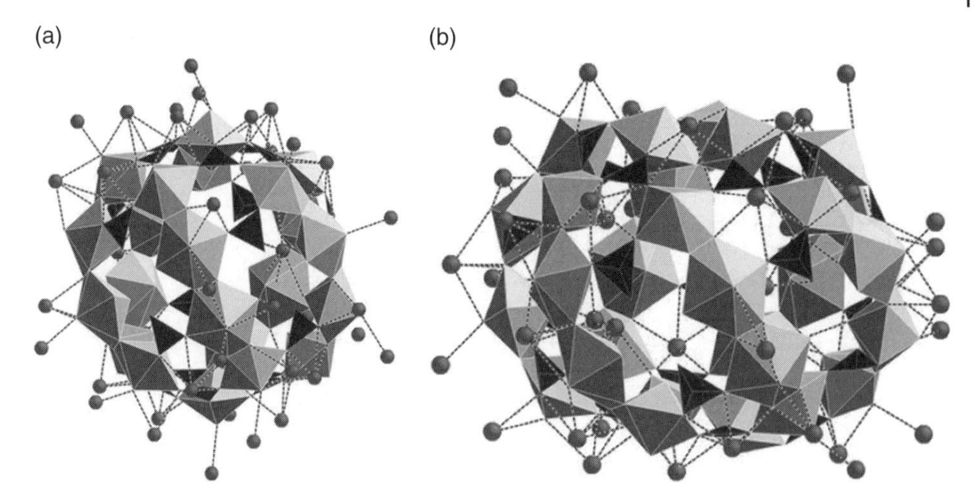

Figure 2.18 Polyhedral representation of (a) $K_{14.73}[(UO_2)_{22}(O2)_{15}(HPO_3)_{20}(H_2O)_{10}]^{11-}$ (**U_{22}**), and (b) $K_{24}[(UO_2)_{28}(O_2)_{20}(HPO_3)_{24}(H_2O)_{12}]^{8-}$ (**U_{28}**). Uranyl centers, phosphite bridges, and potassium ions are represented by light gray polyhedra, black tetrahedra, and dark gray spheres, respectively. Oxygen and hydrogen atoms are not shown. Image reproduced with permission from Qiu J, Nguyen K, Jouffret L, Szymanowski JES, Burns PC. Time-resolved assembly of chiral uranyl peroxo cage clusters containing belts of polyhedra. *Inorganic Chemistry*. 2013;52(1):337–345. Copyright 2013 American Chemical Society.

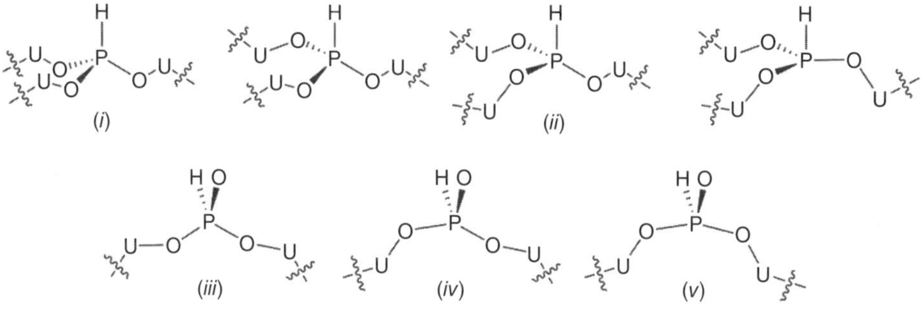

Figure 2.19 Possible conformation of phosphite (HPO_3^{2-}) bonding in the uranyl cluster. Structures noted with Roman numerals have been identified in the solid-state structure of **U_{22}** and **U_{28}**. Image adapted with permission from Oliveri AF, Pilgrim CD, Qiu J., Colla CA, Burns PC, Casey WH. Dynamic phosphonic bridges in aqueous uranyl clusters. *European Journal of Inorganic Chemistry*. 2016;2016(6):797–801. Copyright 2016 Wiley.

diffusion coefficient (D) is represented on the vertical axis and is plotted against a one-dimensional spectrum, most often ^1H. By experimentally calculating the viscosity of the solution, an accurate value for the diffusion coefficient can be calculated which is related directly to the hydrodynamic radius of the dissolved species according to the Stokes–Einstein equation (65). When this experiment was run on solutions of the **U_{22}** and **U_{28}** clusters, the spectra shown in Figure 2.21 were generated.

The selected data points of the cluster (in this instance, protons of the bridging phosphite) are used in the calculation of the diffusion coefficient, and subsequently, the

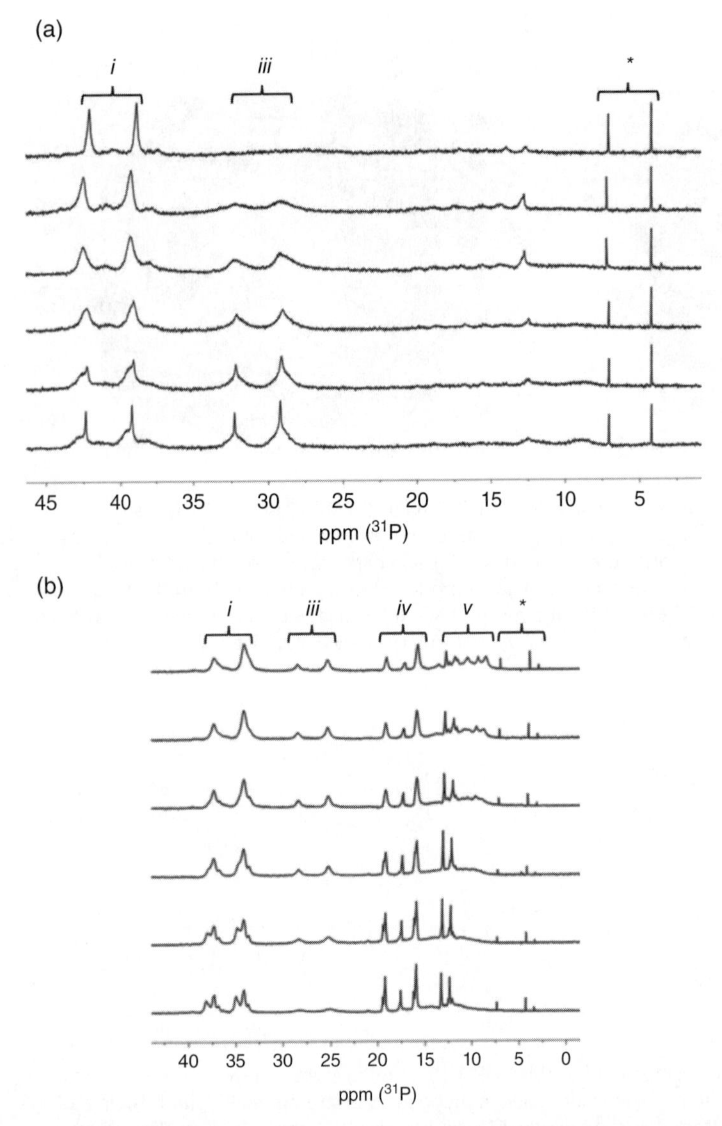

Figure 2.20 ^{31}P-NMR spectra of (a) $\mathbf{U_{22}}$, and (b) $\mathbf{U_{28}}$. Roman numerals correspond to conformation of phosphite (HPO_3^{2-}) shown in Figure 2.19. Asterisks denote free H_3PO_3. Image adapted with permission from Oliveri AF, Pilgrim CD, Qiu J, Colla CA, Burns PC, Casey WH. Dynamic phosphonic bridges in aqueous uranyl clusters. *European Journal of Inorganic Chemistry.* 2016;2016(6):797–801. Copyright 2016 Wiley.

hydrodynamic radius of the cluster. The acquired data in this study showed D values for $\mathbf{U_{22}}$ and $\mathbf{U_{28}}$ of $0.90(16) \times 10^{-10}$ and $1.23(7) \times 10^{-10}$ m^2s^{-1}, respectively; calculating the hydrodynamic radii from the experimental diffusion coefficients yielded values of 9.8(4) and 12.3(5) Å for $\mathbf{U_{22}}$ and $\mathbf{U_{28}}$, respectively. These numbers are in excellent agreement with previous studies that investigated the size of these clusters by single-crystal X-ray diffraction studies and small-angle X-ray scattering (SAXS) (63, 66).

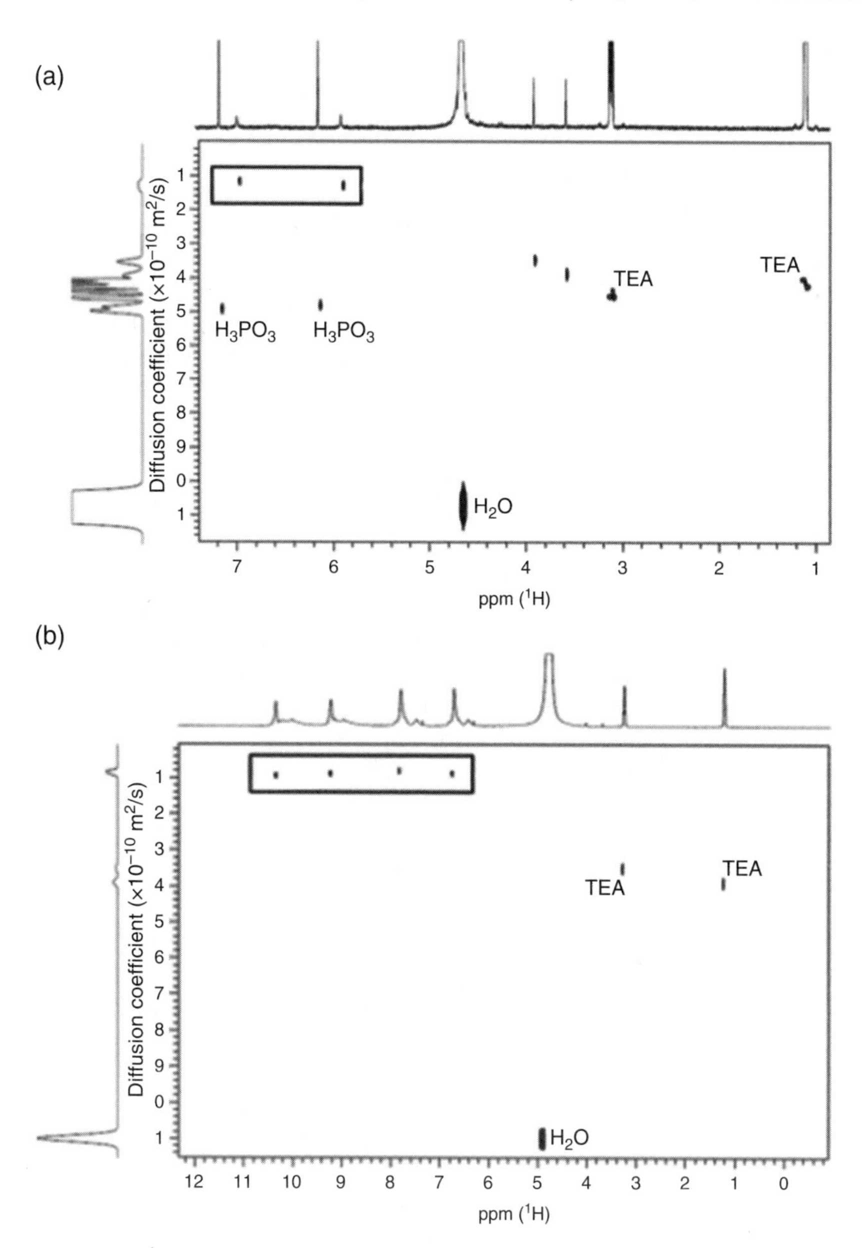

Figure 2.21 ^{1}H-DOSY spectra of (a) $\mathbf{U_{22}}$ cluster, and (b) $\mathbf{U_{28}}$ cluster. The proton signals of the phosphites are enclosed within the inlay box of each spectrum. These data points are used in the calculation of the hydrodynamic radius. Image adapted with permission from Oliveri AF, Pilgrim CD, Qiu J., Colla CA, Burns PC, Casey WH. Dynamic phosphonic bridges in aqueous uranyl clusters. *European Journal of Inorganic Chemistry.* 2016;2016(6):797–801. Copyright 2016 Wiley.

2.4.5 Overview of NMR Spectroscopy in Aqueous Actinide Chemistry

As is illustrated in this brief summary of the analysis of aqueous actinide compounds and clusters, NMR is a powerful experimental technique that may be used for the elucidation of the structure and solution behavior of dissolved inorganic species. Ligand binding, pH, and temperature effects are readily studied using this technique, making it particularly advantageous for dynamic studies including kinetic and thermodynamic investigations. In addition, 118 isotopes are detectable by NMR, providing a range of possible experiments for the study of actinide compounds. Despite the attractive properties of NMR, the analytical method suffers from certain disadvantages. First, it is important to note that NMR does not achieve direct detection of the actinide center, but rather indirect measurements of the ligand environment provide information that are correlated to metal speciation. In addition, relatively high sample concentrations are needed for data acquisition (1 mM); this is especially problematic when working with the heavier synthetic actinide elements and the small sample sizes that are usually employed. When detecting nuclei of low natural abundance, such as deuterium or oxygen-17, another challenge presents in the necessity to perform isotopic labeling experiments that can often prove tedious. Finally, spectral crowding of signals in similar chemical environments can make data analysis difficult or impossible using classical one-dimensional NMR analysis. Even with these challenges, though, a suite of heteronuclear and two-dimensional methods is readily available which provide several options to circumvent potential deficiencies. Such approaches have been demonstrated extensively in modern aqueous actinide chemistry, and improvements in instrumental technologies and spectral resolution aid in resolving complicated data.

2.5 Raman Spectroscopy

Raman scattering was first discovered by Sir Chandrasekhra Venkata Raman in 1928, and since then has emerged as a powerful analytical tool across a number of disciplines, including actinide chemistry, for species identification and quantitative analysis (67). First applied to UO_2^{2+}-containing systems in 1938, Raman spectroscopy is now routinely used for the characterization of solid-state compounds and is increasingly applied to the detection and measurement of An species formation and indeed, identification, in solution (57, 68–76). Raman spectroscopy, as well as providing the means to measure the formation and stability of complexes and clusters, can yield basic information about the structure and bonding of a complex. As we will discuss in this section, Raman spectroscopy has found significant utility for pentavalent and hexavalent actinide ions (U, Np, Pu, and Am), which are characterized by the linear dioxo actinyl, $O=An=O^{+/2+}$. The Raman-active symmetric ν_1 mode of the actinyl ions is highly sensitive to the chemical environment about the metal center, particularly the inner coordination sphere, and as we will see through examples highlighted in this section, the position of this peak can be used for fingerprint identification of actinyl species in solution; the symmetric vibrations are known to vary on the basis of ligand identity and metal ion coordination number, and the frequencies can be used to probe actinide ligand interactions as well as to determine actinide speciation in solution (77, 78). Furthermore, spectral deconvolution can provide quantitative information about solution species (71).

Though Raman has arguably been most successfully applied to the pentavalent and hexavalent actinides that adopt a "yl," Raman has also been used to interrogate speciation of tetravalent Th-Pu and pentavalent Pa systems, using the position of metal–ligand vibrational modes as characteristic signatures from which speciation may be inferred (72, 79). Yet by comparison to the higher-valent oxidation states, the application of Raman to these non-"yl" moieties is fairly limited, particularly for those systems in which the ligand itself exhibits a complex Raman spectrum that may yield overlapping modes which would otherwise be used for species identification (80–82). In these instances, computational studies have proved quite valuable for spectral analysis (82). Computation has also played an important role in attributing the observed bands to Raman-active vibrational modes (83).

It is also worth mentioning that Raman spectroscopy offers many advantages over other spectroscopic techniques, including infrared spectroscopy, such as minimal to no sample preparation; small sample sizes; the ability to measure through glass; and easy sample containment, which is attractive for minimizing radiological risks and maintaining the integrity of the sample. Moreover, Raman is particularly well suited for examining aqueous solutions as water is a weak Raman scatterer, and signals resulting from the solvent do not confound the spectrum. Nevertheless, there are also inherent disadvantages of Raman; for example, during the data collection, a laser is used to observe weak Raman scatters. This can cause local heating, decomposition, and/or oxidation of the sample. Moreover, some compounds fluoresce when irradiated by the laser beam and thus interfere with vibrational signals. Despite these limitations, Raman spectroscopy remains a technique that is routinely used to characterize the speciation of actinides, both in solution and solid-state samples (50, 72, 74, 75, 79).

2.5.1 Cluster Formation and Assembly

Raman has been applied toward understanding the speciation and self-assembly of uranyl-peroxo clusters with considerable success (73, 76, 84). One exemplary study followed the speciation of uranyl peroxo units as a function of LiOH/U ratios as well as the evolution of lower-order monomeric units to form higher-order oligomers as a function of time (84). Importantly, such interrogations of solution speciation are largely facilitated by the observation that different peroxo species have characteristic Raman signatures that may be used for the identification of a particular species in solution. For example, as shown in Figure 2.22, the Raman spectrum of crystalline sodium uranyl triperoxide built from uranyl triperoxide monomers (UTs) is characterized by a broad peak at 716 cm^{-1}, as well as bands at 814, 822, and 844 cm^{-1}. Notably, all compounds built from UT structural units characteristically exhibit a broad peak centered at 710–725 cm^{-1} that can be used for its identification in solution. By contrast, this mode is absent from spectra of U24- and U28-peroxo cluster-containing compounds, rather exhibiting bands at 814 and 845 cm^{-1} for the lithium salts of the U24 cluster (Figure 2.22) and 808 and 836 cm – ^{1} for the lithium salt of the U28 cluster. Though peaks were only recently definitively attributed to certain vibrations, differences in the signatures for the monomeric, U24, and U28-peroxo clusters have been used for the identification of these species in solution, as illustrated by the following examples.

Falaise and Nyman, for example, recently examined the formation of U24 and U28 clusters from monomeric uranyl triperoxide, $UO_2(O_2)_3$, units using SAXS and Raman

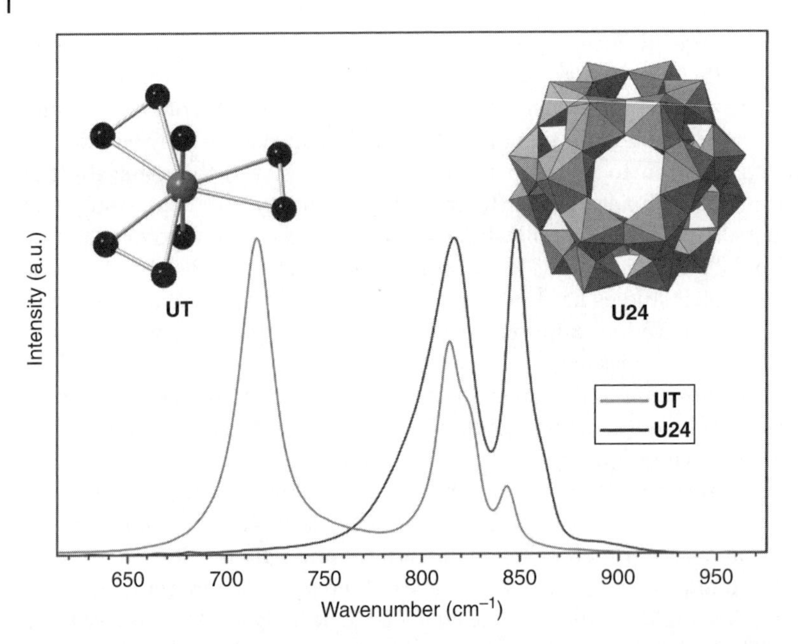

Figure 2.22 Raman spectra for crystalline samples of the sodium salt of the uranyl triperoxide monomer (UT) and Li@U24 cluster, highlighting the signatures that can be used for their identification in solution. Notably, the Raman spectra of UT salts characteristically exhibit a band centered near 710–722 cm^{-1} that is absent from the spectra of larger oligomers including the U24- and U28-peroxide clusters. Adapted with permission from Dembowski M, Bernales V, Qiu J, Hickam S, Gaspar G, Gagliardi L, Burns PC. *Inorganic Chemistry.* Copyright 2017 American Chemical Society.

spectroscopy (84). Using synthetic techniques established in the literature, the authors first isolated a series of compounds including $Li_4[UO_2(O_2)_3]$, $Li_{24}[UO_2(O_2)(OH)]_{24}{\cdot}nH_2O$ ($Li@U_{24}$), and $Li_{28}[UO_2(O_2)_{1.5}]_{28}{\cdot}nH_2O$ ($Li@U_{28}$). Raman spectra were collected for crystalline samples of the compounds and shown to exhibit different vibrational signatures that could be used for the subsequent identification of UT-, U24-, and U28-peroxo units in solution (84). Using the different signatures characteristic of each of these uranyl-peroxo units, the authors examined the product evolution from reaction solutions as a function of LiOH/U ratios. These data are shown in Figure 2.23a. In the spectrum, obtained at a LiOH/U ratio of 2.7, the peaks at 809 and 836 cm^{-1} index to the U28 peroxo cluster, while the peak at 874 cm^{-1} can be attributed to free peroxide. With increasing LiOH/U ratios (>3.5), however, a broad peak is observed at 710 cm^{-1}, consistent with a UT species. It is worth noting that complementary SAXS experiments on these solutions showed no discernible scattering after 2 days, yet after 10 days, solutions with LiOH/U ratios of 4 and 5 showed scattering consistent with the U28 cluster, suggesting that the monomeric UT units assembled into the U28 cluster over time. Solutions with LiOH/U = 10 initially showed peaks at 709, 816, 846, and 874 cm^{-1}, with the first three peaks correlating to the UT, and the peak at 874 arising from free peroxide. After two days, only those bands consistent with the UT monomers were present (Figure 2.23). Interestingly, Raman spectra collected for the aged solutions (Figure 2.23b) showed decreased intensity of the peak at 710 cm^{-1}, while bands at 814 and 846 cm^{-1} associated with the U24 peroxo cluster grew in intensity (84). This evolution in the Raman spectrum (and supported by SAXS data) suggests that U24 assembly occurs

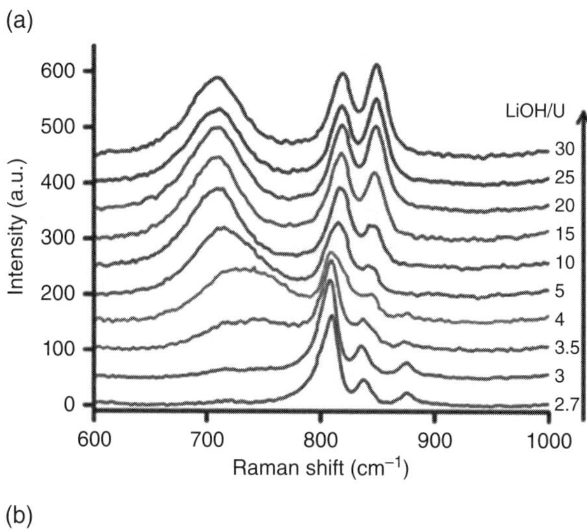

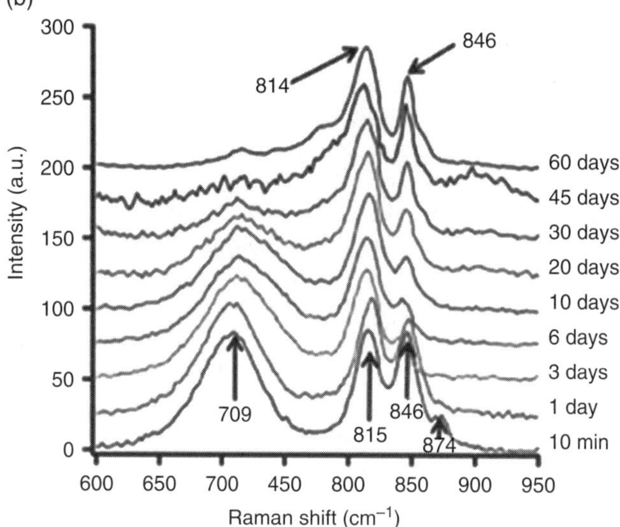

Figure 2.23 (a) Raman spectra of aqueous solutions of U(VI)/H_2O_2/LiOH at variable LiOH/U ratios, after two days. (b) Evolution of Raman spectra of the solution with an LiOH/U ratio of 10. A slow conversion of the monomers to U24 occurs at high hydroxide concentrations over time. Adapted with permission from Falaise C, Nyman M. The key role of U28 in the aqueous self-assembly of uranyl peroxide nanocages. *Chemistry – A European Journal*. 2016;22(41):14678–14687. Copyright 2016 Wiley.

through the linking of UT units. As compared to the solution at LiOH/U = 10, those at higher LiOH/U ratios showed slower onset of U24 cluster formation, with the rate of cluster assembly slowed at higher pH.

2.5.2 Spectral Deconvolution of Raman Data to Yield Speciation

In a monodisperse solution, the assignment of peaks in the Raman vibrational spectrum can be fairly straightforward. Especially when considering the solution-state behavior and bonding of the actinides, various discreet spectral features may exist which provide an added level of convenience in speciation assignments using such spectroscopic methods. However, the aqueous reactivity of the actinides is often more complex and accompanied by the presence of several species in solution. Each of these bands can be considered to produce a single vibrational signal which result in a complex spectral signature of overlapping bands, making the ability to determine even basic qualitative relationships extremely difficult. Recently, Kim et al. investigated the solution speciation and stability of a uranyl peroxocarbonate product over a broad pH range (71). The solutions were prepared by reacting UO_2 in aqueous Na_2CO_3 and H_2O_2, and the pH was adjusted by the addition of HNO_3 or NaOH solutions (71). Under basic conditions (pH 8.5), a variety of uranyl carbonato and peroxo compounds were found to initially exist in solution and generated the Raman spectrum in Figure 2.24a. From a careful inspection of the corresponding Raman spectrum, the presence of particular compounds and ions is apparent; however, the initial solution is seen to possess a complex series of overlapping bands centered at approximately 800 cm^{-1}.

As depicted, the species in solution which give rise to the complex bands seen in the spectrum of the initial solution would be nearly impossible to decipher. However, using spectral deconvolution, the individual bands corresponding to each uranyl compound in solution can be easily derived. Spectral deconvolution refers to the application of a mathematical algorithm that in effect isolates the individual vibrational modes which comprise a more complex overlapping band; one of the most common of these algorithms is the Fourier self-deconvolution (85). Such a transformation was applied to the spectrum generated by Kim and coworkers and revealed the origin of the signals shown in Figure 2.24b. Application of a deconvolution algorithm deconstructs the complex Raman band at 800 cm^{-1} into its constituent peaks and their relative intensities, and provides a more convenient spectrum through which the speciation of the uranyl compounds in solution can be found. Using this data, it is found that a combination of five uranyl carbonate, peroxocarbonate, and peroxide species is initially present at pH 8.5. However, upon reacting for five days, equilibrium is achieved, and the resulting spectrum reveals small amounts of $UO_2(CO_3)_x(OH)_y^{2-2x-y}$ and $UO_2(O_2)_2^{2-}$ with $UO_2(CO_3)_3^{4-}$ as the major product. The ability to perform spectral deconvolution has been improved by the integration of user-friendly operations in most modern spectral software. Moreover, it is noteworthy that these transformations are not only limited to Raman spectroscopy, but can also be easily applied to UV/vis, fluorescence, and FTIR spectra, making the analysis of complex data achievable even by novice users.

2.5.3 Identifying the Nature of Cation–Cation Interactions in Solution

As discussed in the previous examples, dioxo actinyl units exhibit characteristic and well-defined vibrational bands that are sensitive to changes in the coordination environment about the metal center. This feature makes Raman spectroscopy a useful tool for examining the nature and identity of cation–cation interactions (CCIs) in solution as it has been found that there is a strong correlation between the vibrational energy of the actinyl unit and the nature of CCIs (69, 70). First reported by Sullivan in 1961 (86), cation–cation interactions are a fundamental and indeed prevalent feature in higher-valent actinyl compounds, particularly those of neptunium(V), and refer to an interaction wherein the actinyl oxygen atoms of one $AnO_2^{+/2+}$ unit occupy the equatorial girdle of another actinyl unit as illustrated in the inset in Figure 2.25a. Whereas early work

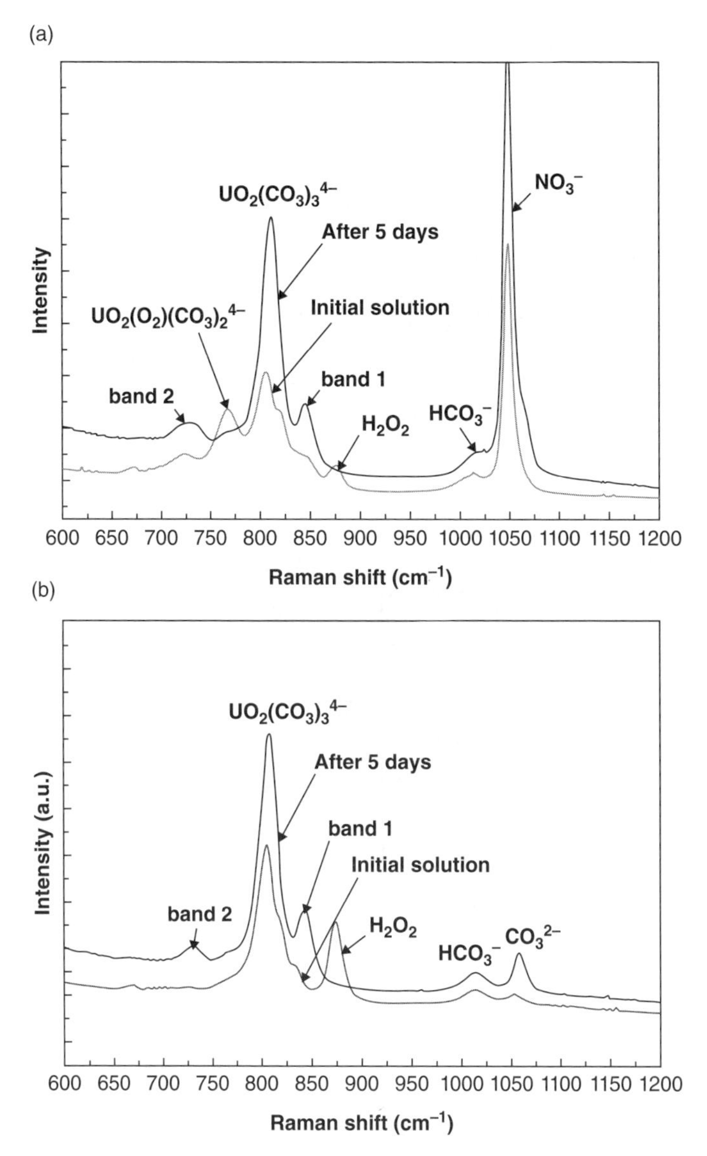

Figure 2.24 (a) Change in Raman spectrum of the reaction containing UO_2, Na_2CO_3, and H_2O_2 after adjusting to pH 8.5 with time. (b) Deconvolution of the initial Raman spectrum shown in (a). Image adapted with permission from Kim K-W, Jung E-C, Lee K-Y, Cho H-R, Lee E-H, Chung D-Y. Evaluation of the behavior of uranium peroxocarbonate complexes in Na–U(VI)–CO_3–OH–H_2O_2 solutions by Raman spectroscopy. *Journal of Physical Chemistry. A* 2012;116(49):12024–12031. Copyright 2012 American Chemical Society.

using Raman spectroscopy provided evidence for the existence of such CCI complexes in solution, details of their structure remained somewhat limited (69). For example, early studies examining the association of Np(V) in $HClO_4$ solutions at various concentrations established characteristic Raman signatures for monomeric and dimeric NpO_2^+ complexes at 765 cm^{-1} and 739 cm^{-1}, respectively (69). From approximately 0.01–0.1 M NpO_2^+, bands consistent with the "yl" symmetric vibration of a monomeric complex

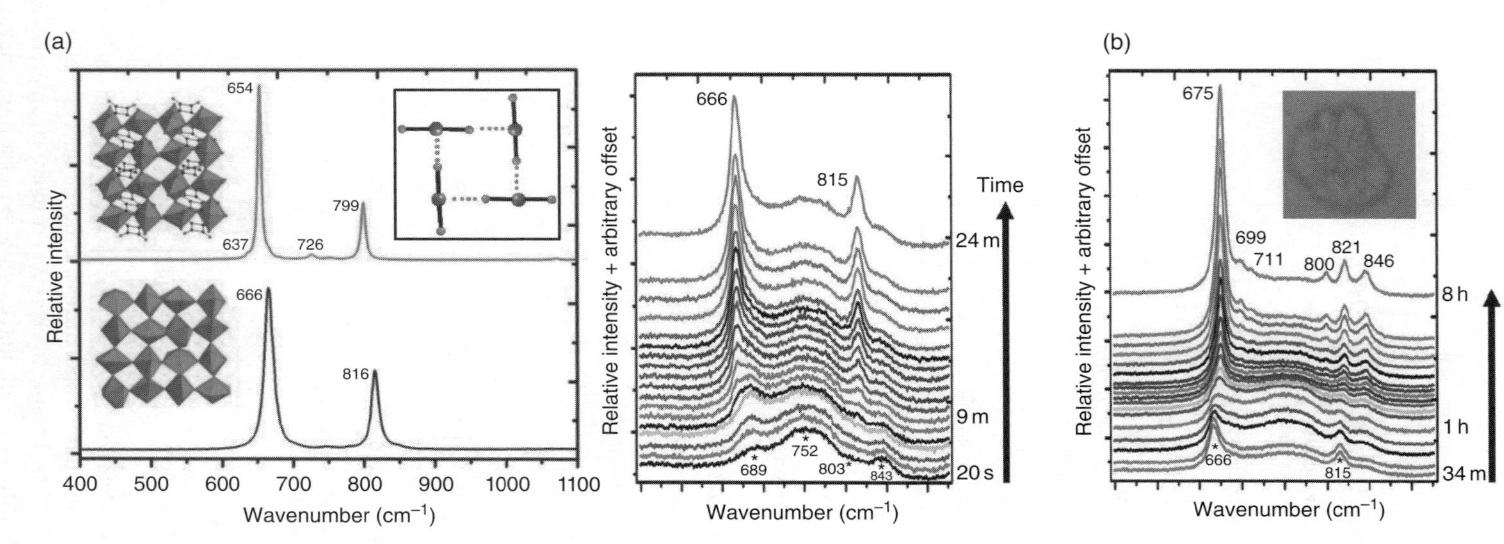

Figure 2.25 (a) Raman spectra of $(NpO_2)_2(C_4O_4)(H_2O)$ (top) and $(NpO_2)_4Cl_4(H_2O)_6{\bullet}(H_2O)_3$ (bottom), both of which consist of square arrangements of cation–cation interactions (inset). Peaks at 654 and 666 cm^{-1} are attributed to the υ_1 mode of the NpO_2^+, and the peaks at 799 and 816 cm^{-1} are consistent with the υ_3 mode. These bands characterize square arrangements of NpO_2^+ CCIs as found in the crystal structures of the two compounds. (b) In situ Raman spectra obtained over the laser-induced slow evaporation of a 1.5 M Np(V) solution. After 8 h, upon evaporation to dryness, a precipitate $(NpO_2)Cl(H_2O)_2$ forms. Single-crystal X-ray diffraction studies shows that the structure consists of 3-D networks of NpO_2^+ cations, whereas neighboring NpO_2^+ adopt a square arrangement of CCIs such as that shown in the inset in (a). Figures adapted/reproduced with permission from Jin GB Three-dimensional network of cation–cation-bound neptunyl(V) squares: Synthesis and in situ Raman spectroscopy studies. *Inorganic Chemistry.* 2016;55(5):2612–2619. Copyright 2016 American Chemical Society.

were observed, and with increasing NpO_2^+ concentrations, from 0.1–1M, bands consistent with self-association or dimerization were discerned. In solutions of 1 M-3M NpO_2^+, bands indicative of higher degrees of association or oligomerization were also observed (69). Unfortunately, the data were insufficient to definitively identify a structure. More recently, a number of solid-state structures containing CCIs have been isolated and structurally characterized. This solid-state data has shed light on the nature of CCIs as they exist in crystalline materials. Moreover, pairing of this structural data with spectroscopic studies of solids as well as solution samples has emerged as an effective technique for investigating CCIs in solution, in addition to the conditions over which they form and persist. To highlight one recent example, Jin used Raman to follow CCI speciation and precipitate formation, using a Raman laser for the purpose of both data collection and controlling the evaporation rate of an aqueous solution (70). Two stock solutions with $[NpO_2^+] = 1M$ and 1.5M were prepared, and Raman spectra were collected on these bulk samples. The 1M solution exhibited peaks at 739 and 765 cm^{-1}, consistent with NpO_2^+ monomers and dimers as had been established by the work of Guillaume et al. (69). Four overlapping peaks were present in the spectrum of the 1.5M solution, which the author attributed to the presence of higher-order oligomers. Such an assignment is consistent with earlier spectroscopic studies on NpO_2^+ solutions with concentrations greater than 1M.

Using the Raman laser as a means of controlling evaporation, the authors followed the formation of CCI networks upon exposure (and evaporation) of the 1.5M solution over time (Figure 2.25). After 24 min, peaks at 666 cm^{-1} and 815 cm^{-1} were observed. Notably, such bands are consistent with those observed in the spectra of solid samples of $(NpO_2)_2(C_4O_4)(H_2O)$ (top) and $(NpO_2)_4Cl_4(H_2O)_6{\bullet}(H_2O)_3$, compounds that consist of extended networks of NpO_2^+ CCIs (Figure 2.25). Importantly, the solid-state structures of these compounds can be described as extended networks formed through the propagation of square arrangements of NpO_2^+ centers that interact through CCIs; such compounds characteristically exhibit bands at approximately 650–670 cm^{-1} and ~800–850 cm^{-1} that are attributed to the υ_1 and υ_3 modes of the NpO_2^+, respectively. The peaks observed in the Raman spectra obtained for the solution after 24 min are then consistent with the presence of higher-order oligomers (square nets of NpO_2^+ CCIs) in solution. Further evaporation of the solution resulted in the formation of a crystalline solid that was shown to be $(NpO_2)Cl(H_2O)_2$, the structure of which can be described as a 3D network of square arrangements of NpO_2^+ CCIs (70).

2.5.4 In the Absence of an "yl": Pa(V) Speciation in HF Solutions

The identification of solution species of those actinide complexes that do not present as an "yl" is not as straightforward. Yet, comparison of the Raman spectra of single crystals of known composition with the solution spectra has been used to identify solution species. For example, the speciation of homoleptic fluoride complexes of Pa(V) in concentrated HF was determined by De Sio and Wilson through the correlation of the Raman spectra obtained for Pa(V) solid-state compounds with those spectra obtained for aqueous Pa(V)/hydrofluoric acid solutions (79). The authors reported the isolation of seven fluoroprotactinate compounds using ammonium, tetramethylammonium

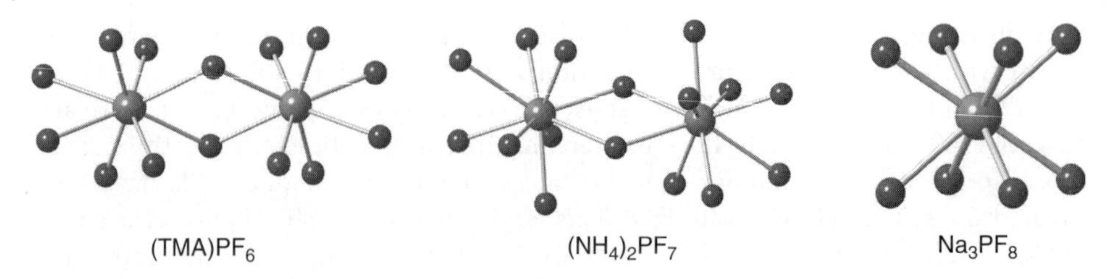

Figure 2.26 Ball-and-stick representation of fluoroprotactinate structural units observed in the solid state by De Sio and Wilson. (79)

(TMA), and various alkali metal cations. The structures were composed of PaF_6^-, PaF_7^{2-}, and PaF_8^{3-} structural units (Figure 2.26) with the Pa(V) in the PaF_7^{2-} and PaF_8^{3-} compounds adopting an eight-coordinate complex, and in PaF_7^{2-}, a nine-coordinate environment. Noteworthy are the structures of PaF_6^- and PaF_7^{2-}, which exist as one-dimensional chains with the Pa(V) metal centers bridged by μ_2-fluorides, whereas PaF_8^{3-} exists in the solid state as an isolated unit.

The Raman spectra collected on single crystals of the fluroroprotactinate compounds are presented in Figure 2.27a. An intense band centered over 545–590 cm^{-1} is assigned to the symmetric Pa-F vibration, characteristic in all of the spectra irrespective of the formula. As seen in the spectra, this band undergoes a slight bathochromic shift in the isostructural series K_2PaF_7, Rb_2PaF_7, and Cs_2PaF_7 as the size of the counterion increases from K^+ to Cs^+ (79). By comparison, the symmetric stretching Pa-F vibration in the Raman spectra of Na_3PaF_8 and $(TMA)(H_3O)PaF_8$ is centered at 557 cm^{-1} and 555 cm^{-1}, respectively. These values are blue shifted as compared to the vibrations of the corresponding PaF_7^{2-}-containing compounds, consistent with the general expectation that higher coordination numbers (nine in PaF_7^{2-}, and eight in PaF_8^{3-}) should lower the vibrational frequency. The highest observed frequency for the Pa-F stretching corresponds to $(TMA)PaF_6$, observed at 581 cm^{-1}. The significant hypsochromic shift as compared to PaF_8^{3-}, both with eight-coordinate Pa(V), is less straightforward, presumably due to the infinite chains present in the structure of $(TMA)PaF_6$ as compared to the isolated monomeric units in Na_3PF_8. In addition to collecting the Raman spectra for the solid-state compounds, the authors proceeded to collect the solution spectra of 1 M Pa(V) in 48% HF, as shown in Figure 2.27b. The broad band observed at 563 cm^{-1} is attributed to the Pa-F symmetric stretching vibration, and the remaining bands are assigned to the sapphire cell used for sample containment. Using plane-polarized light, the authors further demonstrated that the band at 563 cm^{-1} was fully polarized as expected. Although the authors note that the spectra presented for the solid-state compounds and the solution do not exhibit an exact match, the closest frequencies observed in the solid state are those of the PaF_8^{3-} structural units at 557 and 555 cm^{-1}. On the basis of these results as well as observations in *d*-block Nb and Ta fluoride systems, it is evident that as the coordination number about the metal center increases, the vibrational frequency decreases as discussed earlier. The authors hypothesized that in solution, the PaF_7^{2-} and PaF_6^- molecular anions should exhibit bands higher than 590 cm^{-1}, which would correspond to the octacoordinate crystal, $RbPaF_6$, and hence, the dominant solution species PaF_8^{3-}. The authors' findings are supported by previous work on

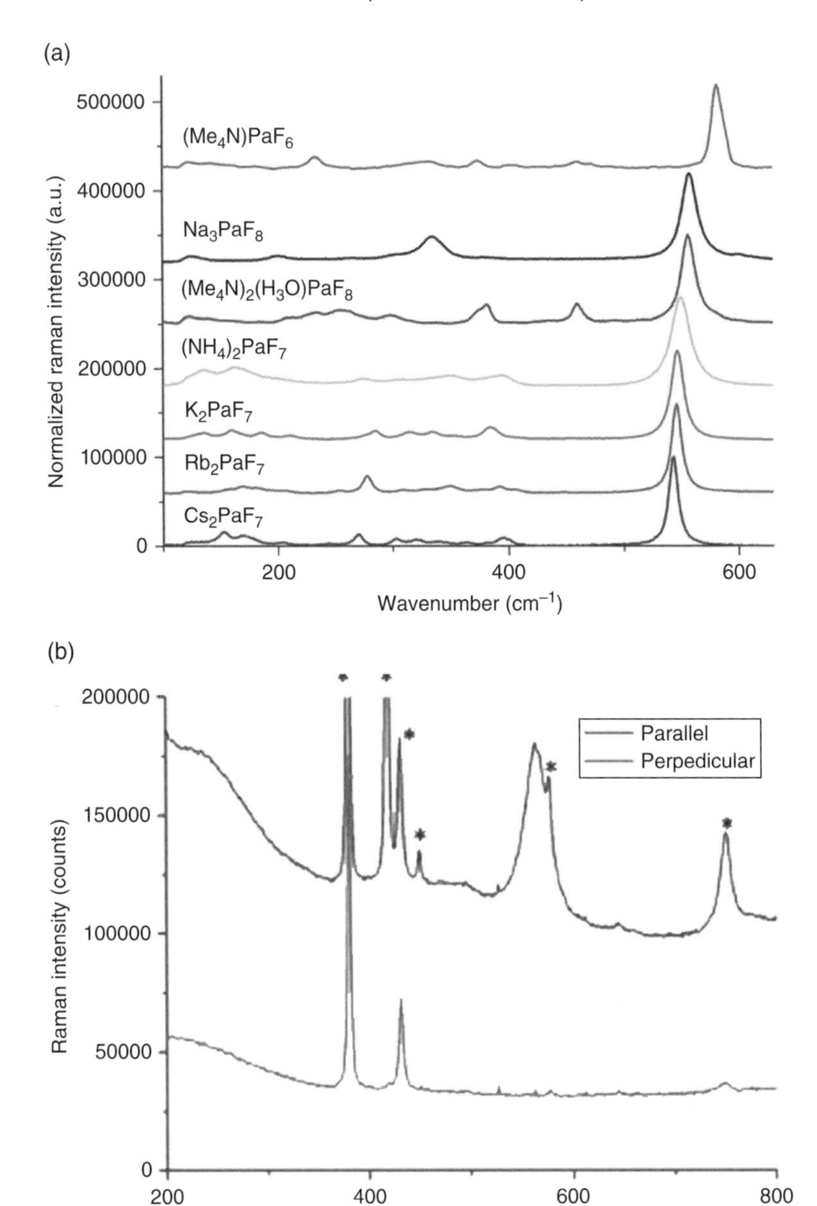

Figure 2.27 Raman spectra collected for (a) solid-state crystalline samples of known composition and structure, and (b) a solution of 1 M Pa in 48% HF using plane-polarized light. Asterisks in (b) indicate sapphire bands. Reproduced with permission from De Sio SM, Wilson RE. Structural and spectroscopic studies of fluoroprotactinates. *Inorganic Chemistry.* 2014;53(3):1750–1755. Copyright 2014 American Chemical Society.

Pa(V) in 0.1–1.42M HF that suggested an equilibrium between seven- and eight-coordinate complexes. More recently, the authors interrogated these systems with the additional use of EXAFS, and the data supported the observation of PaF_8^{3-} under high concentrations of fluoride (87).

2.5.5 Computational Assignment of Vibrational Spectra

The synergy between experiment and theory was illustrated in a study by Dembowski et al. that investigated the characteristic Raman vibrations of uranyl trisperoxide complexes (see Section 2.5.1) using DFT studies (83). B3LYP-level calculations were performed to model the Raman-active vibrations of UT bearing potassium counterion, and the calculated spectrum compared to that of the experimentally determined data (Figure 2.28).

According to the previous studies, the lower energy signal ($720\,cm^{-1}$) was elusive, but known to originate from the mononuclear UT unit. However, the quantum mechanical calculations which provided the predicted Raman spectrum above revealed that the signal of interest actually corresponded to the "yl" moiety of UT. Given that this signal is typically expected to shift a few wavenumbers in different chemical environments, it was especially surprising to see the approximately $100\,cm^{-1}$ red shift of this vibrational mode. The use of DFT in this study succeeded in providing information which was otherwise unable to be determined using empirical methods, and exemplifies the use of in silico experiments in the analysis of spectral data.

2.5.6 Overview of Raman Spectroscopy

One of the advantages of Raman spectroscopy as compared to other techniques is the ease of sample containment, thereby minimizing the risk of radiological exposure; this is particularly attractive when handling radioactive samples. Moreover, Raman is especially well suited for aqueous solutions. As is highlighted in this work, it is useful to draw on known trends in the vibrational frequencies as a function of coordination number to deduce speciation in solutions of unknown composition. Yet, it is also important to note that while this method works well for simple ligand systems, increased complexity in the supporting ligand carries over to the Raman spectrum, thus making data interpretation more challenging. Furthermore, the difficulty in probing solution speciation is perhaps best exemplified by studies examining tetravalent actinide-carboxylates in aqueous solution. In these cases, DFT calculations become quite critical for interpreting the Raman spectra as illustrated in several instances. Though DFT calculations are useful in assigning modes, overlapping bands characteristic of a particular actinide complex or cluster with those of the ligand often confound these systems and make the interpretation of the spectra challenging. Mathematical algorithms have been developed to circumvent these shortcomings and provide a convenient means to resolve such overlapping data, not least of which being spectral deconvolution. As a stand-alone analytical technique, Raman spectroscopy provides a somewhat limited scope of information regarding solution-state speciation. However, when paired with other structural probes and computational models, it has proved to be a powerful technique that can be used for both species identification and quantitative analysis (68).

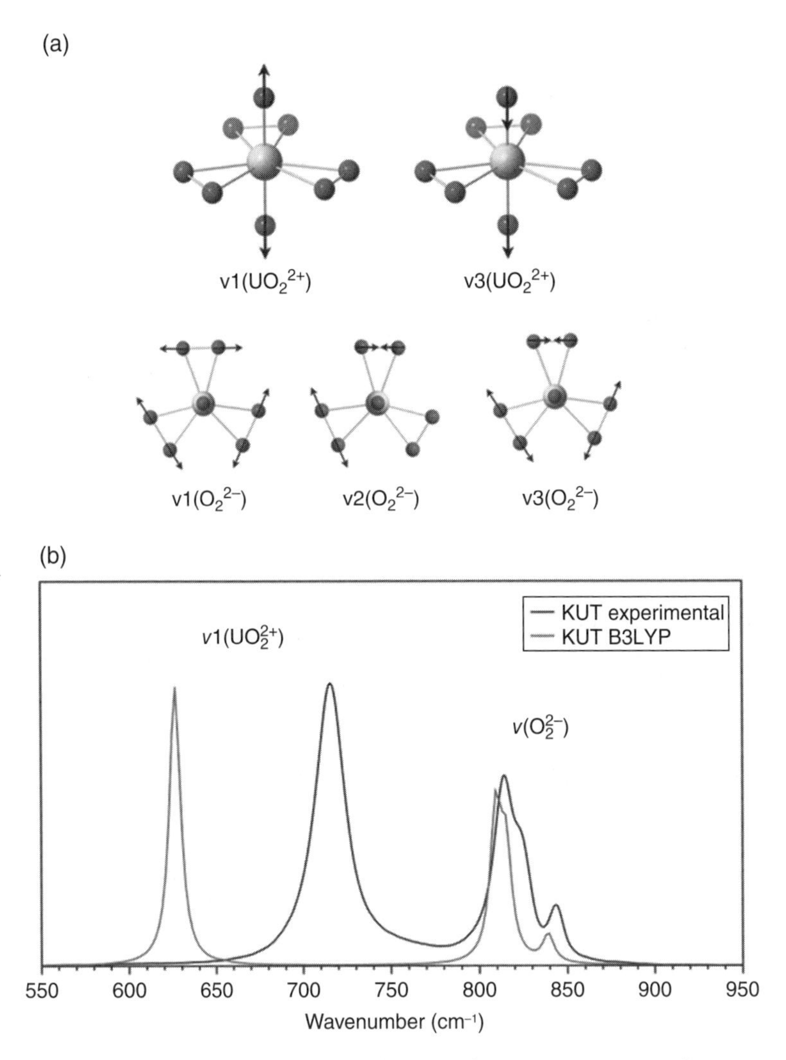

Figure 2.28 (a) Raman- and IR-active vibrations of UT anion. Uranium and oxygen are light and dark spheres, respectively. (b) Experimental and calculated Raman spectra of KUT. Reproduced with permission from Dembowski M, Bernales V, Qiu J, Hickam S, Gaspar G, Gagliardi L, *et al.* Computationally-guided assignment of unexpected signals in the Raman spectra of uranyl triperoxide complexes. *Inorganic Chemistry.* 2017. Copyright 2017 American Chemical Society.

2.6 X-ray Absorption Spectroscopy

X-ray absorption spectroscopy (XAS) is a well-established analytical tool that provides a molecular-level probe of the ion in question. XAS methods are unique when compared to the previously discussed instrumental techniques as it is a single-ion probe that is sensitive to the oxidation state and coordination environment, and can be used on both solid- and solution-state samples (88). The data acquired using modern XAS methods are divided into extended X-ray absorption fine structure (EXAFS) and X-ray absorption near edge structure (XANES); both are common methods to investigate the

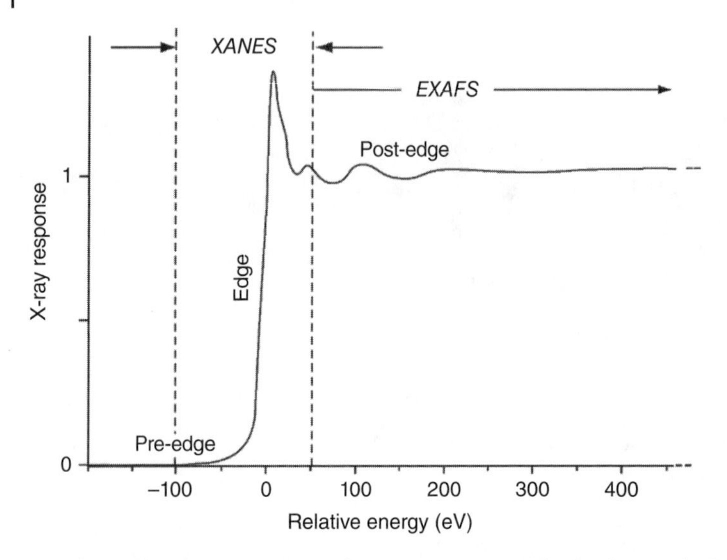

Figure 2.29 Representative X-ray absorption spectrum after background subtraction and normalization. XANES data are typically acquired in the range of 100 eV below the edge up to 50 eV beyond the edge, and the data above this threshold correspond to EXAFS data. Image reproduced with permission from Antonio MR, Soderholm L. X-ray absorption spectroscopy of the actinides. In: Morss LR, Edelstein NM, Fuger J, editors. *The Chemistry of the Actinide and Transactinide Elements.* Dordrecht: Springer Netherlands; 2006. pp. 3086–3198. Copyright 2006 Springer.

X-ray absorption coefficient of a material as a function of energy; however, the latter operates at lower X-ray energies at the absorption edge to elucidate the oxidation state, whereas the former method operates post-edge and provides higher energy data which is directly related to short-range ligand interactions (Figure 2.29). In this section, EXAFS will be discussed as a useful solution-state tool for speciation of the actinides. While XANES is a prominent technique used for oxidation state assignment of the actinides, this particular analysis falls outside the purview of this discussion and is not presented herein.

2.6.1 EXAFS

EXAFS is an incredibly versatile tool for the determination of the speciation of metal ions; this holds true for actinide ions as is evident by the extensive use of this mode of analysis in the characterization of actinide speciation. Owing to the breadth of information that may be obtained, EXAFS has emerged as a widely used technique in the characterization of actinide complexes in solution. Although the introduction of EXAFS occurred nearly a century ago, its true power was realized only recently in the past 40 years. In a typical experiment, synchrotron radiation is used to excite core electrons of the actinide ion, and the ligand electron density further scatters the X-ray photoelectron. These detected photochemical processes provide information regarding the absorbing atom's coordination number, types of coordinating atoms, and metal–ligand bond distances. From the raw data acquired using this analytical method, a Fourier transform is utilized to process the post-edge oscillations (Figure 2.29) into a radial distribution function wherein the observed peaks inform of the interatomic distances to neighboring atoms. However, a limitation of EXAFS is the lack of information on long-range order of the dissolved complex. A detailed description of practical EXAFS acquisition as well as a theoretical background has been summarized in numerous works that describe the aforementioned manipulations (89–96). In the following sections, representative examples of EXAFS are presented which are used to determine speciation of aqueous actinide compounds.

2.6.2 Actinide Solution Speciation by EXAFS

Within the realm of actinide solution chemistry, EXAFS is the technique of choice due to its ability to elucidate solute–solute and solute–solvent interactions. Numerous works have highlighted the utility of EXAFS in studies of the aqueous speciation of actinide ions and yielded data not otherwise accessible using indirect molecular probes as described earlier. In a seminal study by Banik et al., the periodicity of the actinide aquo ions up to berkelium was systematically probed using EXAFS (97). The protactinium(IV) ion was investigated using X-ray absorption studies to determine the characteristics of the An-O bond, and subsequently compared to experimentally determined metal-oxygen bonds of the tetravalent actinide ions (Figure 2.30).

The study revealed that as the ionic radii of An^{4+} decreased, dehydration occurred in moving from nonacoordinate ions (Th^{4+}-Am^{4+}) to octacoordinate ions (Cm^{4+}-Bk^{4+}). Computational analyses using second-order Møller–Plesset perturbation theory (MP2) corroborated these results and were in good agreement with experimentally determined coordination numbers for the An^{4+} aquo ions, wherein increased water coordination was accompanied by bond lengthening (97). These findings provided ample evidence supporting the theory of a "curium break," characterized by a deviation in the electronic and structural behavior of the actinides beyond americium. This study unambiguously showed the power of EXAFS in actinide speciation, and was only enriched through the use of quantum chemical calculations.

In an additional study described earlier in this chapter, the structures of protactinium(V)-fluoride complexes were explored as a function of HF concentration (87). EXAFS again served as an excellent tool for determining the local coordination environment in these complexes. Previous studies using EXAFS and DFT calculations noted the presence of PaF_7^{2-}, but no data supported the presence of other species such as PaF_6^- or $PaF_6(H_2O)^-$ (98). Past studies using Raman spectroscopy suggested the presence of PaF_8^{3-} in solution at very high concentrations of HF, yet these studies remain unsubstantiated up to this point (79). Two fitting models from different single-crystal structures of PaF_8^{3-} (single-shell model and split-shell model) enabled the authors to conclude that at high concentration, protactinium was in fact bound to eight fluorine anions (87). In contrast, low concentrations of HF resulted in a dynamic system in which PaF_8^{3-} and PaF_7^{2-} existed in equilibrium. However, if the HF concentration was reduced even further, a hydrated $PaF_7(H_2O)^{2-}$ species appeared to be present in solution. The combination of Raman (see above) and EXAFS studies in this investigation illustrates the synergistic relationship between otherwise divergent spectroscopic methods to identify species in solution.

The information on short-range interactions that EXAFS can provide is unparalleled in other solution-state studies, and has been vital to the speciation of environmental actinide contaminants. One such work by Hennig surveyed sources of contamination

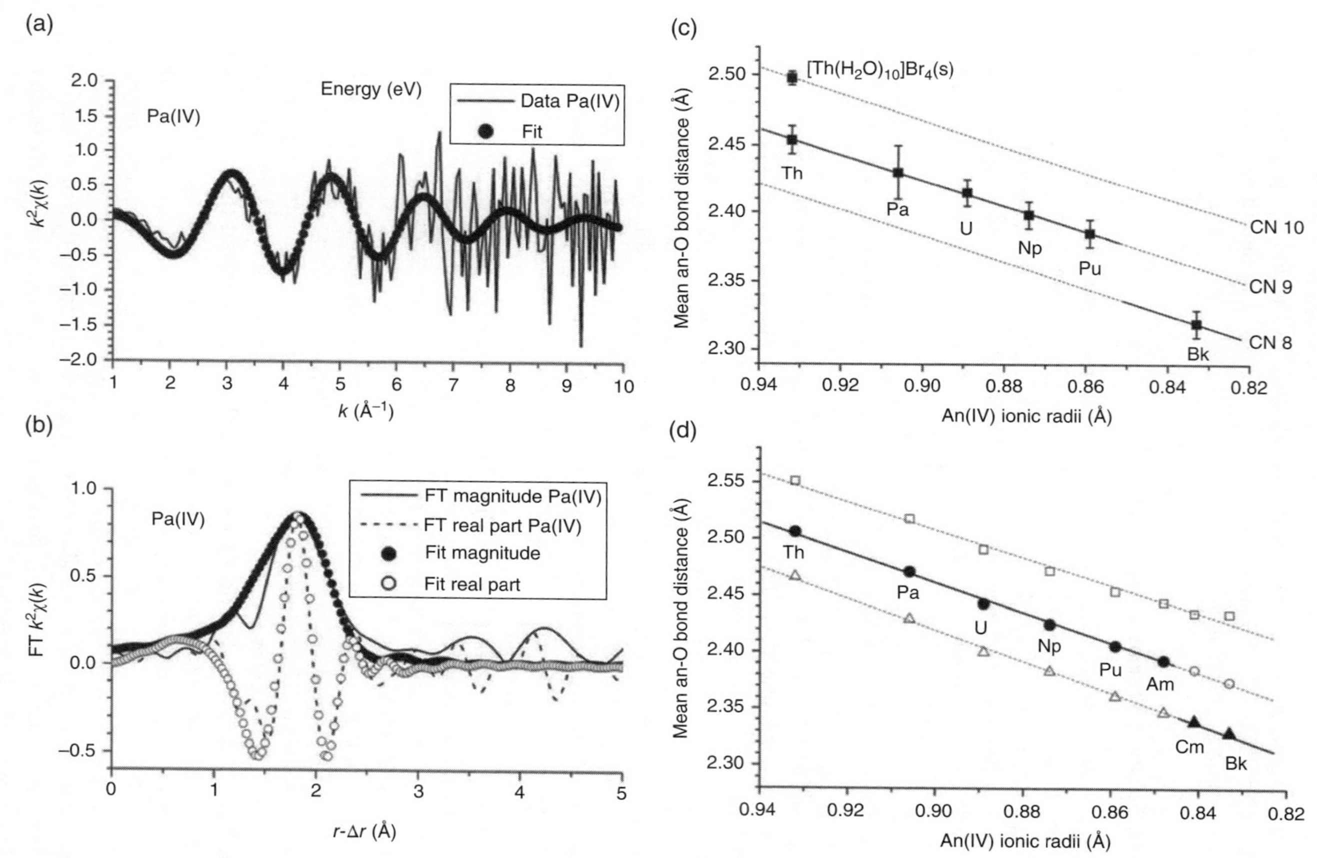

Figure 2.30 (a) Weighted EXAFS data with fit; (b) Fourier-transformed data and fit of 0.3 mM Pa(IV) in 6 M HCl; (c) Experimental average metal-oxygen bond lengths versus tetravalent actinide ionic radii; (d) Average metal-oxygen bond lengths versus tetravalent actinide ionic radii gas-phase optimized 10-, 9-, and 8-coordinate actinide clusters according to quantum chemical calculations. $[An(H_2O)_{10}]^{4+}$ (□), $[An(H_2O)_9]^{4+}\cdot H_2O$ (●), and $[An(H_2O)_8]^{4+}\cdot(H_2O)_2$ (△). Solid symbols represent the most stable clusters, and open symbols correspond to clusters with lower stability. Images adapted with permission from Banik NL, Vallet V, Réal F, Belmecheri RM, Schimmelpfennig B, Rothe J, *et al*. First structural characterization of Pa(IV) in aqueous solution and quantum chemical investigations of the tetravalent actinides up to Bk(IV): the evidence of a curium break. *Dalton Transactions*. 2016;45(2):453–457. Copyright 2016 Royal Society of Chemistry.

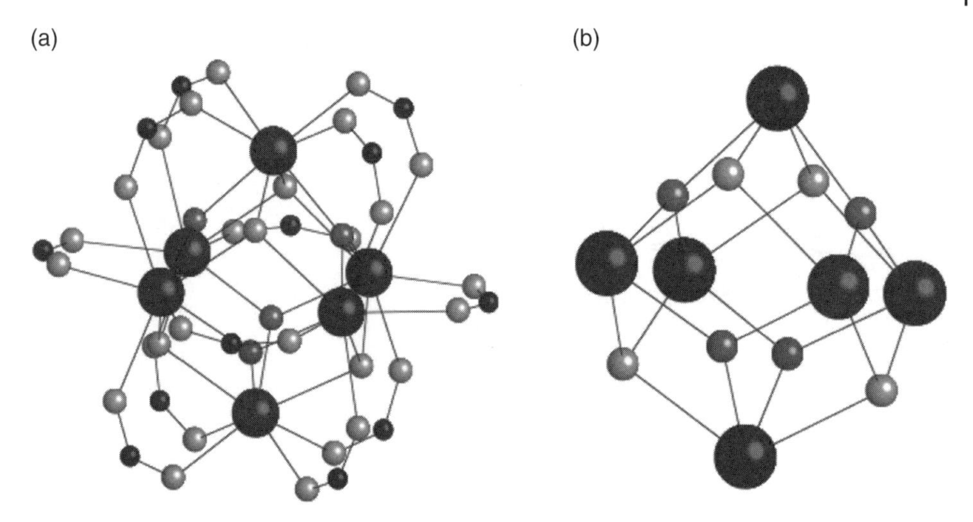

Figure 2.31 (a) Ball-and-stick diagram of Th^{IV}-formate hexanuclear cluster, $[Th_6(\mu_3\text{-}O)_4(\mu_3\text{-}OH)_4(HCOO)_{12}(H_2O)_6]Na_3(ClO_4)_{3.5}(H_2O)_{5.5}(H_3O)_{0.5}$ (Th = charcoal, O = silver, C = black). (b) Ball-and-stick representation of $Th_6(\mu_3\text{-}O)_4(\mu_3\text{-}OH)_4$ core (Th = charcoal, μ_3-O = gray, μ_3-OH = silver). Adapted with permission from Takao S., Takao K., Kraus W., Ernmerling F., Scheinost AC, Bernhard G., Hennig C. First Hexanuclear U-IV and Th-IV formate complexes – Structure and stability range in aqueous solution. *European Journal of Inorganic Chemistry.* 2009, 4771–4775. Copyright 2009 Wiley.

and produced an extensive library detailing the speciation of various actinides in the environment. In many of these works, EXAFS effectively demonstrated the presence of actinide complexes and described the local coordination environment about the metal ion. Emerging from these studies was the first isolation of actinide(IV) hexanuclear clusters bearing formate ligation (99). By utilizing carboxylate groups to cap the metal center and inhibit undesired U(IV) oxidation or colloid formation, discrete hexanuclear clusters were formed and analyzed by EXAFS. Clusters of the formula $[U_6(\mu_3\text{-}OH)_4(\mu_3\text{-}O)_4(HCOO)_{12}(H_2O)_6](N_2H_5)_2(ClO_4)_2(H_2O)_{12}$ and $[Th_6(\mu_3\text{-}O)_4(\mu_3\text{-}OH)_4(HCOO)_{12}(H_2O)_6]Na_3(ClO_4)_{3.5}(H_2O)_{5.5}(H_3O)_{0.5}$ were isolated by slow evaporation of An(IV) solutions with excess HCOOH, yielding single crystals which revealed the solid-state structure. The Th^{IV}-formate cluster is shown in Figure 2.31.

Each metal is nonacoordinate and bridged by $\mu_2\text{-}HCOO^-$ moieties. In order to determine the stability and presence of the clusters in solution, EXAFS was performed on aqueous samples. The sample concentrations for both U^{IV} and Th^{IV} were 1.5×10^{-2} M with formic acid concentrations of 1.0 M.(99) Analysis of the hexanuclear clusters was performed recording the k^3-weighted An L_{III}-edge spectra and performing a Fourier transform (FT) on the acquired raw data shown in Figure 2.32 (99).

The resulting FT of the absorption coefficient data for U^{IV} (Figure 2.32b) revealed U···U backscattering at 3.81 Å corresponding to the distance between neighboring U(IV) ions within the cluster, in addition to U-O distances of 2.26, 2.40, 2.50, and 2.89 Å for the μ^3-O, μ^2-HCOO, μ^3-O, and H_2O moieties, respectively. Further inspection of the radial distribution for the U^{IV} reaction at variable pH values revealed initial formation of the hexanuclear cluster beginning at pH = 1.32 and increasing in concentration up to pH 3.25 (99). In observing the corresponding data for the thorium(IV) analogue

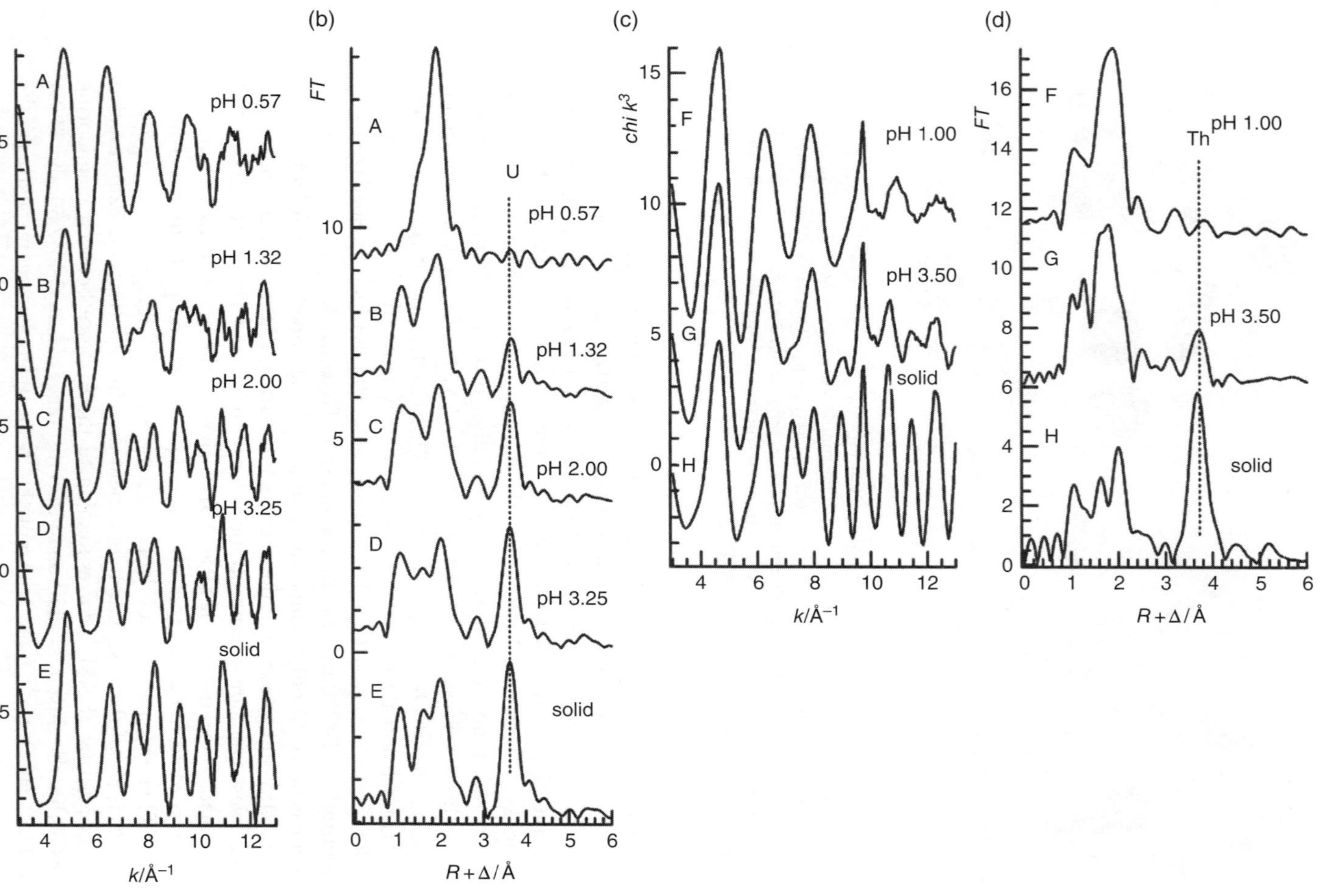

Figure 2.32 (a) EXAFS spectrum and b) Fourier transforms of U^{IV}-formate hexanuclear cluster, $[U_6(\mu_3\text{-}OH)_4(\mu_3\text{-}O)_4(HCOO)_{12}(H_2O)_6](N_2H_5)_2(ClO_4)_2(H_2O)_{12}$. A–D = solution characterization, E = solid state. c) EXAFS spectrum and d) Fourier transforms of Th^{IV}-formate hexanuclear cluster, $[Th_6(\mu_3\text{-}O)_4(\mu_3\text{-}OH)_4(HCOO)_{12}(H_2O)_6]Na_3(ClO_4)_{3.5}(H_2O)_{5.5}(H_3O)_{0.5}$. F–G = solution characterization, H = solid state. Reprinted with permission from Takao S, Takao K, Kraus W, Ernmerling F, Scheinost AC, Bernhard G, Hennig C. First hexanuclear U-IV and Th-IV formate complexes – Structure and stability range in aqueous solution. *European Journal of Inorganic Chemistry.* 2009, 4771–4775. Copyright 2009 Wiley.

(Figure 2.32d), Th···Th interactions are apparent at 3.70 Å, but are found to be much lower in intensity as compared to the solid, suggesting that the Th^{IV} hexanuclear cluster is not as stable in solution (99). Moreover, the thorium hexanuclear cluster is only seen to form in solution within a much more narrow pH range and is present upon titration up to pH 3.50, informing of the inherent solution-state instability of this cluster. While this technique allowed for the identification of polynuclear actinide species in aqueous solution, one of the stark limitations of this analytical method is made apparent in observing the processed data. While the interatomic distances between adjacent actinide(IV) centers was clearly evident in the Fourier-transformed EXAFS data (Figure 2.32b,d), the interatomic distances between terminal U(IV) and Th(IV) ions (5.39 and 5.23 Å, respectively) could not be visualized in this method due to the aforementioned instrumental detection limits that only allows reliable determination of short-range order. Despite this intrinsic weakness of EXAFS analysis, the study described above as well as numerous others probing similar actinide speciation behavior have been performed and provide invaluable data which aids in nuclear waste management and actinide ion separations (100–115).

2.6.3 EXAFS Structural Comparison of Complexes with Varying Oxidation States and Geometries

In addition to identifying the coordination number of actinide complexes in solution, EXAFS has also be used to probe changes in geometry that occur due to differing valences or oxidation states of the metal center. For example, a study performed by Ikeda and coworkers investigated hexavalent and pentavalent uranyl carbonato complexes (116). Due to the many forms of An-carbonato compounds found in groundwater and their recognized environmental mobility, an extensive exploration was sought to characterize the conditions through which actinide-carbonato species form. Although uranium has many accessible oxidation states (III – VI), U(VI) is by far the most prevalent and most readily forms interactions with the carbonate ion (116). U(V), on the other hand, has not been extensively studied, which is attributed largely to its transience and propensity to rapidly oxidize to the hexavalent form. In order to better understand these transformations, an investigation into the redox chemistry of the U(VI)/UO_2^{2+} and U(V)/UO_2^{+} couples was performed. EXAFS was used in combination with electrochemical methods to gain a fundamental understanding of the redox and complexation properties of U(VI)- and U(V)-carbonate compounds.

To form the model compound, $Na_4[UO_2(CO_3)_2]$ was dissolved in an aqueous Na_2CO_3 solution to generate a final uranium concentration of 50 mM. To prepare the pentavalent uranium species, bulk electrolysis of U(VI) was used to achieve reduction (116); the k^3-weighted EXAFS and corresponding FT spectra are shown in Figure 2.33a–b. The oscillation pattern of U(V) clearly differs from that of U(VI), yet the FT spectra still bear striking similarities. It was speculated that this similarity could lead to the conclusion that the coordination geometry remained unchanged between the two redox states. To process the EXAFS data, curve fits were modeled according to established coordination numbers of these complexes. Bond distances for U(VI)-carbonato complexes have been previously reported: U-O_{ax} = 1.76 Å and U-O_{eq} = 2.41–2.42 Å. Experimentally, U(V) was found to have metal-oxygen bond lengths of U-O_{ax} = 1.81 Å and U-O_{eq} = 2.44 Å (116). It is clear that in both U-O bonds, a lengthening occurs from U(VI) to U(V). DFT

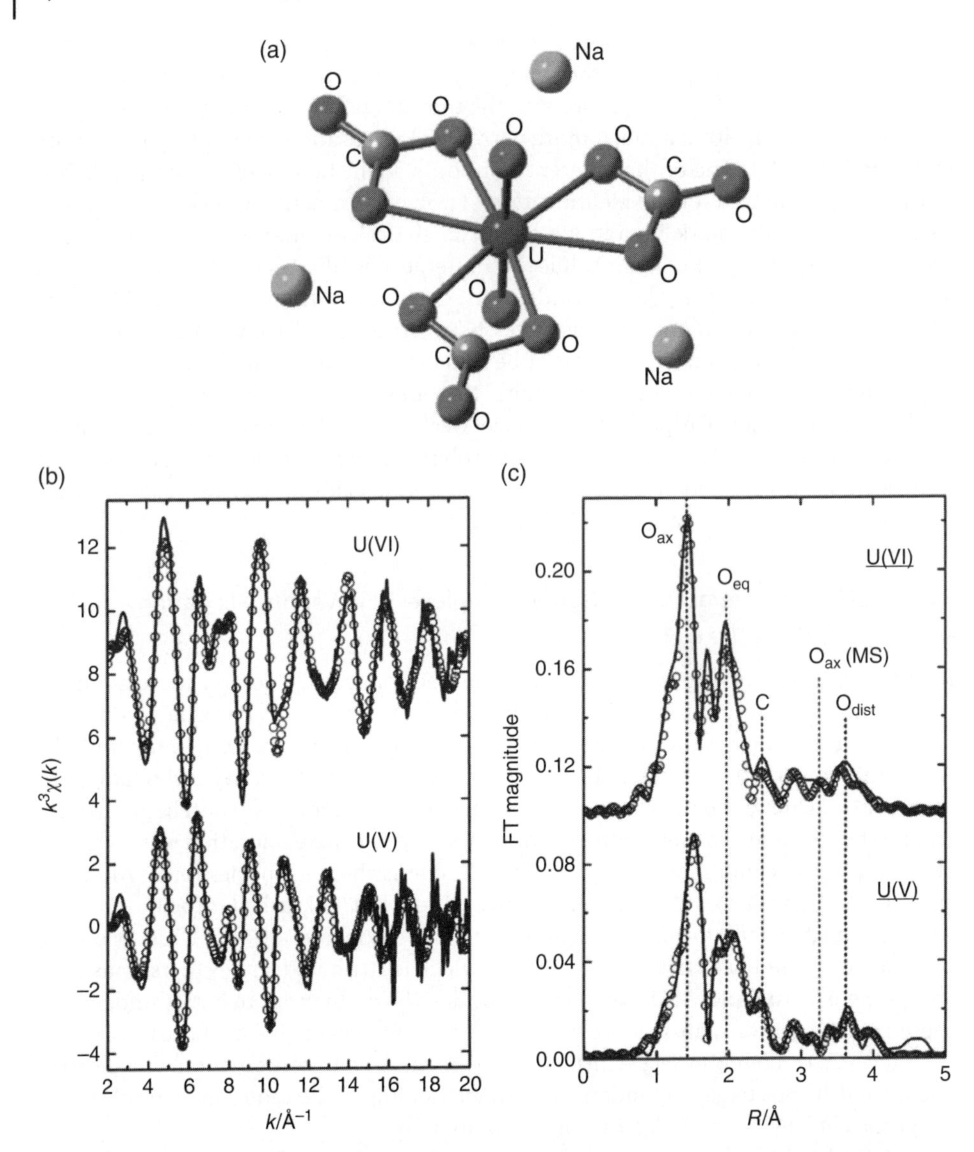

Figure 2.33 (a) Uranyl tricarbonato complex with charge-balancing sodium atoms $[Na_3UO_2(CO_3)_3]^{n-}$ (n = 4 for U(VI), n = 5 for U(V) in optimized geometry. Calculated from B3LYP level in an aqueous phase (Individual atoms are labeled). b) EXAFS spectrum and c) Fourier transforms of U^{VI}- and U^{V}-carbonato complexes. Reprinted with permission from Ikeda A, Hennig C, Tsushima S, Takao K, Ikeda Y, Scheinost AC, Bernhard G. Comparative study of uranyl(VI) and -(V) carbonato complexes in an aqueous solution. *Inorganic Chemistry.* 2007;46:4212–4219. Copyright 2007 American Chemical Society.

calculations attributed this lengthening to increased electron density in the non-bonding orbitals of the uranium center. Although the covalency is seemingly unchanged, the effective charge on the U metal center decreased, which typically is accompanied by an increase in bond length. DFT studies, optimized for the U-triscarbonato structure seen in Figure 2.33a, were also used to validate EXAFS results and ensure appropriate curve fitting with U(VI)- and U(V) carbonato complexes with and without sodium counterions. DFT and EXAFS agree within 0.02 Å when comparing bond distances of U(VI)-carbonato. U(V)-carbonato comparisons were less accurate, evident by the poor agreement between computational and experimental values. While no definitive explanation for this disparity is known, it is predicted to arise from the molecular model used for in silico studies (116). The authors hypothesized, rather, that "apical" water molecules might reside in close proximity to axial oxygens, which may result in the discrepancies observed. This study used EXAFS to determine the isostructural nature of U(VI)-and U(V)-carbonate complexes with observed bond lengthening in moving to the reduced complex, and the power of EXAFS to uncover variations of actinide geometries and local coordination environment based on the oxidation state of the metal.

In a similar manner, EXAFS was used to examine the chemical speciation of neptunium(VI) compounds under highly alkaline environments as variations in conditions supported the presence of more than one hexavalent species (117). From the solutions that were examined, only $[NpO_2(OH)_4]^{2-}$ was observed in the solid state. Alternatively, investigations of speciation in the solution from which the compound crystallized using EXAFS and Raman spectroscopy showed evidence for an increase in the number of equatorial ligands with increasing OH^- concentration, consistent with the presence of $[NpO_2(OH)_5]^{2-}$ in solution. These results pointed toward a heterogeneous mixture of the two species, $[NpO_2(OH)_4]^{2-}$ and $[NpO_2(OH)_5]^{2-}$. Despite the suspected increase in the coordination number, drawn from shifts in the Raman spectra and fits of the EXAFS data, the bond lengths determined from EXAFS of both Np = O and Np-OH remained virtually unchanged; in this case, EXAFS was not able to conclusively discriminate between the $[NpO_2(OH)_4]^{2-}$ and $[NpO_2(OH)_5]^{2-}$, leaving the speciation under these conditions somewhat ambiguous.

2.6.4 Overview of EXAFS

Given that solution speciation plays a major role in many synthetic and environmental processes, the ability to determine short-range information about a specific atom is extremely valuable. EXAFS provides a convenient tool to understand the behavior of actinides in aqueous systems. Many authors have convincingly utilized EXAFS with great success in order to elucidate the coordination environments about the actinide center and hence the nature of the dissolved metal ion. The ability to probe actinide speciation in aqueous samples applies not only to small molecule complexes but has the potential to help investigators understand species that may persist in more complex environmental systems related to waste management. Other examples of unique actinide species that have been identified by EXAFS analysis include plutonium(VII) complexes; similar advances have expanded our knowledge of the participation, energetics, and periodicity between *d* and *f* orbitals (118).

EXAFS, although an invaluable tool to determine the local environment of actinides in aqueous solution, fails to contribute information regarding long-range order within mononuclear or polynuclear systems, showing one of the limitations of this mode of analysis. As a result, only information regarding the first coordination sphere can be attained up to a distance of 4–5 Å (119); interactions much beyond this limit, such as ion–ion interactions, are not observed in the EXAFS spectrum. Although incredibly informative, the inherent instrumental error associated with this absorption technique is rather sizable and can generate data that deviates up to 10% from true values when probing features such as the coordination number. Fortunately, a host of additional experiments are known and well developed which augment data obtained through EXAFS, allowing for a more detailed description of the inner coordination sphere, ligand dynamics, and several other chemical features. These alternative or supplementary analytical methods are described in further detail in the other sections of this chapter.

2.7 Small-Angle X-ray Scattering (SAXS)

SAXS is a powerful analytical method for investigating the structure, size, and morphology of small clusters and nanostructures on the order of ~0.5–100 nm. For these reasons, it has been applied across a number of fields ranging from materials science to soft matter to biophysics (104). The utility of this technique as applied to understanding solution speciation and the processes that occur therein lies in variations in a sample's electron density or scattering differences between dissolved clusters or particles and the solvent. Incident X-rays scattered by the sample are recorded as the scattering intensity as a function of the scattering vector, q. Readers are encouraged to refer to several recent works that provide detailed accounts of the practical use of small-angle X-ray scattering as well as more in-depth descriptions of data measurement and processing (13, 74, 75, 79, 104–108, 120). Several pieces of information regarding the analyte can be extracted from the raw data that together with modeling of the refined data yields a full suite of information regarding the size, shape, dispersity, and aggregation interactions of particles in solution.

Within the field of actinide chemistry, there is much interest in understanding the assembly, stability, and reactivity of actinide clusters in solution. Toward this end, both laboratory- and synchrotron-based SAXS experiments have been used to elucidate the speciation and dynamics of actinide clusters in solution. Much of this work has focused on uranyl-peroxo clusters as they are arguably the most well-developed class of actinide clusters, yet heterometallic polyoxometalates incorporating the actinides have also been examined with great success using SAXS experiments. Some of the larger-order hydroxo/oxo oligomers formed through hydrolysis and condensation reactions likewise lend themselves to analysis via this experimental technique owing to the large electron density gradient that exists between the dissolved nanoclusters and the surrounding solvent, yet surprisingly few studies have harnessed SAXS toward understanding the formation and speciation of condensation products in solution. Nevertheless, this technique is finding increased application in the study of actinide cluster chemistry in solution. Discussed in this section are recent examples that highlight the utility of SAXS in determining actinide speciation and dynamics, namely, in identifying the presence of uranyl-peroxo and metal-oxide clusters in solution, discriminating between clusters of varying nuclearity, examining cluster persistence and degradation, and monitoring the assembly of clusters into higher-order networks or a precipitated phase.

2.7.1 Structure Elucidation by SAXS

Our understanding of the self-assembly processes and solution behavior of metal-oxo clusters, particularly those of the high-valent d-block metals, including molybdenum, tungsten, and vanadium, has developed considerably through the application of SAXS

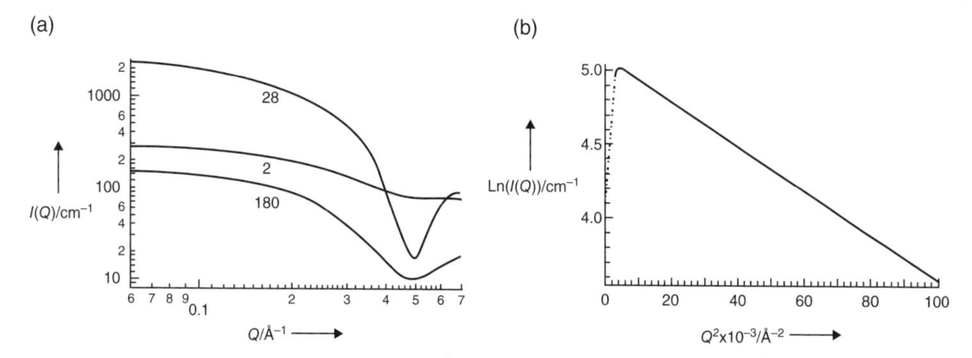

Figure 2.34 (a) SAXS data of mother liquor that produced U24 cluster at 2, 28, and 180 days (numbered on plot area). At high Q, curves broaden and deepen with time, indicating a change in morphology and size of aggregates in solution. b) Guinier plot of U24 mother liquor after 180 days. Linearity indicates monodispersion of clusters in aqueous solution. Reprinted with permission from Burns PC, Kubatko K-A, Sigmon G, Fryer BJ, Gagnon, JE., Antonio MR, Soderholm L. Actinyl peroxide nanospheres. *Angewandte Chemie International Edition Engl.* 2005;44:2135–2139. Copyright 2005 Wiley.

to these systems (12, 120). The recent application of SAXS to investigations of actinide cluster assembly, stability, and aggregation has likewise contributed to our knowledge of the aqueous behavior of actinide cluster solution chemistry. In fact, with the recent development of actinyl peroxo clusters as a new family of actinide polyoxometalates (POMs) (121), SAXS has emerged as a useful technique for understanding the processes that govern cluster formation, speciation, stability, and aggregation into a precipitated phase.

Actinyl peroxo clusters represent a new family of polyoxometalates, and SAXS provides a reliable means to identify the persistence of a cluster and define its size. For example, Burns et al. investigated the mother liquor of a reaction that produced U24 nanospheres after 2, 28, and 180 days (Figure 2.34). Data were collected using synchrotron radiation and processed using the sphere-shell model, which was found to provide the best fit. From this method, the diameter of the U24 cluster was found to be 16.2 Å. A Guinier plot of data after 180 days compared the intensity of the scattered radiation (I(Q)) versus the scattering vector (q), producing the plot shown in Figure 2.34. The linear fit of the Guinier plot indicates an excellent fit over a broad range of scattering vectors (Figure 2.34b). The low-*q* SAXS data obtained were similarly subjected to Guinier analysis and revealed a radius of gyration (R_g) in agreement with solid-state structural parameters obtained from single-crystal X-ray diffraction studies of the U24 cluster (121). In addition to the conclusions stated above, it was found that nanosphere organization occurred in solution as early as two days after sample preparation. Allowing the reaction to proceed further resulted in the crystallization and deposition of the U24 nanospheres (121).

Since the discovery of the U24 actinyl peroxide cluster and its closely related structures, Burns and coworkers have expanded their efforts toward self-assembled derivatives of uranyl-POMs. Only in recent years was an extensive library of uranyl peroxide clusters prepared. This class of compounds has yielded highly variable structures ranging from 1.5 to 4 nm in diameter, and containing as many as 124 uranyl ions (122).

Within this large family of compounds, many assume a cage-like structure with fullerene topology dominating (121, 123–125); however, uranyl bowls and crowns have also been isolated (126). Recent efforts have also targeted the use of alternative bridging ligands which deviate from the classical oxo or hydroxo moieties; it has been shown that incorporation of various anions into the clusters can produce a significant impact on the extension and formation of these clusters. Some prominent examples of this include the use of nitrate (127), oxalate (127, 128), and phosphates/phosphonates such as pyrophosphate or methylenediphosphonate (39, 129). In the analysis of the solution phase behavior of these clusters, SAXS is commonly used as a nondestructive means of studying the size and shape of the actinide clusters in aqueous solution as a function of time. SAXS provides a useful experimental handle to elucidate the persistence of clusters in solution, accurately measure the diameter of clusters, and draw informative parallels between the self-assembly of molecules and the solid-state products.

2.7.2 SAXS Analysis of Cluster Evolution

It is generally understood that the identity of actinide species in solution is crucial to our understanding of numerous chemical and biological processes. No less important, though, is our understanding of the dynamic behavior of such species over time. The transformation from one actinide moiety to another naturally affects the chemical behavior and fate of that species, and to this effect, SAXS proves useful as a probe of species evolution in dynamic systems, whether this occurs due to aging of the system, catalytic activity, or altered solution conditions (120). In the field of actinide peroxide clusters, Burns et al. have employed SAXS to determine the size and differences in cluster aggregation as a result of using various anionic units as a function of time (121). Another example of such studies is seen by Nyman et al., wherein the self-assembly of polynuclear actinide species has been investigated. Uranyl peroxide nanocages were similarly prepared and studied in this research. However, the variations employed focused on the effects of various synthons on the identity of the evolved species. Small-Angle X-ray Scattering provides an appropriate analytical technique in this study to yield fundamental information on the formation of clusters in solution (130).

Recently, a study by Falaise and Nyman probed an integral step in the formation of a U28 peroxide cluster (84). On the basis of the previous synthesis by Burns et al., the $UO_2^{2+}/H_2O_2/LiOH$ aqueous system was used. However, the synthetic methodology employed provides little selectivity or control, forming clusters with low yield as well as several impurities. This was thought to occur due to the long crystallization times, which lead to local concentration gradients, pH changes, and variations in carbonate levels. Despite these shortcomings, self-assembly is still operative. However, conversion between the cluster geometries is common in the reaction mixtures. Both experimental findings and quantum chemical calculations have indicated that cluster formation is largely driven by alkali metal counterions (84). The investigation sought to answer questions of the thermodynamic and kinetic influences on these transformations. The results suggested that the solution pH in addition to the identity of the alkali metal are of considerable consequence in the observed dynamic behavior. With this information in hand, the research proceeded to investigate the effect of variable ratios of LiOH and uranyl cation on product formation. The resulting products that form are summarized in Figure 2.35.

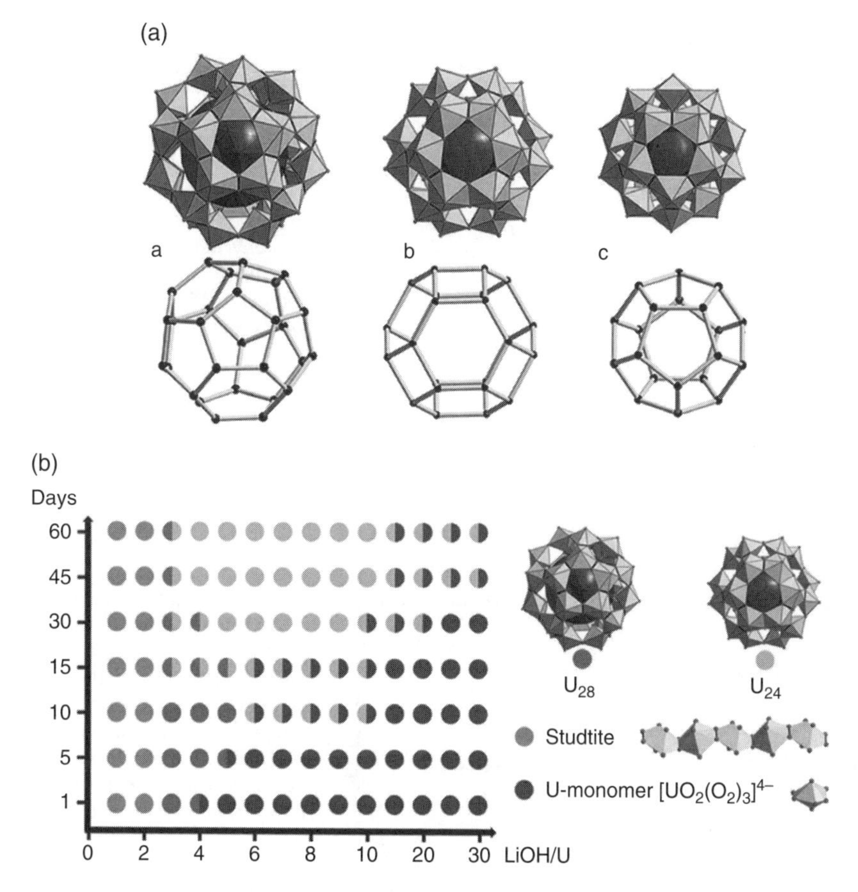

Figure 2.35 (a) Uranyl peroxide clusters studied, including U28, U24, and U20. (b) Schematic representation of persistence of uranyl peroxide clusters over increasing LiOH/U ratios. Reprinted with permission from Falaise C, Nyman M. The key role of U28 in the aqueous self-assembly of uranyl peroxide nanocages. *Chemistry – A European Journal.* 2016;22:14678–14687. Copyright 2016 Wiley.

The clusters of interest shown in Figure 2.35 include U20, U24, and U28. As the clusters are clearly variable in their sizes, SAXS was used as the obvious choice for analysis. Solutions of uranyl nitrate in water, H_2O_2, and lithium hydroxide were combined in various ratios, and the solutions were analyzed at specific time intervals (2, 10, 20, 30, 45, and 60 days). The FTs of the scattering curves gave pair-distance distribution functions (PDDF) indicative of particle shape, and the radius of gyration (R_g) was determined for each cluster. SolX software calculated simulated scattering curves from single-crystal X-ray diffraction data, represented by dotted black lines in Figure 2.36 and Figure 2.37.

The persistence of the U28 cluster was studied at a LiOH/U ratio of 2.7, and subsequent conversion to U24 was observed as a function of time. The ideal base/uranium ratio used in these reactions was derived from the data in Figure 2.35. Further SAXS data collected as a function of time (Figure 2.37) revealed that the U28 polynuclear species formed in solution after 2 days, and from 10 to 60 days, no obvious changes were observed. The experimentally determined R_g value from PDDF analysis was in

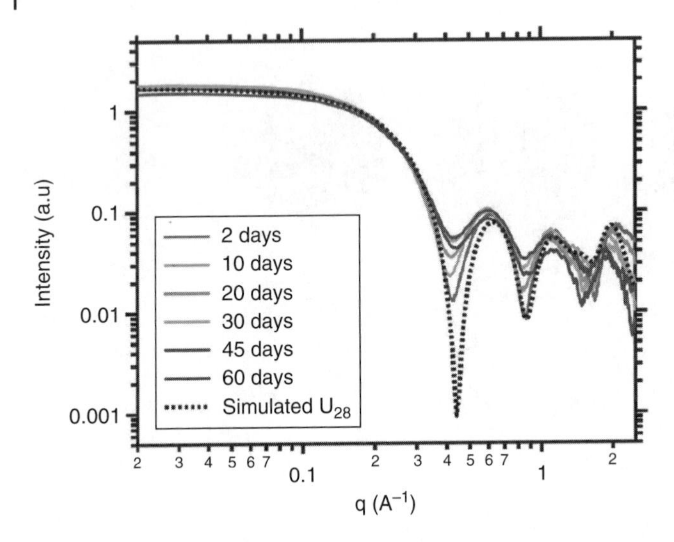

Figure 2.36 Scattering curve of SAXS data showing formation of the U28 cluster in aqueous solution over 2 to 60 days. Reprinted with permission from Falaise C, Nyman M. The key role of U28 in the aqueous self-assembly of uranyl peroxide nanocages. *Chemistry – A European Journal*. 2016;22:14678–14687. Copyright 2016 Wiley.

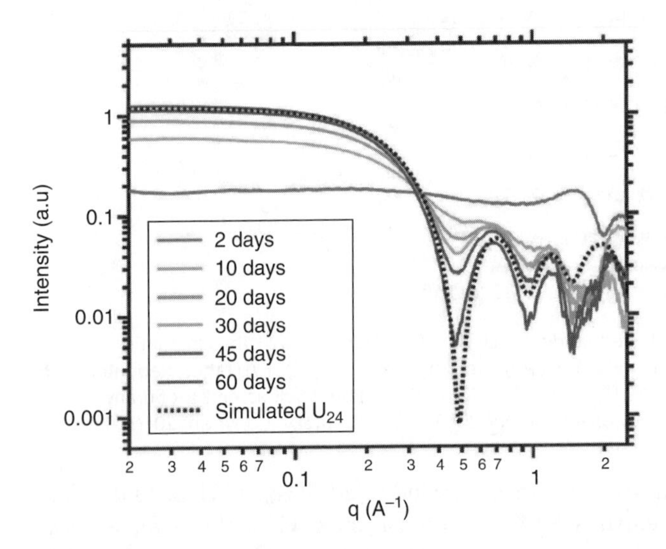

Figure 2.37 Scattering curves of SAXS data over time with a LiOH/U ratio of 10. As the solution ripens, the scattering curve at high q deepens. Day 2 is an exception. Reprinted with permission from Falaise C, Nyman M. The key role of U28 in the aqueous self-assembly of uranyl peroxide nanocages. *Chemistry – A European Journal*. 2016;22:14678–14687. Copyright 2016 Wiley.

good agreement with the simulated pattern and were determined to be 7.52(1)–7.82(1) Å, equal to core-shell clusters 16.99(1)–17.68(1) Å in diameter. With elapsed time, a slight shift is observed corresponding to a decrease in R_g, which consequently results in decreased scattering and lower resolution of the scattering curve. This is attributed to conversions in the geometry of the U28 cluster as determined by SAXS analysis (84).

Next, by increasing the LiOH/U ratio to 10, the evolution of the uranyl peroxide polynuclear species was examined. The SAXS data shown in Figure 2.37 illustrated that after 2 days, a uranyl-peroxo monomer formed, but no evidence of cluster aggregation was present according to the relatively flat scattering curve. At 60 days, PDDF analysis yielded an R_g of 6.88 Å, which is in good agreement with the simulated R_g of 6.58 Å for the U24 cluster (84). Again, the conversion of U28 to U24 was shown to be base-dependent. Furthermore, at very high Li/U ratios, U24 cluster formation would occur after 60 days. However, the concentration of this species was seen to decrease as the LiOH loading was increased, accompanied by increased alkalinity in the reaction solution.

Lastly, the U28 cluster's ability to form with counterions other than Li^+ was explored, and mixtures of clusters were observed in solution via SAXS, which was further supported by Raman spectroscopy. It was found that the U28 cluster was the favored product in the presence of small amounts of peroxide and base and in the absence of any strong templating cation. This finding not only revealed the optimized conditions for selective formation of the uranyl clusters, but also underscored the importance of templating cations to facilitate aggregation (84). The determination of these data clearly illustrates the importance of SAXS analysis in studying the solution-state dynamics of aqueous actinide systems, such as the uranyl peroxide clusters presented herein.

2.7.3 Understanding Self-Assembly Processes by SAXS

It is evident from experiments described thus far that SAXS exhibits promising utility in aqueous actinide cluster chemistry. This is especially true when crystallization of the cluster is not possible, leaving solution-state studies as a means of revealing the nature of the dissolved actinide species. Such studies prove useful for determining a mechanistic understanding that underpins self-assembly and precipitation of discrete clusters. The nature of these mechanisms still remains hotly debated in the scientific community; however, there is ample research that supports the claim that polynuclear species are formed by the aggregation of prenucleation clusters rather than through an atom-by-atom approach (131). SAXS serves an important role in the elucidation of the operative chemical steps in self-assembly mechanisms. POMs have been at the forefront of such self-assembly studies, and this has only increased as a renewed interest in hybrid uranyl peroxide POMs containing transition metals such as those in groups V or VI has become apparent (132–134). Hypervalent hybrid-uranyl-POMs have received considerable attention; however, similar materials containing metals in the lower oxidation states are less studied. One such exception to this is a study performed by Deb et al. that explored tetravalent uranium incorporated into polyoxomolybdates. In this study, the formation of $NaUMo_6$ and $NaMo_6$ clusters was reported (135). $NaUMo_6$ contains chains of Mo_6P_4 clusters linked by U^{4+} and Na^+ cations of the formula $[UNa(Mo_6P_4O_{31}H_7)_2]\cdot5\ Na\cdot(H_2O)_n$ (Figure 2.38). Conversely, in the absence of tetravalent uranium $NaMo_6$ formed wherein dimers linked by sodium cations in $[Na(Mo_6P_4O_{31}H_{10})_2]\cdot5Na\cdot(H_2PO_4)\cdot(H_2O)_n$ (Figure 2.39).

The general structure of $NaUMo_6$ shows each Mo_6P_4 structural unit connected to two other Mo_6P_4 polyhedra by both sodium and uranium cations to form a 1-D network (Figure 2.38b). By comparison, $NaMo_6$ contains two Mo_6P_4 clusters connected by one sodium cation, to form a sandwich-like POM (Figure 2.39) (135).

As the $NaUMo_6$ compound would not readily dissolve in aqueous solution, SAXS studies were performed in 1M LiCl solution. SAXS yielded information about the self-assembly and linkage of both clusters in aqueous solution. $NaMo_6$ was found to exist as dimers in solution as supported by a rough spherical geometry observed using PDDF analysis. This is in agreement with simulated scattering curves, which also were consistent with dimeric rather than hexameric molybdenum species (135). Furthermore, the experimental data agree with the theoretical predictions of R_g values. $NaUMo_6$, on the other hand, proved more complex to analyze because of the LiCl needed to achieve

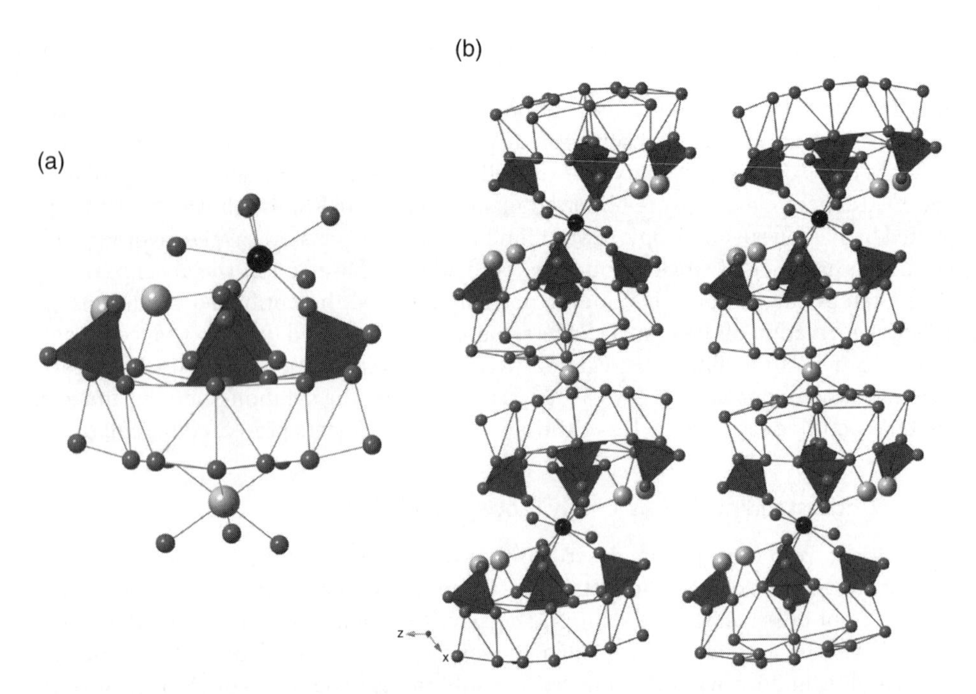

Figure 2.38 (a) Asymmetric unit of $[UNa(Mo_6P_4O_{31}H_7)_2]\cdot 5Na\cdot (H_2O)_n$ ($NaUMo_6$), displaying the Mo_6P_4 cluster and linking sodium and uranium cations. (b) Polyhedral representation of extended network of $NaUMo_6$. Sodium and water atoms in the lattice have been removed for clarity. (MoO_6 = white, U = black, O = gray, Na = silver, PO_4 = charcoal). Adapted with permission from Deb T, Zakharov L, Falaise C, and Nyman M. Structure and solution speciation of U^{IV} linked phosphomolybdate (Mo^{V}) clusters. *Inorganic Chemistry.* 2016;55:755–761. Copyright 2015 American Chemical Society.

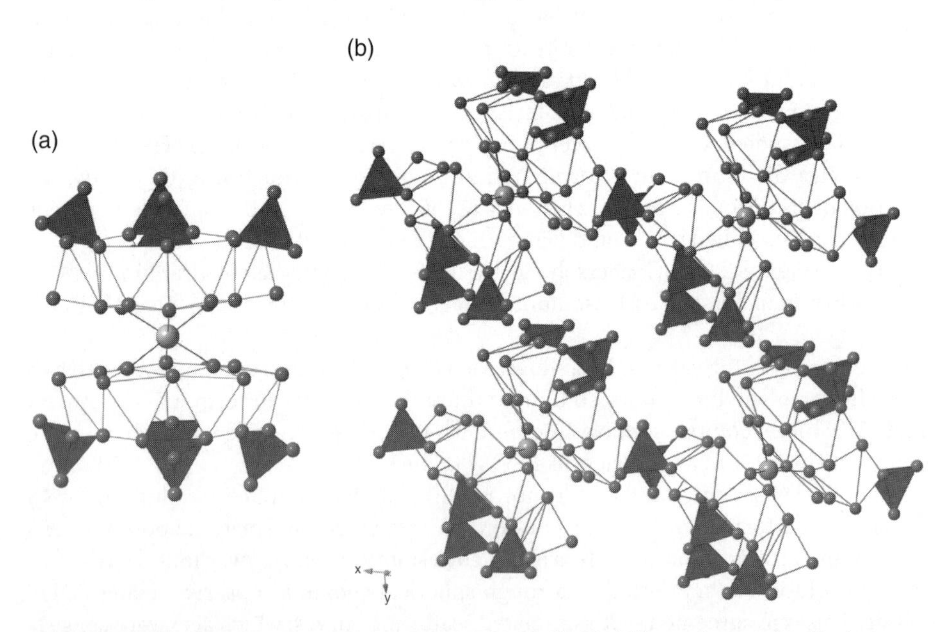

Figure 2.39 (a) $[Na(Mo_6P_4O_{31}H_{10})_2]]\cdot 5Na\cdot (H_2PO_4)\cdot (H_2O)_n$ ($NaMo_6$), illustrating two Mo_6P_4 clusters linked by a sodium cation. (b) Polyhedral representation of extended network of $NaMo_6$. Sodium, H_2PO_4, and water molecules in the lattice have been removed for clarity. (MoO_6 = white, U = black, O = gray, Na = silver, PO_4 = charcoal). Adapted with permission from Deb T, Zakharov L, Falaise C, and Nyman M. Structure and solution speciation of U^{IV} linked phosphomolybdate (Mo^{V}) clusters. *Inorganic Chemistry.* 2016;55:755–761. Copyright 2015 American Chemical Society.

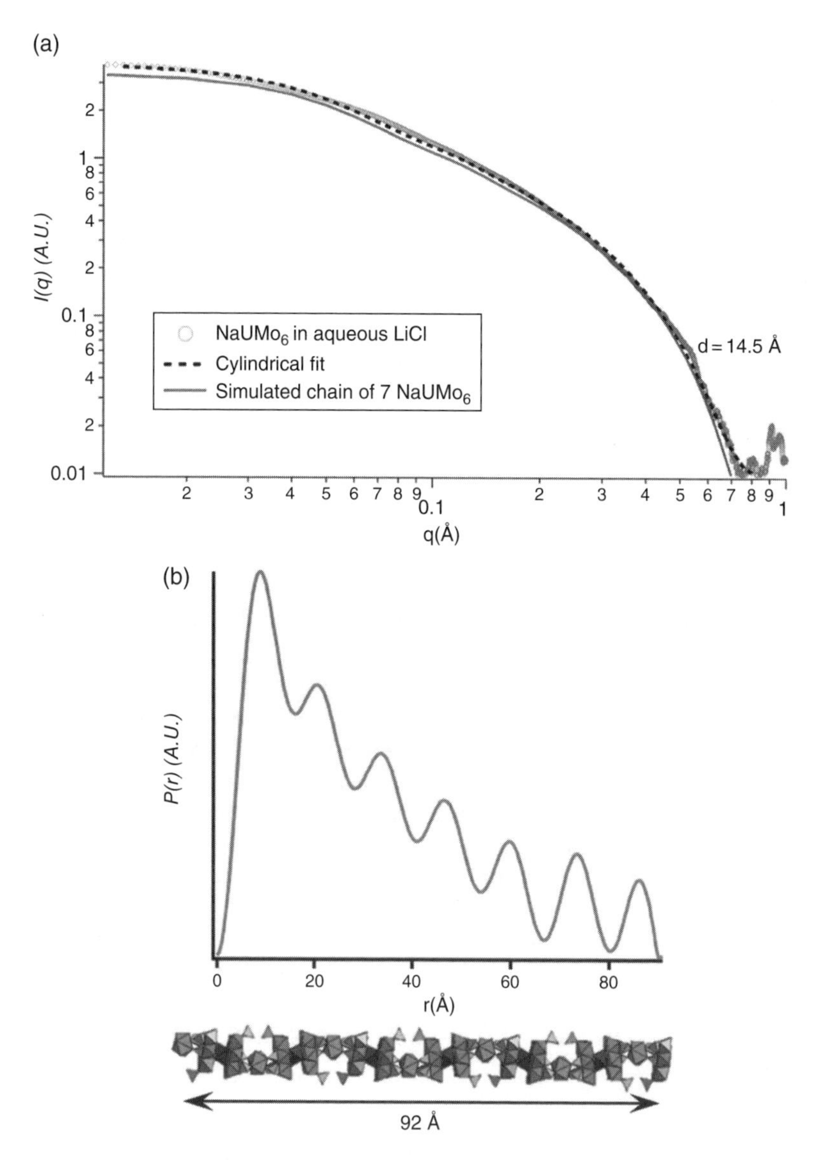

Figure 2.40 (a) Scattering curve of $NaMo_6$ in 1M LiCl (thick gray line). Dotted line illustrates the modeled cylindrical fit, and the thin gray line shows the simulated chain of 7 $NaUMo_6$, both of which are in good agreement with the experimental data. (b) Pair-distance distribution function where each oscillation represents a link in the chain, corresponding to the graphic shown below the PDDF. Reprinted with permission from Deb T, Zakharov L, Falaise C, and Nyman M. Structure and solution speciation of U^{IV} linked phosphomolybdate (Mo^{V}) clusters. *Inorganic Chemistry.* 2016;55:755–761. Copyright 2015 American Chemical Society.

dissolution. As a result, it was found that Li^+ replaced some of the Na and U atoms; however, some sites remained unchanged as indicated by shorter chains detected by SAXS (135). The scattering curve of $NaUMo_6$ suggests linearity, which was found to be in agreement with a cylindrical model fit (Figure 2.40). PDDF analysis predicted a 92 Å maximum linear relationship as shown in Figure 2.40b. The oscillations in the PDDF are attributed to chains corresponding to the aggregation of seven monomer units that were found to persist for two weeks in solution. It is interesting to note that despite dissolution in aqueous LiCl, linkages between the clusters remained intact.

2.7.4 Overview of SAXS

In the short time since SAXS was first introduced, it has already been established as an excellent technique for examining the speciation of metal clusters, especially for those of the actinide ions. The information obtained includes cluster size and shape, both of which can be used to monitor changes in geometry and self-assembly processes. Collectively, these examples showcase the successful application of SAXS to important questions regarding the nature of clusters in solution, their stability, reactivity, and transformation into a solid phase.

Despite the clearly stated advantages of SAXS experiments, certain limitations of this technique should be considered. As the examples above have shown, clusters ranging from 0.5 to 100 nm are best suited for SAXS. A cluster which exists outside this range, such as dinuclear or trinuclear species, is often more difficult to analyze using SAXS and may result in unclear, insufficient data for proper modeling and characterization. In addition, if a sample lacks sufficient solubility in the desired medium, SAXS would certainly not be capable of providing the desired information. Another challenge that remains is the direct relationship between the accuracy of data and the experience of the user; unlike other solution-state measurements such as NMR or Raman spectroscopy, which are sufficiently advanced and can utilize a suite of user-friendly software for data processing, the analysis of SAXS data lacks these features. Especially when considering the need to apply multiple modeling methods for accurate structure determination, this technique presents a strong possibility of human error when attempted by a novice user. As such, corroborative spectroscopic data are ideal and important for drawing accurate conclusions from the SAXS data. As shown in the examples presented throughout this section, SAXS data generated from single-crystal X-ray diffraction structural analysis are extremely important in modeling SAXS data. In the absence of such data, simulations from computational models can likewise be used to generate scattering curves and play an important role in SAXS data analysis and interpretation (136).

2.8 High-Energy X-ray Scattering (HEXS)

Potentiometric titrations, Raman, UV-vis, and IR spectroscopies all provide indirect probes of solution speciation, and as illustrated by the examples highlighted in this chapter, provide limited structural information in the absence of complementary spectroscopic, solid-state, or computational data. Alternatively, EXAFS has found great utility among actinide chemists for probing solution speciation, yet it is limited by its ability to determine only the first coordination sphere (typically <5 Å) reliably and the accompanying 10% error typically associated in providing coordination numbers (93, 119). By comparison, High-Energy X-ray Scattering (HEXS) has emerged as a powerful technique for examining solution speciation that can provide information about local and intermediate range structural organization up to 12 Å, and thereby provide new insight into solution speciation. Such data, which are not available through other X-ray techniques, have been used to quantify mononuclear, polynuclear, and higher-order oligomeric species in solution (61, 88, 119, 137, 138).

In brief, HEXS experiments resemble powder X-ray diffraction experiments, wherein the data of interest are the X-ray intensities as a function of momentum transfer, $Q = 4\pi(\sin\theta)/\lambda$, where θ is the scattering angle and λ is the wavelength of radiation (88, 119, 139, 140). Unlike EXAFS, HEXS is a total scattering method, and hence correlations between all atoms in solution contribute to the scattering intensity. While advantageous in several respects, these features require a careful background subtraction that adds to the complexity and time needed for data acquisition. The Fourier transform of these data provides g(r), a pair-correlation or pair distribution function (PDF), as a function of distance, r, with the resolution dependent on the Q range used in the FT (120, 139–141). Peak positions in the PDF represent coordination distances between ions, and background subtraction gives peak positions in the $g\Delta(r)$ versus r plots that correspond to bond lengths and atomic distances between the metal center and other correlated ions in solution. The peak intensities correspond to the number of electrons involved in the correlation and therefore can be related to the coordination number. Data are often fit using Gaussians, which are subsequently integrated to determine the number of electrons contributing to a correlation. The concentration of the correlated pairs can thus be used to extract coordination numbers. For more details regarding data collection and data processing, readers are encouraged to see the work of Soderholm *et al.* (88, 119, 138, 141) and the recent publications by Chapman and Chupas (139, 140).

Given that X-rays scatter off electrons, HEXS is particularly well suited for examining higher Z element speciation in solution, such as the actinides. Related scattering techniques have been reported using in-house instruments, yet the use of synchrotron radiation provides much higher X-ray energy (60–115 keV) and intensity, thereby minimizing absorption and greatly improving the resolution. This technique has been used to probe the coordination environment about a metal center, determine coordination numbers, identify solution species, and understand the relationship between solution- and solid-state structural entities (61, 72, 88, 137, 138, 142–146). With respect to the latter, it is important to note that the interpretation of results, that is, the assessment of the coordination environment about the metal center as well as ligand binding modes through analysis of the observed pair correlations, is largely dependent on structural data available from solid-state examples or computational models.

2.8.1 Determining Coordination Number and Environment about a Metal Center

HEXS has been applied to many systems to understand the coordination environment about a metal center under a given set of conditions (35, 61, 72, 88, 100, 119, 137, 138, 142–147). For example, using HEXS, Soderholm et al. examined the speciation of UO_2^{2+} under slightly acidic conditions using a 0.858 mol % UO_2^{2+} in 1.74 mol % ClO_4 aqueous solution (138, 148). Perchlorate was chosen for this set of reactions as the ion is known to be non-complexing, and therefore allowed the authors to probe the aquo complexes of uranium. The scattering data including the experimental scattering curve, the difference structure $S\Delta(Q)$ obtained for the background-subtracted data, and the FT of the $S\Delta(Q)$ data are shown in Figure 2.41. As shown in the g(r) versus r plot, there are two

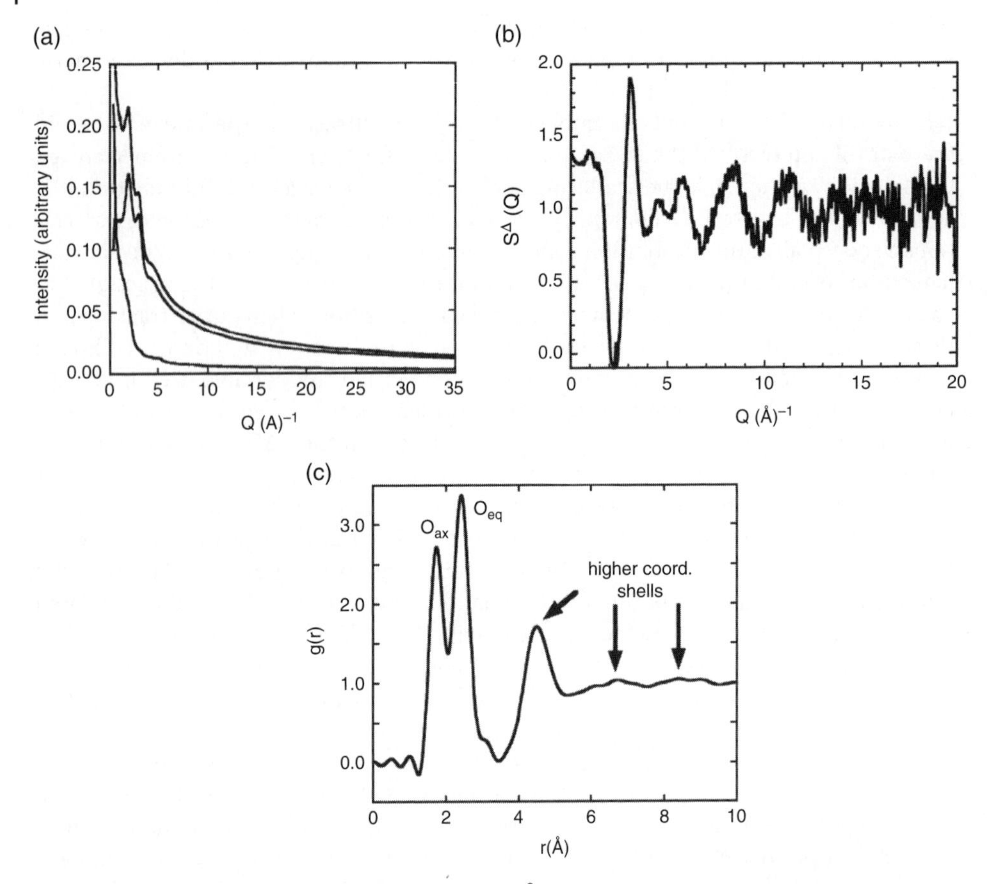

Figure 2.41 (a) Experimental scattering curves for UO_2^{2+}/perchlorate solution (top), background-subtracted UO_2^{2+}/perchlorate solution (middle), and the empty sample holder (bottom). (b) The difference structure factor, SΔ(Q), obtained from the corrected data for which a $LiClO_4$ spectrum has been subtracted – this spectrum represents only those correlations with uranium. (c) FT of the SΔ(Q) data; the peaks in this spectrum represent atomic pair correlations or distances between the metal center and other correlated ions in solution. Reproduced with permission from Soderholm L, Skanthakumar S, Neuefeind J. Determination of actinide speciation in solution using high-energy X-ray scattering. *Analytical and Bioanalytical Chemistry.* 2005;383(1):48–55. Copyright 2005 Springer.

high-intensity peaks at 1.76(2) Å and 2.41(3) Å, which are consistent with the linear dioxo UO_2^{2+} unit and equatorially coordinated ligands, respectively. The correlation at 2.41(3) Å is consistent with EXAFS data, from which it was concluded that under the conditions of the study, five water molecules are bound to the UO_2^{2+} moiety in the equatorial plane. Integration of the peak at 2.41(3) Å yielded 46.1(7) electrons, suggesting a non-integral number of water molecules per UO_2^{2+}. This non-integral value is explained by an equilibrium between the $UO_2(H_2O)_4^{2+}$ and $UO_2(H_2O)_5^{2+}$ complexes, with the $UO_2(H_2O)_5^{2+}$ species reported to account for 86(7)% of the UO_2^{2+} moieties in solution. In addition to these results, the authors were able to determine that the penta-coordinate complex was 1.2(4) kcal/mol more stable than the four-coordinate complex. It is worth noting that the higher *r* correlations at 4.50, 7, and 8.7 Å are attributed to longer, higher-order uranium-solvent correlations (148).

In a related study, the formation of UO_2^{2+}-chloride complexes in solution was examined in an effort to discern whether the UO_2^{2+} cation formed inner- or outer-coordination sphere complexes with the chloride ion, identify the complexes that formed, and determine the stability constants of UO_2^{2+}-Cl complexes (137). Two sets of solutions were prepared. However, this discussion will focus only on those solutions from which stability constants were determined, as the previous case study already highlighted how atomic pair correlations can be used to determine metal ion speciation. Solutions were prepared in which the chloride concentration was varied over a 0–4.8 *M* range at a constant ionic strength (5.3 *M*). The background-subtracted, Fourier-transformed data highlighting the U-O and U-Cl correlations of metal bound water and chloride ligands at 2.40–2.41 Å and 2.72(3) Å, respectively, are shown in Figure 2.42a; note that the FT data of all samples exhibited a peak at 1.76(2) Å attributed to the U-O of the UO_2^{2+} cation that is present in all samples. In each scattering pattern, the chloride peaks at 2.72(3) Å were fit with a Gaussian curve (Figure 2.42), the integration of which is interpreted as the number of electrons corresponding to the peak intensity for chloride ions bound to the uranyl cation in solution. From these data, the average number of chloride ions ($\bar{n}$) bound to the uranyl moiety as a function of free chloride concentration can be determined and used to approximate the stability constants (β_N) for the successive addition of chloride ions to the inner coordination sphere of the uranyl-aqua complex (137). This work yielded values of $\beta_1 = 1.5(10)$ m^{-1}, $\beta_2 = 0.8(4)$ m^{-2}, and $\beta_3 = 0.40(2)$ m^{-3} for 0.5 *M* UO_2^{2+} solutions at an ionic strength of 5.3 *M* (137); these values were further used to construct the speciation diagram shown in Figure 2.42b. As noted by the authors of this work, the stability constants of the U-Cl complexes are very weak, and while the studies confirm the presence of these species in solution, computational studies are required to understand the relative role that UO_2^{2+}-Cl^- bonding, charge transfer, hydrogen bonding, and outer-sphere interactions have on the overall stability of uranyl-chloride complexes.

2.8.2 Deducing Metal–Ligand Coordination Modes

In addition to determining the coordination number of a metal center and speciation under a given set of conditions, HEXS has also been applied toward understanding metal-ligand coordination modes in solution (143). Such work is often carried out with the goal of understanding the correlation between solution species and solid-state structural reports. For example, Knope et al. reported three thorium sulfates isolated from acidic aqueous solution (143). The structure of each of the compounds possessed characteristic monodentate sulfate units. In an effort to understand how the units in solution assembled into the crystalline material, the authors used HEXS to examine the speciation of Th in sulfate solutions. In particular, a solution from which a mononuclear complex $Th(SO_4)_2(H_2O)_7$ was prepared was examined; the PDF obtained from the Th-SO_4 solution after background and solvent-solvent subtraction is shown in Figure 2.43. Interpretation of the PDF suggests that both bidentate- and monodentate-sulfate linkages are present in solution with Th–S correlations at 3.19 Å and 3.75 Å, respectively. Moreover, the HEXS data suggest that a sulfate-bridged polynuclear thorium complex may exist in solution with a peak centered at 5.9 Å, consistent with intermediate range ordering occurring prior to crystallization (143). These results point to differences in the species observed in solution and in the solid state, an observation that could not be rationalized on the basis of the limited dataset and warranted further investigation.

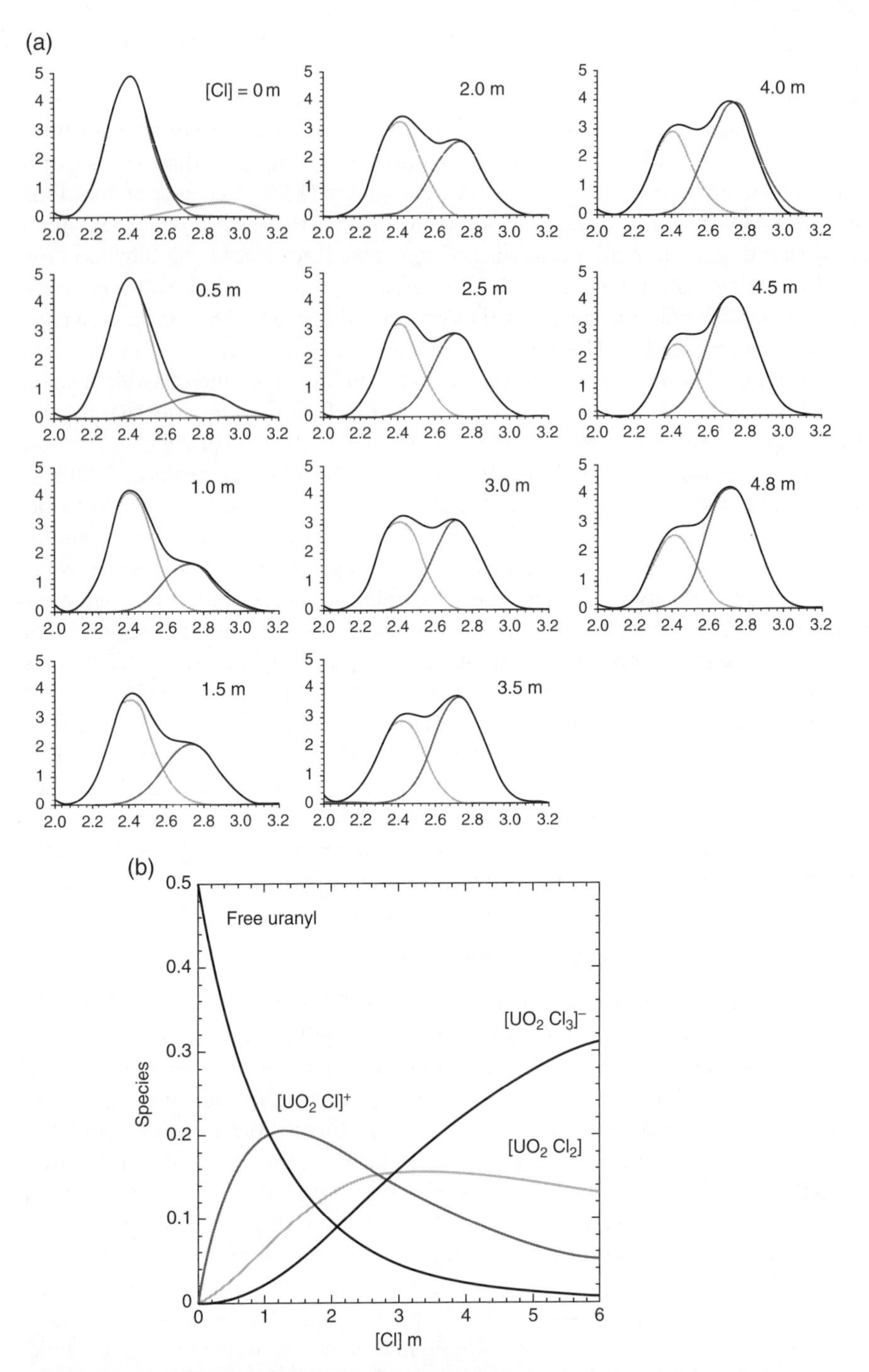

Figure 2.42 (a) FT of the background-subtracted X-ray scattering data collected from aqueous solutions of 0.5 *m* UO_2^{2+} as a function of chloride concentration at constant ionic strength. Visible changes in the pair correlations are attributed to UO_2^{2+}-chloride complexation. The peaks attributed to U-O and U-Cl correlations were fitted with Gaussians centered at 2.41(2) and 2.72(3) Å, respectively. (b) The speciation diagram determined from the stability constants, plotted as a function of total chloride concentration. Reproduced with permission from Soderholm L, Skanthakumar S, and Wilson RE. Structural correspondence between uranyl chloride complexes in solution and their stability constants. *Journal of Physical Chemistry A.* 2011;115(19):4959–4967. Copyright 2011 American Chemical Society.

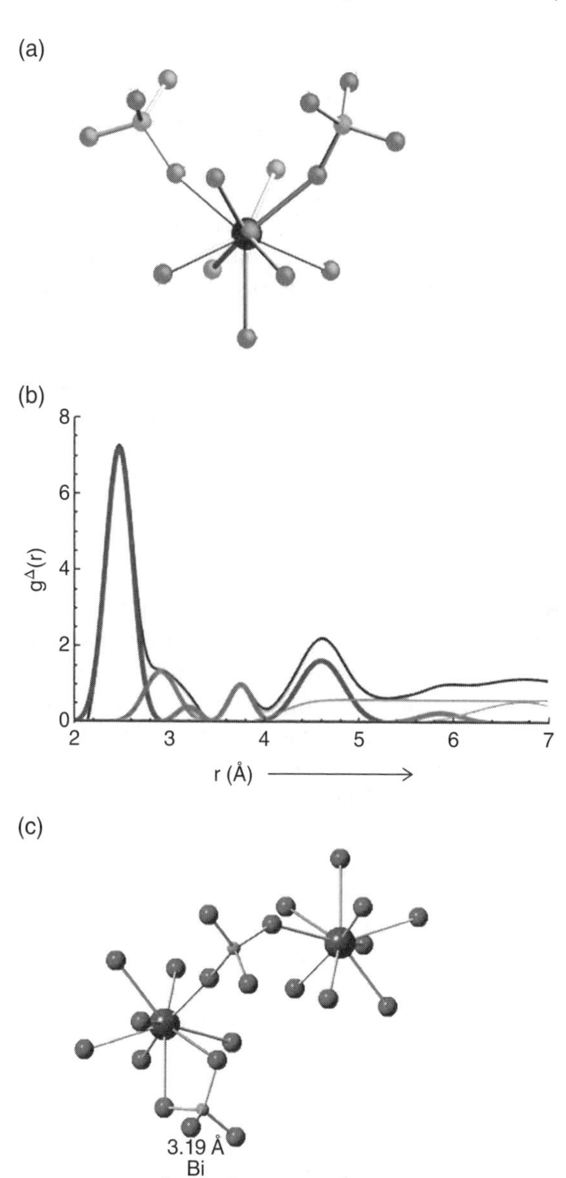

Figure 2.43 (a) Illustration of the structural units observed in the crystal structure of $Th(SO_4)_2(H_2O)_7$. Th is the dark sphere, and sulfate anions are tetrahedral units shown. (b) Fit of the background-subtracted PDF data obtained from a solution from which $Th(SO_4)_2(H_2O)_7$ crystallized. Peaks represent Th correlations; the pattern is fit with Gaussians attributed to Th–O, Th–H, and Th–S, Th–O, and Th–Th interactions that are used to infer the coordination about the Th metal center in solution. (c) Illustration of the proposed correlations that exist in solution. Reproduced with permission from Knope KE, Wilson RE, Skanthakumar S, Soderholm L. Synthesis and characterization of thorium(IV) sulfates. *Inorganic Chemistry.* 2011;50(17):8621–8629. Copyright 2011 American Chemical Society.

2.8.3 Following Oligomer Formation and Stability

HEXS has also yielded important insight into hydrolysis and condensation reactions that are well known to occur for the actinide ions, and an understanding of which is critical to the development of predictive models of actinide chemical behavior. This technique has been applied to monitor the formation of dimeric Th species from mononuclear complexes, ligand-directed assembly of hexanuclear units from lower-order oligomers, and the stability of these and related polynuclear complexes in aqueous solution (61, 149). For example, Hu et al. used HEXS to follow the speciation and evolution of Th complexes in perchlorate solutions as a function of pH and in the presence or absence of a carboxylate donor ligand (61). An aqueous solution containing 0.5 M $Th(ClO_4)_4$ in 2.0 M $HClO_4$ was used as the starting point for electrolysis experiments, and bulk electrolysis was employed to raise the pH of the solution (61). Aliquots were taken from three solutions including (1) the initial solution, (2) immediately prior to the point of bulk thorium precipitation (pH ~2.5), and (3) after the addition of glycine. HEXS data were then collected for these samples, and the FT of the background-subtracted data are shown in Figure 2.44 (61).

The PDF of the initial solution has two main peaks at 2.46 Å and 4.6 Å that correspond to the first and second coordination sphere Th-O(H_2) correlations, respectively, for a monomeric Th species. With increasing pH of the solution, noticeable differences are observed in the scattering data. When bulk electrolysis is halted shortly prior to precipitation (pH 2.5), the resultant PDF, shown in Figure 2.44(a, ii), exhibits correlations consistent with those previously reported for aqueous dimeric hydroxo-bridged Th complexes. Most notable is the appearance of an additional peak at 3.93 Å attributed to Th–Th interactions and consistent with the distances observed for hydroxo-bridged Th in the solid state. The peak at 6.4 Å can be assigned to distal O on the bridged Th (61). The PDF obtained for the solution to which glycine was added (post-electrolysis) also exhibits discernible differences as compared to the monomeric and dimeric solutions. The increase in the intensity of the peak at 3.93 Å in addition to the appearance of a peak centered at 5.56 Å, along with longer-range correlations, are most notable. Single crystals isolated from these solutions show that the structure is built from hexanuclear $[Th_6(OH)_4O_4]^{12+}$ structural units. The Th-Th interatomic distances observed in the crystal structure (3.922 and 5.545 Å) are in excellent agreement with those observed in the PDF obtained for the solution, suggesting that the cluster is present in solution prior to precipitation. Also of note are the relatively low intensity and higher range correlations between 6 and 10 Å. These correlations can be attributed to Th–C correlations from glycine ligands decorating the outside of the thorium hexamer, and provide evidence that not only is the cluster assembled in solution, but also the ligands are well ordered about the cluster core (61).

In addition to providing insight into the hydrolysis and condensation chemistry of metal ions and the ligand-directed assembly of higher-order oligomers, HEXS has also been applied to the examination of the stability of polynuclear species in solution (144). For example, Soderholm et al. isolated and structurally characterized single crystals that formed during the processing of a solution that exhibited the chemical characteristics of plutonium polymer (144). Single-crystal X-ray diffraction of the precipitate showed the structure to be built from polynuclear $[Pu_{38}O_{56}Cl_{54}(H_2O)_8]^{14-}$ structural units that exhibit a core structure related to bulk PuO_2. Dissolution of the compound into aqueous 2M LiCl produced a green solution, the optical spectrum of which was

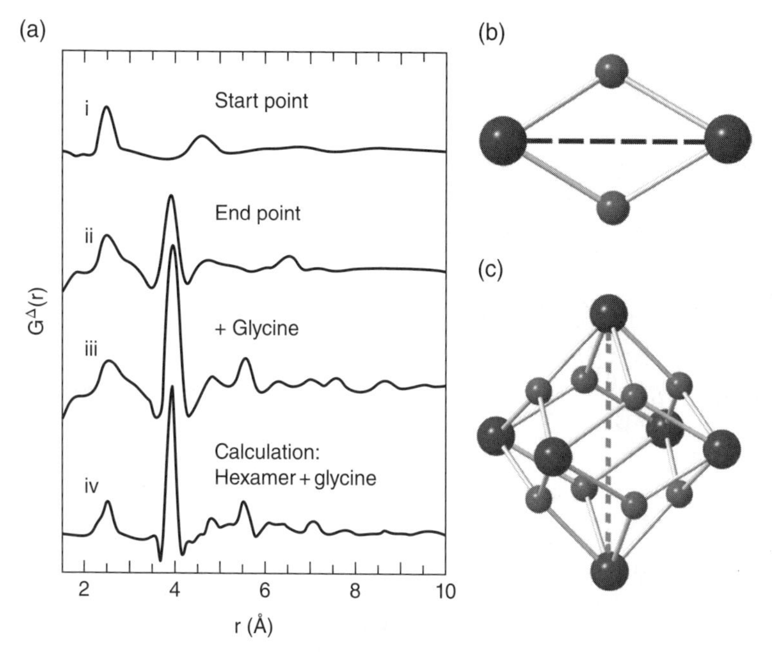

Figure 2.44 (a) PDFs generated from background-subtracted (ii, iii) Fourier-transformed HEXS data: (i) acidic aqueous solution of thorium before bulk electrolysis (2.0 M $HClO_4$), (ii) the solution after bulk electrolysis was stopped, just prior to precipitation (pH ≈ 2.5), (iii) solution PDF after addition of glycine. (iv) Calculated PDF pattern of the Th hexanuclear cluster with glycine ligands as observed in the solid state. The peak at 3.93 Å corresponds to neighboring Th–Th correlations, while the peak at 5.56 Å is assigned to distal Th–Th correlations in the hexameric core. (b) Illustration of a dimeric $Th_2(OH)_2$ structural unit; the dotted black line corresponds to the Th-Th interatomic distance that is reported at 3.998(2)–4.020(2) Å for solid-state $Th_2(OH)_2$ structural units; (c) Representation of the $Th_6(OH)_4O_4$ hexanuclear core that is observed in solid-state structure of crystals which precipitate from solutions to which glycine was added. The dotted gray line highlights the distal Th–Th correlation that has a distance of 5.545(12) Å in the solid state. Reproduced with permission from Hu Y-J, Knope KE, Skanthakumar S, Soderholm L. Understanding the ligand-directed assembly of a hexanuclear ThIV molecular cluster in aqueous solution. *European Journal of Inorganic Chemistry.* 2013, 24:4159–4163. Copyright 2013 Wiley.

consistent with plutonium polymer. In an effort to examine the correlation between the solution species and that observed in the solid state, the authors examined the relationship between the PDF from HEXS data collected for the solution used to obtain the crystal and the PDF calculated from the single-crystal structural analysis; these patterns are shown in Figure 2.45. Notably, the experimental PDF from the solution exhibits Pu–Pu correlations out to 12 Å, which is consistent with the diameter of the cluster core observed in the solid state. Moreover, excellent agreement between the pattern calculated from the crystal structure and the experimentally determined PDF provides evidence that the Pu oligomer exists as a stable, monodisperse solution species (144).

2.8.4 Overview of HEXS

HEXS is emerging as a powerful technique for understanding actinide speciation as demonstrated by the number of recent studies that have used this analytical tool to examine metal ion speciation in solution. As illustrated in the examples above, the

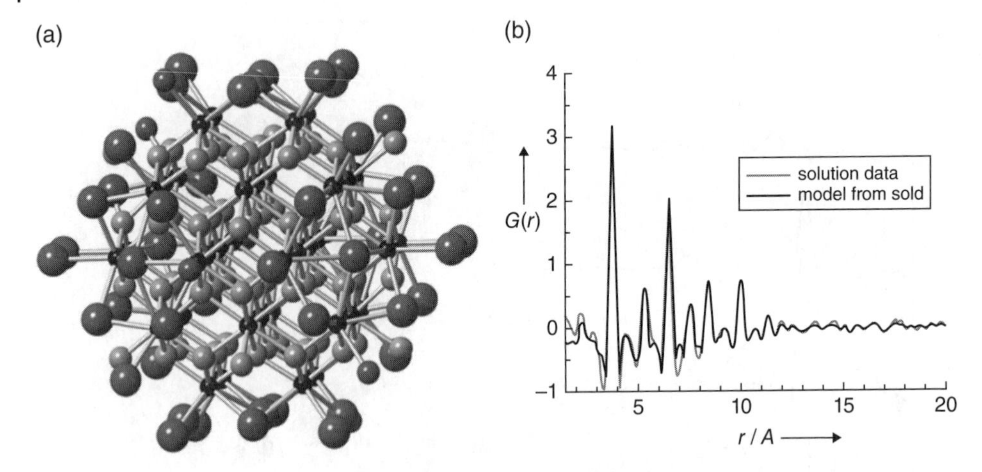

Figure 2.45 (a) Illustration of the $Pu_{38}O_{56}$ core cluster isolated from aqueous solution and (b) the Fourier-transformed HEXS data obtained for the hydrolyzed Pu(IV) solution from which single crystals of $Li_{14}[Pu_{38}O_{56}Cl_{54}(H_2O)_8]$ were isolated (gray) and the PDF calculated from the solid-state structure (black). Reproduced with permission from Soderholm L, Almond PM, Skanthakumar S, Wilson RE, Burns PC. The structure of the plutonium oxide nanocluster $[Pu_{38}O_{56}Cl_{54}(H_2O)_8]$. *Angewandte Chemie International Edition Engl*. 2008;47(2):298–302. Copyright 2008 Wiley.

utility of this analytical technique lies in its ability to directly give information on ligand coordination modes, metal ion speciation, and short- to long-range ordering that can be applied to the examination of the stability and evolution of species over a range of synthetic conditions. Although it is one of the newest analytical methods to be applied in actinide speciation, HEXS arguably provides the most comprehensive view of species in solution, and by virtue of this, reveals chemical reactivity patterns in a much more comprehensive manner. The promise of HEXS experiments in the future of aqueous actinide chemistry is difficult to overstate, and undoubtedly provides the tools to augment our understanding of the 5*f* block as well as the ability to form more accurate predictions of potential reactivity patterns.

References

1 Lütke L, Moll H, Bachvarova V, Selenska-Pobell S, Bernhard G. The U(vi) speciation influenced by a novel Paenibacillus isolate from Mont Terri Opalinus clay. *Dalton Transactions*. 2013;**42**(19):6979–6988.
2 Choppin GR. Actinide speciation in the environment. *Journal of Radioanalytical and Nuclear Chemistry*. 2007;**273**(3):695–703.
3 Choppin GR. Actinide speciation in aquatic systems. *Marine Chemistry*. 2006; **99**(1–4):83–92.
4 Choppin GR, Jensen MP. Actinides in solution: Complexation and kinetics. In: Morss LR, Edelstein NM, Fuger J, editors. *The Chemistry of the Actinide and Transactinide Elements*. Dordrecht: Springer Netherlands; 2011. p. 2524–2621.
5 Rizkalla EN, Choppin GR. Chapter 127 Lanthanides and actinides hydration and hydrolysis. *Handbook on the Physics and Chemistry of Rare Earths*. Volume 18: Elsevier; 1994. p. 529–558.

6 Liu G, Beitz JV. Optical spectra and electronic structure. In: Morss LR, Edelstein NM, Fuger J, Katz JJ, editors. *The Chemistry of the Actinide and Transactinide Elements. 1.* Dordrecht: Springer; 2006. p. 2013–2111.

7 Pilgrim CD, Zavarin M, Casey WH. Pressure dependence of carbonate exchange with $[NpO_2(CO_3)_3]_4$– in aqueous solutions. *Inorganic Chemistry.* 2017;**56**(1):661–666.

8 Liao Z, Ling J, Reinke LR, Szymanowski JES, Sigmon GE, Burns PC. Cage clusters built from uranyl ions bridged through peroxo and 1-hydroxyethane-1,1-diphosphonic acid ligands. *Dalton Transactions.* 2013;**42**(19):6793–6802.

9 Smith PA, Burns PC. Ionothermal effects on low-dimensionality uranyl compounds using task specific ionic liquids. *CrystEngComm.* 2014;**16**(31):7244–7250.

10 Falaise C, Assen A, Mihalcea I, Volkringer C, Mesbah A, Dacheux N, et al. Coordination polymers of uranium(iv) terephthalates. *Dalton Transactions.* 2015;**44**(6):2639–2649.

11 Mei L, Wang L, Yuan LY, An SW, Zhao YL, Chai ZF, et al. Supramolecular inclusion-based molecular integral rigidity: A feasible strategy for controlling the structural connectivity of uranyl polyrotaxane networks. *Chemical Communications.* 2015;**51**(60):11990–11993.

12 Jackson MN, Kamunde-Devonish MK, Hammann BA, Wills LA, Fullmer LB, Hayes SE, et al. An overview of selected current approaches to the characterization of aqueous inorganic clusters. *Dalton Transactions.* 2015;**44**(39):16982–17006.

13 Lehr JH, Lehr JK. *Standard Handbook of Environmental Science, Health, and Technology.* New York: McGraw-Hill; 2000.

14 Bürger S, Banik NL, Buda RA, Kratz Jens V, Kuczewski B, Trautmann N. Speciation of the oxidation states of plutonium in aqueous solutions by UV/Vis spectroscopy, CE-ICP-MS and CE-RIMS. *Radiochimica Acta* 2007. p. 433.

15 Tian G, Rao L, Oliver A. Symmetry and optical spectra: A “silent” 1 : 2 Np(v)-oxydiacetate complex. *Chemical Communications.* 2007(**40**):4119–4121.

16 Weng Z, Wang S, Ling J, Morrison JM, Burns PC. $(UO_2)_2[UO_4(trz)_2](OH)_2$: A U(VI) Coordination intermediate between a tetraoxido core and a uranyl ion with cation–cation interactions. *Inorganic Chemistry.* 2012;**51**(13):7185–7191.

17 Morrison JM, Moore-Shay LJ, Burns PC. U(VI) Uranyl cation – cation interactions in framework germanates. *Inorganic Chemistry.* 2011;**50**(6):2272–2277.

18 Balboni E, Burns PC. Cation–cation interactions and cation exchange in a series of isostructural framework uranyl tungstates. *Journal of Solid State Chemistry.* 2014;**213**:1–8.

19 Roßberg A, Reich T, Bernhard G. Complexation of uranium(VI) with protocatechuic acid—application of iterative transformation factor analysis to EXAFS spectroscopy. *Analytical and Bioanalytical Chemistry.* 2003;**376**(5):631–638.

20 Meinrath G. Chemometric analysis: Uranium(VI) hydrolysis by UV-Vis spectroscopy. *Journal of Alloys and Compounds.* 1998;**275–277**:777–781.

21 Meinrath G. Uranium(VI) speciation by spectroscopy. *Journal of Radioanalytical and Nuclear Chemistry.* 1997;**224**(1):119–126.

22 Tsushima S, Rossberg A, Ikeda A, Müller K, Scheinost AC. Stoichiometry and structure of uranyl(vi) hydroxo dimer and trimer complexes in aqueous solution. *Inorganic Chemistry.* 2007;**46**(25):10819–10826.

23 Duval S, Béghin S, Falaise C, Trivelli X, Rabu P, Loiseau T. Stabilization of tetravalent 4f (Ce), 5d (Hf), or 5f (Th, U) clusters by the $[\alpha\text{-}SiW_9O_{34}]_{10}$– polyoxometalate. *Inorganic Chemistry.* 2015;**54**(17):8271–8280.

24 Falaise C, Volkringer C, Loiseau T. Mixed formate-dicarboxylate coordination polymers with tetravalent uranium: Occurrence of tetranuclear {U4O4} and hexanuclear $\{U_6O_4(OH)_4\}$ Motifs. *Crystal Growth & Design*. 2013;**13**(7):3225–3231.

25 Adelani PO, Cook ND, Burns PC. Use of 2,2-Bipyrimidine for the preparation of UO22+-3d Diphosphonates. *Crystal Growth & Design*. 2014;**14**(11):5692–5699.

26 Falaise C, Volkringer C, Vigier J-F, Henry N, Beaurain A, Loiseau T. Three-dimensional MOF-type architectures with tetravalent uranium hexanuclear motifs (U_6O_8). *Chemistry – A European Journal*. 2013;**19**(17):5324–5331.

27 Falaise C, Volkringer C, Hennig C, Loiseau T. Ex-Situ Kinetic investigations of the formation of the poly-oxo cluster U38. *Chemistry – A European Journal*. 2015;**21**(46):16654–64.

28 Adelani PO, Jouffret LJ, Szymanowski JES, Burns PC. Correlations and differences between uranium(VI) arsonates and phosphonates. *Inorganic Chemistry*. 2012;**51**(21):12032–12040.

29 Diwu J, Wang S, Liao Z, Burns PC, Albrecht-Schmitt TE. Cerium(IV), neptunium(iv), and plutonium(iv) 1,2-phenylenediphosphonates: Correlations and differences between early transuranium elements and their proposed surrogates. *Inorganic Chemistry*. 2010;**49**(21):10074–10080.

30 Geipel G. Some aspects of actinide speciation by laser-induced spectroscopy. *Coordination Chemistry Reviews*. 2006;**250**(7–8):844–854.

31 Edelstein NM, Klenze R, Fanghänel T, Hubert S. Optical properties of Cm(III) in crystals and solutions and their application to Cm(III) speciation. *Coordination Chemistry Reviews*. 2006;**250**(7–8):948–973.

32 Moulin C. On the use of time-resolved laser-induced fluorescence (TRLIF) and electrospray mass spectrometry (ES-MS) for speciation studies. *Radiochim Acta*. 2003;**91**(11):651–657.

33 Barkleit A, Moll H, Bernhard G. Complexation of uranium(VI) with peptidoglycan. *Dalton Transactions*. 2009(**27**):5379–5385.

34 Skerencak A, Panak PJ, Fanghänel T. Complexation and thermodynamics of Cm(III) at high temperatures: The formation of [Cm(SO4)n]3-2n (n = 1, 2, 3) complexes at T = 25 to 200 °C. *Dalton Transactions*. 2013;**42**(2):542–549.

35 Knope KE, Soderholm L. Solution and solid-state structural chemistry of actinide hydrates and their hydrolysis and condensation products. *Chemical Reviews*. 2013;**113**(2):944–994.

36 Meinrath G, Kimura T. Behaviour of U(VI) solids under conditions of natural aquatic systems. *Inorganica Chimica Acta*. 1993;**204**(1):79–85.

37 Eliet V, Grenthe I, Bidoglio G. Time-resolved laser-induced fluorescence of uranium(VI) hydroxo-complexes at different temperatures. *Applied Spectroscopy*. 2000;**54**(1):99–105.

38 Kimura T, Choppin GR. Luminescence study on determination of the hydration number of Cm(III). *Journal of Alloys and Compounds*. 1994;**213/214**:313–317.

39 Adelani PO, Ozga M, Wallace CM, Qiu J, Szymanowski JES, Sigmon GE, et al. Hybrid uranyl-carboxyphosphonate cage clusters. *Inorganic Chemistry*. 2013;**52**(13):7673–7679.

40 Mei L, Wu Q-y, Yuan L-y, Wang L, An S-w, Xie Z-n, et al. An unprecedented two-fold nested super-polyrotaxane: sulfate-directed hierarchical polythreading assembly of uranyl polyrotaxane moieties. *Chemistry – A European Journal*. 2016;**22**(32): 11329–11338.

41 Martin NP, Falaise C, Volkringer C, Henry N, Farger P, Falk C, et al. Hydrothermal crystallization of uranyl coordination polymers involving an imidazolium dicarboxylate ligand: Effect of pH on the nuclearity of uranyl-centered subunits. *Inorganic Chemistry.* 2016;**55**(17):8697–8705.
42 Mihalcea I, Henry N, Loiseau T. Revisiting the uranyl-phthalate system: Isolation and crystal structures of two types of uranyl – organic frameworks (UOF). *Crystal Growth & Design.* 2011;**11**(5):1940–1947.
43 Mihalcea I, Henry N, Volkringer C, Loiseau T. Uranyl–pyromellitate coordination polymers: Toward three-dimensional open frameworks with large channel systems. *Crystal Growth & Design.* 2012;**12**(1):526–535.
44 Mihalcea I, Henry N, Bousquet T, Volkringer C, Loiseau T. Six-fold coordinated uranyl cations in extended coordination polymers. *Crystal Growth & Design.* 2012;**12**(9): 4641–4648.
45 Ling J, Morrison JM, Ward M, Poinsatte-Jones K, Burns PC. Syntheses, structures, and characterization of open-framework uranyl germanates. *Inorganic Chemistry.* 2010;**49**(15):7123–7128.
46 Mihalcea I, Henry N, Clavier N, Dacheux N, Loiseau T. Occurence of an octanuclear motif of uranyl isophthalate with cation–cation interactions through edge-sharing connection mode. *Inorganic Chemistry.* 2011;**50**(13):6243–9.
47 Wåhlin P, Vallet V, Wahlgren U, Grenthe I. Water exchange mechanism in the first excited state of hydrated uranyl(VI). *Inorganic Chemistry.* 2009;**48**(23): 11310–3.
48 Alam TM, Liao Z, Nyman M, Yates J. Insight into hydrogen bonding of uranyl hydroxide layers and capsules by use of 1H magic-angle spinning NMR spectroscopy. *The Journal of Physical Chemistry C.* 2016;**120**(19):10675–10685.
49 Alam TM, Liao Z, Zakharov LN, Nyman M. Solid-state dynamics of uranyl polyoxometalates. *Chemistry – A European Journal.* 2014;**20**(27):8302–8307.
50 Unruh DK, Gojdas K, Flores E, Libo A, Forbes TZ. Synthesis and structural characterization of hydrolysis products within the uranyl iminodiacetate and malate systems. *Inorganic Chemistry.* 2013;**52**(17):10191–10198.
51 Palladino G, Szabó Z, Fischer A, Grenthe I. Structure, equilibrium and ligand exchange dynamics in the binary and ternary dioxouranium(vi)-ethylenediamine-N,N[prime or minute]-diacetic acid-fluoride system: a potentiometric, NMR and X-ray crystallographic study. *Dalton Transactions.* 2006(**43**):5176–5183.
52 Szabó Z, Grenthe I. Reactivity of the "yl"-bond in uranyl(VI) complexes. 1. Rates and mechanisms for the exchange between the trans-dioxo oxygen atoms in $(UO_2)_2(OH)_2^{2+}$ and mononuclear UO_2(OH)n2-n complexes with solvent water. *Inorganic Chemistry.* 2007;**46**(22):9372–9378.
53 McGrail BT, Pianowski LS, Burns PC. Photochemical water oxidation and origin of nonaqueous uranyl peroxide complexes. *Journal of the American Chemical Society.* 2014;**136**(13):4797–4800.
54 Farkas I, Bányai I, Szabó Z, Wahlgren U, Grenthe I. Rates and mechanisms of water exchange of UO_2^{2+}(aq) and UO_2(oxalate)F$(H_2O)_2$-: A variable-temperature 17O and 19F NMR study. *Inorganic Chemistry.* 2000;**39**(4):799–805.
55 Wall TF, Jan S, Autillo M, Nash KL, Guerin L, Naour CL, et al. Paramagnetism of aqueous actinide cations. Part I: Perchloric acid media. *Inorganic Chemistry.* 2014;**53**(5):2450–2459.

56 Miyamoto T, Tsutsui M. Organic derivatives of the *f*-block elements: A quest for *f*-Orbital participation and future perspective. lanthanide and actinide chemistry and spectroscopy. *ACS Symposium Series*. 131: American Chemical Society; 1980. p. 45–58.

57 Allen PG, Bucher JJ, Clark DL, Edelstein NM, Ekberg SA, Gohdes JW, et al. Multinuclear NMR, Raman, EXAFS, and X-ray diffraction studies of uranyl carbonate complexes in near-neutral aqueous solution. X-ray structure of $[C(NH_2)_3]_6[(UO_2)_3(CO_3)_6]$.cntdot.6.5$H_2O$. *Inorganic Chemistry*. 1995;**34**(19):4797–4807.

58 Panasci AF, Harley SJ, Zavarin M, Casey WH. Kinetic studies of the $[NpO_2(CO_3)_3]_4$– ion at alkaline conditions using 13C NMR. *Inorganic Chemistry*. 2014;**53**(8):4202–4208.

59 Johnson RL, Harley SJ, Ohlin CA, Panasci AF, Casey WH. Multinuclear NMR study of the pressure dependence for carbonate exchange in the $UO_2(CO_3)_{34}$ – (aq) ion. *ChemPhysChem*. 2011;**12**(16):2903–2906.

60 Szabó Z, Grenthe I. Potentiometric and multinuclear NMR study of the binary and ternary uranium(VI) – L – fluoride systems, where L Is α-hydroxycarboxylate or glycine. *Inorganic Chemistry*. 2000;**39**(22):5036–5043.

61 Hu Y-J, Knope KE, Skanthakumar S, Soderholm L. Understanding the ligand-directed assembly of a hexanuclear ThIV molecular cluster in aqueous solution. *European Journal of Inorganic Chemistry*. 2013;**2013**(24):4159–4163.

62 Dembowski M, Olds TA, Pellegrini KL, Hoffmann C, Wang X, Hickam S, et al. Solution 31P NMR study of the acid-catalyzed formation of a highly charged $\{U_{24}Pp_{12}\}$ nanocluster, $[(UO_2)_{24}(O_2)_{24}(P_2O_7)_{12}]^{48-}$, and Its structural characterization in the solid state using single-crystal neutron diffraction. *Journal of the American Chemical Society*. 2016;**138**(27):8547–8553.

63 Qiu J, Nguyen K, Jouffret L, Szymanowski JES, Burns PC. Time-resolved assembly of chiral uranyl peroxo cage clusters containing belts of polyhedra. *Inorganic Chemistry*. 2013;**52**(1):337–345.

64 Oliveri AF, Pilgrim CD, Qiu J, Colla CA, Burns PC, Casey WH. Dynamic phosphonic bridges in aqueous uranyl clusters. *European Journal of Inorganic Chemistry*. 2016;**2016**(6):797–801.

65 Oliveri AF, Carnes ME, Baseman MM, Richman EK, Hutchison JE, Johnson DW. Single nanoscale cluster species revealed by 1H NMR diffusion-ordered spectroscopy and small-angle X-ray scattering. *Angewandte Chemie International Edition*. 2012;**51**(44):10992–10996.

66 Oliveri AF, Elliott EW, Carnes ME, Hutchison JE, Johnson DW. Cover picture: Elucidating inorganic nanoscale species in solution: Complementary and corroborative approaches (*ChemPhysChem* 12/2013). *ChemPhysChem*. 2013;**14**(12):2605.

67 Ferraro JR, Nakamoto K, Brown CW. *Introductory Raman Spectroscopy*. San Diego, CS: Elsevier Science; 2003.

68 Lu G, Forbes TZ, Haes AJ. Evaluating best practices in Raman spectral analysis for uranium speciation and relative abundance in aqueous solutions. *Analytical Chemistry*. 2016;**88**(1):773–780.

69 Guillaume B, Begun GM, Hahn RL. Raman spectrometric studies of "cation-cation" complexes of pentavalent actinides in aqueous perchlorate solutions. *Inorganic Chemistry*. 1982;**21**(3):1159–1166.

70 Jin GB. Three-dimensional network of cation–cation-bound neptunyl(V) squares: Synthesis and in situ Raman spectroscopy studies. *Inorganic Chemistry*. 2016;**55**(5): 2612–2619.

71 Kim K-W, Jung E-C, Lee K-Y, Cho H-R, Lee E-H, Chung D-Y. Evaluation of the behavior of uranium peroxocarbonate complexes in Na–U(VI)–CO_3–OH–H_2O_2 solutions by Raman spectroscopy. *The Journal of Physical Chemistry A*. 2012;**116**(49): 12024–12031.
72 Knope KE, Skanthakumar S, Soderholm L. Two dihydroxo-bridged plutonium(IV) nitrate dimers and their relevance to trends in tetravalent ion hydrolysis and condensation. *Inorganic Chemistry*. 2015;**54**(21):10192–10196.
73 Qiu J, Dembowski M, Szymanowski JES, Toh WC, Burns PC. Time-Resolved X-ray scattering and Raman spectroscopic studies of formation of a uranium-vanadium-phosphorus-peroxide cage cluster. *Inorganic Chemistry*. 2016;**55**(14):7061–7067.
74 Rowland CE, Kanatzidis MG, Soderholm L. Tetraalkylammonium uranyl isothiocyanates. *Inorganic Chemistry*. 2012;**51**(21):11798–11804.
75 Wilson RE, Schnaars DD, Andrews MB, Cahill CL. *Supramolecular interactions in PuO2Cl42– and PuCl62– Complexes with protonated pyridines: Synthesis, crystal structures, and Raman spectroscopy. Inorganic Chemistry*. 2014;**53**(1):383–392.
76 Wylie EM, Peruski KM, Weidman JL, Phillip WA, Burns PC. Ultrafiltration of uranyl peroxide nanoclusters for the separation of uranium from aqueous solution. *ACS Applied Materials & Interfaces*. 2014;**6**(1):473–479.
77 Nguyen Trung C, Begun GM, Palmer DA. Aqueous uranium complexes. 2. Raman spectroscopic study of the complex formation of the dioxouranium(VI) ion with a variety of inorganic and organic ligands. *Inorganic Chemistry*. 1992;**31**(25):5280–5287.
78 Nguyen-Trung C, Palmer DA, Begun GM, Peiffert C, Mesmer RE. Aqueous uranyl complexes 1. Raman spectroscopic study of the hydrolysis of uranyl(VI) in solutions of trifluoromethanesulfonic acid and/or tetramethylammonium hydroxide at 25 °C and 0.1 MPa. *Journal of Solution Chemistry*. 2000;**29**(2):101–129.
79 De Sio SM, Wilson RE. Structural and spectroscopic studies of fluoroprotactinates. *Inorganic Chemistry*. 2014;**53**(3):1750–1755.
80 Knope KE, Vasiliu M, Dixon DA, Soderholm L. Thorium(IV)-selenate clusters containing an octanuclear Th(IV) hydroxide/oxide core. *Inorganic Chemistry*. 2012;**51**(7):4239–4249.
81 Knope KE, Wilson RE, Vasiliu M, Dixon DA, Soderholm L. Thorium(IV) molecular clusters with a hexanuclear Th core. *Inorganic Chemistry*. 2011;**50**(19):9696–9704.
82 Vasiliu M, Knope KE, Soderholm L, Dixon DA. Spectroscopic and Energetic properties of thorium(IV) molecular clusters with a hexanuclear core. *Journal of Physical Chemistry A*. 2012;**116**(25):6917–6926.
83 Dembowski M, Bernales V, Qiu J, Hickam S, Gaspar G, *Gagliardi L, et al.* Computationally-guided assignment of unexpected signals in the raman spectra of uranyl triperoxide complexes. *Inorganic Chemistry*. 2017.
84 Falaise C, Nyman M. The key role of U28 in the aqueous self-assembly of uranyl peroxide nanocages. *Chemistry-A European Journal*. 2016;**22**(41):14678–14687.
85 Griffiths PR, Pariente GL. Introduction to spectra deconvolution. *Trends in Analytical Chemistry*. 1986;**5**(8):209–215.
86 Sullivan JC, Hindman JC, Zielen AJ. Specific interaction between Np(V) and U(VI) in aqueous perchloric acid media1. *Journal of the American Chemical Society*. 1961; **83**(16):3373–3378.
87 De Sio SM, Wilson RE. EXAFS study of the speciation of protactinium(V) in aqueous hydrofluoric acid solutions. *Inorganic Chemistry*. 2014;**53**(23):12643–12649.

88 Skanthakumar S, Soderholm L. Studying actinide correlations in solution using high-energy X-ray scattering. *MRS Proceedings.* 2005;**893**.

89 Koningsberger D, Prins R. *X-Ray Absorption: Principles, Applications, Techniques of EXAFS, SEXAFS, and XANES.* New York: John Wiley & Sons; 1988.

90 Teo B, Joy D. *EXAFS spectroscopy. Techniques and Applications.* New York: Plenum; 1981.

91 Choppin GR, Thakur P, Mathur JN. Complexation thermodynamics and structural aspects of actinide-aminopolycarboxylates. *Coordination Chemistry Reviews.* 2006;**250**(7–8):936–947.

92 Den Auwer C, Simoni E, Conradson S, Madic C. Investigating actinyl oxo cations by X-ray absorption spectroscopy. *European Journal of Inorganic Chemistry.* 2003(**21**):3843–3859.

93 Denecke MA. Actinide speciation using X-ray absorption fine structure spectroscopy. *Coordination Chemistry Reviews.* 2006;**250**(7–8):730–754.

94 Shi W-Q, Yuan L-Y, Wang C-Z, Wang L, Mei L, Xiao C-L, et al. Exploring actinide materials through synchrotron radiation techniques. *Advanced Materials.* 2014;**26**(46):7807–7848.

95 Szabó Z, Toraishi T, Vallet V, Grenthe I. Solution coordination chemistry of actinides: Thermodynamics, structure and reaction mechanisms. *Coordination Chemistry Reviews.* 2006;**250**(7–8):784–815.

96 Tan XL, Fang M, Wang XK. Sorption speciation of lanthanides/actinides on minerals by TRLFS, EXAFS and DFT studies: A Review. *Molecules.* 2010;**15**(11):8431–8468.

97 Banik NL, Vallet V, Réal F, Belmecheri RM, Schimmelpfennig B, Rothe J, et al. First structural characterization of Pa(IV) in aqueous solution and quantum chemical investigations of the tetravalent actinides up to Bk(IV): The evidence of a curium break. *Dalton Transactions.* 2016;**45**(2):453–457.

98 Di Giandomenico MV, Le Naour C, Simoni E, Guillaumont D, Moisy P, Hennig C, et al. Structure of early actinides(V) in acidic solutions. *Radiochimica Acta.* 2009;**97**(7):347–353.

99 Takao S, Takao K, Kraus W, Ernmerling F, Scheinost AC, Bernhard G, et al. First Hexanuclear U-IV and Th-IV formate complexes – Structure and stability range in aqueous solution. *European Journal of Inorganic Chemistry.* 2009(**32**):4771–4775.

100 Hennig C, Ikeda A, Schmeide K, Brendler V, Moll H, Tsushima S, et al. The relationship of monodentate and bidentate coordinated uranium(VI) sulfate in aqueous solution. *Radiochimica Acta.* 2008;**96**(9–11):607–611.

101 Hennig C, Ikeda-Ohno A, Emmerling F, Kraus W, Bernhard G. Comparative investigation of the solution species $U(CO_3)(5)$ (6-) and the crystal structure of Na-6 $U(CO_3)(5)$ center dot 12H(2)O. *Dalton Transactions.* 2010;**39**(15):3744–3750.

102 Hennig C, Ikeda-Ohno A, Tsushima S, Scheinost AC. The sulfate coordination of Np(IV), Np(V), and Np(VI) in aqueous solution. *Inorganic Chemistry.* 2009;**48**(12):5350–5360.

103 Hennig C, Kraus W, Emmerling F, Ikeda A, Scheinost AC. Coordination of a uranium(IV) sulfate monomer in an aqueous solution and in the solid state. *Inorganic Chemistry.* 2008;**47**(5):1634–1638.

104 Hennig C, Panak PJ, Reich T, Rossberg A, Raff J, Selenska-Pobell S, et al. EXAFS investigation of uranium(VI) complexes formed at Bacillus cereus and Bacillus sphaericus surfaces. *Radiochimica Acta.* 2001;**89**(10):625–631.

105 Hennig C, Reich T, Dahn R, Scheidegger AM. Structure of uranium sorption complexes at montmorillonite edge sites. *Radiochimica Acta*. 2002;**90**(9–11): 653–657.

106 Hennig C, Schmeide K, Brendler V, Moll H, Tsushima S, Scheinost AC. EXAFS investigation of U(VI), U(IV), and Th(IV) sulfato complexes in aqueous solution. *Inorganic Chemistry*. 2007;**46**(15):5882–5892.

107 Hennig C, Takao S, Takao K, Weiss S, Kraus W, Emmerling F, et al. Identification of hexanuclear Actinide(IV) carboxylates with Thorium, Uranium and Neptunium by EXAFS spectroscopy. In: Wu ZY, editor. *15th International Conference on X-Ray Absorption Fine Structure*. Journal of Physics Conference Series. 430. Bristol: Iop Publishing Ltd; 2013.

108 Hennig C, Takao S, Takao K, Weiss S, Kraus W, Emmerling F, et al. Structure and stability range of a hexanuclear Th(iv)-glycine complex. *Dalton Transactions*. 2012;**41**:12818–12823.

109 Hennig C, Tutschku J, Rossberg A, Bernhard G, Scheinost AC. Comparative EXAFS investigation of uranium(VI) and -(IV) aquo chloro complexes in solution using a newly developed spectroelectrochemical cell. *Inorganic Chemistry*. 2005;**44**(19):6655–6661.

110 Ikeda-Ohno A, Tsushima S, Takao K, Rossberg A, Funke H, Scheinost AC, et al. Neptunium carbonato complexes in aqueous solution: An electrochemical, spectroscopic, and quantum chemical study. *Inorganic Chemistry*. 2009;**48**(24):11779–11787.

111 Le Naour C, Trubert D, Di Giandomenico MV, Fillaux C, Den Auwer C, Moisy P, et al. First structural characterization of a protactinium(V) single oxo bond in aqueous media. *Inorganic Chemistry*. 2005;**44**(25):9542–9546.

112 Mendes M, Hamadi S, Le Naour C, Roques J, Jeanson A, Den Auwer C, et al. Thermodynamical and structural study of protactinium(v) oxalate complexes in solution. *Inorganic Chemistry*. 2010;**49**(21):9962–9971.

113 Moll H, Reich T, Hennig C, Rossberg A, Szabó Z, Grenthe I. Solution coordination chemistry of uranium in the binary UO22+-SO42- and the ternary UO22+-SO42--OH- system. *Radiochimica Acta*. 2000;**88**(9–11):559–566.

114 Takao K, Takao S, Scheinost AC, Bernhard G, Hennig C. Formation of soluble hexanuclear neptunium(iv) nanoclusters in aqueous solution: Growth termination of actinide(iv) hydrous oxides by carboxylates. *Inorganic Chemistry*. 2012;**51**(3): 1336–1344.

115 Tamain C, Dumas T, Guillaumont D, Hennig C, Guilbaud P. First evidence of a water-soluble plutonium(iv) hexanuclear cluster. *European Journal of Inorganic Chemistry.* 2016(**22**):3536–3540.

116 Ikeda A, Hennig C, Tsushima S, Takao K, Ikeda Y, Scheinost AC, et al. Comparative study of uranyl(VI) and -(V) carbonato complexes in an aqueous solution. *Inorganic Chemistry*. 2007;**46**(10):4212–4219.

117 Clark DL, Conradson SD, Donohoe RJ, Gordon PL, Keogh DW, Palmer PD, et al. Chemical speciation of neptunium(VI) under strongly alkaline conditions. Structure, composition, and oxo ligand exchange. *Inorganic Chemistry*. 2013;**52**(7):3547–3555.

118 Antonio MR, Williams CW, Sullivan JA, Skanthakumar S, Hu YJ, Soderholm L. Preparation, stability, and structural characterization of plutonium(vii) in alkaline aqueous solution. *Inorganic Chemistry*. 2012;**51**(9):5274–5281.

119 Soderholm L, Skanthakumar S, Neuefeind J. Determination of actinide speciation in solution using high-energy X-ray scattering. *Analytical and Bioanalytical Chemistry*. 2005;**383**(1):48–55.

120 Nyman M. Small-angle X-ray scattering to determine solution speciation of metal-oxo clusters. *Coordination Chemistry Reviews*. 2016:Ahead of Print.

121 Burns PC, Kubatko K-A, Sigmon G, Fryer BJ, Gagnon JE, Antonio MR, et al. Actinyl peroxide nanospheres. *Angewandte Chemie International Edition*. 2005;**44**(14): 2135–2139.

122 Peruski KM, Bernales V, Dembowski M, Lobeck HL, Pellegrini KL, Sigmon GE, et al. Uranyl peroxide cage cluster solubility in water and the role of the electrical double layer. *Inorganic Chemistry*. 2017.

123 Forbes TZ, McAlpin JG, Murphy R, Burns PC. Inside cover: Metal–Oxygen isopolyhedra assembled into fullerene topologies (*Angew. Chem. Int. Ed.* 15/2008). *Angewandte Chemie International Edition*. 2008;**47**(15):2710.

124 Sigmon GE, Unruh DK, Ling J, Weaver B, Ward M, Pressprich L, et al. Symmetry versus minimal pentagonal adjacencies in uranium-based polyoxometalate fullerene topologies. *Angewandte Chemie International Edition*. 2009;**48**(15):2737–2740.

125 Qiu J, Ling J, Sui A, Szymanowski JES, Simonetti A, Burns PC. Time-resolved self-assembly of a fullerene-topology core–shell cluster containing 68 uranyl polyhedra. *Journal of the American Chemical Society*. 2012;**134**(3):1810–1816.

126 Sigmon GE, Weaver B, Kubatko K-A, Burns PC. Crown and bowl-shaped clusters of uranyl polyhedra. *Inorganic Chemistry*. 2009;**48**(23):10907–10909.

127 Ling J, Ozga M, Stoffer M, Burns PC. Uranyl peroxide pyrophosphate cage clusters with oxalate and nitrate bridges. *Dalton Transactions*. 2012;**41**(24):7278–7284.

128 Ling J, Wallace CM, Szymanowski JES, Burns PC. Hybrid uranium–oxalate fullerene topology cage clusters. *Angewandte Chemie International Edition*. 2010;**49**(40): 7271–7273.

129 Ling J, Qiu J, Sigmon GE, Ward M, Szymanowski JES, Burns PC. Uranium pyrophosphate/methylenediphosphonate polyoxometalate cage clusters. *Journal of the American Chemical Society*. 2010;**132**(38):13395–13402.

130 Hou Y, Zakharov LN, Nyman M. Observing assembly of complex inorganic materials from polyoxometalate building blocks. *Journal of the American Chemical Society*. 2013;**135**(44):16651–16657.

131 Gebauer D, Kellermeier M, Gale JD, Bergström L, Cölfen H. Pre-nucleation clusters as solute precursors in crystallisation. *Chemical Society Reviews*. 2014;**43**(7):2348–2371.

132 Alizadeh MH, Mohadeszadeh M. Sandwich-type Uranium-substituted of bismuthotungstate: Synthesis and structure determination of [Na (UO_2) 2 (H_2O) 4 (BiW_9O_{33}) 2] 13–. *Journal of Cluster Science*. 2008;**19**(2):435–443.

133 Ling J, Hobbs F, Prendergast S, Adelani PO, Babo J-M, Qiu J, et al. Hybrid uranium–transition-metal oxide cage clusters. *Inorganic Chemistry*. 2014;**53**(24):12877–12884.

134 Mal SS, Dickman MH, Kortz U. Actinide polyoxometalates: Incorporation of uranyl–peroxo in U-shaped 36-tungsto-8-phosphate. *Chemistry-A European Journal*. 2008;**14**(32):9851–9855.

135 Deb T, Zakharov L, Falaise C, Nyman M. Structure and solution speciation of UIV linked phosphomolybdate (MoV) clusters. *Inorganic Chemistry*. 2016;**55**(2):755–761.

136 Qiao B, Ferru G, Olvera de la Cruz M, Ellis RJ. Molecular origins of mesoscale ordering in a metalloamphiphile phase. *ACS Central Science*. 2015;**1**(9):493–503.

137 Soderholm L, Skanthakumar S, Wilson RE. Structural correspondence between uranyl chloride complexes in solution and their stability constants. *Journal of Physical Chemistry A*. 2011;**115**(19):4959–4967.

138 Neuefeind J, Soderholm L, Skanthakumar S. Experimental coordination environment of uranyl(VI) in aqueous solution. *Journal of Physical Chemistry A*. 2004;**108**: 2733–2739.

139 Chapman KW, Chupas PJ, editors. *Pair Distribution Function Analysis of High-Energy X-ray Scattering Data*. Hoboken, NJ: John Wiley & Sons, 2013.

140 Chupas PJ, Chapman KW, Chen H, Grey CP. Application of high-energy X-rays and pair-distribution-function analysis to nano-scale structural studies in catalysis. *Catalysis Today*. 2009;**145**(3–4):213–219.

141 Soderholm L, Mitchell JF. Perspective: Toward "synthesis by design": Exploring atomic correlations during inorganic materials synthesis. *APL Materials*. 2016;**4**(5): 053212/1–053212/9.

142 Wilson RE, Skanthakumar S, Sigmon G, Burns PC, Soderholm L. Structures of dimeric hydrolysis products of thorium. *Inorganic Chemistry*. 2007;**46**(7):2368–2372.

143 Knope KE, Wilson RE, Skanthakumar S, Soderholm L. Synthesis and characterization of thorium(IV) sulfates. *Inorganic Chemistry*. 2011;**50**(17):8621–8629.

144 Soderholm L, Almond PM, Skanthakumar S, Wilson RE, Burns PC. The structure of the plutonium oxide nanocluster $[Pu_{38}O_{56}Cl_{54}(H_2O)_8]_{14}$. *Angewandte Chemie International Edition Engl*. 2008;**47**(2):298–302.

145 Wilson RE, Skanthakumar S, Burns PC, Soderholm L. structure of the homoleptic thorium(IV) aqua ion $[Th(H_2O)_{10}]Br_4$. *Angewandte Chemie International Edition*. 2007;**46**(42):8043–8045.

146 Wilson RE, Skanthakumar S, Cahill CL, Soderholm L. Structural studies coupling X-ray diffraction and high-energy X-ray scattering in the UO22+-HBraq system. *Inorganic Chemistry*. 2011;**50**(21):10748–10754.

147 Skanthakumar S, Antonio MR, Wilson RE, Soderholm L. The curium aqua ion. *Inorganic Chemistry*. 2007;**46**(9):3485–3491.

148 Soderholm L, Skanthakumar S, Neuefeind J. Determination of actinide speciation in solution using high-energy X-ray scattering. *Analytical and Bioanalytical Chemistry*. 2005;**383**(1):48–55.

149 Wilson RE, Skanthakumar S, Soderholm L. The structures of polynuclear Th(IV) hydrolysis products. *Materials Research Society Symposium Proceedings*. 2007;986(Actinides 2006 – Basic Science, Applications and Technology):183–188.

3

Complex Inorganic Actinide Materials

Matthew L. Marsh and Thomas E. Albrecht-Schmitt

Florida State University, Tallahassee

3.1 Introduction

When traversing the actinide series, the very diverse behavior of oxidation states and physical traits necessitates a careful description of the types of bonding, structures, and physical phenomena that are presently known. It is self-evident that both a clearer picture and a crisper understanding of these types of functional materials can aid in the broader scope of nuclear-related problems of this and subsequent generations. These problems, for example, pass between improved designs with Generation-IV reactors, advanced fuel fabrication assemblies, reprocessing, and selection and continued monitoring of future nuclear waste repositories. Many of the materials described herein have some of the properties needed for these future developments. Due in part to their exotic nature, scarcity, and lack of equitable coverage compared to other areas of the periodic table, however, much more study of these elements will be needed in order to utilize their potential and solve all of these challenging problems.

Actinides can complex in a wide variety of ways with fluorides, borates, sulfates, and phosphates, which together encompass a large number of known actinide structures. There are, of course, many more, but it is the objective of this chapter to only emphasize the above, which will suffice to highlight the main instructive and thematic points. The topography of the present structures can commonly depend on a wide variety of factors such as the oxidation state, ionic radii, stoichiometry, pH, and temperature. Most of the synthetic reactions are conducted in either the solid state or hydrothermally – in general, this also mimics the expectations for chemistry found in either reactor conditions or in a repository environment. To study the topography and resulting structures in detail, single-crystal x-ray diffraction is considered to be the most robust and reliable technique at hand. Nevertheless, a large number of experimental techniques can be employed to discover structural information, and this will be presented as each individual case studied permits.

Experimental and Theoretical Approaches to Actinide Chemistry, First Edition.
Edited by John K. Gibson and Wibe A. de Jong.

3.2 Fluorides

3.2.1 Trivalent and Tetravalent Fluorides

The beginning of structural actinide chemistry was ushered in by the prolific crystallographer and Manhattan project scientist William Zachariasen. This beginning saw many improvements in the interpretation of crystallographic information, such as weak diffraction lines and absorption corrections. The need to characterize fluoride complexes first came from the ubiquitous use of UF_6 in gas centrifuges for the isotopic enrichment of ^{235}U from ^{238}U. While none of these compounds are necessarily the most complex, they are important to detail as the basic building blocks to higher-order structures.

Crystal parameters for the trivalent and tetravalent fluorides are listed in Table 3.1. The trivalent fluorides crystallize predominantly in the hexagonal crystal system and are isostructural with the LaF_3 structure type. They exhibit the very common behavior of a decrease in overall size due to the decrease in ionic radii across the series.

Depending on the actinide, different treatments to achieve the trivalent fluorides are necessary. For example, since curium in solution already resides preferably in the trivalent state, it can be precipitated out by simply adding HF [1]. However, for an element

Table 3.1 Colors and lattice parameters for some trivalent and tetravalent fluorides [1, 4, 7–11].

Compound	Color	a_0 (Å)	b_0 (Å)	c_0 (Å)	β (°)
AcF_3	White	7.41	7.41	7.55	90
ThF_4	White	13.10	11.01	8.6	126
PaF_4	Brick-red	12.86	10.88	8.54	126.34
UF_3	Black	7.181	7.181	7.348	90
UF_4	Blue-green	12.803	10.792	8.372	126.30
NpF_3	Purple	7.129	7.129	7.288	90
NpF_4	Green	12.67	10.62	8.31	126.16
PuF_3	Purple	7.092	7.092	7.240	90
PuF_4	Dusty-pink	12.59	10.55	8.26	126.16
AmF_3	Pink	7.044	7.044	7.225	90
AmF_4	Rose-tan	12.49	10.47	8.19	126.16
CmF_3	White	6.999	6.999	7.129	90
CmF_4	Yellow-green	12.45	10.45	8.16	126
BkF_3	Yellow-green	6.70	7.09	4.41	90
BkF_3	Yellow-green	6.97	6.97	7.14	90
BkF_4	Yellow-green	12.396	10.466	8.118	126.33
CfF_3	Green	6.653	7.039	4.393	90
CfF_3	Green	6.945	6.945	7.101	90
CfF_4	Green	12.327	10.402	8.113	126.44

like plutonium, it is necessary to maintain a reducing environment with the use of hydroxylamine or ascorbic acid as shown with hydroxylamine in Equation 3.1 [2]:

$$2PuO_2 + 6HF + 2H_3NO \rightarrow 2PuF_3 \cdot 3H_2O + N_2 \tag{3.1}$$

The tetravalent fluorides crystallize in the monoclinic crystal system and are isostructural with the ZrF_4 structure type. UF_4, as a group example, has two distinct uranium sites which are each arranged in a slightly distorted antiprism configuration. There are 12 repeating formula units per unit cell [3]. Curium, which resists oxidation to the tetravalent state, is reacted with purified fluorine gas at 400 °C as follows [4]:

$$2CmF_3 + F_2 \rightarrow 2CmF_4 \tag{3.2}$$

In a different example, dried plutonium oxalate is converted first to plutonium (III) fluoride by a direct hydrofluorination between 120 °C and 130 °C. In the next step, a three hour, 400 °C hydrofluorination follows [5]:

$$4PuF_3 + 4HF + O_2 \rightarrow 4PuF_4 + 2H_2O \tag{3.3}$$

In contrast to Equation 3.1, where a reducing environment is controlled, in this example it is the presence of oxygen that is required to ensure complete conversion from the trivalent to the tetravalent state. Such added oxidation/reducing agents are common when dealing with plutonium chemistry because plutonium can be known to equilibrate up to four oxidation states simultaneously (in solution) [6].

As stated above, most of the trifluorides shown in Table 3.1 exhibit the LaF_3 structure type. However, it should be noted that both berkelium and californium have two entries in this table for the trifluoride. This is because for each of these, the structure type is related to a function of temperature. While both do adopt the LaF_3 structure type at higher temperatures, they assume the YF_3 structure type in the orthorhombic crystal system at 25 °C [9–10].

The number of complex fluorides is vast. A simple step beyond the tetravalent fluorides, for example, is to complex alkali fluorides together with the actinides in order to create a ternary tetravalent fluoride. It just happens that this is very relevant due to the widespread possibility of developing molten salt reactor designs. These reactors can use either eutectic mixtures of either LiF or NaF as the solvent to maintain fuel and fission products for reprocessing. Higher temperatures are possible with these designs, which can allow for more efficient burn-up. However, depending upon the reactor conditions, a spectrum of redox potentials makes for potentially diverse chemical environments. Therefore, various conditions for forming these complexes have been of interest. With the element protactinium, it can be seen that reducing conditions at 400 °C can make tetravalent complexation possible [12]:

$$7NaF + 6PaF_5 + 3H_2 \rightarrow 7NaF \cdot 6PaF_4 + 6HF \tag{3.4}$$

The generic formula of $7AF \bullet 6PaF_4$ above exists for a number of alkali fluorides (A = Na, K, Rb). The lattice type is rhombohedral for this family of compounds. It is also noteworthy that application of fluorine gas to the rubidium compound has a reversible effect on Equation 3.4, and then with reapplication of hydrogen gas, a cyclic type of process can be established. Because protactinium (V) has no valence electron, there is a significantly palpable difference in the absorption spectrum of its tetravalent product.

Meanwhile, for the case of the smaller lithium fluoride, a different, but similar product, can result [13]:

$$LiF + UF_6 + H_2 \rightarrow LiF \cdot UF_4 + 2HF \tag{3.5}$$

This product is tetragonal, and due to a smaller stoichiometric ratio ($AF{:}AnF_4$) has a smaller overall molecular volume. To complete the series, use of ammonium fluoride at lower temperatures result in the largest stoichiometric ratio (4:1) and is monoclinic. Apart from minor differences, uranium also reacts in the same fashion as in Equation 3.4.

Returning to ammonium fluoride as a reagent briefly, it should be mentioned that the monoclinic structure cited above is only for one set of conditions and that many reaction products exist solely for this, depending upon the temperature and initial stoichiometric ratios. For example, the following reaction is for the 2:1 ratio and is set at 130 °C–160 °C. The use of dry gases prevents the plutonium tetrafluoride from hydrolyzing as follows [14]:

$$2NH_4F + PuF_4 \rightarrow (NH_4)_2 PuF_6 \tag{3.6}$$

While anhydrous $(NH_4)_4UF_8$ happens to be the most stable crystalline phase for uranium, the analogous 4:1 plutonium product, a light red to pink-hued crystal, decomposes to the 2:1 product (green) seen above.

3.2.2 Pentavalent and Hexavalent Fluorides

Undoubtedly, the higher oxidation states are represented by a smaller margin of the actinide elements. Nevertheless, because a strong suit of these compounds is represented by uranium and plutonium, their complexes with fluoride are important in a range of applications. Certain designs of the molten salt reactor example given in the preceding section, for example, utilize a purified stream of F_2 to oxidize nuclear fuel into UF_6 and PuF_6 as a purification step for separating out and removing fission products. Besides UF_6 and PuF_6, NpF_6 is possible and shares similar properties as its two neighbors on the periodic table. All of the hexavalent fluorides exist in the tetragonal crystal system and have low-boiling temperatures. Meanwhile, the pentavalent class includes protactinium but loses plutonium. This slight shift to the left of the 5*f*-series also exhibits a significantly lower stability of NpF_5 which disproportionates to NpF_4 and NpF_6 under a majority of circumstances [15]. NpF_5, as do all of the binary pentavalent fluorides, crystallizes in the orthorhombic crystal system. All of the binary species are listed in Table 3.2.

Ammonium fluoride, as well as all of the alkali metals, also forms ternary complexes with pentavalent fluorides. Stoichiometric ratios of 1:1, 2:1, and 3:1 provide, in most cases, distinct products resembling the formulas $AAnF_6$, A_2AnF_7, and A_3AnF_8,where $A = NH_4$, Na, K, Rb, and Cs. Li, regardless of stoichiometric ratio, favors the 1:1 ratio [16]. These reactions only generally apply to the pentavalent actinides listed in Table 3.2, though pentavalent plutonium can also be stabilized by a number of these reactions. Another important observation is that if greater stoichiometric ratios are used, the products remain at the 3:1 ratio. This is unlike the tetravalent fluoride complexes, where several stable and unique compounds can exist with the synthetic use of the 4:1 ratio. For uranium, this tendency can be referred to commonly as the F/U limit. The F/U limit

Table 3.2 Colors and lattice parameters for some pentavalent and hexavalent fluorides [15, 20–22].

Compound	Color	a_0 (Å)	b_0 (Å)	c_0 (Å)
PaF_5	White	11.530	11.530	5.190
α-UF_5	Bluish-white	6.525	6.525	4.472
β-UF_5	Greenish-white	11.473	11.473	5.209
NpF_5	Bluish-white	6.536	6.536	4.456
UF_6	White	9.900	8.962	5.207
NpF_6	Orange	9.910	8.970	5.210
PuF_6	Reddish brown	9.950	9.020	5.260

of 8 is the maximum ratio of fluorine atoms that can structurally coexist per uranium atom in a ternary alkali uranium fluoride complex regardless of the valence state. Therefore, the expectation for complex hexavalent uranium fluorides would be a synthetic limit of the stoichiometric ratio held at 2:1. Such examples include $CsUF_7$ [17], NH_4UF_7 [18], and Na_2UF_8 [19], with other examples seen elsewhere in the literature.

3.2.3 Fluoride Architectures

Continuing along the line of the molten salt reactor that was used to illustrate earlier points, the proposed reactor environments limit O_2 concentrations for several reasons. The first reason is because O_2 has a tendency to promote corrosion due to the ability to react inside the molten salt with the construction alloy. However, another reason is that reducing conditions within the molten salt can promote precipitation out of oxides and oxyfluorides, and this can create a myriad of other operational risks besides just removal of fuel from the salt. Such a risk can also be especially true of thorium in Equation 3.7:

$$ThF_4 + O_2 \rightarrow ThO_2F_2 + F_2 \quad (3.7)$$

Oxyfluorides like the product above are another large class of important ternary fluorides. It is important to note them when building up to more complex architectures, because, as will be seen throughout the rest of the chapter, the An-O bond is the most prevalent bond seen in complex inorganic actinide structures and can assume a variety of forms. Whether capped axially by oxygens in the AnO_2 unit or equatorially by anionic frameworks such as phosphate or borate, the An-O bond forms the basis set of a great topographical variety in structures. While oxyfluorides can be more diverse than what is shown in Equation 3.7, they will not be discussed further here.

Most uranyl structures are represented by alteration or substitution of the equatorially aligned moieties from within the coordination group. This is because the axial bond is several times as strong as the equatorial bond. However, there has been a considerable interest in being able to modify the terminal axial oxygens for the purpose of diversifying 3D-framework topologies. This becomes possible when use of a suitable equatorial binding moiety quenches the charge at the uranium center and causes the movement of electron density, shifting it away from the terminally axial oxygens back onto the uranium center. Moieties that demonstrate a good combination of electrostatic interactions,

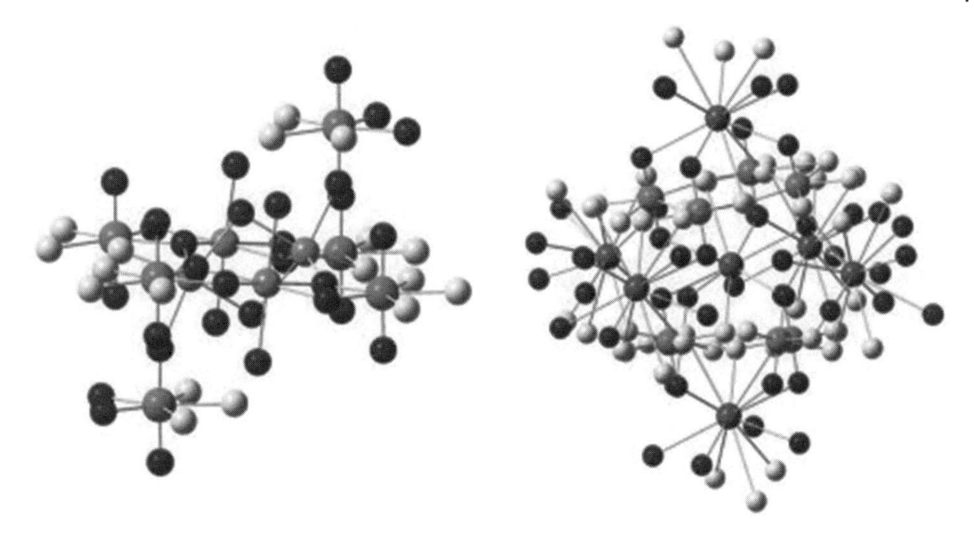

Figure 3.1 On the (left) $Na_3(UO_2)_2F_3(OH)_4(H_2O)_2$. On the (right) $Cs(UO_2)_2F_5$. (*See insert for color representation of the figure.*)

charge-transfer effects, or covalency are best at this destabilization effect and give Lewis basic attributes to the axial oxygens [23]. This, in turn, allows cations to bond with the axial oxygens and form the desired larger frameworks.

Such an example are the complex materials $Na_3(UO_2)_2F_3(OH)_4(H_2O)_2$ and $Cs(UO_2)_2F_5$ that are shown in Figure 3.1 [24]. Synthesizing them hydrothermally, it was found that the extended 3D-structures are due to the critical and abundant interactions of $U^{VI}=O$-Na and $U^{VI}=O$-Cs between the layers in the structures. In the first structure, the uranyl center displays a five-coordinate geometry marked by three U-F and two U-O equatorial bonds. On the other hand, the sodium ions adopt two different coordination environments. One coordination environment displays six Na-O bonds, whereas the other coordination environment has five Na-O bonds and one Na-F bond. In the second structure, the uranyl center is arranged in one layer as a pentagonal bipyramid with bonds to five equatorial fluoride ions. Each layer is then connected via a cesium ion. The cesium ions are 14-coordinate, divided into eight Cs-O axial bonds and six Cs-F bonds.

One important concept that illustrates the difference between the two compounds is the feature of covalency between the axial oxygen and alkali metal. In the case of sodium, molecular orbital plots show that there is a degree of covalent character in the Na-O bond. This contrasts with the case of cesium, which has very poor orbital overlap and hence only shares electrostatic interactions with the axial oxygen. In the case of sodium, the product crystallizes in the monoclinic space group *C*2/*c*, whereas in the case of cesium, the product crystallizes in the orthorhombic space group *Cmcm*. Therefore, besides size differences, fundamental bonding differences in the alkali metals contribute to the structural differences observed.

One remarkable possibility regarding the construction of these topographical features lies in a footnote to the synthesis. In the first synthesis, 1*H*-imidazole-4,5-dicarboxylic acid, an organic ligand with two common functionalities normally observed in extended structures, was absent in the final product. Yet, when the reaction is run without this ligand, no crystalline product results. The same is also found to be true with the second

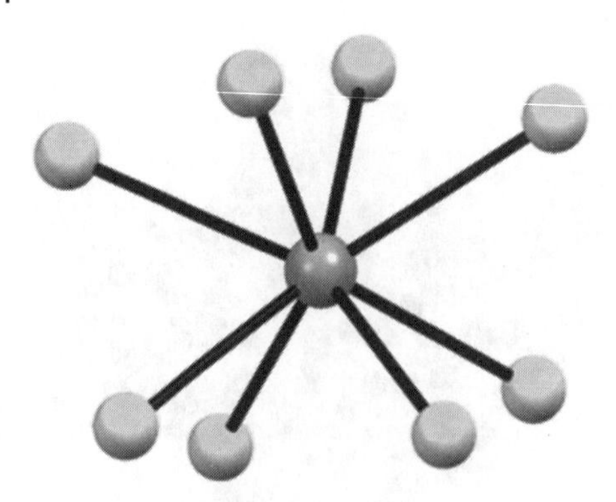

Figure 3.2 Representation of UF_8 dodecahedra in $(NH_4)_4UF_8$.

Table 3.3 Some properties of oxide versus nitride fuels [26].

	Oxides	Nitrides
Heavy metal density (g/cm^3)	9.3	13.1
Melting temperature (K)	3000	3035
Thermal conductivity (W/cm-K)	0.023	0.26
Operating centerline temperature at 40 kW/m (K)	2360	1000

compound, but the ligand exhibited instead is 3,3′,4,4′-benzophenonetetracarboxylic dianhydride. It is likely that these ligands aid in templating the uranyl and various alkali ions into the appropriate layers. Then, when the structure cools down and crystallizes, the organic ligands leave the system as it assembles the various layers together. Such design elements that a chemist can use in the initial synthesis phase are valuable for allowing desirable features and physical properties to be created within future materials.

Returning to the stable compound $(NH_4)_4UF_8$ for a moment, it can be shown that it is important to understand topographical features sometimes for the purpose of improving synthesis. This compound is monoclinic and crystallizes in the space group *C2/c* [25]. The uranium center is bonded to eight fluorine atoms and has a coordination environment that is regarded as a distorted tetragonal antiprism (Figure 3.2). Looking at the ammonium group, there are two distinct nitrogen sites. The first site has eight neighbors with N-H- -F bond distances split between four short and four long. The second site has seven neighbors split between N-H- -F bond distances of four short and three long. In addition, the topographical features have confirmed that the uranium polyhedra arrange into layers.

The possibility of using uranium nitrides as an alternative fuel in advanced reactor types has been a long-studied concept of interest. Table 3.3 compares the properties of oxide and nitride fuels. From the table, it can be gathered that average fuel temperatures will be much lower with nitrides. This could allow for the introduction of simple passive safety features that would otherwise be more difficult to implement in standard designs [26]. One type of synthetic preparation of actinide nitrides is called oxidative ammonolysis. Such a synthesis normally uses tetravalent fluorides as a precursor [27]:

$$UF_4 + 6NH_3 \rightarrow UN_2 + 4NH_4F + H_2 \quad (3.8)$$

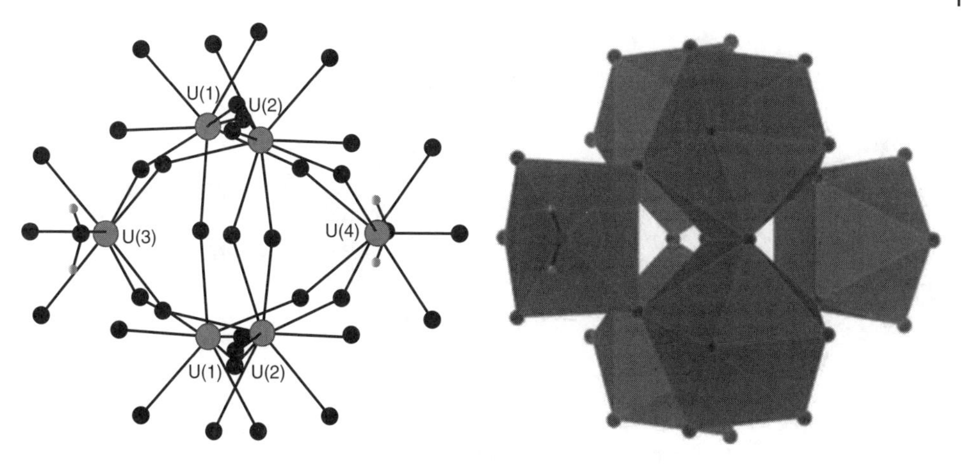

Figure 3.3 Representation of $[U_6F_{33}(H_2O)_2]^{9-}$ found in $U_3F_{12}(H_2O)$. (*See insert for color representation of the figure.*)

UF_4, as discussed earlier, also has a distorted antiprism configuration [3]. However, its structural environment is not as distorted as that of the more complexed fluoride, $(NH_4)_4UF_8$. This slight difference in the structural environment demonstrates that use of the ammonium complexed fluoride as a precursor to a nitride could be feasible, and it might react differently depending upon the conditions. An alternative synthesis, in turn, could improve the nitride/oxide impurity ratio that is formed [27]. Conceptually, it is often not difficult to see the actinide contraction in the 5*f*-series, and it becomes even more definitive when that series is isostructural. For example, $(NH_4)_4NpF_8$ has been confirmed to be isostructural with its uranium analogue [28]. As a result, both a decrease in unit cell volume from 1030.7 Å^3 to 1020.9 Å^3 and a reduction in An-F bond length of ~0.02 Å are observed.

Another fascinating material built up from the UF_4 building unit is $U_3F_{12}(H_2O)$ [29]. Even more unlike UF_4 than the ammonium complex, this hydrated binary fluoride material crystallizes in the separate, noncentrosymmetric space group *Cm*. While UF_4 consists only of corner-shared UF_8 polyhedra, $U_3F_{12}(H_2O)$ can have a variety of different corner and edge-sharing polyhedra such as UF_8, UF_9, and UOF_7. Furthermore, the basic building blocks of the two compounds also are different: $(U_6F_{38})^{14-}$ represents UF_4, whereas $[U_6F_{33}(H_2O)_2]^{9-}$ represents $U_3F_{12}(H_2O)$ (Figure 3.3). In fact, it is these key $[U_6F_{33}(H_2O)_2]^{9-}$ units, connected by fluoride moieties, which allow for the observed unique structural topography.

To get to U^{4+} from UO_2^{2+}, a carefully selected reducing environment is necessary, as was demonstrated in the synthesis of this material. $UO_2(C_2H_3O_2)_2{\bullet}H_2O$ was the starting material and was used because acetate is a common reducing agent in organic chemistry. Furthermore, the use of copper metal as a catalyst was even more critical and prompted the successful complete reduction. This reaction was run hydrothermally at 200 °C in dilute HF.

The desire to discover more complex inorganic materials with U^{4+} comes from the fact that there can be a larger coordination number as well as the fact that there are two unpaired *f*-electrons present. The stability of the U^{4+}-F bond means that fluoride architectures can be very common for this type of chemistry. Moreover, the possibility of altering the structural environment of the two *f*- electrons can create unique electronic and magnetic properties. For example, the material cited above displays semiconducting

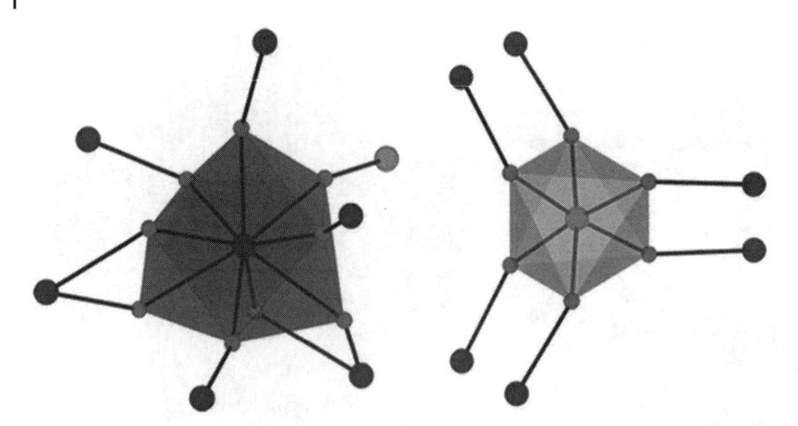

Figure 3.4 The U^{4+} and M^{2+} coordination environments. The U^{4+} cation exhibits a monocapped distorted square antiprism geometry, whereas the M^{2+} cation has a nearly octahedral environment.

behavior and has a representative band gap of 3.8 eV. It also shows deviation away from Curie–Weiss behavior below 100 K due to formation of an A_1 singlet ground state.

Another material, which is slightly different but prepared similarly to the one above, is the quaternary complex fluoride $Na_4MnU_6F_{30}$ [30]. In fact, any of the stable divalent 3*d*-metals can be substituted for Mn^{2+} and will also crystallize in the same trigonal space group $P\bar{3}\ c1$. In this structure, the basic units are corner and edge-sharing UF_9 polyhedra which arrange together into U_2F_{16} dimers. The dimers subsequently arrange into chains through their own corner and edge-sharing vertices. The overall common building block is represented by $U_6F_{30}{}^{6-}$ – these orient together in such a way that both large and small hexagonal channels where sodium ions can reside are formed (Figure 3.5). The coordination geometry of the uranium center is a distorted monocapped square antiprism (Figure 3.4).

For the sister compound $Na_4CuU_6F_{30}$, it is notable that the octahedrally coordinated Cu^{2+} ions are inhibited from adopting their common Jahn–Teller distortions. It is suggested that a possible combination of higher symmetry and rigidity of the lattice underlies the cause of this effect. In addition, there is no evidence of magnetic exchange between the uranium and transition metal centers due to their relative isolation from each other in the structure – the magnetic susceptibilities are observed instead to be a simply additive function between the centers. The magnetic susceptibility loses Curie–Weiss behavior below 150 K, as in the prior structure, due to the development of U (IV)'s singlet ground state.

In conclusion, it can be seen from some of the few examples provided that the number and diversity of actinide fluorides and complex actinide fluoride architectures are impressive. Some crystallographic information for select examples cited throughout are compiled together in Table 3.4. There is much more complexity yet to be explored in the existing architectures, such as with U^{4+} topographies. Even newer topographies such as the modification of the uranyl dication serve to inform the broader audience of new types of chemistry. Both the nuclear and materials industry will need improved technologies to fuel the future, and from the small display of nascent materials presented here, it seems assured that fluoride architectures will continue to remain an important fixture in the foreseeable future.

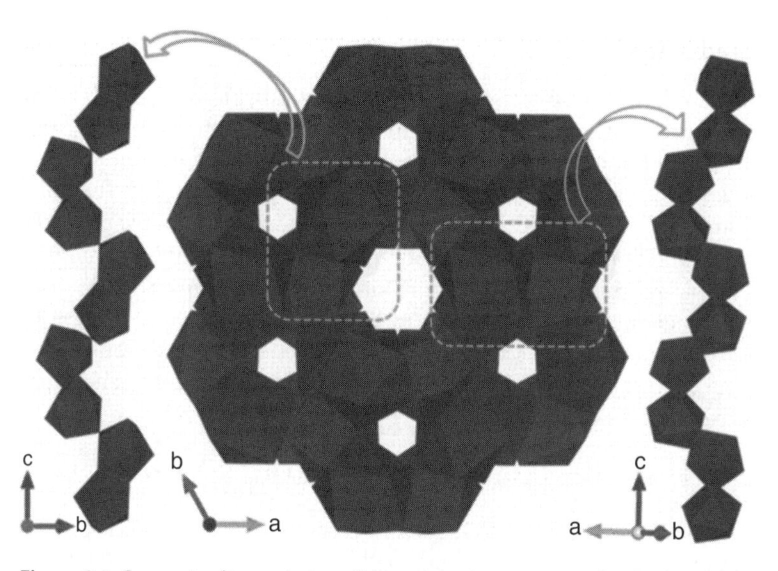

Figure 3.5 Successive linear chains of UF_9 polyhedra come together in the *ab*-plane to structurally produce both large and small hexagonal channels.

Table 3.4 Crystallographic data for select fluoride architectures [3, 23–25, 28–30].

Formula	Crystal system	Space group	a_0 (Å)	b_0 (Å)	c_0 (Å)	β (°)	Volume (Å^3)
$Na_3(UO_2)_2F_3(OH)_4(H_2O)_2$	Monoclinic	*C*2/*c*	15.125	6.941	11.257	94.76	1177.7
$Cs(UO_2)_2F_5$	Orthorhombic	*Cmcm*	12.149	11.862	12.366	90	1782.1
UF_4	Monoclinic	*C*2/*c*	12.73	10.75	8.43	126.2	931.68
$(NH_4)_4UF_8$	Monoclinic	*C*2/*c*	13.126	6.692	13.717	121.19	1030.7
$(NH_4)_4NpF_8$	Monoclinic	*C*2/*c*	13.054	6.681	13.676	121.14	1020.9
$U_3F_{12}(H_2O)$	Monoclinic	*Cm*	12.1594	11.8457	8.5409	128.787	958.92
$Na_4MnU_6F_{30}$	Trigonal	$P\overline{3}$ *c*1	9.9172	9.9172	13.033	90	1110.1

3.3 Borates

Actinide borates are considered to be one of the most successful oxyanions for probing structural variation across the 5*f*-series. This is because borates, most commonly derived from boric acid, can adopt many structure types depending upon only slight changes in reaction conditions. Examples of such conditions include pH, temperature, cation size, stoichiometry, and counterions. A large number of complex topologies exist. Furthermore, borates have become synonymous with the growing evidence that indicates increased participation in bonding by 6*d* and 5*f* orbitals as the series is traversed.

3.3.1 Functionalized Borates

There are several different synthetic strategies for making uranyl borates. In the first strategy, high-temperature melts with B_2O_3 result in structures dominated by UO_6 and UO_7 polyhedra. These reactions were commonly run in the 1980s and exhibited coordination with the borate in the form of BO_3 triangles. One example of this is the noncentrosymmetric complex $Ca(UO_2)_2B_2O_6$ [31]. In the second strategy, slow evaporation with mixtures of UO_2^{2+} and borate at room temperature result in the maintained uranyl core surrounded by isolated clusters of cyclic polyborates. This example was the first uranium borate ever made and was determined to be $K_6[UO_2(B_{16}O_{24}(OH)_8)]{\bullet}12H_2O$ [32]. Finally, the most recently developed boric acid flux reactions are also considered to be an effective way to make borates. Heated with a thorium salt at 200 °C, an inorganic thorium borate complex results [33]:

$$\begin{aligned} Th(NO_3)_4 \cdot 5H_2O + 6H_3BO_3 \rightarrow & \left[ThB_5O_6(OH)_6\right]\left[BO(OH)_2\right] \cdot 2.5H_2O \\ & +4HNO_3 + 5.5H_2O \end{aligned} \tag{3.9}$$

Named NDTB-1, this material crystallizes in the cubic space group *Fd*-3. One of the special attributes of these borates can be seen right away with the ability of separate BO_4 tetrahedra and BO_3 triangles to form within the lattice. This type of borate dichotomy has a multiplicative effect on the potential types of structural topographies that can exist because of the many ways that the two can combine, orient, or segregate within structures. Neither moiety possesses an inversion center, either. Therefore, when studying current actinide borate chemistry, this should be recognized and learned quickly. In the present example, it is the BO_4 tetrahedra which directly chelate the Th^{4+} center, whereas the BO_3 triangles only share vertices with the larger thorium polyhedra. These thorium centers are 12-coordinate and show icosahedral geometry.

This localized geometric progression spires into a much more revealing topography when viewed at a larger scale. NDTB-1 is considered to be a porous supertetrahedral three-dimensional framework and exhibits channels which form along the [110] plane in the cubic environment. The channels further combine at the center of supertetrahedra to form hexagonal cavities – this joint free void space contributed by both the channels and cavities represents 43% of the material's entire volume (Figure 3.6). It was determined that this material might have a significant capability of participating in ion change since the weakly coordinated $H_2BO_2^-$ species can be easily mobilized into vacating the structure due to an equilibrium with its weak acid counterpart, H_3BO_3.

An outstanding property discovered about this material is that it can participate in what is called single crystal to single-crystal ion exchange. This is defined as the ability to engage in ion exchange without exceeding a limiting threshold of perturbations that can disrupt and cause collapse of the crystal structure framework. Most potential ion exchange materials demonstrate collapse because the combination of charge mobility combined with a destabilization of the coordination environment exceeds this limit. NDTB-1, however, resists collapse.

Another type of functionalized material is one with nonlinear optical properties, and the most common type of nonlinear optical application is called second-harmonic generation. To be a strong candidate for second-harmonic generation, a material should crystallize in a noncentrosymmetric space group, show transparency to whatever laser

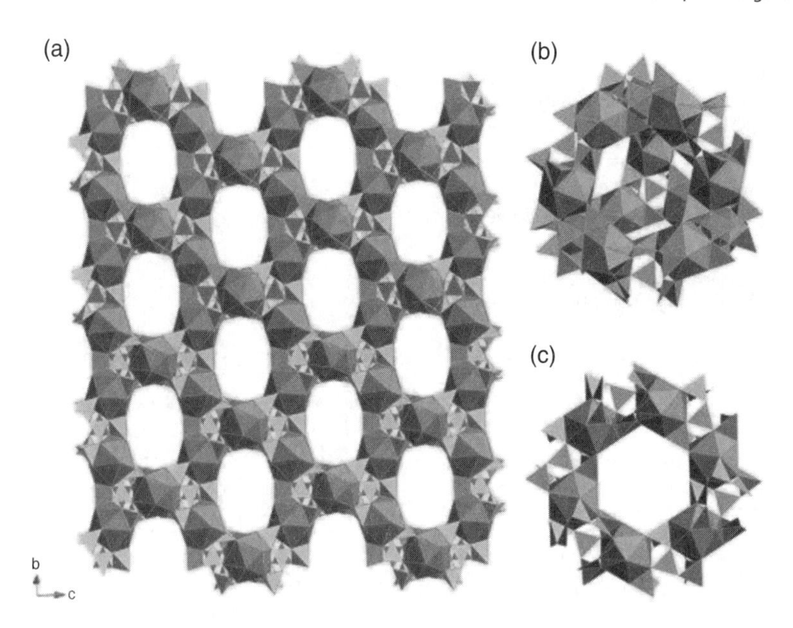

Figure 3.6 (a) The NDTB-1 supertetrahedral framework creates channels, (b) an example of a supertetrahedral cavity, and (c) a better view of a single hexagonal window from (b).

wavelength is both used and exits the material, resist degradation from the intensity of the laser light, and exhibit optimal birefringence. The beta phase of barium borate, for example, displays all of these characteristics and is considered an excellent material for second-harmonic generation. It can also be used to split photon beams for the purpose of generating single photons [34]. The success of this borate material has led to the search and discovery of other nonlinear optical borate materials in the actinide family.

One such material, based on the $Ca(UO_2)_2B_2O_6$ mentioned already, is $Li[(UO_2)B_5O_9]{\bullet}H_2O$ [35]. Similarly to NDTB-1, this material is synthesized in a boric acid flux at 190 °C as follows:

$$\begin{aligned} &UO_2\left(NO_3\right)_2\cdot 6H_2O + LiNO_3 + 5H_3BO_3 \rightarrow Li\left[(UO_2)B_5O_9\right]\cdot H_2O \\ &\quad +3HNO_3 + 11H_2O \end{aligned} \tag{3.10}$$

Crystallizing in the monoclinic and noncentrosymmetric space group *Pn*, this material is characterized by polyborate sheets that are linked together by BO_3 triangles. The uranyl core is maintained and displays UO_8 polyhedra that are each linked by six BO_4 tetrahedra and three BO_3 triangles. The lithium ions are placed between BO_3 triangles and coordinate not only to the polyborate net but also to the uranyl core and to water molecules that reside inside the open cages present within the structure (Figure 3.7a). In general, alkali and alkaline earth metals are favored as counterions of choice when designing nonlinear optical materials because these types of metal-oxygen bonds have no *d-d* electron transitions, and as a result will be more transparent to light in the ultraviolet region. It is also generally uncommon to see uranium (VI) materials without a center of symmetry owing to the uranyl group, which is very symmetric. It was shown that $Li[(UO_2)B_5O_9]{\bullet}H_2O$ doubles the frequency of 1064 nm light, thereby qualifying itself as a second-harmonic generator (Figure 3.7b).

(a)

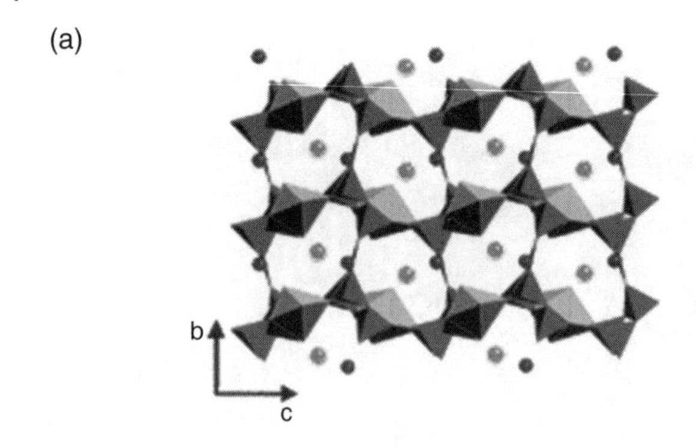

(b)

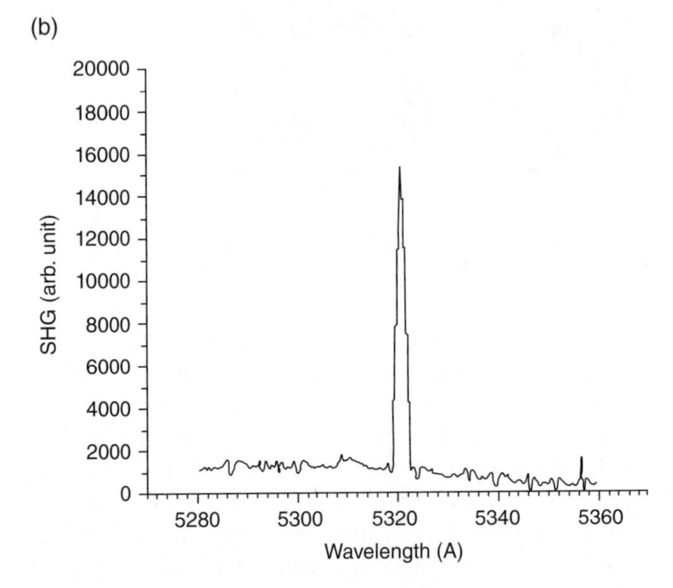

Figure 3.7 (a) Structure of $Li[(UO_2)B_5O_9]{\cdot}H_2O$. The largest polyhedra are UO_8, the smaller polyhedra are BO_3 and BO_4 units, and Na^+ and H_2O are represented by spheres. (b) Second Harmonic Generation (SHG) produces 532 nm laser light from 1064 nm laser light.

Further modifications and functionalization of these material types have arisen. For example, by introducing fluoride into the synthesis of borates, it is possible to obtain materials that incorporate both B-O bonds and B-F bonds. This has two potential advantages when designing nonlinear optical materials. First, because the B-F bond is more polarizable than the B-O bond, a larger polarization density will result and can interact with the electric field to increase the nonlinear optical effect seen in the material. Second, the introduction of other $B\text{-}F_n$ ($n = 1\text{–}4$) moieties alongside the typical BO_3 triangles and BO_4 tetrahedra can introduce lesser symmetry which also favorably affects the anisotropic polarization parameter. An example of this is the complex inorganic actinide material $Tl[(UO_2)B_5O_8(OH)F]$, synthesized by way of a boric acid flux at 220 °C for five days [36]:

$$\begin{aligned} &UO_2(NO_3)\cdot 6H_2O + 5H_3BO_3 + TlF \rightarrow \left[Tl(UO_2)B_5O_8(OH)F\right] \\ &\quad + 2HNO_3 + 12H_2O \end{aligned} \tag{3.11}$$

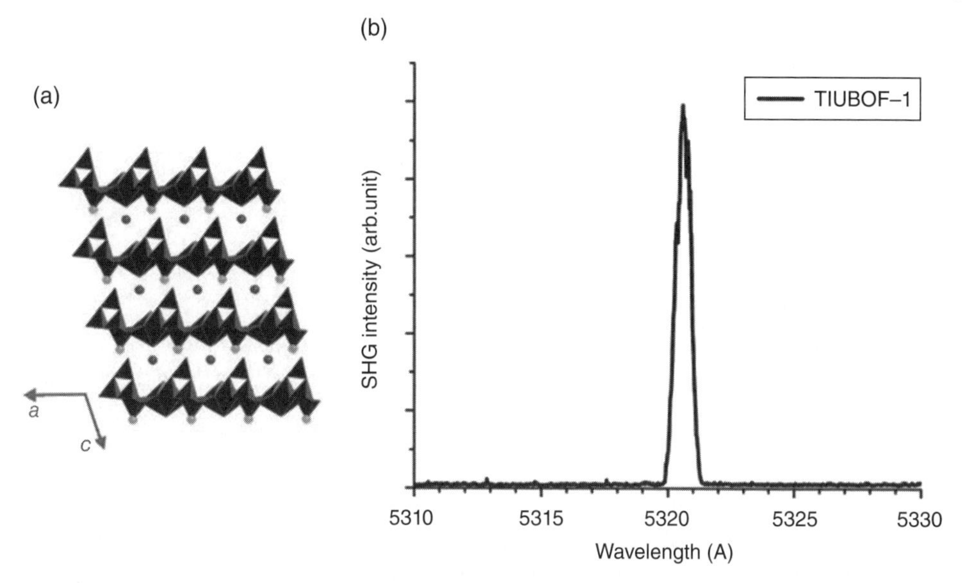

Figure 3.8 (a) The overall topography shows uranium centers linked by BO_3 triangles and BO_4 tetrahedra. F^- are circles that sit at the apex of BO_3 triangles. Tl^+ are circles that reside in the cavities. (b) SHG produces a strong signal at 532 nm from 1064 nm laser light.

This material crystallizes in the triclinic space group $P1$ with the presence of BO_3 triangles, BO_4 tetrahedra, and BO_3F tetrahedra. The topography for this structure is represented by a series of identical layers stacked on top of each other with extra BO_3 triangles angled between the layers. One important difference seen in the topography is that the extra BO_3 triangles assemble on the same face of each layer alongside the substituted BO_3F tetrahedra. In replicate structures without the fluoride moiety, these extra BO_3 triangles are assembled on the opposite face and are linked through bridged cations, while the BO_4 tetrahedra are linked via hydrogen bonding. Because the B-F bond terminates extension of the structure, the extra BO_3 triangles have to switch faces in order to maintain linkage of the layers. The attributes of such a rearrangement in topography are evident in the observation that this functionalized material shows strong second-harmonic generation (Figure 3.8).

Within the context of functionalized architectures, borates represent an impressive diversity of materials. Some crystallographic properties for these materials are summarized in Table 3.5. Yet to be fully explored, there are still many untried possibilities for modifying some of the topologies presented. For example, Chen's anionic group theory argues that a reduction of BO_4 tetrahedra paired with an increase in BO_3 triangles, and especially the creation of the six-membered ring B_3O_6, can lead to a significant increase in second-harmonic generation [37]. This is due to the typically larger response of π-orbitals in planar groups toward polarized light. If some of these advances can be realized in the actinide series, then it could lead to even more impressive and complex materials in the future.

3.3.2 Transuranic Borates

From Section 3.2.2, an example was given of the ability of Np^{5+} to disproportionate into Np^{4+} and Np^{6+}, as seen in NpF_5. This is not an uncommon behavior of Np (V) and is also exhibited in some neptunium borates. Because of the crystalline phases created,

Table 3.5 Uranium and thorium borates [31–33, 35–36].

Compound	Space group	Color	a_0 (Å)	b_0 (Å)	c_0 (Å)	β, α, γ (°)	Volume (Å^3)
$Ca(UO_2)_2B_2O_6$	*C*2	—	16.512	8.169	6.582	96.97	881
$K_6[UO_2(B_{16}O_{24}(OH)_8)]\cdot 12H_2O$	*P*2$_1$/*n*	—	12.024	26.45	12.543	94.74	3975
$[ThB_5O_6(OH)_6][BO(OH)_2]*2.5H_2O$	*Fd*3	Colorless	17.4036	17.4036	17.4036	90.0	5271.3
$Li[(UO_2)B_5O_9]\cdot H_2O$	*Pn*	Yellow-green	6.3783	6.2241	10.5308	89.996	418.06
$Tl[(UO_2)B_5O_8(OH)F]$	*P*1	Light yellow	6.4163	6.4663	7.1357	92.1, 103.3, 119.6	246.67

the disproportionation of neptunium can create rare mixed-valent states that remain trapped within the crystals. Using the standard boric acid flux synthesis at 220 °C for three days, the following reaction was observed [38]:

$$\begin{aligned} &3NpCl_5 + 3Th(NO_3)_4^{*}5H_2O + 2KNO_3 + 13H_3BO_3 \rightarrow \\ &\quad K_2\left[(NpO_2)_3 B_{10}O_{16}(OH)_2(NO_3)_2\right] + Th_3B_3O_{10} + 20H_2O + 15HCl + 12HNO_3 \end{aligned} \tag{3.12}$$

Due to the glassy state of the thorium borate, it is not known with certainty if $Th_3B_3O_{10}$ is the correct composition of the by-product or not. However, it is known with certainty that the crystalline neptunium product is monoclinic and has symmetry of *P*2$_1$/*n*. In addition, through both single-crystal x-ray diffraction and UV-vis-NIR absorption spectroscopy, it is also known with certainty that the neptunium (V) undergoes disproportionation. However, there is only unquestionable certainty for the ingrowth of Np (VI) because the *f-f* transitions seen at 1000 nm and 1200 nm are clear for only Np (V) and Np (VI), respectively. The evidence for Np (IV) is only an inference – unconfirmed to date – from an additional absorption transition seen at 760 nm (Figure 3.9b).

The topography of this structure arranges itself into Np (V) layers which are connected to each other through cation–cation interactions with $NpO_2(NO_3)_2$ moieties. Beyond the rarity of the mixed-valent state is the further infrequence of two different oxidation states of neptunium coordinating each other. The NpO_2^+ centers are coordinated to nine borate anions – five BO_3 triangles and four BO_4 tetrahedra (Figure 3.9a). While differences observed in bond lengths are able to differentiate Np (V) from Np (VI), the lack of site-specific information seen to confirm the presence of Np (IV) makes the intermediate valence of Np (IV) and Np (V) instead of mixed-valence a possibility.

It can be clearly seen from this example that the redox behavior of the actinide site has a much wider movement window due to the versatility of electrochemical potentials in neptunium as opposed to uranium. When implementing the different synthetic approaches to neptunium borates, for instance, it is important to select the proper counterion for the following reasons: (1) an anion such as perchlorate or perbromate at high temperatures will create redox perturbations such that oxidizing conditions and Np (VI) will result while (2) an anion such as bromide or thiosulfate will create weakly reducing conditions such that Np (IV) or Np (III) will result and (3) very weak redox

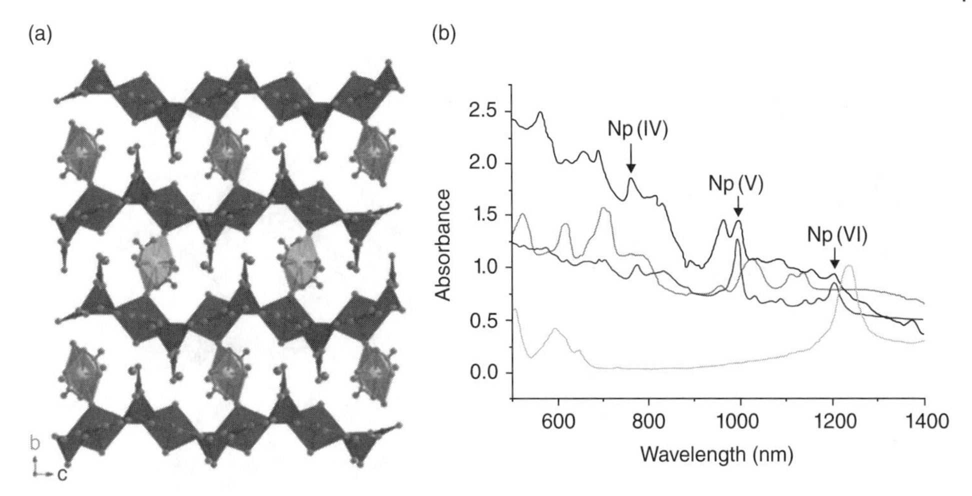

Figure 3.9 (a) The three-dimensional topography of $K_2[(NpO_2)_3B_{10}O_{16}(OH)_2(NO_3)_2]$. Np(V) polyhedra are longer on the *b*-axis (vertical), whereas the Np(VI) polyhedra are longer on the *c*-axis (pink). The BO_3 triangles and the BO_4 tetrahedra bridge between the Np(VI) polyhedra. (b) The UV-vis-NIR spectrum of several compounds. The labeled line is for $K_2[(NpO_2)_3B_{10}O_{16}(OH)_2(NO_3)_2]$ and shows the redox transitions for Np. (*See insert for color representation of the figure.*)

ions such as chloride or nitrate will sustain Np (V) long enough to create conditions for mixed/intermediate valent states. The counterion rules advocated here can also impart a strict judgment on the resulting topographies in these systems because of the lack of redox behavior in borates combined with the redox versatility of the actinide.

Another example for a pair of Np (V) borates synthesized under similar conditions but showing a different structural topology from that above is given by Equations 3.13 and 3.14 [39]:

$$NpCl_5 + KCl + 10H_3BO_3 \rightarrow K\left[(NpO_2)B_{10}O_{14}\left(OH\right)_4\right] + 10H_2O + 6HCl \quad (3.13)$$

$$2NpCl_5 + 2KCl + 16H_3BO_3 \rightarrow K_2\left[\left(NpO_2\right)_2 B_{16}O_{25}\left(OH\right)_2\right] + 17H_2O + 12HCl \quad (3.14)$$

The predominant product is from Equation 3.13. In this neptunium borate structure, sheets are stacked on top of one another with potassium ions retained in the interlayer space. BO_3 units extend out from each side of a given sheet, but the interaction of BO_3 units to former higher-order polyborate moieties is different along each side of the given sheet. Along one side, BO_3 units combine into dimeric $B_2O_5{}^{4-}$ arrangements, whereas on the opposite side they combine into trimeric $B_3O_6{}^{3-}$ arrangements. Each neptunyl center is nine-coordinate with five BO_3 triangles and four BO_4 tetrahedra participating per unit (Figure 3.10). The second product in Equation 3.14, meanwhile, still lacks specific structural information and will not be detailed further here.

Like the product from Equation 3.12, there is also evidence that supports an either mixed or intermediate valence state for this structure. For example, while the crystallographic evidence points only to Np (V), the UV-vis-NIR absorption shows a 736 nm peak characteristic only of Np (IV) species. The possibility of the disproportionation for Np (V) into Np (IV) within the lattice is further compounded by an exhibition of the Alexandrite effect, which is normally seen only in Np (IV) structures. Due to the

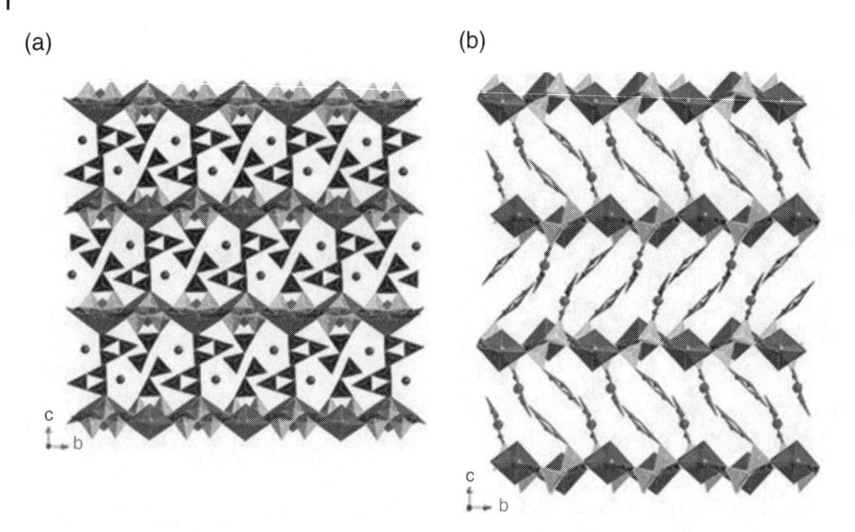

Figure 3.10 (a) A view of the layered structure in the *ac*-plane. (b) A view along the *bc*-plane. Np's have the largest polyhedra followed by BO_4 tetrahedra, BO_3 triangles, and K^+ (circles).

different possible spectral compositions of white light, the Alexandrite effect is commonly observed when there is a color change in the studied crystal due to its movement away from the irradiation of natural white light into the light of a microscope. It is also noteworthy that crystallographic evidence is only a good average of a structure. If there is an intermediate valence state, as the broader evidence could suggest, then such a behavior could be averaged into the structure.

Alternatively, the use of Np (VI) perchlorate via counterion rule number one results in an exclusively Np (VI) borate, as shown by Equation 3.15 [40]:

$$Np(ClO_4)_6 + 6KBO_2 + 2H_3BO_3 \rightarrow NpO_2\left[B_8O_{11}(OH)_4\right] + H_2O + 6KClO_4 \quad (3.15)$$

This neptunium borate crystallizes in the monoclinic and noncentrosymmetric space group *Cc*. The neptunyl core displays an unusual hexagonal bipyramid coordination environment, and each polyhedra is surrounded by nine borate groups divided between two BO_3 triangles and seven BO_4 tetrahedra. Arranged into sheets, this 3-D topology has the unique quality of exhibiting additional BO_4 tetrahedra instead of BO_3 triangles between the sheets (Figure 3.11).

Returning to the unusual hexagonal bipyramidal coordination environment, it is the electron density from an extra equatorial donor group which should alter not only normal crystallographic data but also UV-vis-NIR data, too. Indeed, this extra electron density causes a shift in the observed Np (VI) *f-f* transitions from the ascribed 1200 nm peak normally seen to the blue-shifted 1140 nm peak seen in the present structure.

Meanwhile, plutonium, which is known for even greater redox versatility than neptunium, can display some surprising synthetic behavior in boric acid flux reactions. Hence, while counterions do affect plutonium borate topologies, they do not apply in the same way as the rules for neptunium. For example, a Pu (IV) starting material can oxidize to the Pu (VI) borate without the use of strong oxidizing anions as shown in Equation 3.16 [40]:

$$Pu(NO_3)_4 + 8H_3BO_3 \rightarrow PuO_2\left[B_8O_{11}(OH)_4\right] + 4HNO_3 + 7H_2O + H_2 \quad (3.16)$$

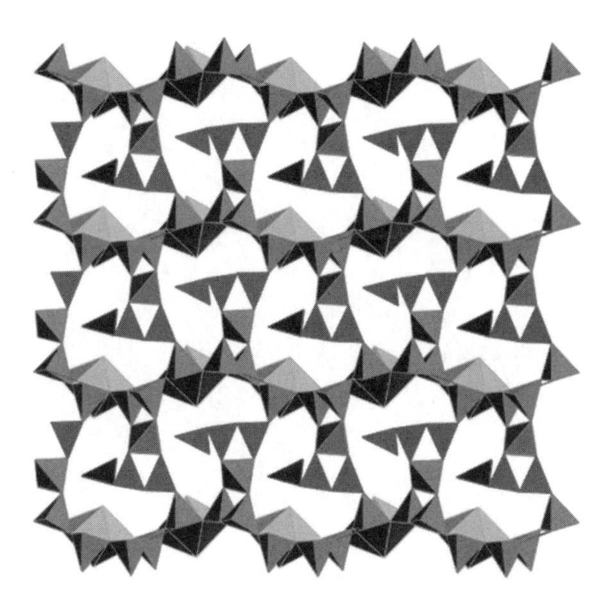

Figure 3.11 Representation of the generic topography for $AnO_2[B_6O_{11}(OH)_4]$ where An = U, Np, Pu. The three-dimensional structure has large AnO_8 polyhedra along with smaller BO_3 triangles and BO_4 tetrahedra.

This plutonium borate is isotypic to the neptunium borate from Equation 3.15. With a $5f^2$ electronic configuration, however, its UV-vis-NIR will be more similar to one of the Np (V) compounds from Equations 3.12–3.14. The main *f-f* transition is a peak at 800 nm, though the absorption properties in this example are also affected by the crystallographic orientation (Figure 3.12). This means that the absorption intensity will vary depending on the path through the lattice that the light travels in order to excite the electronic transitions at each plutonyl site. This property is called pleochroism. In this example, the effect would be noticed by a human observer as the crystals change between a pink and peach color depending upon the crystallographic orientation with respect to the viewer. However, another example where this same effect is not as apparent is the neptunium borate from Equation 3.12. These crystals will remain a pale pink color in any crystallographic orientation. Nevertheless, the absorption intensities will still change.

Plutonium can also be reduced to the trivalent borate under careful conditions. This requires the strict exclusion of oxygen and the use of chloride, bromide, or iodide as counterions. Equations 3.17 and 3.18 show the uses of both bromide and iodide [41–42]:

$$2PuBr_3 + 12H_3BO_3 \rightarrow Pu_2\left[B_{12}O_{18}(OH)_4Br_2(H_2O)_3\right]\cdot 0.5H_2O + 10.5H_2O + 4HBr \quad (3.17)$$

$$PuOI + 8H_3BO_3 \rightarrow Pu\left[B_7O_{11}(OH)(H_2O)_2I\right] + 7H_2O \quad (3.18)$$

The reduced plutonium borates are quite different from any of the previous structures presented. Crystallizing in the noncentrosymmetric, monoclinic space group *Pn*, the first structure is characterized by layers of BO_3 triangles and BO_4 tetrahedra which arrange in a way that resembles lanthanide topographies more so than thorium,

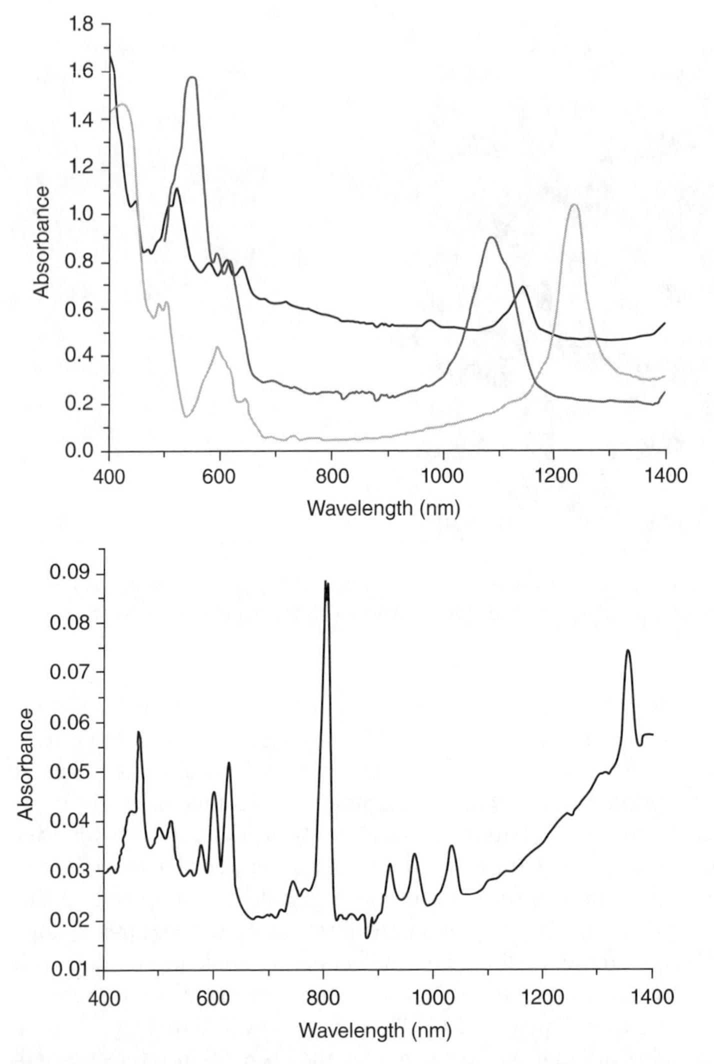

Figure 3.12 (Top) UV-vis-NIR absorption spectra for several Np(VI) compounds. $NpO_2[B_6O_{11}(OH)_4$ is shifted to 1140 nm. For context, $NpO_2(NO_3)_2$•$6H_2O$ is at 1100 nm, while $NpO_2(IO_3)_2(H_2O)$ is at 1220 nm. (bottom) UV-vis-NIR absorption spectra for $PuO_2[B_6O_{11}(OH)_4$ showing strong absorption at 800 nm along the excitation axis. (*See insert for color representation of the figure.*)

uranium, or neptunium. One type of plutonium atom adopts a nine-coordinate, hula-hoop geometry, whereas the other adopts a 10-coordinate geometry defined as a capped triangular cupola (Figure 3.13). All of the plutonium atoms are capped with bromide ions, whereas BO_3 triangles are the primary feature that populates the space between each layer (Figure 3.14).

In the second structure, there is only one unique plutonium center which is 10-coordinate and shows the same capped triangular cupola geometry as the second plutonium center in the bromide structure (Figure 3.15a). While the sheet topology is exactly like the bromide structure, the presence of the larger iodide groups causes the structure to

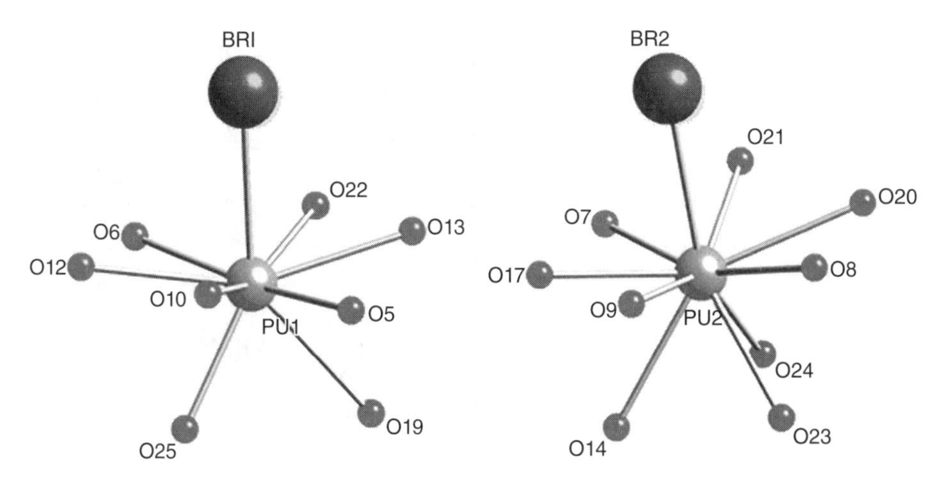

Figure 3.13 (Left) Nine-coordinate Pu_1 in a hula-hoop geometry. (Right) Ten-coordinate Pu_2 in a capped triangular cupola geometry.

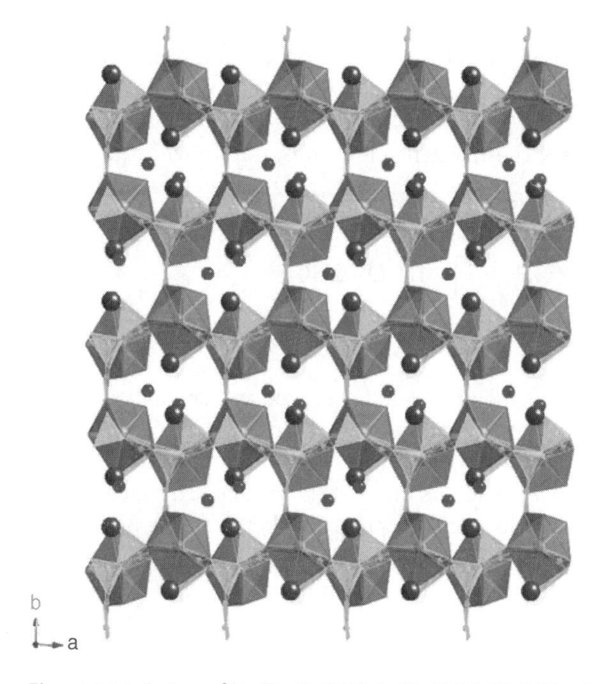

Figure 3.14 A view of $Pu_2[B_{12}O_{18}(OH)_4Br_2(H_2O)_3]\cdot 0.5H_2O$. Sheets are linked together by the BO_3 triangles. The large polyhedra represent Pu coordination, smaller polyhedra represent BO_4, large circles represent Br^- ions, and small circles represent H_2O.

pack less densely. This new size parameter also forces the structure to require an additional BO_3 triangle per plutonium center to connect the layers. The additional BO_3 triangle, in effect then, sums the total to four BO_3 triangles needed to connect the layers together. To clarify this, one layer is connected to another as follows: Pu-BO_3-BO_3-μ_3 BO_4-BO_3-BO_3-Pu (Figure 3.15b). This product crystallizes in the monoclinic space

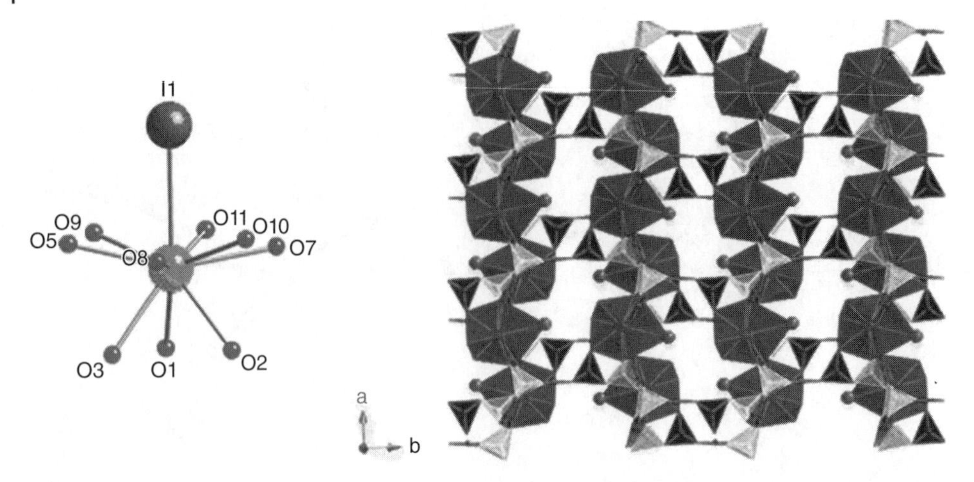

Figure 3.15 (Left) Depiction of the 10-coordinate capped triangular cupola geometry for Pu in $Pu[B_7O_{11}(OH)(H_2O)_2]I$. (Right) Three-dimensional view. Large polyhedra are Pu centers, smaller polyhedra are BO_4 units, BO_3 are triangles, and spheres are I^-.

group $P2_1/n$. One last difference to note about this structure is that because the Pu-I bond is weaker than the Pu-Br bond, disorder accumulates quickly due to fragmentation of the bond by radiolysis.

Finally, it is also relevant to present the trivalent plutonium borate with the use of chloride as the counterion. As it turns out, two products result, which are shown in Equations 3.19–3.20 [43]:

$$PuCl_3 + 4H_3BO_3 \rightarrow Pu\left[B_4O_6(OH)_2Cl\right] + 4H_2O + 2HCl \quad (3.19)$$

$$2PuCl_3 + 13H_3BO_3 \rightarrow Pu_2\left[B_{13}O_{19}(OH)_5Cl_2(H_2O)_3\right] + 12H_2O + 4HCl \quad (3.20)$$

The product from Equation 3.19 is the major product observed and crystallizes in the space group *Cc* [44]. In summary, its sheet topography more closely mimics that of the pentavalent and hexavalent neptunium and plutonium structures cited above rather than the other trivalent halide structures given. Its unique feature that sets it apart from the other structures is the capping chloride ions which bridge between the plutonium centers and thus hold the sheets together (Figure 3.16). Meanwhile, the product from Equation 3.20 is more like the other trivalent halide structures given (Figure 3.17a and Figure 3.18a). For both structures, all of the plutonium centers display the capped triangular cupola geometry that is fashioned in both the bromides and iodides.

As the series progresses to americium and curium, it is important to begin to assess the structural expectations. Looking at the six-coordinate trivalent ionic radii, some of the lanthanides and actinides are listed as follows (in angstroms): Pu = 1.00, Am = 0.975, Cm = 0.970, Ce = 1.01, Pr = 0.99, Nd = 0.98, Sm = 0.958, and Eu = 0.947 [45]. In general, this group of elements (Pu would normally not be trivalent) should behave identically while generating isostructural topographies, especially the ones that are closest in ionic radii such as americium and curium. The first plutonium borate from Equation 3.19, for example, is isostructural with several lanthanide borates. The second structure from Equation 3.20, however, is not. Furthermore, the americium and curium structures, as

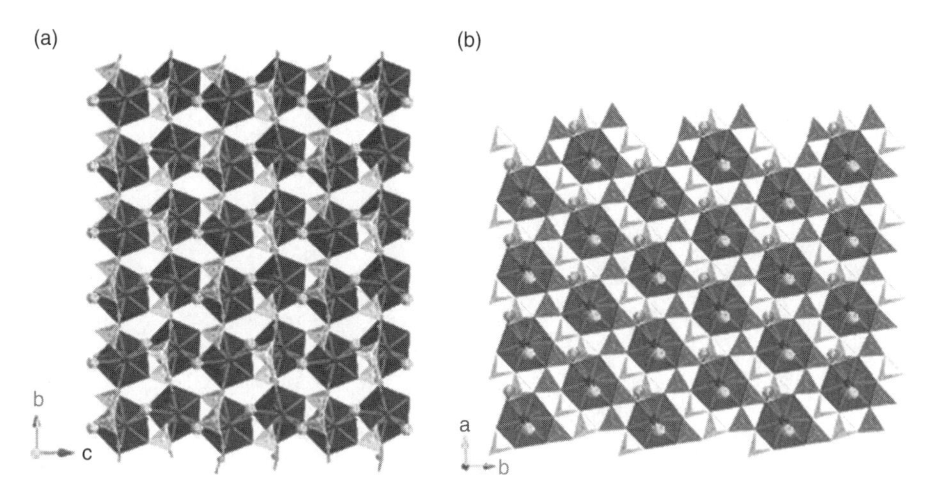

Figure 3.16 (a) Three-dimensional view and (b) sheet topology for $Pu[B_4O_6(OH)_2Cl]$. Large polyhedra are Pu centers, smaller polyhedra are BO_4 units, BO_3 are triangles, and spheres are Cl^-.

will be seen, are isostructural with neither the lanthanides nor with each other. The generic reactions for americium and curium are listed by Equations 3.21–3.22 [43, 46]:

$$AmCl_3 + 9H_3BO_3 \rightarrow Am\left[B_9O_{13}(OH)_4\right]\cdot H_2O + 9H_2O + 3HCl \quad (3.21)$$

$$2CmCl_3 + 14H_3BO_3 \rightarrow Cm_2\left[B_{14}O_{20}(OH)_7(H_2O)_2Cl\right] + 13H_2O + 5HCl \quad (3.22)$$

The americium borate crystallizes in the space group $P2_1/n$ and has a sheet topography that is very similar to the plutonium borate from Equation 3.20 (Figure 3.18b) [44]. Nevertheless, the most drastic changes for the americium structure are the omission of chloride ions as well as the reduction of its coordination number to nine. Its geometry, as a result, is defined to be a hula-hoop (Figure 3.17b) [43]. Instead of a capping chloride, a BO_3 triangle is substituted as the bridging moiety to the next layer. In addition, the protrusion beneath the layer is two oxo atoms from a BO_4 tetrahedra, which is also different. It can be concluded, then, that americium borate adopts some features of both lanthanides and plutonium, yet the structure also maintains its own unique features.

Unexpectedly, curium borate has structural features that borrow from both plutonium and americium (Figure 3.18c) [46]. For example, the curium borate structure utilizes both BO_3 triangles and BO_4 tetrahedra to connect sheet layers as in the americium structure. Yet, unlike americium, the curium borate retains the chloride capping unit that is seen in the plutonium borate. Also, like the plutonium bromide structure, the curium centers adopt two unique geometries that are 9-coordinate hula hoops and 10-coordinate capped triangular cupolas, respectively (Figure 3.17c). However, unlike the bromide structure, which is capped only by bromides, the 10-coordinate feature of curium is capped by a BO_3 triangle. On the basis of these mixed associations, it is concluded that curium borate is an intermediate structure result of both plutonium and americium borates.

The next trivalent borates along the track of the 5*f*-series include those for berkelium and californium. In fact, only at this point in the series is it that two different actinide

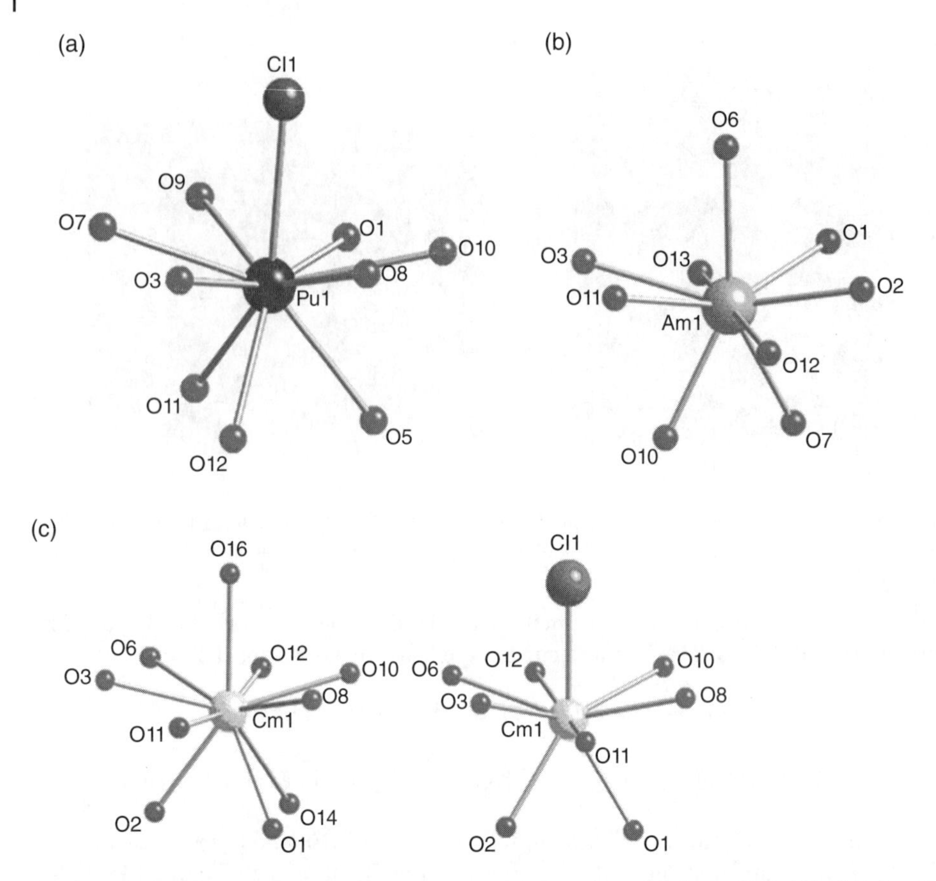

Figure 3.17 Coordination geometries for (a) $Pu_2[B_{13}O_{19}(OH)_5Cl_2(H_2O)_3]$, (b) $Am[B_9O_{13}(OH)_4]\cdot H_2O$, and (c) $Cm_2[B_{14}O_{20}(OH)_7(H_2O)_2Cl]$.

elements are reported to have isotypic borates. Since both reactions are the same, only the californium one is displayed in Equation 3.23 [47–48]:

$$CfCl_3 + 6H_3BO_3 \rightarrow Cf\left[B_6O_8(OH)_5\right] + 5H_2O + 3HCl \tag{3.23}$$

Trivalent berkelium and californium borate crystallize in the triclinic and monoclinic space groups *P*-1 and *C*2/*c*, respectively. Both exhibit a fundamentally different sheet topography from any previous architecture covered. This topography is best described as 1-D chains constructed of four corner-sharing BO_4 tetrahedra and two BO_3 triangles. In addition, due to the actinide contraction, both metals adopt an eight-coordinate square antiprismatic geometry (Figure 3.19a–b). The berkelium and californium centers are coordinated by a combination of BO_4 tetrahedra and μ_3-oxo atoms shared by BO_3 and BO_4 units and are the bridges which tie each chain into larger 2-D sheet topographies (Figure 3.19c). It should be noted that α-decay within the berkelium crystals causes a substantial positive charge build-up which can rupture the crystals after only four days of standing.

Overall, the variation of structures and types of topography shown for all of the actinides have demonstrated that borates are incredibly sensitive to the finer differences of

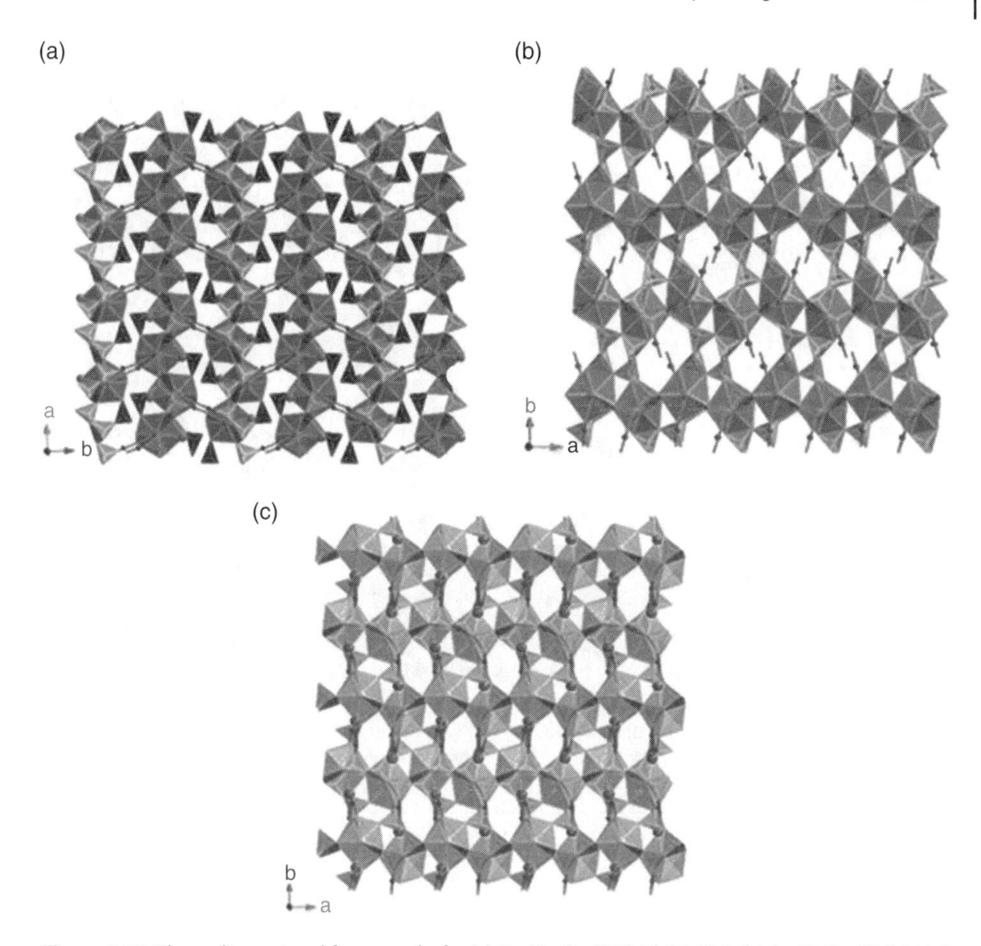

Figure 3.18 Three-dimensional frameworks for (a) $Pu_2[B_{13}O_{19}(OH)_5Cl_2(H_2O)_3]$, (b) $Am[B_9O_{13}(OH)_4]{\bullet}H_2O$, and (c) $Cm_2[B_{14}O_{20}(OH)_7(H_2O)_2Cl]$. The large polyhedra are the actinide centers, smaller polyhedra are BO_4 units, BO_3 are triangles, and spheres are Cl^-.

properties exhibited between each element. For example, it has been shown that there is a significant orbital overlap between the 2*p* orbitals of oxygen atoms and 6*p* orbitals of the metal for the plutonium borate given in Equation 3.19. The americium and curium borates, in a progressive way, show orbital overlap between their 6*d* orbitals with the same oxygen 2*p* orbitals [44]. The degree of covalent bond participation has been too small to clarify with spectroscopic tools in these elements, yet it has been postulated that this plays a role in the deviation from structure-type expectation and prediction. Furthermore, for curium, very strong spin-orbit coupling can also contribute to its selection of structure type [49].

For berkelium, large spin-orbit coupling causes the mixture of several excited states with its ground state. This, as well as orbital mixing, causes a broadening in *f*-*f* transitions which makes it the first 5*f* element in the series to show this kind of spectroscopic behavior [48]. Californium, in line with this trend, not only shows broadened *f*-*f* transitions, but also shows an additional, and abnormally, broad photoluminescence peak originating at 525 nm (Figure 3.20) [47, 50]. This peak has been incorrectly assigned in

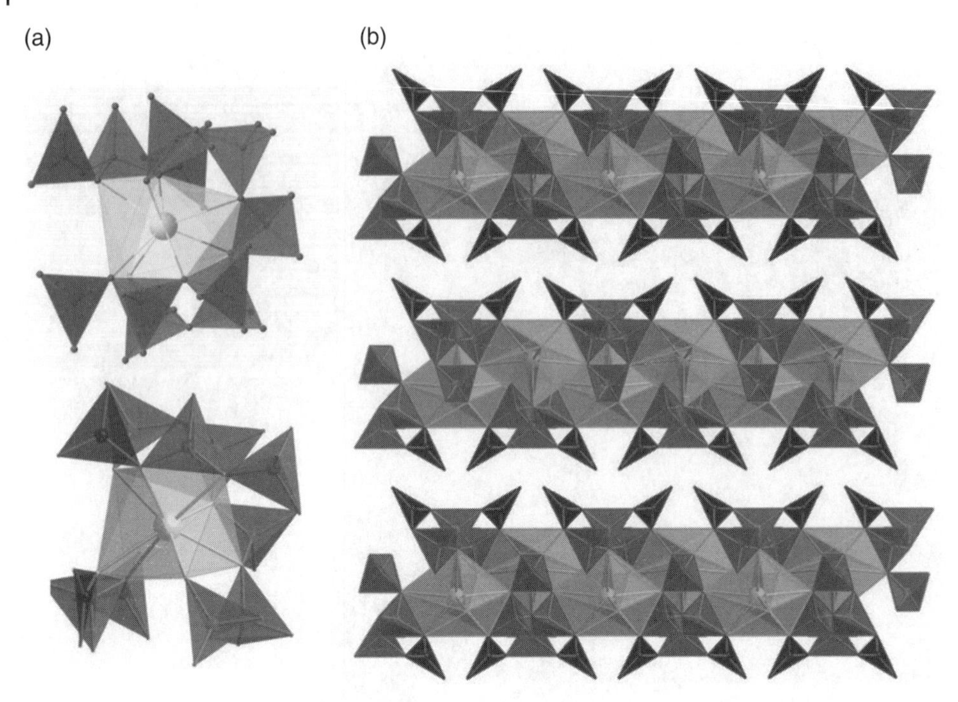

Figure 3.19 (a) $Bk[B_5O_8(OH)_5]$ and (b) $Cf[B_5O_8(OH)_5]$. (c) Two-dimensional sheet topography with the large polyhedra representing an actinide (Bk or Cf) center along with BO_4 tetrahedra and BO_3 triangles.

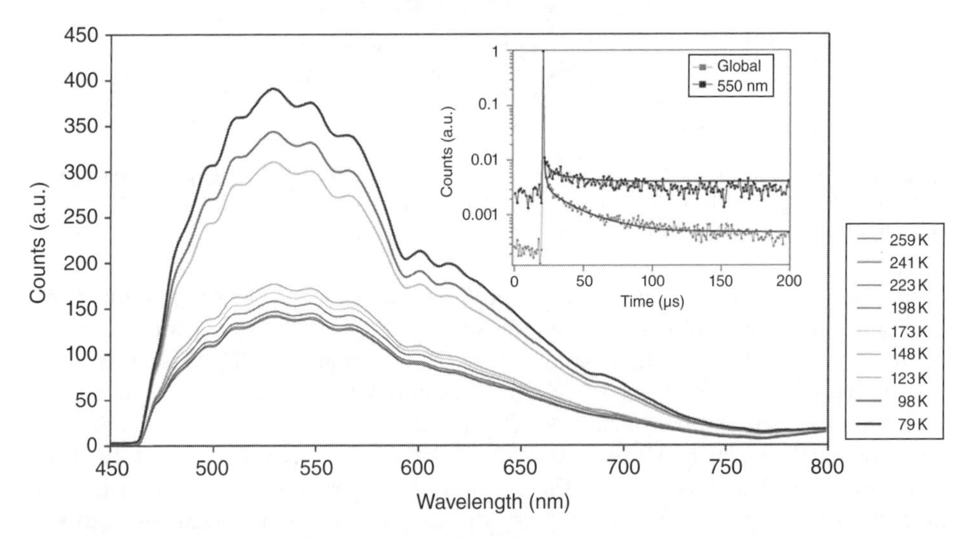

Figure 3.20 Photoluminescence spectra of $Cf[B_5O_8(OH)_5]$ with 420 nm light. Cf(III) emits at 525 nm while its daughter Cm(III) emits at 600 nm. The features of vibronic coupling become more apparent at lower temperatures. The inset shows decay lifetimes of 1.2 ± 0.3 μs for Cf(III) and 20 ± 2 μs for Cm(III). (*See insert for color representation of the figure.*)

the past to be the $J = 5/2 \rightarrow J = 15/2$ *f-f* transition when it is actually more indicative of a ligand-to-metal charge-transfer transition. Theoretically, this assignment first comes into question due to the fact that the calculated emission energy for $J = 5/2$ is energetically too small a value to ascribe to the peak seen. Experimentally, the presence of vibronic coupling also puts this assignment into question and is considered to be an indicator instead of covalent bonding. Quantum mechanical calculations have shown orbital overlap between the $2p$ borate oxygens and both the $6d$ and $5f$ orbitals of the metal. The gain in the divalent redox stability of Cf(II) as well as a reduced magnetic moment also makes the ligand-to-metal charge-transfer assignment practical. Therefore, this spectroscopic transition has been reassigned to the availability of a metastable Cf(II) state.

While some of these structure–property relationships are still being fleshed out, another seemingly unrelated application is growing more imminent. Since the 1950s, it has been proposed that rock salt deposits would be one of the ideal mineralogical underground formations to maintain and store high-level nuclear waste [51]. Such rock salt formations would provide strength and prevent roof collapse, are dry and impervious to groundwater movement, are level, and have higher thermal conductivities which can handle heat loads. In the United States, for instance, large rock salt deposits can be found across the Great Lake boundaries of Michigan, Ohio, Pennsylvania, and New York. Another large deposit extends across the Delaware Basin of New Mexico and Texas, where it is the site of the Waste Isolation Pilot Plant (WIPP) for radiological waste. Such deposits also include relatively higher concentrations of borates, which, in addition to the halides, carbonates, and sulfates, would complex with the actinides in the event of a release [52]. The knowledge of structural actinide borate chemistry then, as related above, is important for designing quick and effective contingency analyses in the event of a future radiological containment breach. In a final summary, some structural information for transuranic borates are listed in Table 3.6.

Table 3.6 Transuranic borates [38–48].

Compound	Space Group	Color	a_0 (Å)	b_0 (Å)	c_0 (Å)	β, α, γ (°)	Volume (Å^3)
$K_2[(NpO_2)_3B_{10}O_{16}(OH)_2(NO_3)_2]$	$P2_1/n$	Pale pink	6.599	16.026	11.053	90.922	1168.7
$K[(NpO_2)B_{10}O_{14}(OH)_4]$	$P2_1/n$	Light-green	9.933	8.1985	21.041	91.302	1713.1
$K_2[(NpO_2)_3B_{16}O_{25}(OH)_2]$	Pn	Light-green	8.3214	15.876	9.889	90.008	1306.4
$NpO_2[B_8O_{11}(OH)_4]$	Cc	Pale pink	6.4426	16.705	10.9843	90.766	1182.1
$PuO_2[B_8O_{11}(OH)_4]$	Cc	Peach pink	6.4391	16.714	10.9648	90.744	1180
$Pu_2[B_{12}O_{18}(OH)_4Br_2(H_2O)_3]$ *$0.5H_2O$	Pn	Navy blue	8.0995	14.635	9.8248	90.028	1164.6
$Pu[B_7O_{11}(OH)(H_2O)_2I]$	$P2_1/n$	Pale blue	8.1103	17.06	9.7923	90.133	1354.9
$Pu[B_4O_6(OH)_2Cl]$	Cc	Blue	6.491	11.184	9.63	105.175	674.7
$Pu_2[B_{13}O_{19}(OH)_5Cl_2(H_2O)_3]$	$P2_1/n$	Light purple	8.0522	14.568	9.82	90.12	1151.8
$Am[B_9O_{13}(OH)_4]\cdot H_2O$	$P2_1/n$	Pink	7.703	16.688	9.872	90.073	1269.1
$Cm_2[B_{14}O_{20}(OH)_7(H_2O)_2Cl]$	$P2_1/n$	Pale yellow	7.9561	14.212	9.836	90.013	1112.2
$Bk[B_6O_8(OH)_5]$	P-1	Gold	6.7994	7.1441	9.9828	73.6, 80.9, 77.5	451.66
$Cf[B_6O_8(OH)_5]$	$C2/c$	Pale green	6.8495	18.809	7.2113	101.364	910.9

3.4 Sulfates

As seen from the example of WIPP, geological repositories can introduce many inorganic groups which can chelate with the actinides. In this section, a brief examination of sulfates will look at a variety of ways that this oxyanion can form complex structures with actinides. Generally, sulfates are considered to only weakly coordinate actinides and are outcompeted by other groups in solution such as carbonates and phosphates. This can sometimes introduce inherent synthetic challenges when trying to crystallize a diverse array of such compounds. In addition, sulfates are weakly reducing oxyanions but often do not cause drastic redox changes to occur on their own. Nevertheless, a variety of approaches for finding useful actinide sulfate architectures have been discovered.

3.4.1 Thorium and Uranium

The simplest thorium sulfates come in a number of hydrated varieties. The first example to show the generic mode of bonding is from that of the octohydrate, $Th(SO_4)_2{\bullet}8H_2O$ [53]. The thorium center has a 10-coordinate bicapped square antiprismatic geometry and coordinates four total oxygens from two sulfate groups with a complement of six waters. It is important to note that the two additional waters do not bind to the thorium center but instead form bridging hydrogen bonds between additional coordinated thorium polyhedra. Due to the anisotropy of coordination, the SO_4 tetrahedra have varying S-O bonds of approximately 0.04 Å difference, with the coordinated bonds being longer than the terminal bonds.

The next example in progression of complexity is that of a ternary thorium sulfate, $Na_2[Th(SO_4)_3(H_2O)_3]{\bullet}3H_2O$ [54]. The thorium center in this example shows a reduction in coordination number with a higher chelating number of sulfate groups involved in the bonding. The division of bonds, as scaled in line with the first example, totals six oxygens from bidentate sulfate groups and three equatorial oxygens from inner-sphere coordinating water molecules, making the thorium center adapt itself to be a nine-coordinate, tricapped trigonal prism. The sulfate groups serve as the bridging moieties between thorium centers, and this extension ultimately creates a chain topography. As before, S-O coordinated bonds are longer than S-O terminal bonds, although the mean difference in bond length is only about 0.03 Å in this example. Furthermore, the equatorial water bond lengths are found to be longer than the bridging sulfate oxygens, which indicates the ease with which such a compound is hydrated.

From these first two examples, it is important to note that inorganic sulfates can be involved in a great deal of hydrogen bonding. Both of these examples also happen to show sulfate acting as a bidentate group that bridges corresponding thorium centers into chains. Sulfates can, however, also adopt monodentate coordination modes and are capable of creating higher-order structures. Such an example is that of $Th_3(SO_4)_6(H_2O)_6{\bullet}H_2O$, which can be made in several ways and is shown by Equation 3.24 [55]:

$$3Th(OH)_4 + 7H_2SO_4 + 2LiOH \rightarrow Th_3(SO_4)_6(H_2O)_6 \cdot H_2O + Li_2SO_4 + 7H_2O \quad (3.24)$$

Unlike the first two structures, which crystallize in the monoclinic space group $P2_1/n$, the observed structure has higher symmetry and crystallizes in the tetragonal $P4_2/nmc$

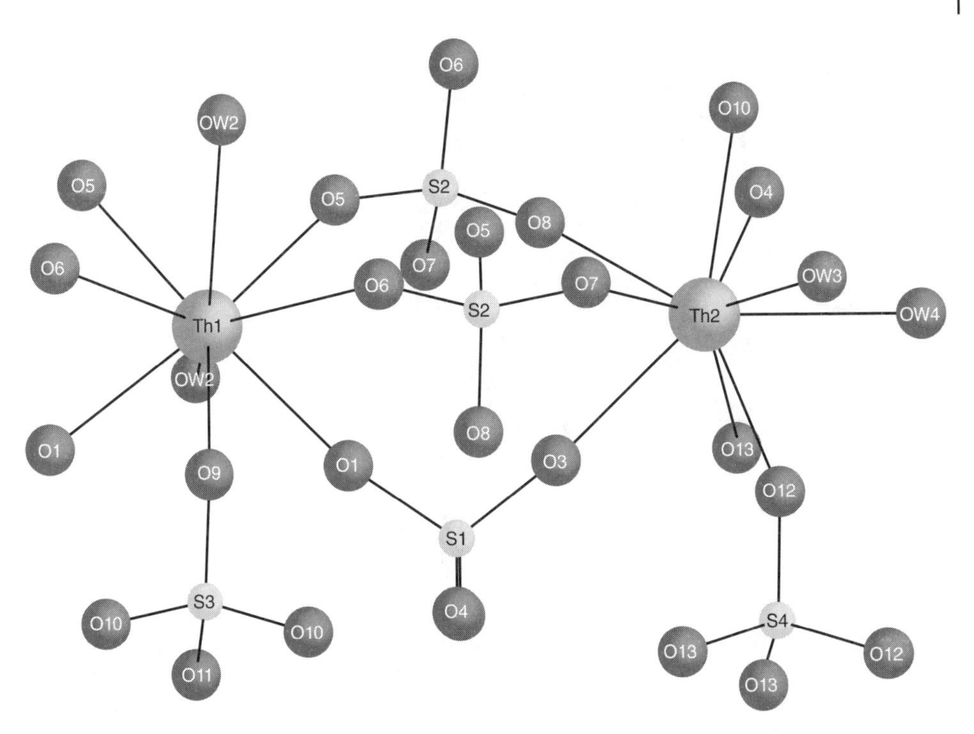

Figure 3.21 Coordination environment showing both unique Th_1 and Th_2 sites. Three sulfate groups are bridging the two metal centers.

space group. The sulfates adopt a monodentate coordination mode and bridge the thorium centers by corner-sharing into chains. The building units are denoted as $[Th_2(SO_4)_3]$.. These chains, however, further intersect into 2D sheets at approximately right angles to form square channels that are 11.5 Å across (Figure 3.22). These channels have also been determined to be vacant with no cations or water molecules occupying the interstitial spaces. As in the second example above, the thorium centers are nine-coordinate. However, these thorium centers diverge into two crystallographically unique sites (Figure 3.21).

One important crystallographic difference between monodentate and bidentate sulfate groups is the experimentally determined bond distance between the An and S atoms. Such bond distances can be as different as ~0.6 Å with those from the bidentate groups being much shorter. The implications of this are clear in that unique crystallographic arrangements and topographies can be achieved when both monodentate and bidentate sulfate groups are present. One example of this, following a different reaction path from that given by Equation 3.24, is the ternary compound $Na_2[Th_2(SO_4)_5(H_2O)_3]\cdot H_2O$ [56]. In this structure, there are two unique thorium sites. The first site is coordinated by one bidentate and six monodentate sulfate tetrahedra, while the second has exclusively seven monodentate binding groups. Since all of the thorium centers are nine-coordinate monocapped square antiprisms, a different number of waters fill the remaining sites for each center. In line with these differences, there are also five unique sulfate group positions. Finally, the topographical features are somewhat similar to $Th_3(SO_4)_6(H_2O)_6\cdot H_2O$ in that they exhibit interconnected sheets with porous channel

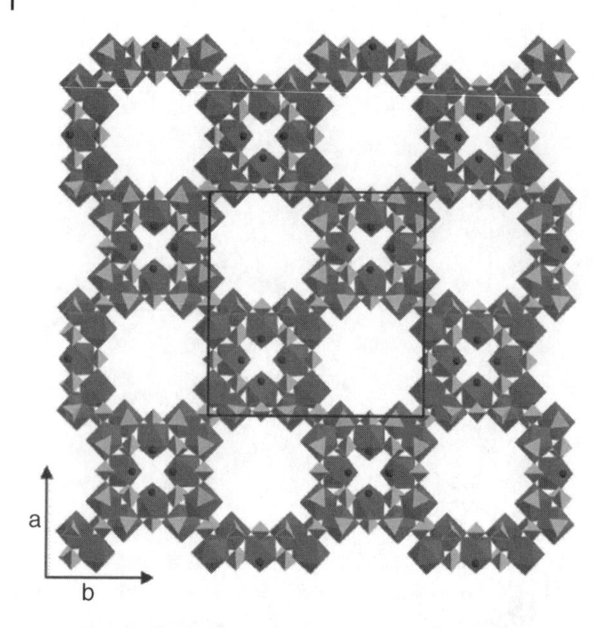

Figure 3.22 A view of $Th_3(SO_4)_6(H_2O)_6{\cdot}H_2O$ along the *ab*-plane showing the 11.5 Å square channels. The large polyhedra are thorium centers, the smaller tetrahedra are sulfates, and the small spheres are unbound waters.

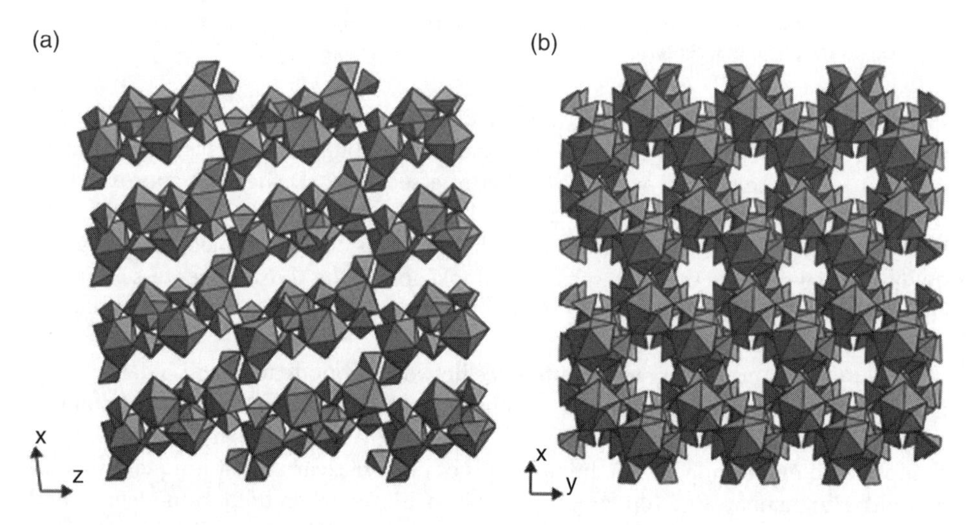

Figure 3.23 Three-dimensional structure of $Na_2[Th_2(SO_4)_5(H_2O)_3]{\cdot}H_2O$ showing (a) the open channels along [010] and (b) the open channels along [001]. The large polyhedra are thorium centers, the smaller tetrahedra are sulfates, and the sodium ions are omitted for clarity.

features (Figure 3.23). These channels, however, are not as large and house charge-balancing sodium cations.

It should be noted that such a large variation of compounds can be produced by simply changing the pH (with acid of varied strength or by using alkali sulfate salts) or changing the reaction temperatures. Another familiar method seen in this chapter,

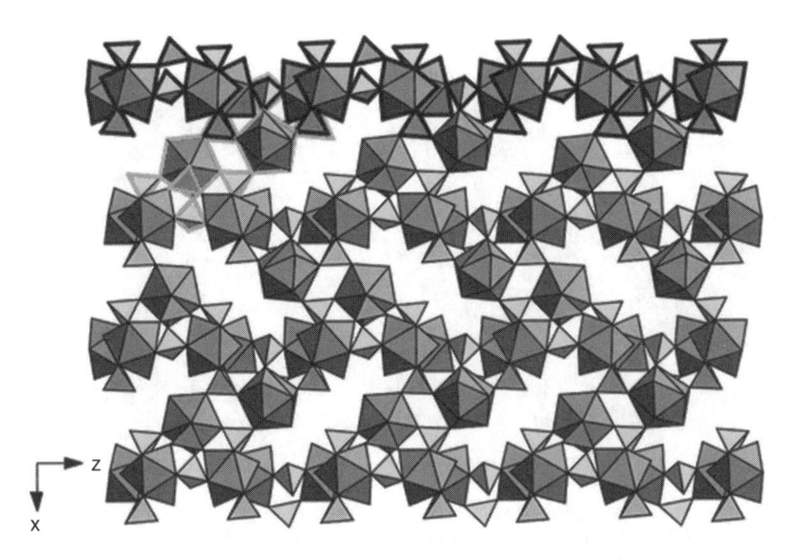

Figure 3.24 Three-dimensional structure of $Th_4(SO_4)_7(H_2O)_7(OH)_2{\cdot}H_2O$ viewed down the [010] plane. A chain of sulfate-bridged Th(2) and Th(3) centers is shown in boldface, while other sulfates bridge to Th(1)-Th(1) dimers. Large polyhedra are thorium centers, and smaller tetrahedra are sulfates.

especially among the borates, is the hydrothermal method. This thorium sulfate, given by Equation 3.25, shows a different product with this method [57]:

$$4Th(OH)_4 + 7H_2SO_4 \rightarrow Th_4(SO_4)_7(H_2O)_7(OH)_2 \cdot H_2O + 6H_2O \tag{3.25}$$

In this structure, all sulfate groups are monodentate and bridge nine-coordinate monocapped square antiprism thorium centers. Sulfate groups also bridge these sheet-arranged ThO_9 polyhedra with Th_2O_{15} dimers to form an overall 3D architecture (Figure 3.24–3.25). In general, with higher dimensionality comes the notion that some sulfate groups can be saturated. This is, in fact, the case as two of the four unique sulfate anions are saturated. There are also three unique thorium centers, and one of the centers is distinguished as coordinating two hydroxides which bridge the same kind of thorium facsimiles. It can be seen once again from this example that minor synthetic conditions can have a very large impact on the resulting structural topography.

A final example of a thorium sulfate demonstrates a hydrothermally in situ method of generating the SO_4 group by means of a redox reaction as shown in Equation 3.26 [58]:

$$\begin{aligned} Th(NO_3)_4 \cdot 5H_2O + 9NaNO_2 + 2HF + CH_3SO_3H &\rightarrow ThF_2(SO_4)(H_2O) \\ + 9NaNO_3 + 5H_2O + CH_3OH + 2N_2 & \end{aligned} \tag{3.26}$$

Furthermore, it should be noted that tripropylamine is a non-stoichiometric addition in the reaction to serve as a templating function (refer to Section 3.2.3). Each thorium has a nine-coordinate tricapped trigonal prism geometry with bonds to four sulfate groups, four fluoride groups, and one water (Figure 3.26a–b). The fluoride groups are responsible for bridging thorium centers into 1-D chains, while the sulfate groups bridge parallel chains into an overall 3D topography (Figure 3.26c–d). Although this example is not emblematic of it, the use of HF is considered to be an excellent functionalizing agent for

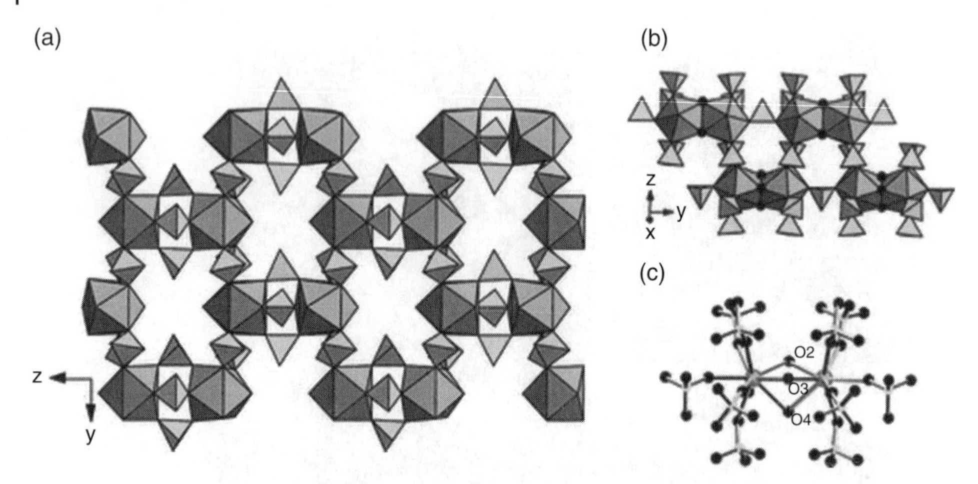

Figure 3.25 (a) Depiction of the 2D topology showing Th(2) and Th(3) centers bridged by sulfate. (b) Th_2O_{15} units linked by sulfates. (c) Coordination environment of the Th_2O_{15} dimers. Large polyhedra are thorium centers, smaller tetrahedra are sulfates, and small spheres are oxygen.

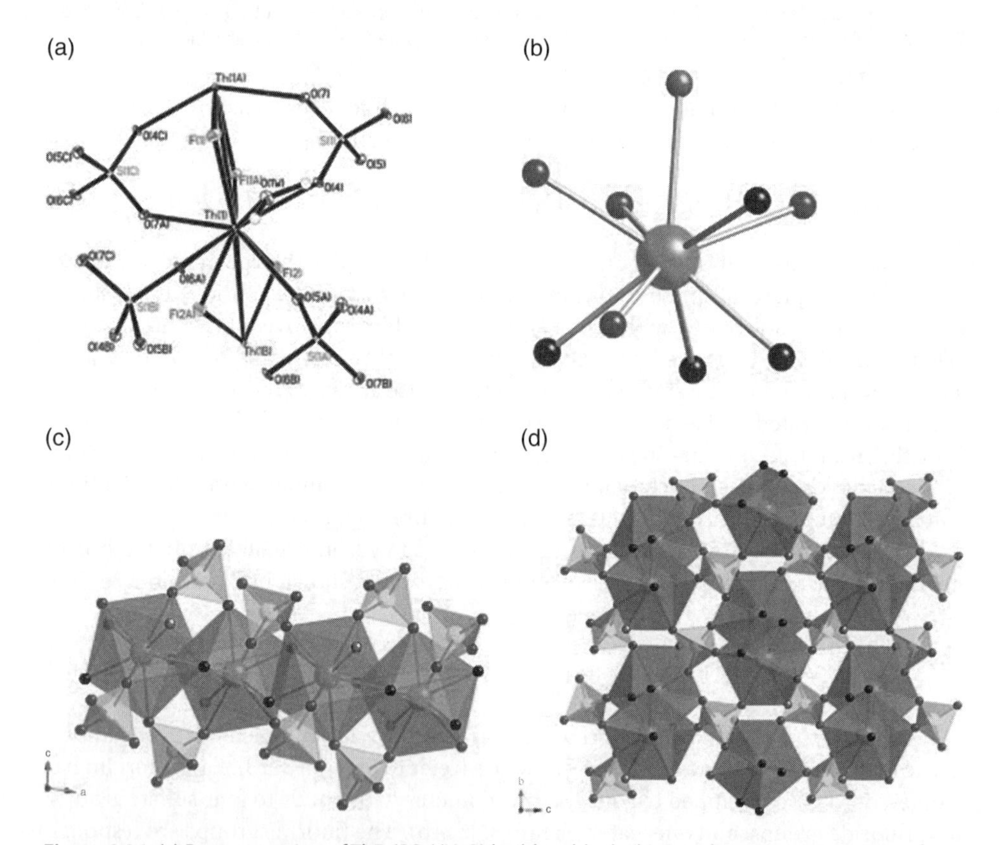

Figure 3.26 (a) Representation of $ThF_2(SO_4)(H_2O)$ building block. (b) Coordination environment of Th^{4+}. (c) ThO_5F_4 polyhedra extended along the *a*-axis. (d) Three-dimensional structure showing successive thorium centers bridged by sulfates. Large polyhedra are thorium centers, small tetrahedra are sulfates, and spheres are F^-, oxygen from sulfate, and oxygen from water.

creating porous structures. Another noteworthy feature of comparison is that while functionalization of borate structures with fluoride leads to terminal B-F bonds, such functionalization in sulfate structures does not lead to S-F bonds but rather An-F bonds. This is due to the order of the bond strength going from B-F > An-F > S-F.

Uranium sulfates, meanwhile, are divided primarily into those with UO_2^{2+} functional groups versus those with U^{4+} groups. The first basic structure worth observing is that of $U(SO_4)_2{\bullet}4H_2O$ [59]. One immediate difference from thorium structures to note is that U^{4+} has a smaller ionic radius and thus will exhibit smaller coordination numbers. In the structure at hand, uranium adopts an eight-coordinate distorted square antiprism geometry. Four monodentate sulfate groups coordinate it while bridging to equivalent uranium centers. Because all of the sulfate groups bridge between metal centers, a topography of successive layers becomes the main feature. Water fills the remaining coordination sites and takes up occupational spaces between the layers. Analogous to the presented thorium structures, a significant quantity of hydrogen bonds decorate the overall lattice.

It is known that in solution the observation of bidentate sulfate linkages is more common in the order $Th^{4+} < U^{4+} < Np^{4+} < Pu^{4+}$ [60, 61]. Bidentate linkages are much less common overall in the solid state, but there is insufficient evidence so far to support whether such a trend carries over from solution. The first identified case of such a bidentate linkage in U^{4+} was found in the compound $Na_{1.5}(NH_4)_{4.5}[U(SO_4)_5{\bullet}H_2O]{\bullet}H_2O$ [62]. This structure crystallizes in the triclinic space group *P*-1 and adopts a higher nine-coordinate monocapped square antiprism geometry. Three sulfate groups offer bidentate linkages, while the two other additional groups offer monodentate linkages. The overall topography arranges itself into zigzag chains via the fundamental unit $[U(SO_4)_5{\bullet}H_2O]^{6-}$ and is bridged by extensive bonding with NH_4^+ cations, while Na^+ is only able to assist in the charge-balancing act (Figure 3.27).

Another important structure is that of $Cs_2U(SO_4)_3{\bullet}2H_2O$, which crystallizes in the monoclinic space group $P2_1/c$ (Figure 3.28) [61]. This structure has similar features to its analogue $Cs_2Th(SO_4)_3{\bullet}2H_2O$ [63] in that both coordinate a total of five sulfate groups and two waters. Four of these sulfate groups are bidentate and bridge to additional metal centers to create topographically layered sheets (Figure 3.29). It is important to note that since uranium has a lower coordination number of nine that the fifth terminal bidentate linkage in the thorium compound becomes a terminal monodentate linkage in the uranium structure.

The most complex forms of inorganic hexavalent uranium sulfate topologies typically involve the use of organic templating agents. The first such structure discovered like this and called MUS-1 was synthesized hydrothermally and is shown by Equation 3.27 [64]:

$$\begin{gathered}6UO_2\left(C_2H_3O_2\right)_2\cdot 2H_2O+2NC_4H_{12}\left(OH\right)\cdot 5H_2O+7H_2SO_4\rightarrow\\ \left[NC_4H_{12}\right]_2\left[\left(UO_2\right)_6\left(SO_4\right)_7\left(H_2O\right)_2\right]+12HC_2H_3O_2+22H_2O\end{gathered}\tag{3.27}$$

This complexed actinide crystallizes in the orthorhombic space group $C222_1$. As expected, the hexavalent uranium centers adopt a seven-coordinate pentagonal bipyramidal geometry. Two of the three distinct uranium sites are saturated with monodentate sulfate groups, while the third site has one equatorial linkage open for a water molecule. On a more panoramic scale, sulfate groups bridge between successive uranyl

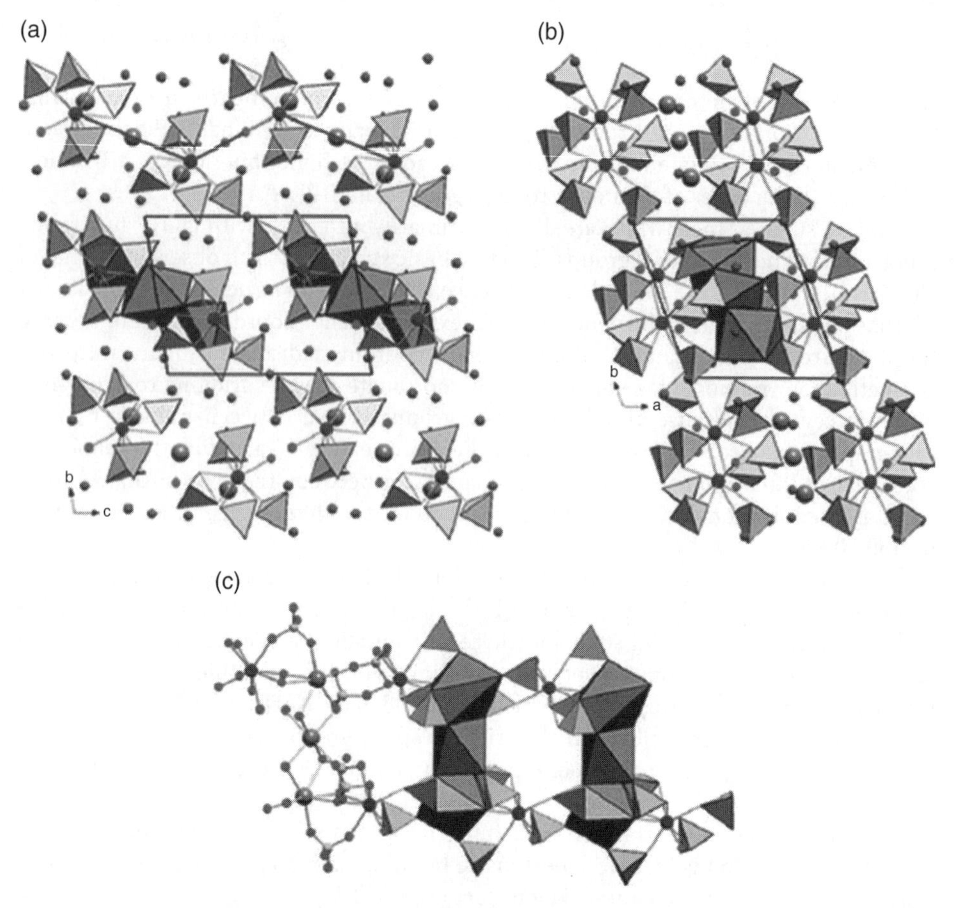

Figure 3.27 Structure of $Na_{1.5}(NH_4)_{4.5}[U(SO_4)_5{\bullet}H_2O]{\bullet}H_2O$ (a) extended along the *a*-axis, (b) extended along the *c*-axis, and (c) two anionic uranium complexes with three sodium units parallel to the *a*-axis. Tetrahedra are sulfates, large polyhedra show the coordination environment of sodium, and spheres represent uranium, sodium, nitrogen, sulfur, and oxygen.

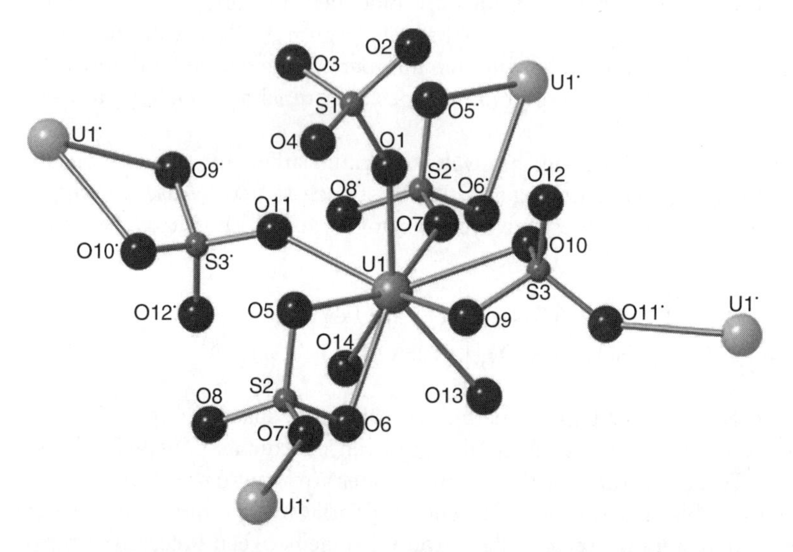

Figure 3.28 Structure showing the coordination environment and bridging in $Cs_2U(SO_4)_3{\bullet}2H_2O$.

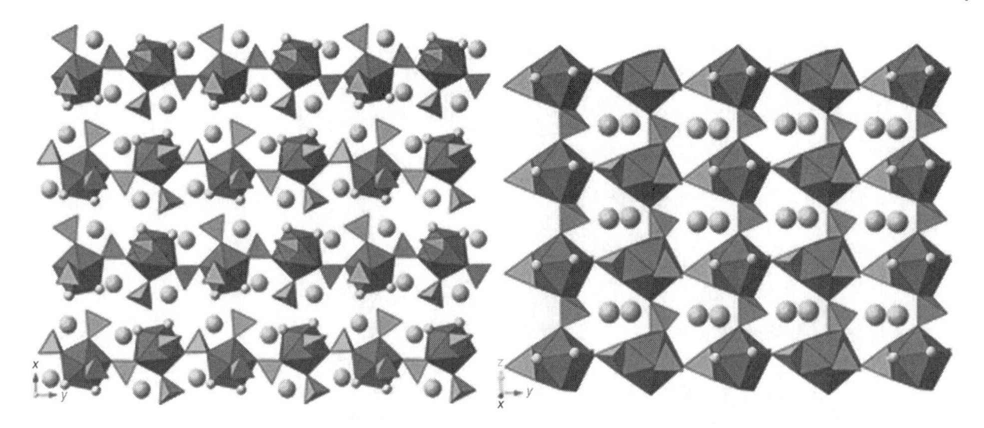

Figure 3.29 Two-dimensional sheets of $Cs_2U(SO_4)_3{\cdot}2H_2O$ (a) observed down the *x*-axis and (b) observed down the *z*-axis. Large polyhedra are uranium centers, smaller tetrahedra are sulfates, and spherical items are water or Cs^+.

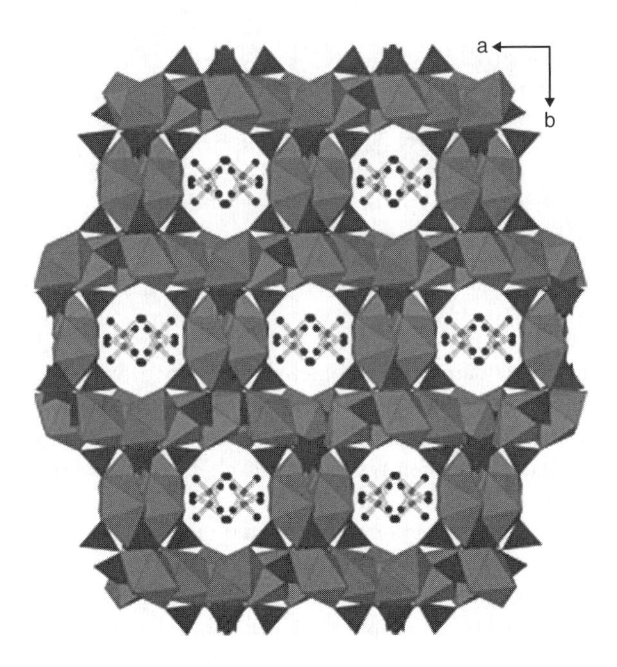

Figure 3.30 Three-dimensional framework of MUS-1 viewed along the [001] plane. Large polyhedra are UO_7 units while smaller tetrahedra are SO_4 units.

cores to create a topographically unique 3D framework (Figure 3.30). A consortium of porous channels is one interesting result of such a framework. These channels are also the sites of charge-balancing tetramethylammonium cations, which are the active templating agents set into the structural design.

A second example of a similarly driven synthetic strategy with the use of the [18]crown-6 ether results in the exceedingly complex structure given by Equation 3.28 [65]:

$$14UO_2(NO_3)_2 \cdot 6H_2O + 19H_2SO_4 + 2[18]crown-6 \rightarrow 28HNO_3 + 49.5H_2O(H_3O)_8[(H_3O)\cdot[18]crown-6]_2[(UO_2)_{14}(SO_4)_{19}(H_2O)_4](H_2O)_{20.5} \quad (3.28)$$

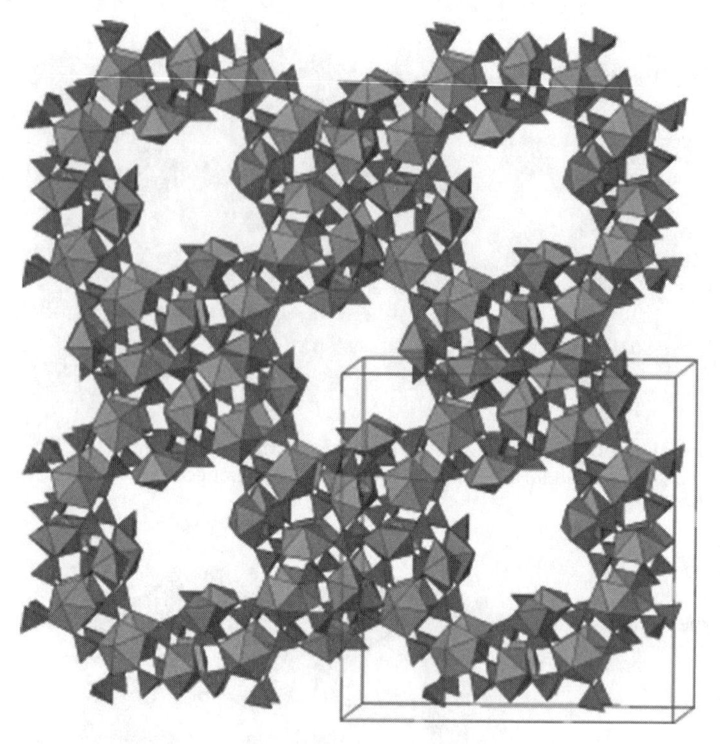

Figure 3.31 Crown ether uranium sulfate complex from Equation 27. Large polyhedra are uranium centers while smaller tetrahedra are SO_4 units.

This complex uranyl sulfate is a highly porous three-dimensional structure which adopts a panoramic pretzel-shaped topography with large channels that are occupied by the bulky hydronium crown ether complexes (Figure 3.31). The sulfate groups bridge from their equatorial linkages and display a significant tolerance toward bond angle deviations in their respective tetrahedra clusters (Figure 3.32).

From these two examples alone, it can be readily seen that uranyl sulfate frameworks can be very complex and depend on the synthesis and directing organic templates used. The limited scope of this section cannot readily describe all of the types of thorium and uranium sulfate frameworks that exist. Nevertheless, a general picture has been given, and many of the same thematic concepts will be applied to the inorganic transuranic sulfate structures, as detailed next.

3.4.2 Transuranic Frameworks

Neptunium, like uranium, can be divided among its many different redox constituents for making sulfate frameworks. Beginning with the most common neptunyl form, NpO_2^+, the first example of a fundamental Np (V) sulfate is $(NpO_2)_2SO_4{\bullet}6H_2O$ [66]. There are two distinct neptunium sites, and of the five available equatorial sites, there is only one sulfate linkage. This is because there is a tendency for the structure to form cation–cation bonds between neptunyl units. Moreover, the hexahydrate adds more water molecules into the coordination sites as opposed to the dihydrate and

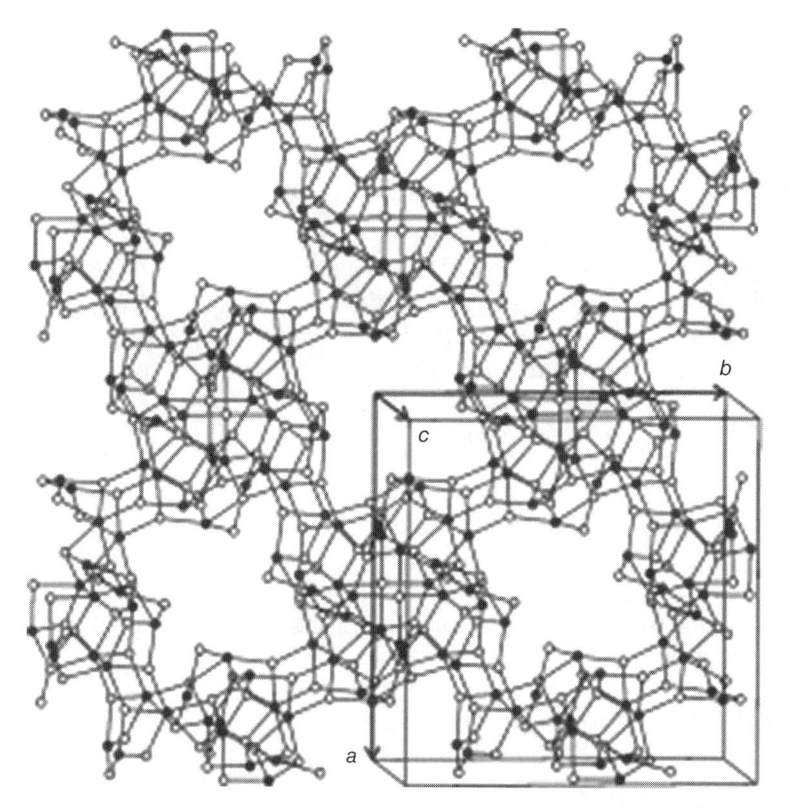

Figure 3.32 Three-dimensional image showing the linkage topology in the crown ether uranium sulfate complex from Equation 3.27.

monohydrate forms [67–68]. Overall, this structure adopts a three-dimensional topography due to a large host of hydrogen bonds that connect the SO_4, H_2O, and NpO_2 units together.

Such neptunyl structures, in general, have a greater likelihood of being constructed under cation–cation interactions, whereas sulfate linkages usually are left only to play a complementary role. It is also unclear if or to what degree sulfate groups can contribute to structures that display magnetic ordering such as $NaNpO_2SO_4H_2O$ which shows ferromagnetic ordering below 7.4 K (Figure 3.33–3.34) [69]. Although the sulfate linkages occupy equatorial positions, it is the neptunyl cation–cation frameworks which are believed to form the basis of the large crystal field splitting parameters and superexchange pathways needed to induce such magnetic responses. Nevertheless, it is possible that the various coordination modes of sulfate could have different impacts on anisotropic symmetry-related factors in related structures.

Next comes a brief check of plutonium. α-$Pu(SO_4)_2{\bullet}4H_2O$ crystallizes in the orthorhombic space group *Fddd*, whereas β-$Pu(SO_4)_2{\bullet}4H_2O$ crystallizes in the space group *Pnma* [70]. Both forms are eight-coordinate square antiprisms which coordinate four monodentate sulfate groups and four water molecules. The sulfate molecules further bridge plutonium centers to form two-dimensional layered topographies. In contrast, $Cs_4Pu(SO_4)_4{\bullet}(H_2O)_2$ crystallizes in the monoclinic space group $P2_1/n$ [71].

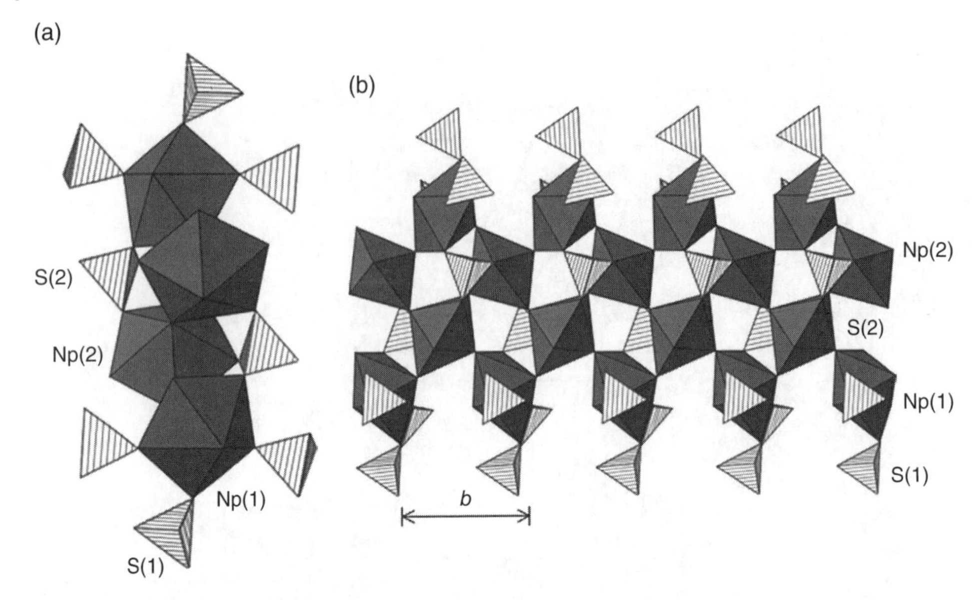

Figure 3.33 (a) Larger neptunyl polyhedra surrounded by sulfate tetrahedra in $NaNpO_2SO_4H_2O$ and (b) same structure extended along the *b*-axis.

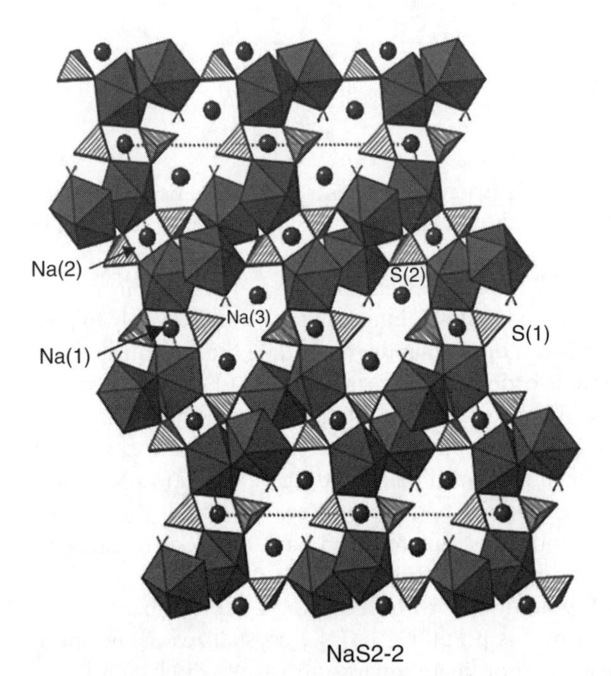

Figure 3.34 Three-dimensional framework for $NaNpO_2SO_4H_2O$. Larger neptunyl polyhedra are bridged by sulfate tetrahedral. Cavity spaces are occupied by Na^+ and H_2O.

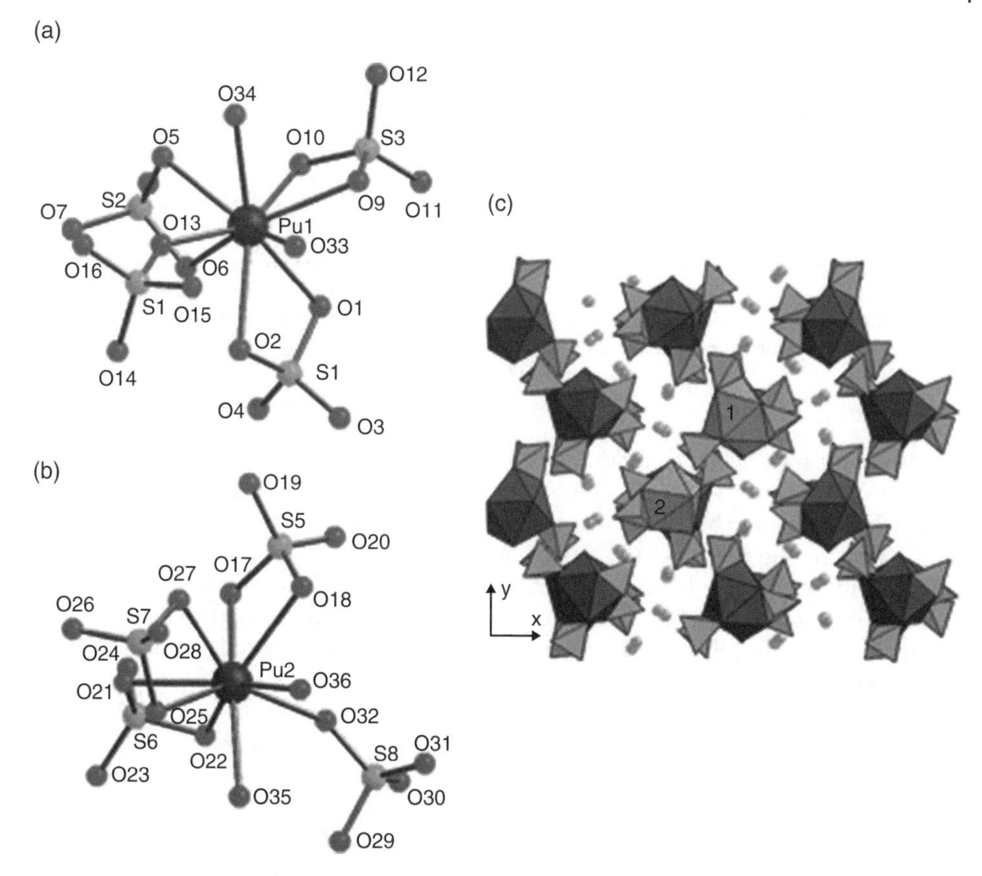

Figure 3.35 Different coordination environments for $Cs_4Pu(SO_4)_4{\cdot}(H_2O)_2$ in (a) Pu(1) and (b) Pu (2). (c) Polyhedral representation with large polyhedra being plutonium centers and tetrahedra being sulfate. The unique Pu(1) and Pu(2) sites are labeled.

This structure features two distinct plutonium sites which are nine-coordinate and confers a preferred privilege on the bidentate sulfate linkages which occupy six total binding sites (Figure 3.35a–b). The other three sites are divided between a single monodentate sulfate group and two water groups. This topography stands apart in that it does not engage in bridging other plutonium centers (Figure 3.35c). Rather, each site maintains a molecular individuality while Cs^+ ions only serve the purpose of conserving charge in the interstitial space.

In fact, it is this ternary cesium actinide sulfate series that has given incredible insight into actinide periodicity. Referring back to the thorium and uranium analogues, it was observed that a decrease in coordination number was paired with a bidentate sulfate switching to a monodentate sulfate. However, when the nine-coordinate $Cs_2Np(SO_4)_3(H_2O){\cdot}H_2O$ is introduced into the series, the monodentate sulfate switches back to a bidentate sulfate at the expense of a water molecule [72]. And now in the cited $Cs_4Pu(SO_4)_4{\cdot}(H_2O)_2$ example, a molecular system prevails instead of an extended structure. It has been proposed that these slight alterations are due to the increasing Lewis

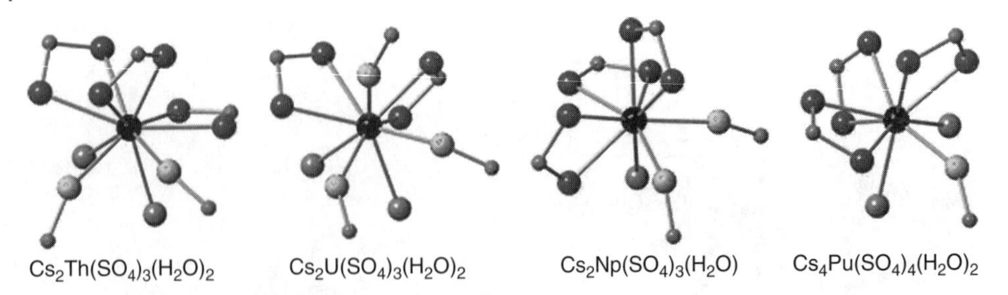

Figure 3.36 Changes in the coordination environment for the cesium actinide sulfate series. Spheres represent An^{4+}, sulfur, oxygen in bidentate linkage, oxygen in monodentate linkage, and water.

acidity of the tetravalent actinides, which compete for a larger share of bidentate sulfate linkages in the order $Th^{4+} < U^{4+} < Np^{4+} < Pu^{4+}$ (Figure 3.36) [61].

The last element to be explored is americium. The first structure which can be definitively represented is that of the americium octahydrate, $Am_2(SO_4)_3{\bullet}8H_2O$ [73]. This compound crystallizes in the monoclinic space group *C2/c*. Similar to plutonium, americium adopts an eight-coordinate distorted square antiprism geometry with four monodentate sulfate linkages and four water groups. Layers held together by extensive hydrogen bonding are brought about by the thematic bridging from the sulfate tetrahedra. From another perspective, a second type of structure for an americium (VI) sulfate was synthesized in a three-step reaction combined as stoichiometric amounts in Equation 3.29 [74]:

$$\begin{aligned} &2Am(NO_3)_3 + 6LiHCO_3 + 2H_2SO_4 + Cs_2SO_4 + O_3 \rightarrow \\ &\quad Cs_2\left[(AmO_2)_2(SO_4)_3\right] + 6CO_2 + 6CO_2 + 6LiNO_3 + 5H_2O \end{aligned} \tag{3.29}$$

This structure at a glance holds some similarities to the uranyl and neptunyl examples presented, and in fact, has isostructural analogues for those elements that were not presented in this section. The linear dioxo americium group adopts a pentagonal bipyramidal geometry with five coordinated equatorial sulfate groups that bridge to other centers. Three of the sulfate groups bridge twice, while two others bridge all of their oxo atom sites. This feature is a unique example that shows a very extensive bridging of the sulfates. Consequently, the larger topographical scene features a series of cationic and anionic layers without the incorporation of water.

In summary, it is quite apparent that actinide sulfate chemistry is both complex and can introduce quite a variety of structural possibilities. For such a variation in potential structures, it has been relatively unexplored in the solid state due to its intermediate complexation ability. This is even more true of the transuranics, with very few known plutonium and americium sulfates and no developed sulfate chemistry past americium. Attaining such structures in the future will likely require new insights of functionalization in order to prevent problems such as radiolysis which could create disorder in any weakly coordinated sulfate structure. Tables 3.7–3.9 summarize crystallographic information for some selected actinide sulfates.

Table 3.7 Hydrated actinide sulfates [53, 59, 66, 69, 73].

Compound	Crystal class	Space group	a_0 (Å)	b_0 (Å)	c_0 (Å)	β (°)	Volume (Å^3)
$Th(SO_4)_2{\cdot}8H_2O$	Monoclinic	*P*2$_1$/*n*	8.51	11.86	13.46	92.65	1357
$U(SO_4)_2{\cdot}4H_2O$	Orthorhombic	*Pnma*	14.674	11.093	5.688	90	925.8
$(NpO_2)_2SO_4{\cdot}6H_2O$	Monoclinic	*P*2$_1$/*c*	7.94	19.03	8.182	96.4	–
α-$Pu(SO_4)_2{\cdot}4H_2O$	Orthorhombic	*Fddd*	26.527	11.995	5.687	90	–
β-$Pu(SO_4)_2{\cdot}4H_2O$	Orthorhombic	*Pnma*	14.544	10.98	5.667	90	–
$Am_2(SO_4)_3{\cdot}8H_2O$	Monoclinic	*C*2/*c*	13.619	6.837	18.405	102.4	–

Table 3.8 Cesium actinide sulfates showing periodicity [61, 63, 71, 72].

Compound	Space group	a_0 (Å)	b_0 (Å)	c_0 (Å)	β (°)	Volume (Å^3)
$Cs_2Th(SO_4)_3{\cdot}2H_2O$	*P*2$_1$/*c*	6.415	15.95	13.078	90.88	1338
$Cs_2U(SO_4)_3{\cdot}2H_2O$	*P*2$_1$/*c*	6.4674	12.782	16.366	102.182	1322.5
$Cs_2Np(SO_4)_3(H_2O){\cdot}H_2O$						
$Cs_4Pu(SO_4)_4{\cdot}(H_2O)_2$	*P*2$_1$/*n*	18.642	10.488	20.445	91.61	3995.9

Table 3.9 Complex actinide sulfates [54–58, 69, 74].

Compound	Space group	a_0 (Å)	b_0 (Å)	c_0 (Å)	β (°)	Volume (Å^3)
$Th_3(SO_4)_6(H_2O)_6{\cdot}H_2O$	*P*4$_2$/*nmc*	25.89	25.89	9.08	90	6086.3
$Th_4(SO_4)_7(H_2O)_7(OH)_2{\cdot}H_2O$	*Pnma*	18.139	11.1729	14.3913	90	2916.7
$ThF_2(SO_4)(H_2O)$	*P*2$_1$/*n*	6.9065	6.9256	10.5892	96.755	502.98
$Na_2[Th(SO_4)_3(H_2O)_3]{\cdot}3H_2O$	*P*2$_1$/*c*	5.567	16.81	15.76	91.925	1474
$Na_2[Th_2(SO_4)_5(H_2O)_3]{\cdot}H_2O$	*C*2/*c*	16.639	9.081	25.078	95.322	3772.8
$NaNpO_2SO_4H_2O$	*C*2/*c*	18.3638	5.6424	23.5512	103.366	2374.2
$Cs_2[(AmO_2)_2(SO_4)_3]$	*P*-421*m*	9.5589	9.5589	8.0105	90	731.94

3.5 Phosphates

Actinide phosphates are another class of widely studied structures. One of the largest hopes with the phosphates is that they will be used as a potential host matrix for nuclear waste disposal. The phosphate anion, like borate and sulfate, is characterized by a non-metallic bond which possesses strong covalent character. In comparison, both the S-O and P-O bonds are more covalent than the B-O bond. However, because PO_4^{3-} holds an extra charge compared to SO_4^{2-}, it prefers insoluble phases when paired with the predominantly trivalent and tetravalent actinides. The number and abundance of these phases and structures has made phosphates a popular subject of study with actinides, and only a brief survey of these will be given.

All of the trivalent orthophosphates from plutonium to einsteinium have been made and characterized [75]. It is notable that the stability of these binary compounds can be increased by producing the monazite crystalline phase typically through heating at very high temperatures. The monazite structure, which is also the monoclinic space group $P2_1/c$, can be found in nature and is one the most predominant forms that lanthanides are found to exist in. It also incorporates both tetravalent uranium and thorium into its structure through a variety of substitutive mechanisms. For example, the most common substitutive mechanism is called cheralitic substitution and is demonstrated by Equation 3.30 [76]:

$$(1-2x)\mathrm{Ln}^{3+}\mathrm{PO}_4 + x\mathrm{An}^{4+} + x\mathrm{M}^{2+} + 2x\mathrm{PO}_4{}^{3-} \leftrightarrow \mathrm{Ln}^{3+}_{1-2x}\mathrm{M}_x{}^{2+}\mathrm{An}_x{}^{4+}(\mathrm{PO}_4) \quad (3.30)$$

The most common divalent substitution partner for an actinide is calcium. The ability of monazite to be able to incorporate nearly all trivalent and tetravalent actinides along with its chemical and radiation stability have made it a promising candidate for a suitable waste form. This is especially true for waste streams generated from reprocessing with molten salt reactors, because the actinide fluorides are much more insoluble and can leach from traditional borosilicate glass waste forms. The monazite structure, as well as other phosphate-based glasses such as sodium aluminophosphate glass, are good alternatives to consider for this [77]. An example of the monazite structure is given in Figure 3.37 for the recently solved PuPO_4 structure which has alluded researchers for some time [78].

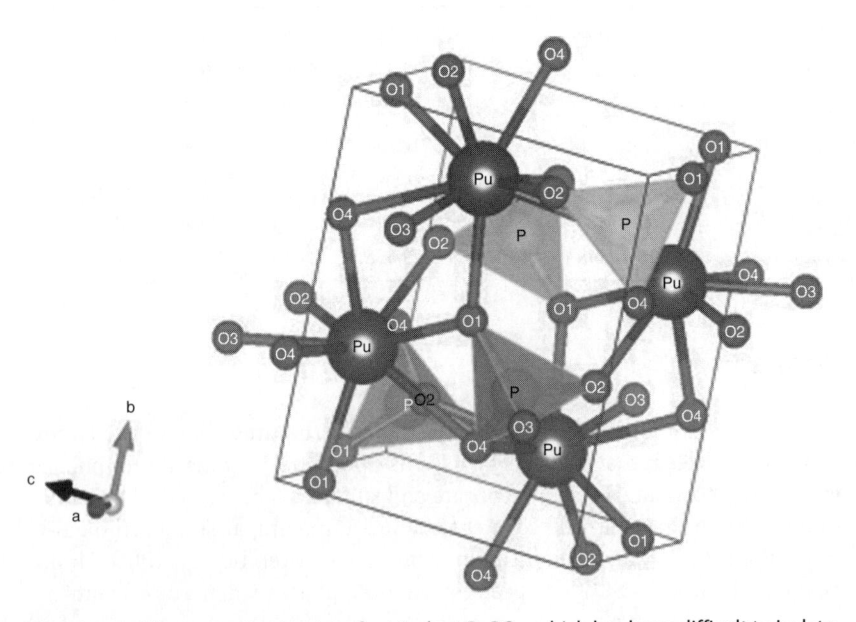

Figure 3.37 The monazite structure for trivalent PuPO_4, which has been difficult to isolate.

Another potential stable waste form is represented by thorium phosphate diphosphate, or TPD, which also requires heating to very high temperatures in a symbolic reaction given by Equation 3.31 [79]:

$$4Th(NO_3)_4 \cdot 6H_2O + 6H_3PO_4 \rightarrow Th_4(PO_4)_4 P_2O_7 + 16HNO_3 + 7H_2O \quad (3.31)$$

This material crystallizes in the orthorhombic space group *Pcam* and features thorium with a coordination of eight surrounded by four monodentate phosphates, one bidentate phosphate, and one bidentate diphosphate. The overall topography is a set of layers that alternate with planes of thorium centers. This material is very insoluble and resists corrosion. Furthermore, it incorporates other tetravalent actinides into its structure by means of a simple substitution with thorium [80].

Uranium also can also adopt structurally complex forms with phosphate. A good example of this in relation to the high-temperature thorium reaction above is the use of high temperatures to get a mixed-valent phosphate, $U(UO_2)(PO_4)_2$. This can be synthesized in a number of ways, but evaporation followed by high temperatures represent one way, as demonstrated by Equation 3.32 [81]:

$$4U(C_2O_4)_2 \cdot 6H_2O + 4H_3PO_4 + 5O_2 \rightarrow 2U(UO_2)(PO_4)_2 + 8CO_2 + 15H_2O \quad (3.32)$$

Crystallizing in the triclinic space group *P*-1, this material has a three-dimensional structure characterized by chains of successive uranium centers bridged by bidentate phosphates. Because of its mixed-valent feature, there are also two distinct uranium sites that have been confirmed spectroscopically.

Another example of an exclusive uranyl phosphate structure is $[(UO_2)_3(PO_4)O(OH)(H_2O)_2](H_2O)$ [82]. In this structure, there are three crystallographically unique uranium centers. Two of them are eight-coordinate hexagonal bipyramids. The first of these has four of its six equatorial positions occupied by two bidentate phosphates, while the other two sites are occupied by O^{2-}. The second site also has two coordinating bidentate phosphates but has two hydroxide groups instead of oxide. The third uranium site is a pentagonal bipyramid which coordinates two bridging monodentate phosphates, one oxide, one hydroxide, and one water. Two pentagonal bipyramids edge-share to form a dimer. This dimer, in turn, is flanked on both sides by hexagonal bipyramid units which extend the units into chains via the bridging phosphates (Figure 3.38). Because of this, the overall topography is able to produce 3.67 Å wide channels that contain water molecules.

As with the sulfates, the search for novel three-dimensional uranyl phosphate frameworks with the use of organic templates has been under way. In fact, it was the initial success with the phosphates which spurred a renewed interest in uranyl sulfate frameworks. The first three-dimensional uranyl phosphate was made hydrothermally with the oxidizing agent diethylhydroxylamine and is represented by Equation 3.33 [83]:

$$\begin{aligned} &10U_3O_8 + 24H_3PO_4 + 12(C_2H_5)_2 NHO \rightarrow \\ &\quad 6[(C_2H_5)_2 NH_2]_2[(UO_2)_5(PO_4)_4] + 30H_2O + O_2 \end{aligned} \quad (3.33)$$

In this structure, there are two distinct uranium coordination environments. In the first case, the common seven-coordinate pentagonal bipyramid geometry persists. It is characterized by three equatorial monodentate phosphates and one bidentate

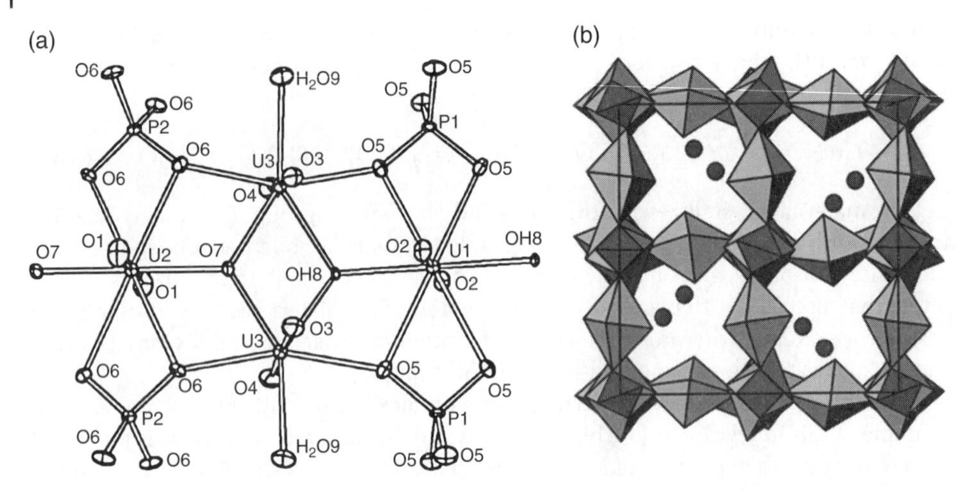

Figure 3.38 (a) Ball-and-stick representation of the three unique uranyl centers and their coordination environments and (b) a view along [001] of the polyhedral arrangement into sheets. Large polyhedra represent UO_7, while tetrahedra represent PO_4.

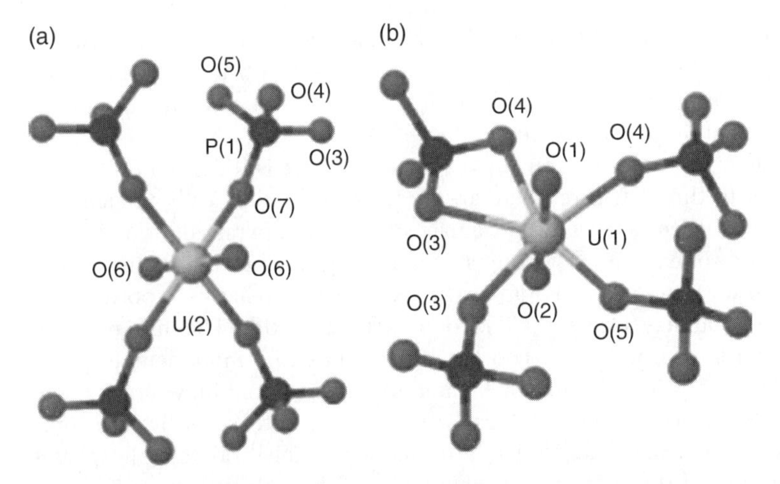

Figure 3.39 (a) The first uranium center which displays pentagonal bipyramidal geometry and (b) the second uranium center which displays a distorted octahedral environment.

phosphate (Figure 3.39a). In the second case, the uranyl unit only has four equatorial bonds which conform its geometry into a distorted octahedron. All of these equatorial bonds are filled by monodentate phosphates (Figure 3.39b). This drives the larger topography to be represented by channels which contain the templating agent $Et_2NH_2^+$. The $UO_2PO_4^-$ units form the sheets, and it is the distorted octahedra which bridge the sheets into the three-dimensional framework observed (Figure 3.40). Finally, it is worth pointing out that this material crystallizes in the centrosymmetric space group *C*2/*m*.

The P-F bond is also weak, and like the S-F bond does not lead to the functionalization of P, enabling the fluoride groups to directly attack the actinide. An example

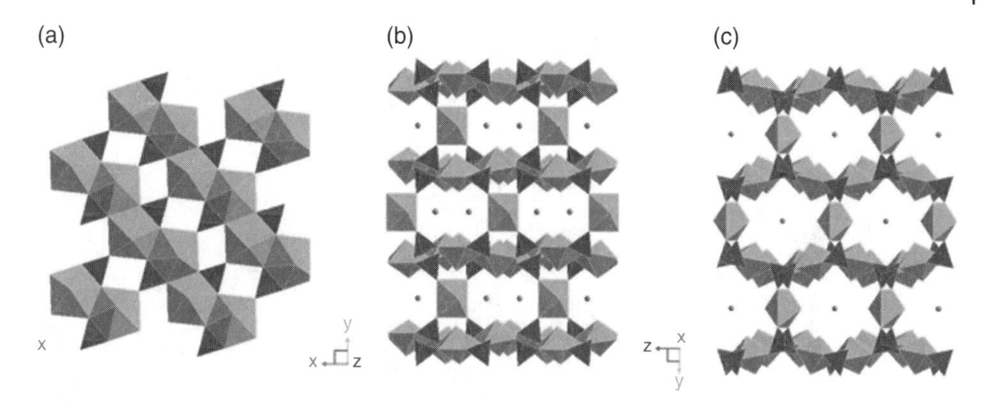

Figure 3.40 A depiction of the three-dimensional structure viewed along the (a) *b*-axis, (b) *c*-axis, and (c) *a*-axis. The large polyhedra represent uranyl centers, the tetrahedra represent phosphate, and the small spheres represent nitrogen atoms from the counterion.

of a functionalized phosphate in this way parallels that of Equation 3.26, except that in this example Dabco is also used as a templating agent. This is represented by Equation 3.34 [84]:

$$6UO_2(C_2H_3O_2)_2 \cdot 2H_2O + 2N_2C_6H_{12} + 3H_4P_2O_7 + 2HF \rightarrow 12HC_2H_3O_2 \\ + 3H_2O + [N_2C_6H_{14}]_2[(UO_2)_6(H_2O)_2F_2(PO_4)_2(HPO_4)_4] \cdot 4H_2O \quad (3.34)$$

This material is called MUPF-1 and crystallizes in the monoclinic space group $P2_1/n$. It has six crystallographically unique uranium centers divided into two different coordination groups. The first has only monodentate PO_4 groups coordinated to equatorial sites. Three of the monodentate oxides that are a part of the PO_4 tetrahedra are shared between two uranium centers. This makes for a complicated bridged design. The second type of coordination group includes three monodentate PO_4 groups, one water, and one fluoride group. There are a number of different ways that these different uranium sites come together into phosphate-bridged dimers which creates parallel chains in the *ac*-plane. These layers are connected into a three-dimensional network through hydrogen bonding and $[UO_5F(H_2O)]$ polyhedra (Figure 3.41). This produces a variety of channel types that host both water molecules and protonated Dabco counterions (Figure 3.42).

Another example of a uranyl phosphate fluoride material is LUPF-1. Instead of using an organic structure directing agent, this material reverts back to the familiar use of alkali metals to serve as charge-balancing ions. In fact, just a simple change like this can drastically change the structural topography. The stoichiometric reaction for LUPF-1 is listed by Equation 3.35 [85]:

$$UO_3 + CsCl + H_3PO_4 + HF \rightarrow Cs(UO_2)F(HPO_4) \cdot 0.5H_2O + HCl + 0.5H_2O \quad (3.35)$$

The uranyl core adopts a distorted pentagonal bipyramidal geometry in this structure, and its five equatorial binding sites are divided between three hydrogen phosphate groups and two fluoride groups. There are two unique uranium centers which are bridged by the fluoride groups, and the fluoride-bridged unit is a dimer denoted by

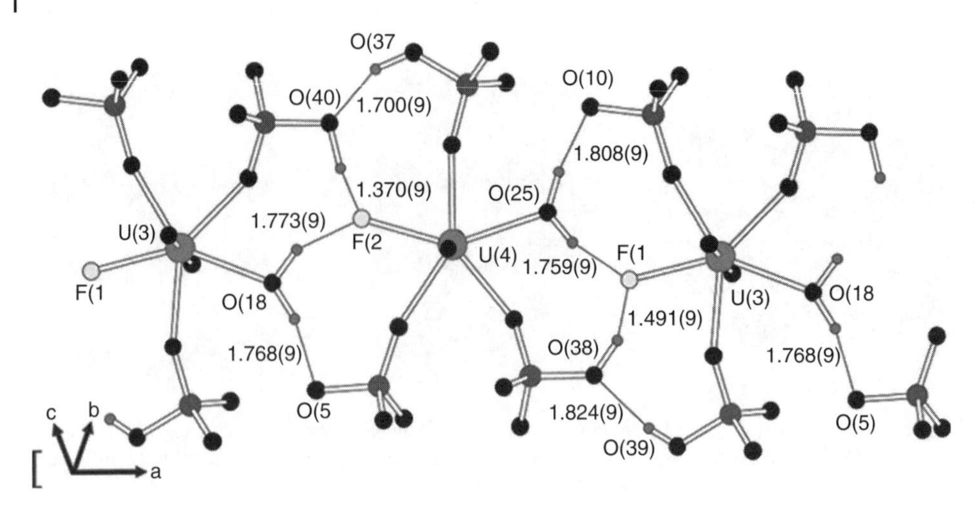

Figure 3.41 Structural representation of bonding in MUPF-1. Note the significance of the hydrogen bonding shown between the phosphate, fluoride, and water groups.

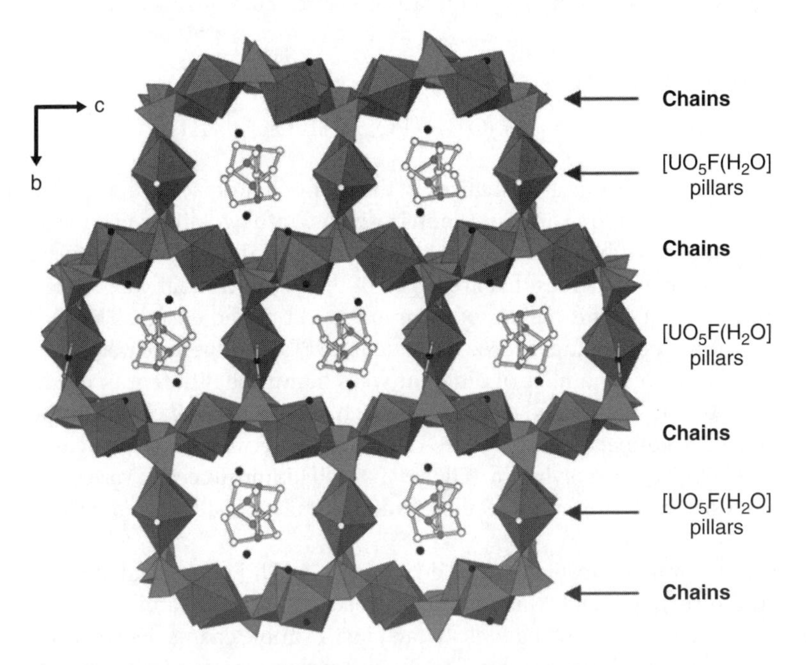

Figure 3.42 Three-dimensional topography of MUPF-1. Large polyhedra are UO_7 units, tetrahedra are PO_4 units, and small spheres are fluoride and water molecules.

$[(UO_2)O_3]_2F_2$. The hydrogen phosphate tetrahedra bridge the dimers into four and five-membered rings (Figure 3.43). The resulting anionic layers are separated by cesium cations and water molecules which provide charge balance and hydrogen bonding (Figure 3.44). Such structural topographies have been looked at for applications in ion

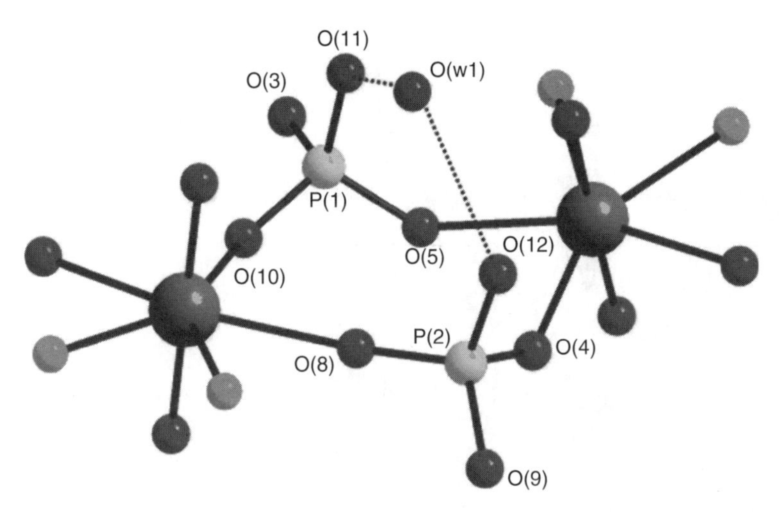

Figure 3.43 A view of how the phosphates bridge the uranyl dimers into 4- and 5-membered rings in LUPF-1. Note also the hydrogen bonding which is engaged between two phosphate groups through a water molecule.

exchange, for example. Furthermore, both the complexity and sensitivity of these structures to the counterion result in quite different topographies for K^+ and Rb^+. This encourages the likelihood of useful structural morphologies to be discovered in this branch of uranyl phosphate fluoride research.

The use of methanesulfonic acid as an *in situ* SO_4 generating agent in Equation 3.26 can also be applied to phosphate research, and furthermore, in the search for more unique phosphate fluoride architectures. Whereas the *in situ* generation of sulfate is derived from the motivation to alter the kinetic relationships of the topographical arrangement process, the *in situ* generation of phosphate is more often derived from the motivation to maintain solubility. One of the main difficulties with functionalizing phosphates is overcoming this challenge. Specifically within the search for phosphate fluorides, use of the hexafluorophosphate anion in hydrothermal syntheses is ideal because both the fluoride and phosphate groups are unavailable until the ion hydrolyzes. An example of such a synthetic use in designing novel actinide materials is given by Equation 3.36 [86]:

$$\begin{aligned}&5UO_2(NO_3)_2\cdot 6H_2O+4\left[(C_4H_9)_4N\right][PF_6]+3H_8C_{10}N_2\rightarrow[H_9C_{10}N_2]_3(UO_2)_5\\&\quad(HPO_4)_3(PO_4)F_4]+20HF+6HNO_3+4\left[(C_4H_9)_4N\right][NO_3]+14H_2O\end{aligned}\qquad(3.36)$$

In addition, the use of 4,4′-bipyridine as a templating agent serves to introduce charge balance between the resulting anionic layers. What is interesting is that a novel pentameric building unit made up of four uranyl pentagonal bipyramids and one uranyl hexagonal bipyramid comes together by bridging PO_4 tetrahedra in the sheets (Figure 3.45). The hexagonal bipyramid has an equatorial representation of two μ_3-fluorides and four phosphates, whereas the pentagonal bipyramid has the same μ_3-fluoride, a μ_2-fluoride, and three phosphates. Two pentagonal bipyramids too are bridged into dimers by the μ_2- and μ_3-fluorides.

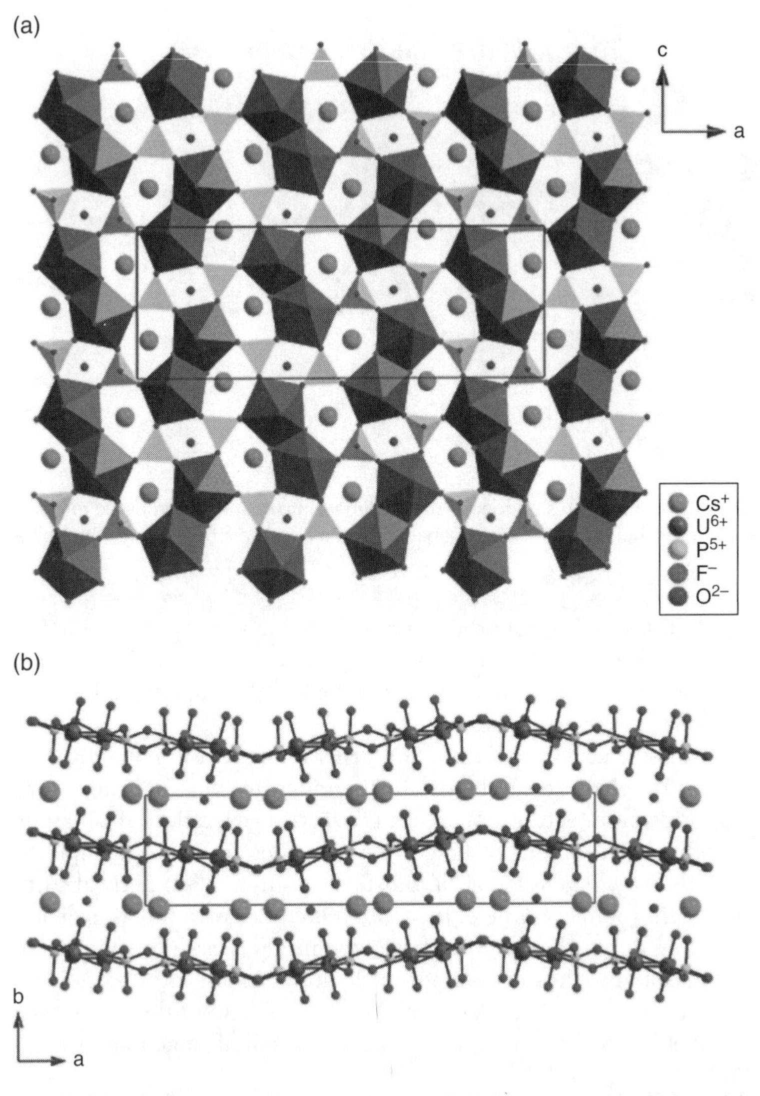

Figure 3.44 The layered architecture of LUPF-1 from the perspective of (a) a polyhedral representation in the *ac*-plane and (b) ball-and-stick representation in the *ab*-plane. Large polyhedra are uranyl centers, and tetrahedra are phosphates. (*See insert for color representation of the figure.*)

From some of these examples, it can be seen that actinide phosphate chemistry, like sulfate chemistry, is dominated by uranium and thorium structures. Other transuranic elements have little developed chemistry with complex phosphate architectures – in general, a survey of the structures only include orthophosphates, diphosphates, and different hydrated permutations of these structural elements. Nevertheless, whereas uranium and thorium chemistry focuses on novel materials with properties for ion exchange, optical materials, and catalysis, complex transuranic phosphate research focuses on characterizing substitution of these elements into geological formations such as monazites. A crystallographic summary of the structures seen in this section is given in Table 3.10.

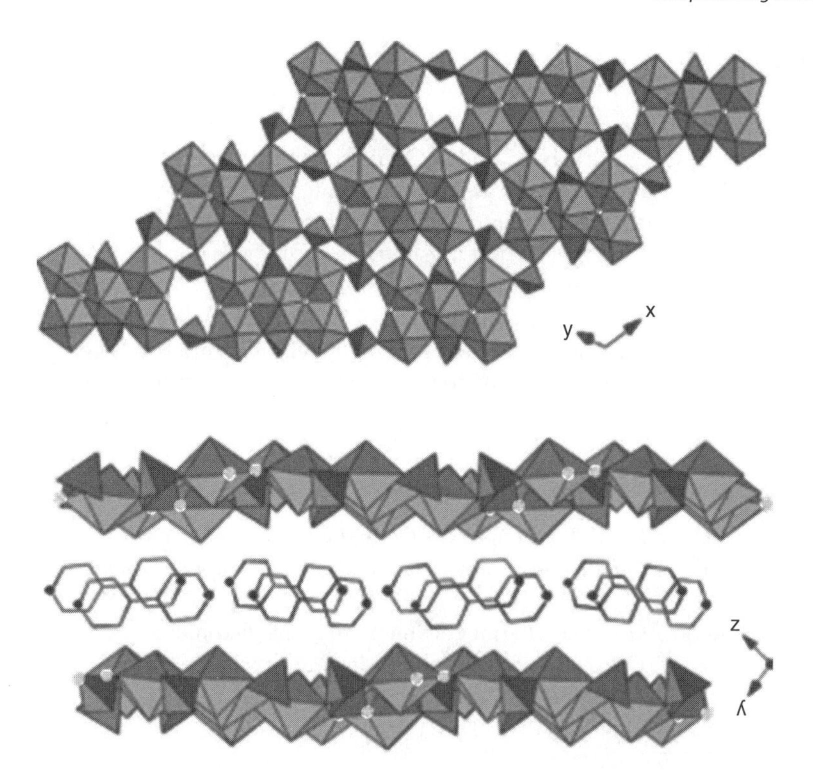

Figure 3.45 (Top) Sheet topology observed with the noticeable pentameric unit building blocks arranged regularly in the plane. (Bottom) A view of the stacking arrangement of protonated 4,4′-bipyridine molecules between the anionic layers. Large polyhedra are uranyl groups, tetrahedra are phosphate groups, and small spheres are fluoride and nitrogen.

Table 3.10 Crystallographic data for some actinide phosphates [78, 80, 82–86].

Compound	Space group	a_0 (Å)	b_0 (Å)	c_0 (Å)	β, α, γ (°)	Volume (Å^3)
$PuPO_4$	$P2_1/n$	6.759	6.98	6.447	103.63	295.6
$Th_4(PO_4)_4P_2O_7$	*Pcam*	12.8646	10.4374	7.0676	90	949.01
$U(UO_2)(PO_4)_2$	*P*-1	8.8212	9.2173	5.4772	97.748, 102.622, 102.459	416.55
$[(UO_2)_3(PO_4)O(OH)(H_2O)_2](H_2O)$	$P4_2/mbc$	14.015	14.015	13.083	90	2575.6
$[(C_2H_5)_2NH_2]_2[(UO_2)_5(PO_4)_4]$	C_2/m	9.4442	15.449	9.5719	93.268	1394.3
$[N_2C_6H_{14}]_2[(UO_2)_6(H_2O)F_2(PO_4)_2(HPO_4)_4]{\bullet}4H_2O$	$P2_1/n$	13.4487	17.921	19.9026	90.9833	4796.1
$Cs(UO_2)F(HPO_4){\bullet}0.5H_2O$	$Pca2_1$	25.656	6.0394	9.2072	90	1426.6
$[H_9C_{10}N_2]_3[(UO_2)_5(HPO_4)_3(PO_4)F_4]$	*P*-1	13.274	14.179	14.437	99.36, 94.727, 114.01	2415.5

3.6 Conclusion

In review, a brief coverage of complex inorganic actinide materials spanning fluorides, borates, sulfates, and phosphates has been undertaken. It is evident from the sequence of material presented that there are many themes that relate one functional group to another. This was done intentionally to demonstrate features that deserve study when approaching other similar systems that are not presented here. It is also hoped that even the general reader will recognize the importance of such structural topologies and the impact of synthesis and design for applications that span a categorical number of nuclear disciplines. Inorganic actinide chemistry has delivered many recent surprises to science, and it is expected that this trend will continue to endure into the foreseeable future.

References

1 Asprey, L., Keenan, T., and Kruse, F. (1965) Crystal structures of the trifluorides, trichlorides, tribromides, and triiodides of americium and curium. *Inorganic Chemistry*, **4** (7), 985–986.
2 Dawson, J. and Elliott, R. (1953) The Thermogravimetry of Some Plutonium Compounds. *Report AERE-C/R-1207*. Great Britain Atomic Energy Research Establishment, Harwell, Oxfordshire, UK.
3 Larson, A., Roof, R., and Cromer, D. (1963) The crystal structure of UF_4. *Acta Crystallographica*. **17**, 555–558.
4 Asprey, L., Ellinger, F., Fried, S., and Zachariasen, W. (1957) Evidence for quadrivalent curium. *Journal of American Chemical Society*. **79**, 5825.
5 Myers, N. (1956) Thermal Decomposition of Plutonium (IV) Oxalate and Hydrofluorination of Plutonium (IV) Oxalate and Oxide. *USAEC Report HW-45128*. Hanford Atomic Products Operation, General Electric Company, Richland, WA.
6 Katz, J., Seaborg, G., and Morss, L. (eds) (1986) *The Chemistry of the Actinide Elements*. Chapman and Hall, London.
7 Zachariasen, W. (1949) Crystal chemical studies of the 5f-Series of elements. *Acta Crystallographica*. **2**, 388.
8 Keenan, T. and Asprey, L. (1969) Lattice constants of actinide tetrafluorides including berkelium. *Inorganic Chemistry*. **8** (2), 235–238.
9 Stevenson, J. and Peterson, J. (1973) The trigonal and orthorhombic crystal structures of CfF_3 and their temperature relationship. *Journal of Inorganic and Nuclear Chemistry*. **35** (10), 3481–3486.
10 Peterson, J. and Cunningham, B. (1968) Crystal structures and lattice parameters of the compounds of berkelium—IV berkelium trifluoride. *Journal of Inorganic and Nuclear Chemistry*. **30** (7), 1775–1784.
11 Haug, H. and Baybarz, R. (1975). Lattice parameters of the actinide tetrafluorides UF_4, BkF_4, and CfF_4. *Inorganic and Nuclear Chemistry Letters*. **11** (12), 847–855.
12 Asprey, L., Kruse, F., and Penneman, R. (1966) Alkali fluoride complexes of tetravalent protactinium. *Inorganic Chemistry*. **6** (3), 544–548.
13 Brunton, G. (1965) Annual report of reactor chemistry division. *ORNL-3913*, 10.
14 Benz, R., Douglass, R., Kruse, F., and Penneman, R. (1962) Preparation and properties of several ammonium uranium (IV) and ammonium plutonium (IV) fluorides. *Inorganic Chemistry*. **2** (4), 799–803.

15 Malm, J. and Carnall, W. (1985) The chemistry of neptunium fluorides in liquid anhydrous hydrogen fluoride. The isolation and characterization of NpF_5. *Journal of Fluorine Chemistry.* **29** (1–2), 26.
16 Penneman, R., Sturgeon, G, and Asprey, L. (1964) Fluoride complexes of pentavalent uranium. *Inorganic Chemistry.* **3** (1), 126–129.
17 Nikolaev, N. and Sukhoverkhov, V. (1961) *Doklady Akademiia Nauk SSSR.* **136**, 621.
18 Volavsek, B. (1961) *Croatica Chemica Acta.* **33**, 181.
19 Malm, J., Selig, H., and Siegel, S. (1966) Complex compounds of uranium hexafluoride with sodium fluoride and potassium fluoride. *Inorganic Chemistry.* **5** (1), 130–132.
20 Asprey, L. and Penneman, R. (1967) Fluorine oxidation of tetravalent uranium and neptunium in the pentavalent state. *Journal of the American Chemical Society.* **89** (1), 172.
21 Baluka, M., Yeh, S., Banks, R., and Edelstein, E. (1980) Preparation and structural characterization of α-NpF_5. *Inorganic and Nuclear Chemistry Letters.* **16** (2), 75–77.
22 Brown, D. (1968) *Halides of the Lanthanides and Actinides*, John Wiley & Sons, Ltd, Bath.
23 Vallet, V., Wahlgren, U., and Grenthe, I. (2012) Probing the nature of chemical bonding in uranyl (VI) complexes with quantum chemical methods. *Journal of Physical Chemistry A.* **116** (50), 12373–12380.
24 Chen, F., Wang, C., Shi, W., *et al.* (2013) Two new uranyl fluoride complexes with $U^{VI}=O$-alkali (Na, Cs) interactions: Experimental and theoretical studies. *CrystEngComm.* **15**, 8041.
25 Rosenzweig, A. and Cromer, D. (1970) The crystal structure of $(NH_4)_4UF_8$. *Acta Crystallographica.* **B26**, 38–44.
26 Kim, T., Grandy, C., and Hill, R. (2009) Carbide and nitride fuels for advanced burner reactor. *International Conference on Fast Reactors and Related Fuel Cycles (FR09) – Challenges and Opportunities.* Kyoto, Japan.
27 Silva, G., Yeamans, C. Ma, L., *et al.* (2008) Microscopic characterization of uranium nitrides synthesized by oxidative ammonolysis of uranium tetrafluoride. *Chemistry of Materials.* **20** (9), 3076–3084.
28 Poineau, F., Silva, C., Yeamans, C., *et al.* (2016) Structural study of the ammonium octafluoroneptunate, $[NH_4]_4NpF_8$. *Inorganica Chimica Acta.* **448**, 93–96.
29 Yeon, J., Smith, M., Sefat, A., *et al.* (2013) $U_3F_{12}(H_2O)$, a noncentrosymmetric uranium (IV) fluoride prepared via a convenient in situ route that creates U^{4+} under mild hydrothermal conditions. *Inorganic Chemistry.* **52**, 8303–8305.
30 Yeon, J., Smith, M., Tapp, J., *et al.* (2014) Application of a mild hydrothermal approach containing an in situ reduction step to the growth of single crystals of the quaternary U (IV)-containing fluorides $Na_4MU_6F_{30}$ ($M = Mn^{2+}$, Co^{2+}, Ni^{2+}, Cu^{2+}, and Zn^{2+}) crystal growth, structures, and magnetic properties. *Journal of the American Chemical Society.* **136**, 3955–3963.
31 Gasperin, M. (1987) Synthèse et Structure du Borouranate de Calcium: $CaB_2U_2O_{10}$. *Acta Crystallographie, Section C.* **43**, 1247–1250.
32 Behm, H. (1985) Hexapotassium (*cyclo*-Octahydroxotetracosaoxohexadecaborato) dioxouranate(VI) Dodecahydrate, $K_6[UO_2(B_{16}O_{24}(OH)_8)]\cdot 12H_2O$. *Acta Crystallographie, Section C.* **41**, 642–645.
33 Wang, S., Yu, P., Purse, B., *et al.* (2012) Selectivity, kinetics, and efficiency of reversible anion exchange with TcO_4^- in a supertetrahedral cationic framework. *Advanced Functional Materials.* **22**, 2241–2250.

34 Nikogosyan, D. (1991) Beta barium borate. *Applied Physics A.* **52** (6), 359–368.
35 Wang, S., Alekseev, E., Stritzinger, J. *et al.* (2010) Structure–property relationships in lithium, silver, and cesium uranyl borates. *Chemistry of Materials.* **22**, 5983–5991.
36 Wang, S., Alekseev, E., Diwu, J., *et al.* (2011) Functionalization of borate networks by the incorporation of fluoride: Syntheses, crystal structures, and nonlinear optical properties of novel actinide fluoroborates. *Chemistry of Materials.* **23**, 2931–2939.
37 Chen, C., Wu, Y., and Li, R. (1989). The anionic group theory of the non-linear optical effect and its applications in the development of new high-quality NLO crystals in the borate species. *International Reviews in Physical Chemistry.* **8** (1), 65–91.
38 Wang, S., Alekseev, E., Depmeier, W., and Albrecht-Schmitt, T. (2010). Further insights into intermediate- and mixed-valency in neptunium oxoanion compounds: Structure and absorption spectroscopy of $K_2[(NpO_2)_3B_{10}O_{16}(OH)_2(NO_3)_2]$. *Chemical Communications.* **46**, 3955–3957.
39 Wang, S., Alekseev, E., Depmeier, W. and Albrecht-Schmitt, T. (2011). New neptunium (V) borates that exhibit the Alexandrite effect. *Inorganic Chemistry.* **51**, 7–9.
40 Wang, S., Villa, E., Diwu, J. *et al.* (2011) Role of anions and reaction conditions in the preparation of uranium (VI), neptunium (VI), and plutonium (VI) borates. *Inorganic Chemistry.* **50**, 2527–2533.
41 Wang, S., Alekseev, E., Depmeier, W., and Albrecht-Schmitt, T. (2011) Surprising coordination for plutonium in the first plutonium (III) borate. *Inorganic Chemistry.* **50**, 2079–2081.
42 Polinski, M., Wang, S., Cross, J. *et al.* (2012) Effects of large halides on the structures of lanthanide (III) and plutonium (III) borates. *Inorganic Chemistry.* **51**, 7859–7866.
43 Polinski, M., Wang, S., Alekseev, E., *et al.* (2011) Bonding changes in plutonium (III) and americium (III) borates. *Angewandte Chemie.* **50**, 8891–8894.
44 Polinski, M., Grant, D., Wang, S., *et al.* (2012) Differentiating between trivalent lanthanides and actinides. *Journal of the American Chemical Society.* **134**, 10682–10692.
45 Shannon, R. (1976) Revised effective ionic radii and systematic studies of interatomic distances in halides and chalcogenides. *Acta Crystallographie, Section A.* **32**, 751–767.
46 Polinski, M., Wang, S., Alekseev, E., *et al.* (2012) Curium (III) borate shows coordination environments of both plutonium (III) and americium (III) borates. *Angewandte Chemie.* **51**, 1869–1872.
47 Polinski, M., Garner, E. III, Maurice, R., *et al.* (2014) Unusual structure, bonding and properties in a californium borate. *Nature Chemistry.* **6**, 387–392.
48 Silver, M., Cary, S., Johnson, J., *et al.* (2016) Characterization of berkelium(III) dipicolinate and borate compounds in solution and the solid state. *Science*, in press.
49 Gouder, T., van der Lann, G., Shick, A., *et al.* (2011) Electronic structure of elemental curium studied by photoemission. *Physical Review B.* **83** (12), 125111.
50 Cary, S., Vasiliu, M., Baumbach, R., *et al.* (2015) Emergence of californium as the second transitional element in the actinide series. *Nature Communications.* **6**, 6827.
51 Heroy, W. (1957) Appendix F, disposal of radioactive waste in salt cavities. *National Research Council (US) Committee on Waste Disposal. The Disposal of Radioactive Waste on Land.* National Academies Press, Washington (DC).
52 Roxburgh, I. (1987) *Geology of High-Level Nuclear Waste Disposal.* Chapman and Hall, New York.
53 Habash, J. and Smith, A. (1983) Structure of thorium sulphate octahydrate, $Th(SO_4){\cdot}H_2O$. *Acta Crystallographica, Section C.* **39**, 413–415.

54 Habash, J. and Smith, A. (1990) Structure of disodium triaquatri-μ-sulfato-thorate(IV) trihydrate. *Acta Crystallographica, Section C.* **46**, 957–960.
55 Wilson, R., Skanthakumar, S., Knope, K., *et al.* (2008) An open-framework thorium sulfate hydrate with 11.5 Å voids. *Inorganic Chemistry.* **47**, 9321–9326.
56 Albrecht, A., Sigmon, G., Moore-Shay, L., *et al.* (2011) The crystal chemistry of four thorium sulfates. *Journal of Solid State Chemistry.* **184**, 1591–1597.
57 Knope, K., Wilson, R., Skanthakumar, S., and Soderholm, L. (2011) Synthesis and characterization of thorium (IV) sulfates. *Inorganic Chemistry.* **50**, 8621–8629.
58 Zhao, Y., Wang, C., Su, J., *et al.* (2015) Insights into the new Th (IV) sulfate fluoride complex: Synthesis, crystal structures, and temperature dependent spectroscopic properties. *Spectrochimica Acta Part A: Molecular and Biomolecular Spectroscopy.* **149**, 295–303.
59 Kierkegaard, P. (1956) The crystal structure of $U(SO_4)_2{\cdot}4H_2O$. *Acta Chemica Scandinavica.* **10**, 599–616.
60 Hennig, C., Ikeda-Ohno, A., Tsushima, S., and Scheinost, A. (2009) The sulfate coordination of Np(IV), Np(V), and Np(VI) in aqueous solution. *Inorganic Chemistry.* **48** (12), 5350–5360.
61 Schnaars, D., and Wilson, R. (2012) Uranium(IV) sulfates: Investigating structural periodicity in the tetravalent actinides. *Inorganic Chemistry.* **51**, 9481–9490.
62 Hennig, C., Kraus, W., Emmerling, F., *et al.* (2008) Coordination of a uranium(IV) sulfate monomer in an aqueous solution and in the solid state. *Inorganic Chemistry.* **47**, 1634–1638.
63 Habash, J. and Smith, A. (1992) Crystal structure of dicesium thorium trisulfate dihydrate, $Cs_2Th(SO_4)_3{\cdot}2H_2O$. *Journal of Crystallographic and Spectroscopic Research.* **22** (1), 21–24.
64 Doran, M., Norquist, A., and O'Hare, D. (2002) $[NC_4H_{12}]_2[(UO_2)_6(H_2O)_2(SO_4)_7]$: The first organically templated actinide sulfate with a three-dimensional framework structure. *Chemical Communications.* 2946–2947.
65 Alekseev, E., Krivovichev, S., and Depmeier, W. (2007) A crown ether as template for microporous and nanostructured uranium compounds. *Angewandte Chemie.* **47** (3), 549–551.
66 Charushnikova, I., Krot, N., and Polyakova, I. (2006) Crystal structure of neptunium(V) sulfate hexahydrate, $(NpO_2)_2SO_4{\cdot}6H_2O$. *Crystallography Reports.* **51** (2), 201–204.
67 Grigor'ev, M., Yanovskiĭ, A., Fedoseev, A., *et al.* (1988) *Doklady Akademiia Nauk SSSR.* **300** (3), 618.
68 Grigor'ev, M., Baturin, N., Budantseva, N., and Fedoseev, A. (1993) *Radiokhimiya.* **35** (2), 29.
69 Forbes, T., Burns, P., Soderholm, L., and Skanthakumar, S. (2006) Crystal structures and magnetic properties of $NaK_3(NpO_2)_4(SO_4)_4(H_2O)_2$ and $NaNpO_2SO_4H_2O$: Cation-cation interactions in a neptunyl sulfate framework. *Chemical Materials.* **18**, 1643–1649.
70 Jayadevan, N., Singh Mudher, K., and Chackraburtty, D. (1982) The crystal structures of α- and β-forms of plutonium(IV) sulphate tetrahydrate. *Zeitschrift für Kristallographie.* **161** (1–2), 7–13.
71 Wilson, R. (2011) Structural periodicity in plutonium(IV) sulfates. *Inorganic Chemistry.* **50**, 5663–5370.
72 Charushnikova, I., Krot, N., and Starikova, Z. (2000) Synthesis, crystal structure, and characteristics of double cesium neptunium(IV) sulfate. *Radiochemistry.* **42** (1), 42–47.

73 Burns, J. and Baybarz, R. (1972) Crystal Structure of americium sulfate octahydrate. *Inorganic Chemistry.* **11** (9), 2233–2237.

74 Budantseva, N., Grigoriev, M., and Fedosseev, A. (2014) Americium dioxocations in solid: Crystal structures and absorption spectra of the new complexes $[Am^{V}O_2(dipy)OOCCH_3(H_2O)]$ and $Cs_2[Am^{VI}O_2(SO_4)_3]$. *Radiochimica Acta.* **102** (5), 377–384.

75 Hobart, D., Begun, G., Haire, R., and Hellwege, H. (1983) Raman spectra of the transplutonium orthophosphates and trimetaphosphates. *Journal of Raman Spectroscopy.* **14**, 59–62.

76 Linthout, K. (2007) Tripartite division of the system $2REEPO_4$-CaTh)$PO_4)_2$-$2ThSiO_4$, discreditation, of brabantite, and recognition of cheralite as the name for members dominated by $CaTh(PO_4)_2$. *The Canadian Mineralogist.* **45**, 503–508.

77 Sun, Y., Xia, X., Qiao, Y., *et al.* (2016) Properties of phosphate glass waste forms containing fluorides from a molten salt reactor. *Nuclear Science and Techniques.* **27**, 63.

78 Popa, K., Raison, P., Martel, L. *et al.* (2015) Structural investigations of Pu^{III} phosphate by X-ray diffraction, MAS-NMR and XANES spectroscopy. *Journal of Solid State Chemistry.* **230**, 169–174.

79 Bénard, P., Branel, V., Dacheux, N., *et al.* (1996) $Th_4(PO_4)_4P_2O_7$, a new thorium phosphate: Synthesis, characterization, and structure determination. *Chemistry of Materials.* **8** (1), 181–188.

80 Brandel, V., Dacheux, N., and Genet, M. (2001) Studies on the chemistry of uranium and thorium phosphates. Thorium phosphate diphosphate: A matrix for storage of radioactive wastes. *Radiochemistry.* **43** (1), 16–23.

81 Bénard, P., and Louër, D. (1994) $U(UO_2)(PO_4)_2$, a new mixed-valence uranium orthophosphate: *Ab Initio* structure determination from powder diffraction data and optical and X-ray photoelectron spectra. *Chemistry of Materials.* **6**, 1049–1058.

82 Burns, P., Alexopoulos, C., Hotchkiss, P., and Locock, A. (2003) An unprecedented uranyl phosphate framework in the structure of $[(UO_2)_3(PO_4)O(OH)(H_2O)_2](H_2O)$. *Inorganic Chemistry.* **43**, 1816–1818.

83 Danis, J., Runde, W., Scott, B., *et al.* (2001) Hydrothermal synthesis of the first organically templated open-framework uranium phosphate. *Chemical Communications.* 2378–2379.

84 Doran, M., Stuart, C., Norquist, A., and O'Hare, D. (2004) $[N_2C_6H_{14}]_2[(UO_2)_6(H_2O)F_2(PO_4)_2(HPO_4)_4]\cdot 4H_2O$: A new microporous uranium phosphate fluoride. *Chemistry of Materials.* **16**, 565–566.

85 Ok, K., Baek, J., Halasyamani, P., and O'Hare, D. (2006) New layered uranium phosphate fluorides: Syntheses, structures, characterizations, and ion-exchange properties of $A(UO_2)F(HPO_4)\cdot xH_2O$ ($A = Cs^+, Rb^+, K^+$; $x = 0$–1). *Inorganic Chemistry.* **45**, 10207–10214.

86 Deifel, N., Holman, K., and Cahill, C. (2008) PF_6^- Hydrolysis as a route to unique uranium phosphate materials. *Chemical Communications.* 6037–6038.

4

Organometallic Actinide Complexes with Novel Oxidation States and Ligand Types

Trevor W. Hayton[1] *and Nikolas Kaltsoyannis*[2]

[1] *Department of Chemistry and Biochemistry, University of California Santa Barbara*
[2] *School of Chemistry, University of Manchester*

4.1 Introduction

The past decade has seen an important expansion of our understanding of actinide–ligand bonding. This expansion has been driven by several factors, including advancements in both characterization techniques and computational methods. Also critical to this improved understanding has been a renewed interest in actinide synthetic chemistry by organometallic chemists, especially younger organometallic chemists (1–4). Perhaps the most important factor in the development of this chemistry, however, has been the close collaboration between synthetic chemists, spectroscopists, and computational chemists during these investigations. These collaborative efforts will be the major focus of this contribution.

This chapter describes advancements in organometallic actinide chemistry from 2006 to 2015. This cut-off was chosen because (1) 10 years is a pleasingly round number and (2) 2006 is the publication year of the *Chemistry of the Actinides and Transactinide Elements,* 3rd edition (5). However, in some sections it will be necessary, and even sometimes desirable, to include earlier papers to tell a more coherent story. For the purposes of this chapter, the definition of an "organometallic" complex will not necessarily be restricted to those that feature an An-C bond. Thus, selected amido, imido, and oxo complexes have been included if they abide by the "spirit" of organometallic chemistry. In addition, it is important to mention that this chapter is not intended to be a comprehensive review, but instead a focused examination of the most important new developments in actinide organometallic chemistry in the opinions of the authors.

4.2 Overview of Actinide Organometallic Chemistry

To get an overview of the scope of actinide organometallic and coordination chemistry, it is useful to explore the Cambridge Structural Database (CSD). As of August 2015, there were 5771 actinide structures listed in the CSD, a number that includes both

Experimental and Theoretical Approaches to Actinide Chemistry, First Edition.
Edited by John K. Gibson and Wibe A. de Jong.

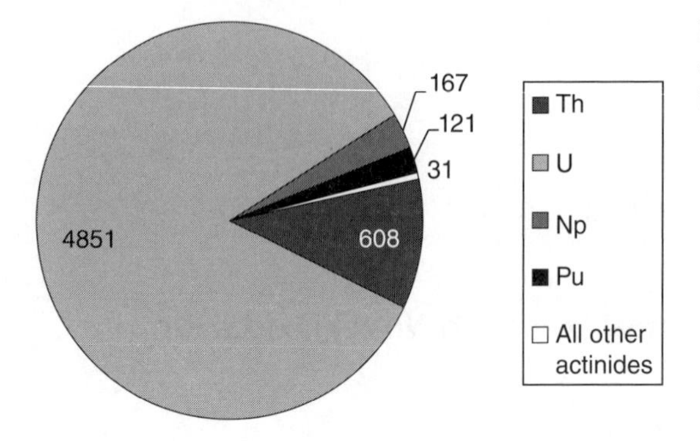

Figure 4.1 Number of actinide complexes in the Cambridge Structural Database (6).

organometallic and coordination complexes. By a large margin, the most abundant actinide element in the CSD is uranium, which features 4851 entries, of which 2626 are of the ubiquitous uranyl ion. In contrast, only 608 thorium-containing structures can be found in the CSD (Figure 4.1) (6), which is surprising considering that the element and its salts are easily acquired and feature low specific activity. Not surprisingly, though, the structural chemistry of Np (167 structures) and Pu (121 structures) is much smaller than that of Th and U; however, the number of Np and Pu structures is a growing – and important – component of actinide structural chemistry. Finally, the structural chemistry of Ac (1 structure), Pa (9), Am (14), Cm (3), Bk (1), and Cf (3) is essentially nonexistent. This observation can be rationalized by the experimental challenges associated with their use, as each is highly radioactive and often difficult to acquire. Comparison of these data with the Group 4 transition metals and the lanthanides is also informative, given that their chemistries often parallel those of An^{4+} and An^{3+}, respectively. As of August 2015, there were 15839 and 28314 Group 4 and lanthanide structures in the CSD, respectively. Thus, the chemistry of the actinide elements still represents a small fraction of coordination chemistry as a whole, although this is beginning to change (see below). Finally, it should be mentioned that this analysis comes with a few caveats. First, some structurally characterized complexes will inevitably be missed during the indexing process, and thus will not be captured by this analysis. Second, complexes that have not been structurally characterized will, of course, not appear in this analysis. Nonetheless, if we assume that indexing errors occur at the same rate for all classes of complexes, then comparisons within the CSD should still give an accurate reflection of emerging trends within the field.

A comparison of the number of actinide structures added to the CSD as a function of the year is also useful. As can be seen in Figure 4.2, the number of new actinide structures added to the CSD *per* year has greatly increased over the past 10 years, essentially doubling over that time period. Importantly, this rate of increase is much greater than the rate of the total number of new structures added to the CSD*. Thus, the pace of progress within actinide chemistry is greater than that found for structural inorganic chemistry as a whole. As a consequence, synthetic actinide chemistry is now approaching the sophistication of its transition metal counterparts, as will become apparent in the sections below.

*The number of new structures added annually to the CSD increased by about 30% from 2005 to 2014.

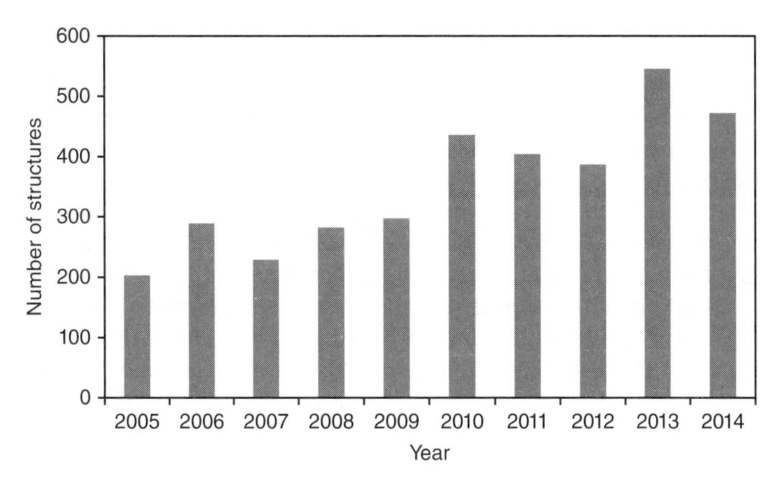

Figure 4.2 Number of new actinide complexes added to the Cambridge Structural Database as a function of year (6).

4.2.1 Overview of Thorium Organometallics

As mentioned above, thorium chemistry is substantially underdeveloped relative to that of uranium. However, several recent advancements in thorium organometallic chemistry may spur greater interest in this element going forward. For example, the past few years have seen a significant expansion of thorium-ligand multiple bond chemistry (7–13). Similarly, recent reports detailing the synthesis of Th(II) and Th(III) complexes, oxidation states that are essentially unknown, suggest another fruitful avenue of investigation (14, 15). Another recent development that may encourage greater interest in the chemistry of thorium is the publication of reliable and convenient procedures for the syntheses of $ThX_4(DME)_2$ (X = Cl, I) by Kiplinger and coworkers (16, 17). Overall, these results, which are discussed in greater detail below, hint at a greater complexity to thorium chemistry than was previously imagined.

Another reason for the growing interest in thorium chemistry is the possible future deployment of $^{232}Th/^{233}U$ nuclear reactors (18, 19). This interest has been motivated by the higher abundance and wider distribution of thorium in Earth's crust, compared with uranium, and the greater proliferation resistance of the thorium fuel cycle (20). Before this reactor type is ever implemented, however, a number of technological problems will need to be overcome, including the development of a thorium-specific fuel reprocessing cycle. This cycle will need to surmount problems not present in uranium fuel reprocessing, such as separation of ^{232}Th from ^{231}Pa ($t_{1/2}$ = 32760 y), which is present in the irradiated fuel, and the dissolution of the highly refractory ThO_2 fuel pellets by acidic media (20). To address these challenges, our fundamental understanding of thorium coordination chemistry will need to be substantially improved. Previous attempts to develop a thorium-specific fuel cycle have focused on modifying the well-understood PUREX process, which led to the development of the THOREX process (21–23). However, it is not clear if tri-*n*-butyl phosphate, the extractant used in both PUREX and THOREX, is the most effective ligand for reprocessing spent thorium fuel, as so little research has been performed in the area.

4.2.2 Overview of Uranium Organometallics

By far the best-understood actinide element, at least with respect to its organometallic chemistry, is uranium. There have been several notable advancements in uranium organometallic chemistry over the past decade, including a substantial expansion of uranium–ligand multiple bond chemistry. Significant examples within this category include the synthesis of the first terminal nitrido complex (24), as well as the synthesis of several actinide carbene complexes (25, 26). Also notable is the isolation of several novel uranyl analogues, such as the *trans*-bis(imido) complexes (27), and more recently, *mer*- and *fac*-tris(imido) complexes (28, 29). Synthetic chemists have also expanded the chemistry of several of uranium's oxidation states, including the +5 and +6 states (29–31), and most notably the +2 state (15, 32). The past decade has also seen the rise of sophisticated new ligands, such as redox-active ligands (29, 33), in uranium organometallic chemistry. In addition, considerable progress has been made in the ligation of soft donor group 16 atoms to uranium, and a large variety of chalcogenide, chalcogenate, and bischalcogenidophosphinate complexes are now known (34–38). These complexes, in particular, have played a central role in the study of covalency in actinide-ligand bonding, as will be described in the sections below. It is also worth noting that several of these advancements have been made possible, in part, by the development of new synthetic protocols that take into account the unique chemical behavior of uranium, such as its proclivity to undergo 1e$^-$ redox transformations (38, 39). Examples include the use of a KC_8 "filtration column" to isolate the first uranium +2 complexes (15, 32), and the use of NO_2^- and $[Ph_3CO]^-$ as novel O-atom transfer agents (38, 39).

4.2.3 Overview of Transuranium Organometallics

The organometallic chemistry of the transuranium elements is not nearly as well developed as that of thorium and uranium, for obvious reasons. Nonetheless, there have been several notable developments over the past ten years, including the isolation of a *trans*-bis(imido) complex of Np (40), and the synthesis of a series of homoleptic soft donor complexes of Np and Pu (34, 41–45). However, recent reports outlining the syntheses of the anhydrous salts, $AnCl_4(DME)_2$ (An = Np, Pu) and $[NBu_4][PuCl_6]$, should make entry into transuranium organometallics much easier (46, 47). It is also notable that these complexes were prepared from An(IV) stock solutions, which are more widely available than alternate An sources, such as Np and Pu metal.

4.3 Overview of Theoretical Methods

One of the most pleasing recent aspects of organoactinide research has been the increasing use of a combined experimental and theoretical approach to tackle key problems in structure and bonding. Reflecting this, and in keeping with the spirit of this volume, the results of computational studies are described throughout this chapter, and hence it is appropriate to provide a short section summarizing the main methods employed in these studies. For more detailed descriptions of modern relativistic computational quantum chemical techniques, the reader is directed to Refs. (48–52).

Modern molecular quantum chemistry is the domain of two distinct theoretical approaches; the ab initio ("from the beginning") Hartree–Fock self-consistent field (HF-SCF) method (and its "post Hartree–Fock" extensions) and density functional theory (DFT). The latter is ubiquitous throughout the periodic table, but the former was, until rather recently, rarely applied to the actinides as it is computationally much more demanding than DFT. However, recent improvements in computational hardware and software have led to an increasing use of post Hartree–Fock methods in molecular actinide chemistry, as we shall see in this chapter.

The HF-SCF method makes very few assumptions, and seeks to solve the electronic Schrödinger equation for a particular geometric arrangement of the nuclei within a molecule. The result of an HF-SCF calculation is the electronic structure of a molecule, expressed in terms of one-electron wavefunctions (molecular orbitals (MOs)) and associated eigenvalues (orbital energies). The MOs are usually broken down into contributions from atom-centered functions that form part of the input to a calculation (the basis set). These basis functions are typically chosen so that they resemble the familiar hydrogenic atomic orbitals, thereby making the results of HF-SCF calculations more chemically accessible.

The HF-SCF method is intrinsically incapable of describing the effects of electron correlation (the tendency of electrons to avoid each other owing to their charged nature) since it uses single-determinantal wavefunctions. The most time-consuming part of *ab initio* calculations then lies in the extension of the HF-SCF method to include these effects. There are many post Hartree–Fock methods, for example, perturbation approaches such as Møller–Plesset theory (the MPn techniques) and techniques based on configuration interaction (CI), which attempt to incorporate correlation by enhancing the wavefunction through additional determinants formed by exciting electrons from the HF solution into virtual orbitals. The complete active space self-consistent field (CASSCF) approach focuses on a partially filled subset of the MOs (the "active space," chosen so as to best reflect the problem at hand). All possible arrangements of the electrons within the active space are considered, with the constraint that they must possess the correct symmetry to contribute to the electronic state under study, with simultaneous optimization of the active space MOs. The resulting CASSCF solution accounts for much of the static electron correlation (that associated with near-degeneracy effects, that is, when different electronic configurations have similar energies), providing that a good choice has been made for the active space orbitals. To recover the dynamic part of the electron correlation (the contribution associated with, for example, two electrons occupying the same spatial orbital), the remaining virtual orbitals are used, as in, for example, the CASPT2 approach of Roos *et al.* (53, 54), in which the multiconfigurational CASSCF wavefunction is used as the starting point for a perturbation treatment.

The great rival to the ab initio family of techniques is DFT (49, 50). The fundamental quantity of DFT is the charge density, from which all of the ground state properties of a system may be derived (55). Kohn and Sham used a mathematical sleight of hand to recast the solution to density functional problems in terms of quantities that are straightforward to evaluate, with the exception of one, namely, the exchange-correlation energy of a system with a given charge density (56). The principal drawback to density functional methods was – and is – that the correct mathematical form of the exchange-correlation energy is not known, although it has been the subject of intense research for many years, and very good approximate formulations are now available. Perdew *et al.*'s engaging review remains a good entry point to these issues (57). One advantage of DFT over HF theory is that even simple formulations of DFT can account for correlation effects in some way (although DFT lacks an analogue of the post-HF "roadmap" to the solution of the Schrödinger equation – full CI within an infinite basis set). Furthermore, most computational implementations of DFT scale somewhere between N^2 and N^3 in the number of basis functions N, making DFT calculations much cheaper computationally than post-HF approaches, which can display scalings of N^7–N^8 for some of the most intensive methods.

The ab initio and DFT methods discussed thus far are non-relativistic. Relativistic analogues of HF theory exist (58), but these four component Dirac–Fock techniques are difficult to implement computationally, are very demanding in terms of computer resources, and are not yet routinely used, especially for large organometallic systems. Rather, the effects of scalar relativity are most often incorporated into ab initio and DFT calculations through relativistic pseudopotentials (or relativistic effective core potentials (RECPs)) and associated valence basis sets. The RECPs replace the computationally demanding, yet chemically unimportant, core electrons by functions designed to mimic their effects on the valence electrons, which are treated explicitly (59). An alternative approach uses a relativistic Hamiltonian such as one based on the Douglas–Kroll transformation (60), which eliminates the coupling between the large and small components of the four component one-electron Dirac Hamiltonian to produce a more tractable two-component method. The zeroth order regular approximation (ZORA) to the Dirac equation is a popular alternative (61–63).

4.4 New Theoretical and Experimental Tools for Evaluating Covalency in the 5f Series

One of the most lively and interesting debates in 5f organometallic chemistry over the last decade has focused on the nature and extent of covalency. As many of the subsequent sections in this chapter discuss aspects of covalency, we here provide an overview and highlight experimental and computational techniques at the forefront of its evaluation. For a recent review, the reader is directed to Ref. (64).

4.4.1 The Quantum Theory of Atoms-in-Molecules

In 1991, Strittmatter and Bursten reported quasi-relativistic Xα-SW (an early form of DFT) calculations on $AnCp_3$ (An = U–Cf; Cp = η^5-C_5H_5) (65). They noted very large 5f contributions to what are expected to be mainly Cp-localized levels, as high as 55% in the Cf compound. A similar conclusion was reached by Ingram *et al.* (44) in a more modern DFT study of lanthanide and actinide imidodiphosphinochalcogenide complexes; the largest 5f contributions to the metal–ligand bonding orbitals were in the Am complexes. These results appear to contradict the traditional view of actinide chemistry (66), which holds that while the early members of the series share certain features with the d transition elements, for example, variable valence and a degree of covalent bonding, as the series is crossed there is a change to more lanthanide-like behavior, with increasing ionicity and a dominant trivalent oxidation state. The answer lies in the behavior of the 5f orbitals as the actinide series is crossed; they become energetically stabilized and radially more contracted. Thus, at a certain point (dependent on the metal and the supporting ligand set) the 5f orbitals become degenerate with the highest-lying ligand-based functions, yet are too contracted for there to be significant spatial overlap. As Strittmatter and Bursten noted, "... the domination of the 5f orbitals in the Cp-bonding orbitals of the later actinides is primarily due to a coincidental energy match of these 5f orbitals with the Cp ... orbitals." Put another way, perturbation theory holds that, to first order, the mixing of MOs ϕ_i and ϕ_j is governed by the mixing coefficient $t_{ij}^{(1)}$ (Eq. (4.1)), where

$$t_{ij}^{(1)} \propto \frac{-S_{ij}}{\varepsilon_i - \varepsilon_j} \tag{4.1}$$

Here, S_{ij} is the overlap between the orbitals, and the denominator is the difference in their energies. Thus, large orbital mixings can arise when ϕ_i and ϕ_j are close in energy, without there necessarily being significant orbital overlap. We must therefore be careful to make the distinction between *energy-driven* covalency (arising from the near degeneracy of metal and ligand orbitals) and the more traditional *overlap-driven* covalency; experimental and theoretical studies show that for actinide-ligand bonds, both types of covalency may be operative for different classes of complexes depending on the oxidation state and nature of the ligand (64).

Energy-driven covalency will not lead to a significant buildup of electron density in the internuclear region, and hence complexes which feature such interactions can, rather engagingly, be simultaneously both highly covalent (in the energy-driven sense) and highly ionic. This has been well illustrated by the recent introduction of the quantum theory of atoms-in-molecules (QTAIM) (67, 68) as a tool for analyzing 5f molecular electron densities. The first extensive use of this technique – which focuses on the properties of the electron density, rather than on orbital structure – targeted $AnCp_4$ (69) and $AnCp_3$ (70), and clearly demonstrated that although the largest 5f/Cp orbital mixings occur in systems toward the center of the actinide series, these compounds also feature the lowest bond critical point electron densities. These studies also showed that all of the An–C bonds are rather ionic, certainly on the basis of QTAIM definitions based on molecules much further up the periodic table, and that the least ionic bonds are found for the U systems. Figure 4.3 shows representations of the pseudo-t_1 valence DFT (PBE functional) MOs of UCp_4 and $AmCp_4$. The Am orbital has about twice the 5f contribution of the U, yet the amplitude is smaller in the internuclear region, leading to a smaller bond critical point electron density. The uranium–ligand bond was also found to be the least ionic in subsequent QTAIM studies of An(^{Ar}acnac)$_4$ (An = Th, U, Np, Pu; ^{Ar}acnac = Ar*N*C(Ph)CHC(Ph)*O*; Ar = 3,5-^{t}Bu$_2$C$_6$H$_3$) (41), and diselenophosphinate complexes of Th–Pu (34).

4.4.2 Ligand K-edge X-ray Absorption Spectroscopy

At about the same time as the 5f elements were being introduced to the QTAIM, a new experimental technique emerged for the study of covalency in the actinides: ligand K-edge X-ray absorption spectroscopy (XAS). This approach was developed by Solomon *et al.* for the study of transition metal inorganic and bioinorganic systems (71), and was

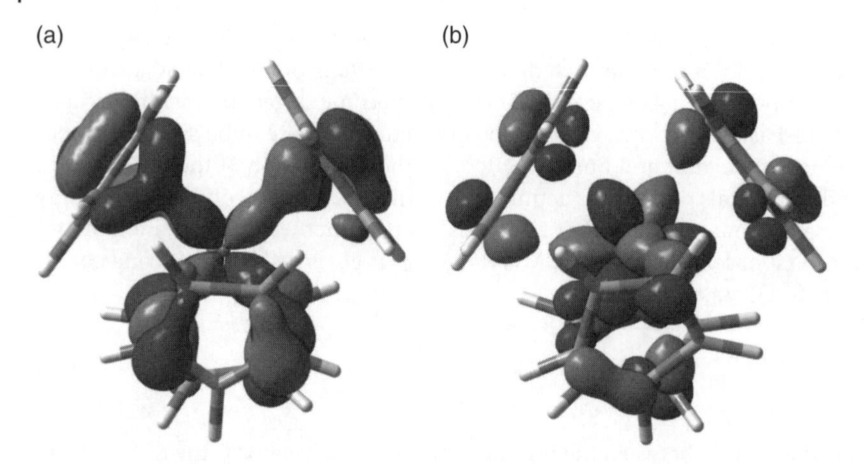

Figure 4.3 Three-dimensional representations of one component of the pseudo-t_1 valence molecular orbitals of (a) UCp_4, (b) $AmCp_4$. The 5f content (Mulliken analysis) of the orbitals are, respectively, 15.4% and 30.9%, and the bond critical point electron densities are 0.034 and 0.029 electron/$bohr^3$. Images and data from Tassell M.J., Kaltsoyannis N. (2010) Covalency in $AnCp_4$ (An = Th–Cm); a comparison of molecular orbital, natural population and atoms in molecules analyses. *Dalton Transactions*, **39** (29), 6719–6725. (http://pubs.rsc.org/en/Content/ArticleLanding/2010/DT/c000704h). Reproduced by permission of The Royal Society of Chemistry. (*See insert for color representation of the figure.*)

translated to the actinides by the Los Alamos team of Clark *et al.*. It focuses on bound-state, electric-dipole-allowed transitions on the low-energy side of the K-edge absorptions of ligating atoms (e.g., Cl, C), and promotes the 1s electrons into partially occupied or unoccupied metal-based MOs, which are usually metal-ligand antibonding. Given the largely atomic-like 1s-np character of the transitions, significant spectral intensity is seen only when there is a ligand np orbital character to the final state; that is, the intensity of the transition quantifies the ligand np content of the metal–ligand antibonding orbitals. The technique thus provides a direct experimental probe of metal–ligand orbital mixing, though it does not distinguish between energy-driven and overlap-driven covalency.

Initial targets for the Los Alamos team were the metallocene dichlorides Cp_2MCl_2 (M = Ti, Zr, Hf, Th, U) (72, 73). Chlorine K-edge XAS and time-dependent DFT calculations were used to determine the Cl 3p character (per M-Cl bond) of the largely metal d-localized M-Cl antibonding orbitals as 25%, 23%, and 22% respectively for M = Ti, Zr, and Hf. Although the spectral resolution was insufficient to obtain useful information for Cp_2ThCl_2, the Cl 3p mixing per U–Cl bond was found to be 9%, of which 4% was attributed to mixing with the 5f orbitals. A more recent application of the technique has been to study $An(COT)_2$ (An = Th, U; COT = η^8-C_8H_8) (74). Covalency in these systems has been widely studied, with unambiguous evidence for both 5f and 6d orbital participation in U–COT bonding coming many years ago from synchrotron radiation-based photoelectron spectroscopy (75). In keeping with earlier ab initio studies (76), the DFT calculations accompanying the C K-edge XAS showed that the C 2p mixing into the 5fδ-based MOs increased from 13% in $Th(COT)_2$ to 28% in $U(COT)_2$, as the 5f orbitals stabilize from Th to U. However, the intensity of the carbon K-edge pre-edge features was larger for $Th(COT)_2$ than $U(COT)_2$, which was interpreted as arising from strong ϕ-type mixing (e_{3u}) between C 2p and Th 5f. As the 5f orbitals stabilize, this metal–ligand interaction diminishes, such that the 5f orbitals are much more metal-localized in $U(COT)_2$. This study nicely highlights how metal–ligand orbital mixings can have rather subtle dependencies on the relative energies of the ligand and metal functions.

4.4.3 Optical Spectroscopy

Optical spectroscopy has proved to be a highly useful means of evaluating bonding and covalency within transition metal coordination complexes (77). However, the ligand field analysis enabled by optical spectroscopy is not as easily transferable to the actinide ions, primarily because most actinide electronic configurations feature a prohibitive number of observed transitions, on account of the larger number of accessible states (78). That said, there is one electronic configuration for which the number of observed optical transitions is manageable, namely, the $5f^1$ configuration in an O_h environment. In this regard, Lukens recently developed a "reduced spin-orbit" molecular orbital model to analyze the bonding in octahedral $5f^1$ complexes (79, 80), which, unlike previous models, takes into account the effect of covalency on the spin-orbit coupling constant. This model was applied to a wide range of octahedral $5f^1$ complexes, including $[PaCl_6]^{2-}$, $[UX_6]^-$ (X = F, Cl, Br, O^{t}Bu, NC_5H_{10}, N = CPh(^{t}Bu), and CH_2SiMe_3), and NpF_6 (Figure 4.4) (79, 80), and also the U(V/IV) bridged oxo complex, $[\{((^{nP,Me}ArO)_3tacn)U^{V/IV}\}_2(\mu\text{-}O)_2]^-$ (81). Using a combination of optical and EPR spectroscopic data, this model predicts modest levels of 5f orbital participation in the t_{1u}-symmetry An-L bonds. For example, $[PaCl_6]^{2-}$ is predicted to feature 6% 5f character in this orbital manifold. This value increases to 19% for $[UCl_6]^-$, consistent with the anticipated drop in 5f orbital energy on moving across the actinide series. Within the $[UX_6]^-$ (X = F, Cl, Br) series, the bromide analogue features the largest amount 5f orbital participation (24%), which is likely due to the lower electronegativity of Br. As a result, its frontier orbitals are a better energy match to those of uranium, which increases covalency via energy-driven overlap (82). Of the organometallic complexes studied, the ketimide complex, $[U(N{=}CPh(^tBu))_6]^-$, features the greatest 5f covalency (16%), which is consistent with our emerging understanding of An–ketimide interactions (see below for more discussion) (30, 83–86). Surprisingly, the U(V) alkyl complex, $[U(CH_2SiMe_3)_6]^-$ featured limited 5f participation within the t_{1u} manifold; however, the data reported for $[U(CH_2SiMe_3)_6]^-$ exhibited sizable errors, limiting the conclusions that could be drawn from them. In this regard, the analysis of other organometallics $[UR_6]^-$-type complexes would be highly useful for developing a fuller picture of the U-C bond within U^{5+} and U^{6+} complexes.

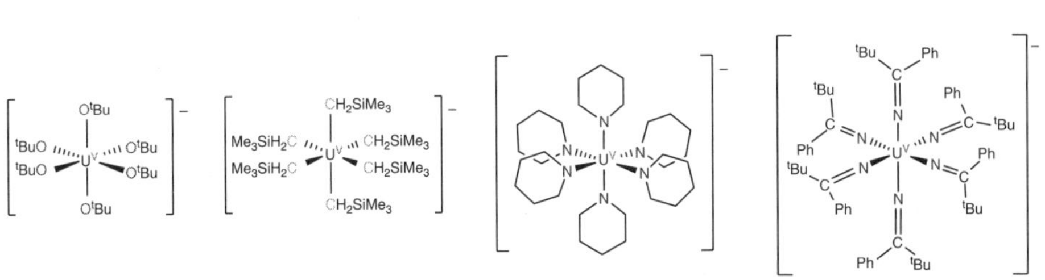

Figure 4.4 Octahedral $5f^1$ complexes analyzed by optical spectroscopy.

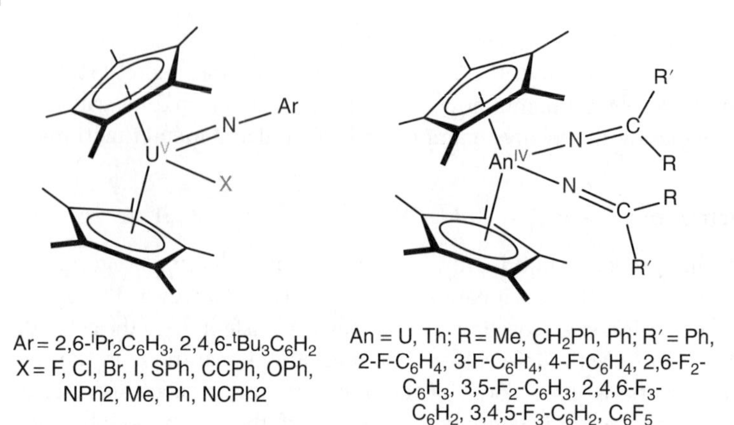

Figure 4.5 Metallocene imide and ketimide complexes characterized by optical spectroscopy.

Kiplinger and coworkers have used optical spectroscopy to evaluate metal–ligand bonding in a series of U(V) and U(IV) metallocene complexes. For example, the optical spectra of the $5f^1$ complexes, $Cp^*_2U(NAr)(X)$ (Ar = 2,6-$^{i}Pr_2C_6H_3$, 2,4,6-$^{t}Bu_3C_6H_2$, X = F, Cl, Br, I, SPh, CCPh, OPh, NPh_2, Me, Ph, $NCPh_2$; Figure 4.5) (30, 83–86), feature transitions in the same region as those measured for $[UX_6]^-$, allowing for a rough estimate of ζ (the spin-orbit coupling constant) to be made (about 2000 cm^{-1}). However, the transitions are much more intense, which may be due to both the lower symmetry of these complexes and also an "intensity stealing" process (87). Of note, similar magnitudes of ζ were reported for $[UX_6]^-$ (X = F, Cl, Br, O^tBu, NC_5H_{10}, N = CPh(^{t}Bu), CH_2SiMe_3) (80). Unfortunately, because of the low local symmetry about the U^{5+} centers in these complexes, a more extensive analysis could not be completed.

The An(IV) ketimido complexes, $Cp^*_2An(N{=}CR(R'))_2$ (An = Th, U; R = CH_3, Ph; R′ = Ph, 2-F-C_6H_4, 3-F-C_6H_4, 4-F-C_6H_4, 2,6-F_2-C_6H_3, 3,5-F_2-C_6H_3, 2,4,6-F_3-C_6H_2, 3,4,5-F_3-C_6H_2, C_6F_5; Figure 4.5), have also been thoroughly investigated by optical spectroscopy (88–93). These complexes exhibit rich UV-vis spectra, characterized by ligand-centered absorptions, LMCT absorptions, and f-f transitions (for the U analogues). This complex spectroscopic behavior is presumably enabled by the strongly σ- and π-donating nature of the ketimido ligand, which results in a relatively covalent An–N interaction. DFT calculations on $Cp^*_2An(N{=}CPh_2)_2$ support this qualitative bonding description. Specifically, a Mulliken population analysis reveals respectable amounts of 5f and 6d orbital participation within the An-L bonds. For example, in the U analogue, the SOMO-1 and SOMO-2 (predominantly the U-N π bonds) feature large contributions from the 5f orbitals (17.7% and 9.9%, respectively) (93). In contrast, for the Th analogue, the HOMO and HOMO-1 (predominantly the Th-N π bonds) feature 9.1% and 3.4% 5f character, respectively. This decrease in 5f contribution in the Th derivative fits with the increase in 5f orbital energy anticipated for Th. However, this decrease in 5f orbital contribution is not compensated for by an increase in 6d contribution, which is minimal for both the HOMO and HOMO-1, respectively (93), revealing that the Th complex features a more ionic An-L interaction than the U analogue.

4.4.4 Nuclear Magnetic Resonance (NMR) Spectroscopy

Although not as widespread as the use of optical spectroscopy, the past decade has seen a significant increase in the use of NMR spectroscopy to probe 5f orbital involvement in actinide-ligand bonding. This research area hinges on the close interplay of theory and experiment, and so is particularly suitable for discussion within this chapter. However, these studies typically require the use of diamagnetic target molecules. As a result, these analyses are restricted to a few select actinide ions, including Th^{4+} and U^{6+}, the latter being somewhat rare in organometallic chemistry. Moreover, the magnetically active nuclei to be probed should be directly bound to the actinide ion. Not surprisingly, it is often an experimental challenge to meet these criteria.

In 2012, Hrobárik and coworkers calculated the ^{1}H NMR chemical shifts for a series of Th^{4+} hydride complexes (94). Significantly, they demonstrated that the large downfield shifts observed for these hydride resonances were the result of large spin-orbit coupling, as a result of 5f orbital contribution to the Th-H bond. For example, the thorium hydride complex, $[Cp^{*}_2Th(H)_2]_2$ features a calculated hydride chemical shift of 17.1 ppm (expt: 19.2 ppm), which is greatly de-shielded relative to most transition metal hydrides. Notably, the spin-orbit contribution to this shift amounts to $\delta_{SO} = 11.8$ ppm. Interestingly, these authors also suggest that the hydride resonance in $[Th(H)(NR_2)_3]$ ($R = SiMe_3$) was mis-assigned in the original report (95), demonstrating the potential benefits of close collaboration between experimentalists and computational chemists in actinide organometallic characterization. Large downfield shifts were also calculated for ^{13}C nuclei directly bonded to the U^{6+} ion. For example, the uranyl carbene complex, $[UO_2(NHC^{C,N})_2]$ (Figure 4.6), was predicted to feature a ^{13}C NMR shift for the carbene carbon at 270.0 ppm, which is pleasingly close to the experimental value (262.8 ppm). More remarkably, the homoleptic U(VI) organometallic, $[U(CH_2SiMe_3)_6]$ was predicted to have a methylene resonance at 529.1 ppm (spin-orbit contribution: $\delta_{SO} = 347.9$ ppm) in its ^{13}C NMR spectrum. For comparison, the experimental spectrum has a resonance at 434.3 ppm, which is not as close as the other predictions; likely on account of the long core tails of the valence 5f orbitals (96). $[U(CH_2SiMe_3)_6]$ is also predicted to feature substantial 5f orbital contribution in its six U-C bonds, in addition to an octahedral geometry (more on this aspect below). In particular, the three t_{1u} symmetry U-C bonds are predicted to feature an amazing 34% 5f character, corroborating the results of an earlier study (31); however, the ability to benchmark these calculations against a measurable spectroscopic quantity makes the more recent calculations more compelling. While this amount of covalency is surprising, it is important to remember that $[U(CH_2SiMe_3)_6]$ should not be compared to other actinide coordination complexes. For one thing, it features relatively electropositive ligands (e.g., C, $\chi_{Pauling} = 2.55$; O,

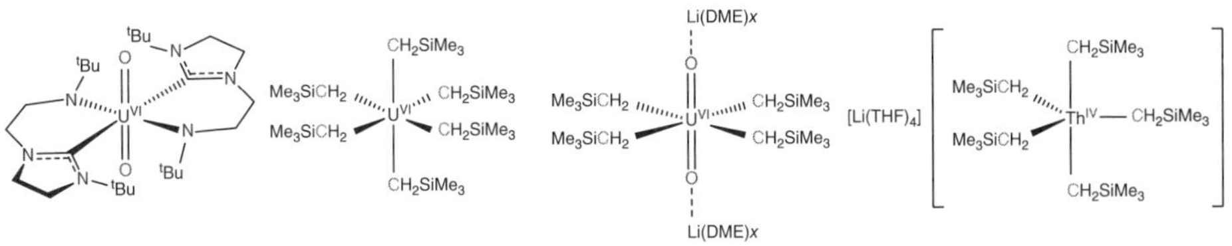

Figure 4.6 Actinide organometallic complexes interrogated by ^{13}C NMR spectroscopy and computational methods.

$\chi_{Pauling} = 3.44$), which decreases the ionicity of the M-L bond; and second, it features uranium in its highest oxidation state (higher-oxidation-state complexes typically feature greater covalency in their M-L bonds) (64). Nevertheless, this result confirms that the actinide ions, at least under certain conditions, can feature levels of covalency similar to those found for the transition metals. It should also be noted that these ^{13}C shifts are among the largest ever reported for diamagnetic molecules, further illustrating the effect that the 5f orbitals can have on spectroscopic properties.

The uranyl alkyl complex, $[Li(DME)_{1.5}]_2[UO_2(CH_2SiMe_3)_4]$ (Figure 4.6), was also investigated in a combined synthetic and computational study (96). Using the PBE0-40HF hybrid functional, this complex was predicted to have a methylene resonance at 250.8 ppm in its ^{13}C NMR spectrum, which is remarkably close to the experimental value (242.9 ppm). Unlike $[U(CH_2SiMe_3)_6]$, however, $[Li(DME)_{1.5}]_2[UO_2(CH_2SiMe_3)_4]$ is only predicted to feature a modest 5f orbital contribution in its four U-C bonds. For example, the two e_u symmetry U-C bonds are predicted to have just 11% 5f character. This reduced 5f contribution, relative to $[U(CH_2SiMe_3)_6]$, is probably not surprising, given that the two oxo ligands in $[UO_2(CH_2SiMe_3)_4]^{2-}$ are anticipated to dominate the metal–ligand bonding manifold, and weaken the U-C bonds as a result. This hypothesis is corroborated by the long U-C bond lengths in $[Li(DME)_{1.5}]_2[UO_2(CH_2SiMe_3)_4]$ (av. U-C = 2.49 Å) relative to those of other high-valent uranium organometallics, such as $[Li(THF)_4][U(CH_2SiMe_3)_6]$ (av. U-C = 2.43 Å) (31).

Finally, the homoleptic thorium alkyl, $[Li(THF)_4][Th(CH_2SiMe_3)_5]$ (97), was interrogated by a similar approach (Figure 4.6) (96). This thermally sensitive complex features a methylene resonance at 83.0 ppm in its ^{13}C NMR spectrum. The calculated value was found to be 79.3 ppm ($\delta_{SO} = 35.6$ ppm,) (96). In this example, the smaller spin-orbit contribution to the overall chemical shift, relative to $[U(CH_2SiMe_3)_6]$, can be rationalized by the lower oxidation state (which affects covalency) and also the small 5f orbital participation in the Th-C bonds (an NLMO analysis shows an average of 2.2% 5f orbital contribution to each Th-C bond). The latter effect is a consequence of the increasing energy of the 5f manifold as the actinide series is traversed from right to left (e.g., from U to Th) (98).

4.4.5 Electrochemistry

There has been a notable increase in the use of electrochemistry to probe the electronic structure of actinide organometallics over the past decade. This work was built on the pioneering studies of Burns and Sattelberger at Los Alamos National Laboratory. In 1990, they reported the U(V/VI) redox potentials for the U(V) imido complexes, U(NR)$(N\{SiMe_3\}_2)_3$ (R = $SiMe_3$, Ph) (Figure 4.7) (99). Both complexes feature U(V/VI) redox potentials at about −0.40 V (vs. Fc/Fc^+). These large negative values are notable because they demonstrate easy access to the highly charged U^{6+} ion, an observation that can be rationalized by the relatively covalent uranium–imido and uranium–amido interactions, which stabilize the 6+ charge of the uranium center.

More recently, Kiplinger and Morris have systematically explored the electrochemical properties of a large series of uranium ketimide complexes, $Cp^*_2U(N=CR(R'))_2$ (R = Ph, Me; R′ = Ph, 2-F-C_6H_4, 3-F-C_6H_4, 4-F-C_6H_4, 2,6-F_2-C_6H_3, 3,5-F_2-C_6H_3, 2,4,6-F_3-C_6H_2, 3,4,5-F_3-C_6H_2, and C_6F_5) (84, 89–91). The prototypal member of this series, $Cp^*_2U(N=CPh_2)_2$ (Figure 4.5), features a reversible U(IV/V) couple at −0.48 V (vs. Fc/Fc^+).

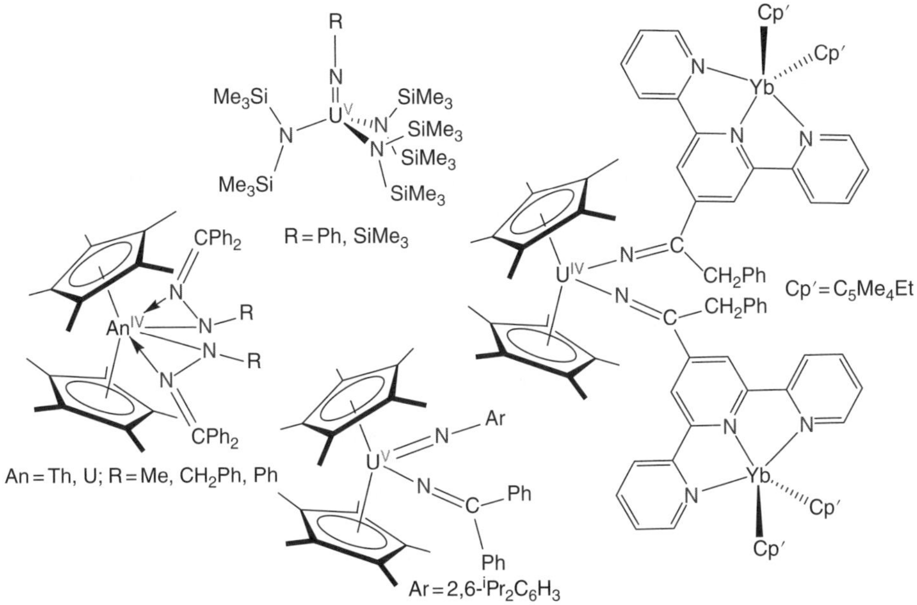

Figure 4.7 Imido, ketimide, and hydrazido complexes characterized by electrochemistry.

The other members of this series have similar values for their U(IV/V) couples. These relatively low values are seen as evidence of strong σ- and π-donation from the ketimide ligands, a hypothesis which has been corroborated by DFT calculations. Kiplinger and Morris also explored the redox chemistry of the U(V) imido complexes, $Cp^*_2U(NAr)(X)$ ($Ar = 2,6\text{-}^iPr_2C_6H_3$; X = OTf, I, Br, Cl, SPh, CCPh, F, Me, OPh, $N = CPh_2$) (Figure 4.7) (84, 87). Within this series, the U(V) ketimide complex, $Cp^*_2U(NAr)(N = CPh_2)$, featured the lowest U(V/VI) redox potential (i.e., it was easiest to oxidize to U(VI)). This result was explained by arguing that the ketimide ligand was the strongest overall donor (both σ and π), and thus was best at stabilizing the U^{6+} ion. These complexes were also characterized by optical spectroscopy (see Section 4.4.3 for more details). Further evidence for covalency within the actinide–ketimide interaction comes from the synthesis of multi-metallic mixed metal uranium-lanthanide complex, $Cp^*_2U(N = C(CH_2Ph)(tpy\text{-}Yb\{C_5Me_4Et\}_2))_2$ (Figure 4.7) (100). Specifically, electrochemical measurements reveal strong electronic coupling between the Yb ions, comparable in magnitude to that exhibited by the archetypal Creutz–Taube ion. This strong coupling is thought to be facilitated by good orbital overlap between U and the ketimide moiety.

A series of An(IV) bis(hydrozonato) complexes, $Cp^*_2An(\eta^2\text{-}(R)N\text{-}N = CPh_2)_2$ (An = Th, U; R = Me, CH_2Ph, Ph), were also interrogated by cyclic voltammetry (Figure 4.7) (101). For instance, $Cp^*_2U(\eta^2\text{-}(Me)N\text{-}N = CPh_2)_2$ features a reversible U(IV/V) redox couple at −0.68 V (vs. Fc/Fc^+), similar to those of the closely related bis(ketimide) complexes (90). These complexes are also noteworthy because they feature four bonded atoms within the metallocene wedge. This results in a unique set of symmetry considerations, which forces a 5f orbital to participate in the An-L bonds. Despite this requirement, DFT calculations reveal that the An-L bonds within the metallocene wedge are predominantly ionic and contain only modest amounts of 6d and 5f character.

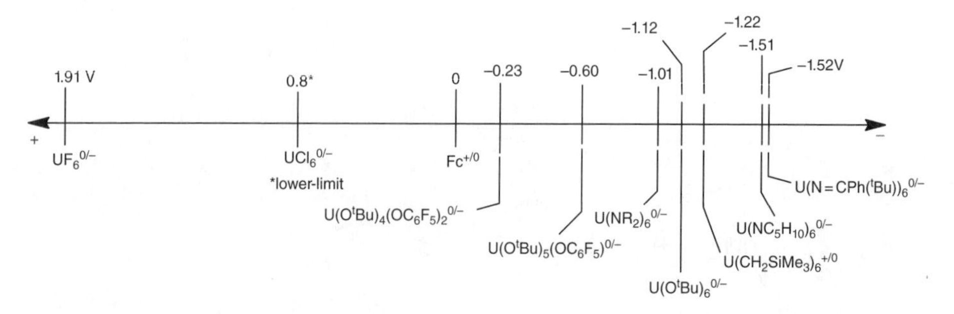

Figure 4.8 U(V/VI) reduction potentials (vs. Fc/Fc$^+$) for selected U(V) and U(VI) complexes. Figure adapted from Ref. (106). Reduction potential for UF_6 taken from Ref. (107). Estimated oxidation potential for $[UCl_6]^-$ taken from Refs. (108, 109). Reduction potential for $U(O^tBu)_6$ taken from Ref. (103). Reduction potentials for $U(O^tBu)_{6-n}(OC_6F_5)_n$ ($n = 1, 2$) taken from Ref. (102). Oxidation potentials for $U(NR_2)_6^-$ (HNR_2 = 2,3:5,6-dibenzo-7-azabicyclo[2.2.1]hepta-2,5-diene) taken from Ref. (105). Oxidation potential for $[U(N{=}CPh(^tBu))_6]^-$ taken from Ref. (79). Oxidation potential for $[U(NC_5H_{10})_6]^-$ taken from Ref. (104). Oxidation potential for $[U(CH_2SiMe_3)_6]^-$ taken from Ref. (31).

Several $[UX_6]^{0/-}$-type complexes (X = F, Cl, O^tBu, NC_5H_{10}, $N{=}CPh(^tBu)$, CH_2SiMe_3, NR_2; HNR_2 = 2,3:5,6-dibenzo-7-azabicyclo[2.2.1]hepta-2,5-diene) have also been studied by electrochemistry (Figure 4.8) (31, 79, 102–105). Within this series, $[U(NC_5H_{10})_6]^-$ and $[U(N{=}CPh(^tBu))_6]^-$ feature the most negative U(V/VI) couples, consistent with the strongly σ- and π-donating character of the amide and ketimide ligands, respectively, which is discussed at several points throughout this chapter. The other organometallic-type complexes in this series feature similarly low U(V/VI) potentials.

Many other high-valent uranium organometallics have also been interrogated by electrochemistry. For example, $[PPh_4][U(O)(NR)(NR_2)_3]$ (R = $SiMe_3$) has an U(VI/V) redox potential at −2.34 V vs. Fc/Fc$^+$ (110). This rather large reduction potential can be rationalized by the negative charge of the complex, as well as by the strong σ- and π-donation of the oxo and imido ligands to the U^{6+} center. Similarly, $[U(O)(Me)(NR_2)_3]$ (R = $SiMe_3$) exhibits a U(V/VI) redox potential at −0.68 V vs. Fc/Fc$^+$ (111). Not all high-valent uranium organometallics have such low U(V/VI) redox potentials, however. For example, $[U(F)_2(NR_2)_3]$ features a U(V/VI) redox potential of 0.98 V vs. Fc/Fc$^+$, a large and positive value consistent with its ionic fluoride ligands (112). The simple coordination complexes $[UX_6]^-$ (X = F, Cl) also feature large and positive U(V/VI) potentials (Figure 4.8), suggesting that the U-X bond within these complexes is weak. This interpretation is also supported by the "reduced spin-orbit" molecular orbital model developed for these materials (see Section 4.4.3 for more details) (80).

4.5 Notable Discoveries in Actinide-Carbon Chemistry

Over the last decade, several significant discoveries in actinide-carbon chemistry have been made. Within this category we have included the following topics: (1) the synthesis of the first authentic An(II) complexes for the early actinides; (2) an improved understanding of actinide–π-acceptor interactions; (3) an expansion of the chemistry of inverted arene sandwich complexes; (4) the synthesis and characterization of phosphorano-stabilized carbene complexes; and (5) the synthesis of homoleptic aryl and alkyl

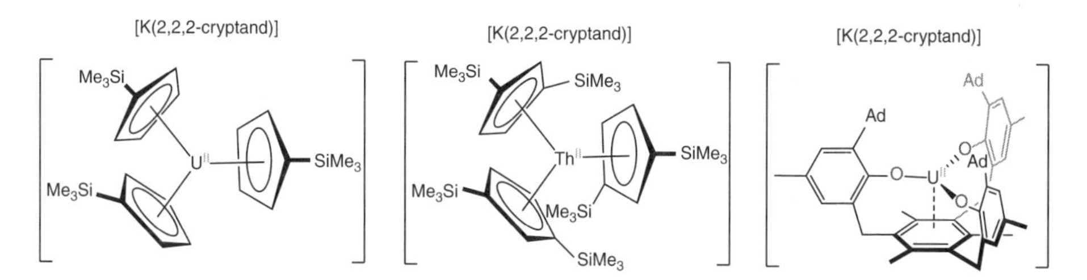

Figure 4.9 Low-valent actinide organometallics.

complexes. Each of these topics is connected by the novel use of an organometallic ligand to produce an unusual property (such as geometry or oxidation state) or the attachment of a novel organometallic ligand to an actinide ion (such as carbene or arene). These areas are discussed in further detail in the next few sections.

4.5.1 An(II) Complexes

In an exciting development, the first authentic U(II) and Th(II) complexes, [K(2,2,2-cryptand)][Cp′$_3$An] (An = U, Cp′ = $C_5H_4(SiMe_3)$; An = Th, Cp′ = C_5H_3-1,3-$(SiMe_3)_2$), were recently isolated by Evans and coworkers (Figure 4.9) (15, 32). DFT calculations suggest that the U(II) complex features a $5f^36d^1$ electronic configuration, whereas the Th(II) analogue has a diamagnetic $5f^06d^2$ electronic configuration (15). The diamagnetism was confirmed experimentally. In both examples, the d electrons occupy the $6d_{z2}$ orbital. In addition, both configurations differ from those determined for the gas phase An^{2+} ions (i.e., Th: $5f^16d^1$; U: $5f^4$). Meyer and coworkers have also successfully isolated a U(II) complex, namely, [K(2,2,2-cryptand)][((^{Ad,Me}ArO)$_3$mes)U] (Figure 4.9) (113). DFT calculations predict a $5f^4$ electronic configuration for this complex. Two of the partially filled 5f orbitals feature substantial δ-backbonding with the ligand's arene π* orbitals. No doubt, this interaction helps stabilize the normally inaccessible U^{2+} ion. The presence of the δ-backbonding interaction is also supported by X-ray crystallography, as [((^{Ad,Me}ArO)$_3$mes)U]$^-$ features a shorter U-C(centroid) distance than its neutral U(III) counterpart (2.18 vs. 2.35 Å). This complex is thermally unstable, and upon warming it undergoes a H-migration that results in partial hydrogenation of the arene ring and oxidation of the U(II) center to U^{4+} (114).

4.5.2 π-Acceptor Ligand Complexes

π-acceptor ligands, such as CO, phosphines, and alkenes, readily coordinate to the transition metals. Indeed, some of the first organometallic complexes ever isolated were transition metal carbonyl or η^2-alkene complexes (115). By contrast, actinide complexes containing these ligands are exceptionally rare. To our knowledge, no simple alkene complexes are known for the actinides (116), while even phosphine and alkyne complexes are quite rare (117–120). This, of course, is due to the contracted nature of the 5f orbitals, which renders them unable to interact with acceptor orbitals on the π-acidic ligand, resulting in a weak An-L bond. That said, there are a few instances where these strongly π-accepting ligands can bind to an actinide ion. For example, reaction of Cp′$_3$U (Cp′ = $C_5H_4SiMe_3$, C_5Me_4H, C_5Me_5) with CO results in formation of the adducts,

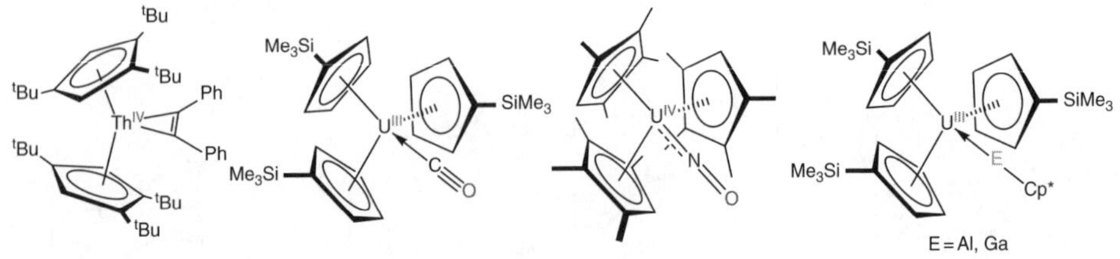

Figure 4.10 π-Acceptor ligand complexes of the actinides.

$Cp'_3U(CO)$ (Figure 4.10) (121–123). As noted by Andersen, the ν(CO) bands in these complexes display an unusual dependence on the identity of the ancillary ligand (124). For example, $(C_5H_4SiMe_3)U(CO)$ features a ν(CO) band at $1976\,cm^{-1}$ in hexane (123), whereas $(C_5Me_4H)_3U(CO)$ has a ν(CO) band at $1900\,cm^{-1}$ in petroleum ether (121). These observations suggest that the Dewar–Chatt–Duncanson model is not appropriate for describing the U-CO interaction in this series of complexes, as the CO stretch should not be so sensitive to the ligand identity. Instead, DFT calculations suggest that the π-backbonding interaction to CO involves donation from filled Cp′-based orbitals of π-symmetry into the CO π^* orbitals. Carbon monoxide also reacts with $[((^{tBu}ArO)_3tacn)U]$ to form a bridged carbonyl complex, $[((^{tBu}ArO)_3tacn)U]_2(\mu,\eta^1{:}\eta^1\text{-}CO)$ (125). In this example, the CO ligand is better described as $[CO]^-$. Accordingly, the strength of the An-CO bond is mediated more by electrostatics, and a π-backbonding interaction need not be invoked to rationalize the stability of this complex. Similarly, carbon monoxide reacts with $[U(\eta^8\text{-}1,4\text{-}C_8H_6(Si^iPr_3)_2)(Cp^*)(THF)]$ to form a series of bridged oxocarbon complexes, including those containing the ynediolate, deltate, and squarate dianions (126–128). The first step in the CO oligomerization process is the formation of a U(III) carbonyl complex, $[U(\eta^8\text{-}1,4\text{-}C_8H_6(Si^iPr_3)_2)(Cp^*)(CO)]$, which has not been isolated but has been observed with IR spectroscopy (126). On the basis of its ν(CO) band ($1920\,cm^{-1}$) this complex is probably electronically similar to the other known U(III) carbonyl complexes.

Several closely related actinide dinitrogen complexes are also known. For example, Cp^*_3U is known to reversibly coordinate N_2, forming $Cp^*_3U(N_2)$ (129). The U-N_2 bond is quite weak, as evidenced by the minimal N_2 activation upon coordination ($\nu(NN) = 2207\,cm^{-1}$; free N_2: $\nu(NN) = 2331\,cm^{-1}$). As a result, this complex is stable only under a pressurized N_2 atmosphere (80 psi). Although calculations have not been performed on this complex, it is likely that the U-N_2 interaction features very little π-backbonding. This is in marked contrast to the family of molecules in which N_2 bridges two uranium centers, which DFT has shown to feature significant metal $\rightarrow N_2$ backdonation (130–133). This leads to substantial lengthening and weakening of the N_2 bond, though X-ray crystallography and theory differ as to the extent of this effect, the latter suggesting greater N–N lengthening than seen experimentally. In 2011, Mansell *et al.* were able to record, for the first time, the Raman spectrum of one of these systems, $[U(OAr)_3]_2(\mu^2\text{-}\eta^2{:}\eta^2\text{-}N_2)$ ($Ar = 2,4,6\text{-}^tBu_3C_6H_2$) (133). Agreement between theory and experiment for ν(NN) was very good ($1486\,cm^{-1}$ vs. $1451\,cm^{-1}$), and it was concluded that the dinitrogen stretching frequency is a better measure of dinitrogen reduction than is the N–N bond length, which X-ray crystallography struggles to determine correctly. More recent computational work has extended the study of these systems to include thorium, protactinium, neptunium, and plutonium analogues, concluding that protactinium activates the N_2 ligand the most in this structural motif, as a result of the key metal → ligand backbonding orbitals being almost exclusively localized on the $Pa–N_2–Pa$ core of the molecules (134).

Although an η^2-alkene complex of the actinides has yet to be reported, Zi and coworkers recently described the first actinide η^2-alkyne complex, $Cp'_2Th(\eta^2\text{-}C_2Ph_2)$ ($Cp' = 1,2,4\text{-}C_5H_2{}^tBu_3$) (Figure 4.10) (135). The C-C distance of the alkyne ligand (1.343(4) Å) is indicative of a double bond, and suggests that this complex is best described as a Th(IV) metallacyclopropene complex. DFT calculations confirm this hypothesis and also reveal that the Th-C bonds are highly ionic (11% total Th character), with the 6d orbitals making the largest contribution from Th. This complex features greater thermal stability than previously observed actinide η^2-alkyne complexes (120), yet still undergoes insertion chemistry with a variety of unsaturated substrates, including ketones, nitriles, and carbodiimides.

Recently, Evans and coworkers synthesized the first f-element nitrosyl complex, $(C_5Me_4H)_3U(NO)$ (Figure 4.10) (136). Although not a carbon-based ligand in the strictest sense, nitric oxide is a strong π-acceptor ligand, like CO, and so is appropriate for inclusion in this discussion. This remarkable complex was synthesized by direct reaction of $(C_5Me_4H)_3U$ with 1 equiv of NO gas. The ν(NO) band for $(C_5Me_4H)_3U(NO)$ appears at about $1440\,cm^{-1}$ but is partially obscured by ligand vibrations. This low NO stretch is consistent with the presence of a nitroxyl ligand, that is, $[NO]^-$. However, the U-N bond is very short (2.013(4) Å), which is paradoxically suggestive of the presence of a U = N double bond. In addition, SQUID magnetometry measurements reveal that $(C_5Me_4H)_3U(NO)$ is only weakly paramagnetic, which is consistent with the presence of temperature-independent paramagnetism and an overall singlet ground state. DFT calculations, using the TPSS functional, suggest that the single and triplet grounds states are only about 2 kcal/mol apart in energy. These correspond to the oxy-imido zwitterion form, for example, $(C_5Me_4H)_3U(\equiv N^+\text{-}O^-)$, and the U^{4+}/NO^- form, for example, $(C_5Me_4H)_3U\text{-}N = O$, respectively. More recent CASSCF calculations by Autschbach and coworkers are broadly consistent with this bonding picture (137), but suggest that a multi-configurational description is more accurate, wherein the antibonding orbitals formed by the U5f-NOπ^* bonding interactions are partially populated. Interestingly, this interaction results in a decrease of the N-O bond order to 1.3, which is close to that predicted for the oxy-imido electronic structure description.

The synthesis of $(C_5H_4SiMe_3)_3U(ECp^*)$ (E = Al, Ga), by Arnold and coworkers, is also relevant to this discussion (Figure 4.10) (119, 138, 139). Cp*E is isoelectronic with CO and, in principle, can act as a π-acceptor ligand, but spectroscopic characterization suggests that very little π-backdonation is occurring in these complexes. For example, U and Al XANES data collected for $(C_5H_4SiMe_3)_3U(AlCp^*)$ support the presence of U(III) and Al(I), consistent with the presence of a dative interaction between U and Al. Moreover, DFT calculations on $(C_5H_4SiMe_3)_3U(ECp^*)$ suggest the U-E interaction features only σ-donation, while the σ acceptor orbital on uranium has predominant d character with smaller amounts of s and f character (e.g., for E = Ga: 67% d, 15% f, 18% s). Finally, a competition experiment between $(C_5H_4SiMe_3)_3U$ and $(C_5H_4SiMe_3)_3Nd$ with Cp*Ga provides a separation factor of $S_{U/Nd} = 14$ at room temperature. The authors argue that this selectivity is due to the rather diffuse nature of the Cp*Ga σ-orbital, which enables its interaction with the relatively compact 5f orbitals, strengthening the

U-Ga bond. This interaction, of course, is not possible with the 4f orbitals, which are too contracted to interact with even the diffuse Cp*E σ-orbitals. $(C_5H_4SiMe_3)_3U(ECp^*)$ are also of interest in the context of direct actinide-metal bond chemistry, a topic that has been recently reviewed and will not be discussed here (140).

4.5.3 (Inverted) Arene Sandwich Complexes

Over the past decade, several groups have reported the isolation of inverted arene sandwich complexes for the actinides (Figure 4.11) (141–147). Typically, these complexes consist of two U(III) centers bridged by a dianionic $[arene]^{2-}$ moiety, although there are a few exceptions (142). The most notable of these are $(\mu\text{-}\eta^6{:}\eta^6\text{-toluene})[U(OSi(O^tBu)_3)_3]_2$ and $(\mu\text{-}\eta^6{:}\eta^6\text{-toluene})[U(HC(SiMe_2(NAr))_3]_2$ ($Ar = 3,5\text{-}Me_2C_6H_3$) (Figure 4.11), which are both proposed to feature U(V) centers bridged by a tetra-anionic $[arene]^{4-}$ moiety (145, 146, 148). This oxidation state assignment is supported by calculations and structural studies. Of particular interest is the $[arene]^{4-}$ moiety, which is an aromatic 10π

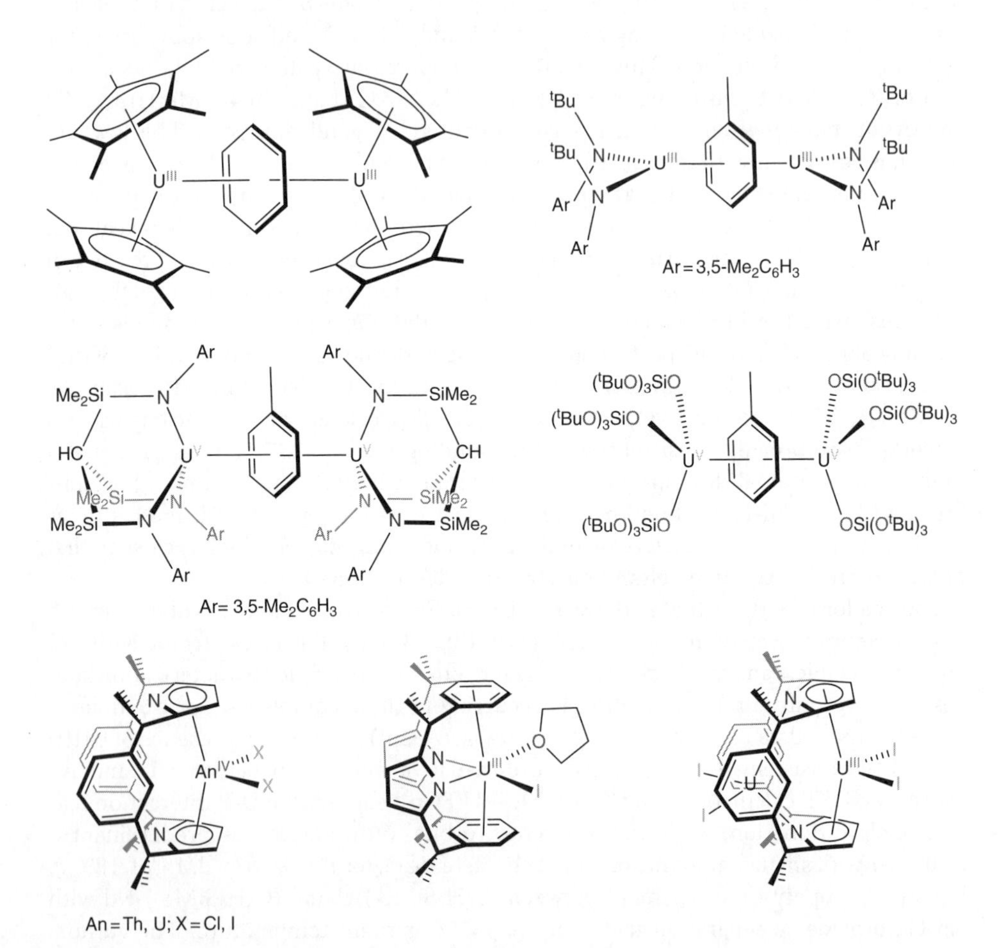

Figure 4.11 Inverted arene sandwich complexes of the actinides, and systems featuring the macrocyclic *trans*-calix[2]benzene[2]pyrrolide ligand.

electron system. Also of note is the proximity of a high-valent U^{5+} center next to the (presumably) highly reducing $[arene]^{4-}$ anion; however, it is likely that aromaticity of the $[arene]^{4-}$ fragment provides some of the driving force to maintain this large charge difference (147).

Determining the electronic structure in these complexes is often a challenge, as the C-C bond lengths within the reduced arene moiety are not necessarily a reliable indicator of charge (148). The first attempt to address the oxidation state ambiguity in this class of complexes with XANES was reported by Diaconescu, Cummins, and coworkers. They performed a thorough electronic structure investigation of the inverted sandwich complex, $(\mu\text{-}\eta^6\text{:}\eta^6\text{-toluene})[U(N[^tBu]Ar)_3]_2$ ($Ar = 3,5\text{-}Me_2C_6H_3$) (Figure 4.11) (144). This complex features a L_3 absorption edge similar to those of other U(III) complexes, according to XANES measurements. Moreover, CASSCF/CASPT2 calculations reveal two δ backbonding interactions between the uranium 5f orbitals and the toluene π^* orbitals. It has been argued that these δ backbonding interactions, which are possible only for metals with valence f orbitals, explain the prevalence of inverted arene sandwich complexes in uranium chemistry (144).

Variable, high-pressure single-crystal X-ray crystallography has very recently been used to study the inverted arene sandwich $[U(NR_2)_2]_2(\mu\text{-}\eta^6\text{:}\eta^6\text{-}C_6H_6)$ ($R = SiMe_3$) (149). Pressurization was found to shorten both the U-N and U-arene bonds, and significantly allowed the formation of extremely close U–CH interactions, but the most significant structural changes were observed in the close contacts between different ligand C-H bonds and the U centers. Although distances between the U and ligand peripheral C-H groups were suggestive of agostic interactions, QTAIM computational analysis of the hybrid DFT electron density revealed no such interaction at ambient pressure. However, at the highest pressure studied (3.2 GPa), pressure-induced agostic interactions to the uranium centers were confirmed by QTAIM and natural bond order (NBO) calculations.

Compared with the d-block, and indeed the inverted arene systems discussed above, the study of regular arene complexes of the actinides remains largely unexplored. In progress toward further understanding this motif, Arnold *et al.* (150) recently generated conformationally restricted Th^{IV} and U^{IV} complexes, $[ThCl_2(L)]$ and $[UI_2(L)]$, of the small-cavity, dipyrrolide, dianionic macrocycle *trans*-calix[2]benzene[2]pyrrolide $(L)^{2-}$, which feature unusual $\kappa^5\text{:}\kappa^5$ binding in a bent metallocene-type structure (Figure 4.11). Single-electron reduction of $[UI_2(L)]$ yields [UI(THF)(L)] and a switch in ligand binding from κ^5-pyrrolide to η^6-arene sandwich coordination (Figure 4.11), demonstrating the preference for arene binding by the electron-rich U^{III} ion. The solvent can be removed under vacuum to produce [UI(L)], which can incorporate a further U^{III} equivalent, UI_3, to form the dinuclear complex $[U_2I_4(L)]$ in which the single macrocycle adopts both $\kappa^5\text{:}\kappa^5$ and $\eta^6\text{:}\kappa^1\text{:}\eta^6\text{:}\kappa^1$ binding modes in the same complex (Figure 4.11). Hybrid DFT (PBE0 functional) calculations found increased contributions to the covalent bonding in $[U_2I_4(L)]$ versus [UI(L)], and similar U–arene interactions in both. Analysis of the Kohn–Sham MOs, and QTAIM calculations, revealed minimal U-U interaction in $[U_2I_4(L)]$.

4.5.4 Phosphorano-Stabilized Carbene Complexes

The 1990s saw major developments in the field of organometallic *N*-heterocyclic carbene chemistry, and this was extended to the 5f series shortly after the turn of the century, with notable contributions, just prior to our time window, coming from the

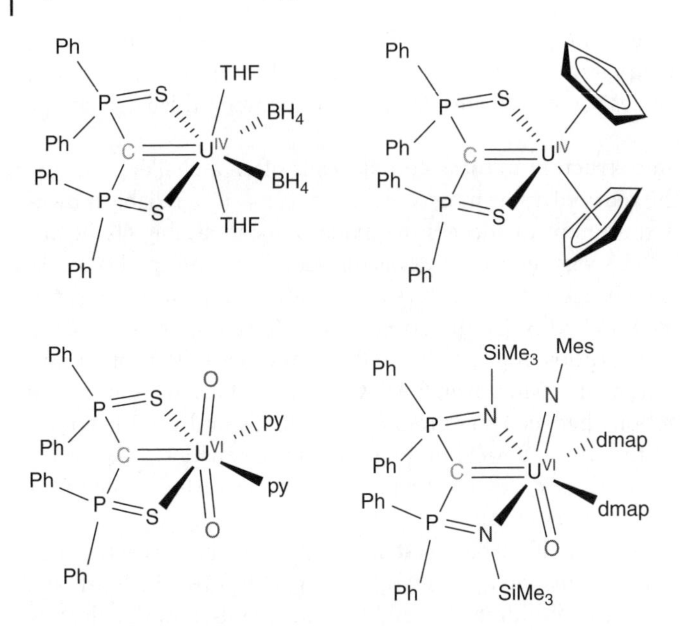

Figure 4.12 Bis-phosphorano stabilized actinide carbene complexes.

groups of Meyer (e.g., $[U(C(NMeCMe)_2)\{N(SiMe_3)_2\}_3]$ and $[U\{tacn(CH_2C_6H_2\text{-}2\text{-}O\text{-}3\text{-}Ad\text{-}5\text{-}Bu^t)_3\}\{C(NMeCMe)_2\}]$)(151) and Arnold (e.g., $[U\{OCH_2CMe_2C(NCHCHNPr^i)\}_4]$) (152). More recently, several U(IV) phosphorano-stabilized carbene complexes have been synthesized by Ephritikhine *et al.*, of the form $[U(SCS)L_2S_n]$ ($L = BH_4$, S = THF, $n = 2$; L = Cp, $n = 0$; Figure 4.12) (153, 154), and the related U(VI) complex $[UO_2(SCS)(py)_2]$ has also been reported (155). The strong bonding in the $[O=U=O]^{2+}$ moiety in the latter leads to a very long U=C bond, 2.430(6) Å, and the dominance of the *trans* $[MesN=U=O]^{2+}$ unit in the related carbene imido oxo system $[U(BIPM^{TMS})(NMes)(O)(dmap)_2]$(156) ($BIPM^{TMS} = C(PPh_2NSiMe_3)_2$, dmap = 4-*N,N'*-dimethylaminopyridine), Figure 4.12) was noted by Liddle *et al.* to also produce long and weak U=C bonding (r(U=C) = 2.383 Å). DFT bond orders are 1.23, 2.34, and 2.68 for the U=C, U=N and U=O bond, respectively, at the Nalewajski-Mrozek/BP86 level, and the U=C QTAIM ellipticity (0.21) is similar to that of the C–C bonds in benzene.

The $BIPM^{TMS}$ ligand was also employed by Liddle *et al.* in the synthesis of the U(IV) carbene complex $[U(BIPM^{TMS})(Cl)_3Li(THF)_2]$ (157). Oxidation by I_2 yields the U(V) complex, $[U(BIPM^{TMS})(Cl)_2(I)]$ (157), and oxidation by 4-morpholine *N*-oxide produces the U(VI) complex, $[U(BIPM^{TMS})(Cl)_2(O)]$ (158), which contains a *trans* $[R_2C=U=O]^{2+}$ unit analogous to uranyl (Figure 4.13). The increase in oxidation state is accompanied by a pleasing reduction in r(U=C) from 2.310(4) to 2.268(10) and 2.183(3) Å, and NBO analyses of the DFT (BP86) electronic structures reveal polarized covalent U=C bonds, with the metal contributions to the σ and π components dominated by the 5f orbitals.

The U(IV) carbene complex $[U\{N(SiMe_3)_2\}_3(CHPPh_3)]$ was prepared by Hayton *et al.* in 2011 (Figure 4.14) (159). The carbene ligand in this complex is notable for only possessing one stabilizing phosphorano moiety. As a result, at 2.278(8) Å, its r(U=C) is short, only 0.01 Å longer than in the U(V) complex, $[U(BIPM^{TMS})(Cl)_2(I)]$, discussed above (157). NBO analysis of the DFT (B3LYP) electronic structure again reveals metal–carbon

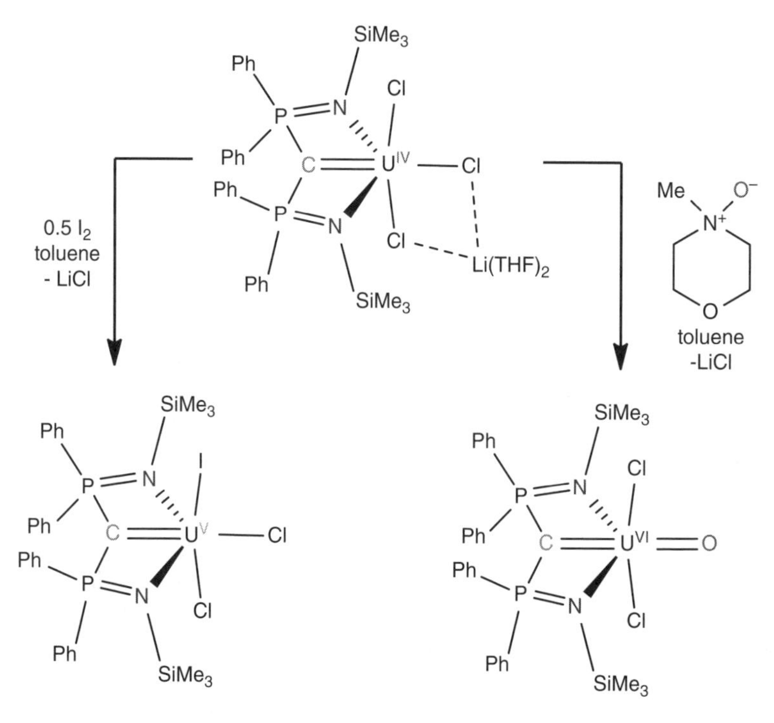

Figure 4.13 High-valent actinide carbene complexes.

bonding strongly polarized toward the carbon but, by contrast to the Liddle systems, there is a significant 6d component to the U contribution, dominantly so in the π bond.

4.5.5 Homoleptic Alkyl and Aryl Complexes

The past decade has seen considerable interest in the synthesis of homoleptic actinide organometallics (2, 97, 160). This work has been motivated by a variety of considerations. For one, these complexes are anticipated to be good starting materials (160), as ligand exchange via protonation should allow for facile entry into a wide variety of structure types (161, 162). Two, their high symmetry simplifies their spectroscopic characterization (see Section 4.4.3 for more details). Three, $[AnR_6]^n$-type complexes (where $n \leq 2$, R = alkyl, aryl) are predicted to feature a trigonal prismatic geometry, and not the more common octahedral geometry. This geometry change occurs because it permits greater d-orbital participation in the M-C bonds, thereby strengthening the M-C interaction (163, 164). As a result, the structural characterization of these complexes allows for an estimation of the amount of 6d participation in the An-C bond.

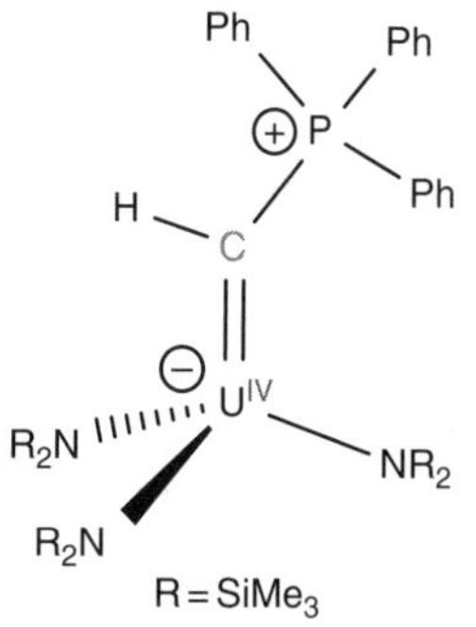

Figure 4.14 A mono-phosphorano stabilized carbene complex. (*See insert for color representation of the figure.*)

Thus far, only five structurally characterized actinide alkyls meet the $[AnR_6]^n$ criterion, including $[K(THF)_x]_2[An(CH_2Ph)_6]$ (An = Th, x = 1; An = U, x = 1.25), $[Li(TMEDA)]_2[UMe_6]$,

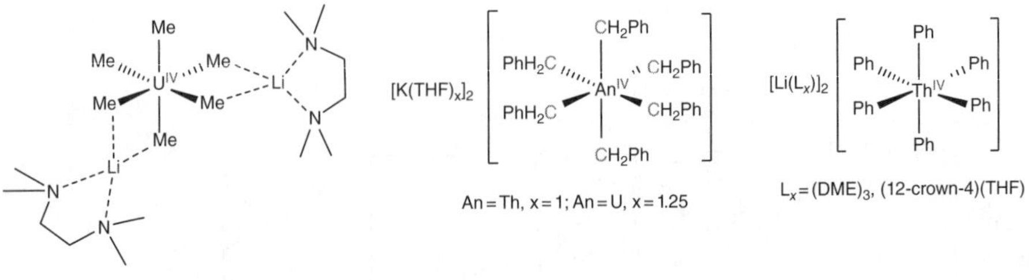

Figure 4.15 Octahedral homoleptic actinide organometallics.

$[Li(THF)_4][U(CH_2SiMe_3)_6]$, and $[ThPh_6]^{2-}$ (Figure 4.15) (31, 97, 165, 166). The uranium complexes $[K(THF)_{1.25}]_2[U(CH_2Ph)_6]$, $[Li(TMEDA)]_2[UMe_6]$, and $[Li(THF)_4]$ $[U(CH_2SiMe_3)_6]$ all feature octahedral geometries (31, 165). This observation can be rationalized by invoking enough 5f orbital participation in their U-C bonds to enforce C-An-C angles of 90° and 180°. The situation is more complicated for $[ThPh_6]^{2-}$. This complex has been structurally characterized as both a $[Li(DME)_3]^+$ salt and a [Li(12-crown-4) $(THF)]^+$ salt. When crystallized as a $[Li(DME)_3]^+$ salt, it features an octahedral geometry. However, when crystallized as a [Li(12-crown-4)(THF)]$^+$ salt, it adopts a severely distorted octahedral geometry. In addition, this form of $[ThPh_6]^{2-}$ has four Th···H_{ortho} anagostic interactions in the solid state. Calculations suggest that the octahedral and non-octahedral structures are close in energy (within $18\,kJ \cdot mol^{-1}$), with the non-octahedral structure being the ground state. While it is tempting to ascribe the non-octahedral ground state in $[Li(12\text{-crown-}4)(THF)]_2[ThPh_6]$ to 6d orbital involvement in the Th-C bonds, the presence of the Th···H_{ortho} anagostic interactions, which are primarily driven by C-H → 5f interactions, complicates this picture. Finally, it is interesting to note that $[ThMe_6]^{2-}$ is also predicted to feature a non-octahedral D_{3h} geometry, which is calculated to be 79 kJ/mol lower in energy than the octahedral geometry (97), a much larger difference than that calculated for $[ThPh_6]^{2-}$. Despite several attempts at its synthesis, however, the isolation of $[ThMe_6]^{2-}$ has remained elusive (167). $[ThH_6]^{2-}$ is also predicted to feature a trigonal prismatic geometry in the ground state (but only 12.6 kJ/mol lower in energy than O_h) (168).

Finally, the U(VI) homoleptic alkyl complex, $[U(CH_2SiMe_3)_6]$ (Figure 4.6), is worthy of further comment. This complex is the only known actinide analogue of the groundbreaking transition metal organometallic, WMe_6, which was first reported by Wilkinson and coworkers in 1973 (169). WMe_6 is notable for its trigonal prismatic geometry (163, 170, 171). While $[U(CH_2SiMe_3)_6]$ has not yet been structurally characterized, it is predicted to feature an octahedral geometry, unlike that exhibited by WMe_6. Similarly, UH_6 is predicted to be octahedral (168). This geometry change is thought to be the result of substantial 5f orbital participation in the U-C bond (see Section 4.4.4 for further discussion), as the t_{1u} set of 5f orbitals, which has the same symmetry as the p orbital manifold, is able to enforce the O_h geometry.

4.6 Single and Multiple Bonding between Uranium and Group 15 Elements

Terminal uranium nitrides were highly sought-after synthetic targets for many years. Despite much elegant chemistry, notably from the groups of Evans (172–174), Cummins (175, 176), Hayton (177), Kiplinger(178), and Mazzanti (179, 180), they eluded isolation

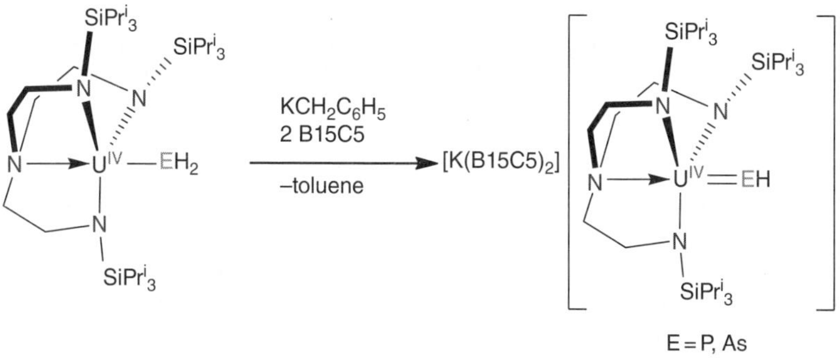

Figure 4.16 Preparation of $[TREN^{TIPS}UEH]^-$ from $TREN^{TIPS}UEH_2$ (E = P, As).

until the recent work of Liddle and coworkers, who developed syntheses based on the $TREN^{TIPS}$ ligand ($TREN^{TIPS} = N(CH_2CH_2NSiPr^i_3)_3$), which they employed to generate single and multiple bonds between U and the group 15 elements N, P, and As. Thus, reaction of alkali metal salts of $[EH_2]^-$ with either $TREN^{TIPS}UCl$ (E = N) (181) or $[TREN^{TIPS}U(THF)][BPh_4]$ (E = P, As) (182, 183) affords U(IV) $TREN^{TIPS}UEH_2$ in high yield. Subsequent reaction with $KCH_2C_6H_5$ in the presence of a crown ether affords $[TREN^{TIPS}UEH]^-$ as U(IV) anions with $K(L)_2^+$ counterions (L = 15-crown-5 for E = N and benzo-15-crown-5 (B15C5) for E = P, As) (Figure 4.16).

The bonding in these systems has been studied by DFT in conjunction with the NBO and QTAIM approaches; key data are summarized in Table 4.1. Agreement between theory and experiment for the U–E bond lengths is satisfactory; in general, the DFT data are longer than experiment, which is typical of generalized gradient approximation (GGA) functionals in this part of the periodic table. The Mayer bond orders (MBOs) are consistent with a single bond formalism for the EH_2 systems, and increase significantly in the EH molecules. For E = P, the QTAIM data also support the single versus double bond description; in particular, the increase in the ellipticity from $TREN^{TIPS}UPH_2$ to $[TREN^{TIPS}UPH]^-$ shows the expected significant deviation from cylindrical symmetry in the latter. NBO finds that the σ bond becomes much less P-localized on going from $TREN^{TIPS}UPH_2$ to $[TREN^{TIPS}UPH]^-$, and also that the relative roles of the 5f and 6d orbitals alter such that there is a much greater 5f bias in $[TREN^{TIPS}UPH]^-$, also seen in the π component. The NBO data for $TREN^{TIPS}UAsH_2$ are similar to the P analogue, but for $[TREN^{TIPS}UAsH]^-$ suggest that both the σ and π components are much more pnictogen-localized, although the greater role of the 5f versus the 6d orbitals is still evident. Despite the low MBO, the NBO and QTAIM data for $TREN^{TIPS}UNH_2$ are not what might be expected of a U–N single bond. The NBO σ component is returned as an N lone pair, and a highly N-localized π bond is found. ε is typical of a double bond. By contrast, that for $[TREN^{TIPS}UNH]^-$ is indicative of cylindrical symmetry, and NBO analysis yields σ + 2π orbitals. As with E = P and As, U 5f dominates over 6d in the π bonding.

Liddle *et al.* also showed that reaction of NaN_3 with $TREN^{TIPS}U$ gives the bimetallic U(V) nitride $[\{TREN^{TIPS}U\}(NNa)\}_2]$, from which the Na cations may be abstracted with 12-crown-4 (12C4) to yield a terminal uranium nitride $[TREN^{TIPS}UN][Na(12C4)_2]$; see Figure 4.17 (24). The pentavalent nature of this species was evidenced by UV/Vis/nIR spectroscopy and SQUID magnetometry, and computational analysis yielded the data

Table 4.1 Experimental and calculated U–E distances, and calculated Mayer Bond Orders, Natural Bond Orbitals and QTAIM bond critical point metrics for $TREN^{TIPS}UEH_2$, $[TREN^{TIPS}UEH]^-$ (E = N, P, As), $[TREN^{TIPS}UN]^-$, $TREN^{TIPS}UN$ and $[\{U(TREN^{TIPS})(AsK_2)\}_4]$. Data from Refs. (24, 181–183). The NBO π data for $[TREN^{TIPS}UNH]^-$, $[TREN^{TIPS}UN]^-$, $TREN^{TIPS}UN$ and $[\{U(TREN^{TIPS})(AsK_2)\}_4]$ are the average of two π orbitals.

	r(U–E)/Å		MBO	NBO σ			NBO π			QTAIM		
	exp	calc		%E	%U	5f:6d	%E	%U	5f:6d	ρ	H	ε
$TREN^{TIPS}UNH_2$	2.228	2.244	0.64	100			89.6	10.4	70:30	0.10	−0.03	0.38
$TREN^{TIPS}UPH_2$	2.883	2.925	0.84	87.4	12.6	45:51				0.05	−0.01	0.01
$TREN^{TIPS}UAsH_2$	3.004	3.060	0.69	86.7	11.3	48:47						
$[TREN^{TIPS}UNH]^-$	2.034	2.002	1.75	90.3	9.7	36:59	82.9	17.2	63:37	0.16	−0.08	0.04
$[TREN^{TIPS}UPH]^-$	2.613	2.621	1.92	75.9	24.1	80:20	72.1	27.9	69:30	0.08	−0.03	0.20
$[TREN^{TIPS}UAsH]^-$	2.716	2.754	1.62	88.8	11.2	61:37	87.3	12.7	66:33			
$[TREN^{TIPS}UN]^-$	1.825	1.810	2.91	68	32	44:47	73	27	72:28			
$TREN^{TIPS}UN$	1.799	1.780	2.92	59	41	89:9	70	30	81:19	0.39	−0.30	
$[\{U(TREN^{TIPS})(AsK_2)\}_4]$	2.731		1.75	82.9	17.1	57:40	82.3	17.7	62:37			
	2.743		1.48	84.3	15.7	73:24	86.7	13.9	59:41			
	2.757		1.40	85.6	14.4	70:27	86.2	13.3	64:36			
	2.774		1.38	86.9	13.1	70:26	87.4	12.6	60:39			

collected in Table 4.1. The significantly shorter U–N distance versus $TREN^{TIPS}UNH_2$ and $[TREN^{TIPS}UNH]^-$ is accompanied by a much higher MBO, and the NBO analysis indicates a σ + 2π system. As with the singly and doubly bonded molecules, there is greater 6d contribution than 5f to the σ bond, but 5f dominates the metal character of the π interactions. Oxidation of $[TREN^{TIPS}UN][Na(12C4)_2]$ with I_2 generates the neutral, diamagnetic U(VI) terminal nitride $[TREN^{TIPS}UN]$ (184). The U–N bond length (Table 4.1) is not statistically shorter than that in $[TREN^{TIPS}UN][Na(12C4)_2]$, in agreement with the removal of a non-bonding $5f^1$ electron, and the MBO is also essentially unaltered. In both the U(V) and U(VI) systems, the σ- and π-bonding NBOs are less N-based than for the $-NH_2$ and $=NH$ molecules. Unusually, however, 5f dominates 6d for both σ and π in the neutral nitride. In this system, the U-amine bond is unusually short (2.465 Å, versus 2.6–2.7 Å more generally), perhaps a consequence of the inverse *trans* influence (ITI).

In an attempt to remove the remaining arsinidene hydrogen atom, Liddle *et al.* treated $TREN^{TIPS}UAsH_2$ with two equivalents of benzyl potassium, isolating $[\{U(TREN^{TIPS})(AsK_2)\}_4]$ in 50% yield (Figure 4.18) (183). In the solid state, this system has a tetrameric structure, with an As_4K_6 adamantane-type core and an interstitial K atom. The

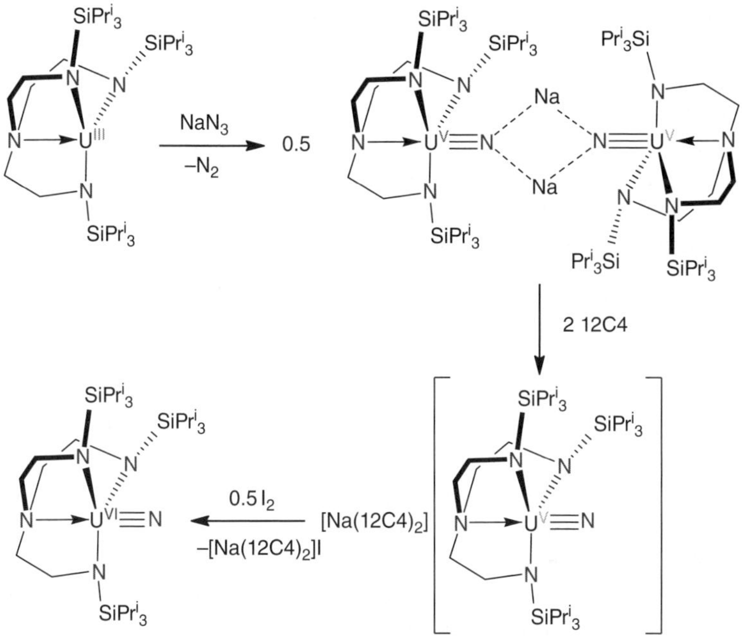

Figure 4.17 Preparation of U(V) and U(VI) terminal nitrides $[TREN^{TIPS}UN][Na(12C4)_2]$ and $[TREN^{TIPS}UN]$.

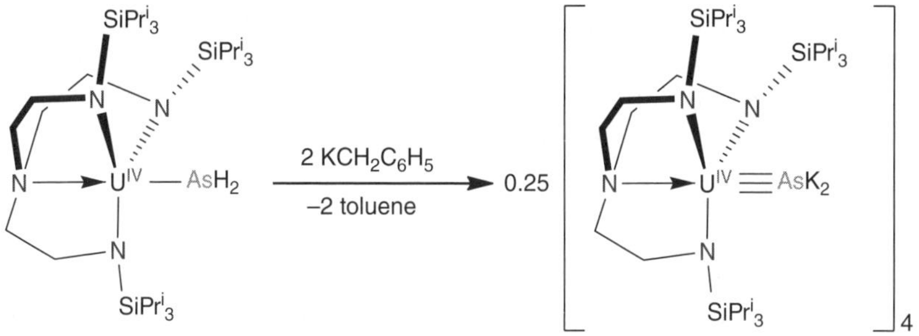

Figure 4.18 Preparation of $[\{U(TREN^{TIPS})(AsK_2)\}_4]$ from $TREN^{TIPS}UAsH_2$.

remaining K is attached to one face of the core to yield a $\{As_4K_8\}^{4-}$ polyhedron, and each As coordinates to a $[U(TREN^{TIPS})]^+$ fragment. Structural and computational data for $[\{U(TREN^{TIPS})(AsK_2)\}_4]$ are given in Table 4.1. The removal of the remaining As-bound H atom allows in principle a U–As triple bond, although Liddle *et al.* are careful not to describe $[\{U(TREN^{TIPS})(AsK_2)\}_4]$ in these terms, preferring "threefold U–As bonding interaction," as the metal–pnictogen bond is attenuated by coordination to multiple K ions. Although NBO analysis of the electronic structure at the crystallographically determined geometry finds $\sigma + 2\pi$ orbitals (with 5f once more dominating 6d), the MBOs are between 1.38 and 1.75, similar to that in $[TREN^{TIPS}UAsH]^-$ and much lower than those in the genuinely terminal nitride analogues.

4.7 Complexes with Group 16 Donor Ligands

In the past 10 years, considerable progress has been made in the synthesis of actinide complexes with group 16 donor ligands. Within this category, two areas stand out in particular: (1) the synthesis and characterization of terminal mono-oxo complexes of the actinides and (2) the synthesis and study of soft donor (S, Se, Te) complexes. Both areas are discussed in detail in the following sections. Note that because of its unique electronic structure, as well as its historical importance to actinide chemistry, recent developments of actinyl chemistry will be considered separately from the other group 16 complexes (see Section 4.8).

4.7.1 Terminal Mono-oxo Complexes

It seems counterintuitive perhaps, given the ubiquity of the actinyl fragment, but terminal mono-oxo complexes of the actinides are actually quite rare. Before 2005, only a few were known (185). However, the past decade has seen considerable progress in their synthesis and study, which parallels the progress seen in group 15-actinide multiple bond chemistry (see Section 4.6). In this section, we will discuss the synthesis of several new U(IV) and U(V) terminal mono-oxo complexes. Omitted from this discussion are mono-oxo complexes that are better described as actinyl analogues or those that manifest the ITI. These are instead discussed in Section 4.8.

In 2005, Andersen and coworkers reported the synthesis of $Cp'_2U(O)(dmap)$ (dmap = 4-N,N′-dimethylaminopyridine; $Cp' = 1,2,4\text{-}^tBu_3C_5H_2$) (Figure 4.19) (186), which represented one of the first structurally characterized terminal mono-oxo complexes of U(IV). This complex features a U-O bond length of 1.860(3) Å, which is only slightly longer that the U-O bond lengths observed in the ubiquitous uranyl fragment (1.76–1.80 Å), and is suggestive of considerable multiple bond character. However, a subsequent computational analysis suggests that the oxo ligand in this species is best described as a highly polar single bond that features negligible π-character (187). Consequently, the short bond length observed in the solid state can be ascribed primarily to Coulombic attraction. This bonding description also rationalizes its facile reactivity with electrophiles, such as SiF_4, TMSCl, and $TMSN_3$. Interestingly, its isoelectronic thorium analogue was recently prepared, namely, $Cp'_2Th(O)(dmap)$ (7). This complex is significant because thorium multiple bonds are more reactive than their uranium

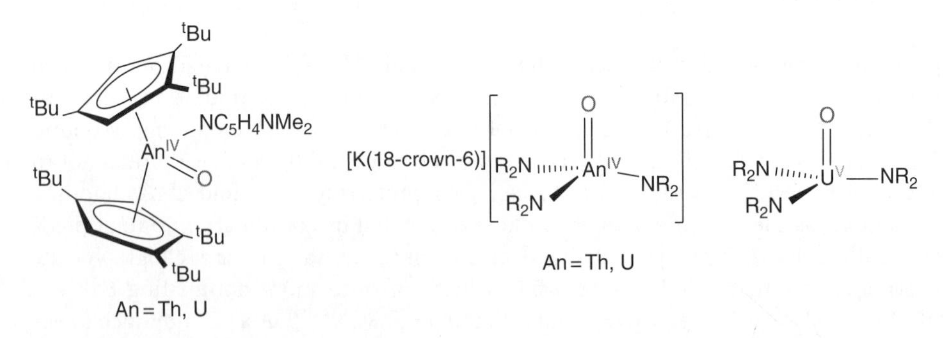

Figure 4.19 Recently prepared terminal mono-oxo complexes of uranium and thorium.

counterparts, which makes their preparation considerably more challenging. This complex features a Th-O bond length of 1.929(4) Å. DFT calculations are also suggestive of minimal π-character in the Th-O bond (7), and like its uranium analogue, $Cp'_2Th(O)$(dmap) readily reacts with electrophiles. A few other An(IV) terminal mono-oxo complexes are also known, including [K(18-crown-6)][An(O)(NR_2)$_3$] (An = Th, U) (9, 38). A DFT analysis of these two complexes suggests the presence of a small but significant π-component in the An-O interaction (11%–15% An character) (9). Thus, their An-O bonds are best described as triple bonds (albeit highly polar). Perhaps more significantly, calculations suggest that the Th-O triple bond is more ionic, and has more 6d character, than the U-O triple bond. This trend in Th versus U bonding is observed more generally throughout this chapter.

Several U(V) terminal mono-oxo complexes have also been synthesized recently, including [Ph_3PCH_3][U(O)($CH_2SiMe_2NSiMe_3$)(NR_2)$_2$] (R = $SiMe_3$) and ((^{Npt,Me}ArO)$_3$tacn)U(O)(py-NO) (81, 188). Of particular note is U(O)(NR_2)$_3$ (R = $SiMe_3$) (Figure 4.19) (189). This four-coordinate complex has an unusual trigonal pyramidal geometry, as exemplified by its N-U-N angles (av. N-U-N = 119°). DFT calculations suggest that this geometry is not a consequence of sterics, but is instead orbitally driven. Like their An(IV) counterparts, the U-O bonding orbitals in U(O)(NR_2)$_3$ are polarized toward oxygen. However, the Kohn–Sham bonding orbitals are delocalized over O, U, and N, making it a challenge to accurately measure the overall O and An contributions to the An-O bonds. Nonetheless, the Gopinathan–Jug bond order (2.34) does reveal significant multiple bond character. Another notable U(V) terminal mono-oxo complex is [U(O){N($CH_2CH_2NSi^iPr_3$)$_3$}] (190), which exhibits single-molecule magnetism and is discussed in greater detail in Section 4.9.

4.7.2 Complexes with Heavy Chalcogen (S, Se, Te) Donors

A key driver behind the move away from O-donor ligands in the search for efficient minor actinide extractants is the idea that ligands based on softer donors will be better able to exploit the greater radial extent of the 5f orbitals versus the 4f, and hence form in some sense more covalent, and thus stronger, bonds with actinide atoms. Thus motivated to explore the bonding between group 16 elements and the f elements, Gaunt *et al.* have synthesized and characterized a range of actinide complexes featuring heavy chalcogen donors, and compared their geometric and electronic structures with analogous 4f systems. Oxidation of uranium metal with, respectively, PhEEPh (E = S, Se) and pySSpy, led to U(IV) sulfur and selenium complexes as shown in Figure 4.20 (left and center) (35). The project was subsequently expanded to focus on Ln(III) and An(III) imidodiphosphinochalcogenide systems, leading to the first structurally characterized examples of An–Te bonds, for An = U, Pu (Figure 4.20 right) (44, 191, 192). Comparison of the experimentally determined imidodiphosphinochalcogenide M-E bond lengths between analogous 4f and 5f systems showed that the An-E bond lengths are shorter than the Ln-E bond lengths for metal ions of similar ionic radii, which was interpreted on the basis of an increase in covalent interactions in the actinide-ligand bonding relative to the lanthanide counterparts. These differences were found to be slightly larger with increasing softness of the chalcogen donor atom. DFT calculations on model systems indicated enhanced metal d-orbital participation in the M-E bond as group 16 is descended, with a larger f-orbital participation in An-E bonds compared with Ln-E.

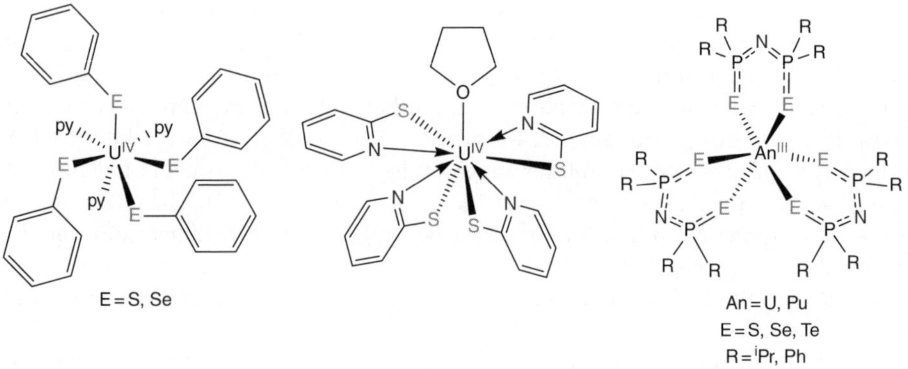

Figure 4.20 Imidodiphosphinochalcogenide, chalcogenate, and pyridylthiolate complexes.

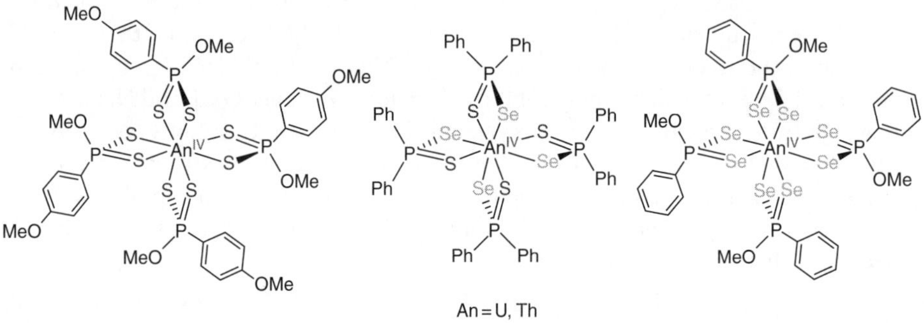

Figure 4.21 Dithiophosphonate, thioselenophosphinate, and diselenophosphonate complexes of Th and U.

More recent work from Gaunt's lab has focused on f-element diselenophosphinate complexes, that is, Ln/An(III) and An(IV) systems featuring the $[Se_2PPh_2]^-$ anion (34). For the trivalent systems, the general synthetic procedure followed was to react the diselenophosphinate dimer $[K(Se_2PPh_2)]_2$ with lanthanide or actinide triflates or halides. $U(Se_2PPh_2)_4$ was produced serendipitously from reaction of $[K(Se_2PPh_2)]_2$ with $UI_3(THF)_4$, while the Np analogue was intentionally synthesized from $[Ph_4P]_2[NpCl_6]$. As with the imidodiphosphinochalcogenide systems, comparison of X-ray crystallographically determined Pu–Se distances with Ce–Se analogues revealed the 5f bonds to be the shorter, consistent with the postulation of a modest enhancement of covalency in Pu–Se versus Ce–Se bonding, a suggestion supported by QTAIM analysis of the bond critical point (BCP) metrics. The computational work further concluded, from analysis of model An(IV) systems $An(Se_2PMe_2)_4$ (An = Th–Pu), that although all of the An–Se bonds are rather ionic, the U–Se bond is the least so.

Simultaneously, work from the Walensky group focused on systems very similar to those of Gaunt *et al.*, specifically dithiophosphonate, thioselenophosphinate, and diselenophosphonate complexes of Th and U (Figure 4.21) (36). These complexes were extensively characterized by NMR spectroscopy, including the first use of ^{77}Se NMR on a compound featuring a direct Th-Se bond. QTAIM analysis was used to show that the actinide–selenium bond has increased covalent character in comparison with the actinide–sulfur bond.

More recent soft-donor work from Bart and Walensky targeted the phenylchalcogenide species, Tp^*_2UEPh (Tp^* = hydrotris(3,5-dimethylpyrazolyl)borate, E = O, S, Se, Te) (193). This study provided a rare opportunity to probe members of an analogous series in which the steric and electronic properties of all the other components in the system bar the chalcogen are held constant, allowing a true comparison of the role of the group 16 element to be made. An extensive range of structural and spectroscopic characterization techniques were employed, together with computational analysis, to confirm that the U-E bond is primarily ionic with a small covalent contribution which increases as group 16 is descended.

A unique subset of actinide chalcogen complexes is those that feature actinide–chalcogen multiple bonds. Terminal actinide–chalcogen multiple bonds make attractive probes for studying covalency because the actinide–chalcogen orbital interactions are not complicated by chalcogen–substituent orbital interactions, which are found in $[EPh]^-$ and $[E_2PPh_2]^-$ complexes. These interactions have previously proved to be a complicating factor in the interpretation of sulfur K-edge XANES, for example (194). Access to this class of materials was first achieved by Ephritikhine and coworkers via a unique synthetic route (195). Specifically, they subjected $Cp^*_2U(S^tBu)_2$ to reducing conditions (Na/Hg), which resulted in formation of [Na(18-crown-6)][$Cp^*_2U(S)(S^tBu)$] via loss of the *tert*-butyl radical (Figure 4.22). No doubt, the relatively high stability of the *tert*-butyl radical plays a role in determining the outcome of the reaction. More recently, several other terminal chalcogenides of uranium and thorium have been isolated, including [K(18-crown-6)][$An(S)(NR_2)_3$] (An = Th, U; R = $SiMe_3$) (Figure 4.22) (9, 38). Both complexes were structurally characterized. The most notable metrical parameters are the An = E distances, which are significantly shorter than those expected for An-E single bonds. For example, the Th-S distance in [K(18-crown-6)][$Th(S)(NR_2)_3$] is ca. 2.52 Å, which is much shorter than a typical Th-S single bond (2.74 Å). DFT calculations on this series of complexes support the presence of a polarized $\sigma + 2\pi$ An-E triple bond. For example, the π orbitals in [K(18-crown-6)][$Th(S)(NR_2)_3$] feature 82% S character and 17% Th character (61% d, 38% f). Several trends can be discerned from a comparison of these two complexes with their oxo congeners (see Section 4.7.1). For one, the sulfur analogues feature greater covalency than the oxygen analogues. Likewise, the uranium complexes have greater covalency than the Th complexes, whereas uranium also features more f-orbital involvement relative to thorium. These trends fit within the developing picture of actinide covalency that is presented within this chapter.

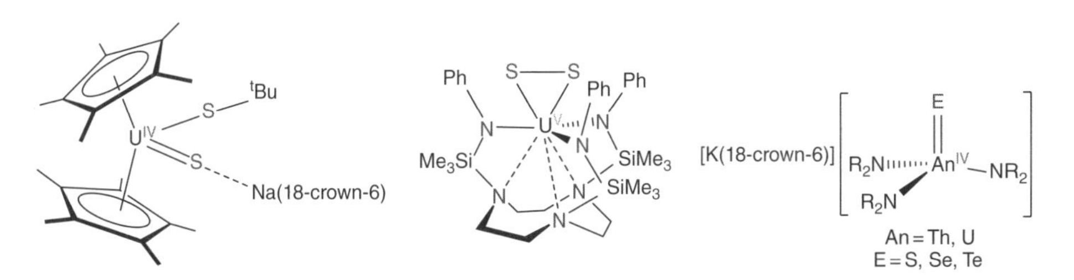

Figure 4.22 Chalcogenide complexes of the actinides.

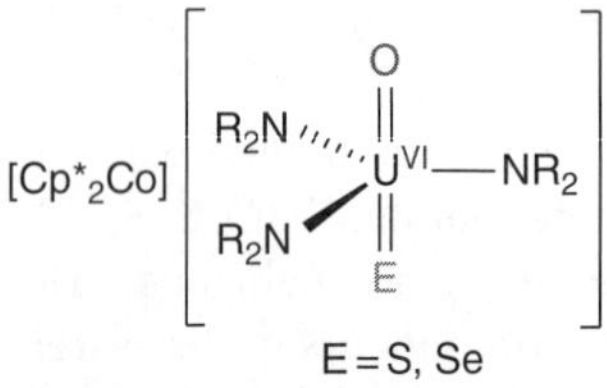

Figure 4.23 Chalcogenido-substituted analogues of the uranyl ion.

Also notable is the isolation of a U(V) persulfide complex, $[U(PhNSiMe_2)_3tacn)(\eta^2\text{-}S_2)]$ (Figure 4.22) (196). This complex is unusual because it contains the redox-active persulfide ligand within the coordination sphere of a U(V) ion, which is normally a good oxidant. Its S-S distance is 2.050(8) Å, which supports the presence of an S_2^{2-} moiety, while the U-S distance (2.66(2) Å) is shorter than those seen for U(IV) persulfides, supporting the presence of a U^{5+} ion. DFT calculations on $[U(PhNSiMe_2)_3tacn)(\eta^2\text{-}S_2)]$ suggest that the HOMO is delocalized over both U and the S_2 unit, which can be taken as evidence for covalency or for partial U(IV)-supersulfide character. However, calculations suggest that the U(V)-persulfide formulation is 8.8 kcal/mol lower in energy that the U(IV)-supersulfide formulation. Several solid-state materials containing high-valent uranium with all-sulfur coordination environments are also known (197). Overall, the observation of high-valent uranium in an easily oxidizable sulfur donor environment can be rationalized by the presence of An-S covalency, which helps satisfy the 5+ charge on the uranium ion and decreases its oxidizing power.

The chalcogenido-substituted analogues of the uranyl ion, $[Cp^*_2Co][U(O)(E)(NR_2)_3]$ (E = S, Se; R = $SiMe_3$), are also worthy of comment (Figure 4.23) (198). Their uranyl "character" is revealed by their solid- state molecular structures, which demonstrate a *trans* arrangement of the oxo and chalcogenido groups (e.g., O-U-S = 177.4(7)°), and short U-E bonds (e.g., U-S = 2.390(8) Å). Notably, this distance is about 0.1 Å shorter than that observed for the related U(IV) sulfides. This large difference is not solely due to the smaller radius of the U^{6+} ion and may be a consequence of the ITI (see Section 4.8.1 for more details). DFT calculations on this series were also very informative. As anticipated, the U = E interaction is best described as a polarized $\sigma + 2\pi$ An-E triple bond. However, the degree of polarization is much smaller than that observed for $[K(18\text{-crown-6})][An(S)(NR_2)_3]$ (An = Th, U). For example, the U-S σ orbital in $[Cp^*_2Co][U(O)(S)(NR_2)_3]$ features 52% S character and 40% U character (mostly f). Surprisingly, though, the U-E bonds become less covalent on going from O to S to Se, which contradicts the findings of other studies. Perhaps more importantly, the isolation of authentic U(VI) chalcogenide complexes is further evidence of the stabilizing ability of the uranyl electronic structure, as well as providing excellent examples of actinide covalency.

4.8 Actinyl and Its Derivatives

There have been several exciting developments in actinyl chemistry that are relevant to the focus of this chapter. For example, the past decade has seen considerable success in the synthesis of new analogues of uranyl, including the *trans*-bis(imido) complexes. Several research groups have also made significant progress toward our understanding of the Inverse *Trans* Influence (ITI), and although a true *cis*-uranyl complex remains elusive, there has been a notable recent development in this arena as well. These advancements are the focus of the next section.

4.8.1 Inverse *Trans* Influence (ITI)

Over the past decade, considerable progress has been made toward our understanding of the ITI. The ITI has been reviewed several times (199, 200); of particular note is the excellent 2011 review by Meyer and La Pierre (201). Briefly, the ITI refers to the shortening of an An-X_{trans} bond relative to an An-X_{cis} bond within the same ligand sphere, such as in the C_{4v} U(VI) complex, $[PPh_4][UOCl_5]$, where the U-Cl_{trans} distance is about 0.1 Å shorter than the U-Cl_{cis} distance (202). This observation was initially quite surprising, as the opposite effect is observed in the transition metals. This led to speculation that the involvement of the 5f orbitals in An-L_{trans} bonds, which of course is not possible in transition metal–ligand bonding, is a key driver of the ITI. Critical to our understanding of the ITI were the seminal contributions of Robert Denning, who was one of the first researchers to recognize the unique nature of the ITI and its potential to provide insight into 5f participation in An-L bonding (203).

The presence of a strong field ligand, such as oxo or imido, within the uranium coordination sphere is thought to be necessary to observe the ITI. Accordingly, it is not surprising that most attempts to study the ITI have focused on the inclusion of these ligands within the actinide coordination sphere. For example, the Schelter group has synthesized a large series of U(VI) oxo and imido complexes, including $U(O)(X)(NR_2)_3$ ($R = SiMe_3$, X = F, Cl, Br, Me, CCPh) (39, 110, 111). Of these, the most notable are the methyl and phenylacetylide derivatives (Figure 4.24), which represent rare examples of stable U(VI) hydrocarbyl complexes. DFT calculations suggest that both complexes manifest the ITI. Specifically, calculations reveal that their O-U-C σ^* interactions feature about 45% $5f_{z3}$ and about 12% $6p_z$ character. The mixing of the "semi-core" $6p_z$ orbital into the $5f_{z3}$ orbital is a hallmark of the ITI and results in a strengthening of the O-U-C bonding interaction (with a concomitant destabilization of the σ^* interaction). These findings, coupled with prediction that $U(O)(H)(NR_2)_3$ should feature comparable levels of $5f_{z3}/6p_z$ mixing, further solidify the observation that the ITI is predominantly a σ-bonding effect and not a π-bonding effect, although there is some debate on this topic (204). The U(V) analogue of $U(O)(CCPh)(NR_2)_3$, namely, $[NEt_4][U(O)(CCPh)(NR_2)_3]$, is also known (112). Not surprisingly, this complex features a large and negative U(V/VI) redox potential (−0.59 V vs. Fc/Fc^+). This low couple is consistent with presence of many strong σ- and π-donor ligands within its coordination sphere (see Section 4.4.5 for further discussion of the electrochemical properties of actinide organometallics). Schelter and coworkers have also isolated a series of 5-coordinate U(V) pseudohalide complexes, $U(X)_2(NR_2)_3$ ($R = SiMe_3$, X = F, Cl, Br, N_3, CN, NCS) (112, 205).

Figure 4.24 Stable U(VI) hydrocarbyl complexes.

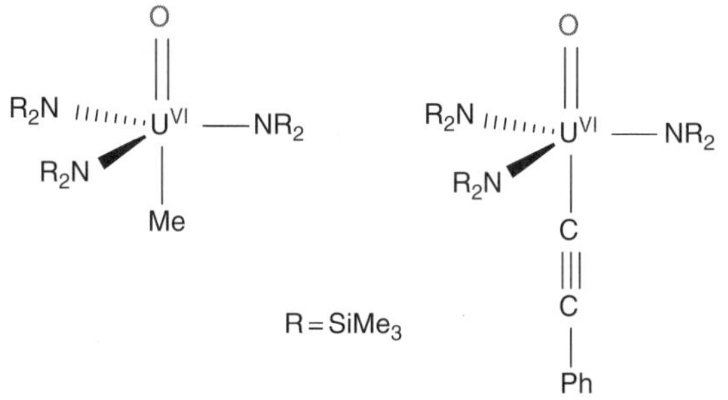

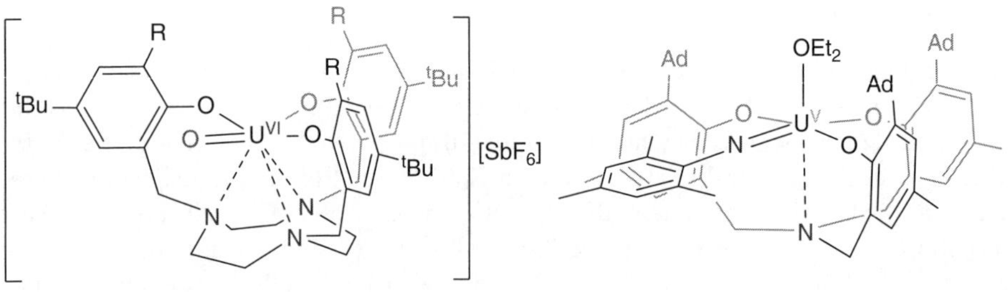

Figure 4.25 U(VI) oxo and U(V) imido complexes that feature the ITI.

DFT calculations on these complexes reveal that the ITI is also observable in the U^{5+} ion, although to a lesser extent than that observed for U^{6+}. For example, the Cl-U-Cl σ bonding interaction in $U(Cl)_2(NR_2)_3$ features 16% $5f_{z3}$ and only 4% $6p_z$ character, which is a smaller degree of $6p_z$ character than that seen for $U(O)(X)(NR_2)_3$.

Meyer and coworkers have also made several notable contributions toward our understanding of the ITI. Specifically, they found that the oxo ligand in $[((^{tBu}ArO)_3tacn)UO][SbF_6]$ occupies an equatorial coordination site that is *trans* to an aryloxide group (e.g., O(aryloxide)-U-O(oxo) = 148.6(2)°) (Figure 4.25) (204, 206), despite the fact that the bulky tripodal $[(^{tBu}ArO)_3tacn]^{3-}$ ligand tends to enforce *axial* coordination of a seventh donor atom (207, 208). This *axial/equatorial* isomerization was rationalized by the presence of the ITI (199–201). Another point of interest is the U-O-C(ipso) angle of the *trans*-aryloxide (157.7(2)°), which is more obtuse than those of the *cis*-aryloxides. It was argued that this difference is evidence for π-donation from the *trans*-aryloxide ligand, and suggests that π-bonding may also contribute to the ITI. The U(V) imido complex, $((^{Ad}ArO)_3N)U(Et_2O)(=NMes)$, also features a multiply-bonded ligand in an equatorial coordination site (Figure 4.25) (209). The *trans* U-O(aryloxide) bond in $((^{Ad}ArO)_3N)U(Et_2O)(=NMes)$ is shorter (by 0.03 Å) than the *cis* U-O(aryloxide) bonds, indicating the presence of a weak ITI. The smaller magnitude of ITI in $((^{Ad}ArO)_3N)U(Et_2O)(=NMes)$ vs $[((^{tBu}ArO)_3tacn)UO][SbF_6]$ may be due to the U(V) oxidation state in the former, and is consistent with the findings of others (112). Computational analyses were performed for both complexes, but these do not mention any evidence for 5f/6p mixing within the O(oxo)-U-O(aryloxide) σ-bonding framework.

4.8.2 Imido-Substituted Analogues of Uranyl

The surprising isolation of the *trans*-bis(imido) and *trans*-oxo-imido analogues of uranyl provided a fruitful avenue for the investigation into actinide-ligand bonding (27, 210, 211). The *trans*-bis(imido) fragment can be conveniently synthesized in a one-pot procedure by oxidation of uranium metal with three equivalents of iodine, in the presence of six equivalents of *tert*-butylamine, in THF (Figure 4.26) (27). Synthesized in this fashion, it is isolated as a bis(THF) adduct, $U(N^tBu)_2(I)_2(THF)_2$. The THF ligands can be readily displaced by other Lewis bases, while the iodide ligands undergo metathesis chemistry (211, 212). Thus, this complex can be easily and quickly derivatized to generate a wide variety of structure types. Interestingly, under most circumstances, iodine is not a sufficiently powerful oxidant to convert uranium metal to U^{6+}; however, the formation of two strong U = N multiple bonds seemingly provides the required driving

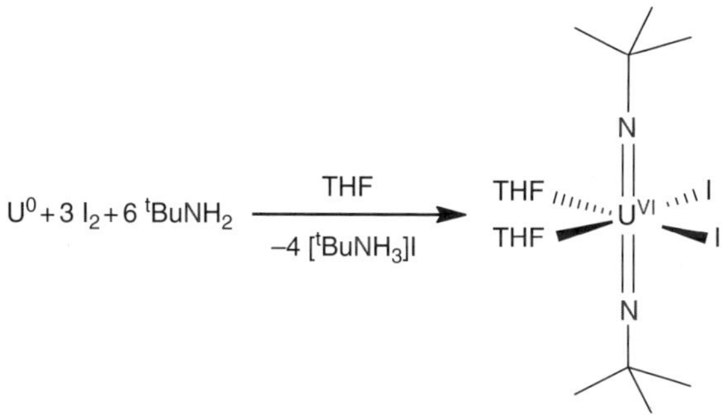

Figure 4.26 Synthesis of $U(N^tBu)_2(I)_2(THF)_2$.

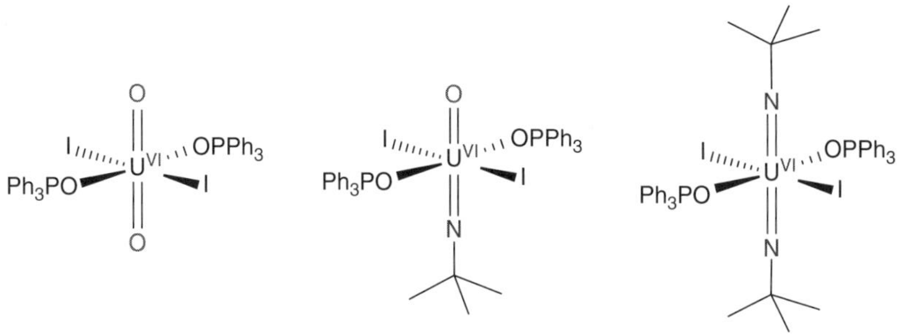

Figure 4.27 *trans*-bis(imido) and *trans*-oxo-imido analogues of uranyl.

Table 4.2 Metrical parameters for $U(O)(E)(I)_2(Ph_3PO)_2$ ($E = O, N^tBu$).

Complex	U-O, Å	U-E, Å	E-U-E, deg	U-N-C, deg	Reference
$UO_2I_2(Ph_3PO)_2$	1.760(4)	1.760(4)	180	—	(213)
$U(N^tBu)(O)I_2(Ph_3PO)_2$	1.821(7)	1.764(5)	178.4(3)	172.3(7)	(210)
$U(N^tBu)_2I_2(Ph_3PO)_2$	—	1.840(3), 1.839(3)	177.8(2)	171.5(3), 170.4(3)	(211)

force to overcome this limitation. It is also noteworthy that the *trans*-bis(arylimido) derivatives are also isolable via a slightly modified synthetic procedure (211).

Conveniently, the triphenylphosphine oxide (TPPO) adducts of uranyl, *trans*-oxo-imido, and *trans*-bis(imido) fragments have all been structurally characterized (Figure 4.27), which renders their structural comparison relatively straightforward (Table 4.2). As is observed for uranyl, both the *trans*-oxo-imido, and *trans*-bis(imido) fragments feature a *trans* arrangement of their multiply-bonded ligands. In addition, their U-O and U-N bond lengths are very short, while the U-N-C angles are nearly linear, which is suggestive of sp hybridization at the nitrogen center. Both features are indicative of multiple bond character in the U–N interactions. Moreover, these structural similarities suggest that the underlying electronic structures of uranyl, *trans*-oxo-imido and *trans*-bis(imido), are closely related. Comparison of computational results for the *trans*-oxo-imido and *trans*-bis(imido) with those collected for uranyl demonstrate that this hypothesis is true. Like uranyl, the *trans*-oxo-imido and

Table 4.3 Mulliken and NBO charges for the U atom in $U(N^tBu)_2I_2(THF)_2$, $U(N^tBu)(O)I_2(Ph_3PO)_2$ and $[UO_2]^{2+}$.

Complex	Mulliken	NBO	Reference
$U(N^tBu)_2I_2(THF)_2$	+1.50	+1.27	(211)
$U(N^tBu)(O)I_2(Ph_3PO)_2$	+1.60	+1.45	(210)
$[UO_2]^{2+}$	+2.73	+2.84	(211)

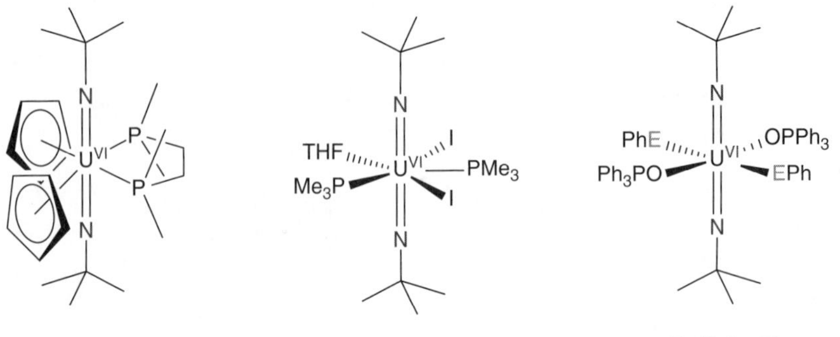

E = S, Se, Te

Figure 4.28 Soft donor *trans*-bis(imido) complexes.

trans-bis(imido) fragments feature four π bonds and two σ bonds. In addition, the bonding within both fragments has considerable 5f and 6d character. For example, the fσ orbital (HOMO-12) in $U(N^tBu)_2I_2(THF)_2$ features 24% $5f_{z3}$ character and 40% total uranium character, demonstrating nearly unprecedented levels of covalency for an An–L interaction (27, 211). Also notable is the $6p_z$ character in this orbital (8%), which further highlights the similarities between the bis(imido) and uranyl fragments. The fσ orbital in the *trans*-oxo-imido fragment also has considerable $5f_{z3}/6p_z$ mixing, with 6.5% overall $6p_z$ character (210).

Interestingly, atomic partial charge analysis (Table 4.3) suggests that the bonding within the *trans*-bis(imido) fragment is more covalent than that within the parent uranyl, while the bonding within the *trans*-oxo-imido is intermediate between the two. For example, the Mulliken charge calculated for the U atom in $U(N^tBu)_2I_2(THF)_2$ is +1.50, whereas that in $U(N^tBu)(O)I_2(Ph_3PO)_2$ is slightly larger (+1.60), and that within the naked $[UO_2]^{2+}$ fragment is larger still (+2.73). Further support for greater covalency within the *trans*-bis(imido) fragment comes from its synthetic chemistry, as the *trans*-bis(imido) fragment is known to bind soft donor ligands (211, 212, 214), such as PMe_3, dmpe, and $[EPh]^-$ (E = S, Se, Te) (Figure 4.28). Complexes of this type are essentially unknown for the uranyl ion (215), suggesting that the $[U(NR)_2]^{2+}$ fragment is a softer Lewis acid than $[UO_2]^{2+}$.

More recently, a tris(imido) complex of uranium, namely, $(^{Mes}PDI^{Me})U(NMes)_3$ (Mes = 2,4,6-$Me_3C_6H_2$), was isolated by Bart and coworkers (Figure 4.29) (29). Notably, this complex features a T-shaped *meridional* arrangement of its three imido ligands. This is significant because gaseous UO_3 also has a T-shaped geometry, which may be a

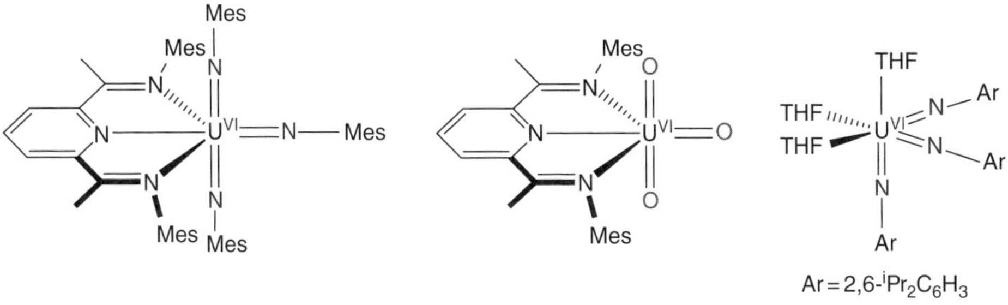

Figure 4.29 Structure of tris(imido) complexes and its oxo analogue.

consequence of the ITI (216–218). Thus, this observation is further evidence to support the similarity of the *trans*-bis(imido) and uranyl electronic structures. However, it could be argued that the multi-dentate $^{Mes}PDI^{Me}$ co-ligand enforces the *meridional* arrangement in this complex. Interestingly, the axial U-N bond lengths in this complex are about 0.1 Å longer than those typically observed for the *trans*-bis(imido) fragment, suggesting that the presence of the equatorial imido group weakens the axial U = N bonds. This suggestion is supported by DFT calculations, which also reveal $5f_{z3}/6p_z$ mixing within the fσ orbital, as expected for a uranyl analogue. The tris(oxo) analogue of ($^{Mes}PDI^{Me}$)U(NMes)$_3$, namely, ($^{Mes}PDI^{Me}$)U(O)$_3$, was also studied by DFT (Figure 4.29). The Nalewajski–Mrozek bond indices suggest that the U-O bonds within ($^{Mes}PDI^{Me}$) U(O)$_3$ are more covalent than those in the tris(imido) complex, in contrast to difference observed for $[UO_2]^{2+}$ versus $[U(NR)_2]^{2+}$. The origin of this inversion is not clear.

Bart and coworkers have also synthesized a uranium tris(imido) complexes that features *facial* stereochemistry (28). Careful reaction of $UI_3(THF)_4$ with ArN_3 (Ar = 2,6-$^iPr_2C_6H_3$) and KC_8 results in formation of $U(NAr)_3(THF)_3$ in high yields (Figure 4.29). Its U-N_{imido} distances range from 1.97(1) to 2.01(2) Å, which are longer than those observed for the *trans*-bis(imido) complexes, but similar to those observed for ($^{Mes}PDI^{Me}$)U(NMes)$_3$. The *fac* arrangement of imido ligands is notable and argues that the ITI is not as strong in the imido complexes as it is in the oxo analogues.

Arnold and coworkers have reported the isolation of the first bis(imido) complex of thorium, $[K][K(THF)_2][Th(NAr)_2(NR_2)_2]$ (Ar = 2,6-$^iPr_2C_6H_3$, R = $SiMe_3$) (Figure 4.30) (10). The two imido ligands in this complex feature *cis* stereochemistry (N_{imido}-Th-N_{imido} = 97.3(1)°), while the Th-N_{imido} distances are 2.165(3) Å, which are significantly longer than the U-N_{imido} distances in related uranium *cis*-bis(imido) complexes (ca. 1.92 Å);(212) likely a function of the larger radius of Th^{4+} versus U^{6+}. DFT calculations on this complex reveal that the Th-N_{imido} σ-bond is quite polarized, featuring only about 10% Th character (65% d, 35% f), which is perhaps not surprising given the +4 oxidation state of the Th ion. Interestingly, a DFT optimization of the *trans* imido geometry results in convergence to the experimentally observed *cis* geometry. This may be due to a reduced strength of the ITI in Th^{4+} versus U^{6+}, which destabilizes the *trans* isomer.

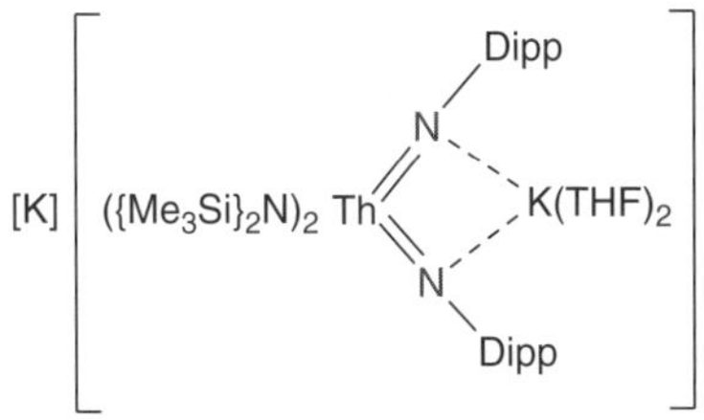

Figure 4.30 Structure of a *cis*-bis(imido) complex of thorium.

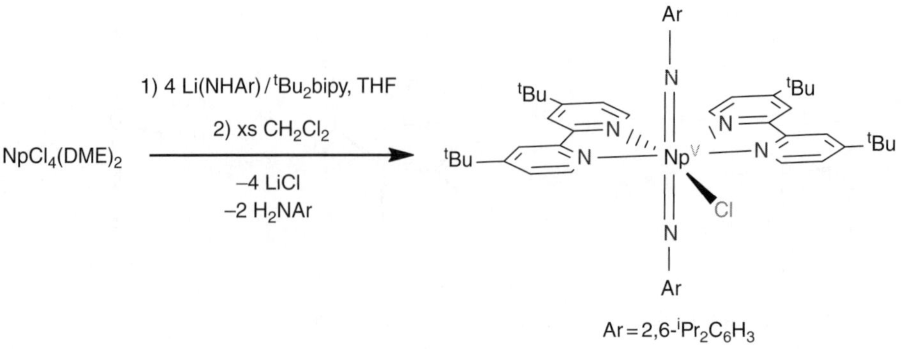

$NpCl_4(DME)_2$ → 1) 4 Li(NHAr)/$^tBu_2$bipy, THF; 2) xs CH_2Cl_2; −4 LiCl; −2 H_2NAr → $Np^V(NAr)_2(^tBu_2bipy)_2Cl$

Ar = 2,6-$^iPr_2C_6H_3$

Figure 4.31 Synthesis of *trans*-bis(imido) Np complex.

In 2015, Gaunt and coworkers reported the synthesis of a *trans*-bis(imido) complex of Np(V), $Np(NAr)_2(^tBu_2bipy)_2Cl$ (Ar = 2,6-$^iPr_2C_6H_3$) (40), which represented the first *trans*-bis(imido) complex reported for an actinide other than uranium (Figure 4.31). This complex was synthesized by reaction of $NpCl_4(DME)_2$ with four equiv of Li(NHAr) and two equiv of $^tBu_2$bipy, followed by addition of excess CH_2Cl_2, and was isolated in 17% yield. Note that its uranium analogue had been previously prepared by a similar synthetic route (219), which allows for convenient comparison of their metrical parameters and electronic structures. Consistent with other complexes in this class, the N_{imido}-Np-N_{imido} angle is 179.3(2)°, while the Np-N_{imido} distances are quite short (1.960(3) and 1.961(3) Å), and are indicative of multiple bond character. According to an NBO analysis, the Np-N bonding orbitals in this complex feature about 30% Np character (40% d, 60% f). For comparison, the isostructural U complex has only about 24% U character (45% d, 55% f). The larger degree of 5f participation in Np versus U, despite the contraction of the 5f subshell, is attributed to a better energy match of 5f orbitals with the imido frontier orbitals (i.e., energy-driven covalency). Perhaps most importantly, this work demonstrates that actinide-ligand multiple bonding can be extended beyond uranium, giving experimentalists and theoreticians even more systems with which to study covalency, and especially 5f orbital participation in bonding.

4.8.3 Progress Toward the Isolation of a cis-Uranyl Complex

The prospect of isolating a *cis*-uranyl complex has intrigued synthetic actinide chemists for several decades (220), as its isolation would provide further insight into f-orbital covalency, as well as the ITI. Although this fragment has yet to be isolated, there have been several unsuccessful attempts made toward its synthesis (220). For instance, several attempted preparations of *cis*-Cp'_2UO_2 have resulted in unwanted ligand oxidation (221, 222), or formation of a uranium(IV) oxide cluster (223). A 2007 report on the isolation of *cis*-uranyl(224) has since come under scrutiny (225), and is likely to be incorrect. Similarly, a 2015 report of a *cis*-uranyl coordination polymer is also likely incorrect (226). Intriguingly, however, the *cis*-uranyl isomer of $[UO_2(OH)_4]^{2-}$ has been studied by DFT methods. These calculations suggest that a *cis*-uranyl species, although 18 kcal/mol higher in energy than the *trans* isomer, represents a local minimum on the energy landscape (227). Similarly, Schelter and coworkers calculated that the *cis* isomer

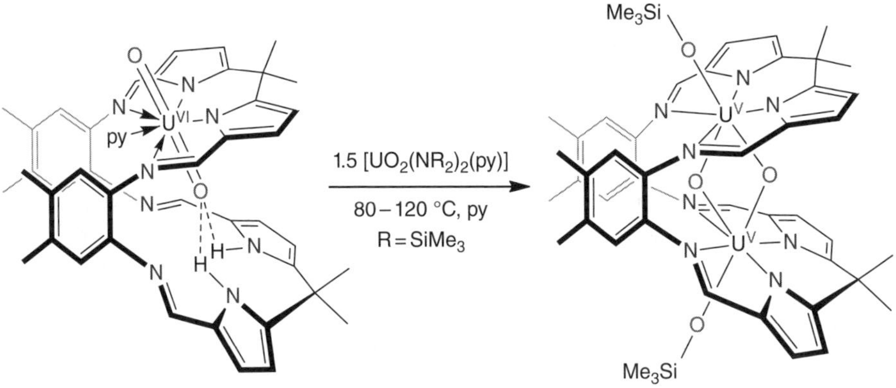

Figure 4.32 Formation of the butterfly-shaped U(V) oxo dimer.

of $[UO_2(N(SiH_3)_2)_3]^-$ is 31 kcal/mol higher in energy than the *trans* isomer, which is substantially higher in energy than the *trans*/*cis* isomerization of the imido analogue, $[U(NSiMe_3)_2(N(SiH_3)_2)_3]^-$ (14 kcal/mol) (110). The *cis*/*trans* isomerization of the bis(imido) complex, $U(NMe)_2(I)_2(THF)_2$ was also studied by DFT methods (211). In this example, the *cis* isomer was found to be about 15 kcal/mol higher in energy than the *trans* isomer. The reduced energy penalty for *cis* isomerization in the *trans*-bis(imido) case is also reflected in its synthetic chemistry, as several *cis*-bis(imido) complexes have actually been isolated (212), unlike the case for uranyl.

Although a true *cis*-uranyl complex has remained elusive, elegant work by Arnold and coworkers has shown that migration of a uranyl *trans* oxo ligand to a *cis* position is possible under certain conditions (228). Specifically, thermolysis of a mixture of $UO_2(NR_2)_2(py)_2$ (R = $SiMe_3$) and $UO_2(py)(H_2L)$ (L = polypyrrole macrocycle) results in formation of $[(ROUO)_2(L)]$ in low yield (Figure 4.32), which is formally generated by coupling of a *trans*-$[U(O)(OSiMe_3)]^{2+}$ fragment with a *cis*-$[U(O)(OSiMe_3)]^{2+}$ fragment. In this example, the large energy penalty required for *trans*/*cis* isomerization is likely lessened by silylation of the "exo" oxo ligands and by reduction of the two uranium ions to U^{5+}, which weakens the uranyl U-O π- and σ-bonds. Interestingly, this complex features substantial magnetic coupling between the U(V) centers, according to SQUID magnetometry. In particular, it has a Néel temperature of 17 K and an exchange constant of $J_{ex} = -33\,cm^{-1}$, which is among the largest antiferromagnetic exchange interactions observed for an actinide complex.

4.9 Organoactinide Single-Molecule Magnets

While the field of single-molecule magnetism is more than 20 years old, it has only recently been extended to the 5f series, and the electronic structure of the actinides is such that intriguing additions to the area have already emerged. Single-molecule magnets (SMMs) are d- and f-block complexes which retain their magnetization for long periods in the absence of an applied magnetic field, albeit at cryogenic temperatures. SMM behavior arises from anisotropic electronic structure, such that the magnetic moment of the individual molecules has a preferred orientation (along the so-called easy axis), leading to bulk magnetization. One of the characteristic properties of SMMs is the anisotropy barrier – the energy barrier which must be overcome to thermalize the individual magnetic moments. This is typically obtained from magnetic susceptibility measurements by analyzing the linear section of the Arrhenius plot of ln τ versus $1/T$, where $\tau = 1/(2\pi\nu)$, ν being the frequency of the applied alternating current magnetic field. A second key property of SMMs is the blocking temperature, which is the highest temperature at which the magnetization displays hysteresis, and a major challenge for the field is to raise blocking temperatures to levels at which device fabrication becomes more feasible.

The first SMMs were polymetallic d-block cage systems (229), and it was established that the anisotropy barriers are functions of the axial zero-field splitting D (a measure of the anisotropy) and the ground spin state S. To maximize the anisotropy barriers, both of these parameters should be as large as possible. Initially, the focus was on generating systems with very large values of S, but recognition that efforts should be targeted at greater anisotropy led to the extension of the field into the lanthanide elements, and SMM behavior has now been observed in many 4f systems containing only one metal center (230). Several Ln elements (especially Tb, Dy, Ho, and Er) have large magnetic moments in their ground electronic multiplets, arising from significant orbital contributions that are not quenched by the weak effect of the ligand fields on the radially contracted 4f orbitals. Spin-orbit coupling is, of course, much larger in the actinides than the lanthanides, and hence greater anisotropies and magnetic moments may be expected, although the larger effects of the ligand fields on the radially more diffuse 5f orbitals may result in a greater quenching of orbital contributions. That said, enhanced SMM behavior may be expected in polymetallic An systems precisely because the 5f orbitals are more extended, leading to greater metal–ligand orbital overlap and stronger exchange couplings between metal centers.

A complexity often seen in f-element SMMs, which distinguishes them from d-block systems, is that an Arrhenius plot of ln τ versus $1/T$ is rarely a straight line. This points to there being contributions to the relaxation of the magnetic moment from mechanisms other than simply overcoming the thermal barrier to $m_j = +J \rightarrow -J$ conversion. Spin-lattice relaxation, in which energy is exchanged with the surroundings (lattice), and quantum tunneling effects, can operate (often in concert) in f-element SMMs, leading to complicated magnetic behavior which must be addressed on a case-by-case basis.

The majority of the known actinide SMMs are f^3 U(III) systems. The most widely studied of these contain pyrazolylborate ligands (231–235), with more recent examples focusing on ligand environments containing iodide and/or N-donors such as BIPM^{TMS} (146, 236). A number of themes have emerged from work on these complexes. All feature anisotropy barriers between about 6 and 31 K, and the SMM properties are not strongly dependent upon either the symmetry of the ligand field or the nature of the donor atoms. It has been suggested that the Orbach mechanism of spin-lattice relaxation does not play a major role in these systems. The Orbach process, which proceeds via real intermediate states of both the spin and vibrations, should give rise to linear behavior in ln τ versus $1/T$ but, as noted above, this is not generally seen for f-element SMMs, and the U(III) systems are no exception. The only other $5f^3$ SMM thus far characterized is neptunocene ($Np(\eta^8\text{-}C_8H_8)_2$ (237). In a field of 3 T the plot of ln τ versus $1/T$ is linear, giving an anisotropy barrier of 41 K, and hysteresis is observed only at larger

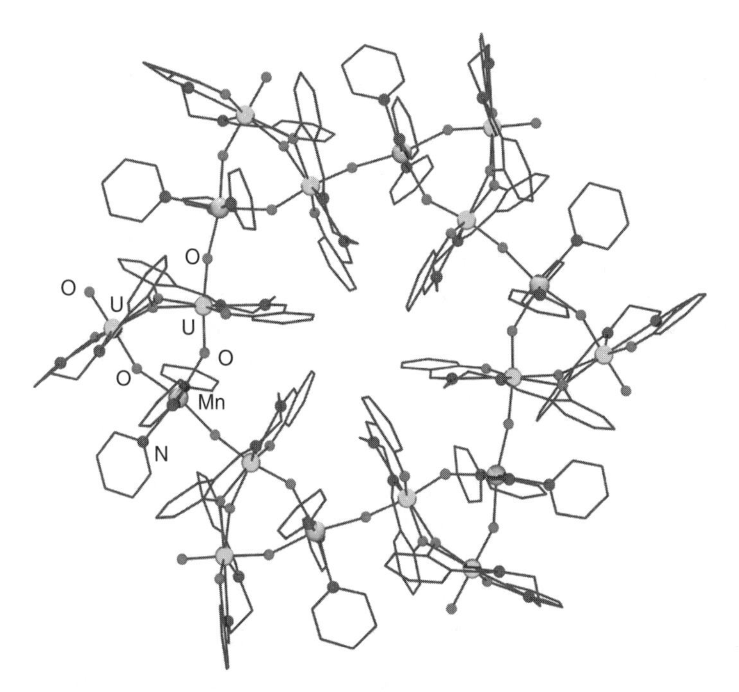

Figure 4.33 Molecular structure of [{[UO_2(salen)]$_2$Mn(py)$_3$}$_6$]. Atomic coordinates taken from Mougel V., Chatelain L., Pécaut J., Caciuffo R., Colineau E., Griveau J.-C., Mazzanti M. (2012) Uranium and manganese assembled in a wheel-shaped nanoscale single-molecule magnet with high spin-reversal barrier. *Nature Chemistry*, **4** (12), 1011–1017. (*See insert for color representation of the figure.*)

fields (>5 T). At low fields (<2 T) the principal relaxation mechanism is nuclear-spin-induced quantum tunneling via the ^{237}Np nuclei, which have $I = 5/2$, while at the high fields required to observe hysteresis, the relaxation mechanism may well involve an Orbach process.

The only known non-f^3 organoactinide SMM is the U(V) f^1 [U(O){N($CH_2CH_2NSi^iPr_3$)$_3$}] (190), which, at 21.5 K, has an anisotropy barrier in the middle of the range displayed by the f^3 systems. Although not organometallic, significantly larger barriers are seen in the U(V)-Mn(II) complexes [{[UO_2(salen)]$_2$Mn(py)$_3$}$_6$](238) and [{[UO_2(salen)(py)] [Mn(py)$_4$](NO_3)}]$_n$ (Figure 4.33) (239), 142 K and 134 K respectively, and the former, rather beautiful macrocyclic system shows significant hysteresis, with open loops below 4 K. The striking magnetic behavior displayed by these 5f/3d compounds is clearly a promising route for future developments of SMMs featuring actinide centers.

4.10 Future Work

Looking toward the future, there are a few obvious organometallic target molecules whose isolation would be of interest to a broad constituency of synthetic chemists, spectroscopists, and computational chemists. One such target is a carbene complex whose An = C bond is not stabilized by a heteroatom substituent (i.e., a "Schrock-type" alkylidene). The An = C bond in such a complex is likely to be stronger, and less polarized, than those of the phosphorano-stabilized carbenes isolated up to this point, making an actinide alkylidene an important goal. Along the same lines, the isolation of an An≡C triple bond (alkylidyne) would also be of significant interest. Further expansion of the chemistry of the An(II) ions is also important, as not only could their study lead to unprecedented modes of small molecule activation, but also because of the fundamental insights into actinide electronic structure and magnetic properties that such studies would furnish. The isolation of an authentic *cis*-uranyl complex is also a key challenge; such a complex should provide deeper understanding of the ITI and, correspondingly, of the uranyl ion itself. The synthesis of new uranyl analogues, such as those containing the heavier chalcogens or the pnictogens, is of interest for the same reason.

From a spectroscopic perspective, an expanded implementation of ligand K-edge XAS, especially for the lighter main group elements (e.g., C and N), would be of significant benefit, in part because many of the most interesting organometallic actinide complexes feature An-C and An-N bonds and also because such analyses have been somewhat rare up to this point. The expanded use of optical and NMR spectroscopies to evaluate An-L bonding is also desirable. Indeed, the deployment of any spectroscopic technique that quantifies actinide-ligand bonding should be pursued to a greater degree than in the past. As has been demonstrated by this chapter, the deepest insights into actinide-ligand bonding are most often achieved with a combined synthetic/spectroscopic/computational approach. Regarding the latter, the use of DFT to study the structure and bonding of organoactinide complexes has become widespread. However, there is no standard approach as yet, and it would be beneficial, particularly when discussing covalency, for further exploration of the conclusions from, particularly, GGA versus hybrid density functionals. The differing localization properties of the 5f orbitals from the two approaches will benefit from additional comparisons with spectroscopic measures. A further area of expansion for computational actinide chemists is enhanced use of post Hartree–Fock methods for the study of optical and magnetic properties. Such approaches are featured on several occasions throughout this chapter, and there is no doubt that they will see increased use in tackling the many problems for which DFT is unsuitable.

Acknowledgments

T.W.H. gratefully acknowledges support from the US Department of Energy, Office of Basic Energy Sciences, Chemical Sciences, Biosciences, and Geosciences Division under Contract no. DE-SC-0001861.

References

1 Ephritikhine M. (2006) The vitality of uranium molecular chemistry at the dawn of the XXIst century. *Dalton Transactions*, (**21**), 2501–2516.
2 Liddle S.T. (2015) The renaissance of non-aqueous uranium chemistry. *Angewandte Chemie International Edition*, **54** (30), 8604–8641.
3 Hayton T.W. (2010) Metal-ligand multiple bonding in uranium: structure and reactivity. *Dalton Transactions*, **39** (5), 1145–1158.

4 Hayton T.W. (2013) Recent developments in actinide-ligand multiple bonding. *Chemical Communications*, **49** (29), 2956–2973.

5 Morss L.R., Edelstein N.M., Fuger J., editors. *The Chemistry of the Actinide and Transactinide Elements*. 3rd ed. Dordrecht, the Netherlands: Springer; 2006.

6 Cambridge Structural Database. (2014) version 1.17.

7 Ren W., Zi G., Fang D.-C., Walter M.D. (2011) Thorium oxo and sulfido metallocenes: Synthesis, structure, reactivity, and computational studies. *Journal of the American Chemical Society*, **133** (33), 13183–13196.

8 Ren W., Zi G., Fang D.-C., Walter M.D. (2011) A base-free thorium–terminal-imido metallocene: Synthesis, structure, and reactivity. *Chemistry – A European Journal*, **17** (45), 12669–12682.

9 Smiles D.E., Wu G., Kaltsoyannis N., Hayton T.W. (2015) Thorium-ligand multiple bonds via reductive deprotection of a trityl group. *Chemical Science*, **6** (7), 3891–3899.

10 Bell N.L., Maron L., Arnold P.L. (2015) Thorium mono- and bis(imido) complexes made by reprotonation of cyclo-metalated amides. *Journal of the American Chemical Society*, **137** (33), 10492–10495.

11 Zi G. (2014) Organothorium complexes containing terminal metal-ligand multiple bonds. *Science China Chemistry*, **57** (8), 1064–1072.

12 Ren W., Zhou E., Fang B., Zi G., Fang D.-C., Walter M.D. (2014) Si-H addition followed by C-H bond activation induced by a terminal thorium imido metallocene: a combined experimental and computational study. *Chemical Science*, **5** (8), 3165–3172.

13 Ren W., Zhou E., Fang B., Hou G., Zi G., Fang D.-C., Walter M.D. (2014) Experimental and computational studies on the reactivity of a terminal thorium imidometallocene towards organic azides and diazoalkanes. *Angewandte Chemie International Edition*, **53** (42), 11310–11314.

14 Walensky J.R., Martin R.L., Ziller J.W., Evans W.J. (2010) Importance of energy level matching for bonding in Th^{3+}-Am^{3+} actinide metallocene amidinates, $(C_5Me_5)_2[^iPrNC(Me)N^iPr]An$. *Inorganic Chemistry*, **49** (21), 10007–10012.

15 Langeslay R.R., Fieser M.E., Ziller J.W., Furche F., Evans W.J. (2015) Synthesis, structure, and reactivity of crystalline molecular complexes of the $\{[C_5H_3(SiMe_3)_2]_3Th\}^{1-}$ anion containing thorium in the formal +2 oxidation state. *Chemical Science*, **6** (1), 517–521.

16 Cantat T., Scott B.L., Kiplinger J.L. (2010) Convenient access to the anhydrous thorium tetrachloride complexes $ThCl_4(DME)_2$, $ThCl_4(1,4\text{-dioxane})_2$ and $ThCl_4(THF)_{3.5}$ using commercially available and inexpensive starting materials. *Chemical Communications*, **46** (6), 919–921.

17 Travia N.E., Monreal M.J., Scott B.L., Kiplinger J.L. (2012) Thorium-mediated ring-opening of tetrahydrofuran and the development of a new thorium starting material: preparation and chemistry of $ThI_4(DME)_2$. *Dalton Transactions*, **41** (48), 14514–14523.

18 Jacoby M. Reintroducing thorium. *Chemical & Engineering News*. 2009 November 16, 2009:44–46.

19 Niiler E. Nuclear power entrepreneurs push thorium as a fuel. *The Washington Post*. 2012 February **20**, 2012.

20 Sokolov F., Fukuda K., Nawada H.P. Thorium Fuel Cycle — Potential Benefits and Challenges. Vienna, Austria: International Atomic Energy Agency, 2005 Contract No.: IAEA-TECDOC-1450.

21 Rainey R.H., Moore J.G. Laboratory Development of the Acid Thorex Process for the Recovery of Consolidated Edison Thorium Reactor Fuel. ORNL-3155. Oak Ridge National Laboratory, 1962 ORNL-3155 Contract No.: ORNL-3155.

22 Benedict G.E. Improvements in the Thorium-Uranium Separation in the Acid-Thorex Process. In: Navratil J.D., Schulz W.W., editors. *Actinide Separations.* ACS symposium series. **117**. Washington D.C.: ACS; 1980.
23 Grant G.R., Morgan W.W., Mehta K.K., Sargent F.P. Heavy element separation for thorium-uranium-plutonium fuels. In: Navratil J.D., Schulz W.W., editors. *Actinide Separations.* ACS symposium series. **117**. Washington D.C.: ACS; 1980.
24 King D.M., Tuna F., McInnes E.J.L., McMaster J., Lewis W., Blake A.J., Liddle S.T. (2012) Synthesis and structure of a terminal uranium nitride complex. *Science,* **337** (6095), 717–720.
25 Ephritikhine M. (2013) Uranium carbene compounds. *Comptes Rendus Chimie,* **16** (4), 391–405.
26 Gregson M., Wooles A.J., Cooper O.J., Liddle S.T. (2015) Covalent uranium carbene chemistry. *Comments on Inorganic Chemistry,* **35** (5), 262–294.
27 Hayton T.W., Boncella J.M., Scott B.L., Palmer P.D., Batista E.R., Hay P.J. (2005) Synthesis of imido analogs of the uranyl ion. *Science,* **310** (5756), 1941–1943.
28 Anderson N.H., Yin H., Kiernicki J.J., Fanwick P.E., Schelter E.J., Bart S.C. (2015) Investigation of uranium tris(imido) complexes: Synthesis, characterization, and reduction chemistry of [U(NDIPP)$_3$(thf)$_3$]. *Angewandte Chemie International Edition,* **54** (32), 9386–9389.
29 Anderson N.H., Odoh S.O., Yao Y., Williams U.J., Schaefer B.A., Kiernicki J.J., Lewis A.J., Goshert M.D., Fanwick P.E., Schelter E.J., Walensky J.R., Gagliardi L., Bart S.C. (2014) Harnessing redox activity for the formation of uranium tris(imido) compounds. *Nature Chemistry,* **6** (10), 919–926.
30 Graves C.R., Kiplinger J.L. (2009) Pentavalent uranium chemistry – synthetic pursuit of a rare oxidation state. *Chemical Communications,* (**26**), 3831–3853.
31 Fortier S., Walensky J.R., Wu G., Hayton T.W. (2011) High-valent uranium alkyls: Evidence for the formation of $U^{VI}(CH_2SiMe_3)_6$. *Journal of the American Chemical Society,* **133** (30), 11732–11743.
32 MacDonald M.R., Fieser M.E., Bates J.E., Ziller J.W., Furche F., Evans W.J. (2013) Identification of the +2 oxidation state for uranium in a crystalline molecular complex, [K(2.2.2-Cryptand)][(C$_5$H$_4$SiMe$_3$)$_3$U]. *Journal of the American Chemical Society,* **135** (36), 13310–13313.
33 Anderson N.H., Odoh S.O., Williams U.J., Lewis A.J., Wagner G.L., Lezama Pacheco J., Kozimor S.A., Gagliardi L., Schelter E.J., Bart S.C. (2015) Investigation of the electronic ground states for a reduced pyridine(diimine) uranium series: Evidence for a ligand tetraanion stabilized by a uranium dimer. *Journal of the American Chemical Society,* **137** (14), 4690–4700.
34 Jones M.B., Gaunt A.J., Gordon J.C., Kaltsoyannis N., Neu M.P., Scott B.L. (2013) Uncovering f-element bonding differences and electronic structure in a series of 1 : 3 and 1 : 4 complexes with a diselenophosphinate ligand. *Chemical Science,* **4** (3), 1189–1203.
35 Gaunt A.J., Scott B.L., Neu M.P. (2006) U(IV) chalcogenolates synthesized via oxidation of uranium metal by dichalcogenides. *Inorganic Chemistry,* **45** (18), 7401–7407.
36 Behrle A.C., Barnes C.L., Kaltsoyannis N., Walensky J.R. (2013) Systematic investigation of thorium(IV)– and uranium(IV)–ligand bonding in dithiophosphonate, thioselenophosphinate, and diselenophosphonate complexes. *Inorganic Chemistry,* **52** (18), 10623–10631.

37 Smiles D.E., Wu G., Hayton T.W. (2014) Synthesis of terminal monochalcogenide and dichalcogenide complexes of uranium using polychalcogenides, $[E_n]^{2-}$ (E = Te, n = 2; E = Se, n = 4), as chalcogen atom transfer reagents. *Inorganic Chemistry*, **53** (19), 10240–10247.

38 Smiles D.E., Wu G., Hayton T.W. (2014) Synthesis of uranium–ligand multiple bonds by cleavage of a trityl protecting group. *Journal of the American Chemical Society*, **136** (1), 96–99.

39 Lewis A.J., Carroll P.J., Schelter E.J. (2013) Reductive cleavage of nitrite to form terminal uranium mono-oxo complexes. *Journal of the American Chemical Society*, **135** (1), 511–518.

40 Brown J.L., Batista E.R., Boncella J.M., Gaunt A.J., Reilly S.D., Scott B.L., Tomson N.C. (2015) A linear trans-bis(imido) neptunium(V) actinyl analog: $Np^V(NDipp)_2({}^tBu_2bipy)_2Cl$ (Dipp = 2,6-$^iPr_2C_6H_3$). *Journal of the American Chemical Society*, **137** (30), 9583–9586.

41 Schnaars D.D., Gaunt A.J., Hayton T.W., Jones M.B., Kirker I., Kaltsoyannis N., May I., Reilly S.D., Scott B.L., Wu G. (2012) Bonding trends traversing the tetravalent actinide series: Synthesis, structural, and computational analysis of $An^{IV}({}^{Ar}acnac)_4$ complexes (An = Th, U, Np, Pu; ^{Ar}acnac = ArNC(Ph)CHC(Ph)O; Ar = 3,5-$^tBu_2C_6H_3$). *Inorganic Chemistry*, **51** (15), 8557–8566.

42 Schnaars D.D., Batista E.R., Gaunt A.J., Hayton T.W., May I., Reilly S.D., Scott B.L., Wu G. (2011) Differences in actinide metal-ligand orbital interactions: comparison of U(IV) and Pu(IV) β – ketoiminate N,O donor complexes. *Chemical Communications*, **47** (27), 7647–7649.

43 Ingram K.I.M., Tassell M.J., Gaunt A.J., Kaltsoyannis N. (2008) Covalency in the f element-chalcogen bond. Computational studies of $M[N(EPR_2)_2]_3$ (M = La, Ce, Pr, Pm, Eu, U, Np, Pu, Am, Cm; E = O, S, Se, Te; R = H, ^{i}Pr, Ph). *Inorganic Chemistry*, **47** (17), 7824–7833.

44 Gaunt A.J., Reilly S.D., Enriquez A.E., Scott B.L., Ibers J.A., Sekar P., Ingram K.I.M., Kaltsoyannis N., Neu M.P. (2008) Experimental and theoretical comparison of actinide and lanthanide bonding in $M[N(EPR_2)_2]_3$ complexes (M = U, Pu, La, Ce; E = S, Se, Te; R = Ph, ^{i}Pr, H). *Inorganic Chemistry*, **47** (1), 29–41.

45 Ingram K.I.M., Kaltsoyannis N., Gaunt A.J., Neu M.P. (2007) Covalency in the f-element–chalcogen bond: Computational studies of $[M(N(EPH_2)_2)_3]$ (M = La, U, Pu; E = O, S, Se, Te). *Journal of Alloys and Compounds*, **47** (17), 369–375.

46 Reilly S.D., Brown J.L., Scott B.L., Gaunt A.J. (2014) Synthesis and characterization of $NpCl_4(DME)_2$ and $PuCl_4(DME)_2$ neutral transuranic An(iv) starting materials. *Dalton Transactions*, **43** (4), 1498–1501.

47 Reilly S.D., Scott B.L., Gaunt A.J. (2012) $[N(n\text{-}Bu)_4]_2[Pu(NO_3)_6]$ and $[N(n\text{-}Bu)_4]_2[PuCl_6]$: Starting materials to facilitate nonaqueous plutonium(IV) chemistry. *Inorganic Chemistry*, **51** (17), 9165–9167.

48 Szabo A., Ostlund N.S. *Modern Quantum Chemistry*. New York: McGraw-Hill; 1989.

49 Jensen F. *Introduction to Computational Chemistry*. Second edition. Chichester: Wiley; 2007.

50 Parr R.G., Yang W. Density-Functional Theory of Atoms and Molecules. Oxford: Oxford University Press; 1989.

51 Dyall K.G., Faegri Jr K. *Introduction to Relativistic Quantum Chemistry*. Oxford: Oxford University Press; 2007.

52 Reiher M., Wolf A. *Relativistic Quantum Chemistry: The Fundamental Theory of Molecular Science*. Second edition. Wienheim, Germany: Wiley-VCH Verlag GmbH & Co; 2015.

53 Roos B.O., Malmqvist P.A. (2004) Relativistic quantum chemistry: The multiconfigurational approach. *Physical Chemistry Chemical Physics*, **6** (11), 2919–2927.

54 Gagliardi L., Roos B.O. (2007) Multiconfigurational quantum chemical methods for molecular systems containing actinides. *Chemical Society Reviews*, **36** (6), 893–903.

55 Hohenberg P., Kohn W. (1964) Inhomogeneous electron gas. *Physical Review B*, **136** (3B), 864–871.

56 Kohn W., Sham L.J. (1965) Self-consistent equations including exchange and correlation effects. *Physical Review A*, **140** (4A), 1133–1138.

57 Perdew J.P., Ruzsinszky A., Tao J.M., Staroverov V.N., Scuseria G.E., Csonka G.I. (2005) Prescription for the design and selection of density functional approximations: More constraint satisfaction with fewer fits. *Journal of Chemical Physics*, **123** (6), 062201.

58 Visscher L. Post Dirac-Hartree-Fock methods – Electron correlation. In: Schwerdtfeger P, editor. *Relativistic Electronic Structure Theory Part I Fundamentals*. Amsterdam: Elsevier; 2002. p. 291–331.

59 Dolg M., Cao X.Y. (2012) Relativistic pseudopotentials: Their development and scope of applications. *Chemical Reviews*, **112** (1), 403–480.

60 Douglas N., Kroll N.M. (1974) Quantum electrodynamical corrections to fine-structure of helium. *Annalen der Physik (Leipzig)*, **82** (1), 89–155.

61 Van Lenthe E., Baerends E.J., Snijders J.G. (1993) Relativistic regular 2-component Hamiltonians. *Journal of Chemical Physics*, **99** (6), 4597.

62 Van Lenthe E., Baerends E.J., Snijders J.G. (1994) Relativistic total-energy using regular approximations. *Journal of Chemical Physics*, **101** (11), 9783.

63 van Lenthe E., van Leeuwen R., Baerends E.J., Snijders J.G. (1996) Relativistic regular two-component Hamiltonians. *International Journal of Quantum Chemistry*, **57** (3), 281–293.

64 Neidig M.L., Clark D.L., Martin R.L. (2013) Covalency in f-element complexes. *Coordination Chemistry Reviews*, **257** (2), 394–406.

65 Strittmatter R.J., Bursten B.E. (1991) Bonding in tris(η^5-cyclopentadienyl) actinide complexes. 5. A comparison of the bonding in Np, Pu, and transplutonium compounds with that in lanthanide compounds and a transition-metal analogue. *Journal of American Chemical Society*, **113** (2), 552–559.

66 Kaltsoyannis N., Scott P. *The f Elements*. Oxford: Oxford University Press; 1999.

67 Bader R.F.W. *Atoms in Molecules: A Quantum Theory*. Oxford: Oxford University Press; 1990.

68 Matta C.F., Boyd R.J. An introduction to the quantum theory of atoms in molecules. In: Matta C.F., Boyd R.J., editors. *The Quantum Theory of Atoms in Molecules*. Weinheim: Wiley-VCH; 2007. p. 1–34.

69 Tassell M.J., Kaltsoyannis N. (2010) Covalency in $AnCp_4$ (An = Th–Cm); a comparison of molecular orbital, natural population and atoms in molecules analyses. *Dalton Transactions*, **39** (29), 6719–6725.

70 Kirker I., Kaltsoyannis N. (2011) Does covalency *really* increase across the 5f series? A comparison of molecular orbital, natural population, spin and electron density analyses of $AnCp_3$ (An = Th-Cm; Cp = η^5-C_5H_5). *Dalton Transactions*, **40** (1), 124–131.

71 Solomon E.I., Hedman B., Hodgson K.O., Dey A., Szilagyi R.K. (2005) Ligand K-edge X-ray absorption spectroscopy: Covalency of ligand-metal bonds. *Coordination Chemistry Reviews*, **249** (1–2), 97.

72 Kozimor S.A., Yang P., Batista E.R., Boland K.S., Burns C.J., Christensen C.N., Clark D.L., Conradson S.D., Hay P.J., Lezama J.S., Martin R.L., Schwarz D.E., Wilkerson M.P., Wolfsberg L.E. (2008) Covalency trends in group IV metallocene dichlorides. Chlorine K-edge X-ray absorption spectroscopy and time-dependent density functional theory. *Inorganic Chemistry*, **47** (12), 5365–5371.

73 Kozimor S.A., Yang P., Batista E.R., Boland K.S., Burns C.J., Clark D.L., Conradson S.D., Martin R.L., Wilkerson M.P., Wolfsberg L.E. (2009) Trends in covalency for d- and f-element metallocene dichlorides identified using chlorine K-edge X-ray absorption spectroscopy and time-dependent density functional theory. *Journal of the American Chemical Society*, **131** (34), 12125–12136.

74 Minasian S.G., Keith J.M., Batista E.R., Boland K.S., Clark D.L., Kozimor S.A., Martin R.L., Shuh D.K., Tyliszczak T. (2014) New evidence for 5f covalency in actinocenes determined from carbon K-edge XAS and electronic structure theory. *Chemical Science*, **5** (1), 351–359.

75 Brennan J.B., Green J.C., Redfern C.M. (1989) Covalency in bis([8]annulene)uranium from photoelectron spectroscopy with variable photon energy. *Journal of the American Chemical Society*, **111** (7), 2373–2377.

76 Kerridge A., Kaltsoyannis N. (2009) Are the ground states of the later actinocenes multiconfigurational? All-electron spin-orbit coupled CASPT2 calculations on $An(\eta^8\text{-}C_8H_8)_2$ (An = Th, U, Pu, Cm). *Journal of Physical Chemistry A*, **113** (30), 8737–8745.

77 Figgis B.N., Hitchman M.A. *Ligand Field Theory and Its Applications*. New York: Wiley-VCH; 2000.

78 Cotton S. *Lanthanide and Actinide Chemistry*. West Sussex, England: John Wiley & Sons; 2006.

79 Seaman L.A., Wu G., Edelstein N.M., Lukens W.W., Magnani N., Hayton T.W. (2012) Probing the 5f orbital contribution to the bonding in a U(V) ketimide complex. *Journal of the American Chemical Society*, **134**, 4931–4940.

80 Lukens W.W., Edelstein N.M., Magnani N., Hayton T.W., Fortier S., Seaman L.A. (2013) Quantifying the σ and π interactions between U(V) f orbitals and halide, alkyl, alkoxide, amide and ketimide ligands. *Journal of the American Chemical Society*, **135** (29), 10742–10754.

81 Schmidt A.-C., Heinemann F.W., Lukens W.W., Meyer K. (2014) Molecular and electronic structure of dinuclear uranium bis-μ-oxo complexes with diamond core structural motifs. *Journal of the American Chemical Society*, **136** (34), 11980–11993.

82 Minasian S.G., Keith J.M., Batista E.R., Boland K.S., Clark D.L., Conradson S.D., Kozimor S.A., Martin R.L., Schwarz D.E., Shuh D.K., Wagner G.L., Wilkerson M.P., Wolfsberg L.E., Yang P. (2012) Determining relative f and d orbital contributions to M-Cl covalency in MCl_6^{2-} (M = Ti, Zr, Hf, U) and $UOCl_5^-$ using Cl K-edge X-ray absorption spectroscopy and time-dependent density functional theory. *Journal of the American Chemical Society*, **134** (12), 5586–5597.

83 Graves C.R., Scott B.L., Morris D.E., Kiplinger J.L. (2009) Selenate and tellurate complexes of pentavalent uranium. *Chemical Communications*, (7), 776–778.

84 Graves C.R., Vaughn A.E., Schelter E.J., Scott B.L., Thompson J.D., Morris D.E., Kiplinger J.L. (2008) Probing the chemistry, electronic structure and redox energetics in organometallic pentavalent uranium complexes. *Inorganic Chemistry*, **47** (24), 11879–11891.

85 Graves C.R., Scott B.L., Morris D.E., Kiplinger J.L. (2008) Tetravalent and pentavalent uranium acetylide complexes prepared by oxidative functionalization with CuCCPh. *Organometallics*, **27** (14), 3335–3337.

86 Graves C.R., Scott B.L., Morris D.E., Kiplinger J.L. (2007) Facile access to pentavalent uranium organometallics: One-electron oxidation of uranium(IV) imido complexes with copper(I) salts. *Journal of the American Chemical Society*, **129** (39), 11914–11915.

87 Graves C.R., Yang P., Kozimor S.A., Vaughn A.E., Clark D.L., Conradson S.D., Schelter E.J., Scott B.L., Thompson J.D., Hay P.J., Morris D.E., Kiplinger J.L. (2008) Organometallic uranium(V)-imido halide complexes: From synthesis to electronic structure and bonding. *Journal of the American Chemical Society*, **130** (15), 5272–5285.

88 Da Re R.E., Jantunen K.C., Golden J.T., Kiplinger J.L., Morris D.E. (2005) Molecular spectroscopy of uranium(iv) bis(ketimido) complexes. Rare observation of resonance-enhanced Raman scattering from organoactinide complexes and evidence for broken-symmetry excited states. *Journal of the American Chemical Society*, **127** (2), 682–689.

89 Jantunen K.C., Burns C.J., Castro-Rodriguez I., Da Re R.E., Golden J.T., Morris D.E., Scott B.L., Taw F.L., Kiplinger J.L. (2004) Thorium(IV) and uranium(IV) ketimide complexes prepared by nitrile insertion into actinide-alkyl and -aryl bonds. *Organometallics*, **23** (20), 4682–4692.

90 Morris D.E., Da Re R.E., Jantunen K.C., Castro-Rodriguez I., Kiplinger J.L. (2004) Trends in electronic structure and redox energetics for early-actinide pentamethylcyclopentadienyl complexes. *Organometallics*, **23** (22), 5142–5153.

91 Schelter E.J., Yang P., Scott B.L., Thompson J.D., Martin R.L., Hay P.J., Morris D.E., Kiplinger J.L. (2007) Systematic studies of early actinide complexes: Uranium(IV) fluoroketimides. *Inorganic Chemistry*, **46** (18), 7477–7488.

92 Schelter E.J., Yang P., Scott B.L., Da Re R.E., Jantunen K.C., Martin R.L., Hay P.J., Morris D.E., Kiplinger J.L. (2007) Systematic studies of early actinide complexes: Thorium(IV) fluoroketimides. *Journal of the American Chemical Society*, **129** (16), 5139–5152.

93 Clark A.E., Martin R.L., Hay P.J., Green J.C., Jantunen K.C., Kiplinger J.L. (2005) Electronic structure, excited states, and photoelectron spectra of uranium, thorium, and zirconium bis(ketimido) complexes $(C_5R_5)_2M[-NCPh_2]_2$ (M = Th, U, Zr; R = H, CH_3). *The Journal of Physical Chemistry A*, **109** (24), 5481–5491.

94 Hrobarik P., Hrobarikova V., Greif A.H., Kaupp M. (2012) Giant spin-orbit effects on ^{1}H and ^{13}C NMR chemical shifts in diamagnetic actinide complexes: Guiding the search for uranium(VI) hydrides and σ-bonded organometallics in the right spectral range. *Angewandte Chemie International Edition*, **51** (43), 10884–10888.

95 Turner H.W., Simpson S.J., Andersen R.A. (1979) Hydrido[tris(hexamethyldisilylamido)]thorium(IV) and -uranium(IV). *Journal of the American Chemical Society*, **101** (10), 2782–2782.

96 Seaman L.A., Hrobárik P., Schettini M.F., Fortier S., Kaupp M., Hayton T.W. (2013) A rare uranyl(VI)–alkyl ate complex $[Li(DME)_{1.5}]_2[UO_2(CH_2SiMe_3)_4]$ and Its comparison with a homoleptic uranium(VI)–hexaalkyl. *Angewandte Chemie International Edition*, **52** (11), 3259–3263.

97 Seaman L.A., Walensky J.R., Wu G., Hayton T.W. (2013) In pursuit of homoleptic actinide alkyl complexes. *Inorganic Chemistry*, **52** (7), 3556–3564.

98 Bursten B.E., Strittmatter R.J. (1991) Cyclopentadienyl—Actinide Complexes: Bonding and electronic structure. *Angewandte Chemie International Edition*, **30** (9), 1069–1085.

99 Burns C.J., Smith W.H., Huffman J.C., Sattelberger A.P. (1990) Uranium(VI) organoimido complexes. *Journal of the American Chemical Society*, **112** (8), 3237–3239.

100 Schelter E.J., Wu R., Veauthier J.M., Bauer E.D., Booth C.H., Thomson R.K., Graves C.R., John K.D., Scott B.L., Thompson J.D., Morris D.E., Kiplinger J.L. (2010) Comparative study of f-element electronic structure across a series of multimetallic actinide and lanthanoid-actinide complexes possessing redox-active bridging ligands. *Inorganic Chemistry*, **49** (4), 1995–2007.

101 Cantat T., Graves C.R., Jantunen K.C., Burns C.J., Scott B.L., Schelter E.J., Morris D.E., Hay P.J., Kiplinger J.L. (2008) Evidence for the involvement of 5f orbitals in the bonding and reactivity of organometallic actinide compounds: Thorium(IV) and uranium(IV) bis(hydrazonato) complexes. *Journal of the American Chemical Society*, **130** (51), 17537–17551.

102 Fortier S., Wu G., Hayton T.W. (2009) Synthesis and redox chemistry of high-valent uranium aryloxides. *Inorganic Chemistry*, **48** (7), 3000–3011.

103 Fortier S., Wu G., Hayton T.W. (2008) Synthesis and characterization of three homoleptic alkoxides of uranium: $[Li(THF)]_2[U^{IV}(O^tBu)_6]$, $[Li(Et_2O)][U^V(O^tBu)_6]$, and $U^{VI}(O^tBu)_6$. *Inorganic Chemistry*, **47** (11), 4752–4761.

104 Seaman L.A., Fortier S., Wu G., Hayton T.W. (2011) Comparison of the redox chemistry of primary and secondary amides of U(IV): Isolation of a U(VI) bis(imido) complex or a homoleptic U(VI) amido complex. *Inorganic Chemistry*, **50** (2), 636–646.

105 Meyer K., Mindiola D.J., Baker T.A., Davis W.M., Cummins C.C. (2000) Uranium hexakisamido complexes. *Angewandte Chemie International Edition*, **39** (17), 3063–3066.

106 Fortier S., Hayton T.W. (2010) Oxo ligand functionalization in the uranyl ion (UO_2^{2+}). *Coordination Chemistry Reviews*, **254** (3–4), 197–214.

107 Winfield J.M., Andersen G.M., Iqbal J., Sharp D.W.A., Cameron J.H., McLeod A.G. (1984) Redox reactions involving molybdenum, tungsten, and uranium hexafluorides in acetonitrile. *Journal of Fluorine Chemistry*, **24** (3), 303–317.

108 Bosse E., Den Auwer C., Berthon C., Guilbaud P., Grigoriev M.S., Nikitenko S., Naour C.L., Cannes C., Moisy P. (2008) Solvation of UCl_6^{2-} anionic complex by $MeBu_3N^+$, $BuMe_2Im^+$, and $BuMeIm^+$ cations. *Inorganic Chemistry*, **47** (13), 5746–5755.

109 Nikitenko S.I., Cannes C., Le Naour C., Moisy P., Trubert D. (2005) Spectroscopic and electrochemical studies of U(IV)-hexachloro complexes in hydrophobic room-temperature ionic liquids $[BuMeIm][Tf_2N]$ and $[MeBu_3N][Tf_2N]$. *Inorganic Chemistry*, **44** (25), 9497–9505.

110 Mullane K.C., Lewis A.J., Yin H., Carroll P.J., Schelter E.J. (2014) Anomalous one-electron processes in the chemistry of uranium nitrogen multiple bonds. *Inorganic Chemistry*, **53** (17), 9129–9139.

111 Lewis A.J., Carroll P.J., Schelter E.J. (2013) Stable uranium(VI) methyl and acetylide complexes and the elucidation of an inverse trans influence ligand series. *Journal of the American Chemical Society*, **135** (35), 13185–13192.

112 Lewis A.J., Mullane K.C., Nakamaru-Ogiso E., Carroll P.J., Schelter E.J. (2014) The inverse trans influence in a family of pentavalent uranium complexes. *Inorganic Chemistry*, **53** (13), 6944–6953.

113 La Pierre H.S., Scheurer A., Heinemann F.W., Hieringer W., Meyer K. (2014) Synthesis and characterization of a uranium(II) monoarene complex supported by δ backbonding. *Angewandte Chemie International Edition*, **53** (28), 7158–7162.

114 La Pierre H.S., Kameo H., Halter D.P., Heinemann F.W., Meyer K. (2014) Coordination and redox isomerization in the reduction of a uranium(III) monoarene complex. *Angewandte Chemie International Edition*, **53** (28), 7154–7157.

115 Cotton F.A., Wilkinson G., Murillo C.A., Bochmann M. *Advanced Inorganic Chemistry*. 6th ed. New York: John Wiley & Sons; 1999. 1146–1147.

116 Burns C.J., Andersen R.A. (1987) Preparation of the first η^2-olefin complex of a 4f-transition metal, $(Me_5C_5)_2Yb(\mu\text{-}(C_2H_4)Pt(PPh_3)_2$. *Journal of the American Chemical Society*, **109** (3), 915–917.

117 Edwards P.G., Andersen R.A., Zalkin A. (1981) Tertiary phosphine derivatives of the f-block metals. Preparation of $X_4M(Me_2PCH_2CH_2PMe_2)_2$, where X is halide, methyl or phenoxy and M is thorium or uranium. Crystal structure of tetraphenoxybis[bis(1,2-dimethylphosphino)ethane]uranium(IV). *Journal of the American Chemical Society*, **103** (26), 7792–7794.

118 Edwards P.G., Andersen R.A., Zalkin A. (1984) Preparation of tetraalkyl phosphine complexes of the f-block metals. Crystal structure of $Th(CH_2Ph)_4(Me_2PCH_2CH_2PMe_2)$ and $U(CH_2Ph)_3Me(Me_2PCH_2CH_2PMe_2)$. *Organometallics*, **3** (2), 293–398.

119 Minasian S.G., Krinsky J.L., Arnold J. (2011) Evaluating f-element bonding from structure and thermodynamics. *Chemistry – A European Journal*, **17** (44), 12234–12245.

120 Foyentin M., Folcher G., Ephritikhine M. (1987) Alkyne- and alkyl-tris(cyclopentadienyl) complexes of uranium(III). *Journal of the Chemical Society, Chemical Communications*, (7), 494–495.

121 Del Mar Conejo M., Parry J.S., Carmona E., Schultz M., Brennann J.G., Beshouri S.M., Andersen R.A., Rogers R.D., Coles S., Hursthouse M. (1999) Carbon monoxide and isocyanide complexes of trivalent uranium metallocenes. *Chemistry – A European Journal*, **5** (10), 3000–3009.

122 Evans W.J., Kozimor S.A., Nyce G.W., Ziller J.W. (2003) Comparative reactivity of sterically crowded nf^3 $(C_5Me_5)_3Nd$ and $(C_5Me_5)_3U$ complexes with CO: Formation of a nonclassical carbonium ion versus an *f* element metal carbonyl complex. *Journal of the American Chemical Society*, **125** (45), 13831–13835.

123 Brennan J.G., Andersen R.A., Robbins J.L. (1986) Preparation of the first molecular carbon monoxide complex of uranium, $(Me_3SiC_5H_4)_3UCO$. *Journal of the American Chemical Society*, **108** (2), 335–336.

124 Maron L., Eisenstein O., Andersen R.A. (2009) The Bond between CO and Cp'_3U in $Cp'_3U(CO)$ Involves back-bonding from the Cp'_3U ligand-based orbitals of π-symmetry, where Cp' represents a substituted cyclopentadienyl ligand. *Organometallics*, **28** (13), 3629–3635.

125 Castro-Rodriguez I., Meyer K. (2005) Carbon dioxide reduction and carbon monoxide activation employing a reactive uranium(III) complex. *Journal of the American Chemical Society*, **127** (32), 11242–11243.

126 Frey A.S., Cloke F.G.N., Hitchcock P.B., Day I.J., Green J.C., Aitken G. (2008) Mechanistic studies on the reductive cyclooligomerisation of CO by U(III) mixed sandwich complexes; the molecular structure of $[(U(\eta\text{-}C_8H_6\{Si^iPr_3\text{-}1,4\}_2)(\eta\text{-}Cp^*)]_2(\mu\text{-}\eta^1{:}\eta^1\text{-}C_2O_2)$. *Journal of the American Chemical Society*, **130** (42), 13816–13817.

127 Summerscales O.T., Cloke F.G.N., Hitchcock P.B., Green J.C., Hazari N. (2006) Reductive cyclotrimerization of carbon monoxide to the deltate dianion by an organometallic uranium complex. *Science*, **311** (5762), 829–831.

128 Summerscales O.T., Cloke F.G.N., Hitchcock P.B., Green J.C., Hazari N. (2006) Reductive cyclotetramerization of CO to squarate by a U(III) complex: The X-ray crystal structure of $[(U(\eta\text{-}C_8H_6\{Si^iPr_3\text{-}1,4\}_2)(\eta\text{-}C_5Me_4H)]_2(\mu\text{-}\eta^2{:}\eta^2\text{-}C_4O_4)$. *Journal of the American Chemical Society*, **128** (30), 9602–9603.

129 Evans W.J., Kozimor S.A., Ziller J.W. (2003) A monometallic f element complex of dinitrogen: $(C_5Me_5)_3U(\eta^1\text{-}N_2)$. *Journal of the American Chemical Society*, **125** (47), 14264–14265.

130 Kaltsoyannis N., Scott P. (1998) Evidence for actinide metal to ligand π backbonding. Density functional investigations of the electronic structure of $\{(NH_2)_3(NH_3)U\}_2(\mu^2\text{-}\eta^2{:}\eta^2\text{-}N_2)$. *Chemical Communications*, (**16**), 1665–1666.

131 Roussel P., Errington W., Kaltsoyannis N., Scott P. (2001) Back bonding without sigma-bonding: A unique pi-complex of dinitrogen with uranium. *Journal of Organometallic Chemistry*, **635** (1–2), 69–74.

132 Cloke F.G.N., Green J.C., Kaltsoyannis N. (2004) Electronic structure of $U_2(\mu^2\text{-}N_2)(\eta^5\text{-}C_5Me_5)_2(\eta^8\text{-}C_8H_4(SiPr_3^i)_2)_2$. *Organometallics*, **23** (4), 832–835.

133 Mansell S.M., Kaltsoyannis N., Arnold P.L. (2011) Small molecule activation by uranium tris(aryloxides): Experimental and computational studies of binding of N_2, coupling of CO, and deoxygenation insertion of CO_2 under ambient conditions. *Journal of the American Chemical Society*, **133** (23), 9036–9051.

134 Huang Q.-R., Kingham J.R., Kaltsoyannis N. (2015) The strength of actinide-element bonds from the quantum theory of atoms-in-molecules. *Dalton Transactions*, **44** (6), 2554–2566.

135 Fang B., Ren W., Hou G., Zi G., Fang D.-C., Maron L., Walter M.D. (2014) An actinide metallacyclopropene complex: Synthesis, structure, reactivity, and computational studies. *Journal of the American Chemical Society*, **136** (49), 17249–17261.

136 Siladke N.A., Meihaus K.R., Ziller J.W., Fang M., Furche F., Long J.R., Evans W.J. (2012) Synthesis, structure, and magnetism of an f element nitrosyl complex, $(C_5Me_4H)_3UNO$. *Journal of the American Chemical Society*, **134** (2), 1243–1249.

137 Gendron F., Guennic B.L., Autschbach J. (2014) Magnetic properties and electronic structures of $Ar_3U^{IV}\text{–}L$ complexes with $Ar = C_5(CH_3)_4H^-$ or $C_5H_5^-$ and $L = CH_3$, NO, and Cl. *Inorganic Chemistry*, **53** (24), 13174–13187.

138 Minasian S.G., Krinsky J.L., Williams V.A., Arnold J. (2008) A heterobimetallic complex with an unsupported uranium(III) – aluminum(I) bond: $(CpSiMe_3)_3U – AlCp^*$ $(Cp^* = C_5Me_5)$. *Journal of the American Chemical Society*, **130** (31), 10086–10087.

139 Minasian S.G., Krinsky J.J., Rinehart J.D., Copping R., Tyliszczak T., Janousch M., Shuh D.K., Arnold J. (2009) A comparison of 4*f vs* 5*f* metal-metal bonds in $(CpSiMe_3)_3M\text{-}ECp^*$ (M = Nd, U; E = Al, Ga; $Cp^* = C_5Me_5$): Synthesis, thermodynamics, magentism, and electronic stucture. *Journal of the American Chemical Society*, **131** (38), 13767–13783.

140 Liddle S.T., Mills D.P. (2009) Metal-metal bonds in f-element chemistry. *Dalton Transactions*, (**29**), 5592–5605.

141 Evans W.J., Kozimor S.A., Ziller J.W., Kaltsoyannis N. (2004) Structure, reactivity, and density functional theory analysis of the six-electron reductant, $[(C_5Me_5)_2U]_2(\mu\text{-}\eta^6\text{:}\eta^6\text{-}C_6H_6)$, Synthesized via a new mode of $(C_5Me_5)_3M$ reactivity. *Journal of the American Chemical Society*, **126** (44), 14533–14547.

142 Diaconescu P.L., Cummins C.C. (2012) μ-η^6,η^6-Arene-bridged diuranium hexakisketimide complexes isolable in two states of charge. *Inorganic Chemistry*, **51** (5), 2902–2916.

143 Arnold P.L., Mansell S.M., Maron L., McKay D. (2012) Spontaneous reduction and C–H borylation of arenes mediated by uranium(III) disproportionation. *Nature Chemistry*, **4** (8), 668–674.

144 Vlaisavljevich B., Diaconescu P.L., Lukens W.W., Gagliardi L., Cummins C.C. (2013) Investigations of the electronic structure of arene-bridged diuranium complexes. *Organometallics*, **32** (5), 1341–1352.

145 Mougel V., Camp C., Pécaut J., Copéret C., Maron L., Kefalidis C.E., Mazzanti M. (2012) Siloxides as supporting ligands in uranium(III)-mediated small-molecule activation. *Angewandte Chemie International Edition*, **51** (49), 12280–12284.

146 Mills D.P., Moro F., McMaster J., van Slageren J., Lewis W., Blake A.J., Liddle S.T. (2011) A delocalized arene-bridged diuranium single-molecule magnet. *Nature Chemistry*, **3** (6), 454–460.

147 Liddle S.T. (2015) Inverted sandwich arene complexes of uranium. *Coordination Chemistry Reviews*, **293–294** (0), 211–227.

148 Camp C., Mougel V., Pécaut J., Maron L., Mazzanti M. (2013) Cation-mediated conversion of the state of charge in uranium arene inverted-sandwich complexes. *Chemistry – A European Journal*, **19** (51), 17528–17540.

149 Arnold P.L., Prescimone A., Farnaby J.H., Mansell S.M., Parsons S., Kaltsoyannis N. (2015) Characterizing pressure-induced uranium CH agostic bonds. *Angewandte Chemie International Edition*, **54** (23), 6735–6739.

150 Arnold P.L., Farnaby J.H., White R.C., Kaltsoyannis N., Gardiner M.G., Love J.B. (2014) Switchable π-coordination and C-H metallation in small-cavity macrocyclic uranium and thorium complexes. *Chemical Science*, **5** (2), 756–765.

151 Nakai H., Hu X.L., Zakharov L.N., Rheingold A.L., Meyer K. (2004) Synthesis and characterization of *N*-heterocyclic carbene complexes of uranium(III). *Inorganic Chemistry*, **43** (3), 855–857.

152 Arnold P.L., Blake A.L., Wilson C. (2005) Synthesis and small molecule reactivity of uranium(IV) alkoxide complexes with both bound and pendant *N*-heterocyclic carbene ligands. *Chemistry – A European Journal*, **11** (20), 6095–6099.

153 Cantat T., Arliguie T., Noel A., Thuéry P., Ephritikhine M., Le Floch P., Mezailles N. (2009) The U = C double bond: Synthesis and study of uranium nucleophilic carbene complexes. *Journal of the American Chemical Society*, **131** (3), 963–972.

154 Tourneux J.-C., Berthet J.-C., Cantat T., Thuéry P., Mezailles N., Le Floch P., Ephritikhine M. (2011) Uranium(IV) nucleophilic carbene complexes. *Organometallics*, **30** (11), 2957–2971.

155 Tourneux J.-C., Berthet J.-C., Cantat T., Thuéry P., Mezailles N., Ephritikhine M. (2011) Exploring the uranyl organometallic chemistry: From single to double uranium-carbon bonds. *Journal of the American Chemical Society*, **133** (16), 6162–6165.

156 Lu E., Cooper O.J., McMaster J., Tuna F., McInnes E.J.L., Lewis W., Blake A.J., Liddle S.T. (2014) Synthesis, characterization, and reactivity of a uranium(VI) carbene imido oxo complex. *Angewandte Chemie International Edition*, **53** (26), 6696–6700.
157 Cooper O.J., Mills D.P., McMaster J., Moro F., Davies E.S., Lewis W., Blake A.J., Liddle S.T. (2011) Uranium-carbon multiple bonding: Facile access to the pentavalent uranium carbene $U\{C(PPh_2NSiMe_3)_2\}(Cl)_2(I)$ and Comparison of $U^V=C$ and $U^{IV}=C$ bonds. *Angewandte Chemie International Edition*, **50** (10), 2383–2386.
158 Mills D.P., Cooper O.J., Tuna F., McInnes E.J.L., Davies E.S., McMaster J., Moro F., Lewis W., Blake A.J., Liddle S.T. (2012) Synthesis of a uranium(VI)-carbene: Reductive formation of uranyl(V)-methanides, oxidative preparation of a $R_2C=U=O^{2+}$ analogue of the $O=U=O^{2+}$ uranyl ion ($R=Ph_2PNSiMe_3$), and comparison of the nature of $U^{IV}=C$, $U^V=C$, and $U^{VI}=C$ double bonds. *Journal of the American Chemical Society*, **134** (24), 10047–10054.
159 Fortier S., Walensky J.R., Wu G., Hayton T.W. (2011) Synthesis of a phosphorano-stabilized U(IV)-carbene via one-electron oxidation of a U(III)-ylide adduct. *Journal of the American Chemical Society*, **133** (18), 6894–6897.
160 Johnson S.A., Bart S.C. (2015) Achievements in uranium alkyl chemistry: Celebrating sixty years of synthetic pursuits. *Dalton Transactions*, **44** (17), 7710–7726.
161 Duhovic S., Khan S., Diaconescu P.L. (2010) In situ generation of uranium alkyl complexes. *Chemical Communications*, **46** (19), 3390–3392.
162 Cruz C.A., Emslie D.J.H., Harrington L.E., Britten J.F., Robertson C.M. (2007) Extremely stable thorium(IV) dialkyl complexes supported by rigid tridentate 4,5-bis(anilido)xanthene and 2,6-bis(anilidomethyl)pyridine ligands. *Organometallics*, **26** (3), 692–701.
163 Seppelt K. (2003) Nonoctahedral structures. *Accounts of Chemical Research*, **36** (2), 147–153.
164 Kaupp M. (2001) "Non-VSEPR" structures and bonding in d^0 systems. *Angewandte Chemie International Edition*, **40** (19), 3534–3565.
165 Fortier S., Melot B.C., Wu G., Hayton T.W. (2009) Homoleptic uranium(IV) alkyl complexes: Synthesis and characterization. *Journal of the American Chemical Society*, **131** (42), 15512–15521.
166 Pedrick E.A., Hrobárik P., Seaman L.A., Wu G., Hayton T.W. (2016) Synthesis, structure and bonding of hexaphenyl thorium(iv): observation of a non-octahedral structure. *Chemical Communications*, **52** (4), 689–692.
167 Lauke H., Swepston P.J., Marks T.J. (1984) Synthesis and characterization of a homoleptic actinide alkyl. The heptamethylthorate(IV) ion: A complex with seven metal-carbon σ bonds. *Journal of the American Chemical Society*, **106** (22), 6841–6843.
168 Straka M., Hrobárik P., Kaupp M. (2005) Understanding structure and bonding in early actinide $6d^05f^0$ MX_6^q ($M=Th-Np$; $X=H, F$) Complexes in comparison with their transition metal $5d^0$ analogues. *Journal of the American Chemical Society*, **127** (8), 2591–2599.
169 Shortland A.J., Wilkinson G. (1973) Preparation and properties of hexamethyltungsten. *Journal of the Chemical Society, Dalton Transactions*, (**8**), 872–876.
170 Kleinhenz S., Pfennig V., Seppelt K. (1998) Preparation and structures of $[W(CH_3)_6]$, $[Re(CH_3)_6]$, $[Nb(CH_3)_6]^-$, and $[Ta(CH_3)_6]^-$. *Chemistry – A European Journal*, **4** (9), 1687–1691.

171 Pfennig V., Seppelt K. (1996) Crystal and molecular structure of hexamethyltungsten and hexamethylrhenium. *Science*, **271** (5249), 626–628.
172 Evans W.J., Kozimor S.A., Ziller J.W. (2005) Molecular octa-uranium rings with alternating nitride and azide bridges. *Science*, **309** (5742), 1835–1838.
173 Evans W.J., Miller K.A., Ziller J.W., Greaves J. (2007) Analysis of uranium azide and nitride complexes by atmospheric pressure chemical ionization mass spectrometry. *Inorganic Chemistry*, **46** (19), 8008–8018.
174 Todorova T.K., Gagliardi L., Walensky J.R., Miller K.A., Evans W.J. (2010) DFT and CASPT2 analysis of polymetallic uranium nitride and oxide complexes: How theory can help when X-ray analysis is inadequate. *Journal of the American Chemical Society*, **132** (35), 12397–12403.
175 Fox A.R., Arnold P.L., Cummins C.C. (2010) Uranium nitrogen multiple bonding: Isostructural anionic, neutral, and cationic uranium nitride complexes featuring a linear U = N = U core. *Journal of the American Chemical Society*, **132** (10), 3250–3251.
176 Fox A.R., Cummins C.C. (2009) Uranium-nitrogen multiple bonding: The case of a four-coordinate uranium(VI) nitridoborate complex. *Journal of the American Chemical Society*, **131** (16), 5716–5717.
177 Fortier S., Wu G., Hayton T.W. (2010) Synthesis of a nitrido-substituted analogue of the uranyl ion, $N = U = O^+$. *Journal of the American Chemical Society*, **132** (20), 6888–6889.
178 Thomson R.K., Cantat T., Scott B.L., Morris D.E., Batista E.R., Kiplinger J.L. (2010) Uranium azide photolysis results in C-H bond activation and provides evidence for a terminal uranium nitride. *Nature Chemistry*, **2** (9), 723–729.
179 Camp C., Pécaut J., Mazzanti M. (2013) Tuning uranium-nitrogen multiple bond formation with ancillary siloxide ligands. *Journal of the American Chemical Society*, **135** (32), 12101–12111.
180 Nocton G., Pécaut J., Mazzanti M. (2008) A nitrido-centered uranium azido cluster obtained from a uranium azide. *Angewandte Chemie International Edition*, **47** (16), 3040–3042.
181 King D.M., McMaster J., Tuna F., McInnes E.J.L., Lewis W., Blake A.J., Liddle S.T. (2014) Synthesis and characterization of an f-block terminal parent imido U = NH complex: A masked uranium(IV) nitride. *Journal of the American Chemical Society*, **136** (15), 5619–5622.
182 Gardner B.M., Balazs G., Scheer M., Tuna F., McInnes E.J.L., McMaster J., Lewis W., Blake A.J., Liddle S.T. (2014) Triamidoamine-uranium(IV)-stabilized terminal parent phosphide and phosphinidene complexes. *Angewandte Chemie International Edition*, **53** (17), 4484–4488.
183 Gardner B.M., Balazs G., Scheer M., Tuna F., McInnes E.J.L., McMaster J., Lewis W., Blake A.J., Liddle S.T. (2015) Triamidoamine uranium(IV)-arsenic complexes containing one-, two- and threefold U-As bonding interactions. *Nature Chemistry*, **7** (7), 582–590.
184 King D.M., Tuna F., McInnes E.J.L., McMaster J., Lewis W., Blake A.J., Liddle S.T. (2013) Isolation and characterization of a uranium(VI)-nitride triple bond. *Nature Chemistry*, **5** (6), 482–488.
185 Roussel P., Boaretto R., Kingsley A.J., Alcock N.W., Scott P. (2002) Reactivity of a triamidoamine complex of trivalent uranium. *Dalton Transactions*, (7), 1423–1428.

186 Zi G., Jia L., Werkema E.L., Walter M.D., Gottfriedsen J.P., Andersen R.A. (2005) Preparation and Reactions of Base-Free Bis(1,2,4-tri-*tert*-butylcyclopentadienyl) uranium Oxide, $Cp'_2U=O$. *Organometallics,* **24** (17), 4251–4264.

187 Barros N., Maynau D., Maron L., Eisenstein O., Zi G.F., Andersen R.A. (2007) Single but stronger UO, double but weaker UNMe bonds: The tale told by Cp_2UO and Cp_2UNR. *Organometallics,* **26** (20), 5059–5065.

188 Fortier S., Kaltsoyannis N., Wu G., Hayton T.W. (2011) Probing the reactivity and electronic structure of a uranium(V) terminal oxo complex. *Journal of the American Chemical Society,* **133** (36), 14224–14227.

189 Fortier S., Brown J.L., Kaltsoyannis N., Wu G., Hayton T.W. (2012) Synthesis, molecular and electronic structure of $U^V(O)[N(SiMe_3)_2]_3$. *Inorganic Chemistry,* **51** (3), 1625–1633.

190 King D.M., Tuna F., McMaster J., Lewis W., Blake A.J., McInnes E.J.L., Liddle S.T. (2013) Single-molecule magnetism in a single-ion triamidoamine uranium(V) terminal mono-oxo complex. *Angewandte Chemie International Edition,* **52** (18), 4921–4924.

191 Gaunt A.J., Scott B.L., Neu M.P. (2005) Homoleptic uranium(III) imidodiphosphinochalcogenides including the first structurally characterised molecular trivalent actinide-Se bond. *Chemical Communications,* (**25**), 3215.

192 Gaunt A.J., Scott B.L., Neu M.P. (2006) A molecular actinide-tellurium bond and comparison of bonding in $[M^{III}\{N(TeP^iPr_2)_2\}_3]$ (M = U, La). *Angewandte Chemie International Edition,* **45** (10), 1638–1641.

193 Matson E.M., Breshears A.T., Kiernicki J.J., Newell B.S., Fanwick P.E., Shores M.P., Walensky J.R., Bart S.C. (2014) Trivalent uranium phenylchalcogenide complexes: Exploring the bonding and reactivity with CS_2 in the Tp^*_2UEPh series (E = O, S, Se, Te). *Inorganic Chemistry,* **53** (24), 12977–12985.

194 Daly S.R., Keith J.M., Batista E.R., Boland K.S., Kozimor S.A., Martin R.L., Scott B.L. (2012) Probing $Ni[S_2PR_2]_2$ electronic structure to generate insight relevant to minor actinide extraction chemistry. *Inorganic Chemistry,* **51** (14), 7551–7560.

195 Ventelon L., Lescop C., Arliguie T., Leverd P.C., Lance M., Nierlich M., Ephritikhine M. (1999) Synthesis and X-ray crystal structure of $[Na(18\text{-crown-}6)][U(Cp^*)_2(SBu^t)(S)]$, the first f-element compound containing a metal-sulfur double bond. *Chemical Communications,* (7), 659–660.

196 Camp C., Antunes M.A., Garcia G., Ciofini I., Santos I.C., Pecaut J., Almeida M., Marcalo J., Mazzanti M. (2014) Two-electron versus one-electron reduction of chalcogens by uranium(iii): synthesis of a terminal U(v) persulfide complex. *Chemical Science,* **5** (2), 841–846.

197 Malliakas C.D., Yao J., Wells D.M., Jin G.B., Skanthakumar S., Choi E.S., Balasubramanian M., Soderholm L., Ellis D.E., Kanatzidis M.G., Ibers J.A. (2012) Oxidation state of uranium in $A_6Cu_{12}U_2S_{15}$ (A = K, Rb, Cs) compounds. *Inorganic Chemistry,* **51** (11), 6153–6163.

198 Brown J.L., Fortier S., Wu G., Kaltsoyannis N., Hayton T.W. (2013) Synthesis and spectroscopic and computational characterization of the chalcogenido-substituted analogues of the uranyl ion, $[OUE]^{2+}$ (E = S, Se). *Journal of the American Chemical Society,* **135** (14), 5352–5355.

199 Denning R.G. (2007) Electronic structure and bonding in actinyl ions and their analogs. *Journal of Physical Chemistry A,* **111** (20), 4125–4143.

200 Denning R. (1992) Electronic structure and bonding in actinyl ions. *Structure and Bonding*, **79**, 215–276.
201 La Pierre H.S., Meyer K. (2013) Uranium–ligand multiple bonding in uranyl analogues, $[L=U=L]^{n+}$, and the inverse trans influence. *Inorganic Chemistry*, **52** (2), 529–539.
202 Bagnall K.W., du Preez J.G.H. (1973) $(C_6H_5)_4P[UOCl_5]$: A new type of uranium(VI) oxochloro-complex. *Chemical Communications*, (21), 820–821.
203 Denning R.G., Green J.C., Hutchings T.E., Dallera C., Tagliaferri A., Giarda K., Brookes N.B., Braicovich L. (2002) Covalency in the uranyl ion: A polarized X-ray spectroscopic study. *Journal of Chemical Physics*, **117** (17), 8008–8020.
204 Kosog B., La Pierre H.S., Heinemann F.W., Liddle S.T., Meyer K. (2012) Synthesis of uranium(VI) terminal oxo complexes: Molecular geometry driven by the inverse trans-influence. *Journal of the American Chemical Society*, **134** (11), 5284–5289.
205 Lewis A.J., Nakamaru-Ogiso E., Kikkawa J.M., Carroll P.J., Schelter E.J. (2012) Pentavalent uranium trans-dihalides and -pseudohalides. *Chemical Communications*, **48** (41), 4977–4979.
206 Bart S.C., Anthon C., Heinemann F.W., Bill E., Edelstein N.M., Meyer K. (2008) Carbon dioxide activation with sterically pressured mid- and high-valent uranium complexes. *Journal of the American Chemical Society*, **130** (37), 12536–12546.
207 Bart S.C., Meyer K. (2008) Highlights in uranium coordination chemistry. *Structure and Bonding (Berlin)*, **127**, 119–176.
208 Castro-Rodriguez I., Meyer K. (2006) Small molecule activation at uranium coordination complexes: Control of reactivity via molecular architecture. *Chemical Communications*, (**13**), 1353–1368.
209 Lam O.P., Franke S.M., Nakai H., Heinemann F.W., Hieringer W., Meyer K. (2012) Observation of the inverse trans influence (ITI) in a uranium(V) imide coordination complex: an experimental study and theoretical evaluation. *Inorganic Chemistry*, **51** (11), 6190–6199.
210 Hayton T.W., Boncella J.M., Scott B.L., Batista E.R. (2006) Exchange of an imido ligand in bis(imido) complexes of uranium. *Journal of the American Chemical Society*, **128** (39), 12622–12623.
211 Hayton T.W., Boncella J.M., Scott B.L., Batista E.R., Hay P.J. (2006) Synthesis and reactivity of the imido analogues of the uranyl ion. *Journal of the American Chemical Society*, **128** (32), 10549–10559.
212 Spencer L.P., Gdula R.L., Hayton T.W., Scott B.L., Boncella J.M. (2008) Synthesis and reactivity of bis(imido) uranium(vi) cyclopentadienyl complexes. *Chemical Communications*, (**40**), 4986–4988.
213 Crawford M.-J., Ellern A., Karaghiosoff K., Mayer P., Nöth H., Suter M. (2004) Synthesis and characterization of heavier dioxouranium(VI) dihalides. *Inorganic Chemistry*, **43** (22), 7120–7126.
214 Spencer L.P., Yang P., Scott B.L., Batista E.R., Boncella J.M. (2009) Uranium(VI) bis(imido) chalcogenate complexes: Synthesis and density functional theory analysis. *Inorganic Chemistry*, **48** (6), 2693–2700.
215 Kannan S., Barnes C.L., Duval P.B. (2005) Synthesis and structural characterization of a uranyl(VI) complex possessing unsupported unidentate thiolate ligands. *Inorganic Chemistry*, **44** (25), 9137–9139.

216 Gabelnick S.D., Reedy G.T., Chasanov M.G. (1973) Infrared spectra of matrix-isolated uranium oxide species. II. Spectral interpretation and structure of UO_3. *The Journal of Chemical Physics*, **59** (12), 6397–6404.

217 Pyykko P., Li J., Runeberg N. (1994) Quasirelativistic pseudopotential study of species isoelectronic to uranyl and the equatorial coordination of uranyl. *Journal of Physical Chemistry*, **98** (18), 4809–4813.

218 Kovács A., Konings R.J.M., Gibson J.K., Infante I., Gagliardi L. (2015) Quantum chemical calculations and experimental investigations of molecular actinide oxides. *Chemical Reviews*, **115** (4), 1725–1759.

219 Jilek R.E., Spencer L.P., Lewis R.A., Scott B.L., Hayton T.W., Boncella J.M. (2012) A direct route to bis(imido)uranium(V) halides via metathesis of uranium tetrachloride. *Journal of the American Chemical Society*, **134** (24), 9876–9878.

220 Duval P.B., Burns C.J., Buschmann W.E., Clark D.L., Morris D.E., Scott B.L. (2001) Reaction of the uranyl(VI) ion (UO_2^{2+}) with a triamidoamine ligand: Preparation and structural characterization of a mixed-valent uranium(V/VI) oxo – imido dimer. *Inorganic Chemistry*, **40** (22), 5491–5496.

221 Arney D.S.J., Burns C.J. (1995) Synthesis and properties of high-valent organouranium complexes containing terminal organoimido and oxo functional groups. A new class of organo-f-element complexes. *Journal of the American Chemical Society*, **117** (37), 9448–9460.

222 Cantat T., Graves C.R., Scott B.L., Kiplinger J.L. (2009) Challenging the metallocene dominance in actinide chemistry with a soft PNP pincer ligand: New uranium structures and reactivity patterns. *Angewandte Chemie International Edition*, **48** (20), 3681–3684.

223 Duval P.B., Burns C.J., Clark D.L., Morris D.E., Scott B.L., Thompson J.D., Werkema E.L., Jia L., Andersen R.A. (2001) Synthesis and structural characterization of the first uranium cluster containing an isopolyoxometalate core. *Angewandte Chemie International Edition*, **40** (18), 3357–3361.

224 Vaughn A.E., Barnes C.L., Duval P.B. (2007) A *cis*-dioxido uranyl: Fluxional carboxylate activation from a reversible coordination polymer. *Angewandte Chemie International Edition*, **46** (35), 6622–6625.

225 Villiers C., Thuéry P., Ephritikhine M. (2008) The first *cis*-dioxido uranyl compound under scrutiny. *Angewandte Chemie International Edition*, **47** (32), 5892–5893.

226 Guan Q.L., Bai F.Y., Xing Y.H., Liu J., Zhang H.Z. (2015) Unexpected cis-dioxido uranyl carboxylate compound: Synthesis, characterization and photocatalytic activity of uranyl-succinate complexes. *Inorganic Chemistry Communications*, **59**, 36–40.

227 Schreckenbach G., Hay P.J., Martin R.L. (1998) Theoretical study of stable trans and cis isomers in $[UO_2(OH)_4]^{2-}$ using relativistic density functional theory. *Inorganic Chemistry*, **37** (17), 4442–4451.

228 Arnold P.L., Jones G.M., Odoh S.O., Schreckenbach G., Magnani N., Love J.B. (2012) Strongly coupled binuclear uranium–oxo complexes from uranyl oxo rearrangement and reductive silylation. *Nature Chemistry*, **4** (3), 221–227.

229 Christou G. (2005) Single-molecule magnets: A molecular approach to nanoscale magnetic materials. *Polyhedron*, **24** (16–17), 2065–2075.

230 Layfield R.A. (2014) Organometallic single-molecule magnets. *Organometallics*, **33** (5), 1084–1099.

231 Antunes M.A., Pereira L.C.J., Santos I.C., Mazzanti M., Marcalo J., Almeida M. (2011) $U(TpMe_2)_2(bipy)^+$: A cationic uranium(III) complex with single-molecule-magnet behavior. *Inorganic Chemistry*, **50** (20), 9915–9917.

232 Coutinho J.T., Antunes M.A., Pereira L.C.J., Bolvin H., Marcalo J., Mazzanti M., Almeida M. (2012) Single-ion magnet behaviour in $U(TpMe_2)_2I$. *Dalton Transactions*, **41** (44), 13568–13571.

233 Rinehart J.D., Long J.R. (2009) Slow magnetic relaxation in a trigonal prismatic uranium(III) complex. *Journal of the American Chemical Society*, **131** (35), 12558–12559.

234 Rinehart J.D., Long J.R. (2012) Slow magnetic relaxation in homoleptic trispyrazolylborate complexes of neodymium(III) and uranium(III). *Dalton Transactions*, **41** (44), 13572–13574.

235 Rinehart J.D., Meihaus K.R., Long J.R. (2010) Observation of a secondary slow relaxation process for the field-induced single-molecule magnet $U(H_2BPz_2)_3$. *Journal of the American Chemical Society*, **132** (22), 7572–7573.

236 Moro F., Mills D.P., Liddle S.T., van Slageren J. (2013) The inherent single-molecule magnet character of trivalent uranium. *Angewandte Chemie International Edition*, **52** (12), 3430–3433.

237 Magnani N., Apostolidis C., Morgenstern A., Colineau E., Griveau J.-C., Bolvin H., Walter O., Caciuffo R. (2011) Magnetic memory effect in a transuranic mononuclear complex. *Angewandte Chemie International Edition*, **50** (7), 1696–1698.

238 Mougel V., Chatelain L., Pécaut J., Caciuffo R., Colineau E., Griveau J.-C., Mazzanti M. (2012) Uranium and manganese assembled in a wheel-shaped nanoscale single-molecule magnet with high spin-reversal barrier. *Nature Chemistry*, **4** (12), 1011–1017.

239 Mougel V., Chatelain L., Hermle J., Caciuffo R., Colineau E., Tuna F., Magnani N., de Geyer A., Pécaut J., Mazzanti M. (2014) A uranium-based UO_2^+-Mn^{2+} single-chain magnet assembled trough cation-cation interactions. *Angewandte Chemie International Edition*, **53** (3), 819–823.

5

Coordination of Actinides and the Chemistry Behind Solvent Extraction

Aurora E. Clark[1], *Ping Yang*,[2] *and Jenifer C. Shafer*[3]

[1] *Department of Chemistry, Washington State University*
[2] *Theoretical Division, Los Alamos National Laboratory, New Mexico*
[3] *Department of Chemistry, Colorado School of Mines*

5.1 Introduction

It is indeed a daunting task to adequately discuss the rich coordination chemistry of actinides and the vast science associated with their separations processes; a task that we make no attempt to complete herein. Instead, this chapter relies upon the ample and excellent literature associated with these topics as reviews, book chapters, and complete treatises, and alternatively has a goal of focusing upon advances made within the last decade. New developments in experimental methods (particularly in the areas of X-ray absorption and scattering methods) as well as computational techniques (as in protocols for increased accuracy and extending statistical methods to larger chemical systems and longer timescales) have inspired a new generation of actinide chemists and materials scientists, and now the statement that we are in a renaissance of actinide chemistry appears to be realized. As a matter of practice, actinide coordination chemistry is intricate, possibly more so than any other s-, p-, or d-block element. The chemical and structural transformations across the period are driven by reduction-oxidation (redox) reactions, variations of ionic radii, and the energetic or spatial availability of a multiplicity of atomic orbitals. The speciation of potential actinide complexes orchestrates solution properties of fundamental and practical importance in natural and engineered settings. In terms of the latter, there is none more important than solvent extraction. Actinide ion redox potentials and coordination environments are important for understanding solvent extraction. These factors are impacted by steric demands, complications of denticity, kinetics, accessibility of actinide orbitals for covalent interaction, and so on. At the same time, solvent extraction also relies upon complex interfacial chemistry, where molecular-scale organization and dynamics has an intricate relationship with the solution-phase conditions, and simple macroscopic observations may belie a myriad of cooperative behaviors.

This chapter discusses in context the separations processes of actinides that involve primarily solvent extraction. Further, we regrettably omit important and significant

Experimental and Theoretical Approaches to Actinide Chemistry, First Edition.
Edited by John K. Gibson and Wibe A. de Jong.

advances in other separations techniques associated with actinides; for example, those that invoke magnetism, light, or nanoporous materials. Given the medium of a book chapter, there is no doubt that many challenging concepts are discussed in a somewhat superficial manner. It is our hope that in this work the excitement of the last decade can be conveyed and serve as a basis for future literature discussion. Due to the ample use of acronyms, a list is provided at the end of the chapter.

5.2 Overview of Separations Processes

Separations chemistry has always been intertwined with actinide science, as the discovery of the transuranium elements required immediate separation and purification for their detection. The earliest solvent extraction based separations of uranium used ether as a solvating "extractant." The discovery of plutonium employed bismuth phosphate ($BiPO_4$), and this separation was scaled up eight orders of magnitude to provide the initial $BiPO_4$ separations process used for industrial-scale recovery and purification of Pu. The demands of World War II required rapid scale-up and deployment of a process that could recover Pu. The environmental consequences of this have been significant, and many lessons were learned from the rapid implementation of nuclear separations technology. A brief overview of separations processes utilized in the early days of transuranic separations is provided with an eye toward understanding the lessons learned from these processes. This framework will set a dialogue for currently considered transuranic separations processes and how they depend upon a complex suite of chemistries – from aqueous speciation and actinide coordination, to metal–ligand covalency and interfacial science.

5.2.1 Classic Processes – U/Pu Recovery

The most substantially used processes during early transuranic research were $BiPO_4$, Redox, and PUREX. All of these processes generally manipulate (1) the redox chemistry available to plutonium relative to the redox stability of uranium and (2) the higher charge density of Pu and U relative to the fission products to generate a purified plutonium product. Process advancements were largely driven by a preference to improve plutonium recovery, recover uranium from fission products, or minimize high-level waste production. The PUREX process was developed in response to these separations goals.[1]

The $BiPO_4$ process is a solid-liquid separation process, where $BiPO_4$ is used as a carrier to initially precipitate Pu and the lanthanides from uranium since the uranyl cation does not co-precipitate with $BiPO_4$. After co-precipitation with the lanthanides, plutonium was oxidized using hexavalent chromium, usually $K_2Cr_2O_7$, for selective recovery. While the scale-up factor ($\sim 10^8$) for this process was highly compatible with the rapid development needs of the time, 1% to 3% of the plutonium was not precipitated in a given $BiPO_4$ strike and was sent as waste material with the uranium. This resulted in 540 kg of Pu being sent to waste. Also, since the uranium comprised approximately 98% of the feed stream, and the uranium was treated as waste, significant waste was produced from this process. In 12 years of operation at the Hanford site with 7000 tons of fuel treated, the $BiPO_4$ process developed 100 million gallons of HLW. This

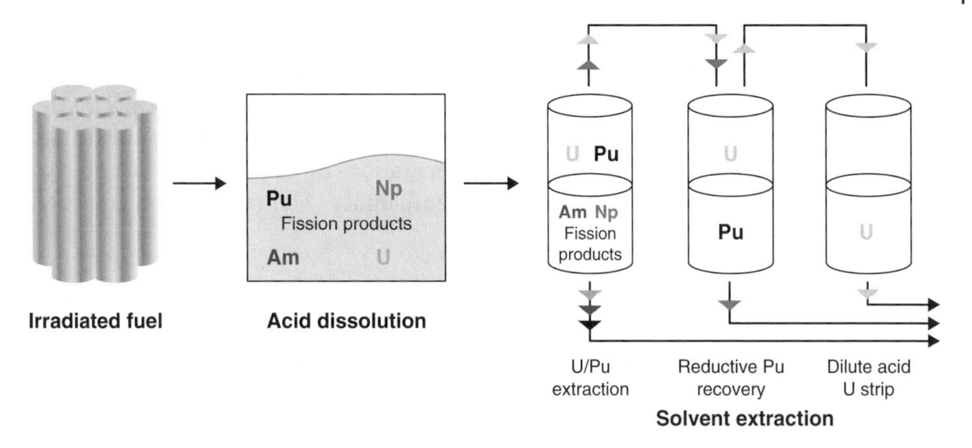

Figure 5.1 A simplified illustration of the PUREX process. (*See insert for color representation of the figure.*)

is comparable to the amount of high-level waste (HLW) produced from the PUREX during operation at Hanford for 50 years and treatment of 80,000 tons of fuel.

The Redox process was the first industrially deployed solvent extraction process for treatment of used nuclear fuel. Methylisobutylketone (MIBK) was used as a solvating organic extractant without dilution. Since this material was quite reactive under higher acid concentrations, the acid concentration had to be capped at 1 M, and additional nitrate necessary for extraction was provided using aluminum nitrate. Plutonium was oxidized to the hexavalent state using potassium chromate and uranium and plutonium were co-recovered by MIBK. The selective recovery of Pu was accomplished by reducing Pu to the trivalent state using ferrous sulfamate. Uranium was recovered by decreasing the acid concentration. The Redox process offered improved recovery of U and Pu (~99.9%) and, as a solvent extraction process, also allowed for continuous processing. However, the MIBK reagent was highly flammable, and the use of aluminum nitrate as a salting agent added significant volume to HLW streams. The Redox process operated for 16 years, processed 19,000 tons of fuel, and generated 40 million gallons of waste.

The PUREX process schematic in Figure 5.1 is the de facto standard process for used nuclear fuel processing. Separations of uranium and plutonium are largely accomplished in a similar fashion to the Redox process, though they are co-extracted by tributyl phosphate (TBP) dissolved in a kerosene diluent which replaces the neat MIBK solvent and consequently removes the need for an aluminum nitrate salting agent. Means of improving the PUREX process have been sought to improve both technological and proliferation concerns. Radiolytic degradation issues with PUREX (discussed in Section 5.4.2) via introduction of monoamide reagents. Monoamides adhere to the CHON (carbon-hydrogen-oxygen-nitrogen) principle developed to minimize low-level waste volumes by using completely incinerable reagents. The PUREX process generates a pure plutonium product stream, which can be a proliferation concern. The oxidation chemistry of the PUREX process can be modified to include co-extraction of neptunium with the uranium and plutonium (the COEX process). The designed co-extraction of other elements that would strip with plutonium has been viewed as a viable means to improve the proliferation resistance of the PUREX process.

The combination of significant process knowledge, a largely technically sound process, and inherent inertia in the nuclear industry has made replacing the PUREX process nontrivial. Monoamide reagents seem to be the most promising extractants for this task, and pilot-scale demonstrations of these reagents have provided reasonable results. Much additional effort has been expended on separations targeting the selective recovery of americium and curium away from the lanthanides. The next section considers those advances in more detail.

5.2.2 Advanced Separation Processes – Am/Cm Recovery

The recovery of americium and curium from the trivalent lanthanides would potentially enable the transmutation of americium in fast neutron spectrum nuclear reactors. The primary curium isotope, Cm-244, produced during reactor operations has a relatively short half-life, $t_{1/2} = 18.1$ years. Most projections suggest that Cm will have largely decayed to Pu-240 by the time fast spectrum transmutation is sought. Since the long-term burden associated with nuclear waste is associated with the decay of the transuranic isotopes, transmutation of these materials could move waste storage timelines from several hundred thousand years to approximately a thousand years.

In general, two different approaches can be envisioned to accomplish the recovery of americium and curium: (1) the use of soft donors, either in the aqueous or organic phase, to drive selective interactions with the trivalent actinides from the lanthanides and (2) the oxidation of americium to the hexavalent state, where it would have chemistry that is similar to the lighter hexavalent actinides, U, Np, and Pu. The United States is currently considering the use of soft donor reagents in the aqueous phase or oxidizing the americium to the hexavalent state. Most European approaches consider the use of nitrogen-containing extractants, such as the BTBP reagents. These are discussed in Section 5.4.

5.2.3 Aqueous-Based Complexants for Trivalent An/Ln Separation

The Trivalent Actinide Lanthanide Separation using Phosphorus Extractants and Aqueous Komplexants (TALSPEAK) process developed at Oak Ridge National Laboratory in the 1960s has been a springboard for development of further advanced processes. TALSPEAK uses an organophosphorus acidic extractant, HDEHP, to recover the lanthanides, while an aqueous soluble holdback reagent, DTPA, is used to selectively retain americium and curium in the aqueous phase.[2, 3] Significant amounts of lactic acid buffer (at least 1 M) have been used to maintain pH where optimal separations are obtained, pH ~ 3.5, and expedite metal phase transfer kinetics. Organic phase aggregation chemistry in TALSPEAK has been shown to exhibit steep pH dependences on metal extraction and, as discussed in Section 5.4.4, the slow extraction kinetics are not compatible with centrifugal contactor technology. In addition, two steps were originally envisioned in the UREX-1a + reprocessing lineup to address americium recovery – a TRansUranic EXtraction (TRUEX) step that would recover trivalent f-elements from the fission products and a TALSPEAK step that would recover the trivalent actinides from the lanthanides. Ultimately, the complicated and slow chemistry of TALSPEAK and additional processing facilities required to complete trivalent f-element separations has encouraged alternative schemes.

A version of Advanced TALSPEAK (sometimes called TALSQuEAK in the literature) substitutes HEH[EHP] for HDEHP and hydroxyethyl-ethylenediaminetriacetic acid (HEDTA) for DTPA.[4, 5] Both the HEH[EHP] and HEDTA reagents are relatively weaker complexants than their respective predecessors and generally provide less complex solution chemistry. Further, being smaller (with presumably faster diffusivity), the HEDTA reagent allows for a modest improvement in phase transfer kinetics. This can allow for lower concentrations of added lactate (employed as a phase transfer catalyst) and ultimately less complex solution chemistry, since organic phase partitioning of lactate is less favored.

Concurrently, compression of the advanced nuclear fuel cycle has been sought by combining the TRUEX extractant, carbamylphosphine oxide (CMPO), with the TALSPEAK organic extractant (HDEHP) in a single organic phase – the TRUSPEAK process.[6] The concept is that trivalent actinides and lanthanides could be co-recovered from other fission products in molar acid media, then the trivalent actinides could be selectively stripped from the lanthanides once contacted with lower acid media. However, advanced fuel cycle separations directly from a PUREX feed are non-ideal for the following reasons: (1) the aqueous feed arising from the raffinate of a PUREX process is in molar nitric acid, and (2) the APC reagents necessary for separation of the f-element groups are non-complexing and insoluble under these conditions. Although the TRUSPEAK process conceptually compresses the fuel cycle, the organic phase aggregation chemistry and slow metal phase transfer kinetics that compromise the TALSPEAK process are also observed in the TRUSPEAK process design. Some success has been observed by exchanging HEH[EHP] for HDEHP in TRUSPEAK,[7] but the most viable combination of extractants has N,N,N′,N′-tetra(2-ethylhexyl)diglycolamide, $T(EH)_2DGA$, and HEH[EHP]. This chemistry is currently described as the Actinide Lanthanide SEParation process (ALSEP).[8]

5.2.4 Recent Trends in Aqueous-Based Trivalent An/Ln Separations

The ALSEP process uses a diglycolamide extractant as the solvating reagent in place of CMPO. Early ALSEP formulations considered N,N,N′,N′-tetraoctyldiglycolamide (TODGA) as the extractant, but the known aggregation chemistry of TODGA discourages use in final ALSEP formulations. The higher americium uptake in molar acid media by diglycolamide reagents has encouraged their use over CMPO. The $T(EH)_2DGA$ does not extract light lanthanides as favorably, and this could be used to improve light lanthanide decontamination from the trivalent actinides. Moreover, the $T(EH)_2DGA$/HEH[EHP] combination provides the best Nd/Am separation factors ($SF_{Nd/Am} \sim 30$), while having a less steep pH dependence than the TODGA/HDEHP combination. Further studies will have to consider the possibility of slow stripping kinetics in the ALSEP process, the decontamination of problematic Sn^{4+}, and process robustness in high radiation fields.

5.2.5 Separation of Hexavalent Actinides (SANHEX) Processes

Americium/curium separations were the initial driver for developing a separation around the barely accessible hexavalent oxidation state of americium. Considering the comparable ionic radii and limited redox chemistry available to these elements, the argument has been presented that the americium/curium separation is *the* most difficult elemental separation on the periodic table. However, the accessibility of hexavalent americium recovery is compelling because a separation could be envisioned where all problematic transuranic elements, U, Pu, Np, and Am, could be co-recovered and selectively stripped as desired in a single, seamless, separation step. This chemistry is available because the "yl" cation coordination geometry is unique to pentavalent and hexavalent actinides. Utilizing the pentavalent oxidation state for separation is less attractive because it is generally non-extractable, and pentavalent uranium is unstable (Section 5.3.1). Under acidic conditions, redox driving forces dictate that hexavalent americium should oxidize water and other organic media. Therefore, the separation challenge in developing hexavalent americium separations is twofold: (1) identifying an oxidant that has sufficient overpotential to oxidize Am under acidic conditions and (2) developing a separations process where kinetics, and not thermodynamics, dictate separations chemistry.

The first report of hexavalent americium recovery using a solvent extraction system was provided by Mincher *et al.* and used sodium bismuthate as the oxidant.[9] The Bi(III/V) redox potential is around 2.0 V, which is sufficient to eclipse that needed to oxidize Am(III) to Am(VI). Early studies used TBP for recovery, though TBP was not a strong enough extractant for significant Am(VI) recovery. The CMPO extractant was considered, but recovered the reduced trivalent bismuth species and therefore was not selective enough for process goals.[10] Diamylamyl phosphonate (DAAP) instead has been identified as a viable extractant that provides sufficient Am(VI) recovery and lacks significant Bi(III) co-extraction.[11] The current issues with americium extraction now relate to the insolubility of the $NaBiO_3$ oxidant and general aversion to solid-liquid separations in nuclear technology arising from the waste legacy associated with the $BiPO_4$ process.

Alternative oxidation techniques and reagents have been considered to oxidize americium. Ozone or noble metal catalyzed ozone americium oxidation have been studied, but the americium oxidation to the hexavalent state only occurs in dilute (<1 M) acid media.[12, 13] Persulfate can oxidize Am to the hexavalent state in dilute acid media. An ITO-functionalized electrode is the first electrode to demonstrate oxidation of macroscopic amounts of americium, but it is only effective in dilute acid media.[14] Currently, the only demonstrated alternative oxidant is copper(III) periodate for applications in molar acid media.[15, 16] Copper periodate is modestly soluble (~300 mg/mL) in aqueous media, but acid-catalyzed dissociation of the higher oxidation state protecting periodate group from the copper metal center enables rapid reduction of Cu(III) to Cu(II). Separation factors for Am from Cm using molar acid media and the DAAP extractant have been competitive with comparable $NaBiO_3$-based systems. One of the biggest disadvantages, or possibly opportunities, associated with the application of copper(III) periodate for americium oxidation and separation is the tendency for periodate to precipitate tetravalent f-elements, such as cerium. If avoiding a solid-liquid separation is the desired advance, a means of reducing cerium to the trivalent, soluble form would need to be added to the separation process.

The overview of separations process schemes described here reflects a tremendously complex chemistry that spans the molecular-scale arrangements of atoms and their electronic structure, to complex self-assembly at interfaces. The rest of this chapter attempts to cover some of these relevant nuances; beginning with coordination chemistry and speciation of actinides under different solution conditions, continuing into ligand design that takes advantage of the geometry and electronic structure of actinides, and finally a detailed discussion regarding the interfacial chemistry that is foundational to an effective transport process during separations.

5.3 Coordination and Speciation of Aqueous Actinides

As described above, separations science relies heavily upon the coordination chemistry of the specific oxidation states involved and their selective manipulation through various chemical mechanisms (i.e., solution-phase conditions). The use of oxidation state control to drive actinide separation is best demonstrated by the original large-scale actinide solvent extraction separation process: PUREX. To enable PUREX processing, spent nuclear fuel is dissolved in 6–11 M nitric acid, diluted to 3–4 M, and nitrite is added to ensure Pu is present as Pu(IV) and uranium as U(VI). The Pu(IV) and U(VI) are complexed by TBP and transported into aliphatic kerosene. Trivalent minor actinides (Am, Cm, etc.) and other trivalent lanthanides remain in the aqueous phase. The Pu is then back-extracted into an aqueous phase by reduction to Pu(III). The U(VI) is not reduced and remains in the organic phase. Further in the process cycle, U(VI) is back-extracted by a dilute nitric acid (0.01 M) aqueous phase and then converted to uranium oxide. The complex redox chemistry within PUREX is utilized because of the stark contrast in the complexation geometries of the high oxidation states (+5, +6) compared to the low oxidation states (+3, +4) of U and Pu (*vide infra*).

As illustrated in the PUREX process, the aqueous solution conditions are essential to oxidation state control, requiring an understanding of the thermodynamically driven speciation favored under differing composition, pH, ionic strength, temperature, and so on. Understanding the redox potentials of actinide-containing species has been a long-standing area of research since the 1950s. Extensive work has been done and summarized in several chapters of *The Chemistry of the Actinide and Transactinide Elements* (CATE).[17] Figure 5.2 shows the redox potentials for the relevant metal ions. Given the

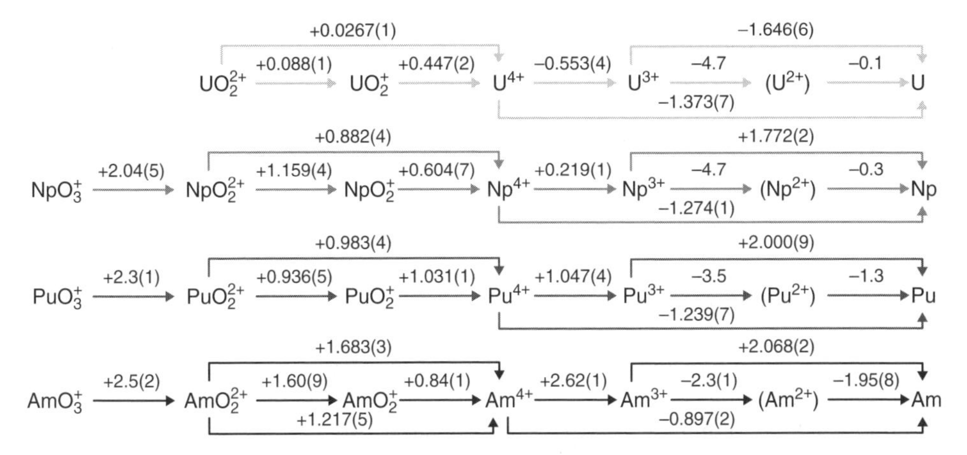

Figure 5.2 Adapted from CATE, p. 1779. Standard reduction potentials diagrams for the actinide ions (in V) versus the standard hydrogen electrode.[17] (*See insert for color representation of the figure.*)

broad overview of the separations processes described above, we now take a bottom-up approach to discuss actinide coordination in solution-phase environments relevant to separations. State-of-the-art methods to probe actinide coordination involve strong coordination between experiments and theory such as X-ray scattering techniques and computational methods. Further discussion will build on this to consider the most fundamental aspects of actinide coordination chemistry – actinide hydration – and progress toward more specific examples of actinide coordination chemistry that involve cation–cation, counterion, and solvent interactions affecting actinide speciation.

Currently, X-ray absorption methods play a major role in clarifying the coordination environment of actinides in solution. These methods have been used to probe coordination environments under relatively low concentrations of actinides, but the role of complex electrolytes, various solvents, and their mixtures are active areas of research. Three main spectral regions exist in X-ray absorption spectroscopy – each of which can contribute to the study of actinide coordination, electronic structure, and bonding. This includes the lower-energy near-edge spectroscopies (metal K-edge, metal L-edge, and also ligand – or solvent – K-edge), and the oft-used X-ray absorption fine structure (EXAFS), which extends from 50 eV to more than 1000 eV above the edge absorption feature. A complementary approach is to use high-energy X-ray scattering (HEXS), an approach particularly suitable for solution that has expanded greatly in the last 10 years.[18]

Though these experimental methods provide bond distances, and sometimes intermolecular angular data within the first coordination shell, they are often coupled with either quantum or statistical mechanical methods to allow further interpretation of the coordination chemistry and dynamic behavior of the solvation environment and to provide energetic information or distributions of configurations and species. As mentioned above, many complementary layers of information are necessary to fully describe actinide coordination chemistry, their dynamics, and how these factors affect separations. Herein, this complexity is illustrated in a few key examples that demonstrate the extent to which close integration between theory and experiment has greatly advanced this area.

In this context, within the realm of electronic structure calculations, you will observe that large clusters of solvated actinide cations are increasingly used as chemical models to account for polarization across solvation shells and to obtain reliable solution-phase thermochemistry using methods like density functional theory, and in a growing literature, correlated wave function methods. Prior work has illustrated that solvating waters donate electron density to highly charged cations in the first shell and that this can propagate to extended solvation shells.[19–21] Solvated clusters may also be embedded in dielectric continuum models to further account for the bulk solution response. Mixed QM/MM approaches are also employed; however, this approach is not as widespread.[22] Complementary statistical mechanics simulations with a broader use of ab initio molecular dynamics paradigms are featured in the recent literature. However, the typical timescales of ab initio molecular dynamics simulations are still limited to the tens of picoseconds and thus apt to the study fast dynamic processes. Care also must be taken to ensure AIMD studies results are independent of starting configurations. Classical molecular dynamics and even Monte Carlo remain a mainstay of this community; however, force field development continues to be an active area of research – with polarizable force fields generally believed to be necessary.[23] Much evidence has illustrated the sensitivity of solvation properties to the specific parameters employed. Therefore, it is appropriate to combine various computational and experimental approaches to obtain a holistic understanding of the solvation properties. In the rest of this discussion, we begin with simple coordination in dilute water systems (hydration), followed by cation–cation interactions in concentrated solution, and then we will discuss the effects of counter ions and mixed solvents.

5.3.1 Actinide Hydration

The predominance of electrostatic interactions in solution mandate that the coordination number (CN) and associated geometries within the primary solvation shell of actinides are largely dictated by the ionic radii, and the CN changes as a function of the oxidation state and position within the period. The crystal ionic radii derived by Shannon[24] are typically employed to derive ion properties in electrolyte solutions. However, these radii have begun to be revised using combined and complementary experimental and theoretical methods.[25]

Depending on the actinide, the An(VI) oxidation states can be observed under a wide range of pH values.[26] The U(V) cations are highly reactive and readily disproportionate to formal IV and VI oxidation states,[26, 27] NpO_2^+ and PuO_2^+ have been observed in acidic conditions. For the actinides in high oxidation states (+5, +6), the cations usually exist as $An^{V}O_2^+$ and $An^{VI}O_2^{2+}$ with a strong covalently bonded linear unit, O = An = O, possessing unique chemical stability. The chemical bonding of this unit is well studied both theoretically[28–30] and experimentally[31] and will thus receive limited discussion. Coordination of solvent or ligands usually occurs in the equatorial plane perpendicular to the linear O = An = O axis with low CNs of four, five, or six. X-ray scattering, electronic structure, and molecular dynamics studies all support five water molecules coordinating the actinyl dications of U, Np, and Pu in dilute solutions to form a dynamic evolving aqua complex.[29, 32–37] The ground states of these species are high-spin states.[29, 37–39] A recent study of $AmO_2^{+/2+}$, using X-ray absorption techniques and first-principles modeling, also suggests five water molecules at the equatorial plane.[40] Somewhat less is known regarding the actinyl monocations; however, experimental studies have indicated that they have 4–6 waters in the first solvation shell.[26] In the gas phase, AnO_2^+ hydrates have been recently produced by electrospray ionization, where only four H_2O were found in the first solvation shell.[31] This was supported by large cluster calculations of solvated actinyl monocations studied with DFT and polarizable continuum model (PCMs).[19]

Regarding the lighter actinides, the trivalent oxidation state increases in stability moving from U to Np to Pu. These trivalent ions are obtained under acidic conditions at pH 0 within non-complexing media.[26] Recent work reveals a strong analogy between the Ln(III) and An(III) series concerning hydration properties.[25] This implies that the two series behave very similarly in water with respect to hydration, structural, and thermodynamic properties, further justifying the use of Ln(III) ions as experimental analogues of An(III) ions under certain circumstances. Numerous studies exist regarding trivalent ion coordination chemistry; however, tetravalent An's have a strong tendency toward hydrolysis in aqueous solution and can undergo polynucleation or colloidal formation, which makes them much more challenging to study.[41] The tetravalent ions of U, Np, and Pu are also easily oxidized to the linear dioxo actinyl species. In contrast with the penta- and hexavalent species, coordination numbers of 8–10 are observed in the tri- and tetravalent spherical ions.[26] The specific CNs of trivalent U, Np, and Pu have been a subject of debate that is still not entirely resolved. Nine to ten solvating waters have been reported for U and Np, and 8–10 waters of hydration for Pu(III). Density functional theory, Møller – Plesset perturbation theory, and single-, double-, and perturbatively treated triple excitations coupled cluster theory (CCSD(T)) calculations have all provided strong evidence that trivalent U, Np, and Pu likely exist with 8 waters in their first solvation shell.[19, 42, 43] A slightly different picture emerges for Cm(III), where simulations from various groups confirmed the experimental results indicating that hydrated Cm(III) exists in an equilibrium between ninefold and eightfold structures,[25, 44–46] with a preference for ninefold coordination.[22]

Yet uncertainty remains in the preferred coordination environment of tetravalent U, Np, and Pu. In the case of U(IV) hydration, the most recent work, including B3LYP DFT and QM/MM studies of hydrated U(IV), have supported CN = 9, and ab initio molecular dynamics have reported an average coordination number of 8.6.[47–49] The most recent tetravalent Np studies have reported coordination numbers of 8 and 9 according to electronic absorption and EXAFS[50, 51] and similarly for Pu(IV).[52, 53] Recent Car – Parrinello molecular dynamics simulations have predicted the octa-aqua species for Pu(IV). Large solvation DFT cluster calculations propose that tetravalent U is proposed to exist in equilibrium between the eight- and nine-coordinate species, with the latter being dominant in solution. Moving to the right of the period, Np(IV) and Pu(IV) exist solely as the $An(H_2O)_8{}^{4+}$ species.[19]

While the geometric structure associated with solvation is important, so too are the thermodynamic stability of different solvation environments and exchange dynamics.[54] One key aspect of the dynamic equilibrium between solvation environments lies in the free energy of solvation and hydration enthalpies. Experimentally determined by calorimetry, the typical uncertainty can be large (~10 kcal/mol). With such a large range of values, theory has played an important role in improving estimates. However, this remains a significant challenge for *ab initio* computational methods. Large changes in the electrostatic interaction between the ion and water necessitate a molecular cluster model with an appropriate number of explicit waters to be paired with a PCM that can adequately describe multiple oxidation states. While prior work has demonstrated that solution-phase cation thermochemistry can depend upon the number of explicit water molecules and the continuum model,[55, 56] it has only been recently that a systematic investigation has examined the physical premise of this behavior or its sensitivity to ion charge. This is particularly relevant as most PCMs have been parametrized in the absence of explicit solvent molecules and the use of ions of low charge.[57–67] A computational approach has been used that encompasses the reactions that represent the hydration reactions of the bare ions. Gas-phase calculations determine the free energy for this process, ΔG_{hyd}, followed by single point PCM calculations that determine the solvation correction (G_{corr}) to the $(H_2O)_n$ and hydrated metal cluster so as to determine the solution-phase free energy for the reaction, which is called the free energy of solvation, ΔG_{solv}, for the ion. Note that the bare ion is left in the gas phase as this is consistent with the experimental calorimetry approximations associated with the heat released from dissolving the ion-containing salt. Further, a cluster of hydrogen-bonded water should be employed to avoid imbalances in the hydrogen bond network of the reactants and products.[56, 68] A "standard state" correction, SS_{corr}, may also be applied; however, it is quite small relative to the other two terms if a hydrogen-bonded cluster is employed.[69] The total solution-phase free energy of solvation is then

$$\Delta G_{solv} = \Delta G_{hyd} + \Delta G_{corr} + SS_{corr} \quad \text{(Eq. 5.1)}$$

The errors associated with this approach are a composite deriving from ΔG_{hyd} as well as ΔG_{corr}. Errors in the gas-phase energetics may derive from either the method or the basis set or other approximations including those associated with relativistic effects. The cluster model employed to define the explicit solvation environment about the ion also influences both the gas- and solution-phase calculations. A variational approach for determining the optimal cluster size has been previously developed.[70] Yet, for clusters of infinite size (the bulk limit) ΔG_{hyd} is equivalent to ΔG_{solv}, and the solvation correction ΔG_{corr} for the reaction should go to zero. Unfortunately, different convergence properties may be observed for the gas-phase energetics and solvation corrections, which may contribute to the aforementioned deviations in ΔG_{solv} as a function of system size and continuum model. To offset this problem, it is often more straightforward to instead try to understand the dynamic equilibrium between different solvation environments through study of water addition solvent configurations.

Initial guesses regarding relevant configurations within theory studies often derive from experimental data. A summary of structural information about trivalent through hexavalent U, Np, and Pu and trivalent Am and Cm can be found in the review article by Knope et al.[26] Notable from this review are the variations in CN and geometry that may be observed as a function of solution-phase conditions, particularly as the ion concentration is increased. Ion–ion interactions are relevant, and counterions, in particular, may coordinate within the first solvation shell, lead to the formation of solvent-separated ion pairs, or can merely act to alter the dynamics and energetics of the solvent as a whole – thereby altering the CN and dynamic properties (solvent exchange). Within the following sections, we provide an overview of relevant changes to speciation that may occur, from the infinite dilution limit to high ionic strength, and mixed solvent conditions.

5.3.2 Cation–Cation Complexes in Separations Solution

Although successful at segregation of UO_2^{2+} and Pu^{4+} from the remainder of the fuel, the PUREX process leaves the minor actinides (Np, Am, and Cm) and other used fuel constituents together. This remaining fuel component is typically at high ionic strength, and has been shown to support the formation of cation–cation complexes. Nagasaki[71] and Guillaume[72] have shown that such interactions impact redox chemistry and thus may influence the chemistry of future nuclear fuel reprocessing efforts. The binuclear actinyl–actinyl complex reported by Sullivan demonstrates the tendency of the NpO_2^+ ion to bind to UO_2^{2+},[73] but a pair of dioxocations is not the only configuration that may form a complex of this type. In 1962,[74] the interactions of a series of trivalent cations ($M^{3+} = Al^{3+}$, Ga^{3+}, In^{3+}, Sc^{3+}, and Fe^{3+}) with NpO_2^+ were examined, and the relative stability of the resulting $[NpO_2 \cdot M]^{4+}$ complexes was determined on the basis of the molar absorptivity of the 980 nm *f-f* NpO_2^+ UV-Vis absorption band. The strength of the complex in perchloric acid media followed the trend: $Fe^{3+} > In^{3+} > Sc^{3+} > Ga^{3+} > Al^{3+}$, and in recent work, the equilibrium constants for $[NpO_2 \cdot M]^{4+}$ ($M = Al^{3+}$, In^{3+}, Sc^{3+}, Fe^{3+}) in $\mu = 10\,M$ nitric acid and $[NpO_2 \cdot Ga]^{4+}$ in $\mu = 10\,M$ hydrochloric acid media have been determined.[75] DFT studies indicate that the NpO_2^+ dioxocation acts as a π-donor with transition metal cations and a sigma donor with Group 13 cations. The small changes in electron donating ability are modulated by the overlap with the coordinating metal ion's valence atomic orbitals.

5.3.3 Counterion Interactions with Aqueous Actinide Ions

The role of counterions cannot be underestimated as they impact the speciation of actinide cations and their reactivity. The extent of ion coordination within the primary coordination sphere or the formation of solvent-separated ion pairs is dictated largely by the specific oxidation state of the actinide and the solution-phase conditions. Consider that in the PUREX process, plutonyl nitrate is extracted with TBP.[76] The bidentate plutonyl(VI) dinitrate, $[PuO_2(NO_3)_2(H_2O)_2]$, has been crystallized and characterized using X-ray diffraction and spectroscopic methods[77]; however, in solution, the dinitrate species is only a minor species in solution, with the mononitrate being dominant. In the case of PuO_2^{2+}, both monodentate and bidentate are equally favored within the mononitrate solution complex. Much effort has focused upon studying the structure and dynamics of uranyl–nitrate complexes as they are also the most probable species during PUREX separation.[36, 78] The formation of 2:1 or 3:1 complexes for uranyl nitrate represents the partition equilibrium for the extraction of uranyl.[79] In the case of U(VI) nitrate complexes, the $UO_2(NO_3)_2$ complex is thermodynamically favored in solution, where in the mononitrate complex, a monodentate mode is preferred to the bidentate mode. This is consistent with the results using molecular dynamics simulations.[80]

For the Pu(IV), two different Pu(IV) solvate adducts, $Pu(NO_3)_4{\bullet}TBP_2$ and $Pu(NO_3)_4{\bullet}TBP_2{\bullet}HNO_3$, were considered as extracted species over a wide range of experimental conditions with their extraction constants determined.[81] Nitrate complexes are also believed to be significant in the solution chemistry of tetravalent ions, like Pu^{4+}, where the dinitrate, tetranitrate, and hexanitrate complexes are known to be present in acidic aqueous nitrate solutions, the latter complexes increasing in prevalence with increasing nitrate concentration. Nitrates are very strongly bonded to the Pu(IV) ion; they do not dissociate easily and prefer to stay in a first coordination shell.[82] A theoretical work using realistic quantum computational methods reported geometries and stabilities of $Pu(NO_3)_6^{2-}$ and $Pu(NO_3)_4{\bullet}TBP_2$ complexes[83] which are in good agreement with experimental results.[84]

Unlike nitrates, chlorides are weakly bonded to Pu(IV) ions, and in chloride media the CN for aquochloro complexes is found to be 8 with three chlorides present in the first shell. In the case of Pu(VI) chlorides, PuO_2^{2+} retains a fivefold CN with two chlorides replacing two water molecules bonded to the Pu(VI) ion. Runde et al.[85] identified the mono- and bis-chloro (and a possible trichloro) complexes formed with the PuO_2^{2+} moiety in sodium chloride solutions. Using EXAFS spectroscopy, formation of $UO_2(H_2O)_4Cl^+$, $UO_2(H_2O)_3\ Cl_2$, and $UO_2(H_2O)Cl_3^-$ complexes were observed as the concentration of chloride ions increases. For U(IV) ions, $U(H_2O)_8Cl_3^+$, $U(H_2O)_{6-7}Cl_2^{2+}$, and $U(H_2O)_5Cl_3^+$ complexes were found to occur.[86] The U-Cl bonding interaction is discussed in Section 5.4.5. Car–Parinnello MD simulations have found that the monochloride $UO_2(H_2O)_4Cl^+$ is five-coordinated, whereas the trichloride complex is four-coordinated. The dichloride complex can exist both as four-coordinated or five-coordinated.

For trivalent Cm^{3+}, Cl^- and Br^- are small anions and form fairly stable hydration shells,[87] whereas ClO_4^- is a large ion and, in general, does not form a strong hydration shell.[88] Ab initio dynamics indicate Cl^- and Br^- make the first coordination sphere more stable, maintaining an eightfold coordination environment. This feature is not observed with perchlorate counterions. Interestingly, trends in the stability of the coordination shell are mirrored by trends in the hydrogen bonds within the first solvation shell – a feature that is mimicked in simulations of a variety of mono and trivalent ions in mixed solutions of water and other polar solvents like methanol.

5.3.4 Changes to Solvation and Speciation in Solvent Mixtures

While ionic strength and the role of counterions should not be ignored in the solution, the composition with respect to the potential for multiple solvents is also relevant. In comparison, much less is understood regarding how co-miscible solvents influence actinide ion speciation. In recent work, Kelley and coworkers have studied water–methanol solutions and their role upon the speciation of Cm^{3+} and lanthanide ions.[89, 90] In addition to the potential effects of methanol on the immediate solvation environment of the ion, the inclusion of methanol as a co-solvent alters the solvent structure of liquid water by interrupting contiguous water–water hydrogen bonding.[91–94] The average number of hydrogen bonds per H_2O decreases by nearly 50% in a 1:3 water/methanol solution, whereas the average hydrogen bond lifetime increases.[95, 96] Overall, this has the effect of creating a more static hydrogen bond network in mixed water–methanol solutions compared to pure solutions of either type. Electronic structure calculations of the Cm^{3+} ion and its first solvation shell have also shown that the solvent dissociation energy of both methanol and water from the solvated ion decreases as more methanol is included in the first solvation shell – by approximately 6 kcal/mol between the $Cm(H_2O)_8(CH_3OH)_1^{3+}$ and $Cm(H_2O)_1(CH_3OH)_8^{3+}$ structures.[89] It appears that two competing forces may alter their solvation dynamics in water–methanol solutions: one based upon changes in the ion-solvent binding energy and the second based upon the perturbation of the solvent dynamics imparted by the methanol co-solvent. In preliminary work, a combination of MP2, classical MD, and ab initio MD was used to demonstrate that the solution dynamics prevail and the ion-solvent dynamic features are retarded as the overall solution hydrogen bond dynamics become more static.

The current discussion regarding An coordination and speciation highlights just a few of the salient complexities associated with the aqueous solution chemistry of the 5f elements. The authors are certain that many other relevant phenomena have been omitted from the discussion; however, at a minimum, the reader should be emboldened to study in detail the expanding literature regarding many of these topics. Of course, the primary relevance to both coordination and speciation lies in the fact that it is these aqueous species to which extracting ligands must bind to change the free energy of aqueous solvation so as to thermodynamically favor transport into the organic phase.

5.4 Ligand Design

Using differences in coordination thermodynamics has been a staple in completing elemental separations of the actinides. When designing selective ligands for separation, it is important to realize the difference between CNs for various oxidation states. Different ligand design principles may be employed depending on the application in mind (nuclear fuel cycle, nuclear forensics, or environmental analysis). Separations related to nuclear fuel cycle applications have generally been completed using solvent extraction since this technique is high throughput, can meet preferred purity specifications (~99% pure material), and can be managed in a largely remote fashion – thus minimizing dose to workers at a reprocessing facility. Nuclear forensics or environmental analysis applications generally utilize some form of chromatography since this technique allows for higher-quality (>99%) separations, and sample throughput demands are less. Ligand design principles for solvent extraction-based separations in the nuclear fuel cycle are generally more restrictive than those considered in chromatography. Therefore, this section will focus most rigorously on ligand design for nuclear fuel cycle applications with the understanding that many design principles may not be as restrictive when completing analytical separations of the actinides or using solvent extraction for matrices other than used nuclear fuel. Generally speaking, ligands used for f-element separations must be compatible with high-radiation environments and acidic conditions. If the ligand needs to be organic soluble, the extractant should be soluble in an aliphatic diluent.

Solvent extraction separations of the actinides will almost always contain an amphiphilic extractant soluble in an organic diluent. For separations related to the nuclear fuel cycle, the organic diluent is frequently preferred to be some sort of aliphatic, long-chain carbon such as kerosene. Many bench top-scale studies will use n-dodecane in place of kerosene to allow for more consistent solution properties. Aliphatic diluents are preferred for many reasons including their low flash point, vapor pressure, density, and toxicity. The most common types of extractants used in nuclear fuel applications targeting recovery of the actinides use solvating extractants, though the additional charge density associated with cation exchange reagents has made them an attractive option for separations targeting trivalent americium and curium.

Lipophilic carbon chains are frequently included in the extractant to encourage solubility in organic diluents. Carbon chains are usually between four to eight carbons long and can include branching. Shorter chain lengths are generally permissible for extractants with less pronounced dipoles. As the strength of the dipole increases, typically more carbon is required to encourage solubility in an aliphatic diluent. Phosphate- or carbamoyl-containing reagents generally have carbon chains four to six carbons long, whereas phosphoric acid or phosphine groups are best solubilized using carbon chain lengths of six to eight carbons.

5.4.1 Solvating Extractants

Solvating extractants are neutral and therefore utilize ion–dipole interactions between the extractant and the metal to drive metal recovery. Dipoles are frequently obtained by using phosphoryl (P = O), carbamoyl (N-C = O), or triazine/pyridine functional groups. Charge-neutralizing anions are co-recovered with the metal to promote solubility of the polar metal with the apolar solvent. A general reaction mechanism for metal recovery by a solvating reagent is provided here:

$$M^{n+} + nA^{-} + \overline{mS} \rightarrow \overline{MA_nS_m} \qquad \text{(Reaction 5.1)}$$

where M is a generic metal ion, A is the charge-neutralizing anion, and S is the solvating extractant.[97] An overbar denotes the species as in the organic phase. The number of extractants included in the extracted metal complex is generally the number required to satisfy the extractant's CN after the addition of the anions. Because actinides have flexible CNs, a result of their contracted 5*f* orbitals, the number of solvating reagents in the metal can increase if a large excess of extractant relative to metal ion is present. A classic example of a solvating reagent is tri-butyl phosphate.[98]

Several properties of solvating reagents have established "benchmarks" for development of next-generation extractants. Since complementary anions are necessary for metal recovery into the neutral organic phase, solvating reagents will generally only target metal ions with CNs of eight or higher. The neutrality of the extractant also limits extractant interaction to trivalent or higher oxidation state metals. In fact, recovery of trivalent f-elements often requires the use of bidentate or higher dipole extractants (such as phosphines) to allow for their recovery. Finally, the relatively small size of the extractant and complementary anions encourages rapid self-assembly of the organic-soluble metal complex. The fast extraction kinetics afforded by solvating extractants allow for the use of high-throughput separations equipment (i.e., centrifugal contactors). While solvating extractants are less commonly used for separations of lighter elements, this combination of reasons provides a basis for their frequent application in solvent extraction separations relevant to the nuclear fuel cycle.

Because of these traits, solvating reagents are most frequently used to recover hexavalent and tetravalent actinides from acidic aqueous media that results from used nuclear fuel dissolution using nitric acid. While tribuylphosphate (TBP) has long demonstrated capability in the recovery of uranium and plutonium, some issues with the use of TBP exist. The hydrolytic and radiolytic degradation products of TBP are dibutyl and monobutyl phosphoric acids.[99] These reagents can recover tetravalent d-block elements (such as zirconium) and trivalent f-block elements (such as the lanthanides or americium and curium). In addition, courtesy their cation exchange extraction behavior (Section 5.6.3), these materials recover metal ions from the low acid (pH ~ 2) media typically used for stripping U and Pu from organic phases. The radiolytic production of dibutylphosphate (DBP) and monobutylphosphate (MBP) requires the inclusion of a sodium hydroxide scrub loop in an engineered-scale separation to remove DBP and MBP. The addition of the scrub step, although an effective engineered solution, can add significant cost to a processing facility.

5.4.2 Recent Trends in Solvating Extractants

Monoamide extractants. Reagents using a monoamide moiety have been recognized as a potentially viable replacement for TBP. Although the carbamoyl dipole is slightly smaller than the TBP dipole, the synthesis of monoamide extractants is straightforward, and the degradation products are aqueous soluble and non-complexing.[100] The good aqueous solubility and non-complexing nature of the monoamide degradation products removes the need for an additional scrub step. In addition, cation exchange reagents, like DBP or MBP, can encourage Chalk River Unidentified Deposits (CRUD) formation. CRUD, when used in the context of solvent extraction systems, is used to describe solid, interfacial precipitates, and its formation can be particularly problematic if a process upset occurs and shutdown of the processing facility is required. Extractant breakdown can be rapid if the organic phase is left to sit in contact with the aqueous phase containing dissolved nuclear fuel. This can ultimately lead to the formation of MBP and DBP stitched together into a metal–polymer interfacial precipitate that can clog the processing equipment. The inherent properties of monoamide reagents could allow advancement beyond these technological issues.

Several recent reviews have captured the current state-of-the-art regarding monoamide reagents.[101–103] Numerous carbon tail structures have been considered to optimize monoamide separation performance. To date, the DEHBA extractant (N,N-di(2-ethylhexyl)butanamide)) is viewed as the most appropriate replacement for TBP.[104] Separation of U and Pu from fission products is viewed as adequate, and contactor tests with irradiated nuclear fuel have been successful. Separations considering U and Pu recovery have largely been led by France. Branching on the monoamide acyl group can encourage selective recovery of hexavalent actinides from tetravalent actinides. Since operation of a Th-based nuclear fuel cycle would most likely require a separation of U from Th, and India has been considering a Th-based cycle, U and Th separations by monoamides have been led by India.[105] For these types of separations, the DEHiBA extractant is currently viewed as most viable.[106]

Diglycolamide extractants. Dyglycolamide extractants have become increasingly popular for trivalent f-element recovery. The most studied reagents are currently TODGA and TEHDGA (N,N,N′,N′-tetra(2-ethylhexyl)diglycolamide).[107, 108] Because of their ability to recover trivalent actinides, this extractant class is most commonly used for advanced nuclear fuel cycle separations targeting the recovery of americium and curium. The additional polarity associated with these reagents, courtesy the presence of two carbamoyl groups, encourages the use of longer, octyl carbon chains to promote their solubility in aliphatic diluents. These reagents can be prone to third phase formation – a phenomenon where the organic phase splits into heavy and light fractions upon high loading of polar analytes. The TODGA extractant is particularly prone to this type of aggregation chemistry, and numerous studies have been performed considering this extractant.[109] The ethyl-hexyl groups of the TEHDGA extractant have a tendency to mitigate extractant aggregation, ultimately making this the preferred reagent for trivalent f-element recovery in the ALSEP process.[110] Density functional theoretical studies have demonstrated the structure, bonding, interaction, and thermodynamic selectivity of hexavalent uranium (UO_2^{2+}) and tetravalent plutonium (Pu^{4+}) ion complexes of tetramethyl diglycolamide (TMDGA).[111] Two TMDGA ligands can bind to uranyl because of the limited space in the equatorial plane, while four TMDGA ligands can bind to Pu^{4+}. A thorough review of the chemistry of diglycoamides as promising extractants for actinide separation was published in 2012.[112]

Bistriazylbipyridine extractants. Bistriazylbipyridine (BTBP) extractants have been the most recent iteration of nitrogen donor extractants developed in Europe for the selective recovery of trivalent actinides. A wonderful review of N-donor extractant developments leading up to the BTBP reagents was provided by Eckberg et al.[113] In short, BTBP reagents were developed to advance upon terpyride and bistriazyl pyridine reagents. Both of these smaller reagents had issues with protonation and aqueous phase partitioning that the BTBP reagents were able to address, though further development of BTBP reagents was necessary to overcome the radiolytic instability of these reagents. The addition of camphor groups on the outer side of the triazyl groups seems to limit alpha-hydrogen extraction and improve the stability of these materials. Since four nitrogens are available to bind from each BTBP molecule, and the stoichiometry of the extracted trivalent actinide complex is generally 1:2 metal–BTBP, the coordination sites of a given metal are nearly saturated by the two BTBP molecules. To provide the necessary charge neutrality, nitrate groups are generally understood to be present in the second coordination (outer) sphere of the extracted metal complex. The presence of outer sphere nitrates makes more polar, alcoholic diluents appropriate for use with BTBP molecules. The significant size of the BTBP reagents can make extraction kinetics slower than classical reagents, such as TBP, but reasonable compatibility with centrifugal contactors has been demonstrated in the recent literature. A recent modification of BTBP reagents includes sulfonating the extractant to allow for an aqueous soluble BTBP.[114] Because of their planer nature and selective interactions with trivalent actinides, the BTBP reagents are natural materials for crystal structure determinations and computational study. Many of the recent reports have focused on considering properties most readily obtained by using this combination of methods.[115–117]

5.4.3 Cation Exchange Reagents

Cation exchange reagents are useful for actinide recovery from dilute acid media. This trait makes them particularly compatible for analytical separations of the actinides, where the constraint of using more concentrated acidic feed streams is preferred due to the use of concentrated nitric acid for dissolution of used nuclear fuel. Cation exchange reagents are also useful for the recovery of less charge dense, trivalent actinides. Interest in examining cation exchange reagents has grown as separations research related to fuel cycle goals has shifted toward improving americium and curium separations from lanthanides to potentially allow for transmutation of americium in fast spectrum reactors.

Metal recovery by cation exchange reagents can be generally described using the following reaction:

$$M^{n+} + n\overline{HA} \rightarrow \overline{MA_n} + nH^+ \qquad \text{(Reaction 5.2)}$$

where M is the metal ion and HA is the acidic extractant, and the overbar indicates the species is in the organic phase.[97] Frequently, cation exchange reagents self-associate in solution. Reaction r2 is therefore modified as follows:

$$M^{n+} + n(\overline{HA})_2 \rightarrow \overline{M(AHA)_n} + nH^+ \qquad \text{(Reaction 5.3)}$$

In the case of MBP and DBP extractants, this reaction design encourages metal recovery from dilute acid media and metal stripping under higher ($>1\,M\,H^+$) acid conditions. The most common cation exchange reagents are bis-2-ethylhexyl phosphoric acid (HDEHP) and (HEH[EHP]).

Recent solvent extraction studies have focused more significantly on the HEH[EHP] extractant relative to HDEHP. The HEH[EHP] reagent seems to be less prone to aggregation chemistry than HDEHP, but the reason for this is not currently obvious. The HDEHP and HEH[EHP] reagent are structurally identical except that HDEHP is a phosphoric acid ($O=P(OR)_3$), whereas HEH[EHP] is a phosphonic acid ($O=P(OR)_2R$). Perhaps the higher symmetry of the HDEHP reagent may encourage better packing and liquid-crystalline-type behavior. Advanced fuel cycle processes currently considered in the United States attempt to use HEH[EHP] instead of HDEHP. Having noted this, the HDEHP and HEH[EHP] reagents are still the most prolific liquid cation exchange reagents with no obvious recent advances that might replace their use.

5.4.4 Aqueous Complexants

Many "next-generation" fuel cycle separations contain an aqueous soluble complexant to provide another level of selectivity. These "holdback reagents" are used to complement the separation afforded by the organic extractant. The most common holdback reagents are (poly)aminopolycarboxylates (APCs).[118, 119] The nitrogen groups of the APC ligand are generally recognized to provide selectivity for the trivalent actinides over trivalent lanthanides, while the acetate groups provide enough coordination "horsepower" to force interaction of the metal ion with the nitrogens in the molecule. The best-known APC is ethylenediaminetetraacetic acid (EDTA), though due to the larger CN of the actinides, diethylenetriamineppentaacetic acid (DTPA) is frequently used for actinide separations.

The APC class of ligands are a favorable ligand design for several reasons. The pre-organization of APC ligands makes them quite strong complexants. Complexation with trivalent metal ions frequently occurs around acid concentrations as high as 0.01 M, but the APC ligand is generally too protonated for trivalent metal complexation at higher acid concentrations. Under acidic conditions, the solubility of APC complexants can generally be limited to several hundred millimolar for small complexants, such as iminodiacetic acid, to less than five millimolar for the pyridine-containing dipicolinic acid. All APC complexants are able to form fully depronated complexes with a metal ion, and some of the larger APC complexants, such as DTPA or triethylenetetraminehexaacetic acid (TTHA) are able to form metal–ligand complexes when only partially deprotonated.

Advancements in APC ligand design currently focus on improving aqueous solubility, metal–ligand complexation kinetics, and metal complexation under more acidic conditions to improve compatibility with nuclear fuel recycling needs.[120, 121] Improvements have been provided by functionalizing the pyridine ring of dipicolinic acid or modifying aliphatic APCs with amide groups. These advances are captured in more detail below.

Dipicolinic acid functionalization. The pre-organization of dipicolinic acid (DPA) makes the 1:3 metal–ligand complex of this reagent particularly strong. The softer, pyridinic nitrogen, relative to the aliphatic nitrogens usually associated with APCs, also appears to make this reagent more selective for the trivalent actinides relative to the lanthanides.[121] While the pyridine group provides more selectivity and pre-organization, this functionality also makes DPA one of the least soluble APC reagents. The limited solubility of DPA nearly eliminated this reagent from consideration for nuclear-fuel-cycle-relevant separations, though functionalization of this ring using a trimethyl ammonium chloride functional group improves DPA solubility by an order of magnitude. The electronic structure of the ring was significantly altered by the presence of the trimethyl ammonium, causing the basicity of the complex to decrease and ultimately resulted in a weaker complex. Further iterations on this design could involve the addition of an additional carbon "spacer" between the pyridine and the trimethyl ammonium group to preserve the electronic properties of the DPA complex.

Amide-modified APC reagents. Recent separations studies have showed that the metal extraction kinetics for systems containing medium- to large-size APC reagents are too slow for use with centrifugal contactors.[122, 123] The slow phase transfer kinetics have been hypothesized to originate from energetics associated with structural reorganization of APC metal complexes at the liquid–liquid interface. Previous studies have suggested that numerous factors influence complexation kinetics, including (1) the number or size of chelate rings, (2) bulk steric effects, (3) formation of ternary complexes, (4) acid-catalyzed dissociation, and (5) substitution of amide groups or longer alcohol chains in place of acetate functionalities.[124–126] The replacement of acetate groups for amide groups or longer alcohol chains to improve phase transfer kinetics while maintaining complex selectivity has recently been considered.[124, 127]

A recent paper from Idaho National Laboratory has considered diethylenetriamine-N,N″-bis(acetylglycine)-N,N′,N″-triacetic acid (DTTA–DAG) as a reagent that could enable trivalent actinide/lanthanide separation while improving phase transfer kinetics by amidating the classical DTPA framework.[124] Comparable stability of DTPA and DTTA–DAG complexes was observed, indicating that the amidification did not produce a complex that might be too weak to be effective in a separations process. Because of the increased basicity of the DTTA–DAG, separations could be completed in higher acid media (0.01 M H^+) and leverage the benefits of acid-catalyzed dissociation that can occur under more acidic conditions. This enabled better phase transfer than the DTPA-based system.

5.4.5 Covalency and Ligand Design

The concept of ligand selectivity described above is often linked to traditional hard-soft-acid-base concepts; however, a more quantitative representation would be from the perspective of chemical bonding in combination with the ligand organization about the An for a specific oxidation state and corresponding CN. For d-block elements, the *d*-orbitals extend well into the periphery of the atom and can interact with valence orbitals of ligand atoms to form covalent chemical bonds. In contrast, the 4f-orbitals of lanthanide elements are very core-like, and their interactions with ligands are of comparatively little chemical consequence. The 5f-elements lie between these two extremes, and there has been much debate over the ability of these elements to use either 5*f*- or 6*d*-orbitals, or both, in chemical bonding interactions. A complete understanding of the covalency of actinide elements will provide guiding principles for novel ligand design. Difference in covalency have also been invoked to rationalize the observation that soft donor S- and N- based ligands are highly selective in separating trivalent actinides from lanthanides in nuclear fuel reprocessing.

There have been great efforts in quantifying the covalency of actinide elements using an integrated theoretical and experimental approach. First-principles calculations are powerful tools to understand the electronic state of f-block compounds. Many studies focus on the chemical stability and bonding properties of f-block complexes using DFT calculations. A recent book discussed in great detail the relevant aspects of theoretical studies of the computational actinide sciences including relativistic effects, basis sets, pseudopotentials, and various methods.[128] Hence, we limit our discussions to the applications of theory in separation sciences. Of experimental approaches to determine covalency, ligand K-edge X-ray absorption spectroscopy (XAS) has emerged as an effective method for measuring orbital mixing in metal–ligand bonds. This technique has recently been extended to the f-block elements.[129–133] These recent works confirm that f-block elements engage in covalent orbital mixing.

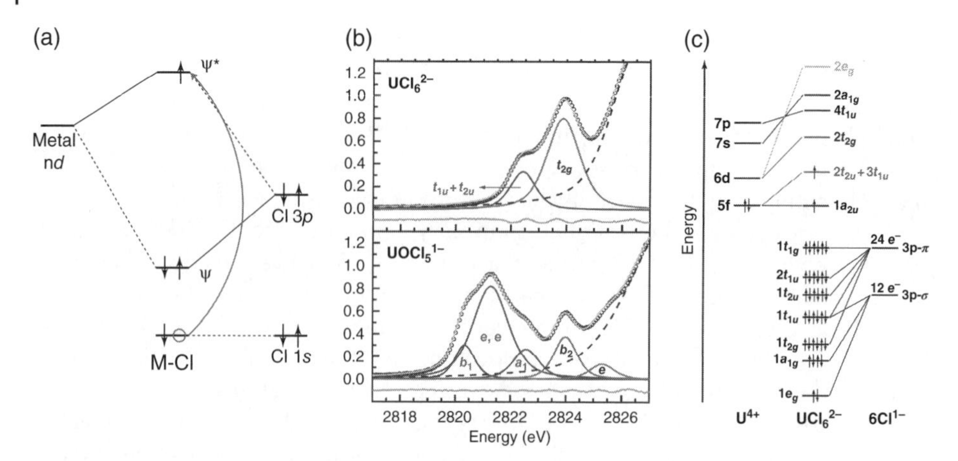

Figure 5.3 (a) Illustration of electronic excitations in a typical ligand K-edge XAS experiment. (b) Cl K-edge fluorescence yield experimental data. Adapted from Ref. 132; (c) qualitative molecular orbital diagram for UCl_6^{2-}. (*See insert for color representation of the figure.*)

In ligand K-edge XAS, bound state transitions can occur on the low-energy side of the absorption edge involving the excitation of 1s electrons (localized on the ligand) into singly occupied or unoccupied acceptor orbitals of the metal complex. The presence of covalent mixing is observed as a pre-edge peak in the ligand K-edge XAS, and can only have transition intensity if the acceptor orbitals (in this case *f* or *d*) contains a significant component of ligand *p* character, Figure 5.3(a). The energies and intensities of these transitions have been simulated by density functional theory using either the dipole transition moment method or time-dependent DFT. Good agreement between theory and experiment has been observed across multiple systems spanning transition metals to lanthanides and actinides. Hence, this technique affords an opportunity to quantify the degree of covalency and assess the relative roles of valence *f*- and *d*-orbitals in An/Ln metal–ligand bonds. Distinct *f*- and *d*-covalent orbital mixing has been observed in $(C_5Me_5)_2MCl_2$ (M = Ti, Zr, Hf, Th, U) series compared to transition metals.[129] Not surprisingly actinide covalency is dependent upon a number of factors, including ligand, oxidation state, and symmetry as demonstrated in the examples of MCl_6^{2-}; $LnCl_6^{x-}$, x = 2, 3; and $UOCl_5^{-}$, shown in Figure 5.3.[130, 132] Figure 5.4 presents the representative virtual Kohn–Sham orbitals for UCl_6^{2-}. Yet it should be noted that the covalency can be tuned by altering the axial atom to imido[133] or by changing ligands.[131] Traversing across the period, the covalency increases within the early actinides due to better overlap with ligand orbitals, as indicated by electron population analysis. It has been reported that as the actinide series is traversed, the energy of 6*d* orbitals stays the same, but the energy for 5*f* orbitals drops due to the relativistic effects (also called actinide contraction) and becomes more degenerate with ligand orbitals, hence promoting more orbital mixing. This trend has been observed for various ligands including those with coordinating oxygen and carbon.[134–136]

There are many examples where the electronic structure information imparted by DFT has played a role in ligand design. As an example, consider the separations of trivalent An from Ln, where it has been hypothesized that trivalent actinides are capable of more covalent bonding compared to the lanthanides. The roles of the 4*f* and 5*f*

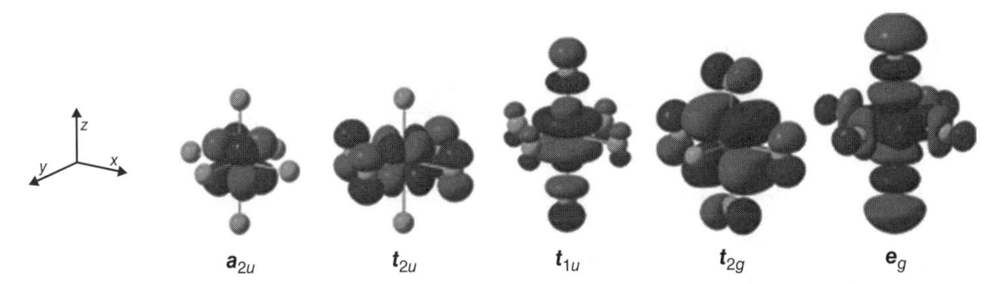

Figure 5.4 Representative virtual Kohn–Sham orbitals for UCl_6^{2-}. Adapted from Minasian, S. G., Keith, J. M., Batista, E. R., *et al.* (2012). Determining relative f and d orbital contributions to M–Cl covalency in MCl_6^{2-}(M = Ti, Zr, Hf, U) and $UOCl_5^-$ Using Cl K-Edge X-ray absorption spectroscopy and time-dependent density functional theory. *Journal of the American Chemical Society*. **134** (12), 5586–5596. (*See insert for color representation of the figure.*)

orbitals in bonding were subsequently studied using isostructural Eu(III) and U(III) complexes by a set of experimental and theoretical tools including synthesis, magneto-chemical, crystal field, DFT, and multi-reference wavefunction methods. The results demonstrated that the 5*f* orbitals, but not the 4*f* orbitals, were significantly involved in bonding to the isocyanide ligand.[137]

5.4.6 Computational Screening of Separation Selectivity

The brief electronic structure studies described above underlie another need within the computational community – the ability to perform computational screening for viable ligand candidates using a high-throughput method. In principle, the efficacy of a ligand for solvent extraction can be calculated using methods based on first principles. Specifically, the free energy of extraction, ΔG_{ext}, can be calculated using the following reaction pathway:

$$M^{3+}{}_{(aq)} + 3H_2O_{(aq)} + 3HL_{(org)} \xrightarrow{\Delta G_{ext}} ML_{3(org)} + 3H_3O^+{}_{(aq)} \qquad \text{(Reaction 5.4)}$$

Reaction 5.4 is the simplest example of solvent extraction by neutral ligands from aqueous solutions via a cation exchange mechanism. The ΔG_{ext} can be calculated using a Born–Haber thermodynamic cycle. However, as described above, it is very challenging to accurately calculate the solvation free energies of each species. Using the assumption that lanthanides and actinides form the same type of complexes in the same environment because the metal ions have similar size, one can instead consider the difference of extraction free energies of two metal ions, a reaction that could result in a more accurate computational value due to error cancellation (Eq. 5.2)

$$\Delta\Delta G_{ext} = \Delta G_{ext}(An) - \Delta G_{ext}(Ln) \qquad \text{(Eq. 5.2)}$$

In Eq. 5.2, the $\Delta\Delta G_{ext}$ term could correspond to the selectivity of An^{3+} over the Ln^{3+} for a particular ligand. Many recent reports used this method when comparing the selectivity of one ligand over multiple metal ions. The ligands being studied include diethylenetriamine-pentaacetic acid (DTPA),[138] tetra-n-octyl diglycolamide (TODGA),[139] tetramethyl diglycolamide (TMDGA), sulfur, nitrogen- and oxygen-donor ligands,[140] and mixed O,N-donor ligands.[141]

$$
\begin{array}{lcl}
An^{3+}_{(aq)} + 3H_2O_{(aq)} + 3\,HL^B_{(org)} & \xrightleftharpoons{\Delta G_{ext}} & AnL^B_{3(org)} + 3\,H_3O^+_{(aq)} \\
-\left(An^{3+}_{(aq)} + 3H_2O_{(aq)} + 3\,HL^A_{(org)}\right. & \xrightleftharpoons{\Delta G_{ext}} & \left. AnL^A_{3(org)} + 3\,H_3O^+_{(aq)}\right) \\
+\left(Ln^{3+}_{(aq)} + 3H_2O_{(aq)} + 3\,HL^A_{(org)}\right. & \xrightleftharpoons{\Delta G_{ext}} & \left. LnL^A_{3(org)} + 3\,H_3O^+_{(aq)}\right) \\
-\left(Ln^{3+}_{(aq)} + 3H_2O_{(aq)} + 3\,HL^B_{(org)}\right. & \xrightleftharpoons{\Delta G_{ext}} & \left. LnL^B_{3(org)} + 3\,H_3O^+_{(aq)}\right) \\
\hline
AnL^{A+}_{3(org)} + LnL^B_{3(org)} & \xrightleftharpoons{\Delta\Delta\Delta G_{sel}} & LnL^A_{3(org)} + AnL^B_{3(org)}
\end{array}
$$

Figure 5.5 Competition reactions between two ligands and two metal ions, generalized from Ref. 142. $\Delta\Delta\Delta G_{sel}$ corresponds to the selectivity difference between two ligands and two metal ions.

As an extension of this concept, to predict relative selectivity of multiple extractants, an even better cancellation of errors can be achieved by considering the competition reactions among all metal–ligand complexes, as proposed in an Am(III) and Eu(III) study,[142] by calculating $\Delta\Delta\Delta G_{sel}$. As shown in Figure 5.5, this method cancels out the potential large uncertainties introduced by the solvation effects for highly charged species. Therefore, this method provides a convenient way to differentiate between the selectivities of different extractants and pave the way for the high-throughput ligand design. A recent publication applied this method to study the separation of Am(III) and Eu(III) using a series of O,N-donor ligands.[141]

These computational treatments have proved very fruitful; however, they are not anticipated to work in all cases. Specifically, they tacitly ignore any perturbations in the interfacial transport that may occur under the various solution-phase conditions that are tuned for specific ligand combinations. Toward that end, the following treatment illustrates many of the complexities associated with the interface as it pertains to solvent extraction.

5.5 Interfacial Chemistry of Solvent Extraction

Though much effort has been dedicated toward the effective manipulation of actinide oxidation state and ligand design, perhaps the heart and soul of liquid-liquid extraction lies within the interfacial chemistry that envelopes the metal–ligand complexation reaction and subsequent transport from the aqueous phase to the organic. The historical literature is impressive in this context and include multiple book chapters and reviews (Refs. 143–145 to name a few). Here, we focus upon recent developments regarding the characterization and structure of liquid–liquid interfaces, changes to speciation and reactivity that may occur at the interface as a function of solution-phase conditions, and discuss the concepts associated with synergism and distinct cooperative phenomena (including aggregation) during extraction conditions.

As it pertains to the properties of the interface and its characterization, many features are relevant, including

- the organization of both solvents, surface-active ligands, and electrolytes
- the speciation of solutes near and within the interface, including the specific solvation environments

- the surface roughness (or capillary wave nature) of the interface and the interfacial width, including phase transitions that may occur as the interface is approached
- the heterogeneity of the interfacial surface
- the length and timescale of interfacial transformations (as it pertains to organizational domains)

To understand these issues, advanced experimental and computational methods have emerged, which are systematically discussed in the following sections.

5.5.1 Properties of the Interface and Its Characterization

Experimental methods. Though once out of reach, the structural characterization of aqueous–organic interfaces has begun in earnest and is a flourishing area of contemporary research. This has been largely enabled by advances in experimental methods, generally X-ray and neutron scattering techniques but also including surface harmonic generation, and vibrational sum-frequency spectroscopies.[146–149] Here we will focus on the scattering methods, where techniques for flattening the interface have simplified these measurements and allowed for precise measurements of the structure of interfaces. Within such studies, liquid–liquid interfaces are configured as in Figure 5.6. The thin film configuration in Fig. 5.6c is preferred for neutron reflection because a neutron beam is strongly attenuated when passing through an aqueous or organic liquid. Although the earliest neutron reflection experiments used the configuration in Fig. 5.6a, the configuration shown in Fig. 5.6c was recently introduced to alleviate the difficulty of maintaining the thin layer as well as to allow the use of non-volatile or non-wetting oils.

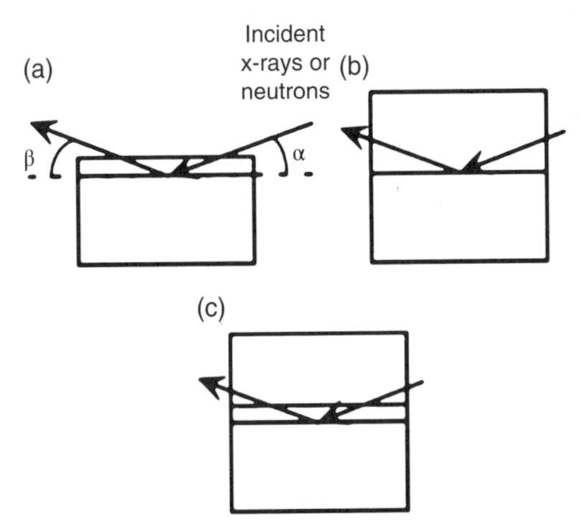

Figure 5.6 Taken from Ref. 147, scattering configurations at liquid–liquid interfaces. Configuration (a) has a thin liquid layer (thickness of nanometers to micrometers) on top of a bulk liquid. Configuration (b) consists of two bulk liquids. Configuration (c) is similar to (a) except for a top single crystal block, either single-crystal silicon or quartz which does not significantly attenuate a neutron beam. Configurations (a) and (c) are used for neutron reflectivity because of the large attenuation of neutrons by aqueous or organic liquids. Appropriate deuteration of the components in the thin film can contrast-match it to the upper vapor phase or solid block. This eliminates reflection from the upper interface, allowing for enhanced sensitivity to reflection from the liquid–liquid interface. Configurations (a) and (b) are used for X-ray reflectivity and off-specular diffuse scattering.

These different configurations also yield somewhat different information. For example, the advantage of configuration (b) in Figure 5.6 is that different thermodynamic states in the binary phase diagram can be reached easily by changing the temperature and re-equilibrating the entire sample. X-ray or neutron specular reflectivity (when $\alpha = \beta$) in Figure 5.6. measures the electron density profile perpendicular to the interface, but averaged over the interfacial plane. When $\alpha \neq \beta$, in-plane inhomogeneities in the electron density profile are revealed. In the case of capillary wave fluctuations of the interface, this diffuse scattering probes the height–height correlation function of the interface.

In recent years, other techniques, such as X-ray reflectivity and X-ray fluorescence have been combined with this approach to gain more insight into ligand and ion distributions at the interface, as described below. X-ray fluorescence near total reflection (XFNTR) has been used study liquid–liquid interfaces, particularly through emission lines of specific ions at the interface, like trivalent Er.[150] This was demonstrated with $ErCl_{3(aq)}$ in contact with dodecane and in the presence of bis(2-ethylhexyl) phosphoric acid (HDEHP) during interfacial CRUD formation.

Statistical mechanical simulations. Theory has been integral to the development of our current understanding of the structure and dynamics of liquid–liquid interfaces. In principle, MD methods can calculate the same or closely related spatial and time correlation functions (TCFs) as measured by experiment, and thus it has been foundational to the interpretation of that data. MD also provides a framework to correlate TCFs to ensembles of molecular configurations that in turn are responsible for the chemical transformations that occur during the complexation and transport process. These can be complemented by computational analysis of the changes to the dipole orientations,[151–155] hydrogen bonding,[154, 156–159] and other structural and dynamic properties.[160–163] In this context, the concept of the Gibbs dividing surface plays an important role as a reference point for defining, in essence, an ideal two-dimensional flat plane that mathematically enables the subsequent definitions of thermodynamic quantities associated solely with molecules at the interface (those adsorbed at the surface) and excess quantities associated with changes in a fluid in the presence of the interface. The Gibbs dividing surface is generally defined as the position in the interface where the density of the solvent is half that in the bulk. In liquid–liquid interfaces, the distance between the two Gibbs dividing surfaces of the two liquids may define the interfacial region (generally termed the intrinsic width[153]). Alternatively, the structural organization and dynamic features of a solvent may be analyzed in consecutive slab layers parallel to the Gibbs dividing surface.[156, 164, 165] The interfacial region is then defined as the region where the properties of solvent have been perturbed relative to the bulk. However, these approaches do not account for the surface roughness of the interface, known as capillary waves, which derives primarily from the surface tension.[166–169] Capillary wave theory mathematically describes these components and provides an alternative mechanism with which to separate the total interfacial width in terms of intrinsic and the capillary wave components.

Complementary to the concepts found in capillary wave theory and the slab-layering analyses based upon the Gibbs dividing surface are recent developments that have focused upon the identification of truly interfacial molecules (ITIM), wherein the molecules that are directly exposed to the opposite phase on the capillary wave front can be elucidated.[155, 163, 170, 171] Recently this method has been used to study the dynamics of the water molecules at the intrinsic water–vapor interface,[171] the width of the water interface with apolar phases,[170] and the surface tension of the liquid–vapor interface.[163]

Capillary wave theory. Much of the experimental observation (from X-ray scattering) and MD study employs capillary wave theory for interpretation.[146] Therein, the total interfacial width consists of two components, that of intrinsic width and that due to capillary waves. From MD simulations, it is approximated that the two contributions can be decoupled so that the total density profile is a convolution of the intrinsic profile $\psi(z)$ and the effect due to capillary waves. The corresponding width parameters can then be obtained by fitting the total density profile to an error function. The density of either liquid is generally fit to

$$\begin{cases} \rho_w(z) = 0.5\rho_w - 0.5\rho_w erf\left(\dfrac{z - z_{0,w}}{\sqrt{2}w_c}\right) \\ \rho_h(z) = 0.5\rho_h - 0.5\rho_h erf\left(\dfrac{z - z_{0,h}}{\sqrt{2}w_c}\right) \end{cases} \qquad \text{(Eq. 5.3)}$$

where ρ_w and ρ_h are the bulk density of the water phase and the *n*-hexane phase, $z_{0,w}$ and $z_{0,h}$ are the Gibbs dividing surfaces for the two liquids, and w_c is the capillary wave width parameter. In this way, the total interfacial width (w), the intrinsic interfacial width (w_0), and the width due to capillary waves (w_c in Eq. (4)) are determined as

$$\begin{cases} w^2 = w_0^2 + w_c^2 \\ w_0 = \left|z_{0,w} - z_{0,h}\right| \\ w_c^2 = \dfrac{k_B T}{2\pi\gamma_e} ln\left(\dfrac{L}{l_b}\right) \end{cases} \qquad \text{(Eq. 5.4)}$$

where L is the length of the cross section of the simulation box $L = L_x = L_y$, γ_e is the surface tension, and l_b is the correlation length.

5.5.2 Current Understanding of Interfacial Structure and Properties under Different Conditions

Using synchrotron X-ray reflectivity, the electron density profile normal to water–organic interfaces has been fit to an error function (from capillary wave theory) to yield the widths of a variety of interfaces. As an example, the water–hexane interfacial width has been measured at 3.3 Å,[172] while water–n-docosane has a measured width of 5.7 Å. Interestingly, the docosane data disagrees with the predictions from capillary wave theory, and the bulk correlation length of docosane must also be incorporated.[173]

As a corollary to the study of liquid–vapor interfaces, and because of the importance of the surface activity of many extracting ligands, significant study has been dedicated to understanding the role of surfactants in the behavior of the interface. Surfactants that self-assemble at the interface between the two bulk liquids are often described or interpreted as being a Langmuir monolayer (or a quasi-two-dimensional thermodynamic system that depends on only two thermodynamic variables like temperature and surface pressure). As such, much of this work focuses on the "phase transition behavior" at which the surfactant transitions from selectively partitioning to the interface and instead migrates into the organic phase. The interfacial monolayer can be spatially homogeneous or inhomogeneous,[174] and depending on the nature of the surfactant and the organic solvent, the phase transition can occur at a wide variety of temperatures.[175] This is quite relevant to separations chemistry due to the necessity of many extracting ligands to have an appreciable residence time in the interface to enable metal–ligand complexation. Indeed, recent work has illustrated the ability to "turn on" or "turn off" extraction through temperature-driven adsorption of an extractant to the interface.[176] The temperature-modulated adsorption has the added advantage of "trapping" a metal–ligand complex in the interface to enable further characterization.[177]

Small changes in the surfactant may produce large changes in the interfacial ordering and speciation. As an example, MD studies of protonated versus neutral forms of heterocyclic N-donor ligands like bistriazinylpyridines have been observed to have large variations in their surface activity (a tendency to be in the interface itself) at the water or acidic water interface with organic solvents.[178] Further, the orientation of these ligands relative to the interface normal changed significantly as a function of their substituents.

The distribution of ligands at the interface, their orientation, and the extent of mixing of water and extractant molecules have in recent years been demonstrated to alter other aspects of the aqueous speciation, including ion distributions near the interface.[149] Indeed, much is left to be learned regarding the condensation of ions, with implications that complex chemistry may lead to non-intuitive behavior. The distribution of ions as a function of formal charge does not appear to follow simple trends or predicted behaviors based on Poisson–Boltzmann statistics applied to planar surfaces (Gouy–Chapman theory and its modifications). This has necessitated combining Poisson's equation with ion potentials of mean force.

In the extreme case, ion condensation at the interface can lead to supersaturation and crystallization, as in the case of a simple solution of NaCl or KCl in the presence of 1-butanol.[179] Recent studies of $ErCl_3$ in HCl in the presence of an interface with $He_{(g)}$ indicate that interfacial hydrolysis reactions may lead to polynuclear cluster motifs in the interfacial region.[180] However, this behavior may be highly sensitive to the conditions employed, as a somewhat different study has instead indicated there is a non-monotonic interfacial distribution of Er(III) that exhibit have a damped, oscillatory behavior in the subsurface in the presence of chloride anions.[181] The total ion concentration (ionic strength) of the solution may influence this behavior, as may the pH of the solution. Co-adsorption of anions may occur with presumed weak adsorption of hydronium ion, which in turn may alter the extent of dissociation of protic extracting ligands and even the distribution or surface activity of ligands near the interface. Recent MD studies of bistriazinyl-phenanthroline ligands have indicated that the nitric acid concentration dramatically impacts the surface activity of these ligands as a function of their substituted functional groups and subsequent total ligand charge.[182]However, the extent of hydronium and hydroxide surface adsorption is still the subject of an active debate, and the impact that this may have upon the protonation state of ampiphilic surface-active extracting ligands and upon electrolyte ion distributions is far from understood.

5.5.3 Synergism and Cooperative Phenomena at Interfaces

Synergism is a deep topic that can include a variety of phenomena depending upon the specific literature examined. The intent of this chapter is not to provide an exhaustive review, but to briefly discuss some important concepts that are relevant when studying the expansive literature and to consider important lessons associated with synergism in communities beyond the separations of actinides. In the actinide separations community, synergism generally refers to the use of more than one extractant to achieve enhanced metal recovery compared to either individual extractant on its own at the same concentration. Given that extracting ligands may be considered in several different classes, the effect of the co-extractant may be due to a number of factors (i.e., modification of interfacial properties, changes to speciation, formation of different metal–ligand complexes, etc.). Thus, the underlying chemistry and physics behind this behavior may be complex and different depending upon the solution conditions employed.

For many decades, increased metal recovery in synergistic systems was thought to occur due to increases in lipophilicity, and therefore solubility, of the extracted metal complex. The increased lipophilicity of the extracted metal species was brought on by swapping polar ligands for more apolar ligands and changing the coordination chemistry of the recovered metal ion. Choppin's review is probably the most direct treatment of how changes in coordination environment can drive increases in metal partitioning to the organic phase.[183] This review divides synergistic systems into two classes:

1) Systems containing an acidic chelant and a neutral (solvating) adduct
2) Systems containing an alkylphosphoric, carboxylic, or sulfonic acid and a neutral (solvating) adduct

In Class I type systems, TBP or trioctylphosphine oxide (TOPO) are frequently used as solvators, and thenoyltrifluoroacetone, HTTA, or other β diketones are used as the acidic chelants. Due to the chelate effect, synergistic effects are generally more pronounced for Class I type systems.

The following mechanisms for increasing metal extraction are also recognized in the Choppin review:

- Opening of one or more of the chelate rings and occupation by solvating molecules in the uncomplexed metal site
- Undersaturating the metal coordination environment, occupying the extra coordination sites with waters, and displacing these waters with a solvating extractant with more lipophilic character
- Increasing the coordination environment of the extracted metal with a solvating extractant, but without displacing waters

Typically, the best way to delineate the particular synergistic mechanism is to assess the system thermodynamics. However, the extraction mechanisms associated with solvating extractants are complicated by the co-extraction of multiple auxiliary species such as water and nitric acid. Since the extraction mechanisms can become difficult to resolve under metal and acid loadings relevant to actinide separations under more applied conditions, thermodynamic assessments using classical methods such as van't Hoff or calorimetric methods can become unresolvable.

The topic of synergism has received renewed interest courtesy the development of TRUSPEAK- and ALSEP-type systems that use combined solvation and cation exchange reagents. From a processing standpoint, synergistic interactions are generally viewed as a disadvantage. Robust chemistry that is insensitive to changes in relative amounts of extractant is generally preferred as this provides more flexibility upon process scale-up and operation. When tons of radioactive material are being processed, buckets are preferred for measurement relative to volumetric flasks.

Untangling synergistic effects has been complicated by trying to capture the role of long-range solution ordering and aggregation in solvent extraction systems. More recent data, using X-ray or neutron scattering, MD simulation, and nuclear magnetic resonance (NMR) spectroscopy, have suggested that aggregation of solvating reagents may be significant under applied conditions when macroscopic amounts of metal are recovered. Currently, the extraction chemistry of long-standing processes, such as PUREX, is under re-examination to delineate if reverse micelle formation, ligand-bridged polymetallic nuclear species, or other extraction mechanisms are relevant. Consideration of the role of the liquid–liquid interface and microemulsion chemistry has provided advances in understanding other potentially aspects of synergistic chemistry.

When considering synergism arising from aggregation, it is useful to consider extraction of organic solutes within microemulsions (e.g., in oil recovery). In the most simplistic sense, synergism in these systems is obtained when the surfactant concentration ensures a minimum in the interfacial tension, thus facilitating transport across the phase boundary. Depending on the system, the drop in interfacial tension has been attributed to major heterogeneities in the surfactant distribution at the interface which cause local disruptions in organization and enhance "mixing" at a molecular level.[184] In the context of actinide solvent extraction, it is also possible that specific extractant mixtures may similarly lead to a broadening of the water–organic interface through a decrease of interfacial tension, in turn decreasing any free energy barriers to transport into the organic phase. Indeed, recent MD studies have indicated that solutes like TBP at low concentrations at the water–organic phase boundary can enhance molecular-scale mixing of both solvents,[159, 164] though current research is determining the extent to which this alters transport of free energy. Other interfacial effects also may be relevant. For example, under certain circumstances, electrostatics may be the source of synergism. Recent MD studies of anionic surfactants (specifically chlorinated cobalt bis(dicarbollide) anions) within water–octanol systems indicate a complex interplay of the octanol itself, and its ability to solvate the anion and the ability of the anion to alter the distribution of metal cations in the interfacial region – thereby increasing cation condensation at the interface and subsequent extraction.[185, 186]

Much of the lore regarding synergism in actinide extraction has not emphasized the important role of modifications to the interfacial chemistry. This is particularly true within the synergism observed for neutral phosphorous-containing extractants with acidic extractants.[187–191] As an example, much work has aimed to study the synergism associated with mixtures of TBP and its radiolysis product di-n-butyl phosphoric acid (HDBP) associated with PUREX separations. There, previous studies have suggested the existence of complexes involving a mixture of ligands, accounting for extraction synergy. At the heart of these studies lies the concept that under the right circumstances synergy may derive from fundamental changes to molecular-scale interactions associated with the metal ions.[97, 188, 192, 193] If different metal–ligand complexes are formed when multiple extracting ligands are present, then perhaps the overall efficacy of the extraction is enhanced (though exactly why the change in metal-ligand (ML) complex distribution would alter the distribution is not well understood). Indeed, recent work has indicated that for the TBP/HDBP explicitly, synergism cannot be related to changes in coordination chemistry in and of itself, and instead begin to invoke changes to the interfacial chemistry imparted by the formation of reverse micelles of extracting ligand.[194] This is particularly interesting in light of other studies that have indicated that formation of hydrocarbon nanoparticles can influence the interfacial thickness (specifically of water–trichloroethylene interfaces).[195] Thus, in some situations, the tide is turning regarding the role of interfacial chemistry in synergism within actinide extraction.

The recent expansion of literature associated with aggregation phenomena (including the formation of reverse micelles described for TBP) at water–organic interfaces, however, goes beyond the PUREX process and may be more generalized to the variety of solvent extraction processes, including TALSPEAK.[196] Further, these aggregation phenomena may have influence beyond behavior like synergism, and their role may impinge upon other areas like third phase formation. Third phase formation occurs when sufficient extracted metal ion and acid concentration exists in the post-contact organic phase. Upon extracting a sufficient quantity of metal ions, or other polar species, the organic phase partitions into two phases: a heavy organic phase laden with the vast majority of the extraction complexes and a light organic phase, containing primarily the organic diluent. The heavy organic phase, or third phase, complicates the extraction system in two ways, both of which require its avoidance to properly operate. First, a partitioned organic phase makes post-extraction recovery of the metal ions difficult as the recovery process relies on a uniform organic phase. Second, the uncontrolled and increased concentration of fissile metal ions in the dense third phase leads to criticality concerns. Depending on the solution-phase conditions, recent SANS and SAXS studies have implicated micelles, reverse micelles, and other aggregate topologies within the formation of the third phase.[197–199]

The third phase may also be thought of as a large "interphase" region that may or may not have a well-defined phase boundary with either water or the lighter organic phase. From a physical chemistry perspective, this can be thought of a phase transition process. Instead of the monolayer of surface-active extractant at the interface, the interface evolves with significant mixing of the organic solvent and water due to the presence of the extractant.[149] The formation of a thick interface or an interphase enriched with extractant molecules over several nanometers can be compared to a surface-induced pre-transitional effect or a critical adsorption already observed for surfactants close to solid or liquid surfaces.[200] A prealignment of the molecules at an interface, or more generally the loss of entropy of the molecules at an interface, can indeed induce such a transition toward a more concentrated phase if it exists in the phase diagram. However, the boundary between what would be considered an "interphase" and a "third phase" is not straightforward.

Several empirical observations can be made regarding third phase formation. Generally, third phase formation is encouraged by the recovery of higher-charge-density metal ions (such as Pu^{4+} or Zr^{4+}), increasing the polarity of a given reagent, or increasing the hydrogen bonding opportunities for a given acid (i.e., $HClO_4 > HNO_3 > HCl$). Many interpretations of third phase formation rely on reverse micelle formation, but a percolation model that accounts for the development of a rich hydrogen bonding network could be another means of obtaining a highly ordered, third phase system. The possibility of an aggregation mechanism only being applicable to a specific aggregation should also be considered when discussing third phase formation for a particular system. Delineating the role of long-range ordering in these systems would also assist in interpretation of synergistic phenomena.

5.6 Concluding Remarks

The future directions of separations processes and the grand challenges underlying solvent extraction extend from basic science to practical implementation. The latter are perhaps more amenable for discussion because the largest demands associated with nuclear fuel cycle separations focus on compressing the number of separations necessary to selectively recover fuel-cycle-relevant actinides (U, Np, Pu, and Am) and upon improving the proliferation resistance of currently utilized separations. In an ideal world, a hexavalent americium separation might be the most feasible means of accomplishing this. Group recovery of actinides would improve both proliferation resistance and, hypothetically, could be accomplished in one separation facility. The key to this type of group actinide separation centers on being able to stabilize hexavalent americium. While possible in basic media – and carbonate-based solid-liquid separations of americium have been demonstrated – the ability to run a solvent extraction process seems to be preferable due to the opportunity for continuous operation. Nevertheless, the following options are under consideration or might be a viable means of achieving hexavalent actinide precipitations:

- Nitrate-based precipitation under acidic conditions[201]
- Making nano-sized solid oxidants that could dissolve upon oxidation
- Use of other higher-oxidation-state transition metals (such as Fe^{4+}, Au^{3+}, or Ni^{4+}) to drive oxidation
- Means of slowing periodate dissociation to improve the lifetime of Cu^{3+} or use/development of other ligand systems with increased solubility and slower dissociation kinetics
- Using the wider electrochemical window of ionic liquids to stabilize Am

It should be noted that using oxidizing americium in an ionic liquid environment has been attempted with limited success, but may be worth reconsidering at a future date.

The future directions are somewhat different for nuclear fuel cycles using a soft donor approach to accomplish americium recovery with less separations steps. For ALSEP-type systems, obstacles still manifest in the metal phase transfer from the aqueous to the organic phase. Some of these challenges could be addressed with ligand design, and that was presented in an earlier section. Alternative contactor designs fabricated using 3D printer technology have been demonstrated as an engineered means of improving phase mixing and kinetics.[202] Recent reports have considered using mixed hard-soft extractants that could compress a fuel cycle in a comparable fashion to that accomplished by a hexavalent americium separation. The mixed hard-soft extractants are a next-generation iteration of group actinide separations that used TBP and BTBP reagents.[203] The biggest obstacle facing these materials is their limited solubility in aliphatic diluents and tendency to hydrolyze in acidic conditions, though completing a group actinide separation using these tactics may be more approachable than oxidizing americium.

Perhaps a final consideration worth mentioning is the use of sulfur-donating reagents to selectively recover trivalent actinides.[204, 205] Their tendency toward radiolysis and acid-catalyzed hydrolysis, as well as a lack of interest in generating significant amounts of a sulfur-bearing waste form, has tabled their use for fuel cycle applications. However, their use in analytical separations may still be attractive for various analytical applications.

As solvent extraction systems are more comprehensively understood in the context of basic science, the ability to tune various ligand properties shift. Certainly the opportunities for accurate computational approaches that accurately describe the physics of several hundreds of atoms in a simulation have made a significant impact on our quantitative understanding of metal–ligand interactions. The ability of high-level methods (beyond DFT) to accurately capture the electronic interaction between a metal and complexant is encouraging more sophisticated considerations than using the empirical Pearson's hard-soft acid-base concepts to discuss actinide/lanthanide selectivity driven by covalency. Further, prior to advances in MD simulations, quantities like entropy and the role of the bulk solvent in these interactions were more difficult to quantify at an atomistic level. Although recognition that bulk solvent properties could impact metal–ligand interactions exists, ligand design is currently centered on direct interactions between a metal center, the ligand, and some qualitative understanding of solubility. The ability to more quantitatively describe the role of bulk solvent properties, and therefore which ligands can manipulate these aspects most effectively, is now a challenge that can be targeted as a part of the design process. In addition, the development of new experimental and computational techniques to probe the portal that enables separation, the liquid–liquid interface, has initiated a new design basis for consideration. The following represent a few of the potential frontiers of the basic science of solvent extraction that will further advance the current state of the art:

- Fully understanding and controlling interfacial activity
- Learning the mechanisms that drive third phase formation and developing extractants resistant to this process
- Decoupling the complex behavior behind synergism and its dependence upon extractant and solution composition
- Developing new materials that can complex actinides in acidic aqueous media
- Utilizing electronic structure to optimize selectivity and solvation chemistry across a broad ligand design space

Acronyms

Acronym	Definition
AIMD	Ab initio molecular dynamics
ALSEP	Actinide Lanthanide SEParation process
APC	(poly)aminopolycarboxylates
BTBP	Bistriazylbypyridine
CATE	The Chemistry of the Actinide and Transactinide Elements

CCSD(T)	Coupled Clusters with Single, Double and perterbatively treated Triple excitations
CHON	Carbon-Hydrogen-Oxygen-Nitrogen concept for incinerable reagents
CMPO	Carbamylphosphine oxide
CN	Coordination Number
COEX	Process of co-extraction of neptunium with the uranium and plutonium
CRUD	Chalk river unidentified deposits, also referring to any precipitates that form in the interface
DAAP	Diamylamyl phosphonate
HCBP/DBP	Dibutyl phosphate radiolysis product of tributyl phosphate
DEHBA	N,N-di(2-ethylhexyl)butanamide
DEHiBA	*N,N*-di-2-ethylhexyl-isobutyramide
DFT	Density functional theory
DPA	Dipicolinic acid
DTPA	Diethylenetriamineppentaacetic acid
DTTA–DAG	Diethylenetriamine-N,N″-bis(acetylglycine)-N,N′,N″-triacetic acid
EXAFS	X-ray absorption fine structure
HDEHP	bis-2-ethylhexyl phosphoric acid
HEDTA	Hydroxyethyl-ethylenediaminetriacetic acid
HEH[EHP]	2-ethylhexyl phosphonic acid mono-2-ethylhexyl ester
HEXS	High-energy X-ray scattering
HLW	High-level waste
ITIM	Identification of truly interfacial molecules
MBP	Monobutyl phosphate
MD	Molecular dynamics
MIBK	Methylisobutylketone
MM	Molecular mechanics
MP2	Second-order Møller–Plesset perturbation theory
PCM	Polarizable continuum model
PUREX	Plutonium Uranium Redox Extraction
QM	Quantum mechanical
SANHEX	Separation of Hexavalent Actinides
TALSPEAK	Trivalent Actinide Lanthanide Separation using Phosphorus Extractants and Aqueous Komplexants
TALSQuEAK	Trivalent Actinide Lanthanide Separation using Quicker Extractants and Aqueous Komplexes
TBP	Tributyl phosphate
TCFs	Time correlation functions
TEHDGA	*N,N,N′,N′*-tetra(2-ethylhexyl)diglycolamide
TODGA	N,N,N′,N′-tetraoctyldiglycolamide
TRUEX	TRansUranic EXtraction
TRUSPEAK	Transuranium Separation by Phosphorous Reagent Extraction from Aqueous Komplexes
TTHA	Triethylenetetraminehexaacetic
XFNTR	X-ray fluorescence near total reflection

Acknowledgments

Aurora Clark acknowledges support from the US Department of Energy, Office of Science, Separations Program, DE-SC-0001815. Jenifer Braley acknowledges support by the US Department of Energy, Office of Science, Office of Basic Energy Sciences, Heavy Elements Chemistry Program at Colorado School of Mines under Award Number DE-SC0012039.Work was supported for Ping Yang under the Heavy Element Chemistry Program at LANL by the Division of Chemical Sciences, Geosciences, and Biosciences, Office of Basic Energy Sciences, US Department of Energy. Los Alamos National Laboratory is operated by Los Alamos National Security, LLC, for the National Nuclear Security Administration of US Department of Energy (contract DE-AC52-06NA25396).

References

1 Lanham, W. B., and Runion, T. C. (1949) *Purex Process for Plutonium and Uranium Recovery* (No. ORNL-479 (Del.)). Oak Ridge National Lab., Tenn., USA.
2 Weaver, B., and Kappelmann, F. A. (1964) TALSPEAK A New Method of Separating Americium and Curium from the Lanthanides by Extraction from an Aqueous Solution of an Aminopolyacetic Acid Complex with a Monoacidic Organophosphate or Phosphonate (No. ORNL-3559). Oak Ridge National Lab., Tenn., USA.
3 Nilsson, M., and Nash, K. L. (2007) Review article: A review of the development and operational characteristics of the TALSPEAK process. *Solvent Extraction and Ion Exchange.* **25** (6), 665–701.
4 Braley, J. C., Carter, J. C., Sinkov, S. I., *et al.* (2012) The role of carboxylic acids in TALSQuEAK separations. *Journal of Coordination Chemistry.* **65** (16), 2862–2876.
5 Braley, J. C., Grimes, T. S., and Nash, K. L. (2011) Alternatives to HDEHP and DTPA for simplified TALSPEAK separations. *Industrial and Engineering Chemistry Research,* **51** (2), 629–638.
6 Lumetta, G. J., Gelis, A. V., Braley, J. C., *et al.* (2013) The TRUSPEAK concept: Combining CMPO and HDEHP for separating trivalent lanthanides from the transuranic elements. *Solvent Extraction and Ion Exchange.* **31** (3), 223–236.
7 Braley, J. C., Lumetta, G. J., and Carter, J. C. (2013) Combining CMPO and HEH [EHP] for separating trivalent lanthanides from the transuranic elements. *Solvent Extraction and Ion Exchange.* **31** (6), 567–576.
8 Lumetta, G. J., Gelis, A. V., Carter, J. C., *et al.* (2014) The actinide-lanthanide separation concept. *Solvent Extraction and Ion Exchange.* **32** (4), 333–346.
9 Mincher, B. J., Martin, L. R., and Schmitt, N. C. (2008) Tributylphosphate extraction behavior of bismuthate-oxidized americium. *Inorganic Chemistry.* **47** (15), 6984–6989.
10 Mincher, B. J., Schmitt, N. C., and Case, M. E. (2011) A TRUEX-based separation of americium from the lanthanides. *Solvent Extraction and Ion Exchange.* **29** (2), 247–259.
11 Mincher, B. J., Martin, L. R., and Schmitt, N. C. (2012) Diamylamylphosphonate solvent extraction of Am (VI) from nuclear fuel raffinate simulant solution. *Solvent Extraction and Ion Exchange.* **30** (5), 445–456.

12 Burns, J. D., Shehee, T. C., Clearfield, A., *et al.* (2012) Separation of americium from curium by oxidation and ion exchange. *Analytical Chemistry.* **84** (16), 6930–6932.
13 Mincher, B. J., Schmitt, N. C., Schuetz, B. K., *et al.* (2015) Recent advances in f-element separations based on a new method for the production of pentavalent americium in acidic solution. *RSC Advances.* **5** (34), 27205–27210.
14 Dares, C. J., Lapides, A. M., Mincher, B. J., *et al.* (2015) Electrochemical oxidation of 243Am (III) in nitric acid by a terpyridyl-derivatized electrode. *Science.* **350** (6261), 652–655.
15 Sinkov, S. I., and Lumetta, G. J. (2015) Americium (III) oxidation by copper (III) periodate in nitric acid solution as compared with the action of Bi (V) compounds of sodium, lithium, and potassium. *Radiochimica Acta.* **103** (8), 541–552.
16 McCann, K., Brigham, D. M., Morrison, S., *et al.* (2016) Hexavalent americium recovery using copper (III) periodate. *Inorganic Chemistry.* **55** (22), 11971–11978.
17 Katz, J. J. (2007) *The Chemistry of the Actinide and Transactinide Elements.* (**1–5**) (Vol. 1). L. R. Morss, N. Edelstein, and J. Fuger (Eds.). Springer Science and Business Media.
18 Soderholm, L., Skanthakumar, S., and Neuefeind, J. (2005) Determination of actinide speciation in solution using high-energy X-ray scattering. *Analytical and Bioanalytical Chemistry.* **383** (1), 48–55.
19 Clark, A. E., Samuels, A., Wisuri, K., *et al.* (2015) Sensitivity of solvation environment to oxidation state and position in the early actinide period. *Inorganic Chemistry.* **54** (13), 6216–6225.
20 Parmar, P., Samuels, A., and Clark, A. E. (2014) Applications of polarizable continuum models to determine accurate solution-phase thermochemical values across a broad range of cation charge–the case of U (III–VI). *Journal of Chemical Theory and Computation.* **11** (1), 55–63.
21 Kuta, J., and Clark, A. E. (2010) Trends in aqueous hydration across the 4f period assessed by reliable computational methods. *Inorganic Chemistry.* **49** (17), 7808–7816.
22 Yang, T., and Bursten, B. E. (2006) Speciation of the curium (III) ion in aqueous solution: A combined study by quantum chemistry and molecular dynamics simulation. *Inorganic Chemistry.* **45** (14), 5291–5301.
23 Marjolin, A., Gourlaouen, C., Clavaguéra, C., *et al.* (2012) Toward accurate solvation dynamics of lanthanides and actinides in water using polarizable force fields: From gas-phase energetics to hydration free energies. *Theoretical Chemistry Accounts.* **131**:1198, 1–14.
24 Shannon, R. D. (1976) Revised effective ionic radii and systematic studies of interatomie distances in halides and chaleogenides. *Acta Crystallographica.* **A32**, 751–766.
25 D'Angelo, P., Martelli, F., Spezia, R., *et al.* (2013) Hydration properties and ionic radii of actinide(III) ions in aqueous solution. *Inorganic Chemistry.* **52**, 10318–10324.
26 Knope, K. E., and Soderholm, L. (2013) Solution and solid-state structural chemistry of actinide hydrates and their hydrolysis and condensation products. *Chemical Reviews.* **113**, 944–994.
27 Steele, H., and Taylor, R. J. (2007) A Theoretical study of the inner-sphere disproportionation reaction mechanism of the pentavalent actinyl ions. *Inorganic Chemistry.* **46**, 6311–6318.
28 Zhang, Z., and Pitzer, R. M. (1999) Application of relativistic quantum chemistry to the electronic energy levels of the uranyl ion. *Journal of Physical Chemistry A.* **103**, 6880–6886.

29 Hay, P. J., Martin, R. L. and Schreckenbach, G. (2000) Theoretical studies of the properties and solution chemistry of AnO_2^{2+} and AnO_2^{+} aquo complexes for An = U, Np, and Pu. *Journal of Physical Chemistry A.* **104**, 6259–6270.

30 Denning, R. G. (2007) Electronic structure and bonding in actinyl ions and their analogs. *Journal of Physical Chemistry A.* **111**, 4125–4143.

31 Rios, D., Michelini, M. C., Lucena, A. F., *et al.* (2012) Gas-phase uranyl, neptunyl, and plutonyl: Hydration and oxidation studied by experiment and theory. *Inorganic Chemistry.* **51**, 6603–6614.

32 Neuefeind, J., Soderholm, L. and Skanthakumar, S. (2004) Experimental coordination environment of uranyl(VI) in aqueous solution. *Journal of Physical Chemistry A.* **108**, 2733–2739.

33 Sémon, L., Boeheme, C., Billard, I., *et al.* (2001) Do Perchlorate and triflate anions bind to the uranyl cation in an acidic aqueous medium? A combined EXAFS and quantum mechanical investigation. *ChemPhysChem.* **2**, 591–598.

34 Dau, P. D., and Gibson, J. K. (2015) Halide abstraction from halogenated acetate ligands by actinyls: A competition between bond breaking and bond making. *Journal of Physical Chemistry A.* **119**, 3218–3224.

35 Dau, P. D., Maurice, R., Renault, E., *et al.* (2016) Heptavalent neptunium in a gas-phase complex: $(Np^{VII}O_3^{+})(NO_3^{-})_2$. *Inorganic Chemistry.* **55**, 9830–9836.

36 Bühl, M., Kabrede, H., Diss, R. *et al.* (2006) Effect of hydration on coordination properties of uranyl(VI) complexes. A first-principles molecular dynamics study. *Journal of the American Chemistry Society.* **128**, 6357–6368.

37 Li, P., Niu, W. and Gao, T. (2017) Systematic analysis of structural and topological properties: New insights into $PuO_2(H_2O)_n^{2+}$ (n = 1–6) complexes in the gas phase. *RSC Advances.* **7**, 4291–4296.

38 Vallet, V., Macak, P., Wahlgren, U., *et al.* (2006) Actinide chemistry in solution, quantum chemical methods and models. *Theoretical Chemistry Accounts.* **115**, 145–160.

39 Kovács, A., Konings, R. J. M., Gibson, J. K., *et al.* (2015) Quantum chemical calculations and experimental investigations of molecular actinide oxides. *Chemical Reviews.* **115**, 1725–1759.

40 Riddle, C., Czerwinski, K., Kim, E., *et al.* (2016) Characterization of pentavalent and hexavalent americium complexes in nitric acid using X-ray absorption fine structure spectroscopy and first-principles modeling. *Journal of Radioanalytical and Nuclear Chemistry.* **309**, 1087–1095.

41 Neck, V. and Kim, J. I. (2001) Solubility and hydrolysis of tetravalent actinides. *Radiochimica Acta*, **89**, 1 – 16.

42 Heinz, N., Zhang, J., and Dolg, M. (2014) Actinoid(III) hydration – First principle Gibbs energies of hydration using high level correlation methods. *Journal of Chemical Theory and Computation.* **10**, 5593–5598

43 Wiebke, J., Moritz, A., Cao, X., *et al.* (2007) *Physical Chemistry Chemical Physics.* **9**, 459 – 465.

44 D'Angelo, P., and Spezia, R. (2012) Hydration of lanthanoids (III) and actinoids (III): An experimental/theoretical saga. *Chemistry – A European Journal.* **18**, 11162–11178.

45 Duvail, M., Martelli, F., Vitorge, P., *et al.* (2011) Polarizable interaction potential for molecular dynamics simulations of actinoids (III) in liquid water. *The Journal of Chemical Physics.* **135** (4), 044503.

46 Hagberg, D., Bednarz, E., Edelstein, N. M., *et al.* (2007) A quantum chemical and molecular dynamics study of the coordination of Cm (III) in water. *Journal of the American Chemical Society.* **129** (46), 14136–14136.
47 Tsushima, S., and Suzuki, A. (2000) Hydration numbers of pentavalent and hexavalent uranyl, neptunyl, and plutonyl. *Journal of Molecular Structure: THEOCHEM.* **529** (1), 21–25.
48 Frick, R. J., Pribil, A. B., Hofer, T. S., *et al.* (2009) Structure and dynamics of the U4+ ion in aqueous solution: An ab initio quantum mechanical charge field molecular dynamics study. *Inorganic Chemistry.* **48** (9), 3993–4002.
49 Atta-Fynn, R., Johnson, D. F., Bylaska, E. J., *et al.* (2012) Structure and hydrolysis of the U (IV), U (V), and U (VI) aqua ions from ab initio molecular simulations. *Inorganic Chemistry.* **51** (5), 3016–3024.
50 Antonio, R., Soderholm, L., Williams, C. W., *et al.* (2001) Neptunium redox speciation. *Radiochimica Acta.* **89** (1), 17–26.
51 Tsushima, S., and Yang, T. (2005) Relativistic density functional theory study on the structure and bonding of U (IV) and Np (IV) hydrates. *Chemical Physics Letters.* **401** (1), 68–71.
52 Odoh, S. O., and Schreckenbach, G. (2011) Theoretical study of the structural properties of plutonium (IV) and (VI) complexes. *The Journal of Physical Chemistry A.* **115** (48), 14110–14119.
53 Horowitz, S. E., and Marston, J. B. (2011) Strong correlations in actinide redox reactions. *The Journal of Chemical Physics.* **134** (6), 064510.
54 D'Angelo, P., Zitolo, A., Migliorati, V., *et al.* (2011) Revised ionic radii of lanthanoid (III) ions in aqueous solution. *Inorganic Chemistry.* **50** (10), 4572–4579.
55 Dinescu, A., and Clark, A. E. (2008) Thermodynamic and structural features of aqueous Ce (III). *The Journal of Physical Chemistry A.* **112** (44), 11198–11206.
56 Gutowski, K. E., and Dixon, D. A. (2006) Predicting the energy of the water exchange reaction and free energy of solvation for the uranyl ion in aqueous solution. *The Journal of Physical Chemistry A.* **110** (28), 8840–8856.
57 Miertuš, S., Scrocco, E., and Tomasi, J. (1981) Electrostatic interaction of a solute with a continuum. A direct utilizaion of AB initio molecular potentials for the prevision of solvent effects. *Chemical Physics.* **55** (1), 117–129.
58 Cramer, C. J., and Truhlar, D. G. (1992) An SCF solvation model for the hydrophobic effect and absolute free energies of aqueous solvation. *Science.* **256** (5054), 213.
59 Miertuš, S., and Tomasi, J. (1982) Approximate evaluations of the electrostatic free energy and internal energy changes in solution processes. *Chemical Physics.* **65** (2), 239–245.
60 Mohan, V., Davis, M. E., McCammon, J. A., *et al.* (1992) Continuum model calculations of solvation free energies: Accurate evaluation of electrostatic contributions. *Journal of Physical Chemistry.* **96** (15), 6428–6431.
61 Still, W. C., Tempczyk, A., Hawley, R. C., *et al.* (1990) Semianalytical treatment of solvation for molecular mechanics and dynamics. *Journal of the American Chemical Society.* **112** (16), 6127–6129.
62 Curutchet, C., Orozco, M., and Luque, F. J. (2001) Solvation in octanol: Parametrization of the continuum MST model. *Journal of Computational Chemistry.* **22** (11), 1180–1193.

63 Giesen, D. J., Hawkins, G. D., Liotard, D. A., *et al.* (1997) A universal model for the quantum mechanical calculation of free energies of solvation in non-aqueous solvents. *Theoretical Chemistry Accounts: Theory, Computation, and Modeling (Theoretica Chimica Acta).* **98** (2), 85–109.
64 Li, J., Zhu, T., Hawkins, G. D., *et al.* (1999) Extension of the platform of applicability of the SM5. 42R universal solvation model. *Theoretical Chemistry Accounts: Theory, Computation, and Modeling (Theoretica Chimica Acta).* **103** (1), 9–63.
65 Luque, F. J., Bachs, M., Alemán, C., *et al.* (1996) Extension of MST/SCRF method to organic solvents: Ab initio and semiempirical parametrization for neutral solutes in CCl4. *Journal of Computational Chemistry.* **17** (7), 806–820.
66 Luque, F. J., Zhang, Y., Alemán, C., *et al.* (1996) Solvent effects in chloroform solution: Parametrization of the MST/SCRF continuum model. *The Journal of Physical Chemistry.* **100** (10), 4269–4276.
67 Sitkoff, D., Ben-Tal, N., and Honig, B. (1996) Calculation of alkane to water solvation free energies using continuum solvent models. *The Journal of Physical Chemistry.* **100** (7), 2744–2752.
68 Bryantsev, V. S., Diallo, M. S., and Goddard Iii, W. A. (2008) Calculation of solvation free energies of charged solutes using mixed cluster/continuum models. *The Journal of Physical Chemistry B.* **112** (32), 9709–9719.
69 Asthagiri, D., Pratt, L. R., and Ashbaugh, H. S. (2003) Absolute hydration free energies of ions, ion–water clusters, and quasichemical theory. *The Journal of Chemical Physics.* **119** (5), 2702–2708.
70 Pliego, J. R., and Riveros, J. M. (2001) The cluster - continuum model for the calculation of the solvation free energy of ionic species. *The Journal of Physical Chemistry A.* **105** (30), 7241–7246.
71 Nagasaki, S., Kinoshita, K., Enokida, Y., *et al.* (1992) Solvent extraction of Np (V) with CMPO from nitric acid solutions containing U (VI). *Journal of Nuclear Science and Technology.* **29** (11), 1100–1106.
72 Guillaume, B., Moulin, J. P., and Maurice, C. (1984) Chemial properties of neptunium applied to neptunium management in extraction cycles of purex process. *Proccess of Extraction.* **88**, 31–45.
73 Sullivan, J. C., Hindman, J. C., and Zielen, A. J. (1961) Specific Interaction between Np (V) and U (VI) in aqueous perchloric acid media1. *Journal of the American Chemical Society.* **83** (16), 3373–3378.
74 Sullivan, J. C. (1962) Complex ion formation between cations. Spectra and identification of a Np (V)-Cr (III) complex. *Journal of the American Chemical Society.* **84** (22), 4256–4259.
75 Freiderich, J. W., Burn, A. G., Martin, L. R., *et al.* (2017) A combined density functional theory and spectrophotometry study of the bonding interactions of $[NpO_2 \cdot M]^{4+}$ Cation-cation complexes. *Inorganic Chemistry.* **56**, 4788–4795.
76 Todd, T. (2008) Spent Nuclear Fuel Reprocessing [PDF Document]. Retrieved from http://www.state.nv.us/nucwaste/library/Reprocessing/NRCseminarreprocessing_Terry_Todd.pdf.
77 Gaunt, A. J., May, I., Neu, M. P., *et al.* (2011) Structural and spectroscopic characterization of plutonyl (VI) nitrate under acidic conditions. *Inorganic Chemistry.* **50** (10), 4244–4246.

78 Ye, X., Smith, R. B., Cui, S., *et al.* (2010) Influence of nitric acid on uranyl nitrate association in aqueous solutions: A molecular dynamics simulation study. *Solvent Extraction and Ion Exchange.* **28** (1), 1–18.
79 Hlushak, S. P., Simonin, J. P., Caniffi, B., *et al.* (2011) Description of partition equilibria for uranyl nitrate, nitric acid and water extracted by tributyl phosphate in dodecane. *Hydrometallurgy.* **109** (1), 97–105.
80 Sahu, P., Ali, S. M., and Shenoy, K. T. (2016) Passage of TBP–uranyl complexes from aqueous–organic interface to the organic phase: Insights from molecular dynamics simulation. *Physical Chemistry Chemical Physics.* **18** (34), 23769–23784.
81 Tkac, P., Paulenova, A., Vandegrift, G. F., *et al.* (2009) Modeling of Pu (IV) extraction from acidic nitrate media by tri-n-butyl phosphate. *Journal of Chemical & Engineering Data.* **54** (7), 1967–1974.
82 Odoh, S. O., and Schreckenbach, G. (2011) Theoretical study of the structural properties of plutonium (IV) and (VI) complexes. *The Journal of Physical Chemistry A.* **115** (48), 14110–14119.
83 Šulka, M., Cantrel, L., and Vallet, V. (2014) Theoretical study of plutonium (IV) complexes formed within the PUREX process: A proposal of a plutonium surrogate in fire conditions. *The Journal of Physical Chemistry A.* **118** (43), 10073–10080.
84 Allen, P. G., Veirs, D. K., Conradson, S. D., *et al.* (1996) Characterization of aqueous plutonium (IV) nitrate complexes by extended X-ray absorption fine structure spectroscopy. *Inorganic Chemistry.* **35** (10), 2841–2845.
85 Runde, W., Reilly, S. D., and Neu, M. P. (1999) Spectroscopic investigation of the formation of PuO_2Cl^+ and PuO_2Cl_2 in NaCl solutions and application for natural brine solutions. *Geochimica et Cosmochimica Acta.* **63** (19), 3443–3449.
86 Hennig, C., Tutschku, J., Rossberg, A., *et al.* (2005) Comparative EXAFS investigation of uranium (VI) and-(IV) aquo chloro complexes in solution using a newly developed spectroelectrochemical cell. *Inorganic Chemistry.* **44** (19), 6655–6661.
87 Kropman, M. F., and Bakker, H. J. (2001) Dynamics of water molecules in aqueous solvation shells. *Science.* **291** (5511), 2118–2120.
88 Skanthakumar, S., Antonio, M. R., Wilson, R. E., *et al.* (2007) The curium aqua ion. *Inorganic Chemistry.* **46** (9), 3485–3491.
89 Kelley, M. P., Yang, P., Clark, S. B., and Clark, A. E. (2016) Structural and thermodynamic properties of the CmIII ion solvated by water and methanol. *Inorganic Chemistry.* **55** (10), 4992–4999.
90 Kelley, M.; Yang, P.; Clark, S. B., Clark, A. E. (2017) Competitive interactions within trivalent ion solvation in binary water/methanol salvations. *Inorganic Chemistry.* Submitted.
91 Morrone, J. A., Haslinger, K. E., and Tuckerman, M. E. (2006) Ab initio molecular dynamics simulation of the structure and proton transport dynamics of methanol - water solutions. *The Journal of Physical Chemistry B.* **110** (8), 3712–3720.
92 Bakó, I., Megyes, T., Bálint, S., Grósz, T., and Chihaia, V. (2008) Water–methanol mixtures: Topology of hydrogen bonded network. *Physical Chemistry Chemical Physics.* **10** (32), 5004–5011.
93 Wang, C. H., Bai, P., Siepmann, J. I., and Clark, A. E. (2014) Deconstructing hydrogen-bond networks in confined nanoporous materials: Implications for alcohol–water separation. *The Journal of Physical Chemistry C.* **118** (34), 19723–19732.

94 Hawlicka, E., and Swiatla-Wojcik, D. (2002) MD simulation studies of selective solvation in methanol – Water mixtures: An effect of the charge density of a solute. *The Journal of Physical Chemistry A.* **106** (7), 1336–1345.
95 Noskov, S. Y., Kiselev, M. G., Kolker, A. M., and Rode, B. M. (2001) Structure of methanol-methanol associates in dilute methanol-water mixtures from molecular dynamics simulation. *Journal of Molecular Liquids.* **91** (1–3), 157–165.
96 Yu, H., Geerke, D. P., Liu, H., and van Gunsteren, W. F. (2006) Molecular dynamics simulations of liquid methanol and methanol–water mixtures with polarizable models. *Journal of Computational Chemistry.* **27** (13), 1494–1504.
97 Rydberg, J., Cox, M., Musikas, C., and Choppin, G. R. (Eds.). (2004) *Solvent Extraction Principles and Practice.* Revised and Expanded.
98 Cotton, S. (2013) *Lanthanide and Actinide Chemistry.* John Wiley & Sons, Chichester, England.
99 Mincher, B. J., Modolo, G., and Mezyk, S. P. (2009) Review article: The effects of radiation chemistry on solvent extraction: 1. Conditions in acidic solution and a review of TBP radiolysis. *Solvent Extraction and Ion Exchange.* **27** (1), 1–25.
100 Musikas, C. (1997) Completely incinerable extractants for the nuclear industry – A review. *Mineral Processing and Extractive Metullargy Review.* **17** (1–4), 109–142.
101 Pathak, P. N. (2014) N, N-Dialkyl amides as extractants for spent fuel reprocessing: An overview. *Journal of Radioanalytical and Nuclear Chemistry.* **300** (1), 7–15.
102 Manchanda, V. K., and Pathak, P. N. (2004) Amides and diamides as promising extractants in the back end of the nuclear fuel cycle: An overview. *Separation and Purification Technology.* **35** (2), 85–103.
103 McCann, K.; Drader, J. A.; Braley, J. C. (2017) Comparisons of actinide recovery between branched and straight-chained monoamide extractants. *Separation and Purification Reviews.* accepted.
104 Ban, Y., Hotoku, S., Tsutsui, N., Suzuki, A., Tsubata, Y., and Matsumura, T. (2016) Uranium and plutonium extraction by N, N-dialkylamides using multistage mixer-settler extractors. *Procedia Chemistry.* **21**, 156–161.
105 Prabhu, D., Mahajan, G., and Nair, G. (1997) Di (2-ethyl hexyl) butyramide and di (2-ethyl hexyl) isobutyramide as extractants for uranium (VI) and plutonium (IV). *Journal of Radioanalytical and Nuclear Chemistry.* **224** (1–2), 113–116.
106 Pathak, P. N., Prabhu, D. R., Ruikar, P. B., and Mancha, V. K. (2002) Evaluation of di (2-ethylhexyl) isobutyramide (D2EHIBA) as a process extractant for the recovery of 233U from irradiated Th. *Solvent Extraction and Ion Exchange.* **20** (3), 293–311.
107 Ansari, S. A., Pathak, P. N., Manchanda, V. K., Husain, M., Prasad, A. K., and Parmar, V. S. (2005) N, N, N′, N′ tetraoctyl diglycolamide (TODGA): A promising extractant for actinide partitioning from high level waste (HLW). *Solvent Extraction and Ion Exchange.* **23** (4), 463–479.
108 Ansari, S. A., Pathak, P., Mohapatra, P. K., and Manchanda, V. K. (2011) Chemistry of diglycolamides: Promising extractants for actinide partitioning. *Chemical Reviews.* **112** (3), 1751–1772.
109 Tachimori, S., Sasaki, Y., and Suzuki, S. I. (2002) Modification of TODGA-n-dodecane solvent with a monoamide for high loading of lanthanides (III) and actinides (III). *Solvent Extraction and Ion Exchange.* **20** (6), 687–699.

110 Deepika, P., Sabharwal, K. N., Srinivasan, T. G., and Vasudeva Rao, P. R. (2010) Studies on the use of N, N, N, N-tetra (2-ethylhexyl) diglycolamide (TEHDGA) for actinide partitioning. I: Investigation on third-phase formation and extraction behavior. *Solvent Extraction and Ion Exchange.* **28** (2), 184–201.

111 Pahan, S., Boda, A., and Ali, S. M. (2015) Density functional theoretical analysis of structure, bonding, interaction and thermodynamic selectivity of hexavalent uranium (UO22+) and tetravalent plutonium (Pu4+) ion complexes of tetramethyl diglycolamide (TMDGA). *Theoretical Chemistry Accounts.* **134** (4), 41.

112 Ansari, S. A., Pathak, P., Mohapatra, P. K., and Manchanda, V. K. (2011) Chemistry of diglycolamides: Promising extractants for actinide partitioning. *Chemical reviews.* **112** (3), 1751–1772.

113 Ekberg, C., Fermvik, A., Retegan, T., Skarnemark, G., Foreman, M. R. S., Hudson, M. J., Englund, S., and Nilsson, M. (2008) An overview and historical look back at the solvent extraction using nitrogen donor ligands to extract and separate An (III) from Ln (III). *Radiochimica Acta.* **96**(4–5), 225–233.

114 Lewis, F. W., Harwood, L. M., Hudson, M. J., Geist, A., Kozhevnikov, V. N., Distler, P., and John, J. (2015) Hydrophilic sulfonated bis-1, 2, 4-triazine ligands are highly effective reagents for separating actinides (III) from lanthanides (III) via selective formation of aqueous actinide complexes. *Chemical Science.* **6**(8), 4812–4821.

115 Bhattacharyya, A., Mohapatra, M., Mohapatra, P. K., Gadly, T., Ghosh, S. K., Manna, D., Ghanty, T. K., Rawat, N., and Tomar, B. S. (2016) An insight into the complexation of trivalent americium vis-à-vis lanthanides with bis (1, 2, 4-triazinyl) bipyridine derivatives. *European Journal of Inorganic Chemistry.* **2017** (4), 820–828.

116 Cao, S., Wang, J., Tan, C., Zhang, X., Li, S., Tian, W., Guo, H., Wang, L., and Qin, Z. (2016) Solvent extraction of americium (iii) and europium (iii) with tridentate N, N-dialkyl-1, 10-phenanthroline-2-amide-derived ligands: Extraction, complexation and theoretical study. *New Journal of Chemistry.* **40** (12), 10560–10568.

117 Fryer-Kanssen, I., Austin, J., and Kerridge, A. (2016). Topological study of bonding in aquo and bis (triazinyl) pyridine complexes of trivalent lanthanides and actinides: Does covalency imply stability?. *Inorganic Chemistry.* **55** (20), 10034–10042.

118 Choppin, G. R., Thakur, P., and Mathur, J. N. (2006) Complexation thermodynamics and structural aspects of actinide–aminopolycarboxylates. *Coordination Chemistry Reviews.* **250** (7), 936–946.

119 Drader, J. A., Luckey, M., and Braley, J. C. (2016) Thermodynamic considerations of covalency in trivalent actinide-(poly) aminopolycarboxylate interactions. *Solvent Extraction and Ion Exchange,* **34** (2), 114–125.

120 Grimes, T. S., Heathman, C. R., Jansone-Popova, S., Bryantsev, V. S., Goverapet Srinivasan, S., Nakase, M., and Zalupski, P. R. (2017) Thermodynamic, spectroscopic, and computational studies of f-element complexation by N-Hydroxyethyl-diethylenetriamine-N, N′, N ″, N ″-tetraacetic acid. *Inorganic Chemistry.***56** (3), 1722–1733.

121 Heathman, C. R. (2013) *Functionalization Characterization and Evaluation of Dipicolinate Derivatives with F-Element Complexes.* Washington State University.

122 Weaver, B., and Kappelmann, F. A. (1964) Talspeak: A New Method of Separating Americium and Curium from an Aqueous Solution of an Aminopolyacetic Acid Complex with a Monoacidic Organophosphate or Phosphonate. Report ORNL-3559; Oak Ridge National Laboratory: TN, USA.

123 Nash, K. L., Brigham, D., Shehee, T. C., and Martin, A. (2012) The kinetics of lanthanide complexation by EDTA and DTPA in lactate media. *Dalton Transactions,* **41** (48), 14547–14556.

124 Heathman, C. R., Grimes, T. S., and Zalupski, P. R. (2016) Coordination chemistry and f-element complexation by diethylenetriamine-N, N″-bis (acetylglycine)-N, N′, N″-triacetic Acid. *Inorganic Chemistry.* **55** (21), 11600–11611.

125 Danesi, P. R., and Vandegrift, G. F. (1981) Kinetics and mechanism of the interfacial mass transfer of Eu/sup 3+/and Am/sup 3+/in the system bis (2-ethylhexyl) phosphate-n-dodecane-NaCl-HCl-water. *Journal of Physical Chemistry (United States).* **85** (24), 969–984.

126 Danesi, P. R., and Cianetti, C. (1982). Kinetics and mechanism of the interfacial mass transfer of Eu (III) in the system: Bis (2-ethylhexyl) phosphoric acid, n-dodecane-NaCl, lactic acid, polyaminocarboxylic acid, water. *Separation Science and Technology.* **17** (7), 969–984.

127 Heathman, C. R., Grimes, T. S., and Zalupski, P. R. (2016) Thermodynamic and spectroscopic studies of trivalent f-element complexation with ethylenediamine-N, N′-di (acetylglycine)-N, N′-diacetic acid. *Inorganic chemistry,* **55** (6), 2977–2985.

128 Batista, E. R., Martin, R. L., and Yang, P. (2015). *Computational Studies of Bonding and Reactivity in Actinide Molecular Complexes* (pp. 375–400). Wiley: West Sussex, United Kingdom.

129 Kozimor, S. A., Yang, P., Batista, E., *et al.* (2009) Trends in covalency for d-and f-element metallocene dichlorides identified using chlorine K-edge X-ray absorption spectroscopy and time-dependent density functional theory. *Journal of the American Chemical Society.* **131** (34), 12125–12136.

130 Löble, M. W., Keith, J. M., Altman, A. B., *et al.* (2015) Covalency in lanthanides. An X-ray absorption spectroscopy and density functional theory study of LnCl6 x–(x = 3, 2). *Journal of the American Chemical Society.* **137** (7), 2506–2523.

131 Minasian, S. G., Keith, J. M., Batista, E., *et al.* (2013). Covalency in metal–oxygen multiple bonds evaluated using oxygen k-edge spectroscopy and electronic structure theory. *Journal of the American Chemical Society,* **135** (5), 1864–1871.

132 Minasian, S. G., Keith, J. M., Batista, E. R., *et al.* (2012). Determining relative f and d orbital contributions to M–Cl covalency in MCl62–(M = Ti, Zr, Hf, U) and UOCl5– Using Cl K-Edge X-ray absorption spectroscopy and time-dependent density functional theory. *Journal of the American Chemical Society.* **134** (12), 5586–5596.

133 Spencer, L. P., Yang, P., Minasian, S. G., *et al.* (2013). Tetrahalide complexes of the [U (NR) 2] 2+ ion: Synthesis, theory, and chlorine K-edge X-ray absorption spectroscopy. *Journal of the American Chemical Society.* **135** (6), 2279–2290.

134 Kaltsoyannis, N. (2012) Does covalency increase or decrease across the actinide series? Implications for minor actinide partitioning. *Inorganic Chemistry.* **52** (7), 3407–3413.

135 Frenking, G., and Shaik, S. (Eds.). (2014). *The Chemical Bond: Chemical Bonding Across the Periodic Table (Vol. 2).* John Wiley & Sons, Weinheim, Germany. 337–353.

136 Prodan, I. D., Scuseria, G. E., and Martin, R. L. (2007) Covalency in the actinide dioxides: Systematic study of the electronic properties using screened hybrid density functional theory. *Physical Review B.* **76** (3), 033101.

137 Lukens, W. W., Speldrich, M., Yang, P., *et al.* (2016). The roles of 4f-and 5f-orbitals in bonding: A magnetochemical, crystal field, density functional theory, and multi-reference wavefunction study. *Dalton Transactions.* **45** (28), 11508–11521.

138 Roy, L. E., Bridges, N. J., and Martin, L. R. (2013) Theoretical insights into covalency driven f element separations. *Dalton Transactions.* **42** (7), 2636–2642.

139 Ali, S. M., Pahan, S., Bhattacharyya, A., *et al.* (2016) Complexation thermodynamics of diglycolamide with f-elements: Solvent extraction and density functional theory analysis. *Physical Chemistry Chemical Physics,* **18** (14), 9816–9828.

140 Kaneko, M., Miyashita, S., and Nakashima, S. (2015) Bonding study on the chemical separation of Am (III) from Eu (III) by S-, N-, and O-donor ligands by means of all-electron ZORA-DFT calculation. *Inorganic Chemistry.* **54** (14), 7103–7109.

141 Bryantsev, V. S., and Hay, B. P. (2015) Theoretical prediction of Am (III)/Eu (III) selectivity to aid the design of actinide-lanthanide separation agents. *Dalton Transactions.* **44** (17), 7935–7942.

142 Keith, J. M., and Batista, E. R. (2011) Theoretical examination of the thermodynamic factors in the selective extraction of Am3+ from Eu3+ by dithiophosphinic acids. *Inorganic Chemistry.* **51** (1), 13–15.

143 Danesi, P. R., Chiarizia, R., and Coleman, C. F. (1980) The kinetics of metal solvent extraction, *CRC Critical Reviews in Analytical Chemistry,* **10**, 1–126.

144 Tarasov, V. V., Yagodin, G. A., and Pichugin, A. A. (1984) Kinetics of extraction of inorganic substances. *Itogi Nauki Tekh., Ser.: Neorg. Khim,* **11** (3), 170.

145 Tarasov, V. V., and Yagodin, G. A. (1988) Interfacial phenomena in solvent extraction, *Ion Exchange and Solvent Extraction.* New York: Marcel Dekker, **10**, p. 141.

146 Schlossman, M. L., Li, M., Mitrinovic, D. M., *et al.* (2000) X-ray scattering from liquid-liquid interfaces. *High Performance Polymers.* **12** (4), 551–563.

147 Schlossman, M. L. (2002) Liquid–liquid interfaces: Studied by X-ray and neutron scattering. *Current Opinion in Colloid & Interface Science.* 7 (3), 235–243.

148 Schlossman, M. L. (2005) X-ray scattering from liquid–liquid interfaces. *Physica B: Condensed Matter.* **357** (1), 98–105.

149 Scoppola, E., Watkins, E., Li Destri, G., *et al.* (2015) Structure of a liquid/liquid interface during solvent extraction combining X-ray and neutron reflectivity measures. *Physical Chemistry Chemical Physics,* **17**, 15093–15096.

150 Bu, W., Hou, B., Mihaylov, M., *et al.* (2011) X-ray fluorescence from a model liquid/liquid solvent extraction system. *Journal of Applied Physics,* **110**, 102214.

151 Wilson, M. A., Pohorille, A., and Pratt, L. R. (1987) Molecular-dynamics of the water liquid vapor interface. *Journal of Physical Chemistry,* **91** (19), 4873–4878.

152 Taylor, R. S., Dang, L. X., and Garrett, B. C. (1996) Molecular dynamics simulations of the liquid/vapor interface of SPC/E water. *Journal of Physical Chemistry,* **100** (28), 11720–11725.

153 Nicolas, J. P., and de Souza, N. R. (2004) Molecular dynamics study of the n-hexane-water interface: Towards a better understanding of the liquid-liquid interfacial broadening. *Journal of Chemical Physics,* **120** (5), 2464–2469.

154 Chowdhary, J., and Ladanyi, B. M. (2006) Water-hydrocarbon interfaces: Effect of hydrocarbon branching on interfacial structure. *Journal of Physical Chemistry B,* **110** (31), 15442–15453.

155 Partay, L. B., Hantal, G., Jedlovszky, P., *et al.* (2008) A new method for determining the interfacial molecules and characterizing the surface roughness in computer simulations. Application to the liquid-vapor interface of water. *Journal of Computational Chemistry.* **29** (6), 945–956.

156 Jedlovszky, P. (2004) The hydrogen bonding structure of water in the vicinity of apolar interfaces: A computer simulation study. *Journal of Physics: Condensed Matter.* **16** (45), S5389–S5402.

157 Liu, P., Harder, E., and Berne, B. J., (2005) Hydrogen-bond dynamics in the air-water interface. *Journal of Physical Chemistry B.* **109** (7), 2949–2955.

158 Walker, D. S., and Richmond, G. L. (2007) Understanding the effects of hydrogen bonding at the vapor-water interface: Vibrational sum frequency spectroscopy of H_2O/HOD/D_2O mixtures studied using molecular dynamics simulations. *Journal of Physical Chemistry C.* **111** (23), 8321–8330.

159 Ghadar, Y., and Clark, A. E. (2014) Intermolecular network analysis of the liquid and vapor interfaces of pentane and water: Microsolvation does not trend with interfacial properties. *Physical Chemistry Chemical Physics.* **16** (24), 12475–12486.

160 Liu, P., Harder, E., and Berne, B. J. (2004) On the calculation of diffusion coefficients in confined fluids and interfaces with an application to the liquid-vapor interface of water. *Journal of Physical Chemistry B.* **108** (21), 6595–6602.

161 Chakraborty, D., and Chandra, A. (2012) A first principles simulation study of fluctuations of hydrogen bonds and vibrational frequencies of water at liquid-vapor interface. *Chemical Physics.* **392** (1), 96–104.

162 Ni, Y. C., Gruenbaum, S. M., and Skinner, J. L. (2013) Slow hydrogen-bond switching dynamics at the water surface revealed by theoretical two-dimensional sum-frequency spectroscopy. *Proceedings of the National Academy of Sciences.* **110** (6), 1992–1998.

163 Sega, M., Fabian, B., Horvai, G., and Jedlovszky, P. (2016) How is the surface tension of various liquids distributed along the interfacial normal. *Journal of Physical Chemistry C.* **120** (48), 27468–27476.

164 Ghadar, Y., Parmar, P., Samuels, A. C., and Clark, A. E. (2015) Solutes at the liquid: Liquid phase boundary-solubility and solvent conformational response alter interfacial microsolvation. *Journal of Chemical Physics.* **142** (10), 104707.

165 Vassilev, P., Hartnig, C., Koper, M. T. M., *et al.* (2001) Ab initio molecular dynamics simulation of liquid water and water-vapor interface. *Journal of Chemical Physics.* **115** (21), 9815–9820.

166 Braslau, A., Deutsch, M., Pershan, P. S., *et al.* (1985) Surface-roughness of water measured by X-ray reflectivity. *Physical Review Letters.* **54** (2), 114–116.

167 Braslau, A., Pershan, P. S., Swislow, G., *et al.* (1988) Capillary waves on the surface of simple liquids measured by X-ray reflectivity. *Physical Review A.* **38** (5), 2457–2470.

168 Sides, S. W., Grest, G. S., and Lacasse, M.-D. (1999) Capillary waves at liquid-vapor interfaces: A molecular dynamics simulation. *Physical Review E.* **60** (6), 6708–6713.

169 Ismail, A. E., Grest, G. S., and Stevens, M. J. (2006) Capillary waves at the liquid-vapor interface and the surface tension of water. *Journal of Chemical Physics.* **125** (1), 014702.

170 Hantal, G., Darvas, M., Partay, L. B., *et al.* (2010) Molecular level properties of the free water surface and different organic liquid/water interfaces, as seen from ITIM analysis of computer simulation results. *Journal of Physics: Condensed Matter.* **22** (28), 284112.

171 Fábián, B., Senćanski, M. V., Cvijetić, I. N., *et al.* (2016) Dynamics of the water molecules at the intrinsic liquid surface as seen from molecular dynamics simulation and identification of truly interfacial molecules analysis. *The Journal of Physical Chemistry C.* **120** (16), 8578–8588.
172 Mitrinovic, D. M., Zhang, Z., Williams, S. M., Huang, Z., *et al.* (1999) X-ray reflectivity study of the water – hexane interface. *The Journal of Physical Chemistry B.* **103** (11), 1779–1782.
173 Tikhonov, A. M., Mitrinovic, D. M., Li, M., *et al.* (2000) An X-ray reflectivity study of the water-docosane interface. *The Journal of Physical Chemistry B.* **104** (27), 6336–6339.
174 Schlossman, M. L. and Tikhonov, A. M. (2008) Molecular ordering and phase behavior of surfactants at water-oil interfaces as probed by X-ray surface scattering. *Annual Review of Physical Chemistry.* **59**, 153–176.
175 Tikhonov, A. M. and Schlossman, M. L. (2007) Vaporization and layering of alkanols at the oil/water interface. *Journal of Physics: Condensed Matter.* **19** (37), 375101.
176 Bu, W., Mihaylov, M., Amoanu, D., *et al.* (2014) X-ray studies of interfacial strontium–extractant complexes in a model solvent extraction system. *The Journal of Physical Chemistry B.* **118** (43), 12486–12500.
177 Bu, W., Yu, H., Luo, G., *et al.* (2014) Observation of a rare earth ion–extractant complex arrested at the oil–water interface during solvent extraction. *The Journal of Physical Chemistry B.* **118** (36), 10662–10674.
178 Benay, G., Schurhammer, R., and Wipff, G. (2010) BTP-based ligands and their complexes with Eu3+ at "oil"/water interfaces. A molecular dynamics study. *Physical Chemistry Chemical Physics.* **12** (36), 11089–11102.
179 Kadota, K., Shirakawa, Y., Matsumoto, I., *et al.* (2007) Formation and morphology of asymmetric NaCl particles precipitated at the liquid–liquid interface. *Advanced Powder Technology.* **18** (6), 775–785.
180 Bera, M. K., Luo, G., Schlossman, M. L., *et al.* (2015) Erbium (III) coordination at the surface of an aqueous electrolyte. *The Journal of Physical Chemistry B.* **119** (28), 8734–8745.
181 Luo, G., Bu, W., Mihaylov, M., *et al.* (2013). X-ray reflectivity reveals a nonmonotonic ion-density profile perpendicular to the surface of ercl3 aqueous solutions. *The Journal of Physical Chemistry C.* **117** (37), 19082–19090.
182 Benay, G. and Wipff, G. (2013) Oil-soluble and water-soluble BTPhens and their europium complexes in octanol/water solutions: Interface crossing studied by MD and PMF simulations. *The Journal of Physical Chemistry B.* **117** (4), 1110–1122.
183 Muller, J. M., Berthon, C., Couston, L., *et al.* (2016) Extraction of lanthanides (III) by a mixture of a malonamide and a dialkyl phosphoric acid. *Solvent Extraction and Ion Exchange.* **34** (2), 141–160.
184 Li, Y., He, X., Cao, X., *et al.* (2007) Molecular behavior and synergistic effects between sodium dodecylbenzene sulfonate and Triton X-100 at oil/water interface. *Journal of Colloid and Interface Science.* **307** (1), 215–220.
185 Chevrot, G., Schurhammer, R., and Wipff, G. (2007) Synergistic effect of dicarbollide anions in liquid–liquid extraction: A molecular dynamics study at the octanol–water interface. *Physical Chemistry Chemical Physics.* **9** (16), 1991–2003.

186 Chevrot, G., Schurhammer, R., and Wipff, G. (2007) Molecular dynamics study of dicarbollide anions in nitrobenzene solution and at its aqueous interface. Synergistic effect in the Eu (III) assisted extraction. *Physical Chemistry Chemical Physics.* **9** (44), 5928–5938.

187 Dyrssen, D. and Kuca, L. (1960) The extraction of uranium (VI) with DBP in the presence of TBP. The synergic effect: Substitution or addition. *Acta Chemica Scandinavica.* **14** (9), 1945–1956.

188 Irving, H. M. N. H. and Edgington, D. N. (1960) Synergic effects in the solvent extraction of the actinides—I uranium (VI). *Journal of Inorganic and Nuclear Chemistry.* **15** (1–2), 158–170.

189 Irving, H. M. N. H. (1968) *Synergism in the Solvent Extraction of Metal Chelates.* Univ. of Leeds, Eng.

190 Shmidt, O. V., Zilberman, B. Y., Fedorov, Y. S., *et al.* (2003) Effect of TBP on extraction of REE and TPE from nitric acid solutions with dibutylphosphoric acid and its zirconium salt. *Radiochemistry.* **45** (6), 596–601.

191 Zangen, M. (1963) Some aspects of synergism in solvent extraction—II: Some di-, tri-and tetravalent metal ions. *Journal of Inorganic and Nuclear Chemistry.* **25** (8), 1051–1063.

192 Kassierer, E. F. and Kertes, A. S. (1972) Displacement of water in synergic extraction systems. *Journal of Inorganic and Nuclear Chemistry.* **34** (2), 778–780.

193 Marcus, Y. and Kertes, A. S. (1969) Ion exchange and solvent extraction of metal complexes. *Science.* **166** (3911), 1391–1392.

194 Braatz, A. D., Antonio, M. R., and Nilsson, M. (2017) Structural study of complexes formed by acidic and neutral organophosphorus reagents. *Dalton Transactions.*

195 Luo, M., Mazyar, O. A., Zhu, Q., *et al.* (2006) Molecular dynamics simulation of nanoparticle self-assembly at a liquid – liquid interface. *Langmuir.* **22** (14), 6385–6390.

196 Grimes, T. S., Jensen, M. P., Debeer-Schmidt, L., *et al.* (2012) Small-angle neutron scattering study of organic-phase aggregation in the TALSPEAK process. *The Journal of Physical Chemistry B.* **116** (46), 13722–13730.

197 Chiarizia, R., Jensen, M. P., Rickert, P. G., *et al.* (2004) Extraction of zirconium nitrate by TBP in n-octane: Influence of cation type on third phase formation according to the "sticky spheres" model. *Langmuir.* **20** (25), 10798–10808.

198 Chiarizia, R., Rickert, P. G., Stepinski, D., *et al.* (2006) SANS study of third phase formation in the HCl-TBP-n-octane system. *Solvent Extraction and Ion Exchange.* **24** (2), 125–148.

199 Chiarizia, R., Stepinski, D. C., and Thiyagarajan, P. (2006) SANS study of third phase formation in the extraction of HCl by TBP isomers in n-octane. *Separation Science and Technology.* **41** (10), 2075–2095.

200 Hirtz, A., Bonkhoff, K., and Findenegg, G. H. (1993) Optical studies of liquid interfaces in amphiphilic systems: How wetting and absorption are modified by surfactant aggregation. *Advances in Colloid and Interface Science.* **44**, 241–281.

201 Burns, J. D. and Moyer, B. A. (2016) Group hexavalent actinide separations: A new approach to used nuclear fuel recycling. *Inorganic Chemistry.* **55** (17), 8913–8919.

202 Brown, M. A., Wardle, K. E., Lumetta, G., and Gelis, A. V. (2016). Accomplishing equilibrium in ALSEP: Demonstrations of modified process chemistry on 3-D printed enhanced annular centrifugal contactors. *Procedia Chemistry.* **21**, 167–173.

203 Hawkins, C. A., Bustillos, C. G., May, I., Copping, R., and Nilsson, M. (2016) Water-soluble Schiff base-actinyl complexes and their effect on the solvent extraction of f-elements. *Dalton Transactions.* **45**, 15415–15426.

204 Daly, S. R., Klaehn, J. R., Boland, K. S., Kozimor, S. A., MacInnes, M. M., Peterman, D. R., and Scott, B. L. (2012) NMR spectroscopy and structural characterization of dithiophosphinate ligands relevant to minor actinide extraction processes. *Dalton Transactions.* **41**, 2163–2175.

205 Peterman, D. R., Greenhalgh, M. R., Tillotson, R. D., Klaehn, J. R., Harrup, M. K., Luther, T. A., and Law, J. D. (2010) Selective extraction of minor actinides from acidic media using symmetric and asymmetric dithiophosphinic acids. *Separation Science and Technology.* **45**, 1711–1716.

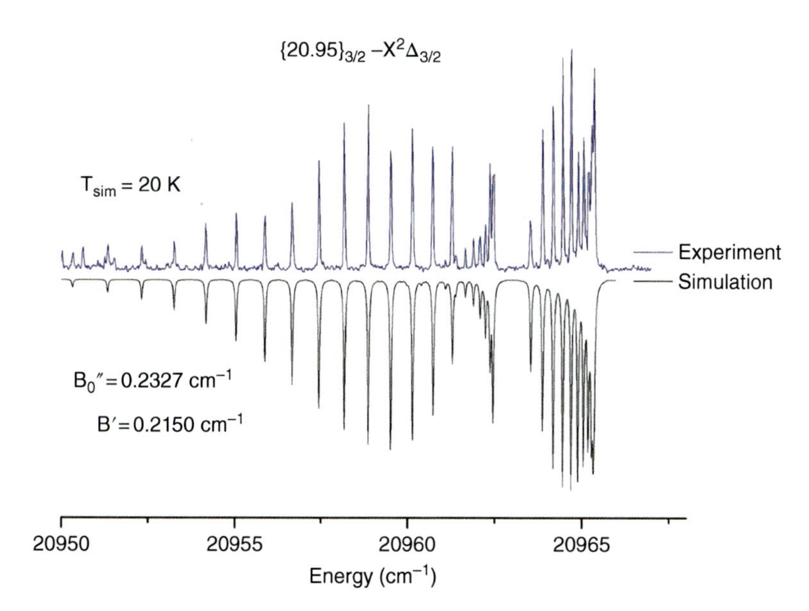

Figure 1.4 Rotationally resolved laser-induced fluorescence (LIF) spectrum of the ThF {20.95}3/2-$X^2\Delta_{3/2}$ band. The downward-going trace is a computer simulation.

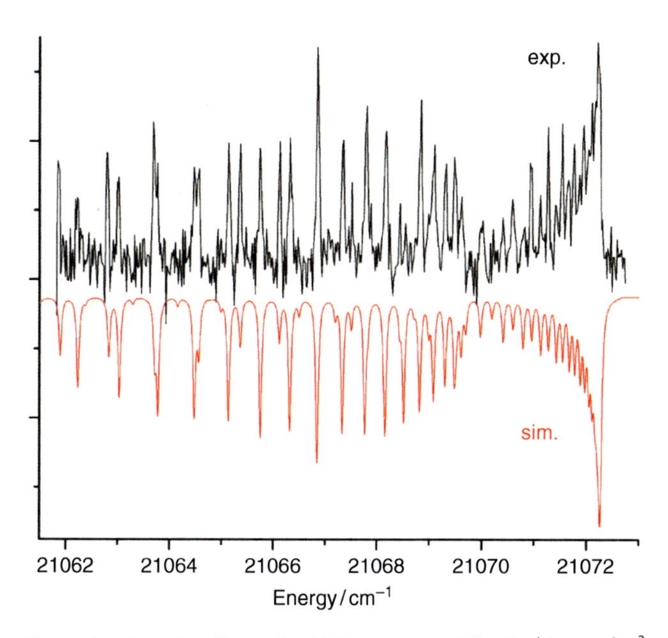

Figure 1.6 Rotationally resolved LIF spectrum of the ThF^+ [21.1]0^+-$X^3\Delta_1$ band. The downward-going trace is a computational simulation of the band.

Experimental and Theoretical Approaches to Actinide Chemistry, First Edition.
Edited by John K. Gibson and Wibe A. de Jong.

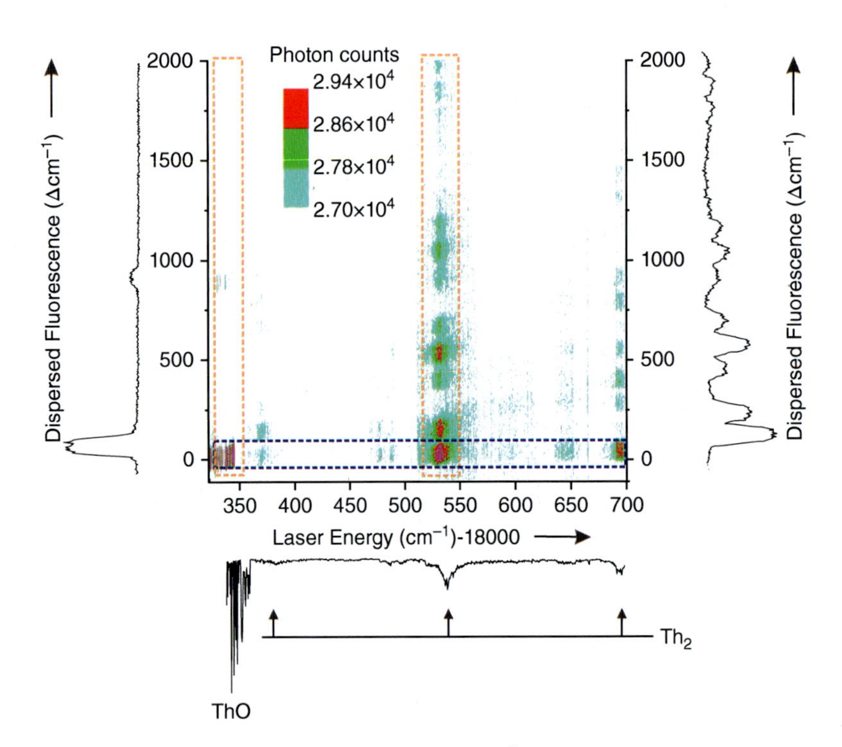

Figure 1.7 2D spectrum over the 18325 – 18700 cm^{-1} laser excitation range of ablated thorium in the presence of a SF6/Ar mixture. Not corrected for laser power variation or the spectral response of the CCD. At the bottom is the on-resonance detected laser excitation spectrum obtained from the vertical integration of the intensities of the horizontal slice marked by the dashed blue rectangle. At the left and right are the dispersed fluorescence spectra resulting from excitation of the (0,0) $F^1\Sigma(+) - X^1\Sigma^+$ band of ThO at 18340 cm^{-1} and Th_2 band at 18530 cm^{-1}, respectively, obtained by horizontal integration of the intensities of the left and right horizontal slices marked in red. (Reproduced with permission from Steimle, T., Kokkin, D.L., Muscarella, S., and Ma, T. (2015) Detection of thorium dimer via two-dimensional fluorescence spectroscopy. *Journal of Physical Chemistry A*, **119**, 9281–9285.)

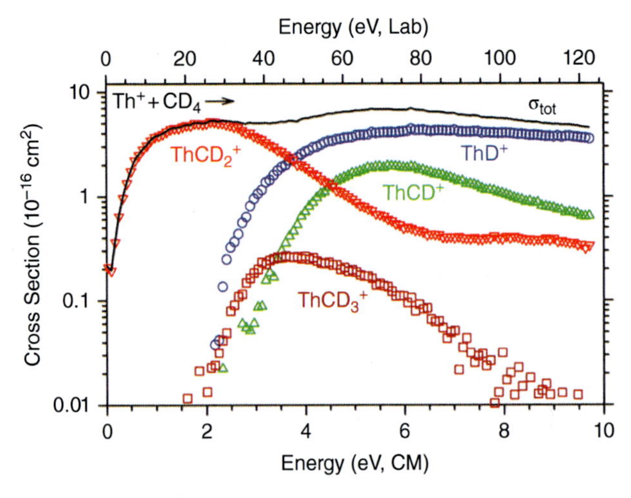

Figure 1.12 Cross sections for the reaction between Th^+ and CD_4 as a function of energy in the center of mass (lower *x* axis) and lab (upper *x* axis) frames of reference. (Reproduced with permission from Cox, R.M., Armentrout, P.B., and de Jong, W.A. (2015) Activation of CH_4 by Th^+ as studied by guided ion-beam mass spectrometry and quantum chemistry. *Inorganic Chemistry*, **54**, 3584–3599.)

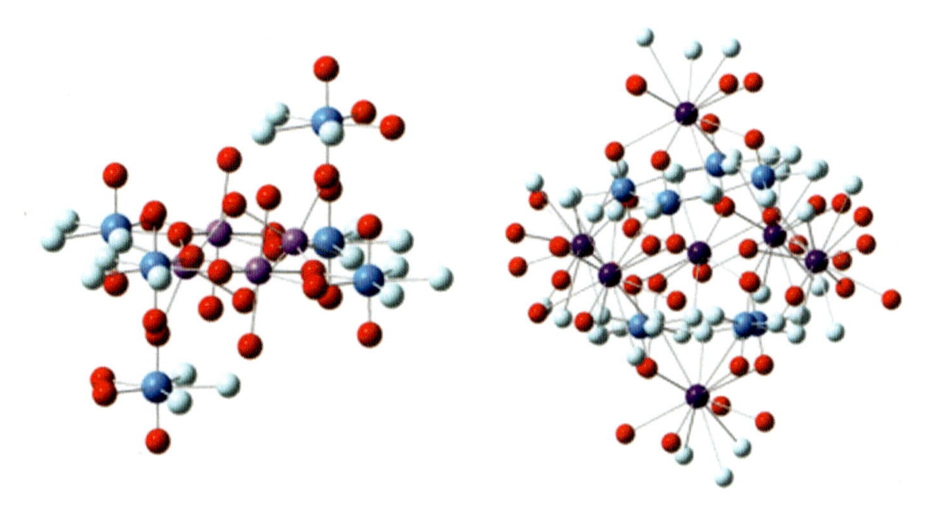

Figure 3.1 On the (left) $Na_3(UO_2)_2F_3(OH)_4(H_2O)_2$. On the (right) $Cs(UO_2)_2F_5$.

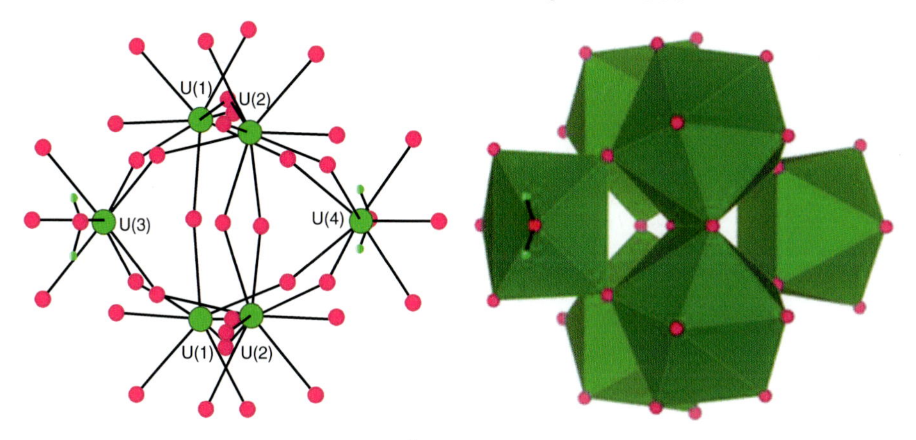

Figure 3.3 Representation of $[U_6F_{33}(H_2O)_2]^{9-}$ found in $U_3F_{12}(H_2O)$.

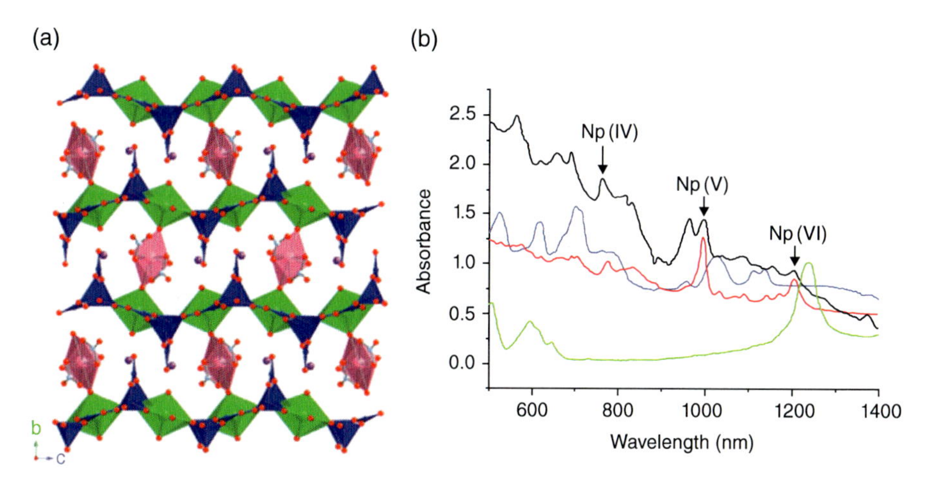

Figure 3.9 (a) The three-dimensional topography of $K_2[(NpO_2)_3B_{10}O_{16}(OH)_2(NO_3)_2]$. Np(V) polyhedra are longer on the *b*-axis (vertical), whereas the Np(VI) polyhedra are longer on the *c*-axis (pink). The BO_3 triangles and the BO_4 tetrahedra bridge between the Np(VI) polyhedra. (b) The UV-vis-NIR spectrum of several compounds. The labeled line is for $K_2[(NpO_2)_3B_{10}O_{16}(OH)_2(NO_3)_2]$ and shows the redox transitions for Np.

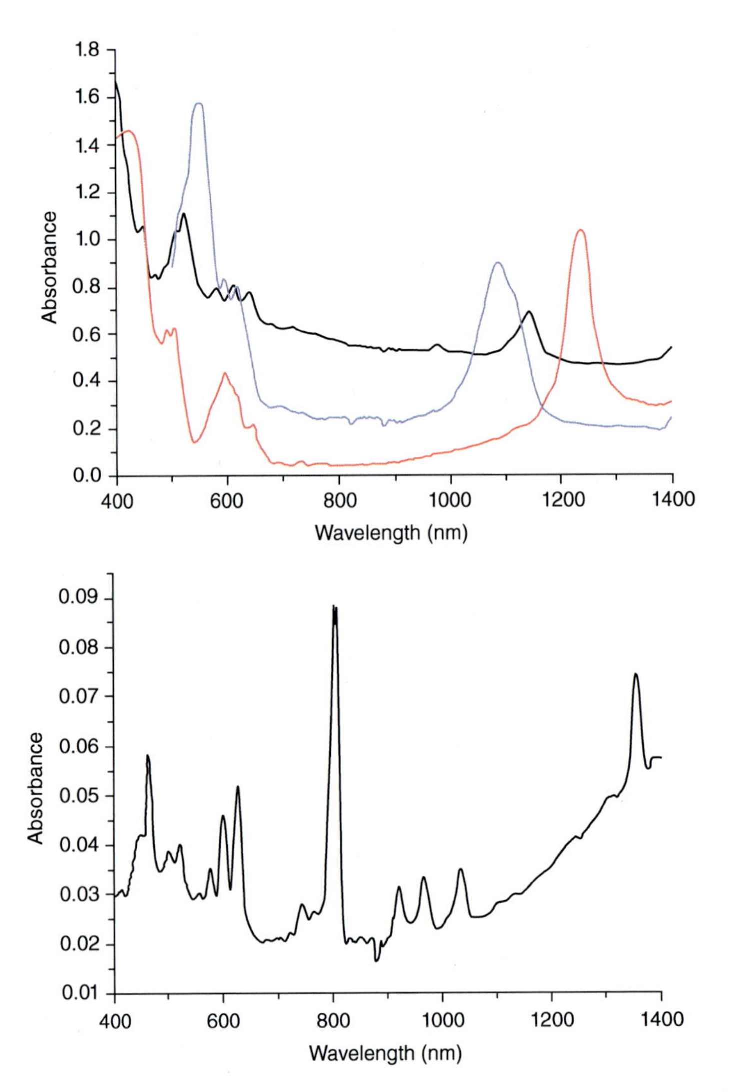

Figure 3.12 (Top) UV-vis-NIR absorption spectra for several Np(VI) compounds. $NpO_2[B_6O_{11}(OH)_4$ is shifted to 1140 nm. For context, $NpO_2(NO_3)_2{\cdot}6H_2O$ is at 1100 nm, while $NpO_2(IO_3)_2(H_2O)$ is at 1220 nm. (Bottom) UV-vis-NIR absorption spectra for $PuO_2[B_6O_{11}(OH)_4$ showing strong absorption at 800 nm along the excitation axis.

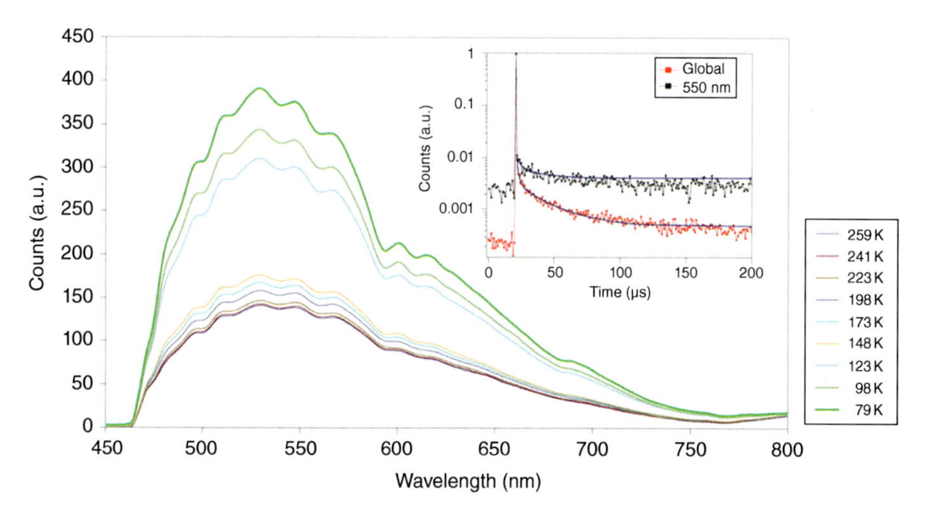

Figure 3.20 Photoluminescence spectra of $Cf[B_5O_8(OH)_5]$ with 420 nm light. Cf(III) emits at 525 nm while its daughter Cm(III) emits at 600 nm. The features of vibronic coupling become more apparent at lower temperatures. The inset shows decay lifetimes of 1.2 ± 0.3 μs for Cf(III) and 20 ± 2 μs for Cm(III).

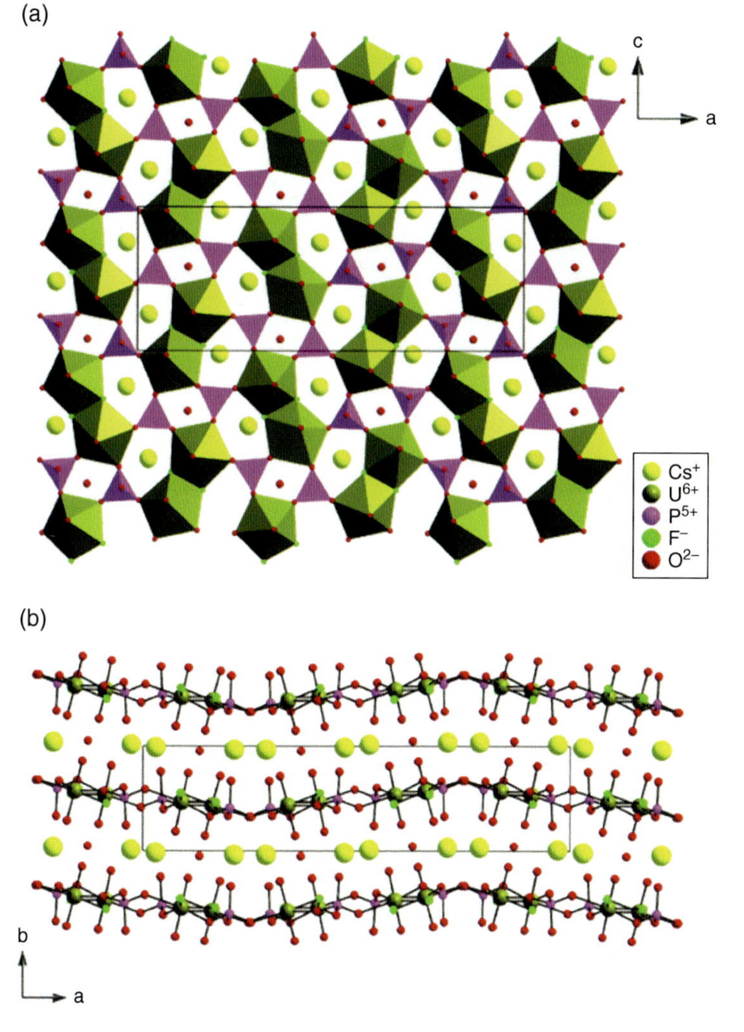

Figure 3.44 The layered architecture of LUPF-1 from the perspective of (a) a polyhedral representation in the *ac*-plane and (b) ball-and-stick representation in the *ab*-plane. Large polyhedra are uranyl centers, and tetrahedra are phosphates.

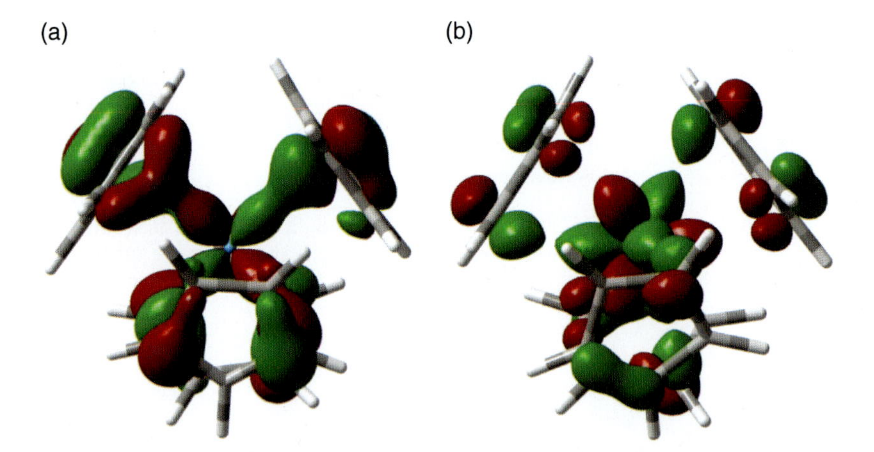

Figure 4.3 Three-dimensional representations of one component of the pseudo-t_1 valence molecular orbitals of (a) UCp_4, (b) $AmCp_4$. The 5f content (Mulliken analysis) of the orbitals are, respectively, 15.4% and 30.9%, and the bond critical point electron densities are 0.034 and 0.029 electron/$bohr^3$. Images and data from Tassell M.J., Kaltsoyannis N. (2010) Covalency in $AnCp_4$ (An = Th–Cm); a comparison of molecular orbital, natural population and atoms in molecules analyses. *Dalton Transactions*, **39** (29), 6719–6725. (http://pubs.rsc.org/en/Content/ArticleLanding/2010/DT/c000704h). Reproduced by permission of The Royal Society of Chemistry.

Figure 4.14 A mono-phosphorano stabilized carbene complex.

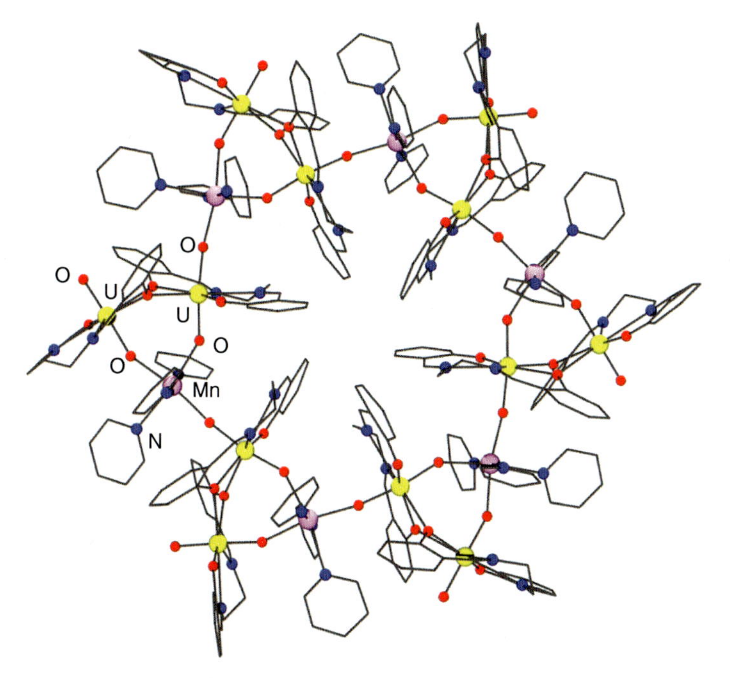

Figure 4.33 Molecular structure of $[\{[UO_2(salen)]_2Mn(py)_3\}_6]$. Atomic coordinates taken from Mougel V., Chatelain L., Pécaut J., Caciuffo R., Colineau E., Griveau J.-C., Mazzanti M. (2012) Uranium and manganese assembled in a wheel-shaped nanoscale single-molecule magnet with high spin-reversal barrier. *Nature Chemistry*, **4** (12), 1011–1017.

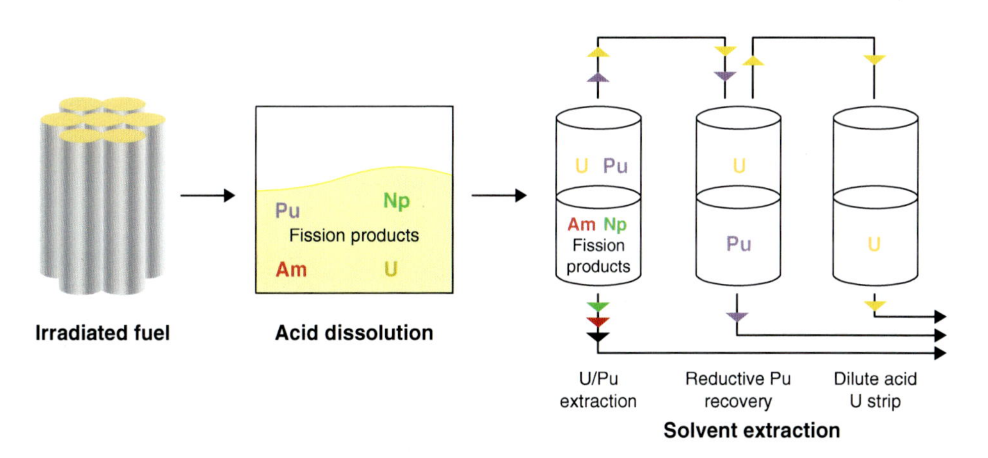

Figure 5.1 A simplified illustration of the PUREX process.

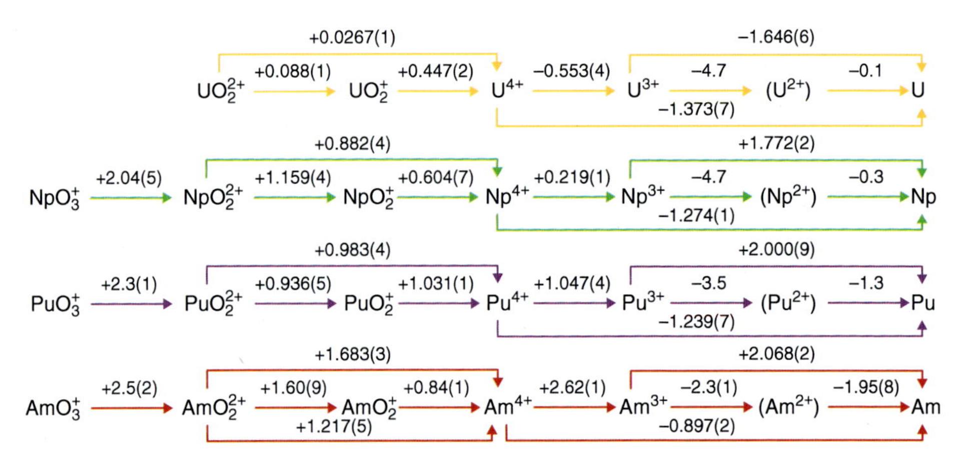

Figure 5.2 Adapted from CATE, p. 1779. Standard reduction potentials diagrams for the actinide ions (in V) versus the standard hydrogen electrode.[17]

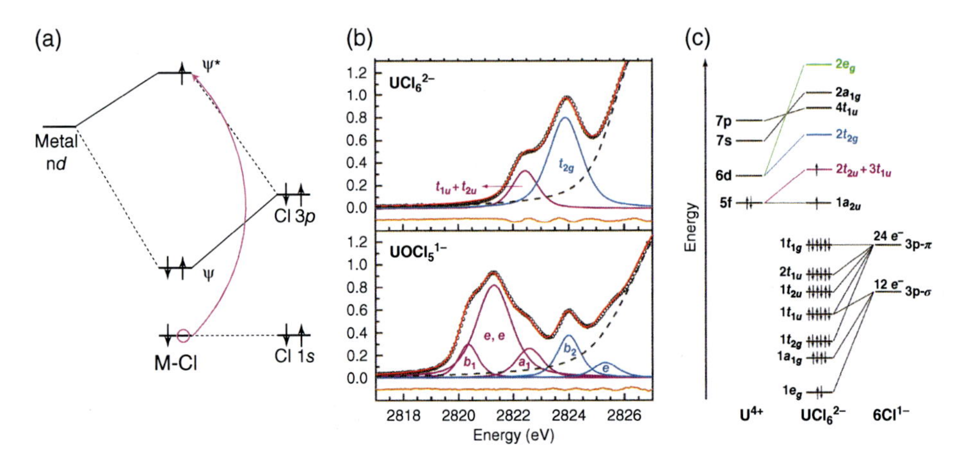

Figure 5.3 (a) Illustration of electronic excitations in a typical ligand K-edge XAS experiment. (b) Cl K-edge fluorescence yield experimental data. Adapted from Ref. 132; (c) qualitative molecular orbital diagram for UCl_6^{2-}.

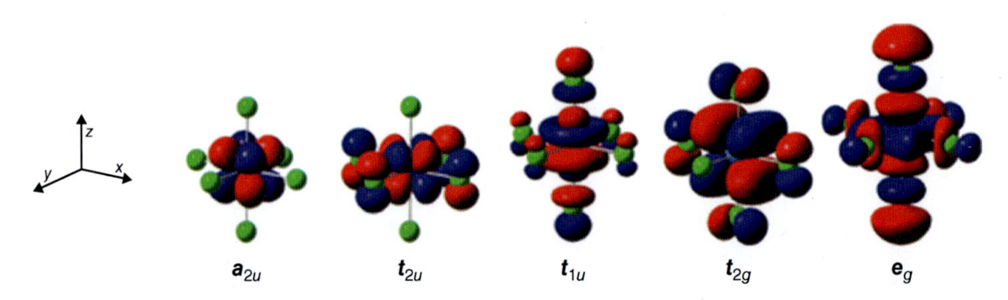

Figure 5.4 Representative virtual Kohn–Sham orbitals for UCl_6^{2-}. Adapted from Minasian, S. G., Keith, J. M., Batista, E. R., *et al.* (2012). Determining relative f and d orbital contributions to M–Cl covalency in MCl_6^{2-}(M=Ti, Zr, Hf, U) and $UOCl_5^-$ Using Cl K-Edge X-ray absorption spectroscopy and time-dependent density functional theory. *Journal of the American Chemical Society*. **134** (12), 5586–5596.

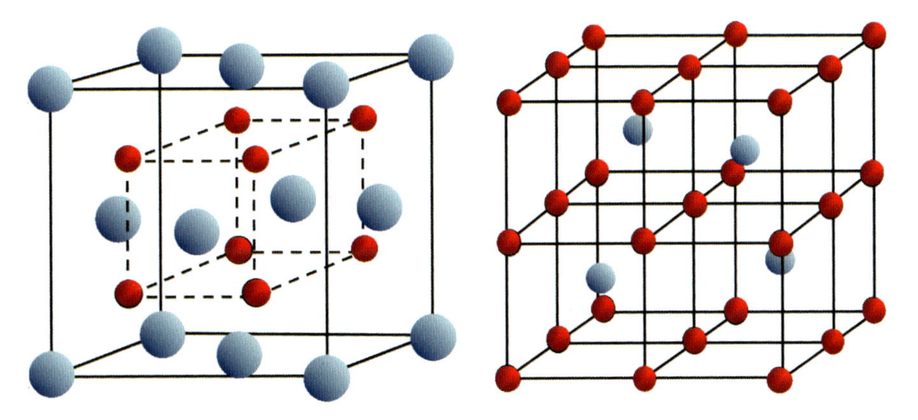

Figure 6.1 The crystal structure of UO_2. The unit cell (left) and the oxygen lattice (right) with the uranium atoms in blue and the oxygen atoms in red.

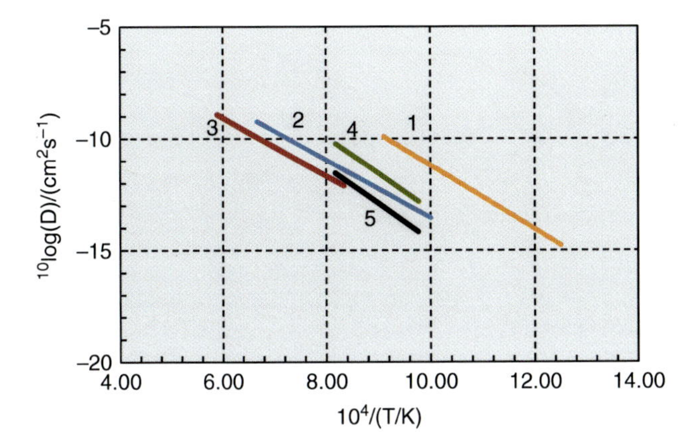

Figure 6.4 The oxygen self-diffusion in UO_2. 1, Auskern and Belle (68); 2, Marin and Contamin (70); 3, Hadari et al. (71); 4, 5, Dorado et al. (78). All curves refer to polycrystalline UO_2, except 5, which refers to single crystal.

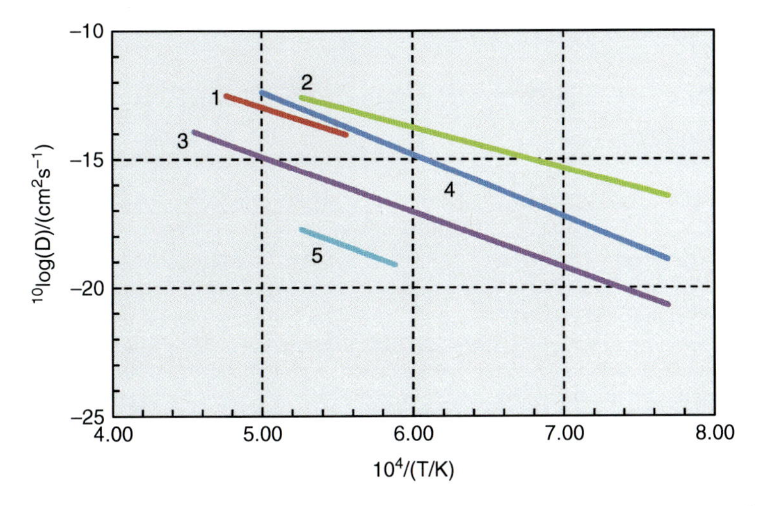

Figure 6.5 Uranium self-diffusion in UO_2. 1, Auskern and Belle (79); 2, Yajima et al. (80); 3, Reimann and Lundy et al. (82); 4, Matzke (83); 5, Sabioni et al. (84).

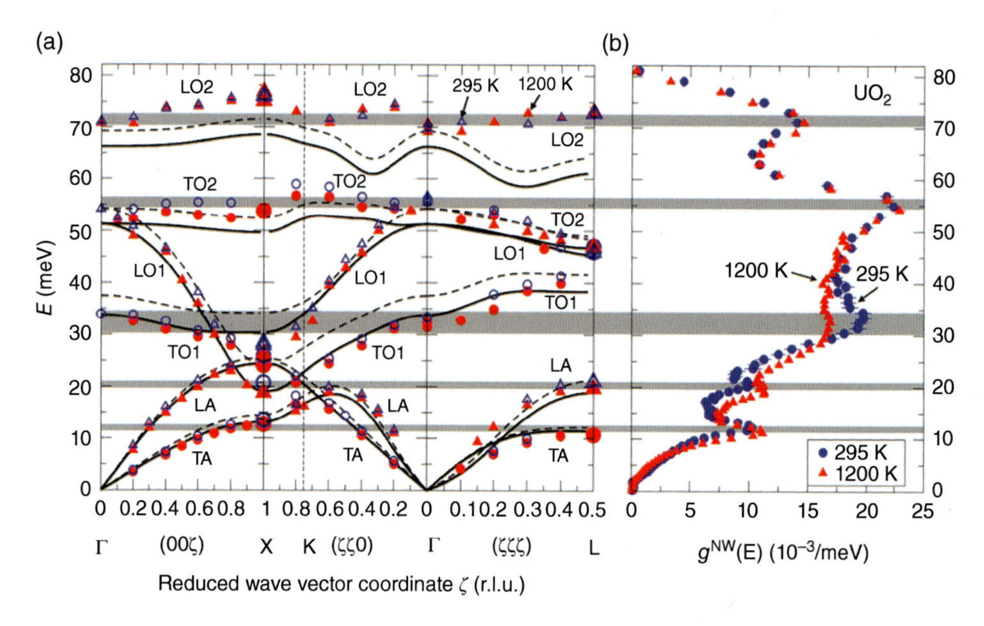

Figure 6.6 The phonon dispersion curves (a) and phonon density of states (b) of UO_2, after Pang JWL, Chernatynskiy A, Larson BC, Buyers WJL, Abernathy DL, McClellan KJ, et al. Phonon density of states and anharmonicity of UO_2. *Phys Rev B*. 2014;89(11):115316. The open symbols refer to T 0 295 K, the solid symbols to T = 1200 K, triangles represent the transverse phonons. The dashed and solid lines show the results of the GGA + U calculations for T = 295 K and T = 1200 K, respectively. Reproduced with permission.

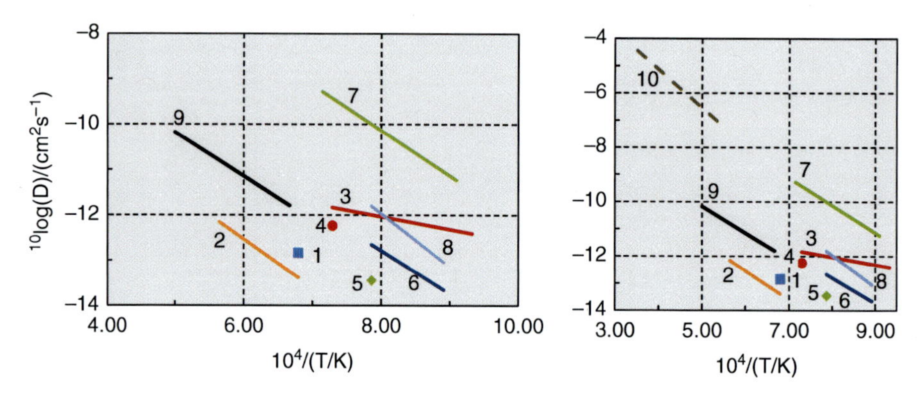

Figure 6.19 The helium diffusion coefficient in UO_2. Curves 1 to 9 show the experimental results. 1, Rufeh (188); 2, Sung (189); 3, Martin et al. (146); 4, Trocellier (209); 5, Guilbert et al. (210), 6, Roudil et al. (207); 7, Ronchi and Hiernaut (208), 8, Pipon et al. (211); 9, Nakajima et al. (190). Note that curve 7 refers to a $(U_{0.9}Pu_{0.1})O_2$ sample. Curve 10 shows the results from the MD simulations by Yakub (212).

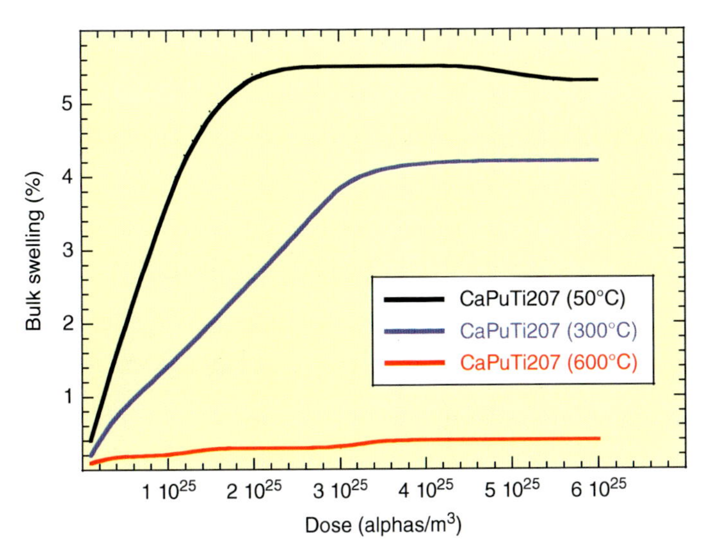

Figure 7.1 Effect of thermal annealing on radiation damage in $CaPuTi_2O_7$ produced by the decay of ^{238}Pu. Samples held at 50 °C and 300 °C become amorphous, but the saturation dose for swelling is increased, and the volume expansion is lower at the higher temperature. When held at 300 °C, the material does not become amorphous, and the bulk swelling saturates at about 0.4 vol% primarily due to the effect of retained Frenkel defects and alpha-recoil collision cascades in the crystalline lattice. Modified from Clinard, F.W., Jr., Peterson, D.E., Rohr, D.L., and Hobbs, L.W. (1984) Self-irradiation effects in 238Pu-substituted zirconolite, I: Temperature dependence of damage. *Journal of Nuclear Materials*, **126**, 245–254.

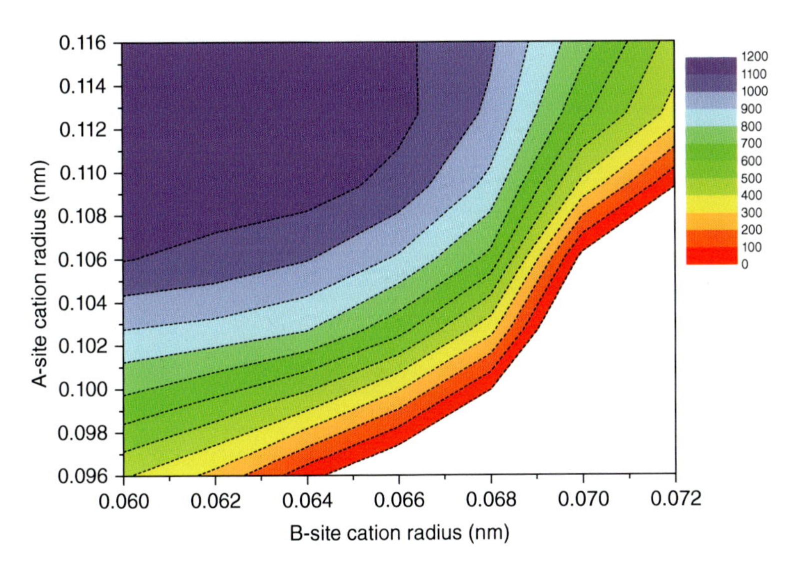

Figure 7.2 Contour map of predicted critical amorphization temperature for pyrochlore and defect fluorite compounds as a function of the radii of the A-site and B-site cations. This map was calculated from the defect energies, lattice parameters, and electronegativity data of the various compounds in this system. It can be used as a predictive tool for testing suitable compositions that are likely to remain crystalline under irradiation. The temperature scale is in degrees Celsius from most tolerant (red, 0–100) to the least tolerant (darker blue, 1100 – 1200) compositions based on $A_2B_2O_7$ stoichiometry. Unpublished data of the author, see references [Lumpkin, G.R., Pruneda, M., Rios, S., et al. (2007) Nature of the chemical bond and prediction of radiation tolerance in pyrochlore and defect fluorite compounds. *Journal of Solid State Chemistry*, **180**, 1512–1518; Lumpkin, G.R., Smith, K.L., Blackford, M.G., et al. (2009) Ion irradiation of ternary pyrochlore oxides. *Chemistry of Materials*, **21**, 2746–2754] for a general discussion and description of methods involved.

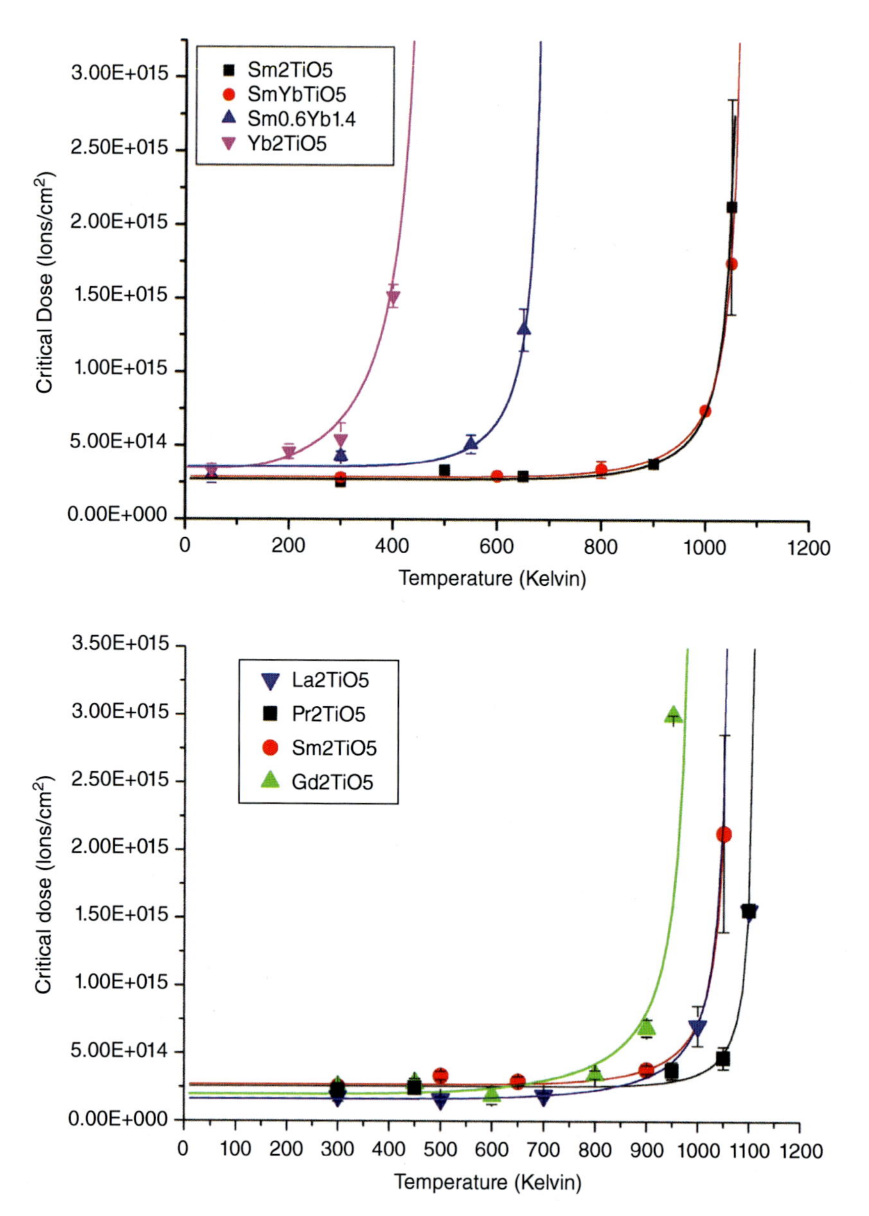

Figure 7.3 In situ ion irradiation results for the Ln_2TiO_5 type compounds. The data were obtained at the IVEM-Tandem Facility at Argonne National Laboratory using 1.0 MeV Kr ions passing through thin TEM samples. Top: Hexagonal SmYb, pyrochlore-like Sm0.6Yb1.4, and fluorite-like Yb irradiation data relative to orthorhombic Sm_2TiO_5. Bottom: Orthorhombic structures with Ln = Gd, Sm, Pr, and La. These figures are adapted from pre-publication files related to references [Aughterson, R.D., Lumpkin, G.R., Ionescu, M., et al. (2015) Ion-irradiation resistance of the orthorhombic Ln_2TiO_5 (Ln = La, Pr, Nd, Sm, Eu, Gd, Tb and Dy) series. *Journal of Nuclear Materials*, **467**, 683–691; Aughterson, R.D., Lumpkin, G.R., De los Reyes, M., et al. (2016) The influence of crystal structure on ion-irradiation tolerance in the $Sm_{(x)}Yb_{(2-x)}TiO_5$ series. *Journal of Nuclear Materials*, **471**, 17–24].

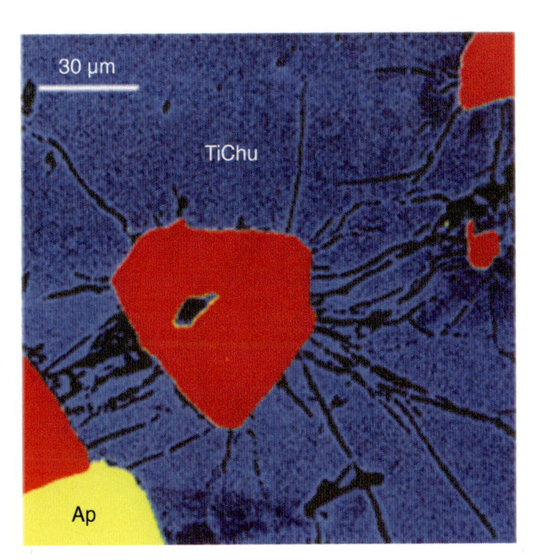

Figure 7.4 Colorized SEM map showing U-pyrochlore (red), apatite (yellow), and titanian clinohumite (blue) in hydrothermal veins from the Adamello massif, northern Italy. The minerals are approximately 40 million years old. Due to the content of ~30 wt% UO_2, the pyrochlore is amorphous due to alpha decay damage. Note the radial cracks in surrounding clinohumite due to radiation-induced volume expansion. Courtesy of Reto Gieré.

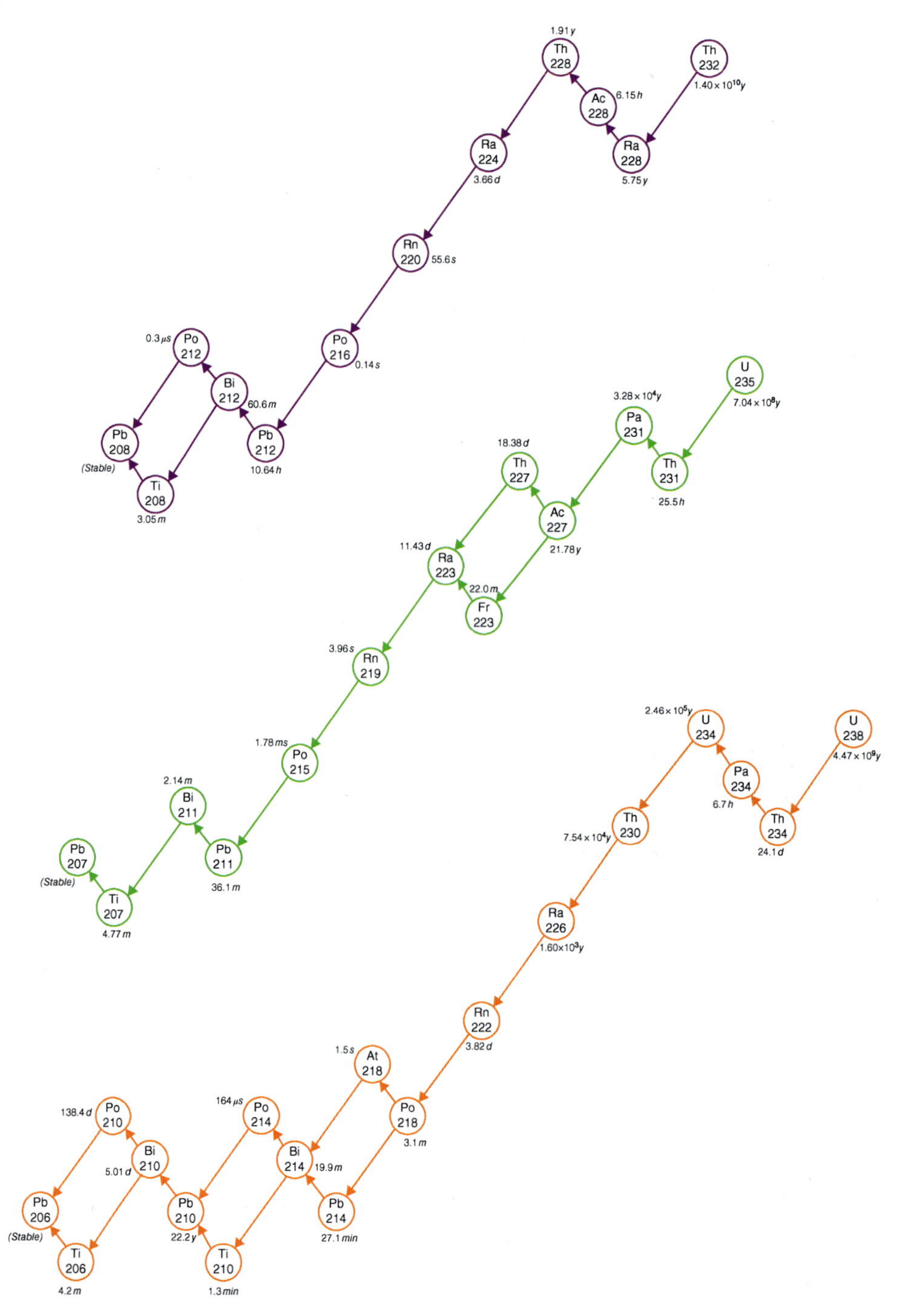

Figure 8.1 Decay series for ^{232}Th, ^{235}U, and ^{238}U, often respectively referred to as the thorium series, the actinium series, and the radium or uranium series, or as the 4n, 4n + 3, and 4n + 2 series (where n is an integer, 4 is the mass change associated with alpha decay and 0, and 3 and 2 are the remaining number of protons and neutrons left over following removal of all possible alpha particles).

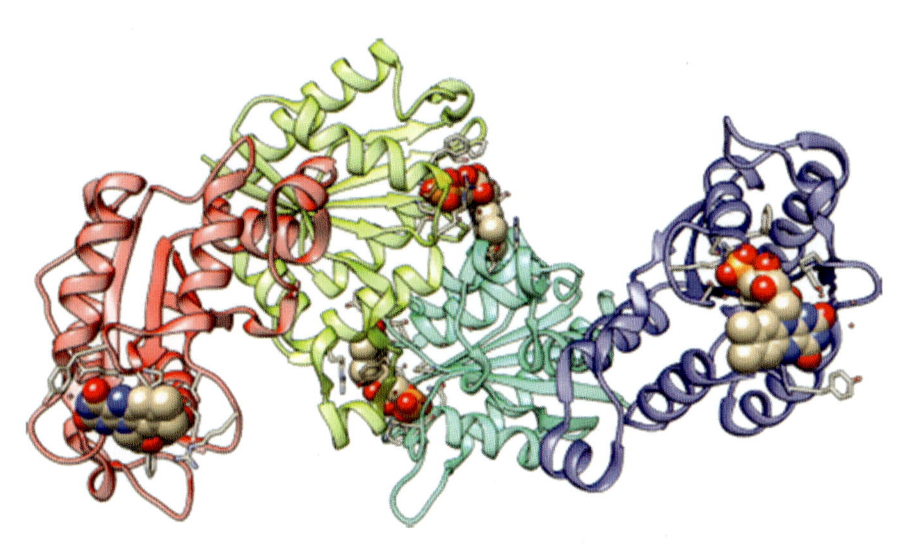

Figure 9.3 Crystal structure of the *G. hansenii* chromate reductase homotetramer (PDB: 3S2Y). Individual monomer chains are differentiated by color, with bound flavin mononucleotide illustrated as spheres.

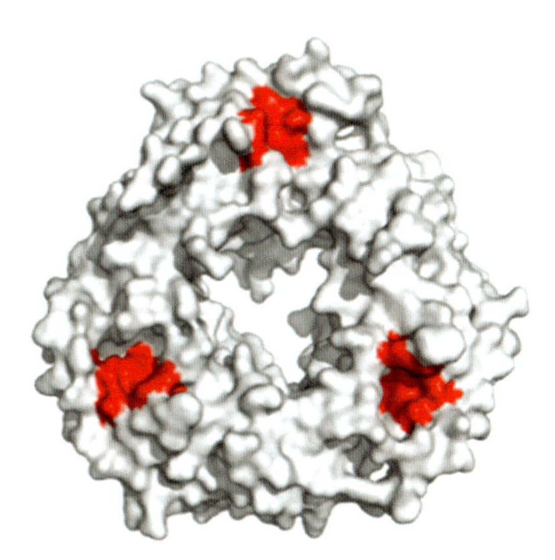

Figure 9.5 Putative Pu^{4+} binding sites on the human p32 trimer (PDB: 1P32). The sites (in red) are formed on the negatively charged side of the trimer from residues E89, E93, L231, D231, and Y268 in each chain.

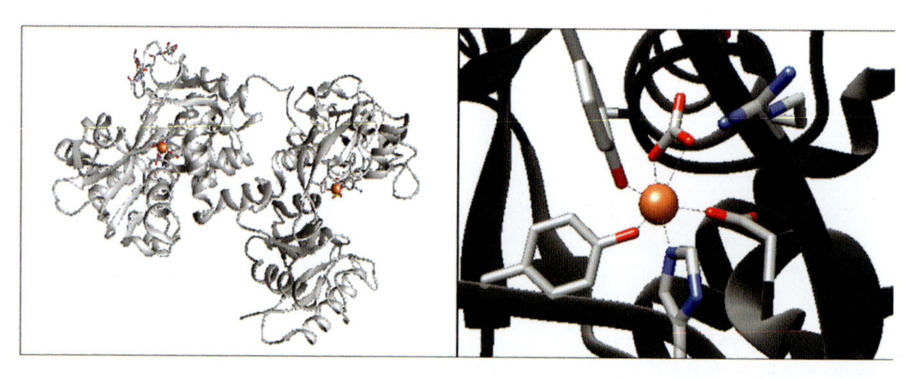

Figure 9.7 Human serum transferrin is a ~80 kDa protein folded into a single polypeptide chain with two homologous lobes (left). Each lobe can tightly coordinate a single ferric ion in a binding pocket, through two tyrosine residues, one histidine, and one aspartate, in the presence of a bidentate synergistic carbonate anion (right).

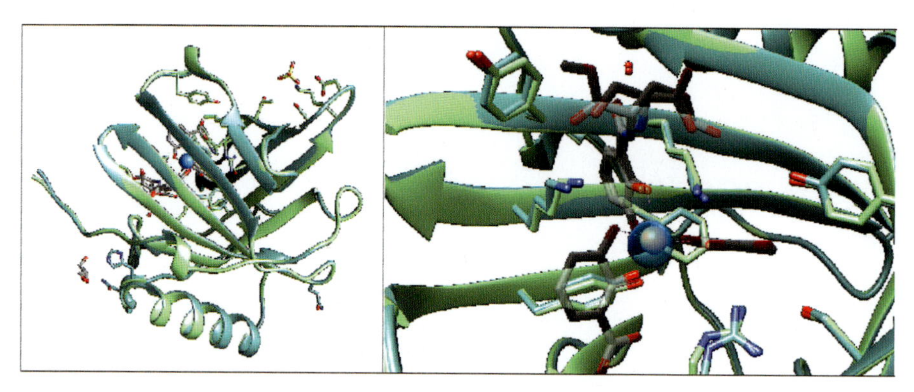

Figure 9.8 Structural superposition of the Scn adducts formed with Pu(IV)-enterobactin and Fe(III)-enterobactin, with a zoom into the Scn calyx (right panel) showing the high rigidity of the protein calyx and the exact same position for the metal ions.

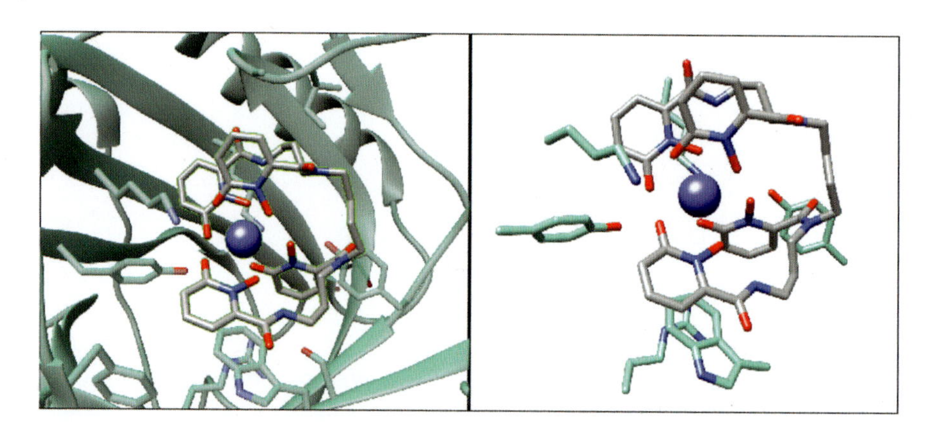

Figure 9.13 Structure of $[Am(3,4,3\text{-}LI(1,2\text{-}HOPO))]^-$ bound in the Scn calyx (PDB 4ZHG). Left: In a QM/MM calculation, the metal and ligand (highlighted with a gray backbone) would be treated using a QM approach, while the protein (aqua) would be treated using MM. Right: The protein structure is removed except for the residues that ligate the complex (same coloring scheme), and the entire model is treated at the QM level.

6

Behaviour and Properties of Nuclear Fuels

Rudy Konings[1] *and Marjorie Bertolus*[2]

[1] *European Commission, Joint Research Centre, Karlsruhe, Germany*
[2] *CEA, DEN, Centre de Cadarache, France*

6.1 Introduction

Nuclear reactor fuel is probably one of the most complex materials in modern technology. This is due to a variety of reasons. First of all, the chemistry of the fuel materials is complex. In compounds, the actinide elements such as uranium and plutonium exhibit various oxidation states depending on the thermodynamic conditions, with concomitant charge compensation taking place through lattice defects (vacancies or interstitials), often without substantially changing the crystal structure but strongly affecting the material properties. Second, during irradiation the fissile uranium (or plutonium) is fissioned, generating a range of fragments that are chemically very different in nature. These fission products are exotic in the crystal lattice of the fuel material and must be accommodated. Third, the fission process and the radioactive decay processes create highly energetic particles (fission fragments, alpha particles) that cause severe damage to the crystalline lattice, leading to lattice displacements, disorder, and defects, with severe impact on the material properties.

Nuclear fuel for the current generation light water reactors (LWRs), which we will consider in this chapter, is a ceramic material of a polycrystalline nature. The fuel is made of cylindrical pellets of uranium dioxide (UO_2) or mixed uranium-plutonium oxide $(U,Pu)O_2$, generally referred to as MOX. The pellets are produced by biaxial pressing of a starting powder, and are enclosed in a tightly closed metallic tube of a zirconium alloy, the fuel pin (1). The fuel has generally an enrichment of 3%–5% ^{235}U or 6%–8% ^{239}Pu, which allows an extent of fission (burnup) of 4–6 atom% of the initial heavy metal content to be reached, equivalent to an energy production of 40 to 55 $MWd \cdot kgU^{-1}$. This corresponds to a residence time in the reactor of up to four years, during which period it must be ensured that no radioactive material is released from the fuel pin. Although the metallic pin acts as the primary barrier between the fuel and the coolant, the fuel matrix itself is designed to retain a large fraction of the fission products. For that reason, the dissolution, diffusion, and transport of the fission product in

Experimental and Theoretical Approaches to Actinide Chemistry, First Edition.
Edited by John K. Gibson and Wibe A. de Jong.

the lattice and along the grain boundaries of the material are key processes that need to be understood.

The situation is further complicated by the fact that the temperature gradient in the fuel pellet is substantial because of the relatively low thermal conductivity of actinide oxides, especially when undergoing changes induced by irradiation: at the pellet rim it is around 650 K, slightly above the temperature of the reactor coolant, and in the centre the temperature can reach values up to 1500 K during normal steady-state operation. It is obvious that thus a variety of thermodynamic and kinetic conditions must be considered when describing the chemical state of the fuel.

During the last 50 years, extensive research has been performed on nuclear fuel materials, leading to a good understanding of its properties and irradiation behaviour at the macroscopic level and of some of the underlying physical, mechanical, and chemical processes (2). It is, however, very difficult to study irradiated nuclear fuel at the (sub-)microscopic level, which is necessary to better understand the fundamental mechanisms controlling radiation effects, matter transport, or secondary phase formation, which influence the bulk material behaviour. This is because observation and measurements are generally made 'post-irradiation' in the absence of the temperature and radiation conditions that are characteristic of the fuel during in-reactor operation. Computational materials science can help increase this understanding, and the advancements in this field have been enormous during the last decade, also for the understanding of nuclear fuel behaviour. In this chapter, we will give an overview of the properties and irradiation behaviour of nuclear fuel, addressing both the results from experimental and computational studies, and demonstrating how macroscopic and microscopic/atomistic information can be merged into a basic understanding of the underlying mechanisms. Because the number of computational studies has increased enormously during recent years, it is impossible to present a comprehensive compilation in this chapter, and only selected examples will be discussed.

6.2 UO_2

6.2.1 Crystal Structure

Uranium dioxide has a face-centred cubic (fcc) crystal structure (space group $Fm\bar{3}m$), isostructural with fluorite, CaF_2. The unit cell contains four molecules of UO_2 (Figure 6.1). It is face-centred with respect to the uranium ions, which occupy the octahedral positions (0,0,0), $\left(\frac{1}{2},\frac{1}{2},0\right)$, $\left(\frac{1}{2},0,\frac{1}{2}\right)$, and $\left(0,\frac{1}{2},\frac{1}{2}\right)$, whereas the oxygen ions occupy the $\left(\frac{1}{4},\frac{1}{4},\frac{1}{4}\right)$ and its equivalent positions (tetrahedrally coordinated by uranium). Interstitial ions may be accommodated at octahedral vacant sites, which is evident from the oxygen sublattice, showing eight cubes of oxygen per unit cell, of which only half are occupied by an U^{4+} ion.

At low temperature, a Jahn–Teller distortion of the oxygen cage of UO_2 occurs due to the strong coupling between the electronic and phononic systems, as demonstrated by elastic constant measurements (3), neutron diffraction, (4) and neutron scattering studies (5, 6). It has been found that the oxygen sublattice is distorted with a displacement of the oxygen atoms of 0.014 Å in the $\langle 111 \rangle$ direction, changing the space group $Pa\bar{3}$ (4, 7, 8). Dorado et al. (9) confirmed the effects of the Jahn–Teller distortion in a

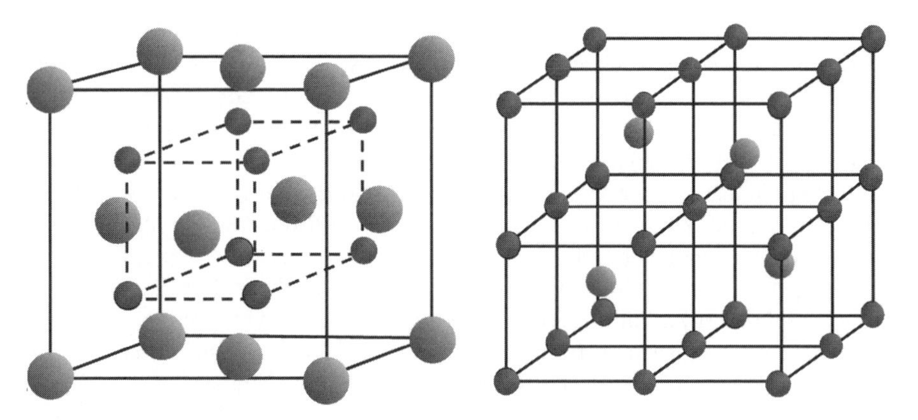

Figure 6.1 The crystal structure of UO_2. The unit cell (left) and the oxygen lattice (right) with the uranium atoms in blue and the oxygen atoms in red. (*See insert for color representation of the figure.*)

computational study of the noncollinear antiferromagnetic order in UO_2 by density functional theory (DFT + U approximation). The distortion caused by the oxygen atom displacement in the $\langle 111 \rangle$ direction stabilises the system by 50 meV/UO_2 compared to the fluorite structure.

At temperatures above about 2400 K, the oxygen sublattice of UO_2 undergoes distortion due to thermally induced disorder as a result of oxygen Frenkel pair formation (strictly speaking, anti-Frenkel defects), and a large fraction of the oxygen atoms is displaced to interstitial positions (10–14). The high-temperature neutron diffraction analysis by Clausen et al. (11) showed that the number of defective anions increases to almost 25% near the melting point. This is evident from an anomalous behaviour of the thermal expansion and the heat capacity. A recent high-temperature synchrotron X-ray study indicated substantial oxygen disorder in UO_2 in this temperature range (15), with increasing U-U distance with temperature, in agreement with lattice expansion, but decreasing U-O distance. It is claimed (14) that the thermal disorder leads to a λ-type transition in UO_2, similar to the anomaly found in CaF_2 (16), which is often referred to as the 'Bredig' or 'superionic' transition. Molecular dynamics (MD) calculations using empirical potentials have been performed to investigate this phenomenon in UO_2 (17–21) and generally reproduce the λ transition. We will discuss this in more detail in Section 6.2.5.

6.2.2 Electronic Structure

The U^{4+} ion has a partially filled 5*f* orbital, and the $5f^2$ states are localised in UO_2, which has a strong influence on its physical and chemical properties. In the cubic crystalline environment, the U^{4+} ions possess a ninefold degenerate $^3H_4$ ground state according to Hund's rules. The crystal field splits the U^{4+} multiplet into a Γ_1 singlet, a Γ_3 doublet, and two triplets, Γ_4 and Γ_5 (22, 23).

A large number of electronic structure calculations in the DFT framework have been performed for UO_2 as discussed in the reviews by Liu et al. (24), Dorado et al. (25), and Bertolus et al. (26). The early works using the standard local density and generalised gradient approximations, LDA, and GGA, respectively, were not able to reproduce the insulator properties of UO_2 and predicted it to be a metal. This is generally attributed to the fact that these levels of theory are not able to describe the electron behaviour in strongly correlated systems, such as uranium compounds. This problem has been overcome by the DFT + U method, in which an intra-atomic interaction parameter (U) is introduced to describe the strong on-site Coulomb interaction of localised electrons. The U value for UO_2 has been fitted to obtain the best results for several physical properties in earlier studies, but recent studies generally use the value 4.5 eV for U extracted from experiments (27, 28) or calculated ab initio (29), and the results reproduce the antiferromagnetic insulator properties of the ground-state electronic structure with a band gap at the Fermi level. However, several authors showed that the DFT + U method results in multiple energy minima on the potential energy curve. This was demonstrated for a variety of materials, such as metallic cerium by Shick et al. (30) and Amadon et al. (31), rare-earth nitrides by Larson et al. (32), plutonium oxide by Jomard et al. (33), and uranium dioxide by Dorado et al. (9, 34). It is now widely recognised that the application of DFT + U method requires a special approach to identify the ground-state configuration, for instance, a systematic occupation matrix control.

UO_2 is an antiferromagnetic insulator with a band gap of the order of about 2 eV, and a Néel temperature of $T_N = 30.44$ K (35). Since the compound exhibits excitations of $f \rightarrow f$ character across the band gap, it can be classified as a strongly correlated Mott-type insulator (36). Below the Néel temperature, UO_2 exhibits a 3**k** noncollinear antiferromagnetic ordering with the moments pointing along the $\langle 111 \rangle$ directions consistent with the oxygen displacement, as shown by the combination of neutron diffraction (37, 38), including under a magnetic field (7), polarised inelastic neutron scattering (5), and nuclear quadrupole resonance experiments (39). LDA + U calculations including spin-orbit coupling by Laskowski et al. (40) yield noncollinear magnetic moments for the low-temperature phase in the $\langle 111 \rangle$-type directions, as shown in Figure 6.2, corresponding to the distortion of the oxygen sublattice in the same direction, in agreement with the experimental observations (see above). This was further confirmed by GGA + U calculations by Gryaznov et al. (41).

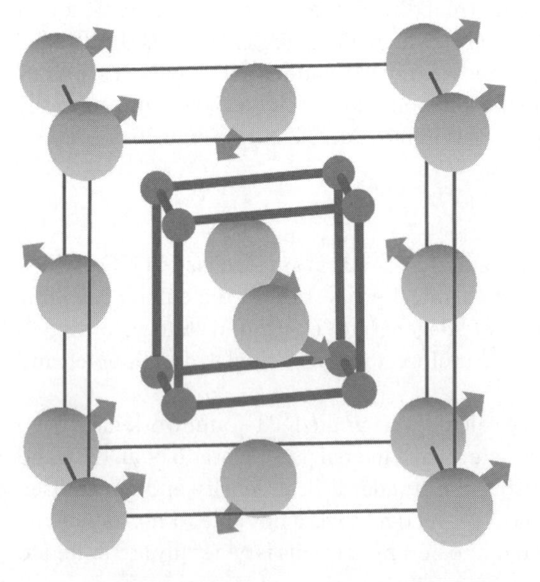

Figure 6.2 The collinear magnetic structure of UO_2 at low temperatures [after Thompson AE, Wolverton C, First-principles study of noble gas impurities and defects in UO_2. *Phys Rev B*. 2011;84(13):134111.]

6.2.3 Defect Chemistry

At low temperatures, UO_2 is a stoichiometric compound, but at high temperatures it shows substantial deviations from stoichiometry. In the hyperstoichiometric range, the interstitial octahedral holes in the lattice are filled with oxygen atoms, compensated by oxidation of the uranium ions from 4+, to formally 5+. In Kröger–Vink notation, this reaction can be expressed as

$$2U^{4+} + \frac{1}{2}O_2(g) = 2U^{5+} + O_i^{2-} \tag{6.1}$$

Above about 2000 K, substantial hypostoichiometry can also occur, which means that oxygen vacancies are formed in the lattice, compensated by reduction of some of the uranium ions from 4+, to formally 3+:

$$2U^{4+} = 2U^{3+} + V_O^{\bullet\bullet} \tag{6.2}$$

Catlow (43, 44) performed empirical potential calculations using rigid-ion pair potentials to study the defect structure of UO_2 as early as 1977. His results showed that Frenkel defects are the predominant form of atomic disorder and that electronic (hole-electron, $\oslash = e' + h^{\bullet}$) disorder is far more extensive than any atomic process, particularly for hyperstoichiometric UO_{2+x}. Grimes and Catlow (45) found that the predominant defects in the fluoride structure are oxygen Frenkel pairs (OFPs) and neutral Schottky trivacancies, which are visualised in Figure 6.3. The formation reactions of these defects can be written as

$$O_O^{\times} = O_i^{''} + V_O^{\bullet\bullet} \tag{6.3}$$

$$\oslash = V_U^{''''} + 2V_O^{\bullet\bullet} \tag{6.4}$$

The predominance of stoichiometric defects in these calculations could be due to the type of potentials used, which favours them significantly over other types of defects. The defects can be charged, as in the above equations, or neutral, and recent DFT + U

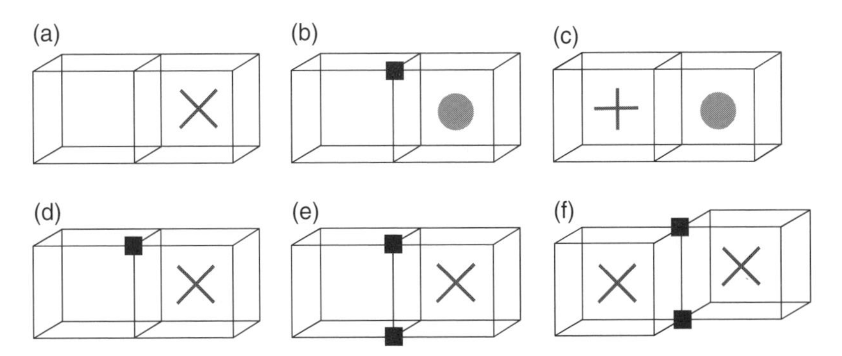

Figure 6.3 Typical point defects in UO_2; ■ oxygen vacancy, • uranium ion, × uranium vacancy, + interstitial site. (a) uranium vacancy; (b) oxygen vacancy; (c) interstitial site; (d) divacancy; (e) trivacancy; (f) tetravacancy [after Grimes RW, Catlow CRA. The stability of fission-products in uranium-dioxide. *Philos T R Soc A*. 1991;335(1639):609–634].

calculations have indicated that the charged defects have much lower formation enthalpy than the neutral ones (46–48). U^{4+} ions are always favoured over U^{3+} and U^{5+}, which partly justifies a posteriori the use of rigid-ion empirical potentials.

Neutron diffraction studies by Willis (49, 50) have shown that for $x > 0.03$ in UO_{2+x}, oxygen interstitials do not occupy the octahedral site at the centre of the oxygen sublattice but are displaced approximately by 1 Å from their normal site along the ⟨110⟩ (O^a) and ⟨110⟩ (O^b) directions. These oxygen interstitials form so-called Willis type clusters by associating with the nearby oxygen vacancies. The most important is the 2:2:2 cluster, formed by two oxygen O^a interstitials, two oxygen O^b interstitials, and two oxygen vacancies, either neutral or charged:

$$2V_i^a + 2V_i^b + 2O_O^{2-} + O_2(g) = \left\{2\left(O_i^a O_i^b V_O\right)\right\} \tag{6.5}$$

$$2V_i^a + 2V_i^b + 2O_O^{2-} + O_2(g) = \left\{2\left(O_i^a O_i^b V_O\right)\right\}^m + mh \tag{6.6}$$

where m indicates the charge of the cluster. Brincat et al. (51) confirmed that the 2:2:2 Willis cluster is a genuine feature of UO_{2+x} using GGA + U calculations, but its formation is dependent on the O/U ratio. Edge-sharing 2:2:2 Willis cluster chains are stable configurations in $UO_{2.25}$ and $UO_{2.125}$. In the latter, the defect cluster remains present in accordance with the original observations of Willis, but for lower O/U ratio ($UO_{2.0625}$) the cluster destabilises and splits in di-interstitial defects. The latter was already predicted in earlier calculations.

Yakub et al. (52) studied defect formation in UO_2 by MD using a partly ionic potential model considering the various oxidation states for uranium cations. Several defect types were considered in the calculations (single interstitial, interstitial dimeric clusters of the type 2:2:2, 2:4:4, and 2:1:1, and interstitial tetra- and pentamer cuboctahedral clusters), and the results showed that the 2:4:4 dimers are more stable than the Willis 2:2:2 clusters, whereby cuboctahedral tetra-and pentamers are more stable than any dimeric cluster at T = 0 K.

Ngayam-Happy et al. (53) performed molecular statics simulations using an extended pair potential model that accounts for a disproportionation mechanism as charge compensation. In that study, the volume changes observed experimentally in non-stoichiometric UO_{2-x} and UO_{2+x} were reproduced, which is not the case for models ignoring disproportionation effects. The model predicts the generation of split-interstitial clusters as the dominant defect type in non-stoichiometric uranium dioxide and progressive aggregation of primitive blocks identified as 1-vacancy split-interstitial clusters as the key mechanism for defect clustering in hyperstoichiometric uranium dioxide.

Experimental data for the formation energies of these defects is limited, and indirect. The values are generally derived from measurements of bulk properties (oxygen diffusion, heat capacity) coupled to an atomistic model assuming that thermodynamic equilibrium conditions are achieved and that diffusion of point defects is the principal mechanism. Matzke (54, 55) derived the formation enthalpy of some defects from defects models applied to oxygen diffusion experiments, and these numbers are generally cited as the experimental values. In contrast, many authors have reported computational studies at the DFT level of the formation energies of these defects using a variety of theoretical approaches and levels of theory, as summarised in Table 6.1. The

Table 6.1 Enthalpy of formation of an oxygen Frenkel pair and Schottky trivacancy defects in UO_2 (in eV).

	Method[a]	OFP	Trivacancy
Clausen et al. (11)	Exp[b]	4.6 ± 0.5	
Matzke (54, 55)	Exp[c]	3.5 ± 0.5	
Murch and Catlow (57)	Exp[c]	4.1	
Konings and Beneš (58)	Exp[d]	3.3	
Crocombette et al. (47, 59)	DFT/LDA	3.9	5.8
Freyss et al. (60)	DFT/GGA	3.6	5.6
Gupta et al. (61)	DFT/GGA + U	3.5–4.0	7.2
Nerikar et al. (46)	DFT/GGA + U	3.95[e]/2.70[f]	7.6
Yu et al. (62)	DFT/GGA + U	2.6[e]	7.0
Yun et al. (63)	DFT/GGA + U	4.5[e]	
Dorado et al. (9)	DFT/GGA + U	6.2[e]	
Crocombette et al. (47)	DFT/GGA + LHFCA	6.4[e]/4.8[f]	9.9
Andersson et al. (64)	DFT/LDA + U	5.3[e]/3.4[f]	10.2[e]/6.4[f]
Vathonne et al. (48)	DFT/LDA + U	2.4–2.6[f,g]	4.2–4.6[f,g]
Ngayam-Happy et al. (53)	EP	2.62 ± 0.27	

a Exp, experimental; DFT, density functional theory; EP, empirical potentials.
b From high-temperature neutron diffraction experiments.
c From oxygen diffusion experiments.
d From high-temperature heat capacity data.
e Neutral defect ($O_O^{\times} + V_O^{\times}$).
f Charged defect ($O_i^{''} + V_O^{\bullet\bullet}$).
g The two values correspond to bound and isolated Schottky defects.

computational results show a wide range of values, for example, for the OFP between 2.6 eV and 6.6 eV. Crocombette et al. (47) and Dorado et al. (9, 56) attributed this to (a) the possible occurrence of multiple minima in earlier calculations in which different initial occupations of the *f* orbitals drive the calculation to different local minima, even with little change in these initial occupations, (b) difference between the numerical approaches, simple LDA and GGA versus the ones with U correction, and general differences between LDA and GGA, and (c) to the chemical potentials taken as reference for the formation energy calculations in the case of non-stoichiometric defects.

In addition, as shown in Table 6.1, the formation energies obtained by DFT + U using charged defects are lower and closer to those obtained from pair potential calculations, which is consistent with the fact that these potentials consider only U^{4+} and O^{2-} ions. These values are also in better agreement with the experimental values, even if the scatter in both calculations and experimental results remains significant. Thus, to obtain the best possible results, special care must be taken to ensure that calculations converge to the system ground state, adapted energy reference states must be chosen, and various charged states must be considered. These conditions are only satisfied in the most recent studies. It must be noted, however, that the energies of charged defects show a strong dependence on the presence of impurities and stoichiometry, and thorough comparison between calculation and experimental results can only be done if the experimental conditions are precisely known (48).

6.2.4 Transport Properties

The atomic transport properties of UO_2 are of key importance for understanding and describing fuel behaviour, in particular, the self-diffusion of oxygen and uranium. Atomic diffusion is a basic mechanism in materials science that influences material behaviour at elevated temperatures, during the fabrication (sintering) and in-pile utilisation (affecting creep, grain growth, etc.). Because diffusion is a thermally activated process, meaning that an activation energy must be overcome for it to become effective, temperature is an important parameter, as is evident from the general Arrhenius-type equation for the diffusion coefficient D:

$$D = D_0 e^{-E_a/(RT)} \tag{6.7}$$

where D_0 is the pre-exponential factor, E_a the activation energy, $\boldsymbol{R}$ the universal gas constant, and T the absolute temperature.

Oxygen is the most mobile species in UO_2, $D^O/D^U > 10^5$, but is strongly dependent on the O/U ratio (54). However, for true mass transport to occur, all atomic species have to be transported, the slowest species thus being rate-controlling. The detailed mechanisms of self-diffusion are not very well known. From an atomistic point of view, diffusion is a stochastic process involving jumping of atoms between vacant and/or interstitial lattice sites and is thus closely related to the defect chemistry. It is not possible to observe it at that level, and the underlying mechanisms must be inferred from indirect observations.

Atomic-scale methods (electronic structure calculations or empirical potential methods) can be used to elucidate the elementary mechanisms involved in the diffusion mechanisms, and diffusion coefficients can be obtained in two ways:

1) Simulate the evolution of a system with time using MD, and determine the time dependence of the mean square displacements of atoms. The diffusion coefficient can then be calculated using the Einstein relation. If diffusion is slow, however, it might be difficult to observe enough events using classical MD, and the application of accelerated dynamics methods might be necessary. In UO_2, this method has so far only been applied in combination with empirical potentials and at high temperatures where diffusion is fast enough so that it can be observe in the time accessible (a few hundreds of nanoseconds).
2) Determine the rates of the elementary mechanisms using a static method, such as the nudged elastic band method (65, 66), and derive the diffusion coefficient using a diffusion model, such as the 5-frequency model (67).

6.2.4.1 Oxygen Diffusion

The oxygen diffusion in stoichiometric UO_2 has been measured by several authors (68–72) with a variety of techniques based on oxygen tracer experiments and electrical conductivity experiments. Since no suitable radioactive tracers exist for oxygen, the measurements are generally based on weight changes or mass spectrometry of the stable

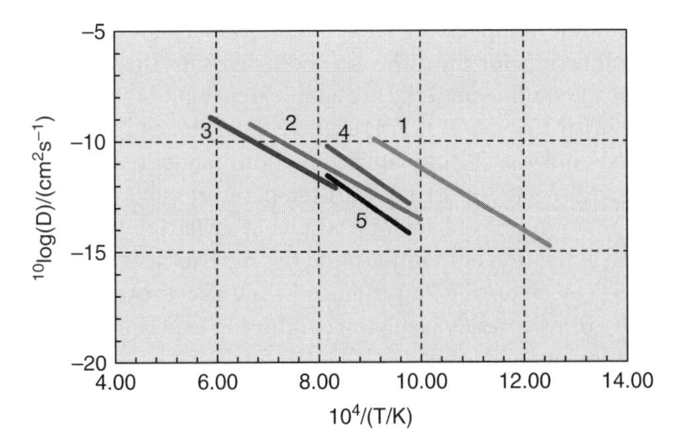

Figure 6.4 The oxygen self-diffusion in UO_2. 1, Auskern and Belle (68); 2, Marin and Contamin (70); 3, Hadari et al. (71); 4, 5, Dorado et al. (78). All curves refer to polycrystalline UO_2, except 5, which refers to single crystal. (*See insert for color representation of the figure.*)

isotope ^{18}O. As discussed by Matzke (54, 55), these are difficult measurements, and depend on the rate at which thermodynamic equilibrium conditions are achieved by diffusion of point defects (oxygen vacancies for UO_{2-x}, oxygen interstitials for UO_{2+x}) in a chemical gradient when the oxygen pressure of the gas atmosphere or the temperature is suddenly changed. Their interpretation requires that oxygen transport across the gas–solid interface is fast and not rate-limiting. Because the equilibrium oxygen pressure plays an important role, the following relation has been suggested by Matzke (54, 55):

$$\tilde{D}^{\circ} = D\frac{(2+x)}{2\boldsymbol{R}T}\frac{d\{\Delta\bar{G}(O_2)\}}{dx} \tag{6.8}$$

where $\tilde{D}^{\circ}$ is the chemical diffusion, D the self-diffusion, $\Delta\bar{G}(O_2)$ the equilibrium oxygen potential, $\boldsymbol{R}$ the universal gas constant, and T the absolute temperature.

Figure 6.4 shows the experimental results for the oxygen diffusion measurements for samples of claimed nominal UO_2 composition, and the agreement is reasonable, with the exception of the early work by Auskern and Belle (68). The variation might be due to possible deviations from stoichiometry and the polycrystalline nature of UO_2. These factors were demonstrated to affect the results considerably in most of the studies. In the work of Dorado et al. (72), the difference between single-crystal and polycrystalline material was measured, yielding slightly higher values (factor 10) for the polycrystalline material, showing the additional effect of grain boundary diffusion.

Yakub et al. (19) performed MD simulations using empirical pair potentials of the oxygen diffusion in UO_2, including the premelting and liquid ranges. They found that the diffusion proceeds via a vacancy or interstitialcy jumping mechanism. They also observed that a strong deviation from the Arrhenius behaviour occurs near the melting point, where the diffusion mechanism becomes more complex due to the formation of defect clusters and cluster chains. Also, Arima et al. (73) studied the oxygen diffusion at high temperatures (2000–5000 K) using MD simulating perfect crystalline UO_2, the same material with Schottky defects, and polycrystalline material. The calculations showed that the predominant mechanism is (1) oxygen vacancy diffusion in the low-temperature range, (2) lattice diffusion in the middle-temperature range, and (3) superionic diffusion just below the melting point. Their results for the diffusion coefficients for the UO_2 cell with Schottky defects and the polycrystalline material are a fair extension of the experimental data at lower temperature. Also, Cooper et al. (74) used a MD approach combined with new empirical potentials beyond the 2-body approximation (Cooper-Rushton-Grimes, CRG) to calculate oxygen diffusivity, obtaining high-temperature values that are in fair agreement with the extrapolation of experimental results at lower temperature. They found a significant change in activation enthalpy around the superionic transition, the values decreasing from about 6–7 eV before the transition, to 1–2 eV after the transition.

Dorado et al. (72) performed an extensive study using tracer diffusion experiments and DFT + U calculations of the activation energies of neutral O interstitials and vacancies, as well as a careful review of previous results on oxygen diffusion. The results showed that for hyperstoichiometric UO_2, bulk oxygen diffusion proceeds via an interstitialcy mechanism with an activation energy of 0.88 eV, which compares very favourably with the value determined experimentally and confirms the result of the empirical potential simulations. For hypostoichiometric conditions, the O vacancy migration in the ⟨100⟩ direction is the energy-lowest mechanism, and the activation energy calculated compares favourably with the value measured by Kim and Olander (75). Recent DFT + U results are available on the migration of charged O interstitials and vacancies as a function of stoichiometry (76), but a detailed comparison with experimental data remains to be done.

Finally, Bai et al. (77) studied the effect of oxygen interstitial clusters of size one to five atoms in UO_2 by temperature-accelerated dynamics simulations. They found that for some clusters, multiple migration paths and barriers exist, and the minimum migration barrier of each oxygen interstitial cluster has the following order: $2O_i < 3O_i < 1O_i < 5O_i < 4O_i$, suggesting that cluster diffusion may be important for the oxygen diffusivity at high deviations from stoichiometry in UO_{2+x}.

6.2.4.2 Uranium Diffusion

The uranium diffusion in UO_2 can be measured from diffusion couples, in which a coating of isotopically enriched UO_2 (^{235}U or ^{233}U) on natural UO_2 is heated at high temperature, after which the surface depletion or the axial profile are measured, for example, by alpha spectrometry. The difficulties of this approach are the control of the sample composition at high temperature, avoiding surface evaporation and grain boundary diffusion, and achieving a sufficient spatial resolution. It is thus not surprising that the experimental studies of the uranium diffusion in UO_2 (70, 79–84) show a wide range of results, as depicted in Figure 6.5. In particular, the most recent study by Sabioni et al. (84) yields diffusion coefficients that are several orders of magnitude lower than those from earlier studies. In that study, single-crystal UO_2 was used with ^{235}U as the tracer, and the diffusion profile was measured by secondary ion mass spectrometry (SIMS). These authors showed in a subsequent study (85) that grain boundary diffusion has a large impact on the measurements, being several orders of magnitude larger than the bulk diffusion (see Section 6.4.5).

Dorado et al. (78) studied uranium self-diffusion in $UO_{2\pm x}$ by first-principles DFT + U calculations for small deviations from stoichiometry. They found the lowest migration barrier to be a vacancy mechanism along the ⟨110⟩ direction, involving a significant contribution from the oxygen sublattice, with several oxygen atoms being displaced from their original position. The ⟨110⟩ vacancy diffusion mechanism was found to have a lower

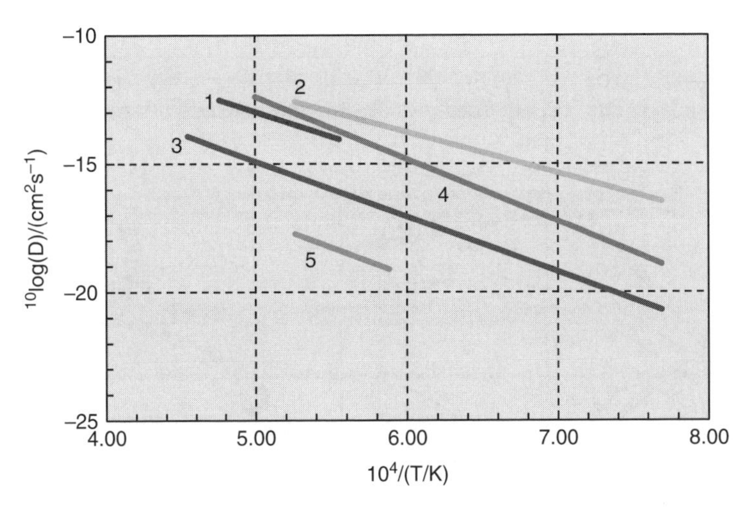

Figure 6.5 Uranium self-diffusion in UO_2. 1, Auskern and Belle (79); 2, Yajima et al. (80); 3, Reimann and Lundy et al. (82); 4, Matzke (83); 5, Sabioni et al. (84). (*See insert for color representation of the figure.*)

activation energy than any of the interstitial mechanisms. As for O diffusion, recent results are available on the migration of charged U interstitials and vacancies as a function of stoichiometry (76), but a detailed comparison with experimental data remains to be done.

Boyarchenkov et al. (86) studied uranium self-diffusion in UO_2 by MD using a pair potential approach. They showed that under periodic boundary conditions (PBCs), the uranium cations diffuse via an exchange mechanism (with the formation of Frenkel defects) with an activation energy of 15–22 eV, while under isolated boundary conditions (IBCs), there is competition between the exchange and vacancy (via Schottky defects) diffusion mechanisms, which give an effective activation energy of 11–13 eV near the melting temperature of the simulated $UO_{2.00}$ nanocrystals. Vacancy diffusion with a lower activation energy of 6–7 eV was dominant in the non-stoichiometric crystals $UO_{2.10}$, $UO_{2.15}$, and $UO_{1.85}$. The energies obtained are very high compared to those derived from DFT + U calculations, and the meaning of the difference between PBC and IBC must be clarified.

6.2.5 Thermophysical Properties

6.2.5.1 Phonon Kinetics

The thermophysical properties of UO_2 are strongly dependent on the phonon kinetics, through the internal vibrational energy. The harmonic approach, which assumes that the equilibrium distance between atoms is independent of temperature, is not suitable for describing these properties, and knowledge of the anharmonicity of the phonon lattice waves is therefore essential. The quasiharmonic approximation, which assumes that the temperature dependence of the phonon frequencies can be described by the dependence on the volume, is generally considered a suitable model.

Neutrons are ideal for studying phonon kinetics as their wavelength can be tuned to be comparable to the interatomic distance of the crystal lattice, and they have energy quanta comparable to the phonon energy (which is not the case for X-rays). Inelastic

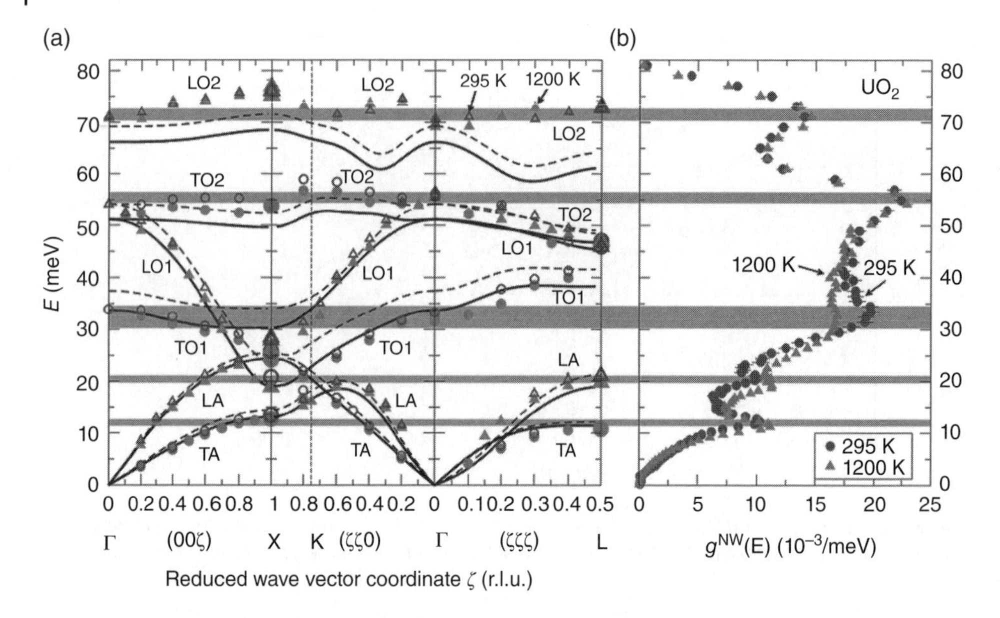

Figure 6.6 The phonon dispersion curves (a) and phonon density of states (b) of UO_2, after Pang JWL, Chernatynskiy A, Larson BC, Buyers WJL, Abernathy DL, McClellan KJ, et al. Phonon density of states and anharmonicity of UO_2. *Phys Rev B*. 2014;89(11):115316. The open symbols refer to T 0 295 K, the solid symbols to T = 1200 K, triangles represent the transverse phonons. The dashed and solid lines show the results of the GGA + U calculations for T = 295 K and T = 1200 K, respectively. Reproduced with permission. (*See insert for color representation of the figure.*)

neutron scattering, in particular, yields essential information on the phonon density of states (PDOS) as demonstrated by Dolling et al. (87) as early as 1965. These authors determined the frequencies of the normal modes of vibration along the ⟨00ζ⟩, ⟨ζζ0⟩, and ⟨ζζζ⟩ direction in single-crystal UO_2 and showed that a rigid-ion model with axially symmetric forces between the nearest neighbouring molecules can provide a qualitative agreement with experiment, but that a shell model including ionic polarizabilities was needed to obtain quantitative agreement.

Yun et al. (88) calculated the phonon spectrum using the DFT/GGA + U approach, and derived the phonon dispersion curves and density of states in the quasiharmonic approximation from the Hellmann–Feynman forces generated by displacing one atom in the supercell (of 96 atoms) from its equilibrium position. The obtained PDOS is in good agreement with the experimental result of Dolling et al. (87)

Pang et al. (89, 90) recently re-measured the phonon spectra of UO_2 at T = 295 and 1200 K with the inelastic neutron scattering technique (Figure 6.6) and combined the measurements with electronic structure calculations of the PDOS. They found a large impact of anharmonicity-induced linewidth broadening on the vibrational spectrum at both ambient and high temperatures.

6.2.5.2 Thermal Expansion

Lattice expansion is related to the pressure and temperature dependence of the phonons and the phonon–phonon interactions. The thermal expansion of UO_2 has been measured by many authors using high-temperature X-ray diffraction (XRD), and these

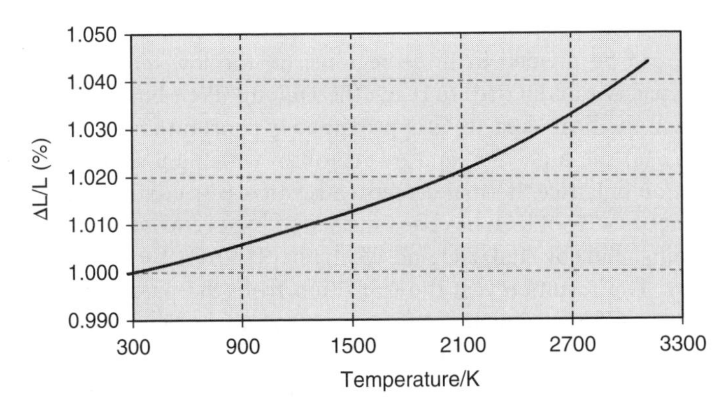

Figure 6.7 The thermal expansion of UO_2, expressed as $\Delta L/L$ according to the recommendation by Martin (91).

data together with the neutron diffraction data were reviewed by Martin (91) and Fink (92), and their recommendations are generally accepted. Recent work by Guthrie et al. (93) using high-energy XRD measurements in combination with containerless laser heating, and thus not subject to sample–containment interaction, give slightly but systematically higher values.

The cubic structure UO_2 expands with temperature in an isotropic manner along the three principal axes. The results show an almost linear increase of the volume up to about 1600 K, above which the volume increase non-linearly up to its melting point, as shown in Figure 6.7.

Willis (94, 95) concluded from neutron diffraction studies of single-crystal and polycrystalline UO_2 from room temperature up to 1373 K that anharmonic vibration of the oxygen atoms must be taken into account in the interpretation of the results, and proposed a (single particle) potential (U) to describe the thermal displacement of the oxygen atoms in ⟨xyz⟩ directions:

$$U(x,y,z) = U_0 + \frac{\alpha_0}{2}\left(x^2 + y^2 + z^2\right) + \beta_0(xyz) \tag{6.9}$$

where the cubic term results from the tetrahedral coordination of the oxygen site and describes the movement of the anions towards adjacent empty cubes (13). The movement of the U cation can be described by a harmonic potential. The Debye Waller coefficients (temperature factors) derived agree well with those derived from the work by Dolling et al. (87). Clausen et al. (11) extended the neutron diffraction measurements up to the melting point and found evidence for OFP disorder in UO_2, as a result of which the lattice expands in a non-linear way above about 2000 K.

The DFT + U study by Yun et al. (88) is in good agreement with experimental data up to 500 K. The deviation between the calculated and measured data slightly increases above 500 K and becomes significant at around 1000 K. This might be due to an increased electronic contribution to the thermal expansion or to an increasing anharmonicity.

Thermal expansion is one of the most frequently simulated properties of UO_2 as it is often used as to derive the empirical potential parameters, and for that reason it makes no sense to compare experiment and empirical potential calculations.

6.2.5.3 Heat Capacity

The heat capacity of UO_2 can be divided in three regions: (a) the low-temperature region in which the heat capacity rapidly rises to reach the Dulong–Petit limit, (b) the intermediate region in which the heat capacity is approximately constant and close to the Dulong–Petit limit, and (c) the high-temperature region in which anharmonicity effects due to the contribution of lattice thermal defects cause an excess contribution.

The low-temperature heat capacity of UO_2 has been measured by several authors, the most authoritative study being made by Hunziker and Westrum (35) from T = 5 to 350 K using adiabatic calorimetry. The results reveal the transition from the paramagnetic state to the low-temperature antiferromagnetic state at T = 30.44 K (see Figure 6.8). In the region below the Néel temperature, the results are in fair agreement with the lattice (phonon) and magnetic contributions derived from the inelastic neutron scattering work of Dolling et al. (87), the discrepancy being likely due to coupling of phonons and magnons (the quanta of magnetisation excitations). The region above the Néel temperature could very well be described by a harmonic contribution of the normal modes, plus an anharmonic contribution and a small electronic contribution resulting from crystal-field states. In this temperature range, UO_2 behaves like a regular Debye solid.

In the intermediate range (300 to 2000 K), the behaviour is still regular, and the heat capacity is close to the Neumann–Kopp rule ($3nR$), as demonstrated by a large number of direct heat capacity and indirect enthalpy increment ($dH/dT = C_p$) measurements (96).

Above about 2000 K, the heat capacity of UO_2 shows a strong increase, which has been attributed to the formation of defects, particularly OFPs. A substantial number of enthalpy increment measurements have been made in this temperature range and give a rather consistent indication that the increase is continuous. It should, however, be realised that measurements in this temperature range are not straightforward and are subject to uncertainties pertaining to the sample composition: as a result of non-congruent vaporisation, the composition of the sample can change, and at temperatures of around T = 3000 K, reactions with the container materials (Ta, W) might take place. Only one study of the heat capacity in the premelting range has been reported: Hiernaut et al. (97) reported the heat capacity derived from laser heating experiments showing an order-disorder (superionic) transition with a peak at (2670 ± 30) K. The heat capacity

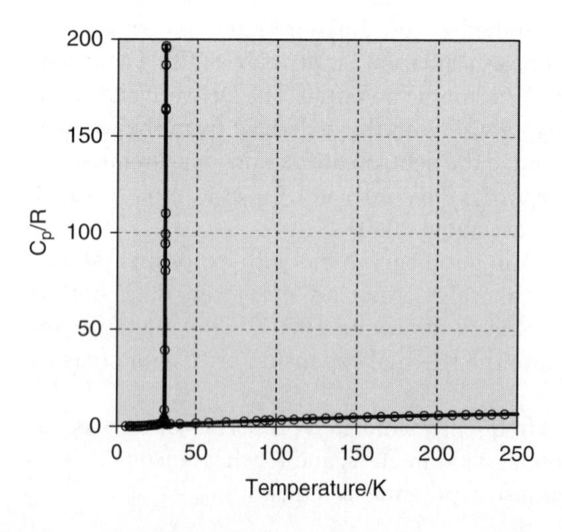

Figure 6.8 The low-temperature heat capacity of UO_2 showing the antiferromagnetic state at T = 30.44 K.

maximum of this peak reaches values around 215 J K^{-1} mol^{-1}, thus contrasting with the enthalpy increment measurements, which give no indication of such a transition.

Szwarc (98) proposed a model for the excess heat capacity resulting from the OFP contribution, which is given by the equation:

$$C_{p,exc} = \frac{\Delta H_{OFP}^2}{\sqrt{2}RT^2}\exp\left(\frac{-\Delta H_{OFP}}{2RT}\right)\exp\left(\frac{\Delta S_{OFP}}{2\boldsymbol{R}}\right) \tag{6.10}$$

where ΔH_{OFP} and ΔS_{OFP} are the enthalpy and entropy of OFP formation, $\boldsymbol{R}$ is the universal gas constant, and T is the absolute temperature. When the excess heat capacity is fitted to this equation, the enthalpy and entropy of OFP formation are 3.3 eV and 0.7 eV, respectively (99). As can be seen in Table 6.1, the enthalpy value is in reasonable agreement with other experimental values and the most recent theoretical values.

Yakub et al. (19) performed an MD study of the superionic transition in UO_2 using a partial ionic model, clearly demonstrating a λ-type anomaly in the heat capacity, in both small and large MD cells, although its magnitude was affected by the cell size. Cooper et al. (20, 21) used the CRG many-body potential model for describing a range of thermophysical properties of actinide dioxides and their solid solutions. They also observed the superionic transition in UO_2 as an inflection in the calculated enthalpy increment data, which results in a λ-type peak in the heat capacity. As discussed by Vălu et al. (96), the discrepancy between experiments and shown in Figure 6.9 may be related to the defect types considered and their energies. It is well known that defect clustering occurs in UO_2 (100) and that the concentration of such clusters increases with temperature where the mobility of point defects increases.

6.2.5.4 Thermal Conductivity

UO_2 has poor thermal conductivity, resulting in a relatively high fuel operating temperature. For that reason, the thermal conductivity of UO_2 is an important engineering property, and many measurements have been reported for UO_2. They were carefully evaluated by an expert group of the IAEA (101), which gave a recommendation for the thermal

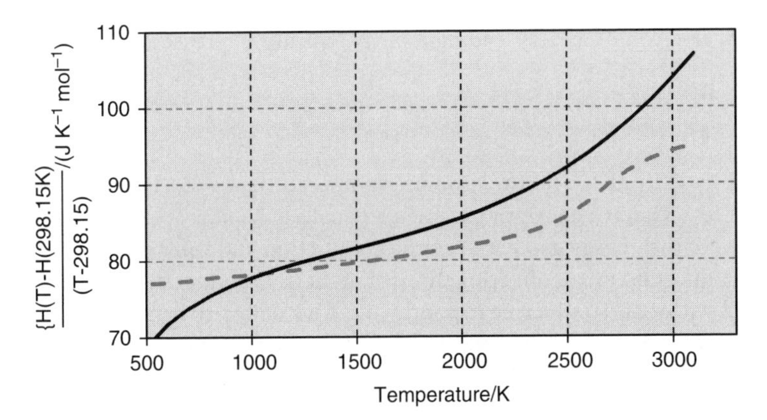

Figure 6.9 The enthalpy increment of UO_2; the solid line shows the values from a critical assessment of the experimental enthalpy increment measurements (99), the broken line shows the results of a many-body potential model calculation (20, 21).

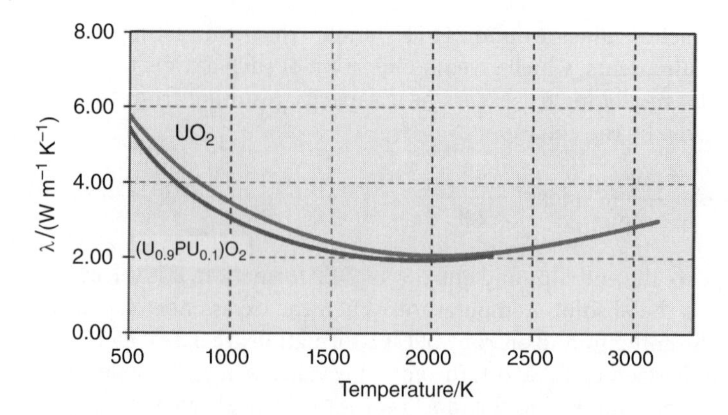

Figure 6.10 The thermal conductivity of UO_2 and $(U_{0.9}Pu_{0.1})O_2$ as a function of temperature.

conductivity of unirradiated (virgin) UO_2. As shown in Figure 6.10, the thermal conductivity of UO_2 has a minimum at approximately 2000 °C. This is related to the fact that heat conduction occurs by phonons and by the kinetic energy of electrons: $\lambda = \lambda_{ph} + \lambda_{el}$.

At temperatures below 1500 K, the phonon contribution predominates. When applying the kinetic gas theory to the propagation of atomic vibrations (phonons) or quasiparticles, it appears that the phonon conductivity in the temperature range of interest can be expressed as

$$\lambda_{ph} = \frac{1}{A + BT} \tag{6.11}$$

where A corresponds to the scattering of phonons by imperfections such as point defects, line and planar defects, voids, pores, and so on. The parameter B corresponds to the scattering by phonon–phonon (Umklapp) interactions. Equation 6.11 also suggests that the thermal conductivity decreases with increasing temperature, which is the case for UO_2 up to about 2000 K.

Above this temperature, the electronic contribution becomes important, as a result of which the thermal conductivity slightly increases again (Figure 6.10). The temperature-dependent creation of electronic carriers leading to λ_{el} is typically expressed as (102):

$$\lambda_{el} = \frac{C}{T^n} \exp\left(-\frac{W}{k_B T}\right) \tag{6.12}$$

where λ is expressed in W/(m·K); T is the absolute temperature; $k_B = 8.6144 \times 10^{-5}$ is Boltzmann's constant (eV/K); n is a constant parameter, usually 2, $\frac{5}{2}$, or 3; $C = 4.715 \times 10^9$; and $W = 1.41$ is related to the energy gap between the conduction and valence bands.

Also, porosity is an important factor affecting the overall thermal conductivity of UO_2. Pores, empty or filled with gas, conduct heat poorly and thus act as thermal barriers. Many formulas have been suggested to correct for this effect, mainly assuming that in highly dense materials the pores have a spherical shape, which is indeed the case for sintered UO_2. One of the most common expressions is the Maxwell–Eucken correction (103):

$$\lambda = \lambda_0 \frac{1 - P}{1 + \beta P} \tag{6.13}$$

Here, λ_0 is the thermal conductivity of the fully (100%) dense material, P is the porosity, and β is a constant, which is unity for perfect spherical pores. For complex pore shapes, other corrections are needed (104).

The thermal conductivity of UO_2 was calculated using empirical potential methods by several authors. Uchida et al. (105) used a non-equilibrium MD approach based on pair potentials, which yielded thermal conductivity values for the perfect crystal that are substantially higher than the experimental values at low temperatures. The authors also studied the effect of lattice defects (0.25%, 0.5%, and 1% Schottky defects in the cell). This strongly decrease the thermal conductivity at low temperatures, which gets close to the experimental values. Nichenko and Staicu (106) performed MD simulations based on pair potentials using the Green–Kubo approach. Their calculated thermal conductivity for perfect stoichiometric UO_2 is in good agreement with the literature data over the temperature range corresponding to heat transfer by phonons (up to 1700 K). The simulation of the effect of non-stoichiometry on the thermal conductivity showed that for the same extent of deviation from the ideal stoichiometry, the effect of oxygen vacancies (hypostoichiometry) is more pronounced than the effect of oxygen interstitials (hyper-stoichiometry). The simulation of the influence of the oxygen Frenkel pairs (0.22%, 0.4%, and 0.6% defects in the cell) showed that OFPs not only lower the thermal conductivity, but also that they recombine progressively and disappear at around 750–800 K. This results in a recovery of the thermal conductivity, as also observed experimentally in a study (107) of the effects of α-damage on the thermophysical properties of UO_2 using samples doped with ^{238}Pu, which will be discussed in further detail in Section 6.4.1.

6.2.6 Melting and the Liquid

Stoichiometric UO_2 melts at 3120 ± 30 K (99). This value is the average of a large number of consistent thermal analysis measurements obtained using traditional heating techniques of encapsulated samples, and by the self-crucible laser heating technique (99). In the latter, the interaction of the sample with the encapsulation material (tantalum, tungsten) is avoided, and due to the short duration (milliseconds to seconds) the compositional changes at high temperatures resulting from non-congruent sublimation and vaporisation can be limited. The agreement between the results shows that these effects are limited for stoichiometric UO_2, but they play a substantial role for PuO_2 and mixed $(U,Pu)O_2$ (109), as will be discussed in Section 6.3.

Arima et al. (108) simulated the melting of UO_2 by a two-phase MD approach, in which the supercell consisted of solid and liquid phases at the initial state, using three different empirical pair potentials. Examples of the structures obtained are shown in Figure 6.11, revealing the progressive lattice disorder with temperature. Their results yielded a melting temperature comparable to the experimental value when using the potential proposed by Yakub (19). The enthalpy of fusion was calculated as $60\,kJ{\cdot}mol^{-1}$, somewhat lower than the recommended value of $70\,kJ{\cdot}mol^{-1}$, which was obtained from high-temperature enthalpy drop measurements. Also, Ghosh et al. (110) performed MD simulation of the melting of UO_2 and found the melting temperature (between 3050 K and 3075 K) to be slightly lower than and the enthalpy of fusion ($58.9\,kJ{\cdot}mol^{-1}$) to be almost identical to the value calculated by Arima et al. (108). Boyarchenkov et al. (111) simulated the melting of uranium dioxide (UO_2) nanocrystals by MD, using a set

Figure 6.11 The structure of UO_2 at room temperature (left), of the solid just below the melting point (middle), and of the liquid, as obtained by molecular dynamics calculations (108); with courtesy of Dr. T. Arima and reproduced with permission.

of widely used pair potentials in the rigid-ion approximation, and extrapolated the results to the bulk. They performed the calculations under PBCs and IBCs. In the PBC case, the melting temperature calculated was more than 600 K above the recommended value, and in the IBC case some potentials yielded melting temperatures very close to the experimental one. Boyarchenkov et al. (111) found the enthalpy of fusion to be substantially lower than the recommended value (by 25–30 kJ/mol). This discrepancy could be due to the (partial) inclusion of the enthalpy of the superionic transition and the cationic Frenkel disordering into the recommended enthalpy of melting.

Yakub et al. (19) simulated liquid UO_2 by MD using a partly ionic model, and compared their results to the experimental results for the heat capacity and density of the liquid. The calculated heat capacity is in good agreement with the experimental results obtained by Ronchi et al. (112). The latter authors used a pulsed-heating laser technique to heat a microsphere of UO_2 up to 8000 K and derived the heat capacity from a model that describes the dynamics and heat losses of heat pulses applied to the specific geometry. Their results indicate a decrease in the heat capacity from the melting point to about 4500 K, and a slight increase above that temperature. There are no experimental data for the density of liquid UO_2, but Yakub found that his results compare well to the experimental results by Breitung and Reil (113), who derived the density and compressibility of $(U_{0.77}Pu_{0.23})O_{2.00}$ from an in-pile test during a power excursion of the reactor, reaching a temperature of about 7500 K. In view of the good agreement with the experimental results for UO_2 just above the melting point, these authors concluded that the thermal expansion of UO_2 and $(U_{0.77}Pu_{0.23})O_{2.00}$ can be described by the same relationship.

6.3 Mixed Oxides

Mixed oxides are binary or ternary mixtures of the actinide oxides that are used as nuclear fuel, and the term is generally used for $(U,Pu)O_2$. In the ideal case, the mixed oxide is a homogeneous material, with the Pu^{4+} ions randomly substituting U^{4+} in the crystal lattice. In practice, this can be obtained, for example, by preparing the material by co-precipitation from an aqueous solution. This was clearly demonstrated for $(U_{0.7}Pu_{0.3})O_{2.0}$ by Vigier et al. (114) using ^{17}O MAS NMR, which showed the presence of the signals for the uranium-plutonium environment of the cationic tetrahedron $O(U)_{4-z}(Pu)_z$ for $0 \le z \le 3$, with no $O(Pu)_4$. However, in mixed $(U,Pu)O_2$ oxide produced according to current industrial procedures based on powder blending (115), the Pu distribution is not homogeneous, but spots high in Pu are present. This is due to the fact that in the industrial process a first mixture containing about 30% PuO_2 is fabricated by ball milling, the primary or master blend, which is then mixed with UO_2 powder to obtain the required Pu concentration. The milling and blending does not yield a homogenous material at the atomic scale, and the industrial material must therefore be treated as a two-phase mixture, consisting of Pu-rich and U-rich zones.

The substitution of Pu^{4+} in the UO_2 lattice will substantially affect the properties of the fuel. The Pu^{4+} ion is slightly smaller than U^{4+}, and substitution leads to a slight contraction of the unit cell volume. Dorado et al. (25) performed DFT + U calculations for $(U,Pu)O_2$ with 12.5 and 25% Pu content and reproduced this lattice contraction for Pu contents up to 25%. They also showed that most of the ground-state electronic properties of UO_2 are affected by the addition of 12.5% Pu but do not change significantly with increasing Pu content up to 25%. They found that the antiferromagnetic order is a very good approximation of the paramagnetic state for energetic calculations, but that the elastic constants depend strongly on the magnetic ordering. Results obtained on ferromagnetic $(U,Pu)O_2$ raise some concerns regarding the electronic structure of the Pu atoms, which should be investigated further in $(U,Pu)O_2$, as well as in PuO_2.

The Pu^{4+} ions act as phonon scattering centres in the crystal lattice, as a result of which the thermal properties, and principally the thermal conductivity of $(U,Pu)O_2$, will be different from those of UO_2. For solid solutions such as $(U_{1-y}Pu_yO_{2\pm x}$, Equation 6.11 can be changed to

$$\lambda = \frac{1}{A_0 + A_1(x,y) + BT} \tag{6.14}$$

where the parameter A_0 describes the temperature-independent phonon scattering in the pure compound. The parameter $A_1(x,y)$ represents the influence of metal substitution (y) or non-stoichiometry (x) on the temperature-independent phonon scattering, and the parameter B is less sensitive to substitutions or other scattering centres. According to the theory developed by Ambegaokar (116) and Abeles (117) the parameter A_1 in Equation 6.14 can be represented by

$$A_1 = \frac{\pi^2 V \Theta}{3v^2 h}\Gamma \tag{6.15}$$

where V is the average atomic volume, Θ is the Debye temperature, v is the average phonon velocity, h is Planck's constant, and Γ is the scattering cross-section parameter of the phonon by the impurity type. For stoichiometric dioxide, the dependence of Γ on the metal substitution (y) is given by

$$\Gamma = y(1-y)\left(\frac{M_1 - M_2}{(1-y)M_1 + M_2}\right)^2 + \in y(1-y)\left(\frac{r_1 - r_2}{(1-y)Mr_1 + yr_2}\right)^2 \tag{6.16}$$

where M_1 is the mass of the host ion, M_2 is the mass of the substituted ion, r_1 is the ionic radius of the host atom, and r_2 is the ionic radius of the impurity atom. $\in$ is a parameter that represents the strain generated in the lattice and is assumed to be about 100, which is obtained from an analysis of various solid solutions by Fukushima et al. (118).

Gibby (119) systematically measured the effect of the Pu content on the thermal conductivity of $(U,Pu)O_2$ solid solutions up to 30 mol% PuO_2, observing a small but systematic decrease with increasing PuO_2 content, which he explained by the fact that the Pu^{4+} ions in the UO_2 lattice act as phonon scattering centres, in line with theory. However, Schmidt (120) found a different dependence on the PuO_2 content, with a peak at about 15 mol% in the thermal conductivity, particularly evident at low temperatures. Beauvy (121) reported similar results, and attributed this to differences in the defect cluster concentrations below and above 12.5 mol% PuO_2. Duriez et al. (122) made systematic measurements of the thermal conductivity of $(U,Pu)O_2$ mixed oxide fuel, for average Pu concentrations from 3 to 15 wt % and in the temperature range 700 K to 2300 K. They used both homogeneous samples and industrial MIMAS samples and found the thermal conductivity to be significantly lower than that of the UO_2, with no dependence on the Pu concentration, nor a difference between the homogeneous and MIMAS samples. Classical MD simulations have shed light on this. Yamada et al. (123), as well as Nichenko et al. (124), using pair potentials based on a partially ionic model, showed that Pu substitution results in a regular trend in line with Equation 6.16: the conductivity shows a minimum at around $x = 0.4$, which becomes less pronounced with increasing temperature, in line with Gibby's observations (see Figure 6.12).

Schmidt (126) studied the effect of the O/M ratio on the thermal conductivity of $U_{0.8}Pu_{0.2}O_{2-x}$ in the temperature range 373 K to 2073 K, observing a strong decrease as a

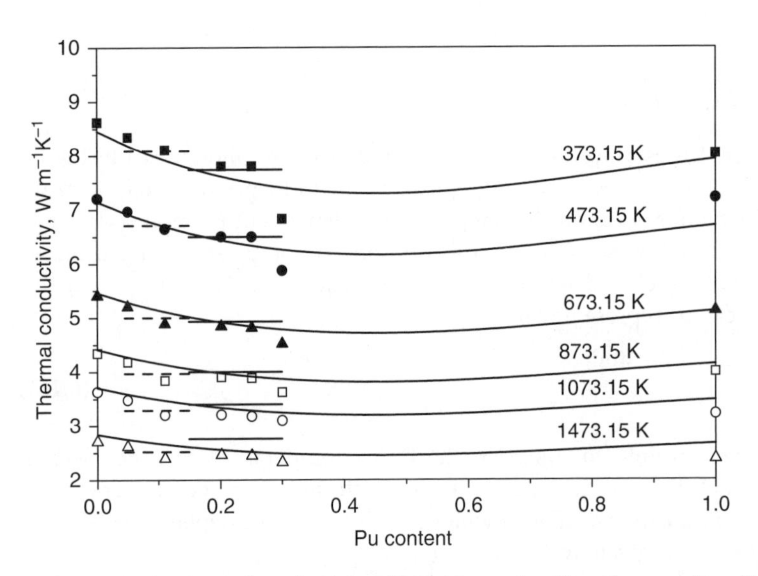

Figure 6.12 The thermal conductivity of $(U,Pu)O_2$ as a function of composition. The continuous lines show the results of the MD simulations by Nicheko et al. (124) for the temperatures indicated. The symbols denote the experimental results by Gibby (119), the broken horizontal line the experimental results by Duriez et al. (122) for average Pu concentrations from 3 to 15 wt%, and the straight horizontal line the experimental results by Philipponeau (125) for average Pu concentrations from 15 to 30 wt% (after Nichenko and Staicu (106)).

function of x, particularly at the lowest temperatures, which is due to the fact that Pu^{3+} and oxygen vacancies act as scattering centres. Schmidt also studied the $U_{0.8}Pu_{0.2}O_{2-x}$ (see Reference (127)), revealing a similar trend as for $UO_{2\pm x}$; that is, the thermal conductivity decreases for positive as well as negative values of x, though it is slightly asymmetric around O/M = 2.00. This is in agreement with the CMD simulations for UO_2 by Nichenko and Staicu (106), discussed above.

The high-temperature heat capacity of $(U,Pu)O_2$ was reviewed and discussed by Vălu et al. (96) The analysis of the experimental data for the whole composition range shows that from room temperature up to about 2000 K the enthalpy increment data do not show significant deviations from the Neumann–Kopp additivity rule, which states that the heat capacity of the solid solution is the sum of the molar fractions of the end members in the whole composition range. Above 2000 K, the effect of OFP formation leads to an excess enthalpy (heat capacity) that can be modelled by Equation 6.10 using the enthalpy and entropy of OFP formation of the end members as derived in (99). Cooper et al. (74) and Vălu et al. (96) also presented results of MD calculations based on the CRG many-body potential model (20, 21), and the simulations indicate the presence of the diffuse Bredig (superionic) transition; this is, however, not found in the experimental enthalpy increment data, as is discussed for UO_2 (see Section 6.2.5.3).

The melting temperature of $(U,Pu)O_2$ has recently been a subject of discussion. The early experimental work (128) suggested a regular decrease of the solidus and liquidus temperatures between the end members UO_2 and PuO_2. However, it was shown later (129, 130) that the commonly accepted melting temperature of PuO_2 was significantly wrong because of interaction of the samples with the containment at the high temperatures and the compositional changes due to the relatively high oxygen potential of Pu-rich samples. These problems could be avoided through the use of a fast and containerless laser melting technique as demonstrated by DeBruycker et al. (130), who showed that the melting temperature of PuO_2 is about 300 K higher than found in the early work. New measurements with the same technique for the UO_2–PuO_2 system have shown that a minimum exists in the liquidus and solidus, around $x(PuO_2) = 0.8$ (109, 131). To the best of our knowledge, no simulation of the melting of $(U,Pu)O_2$ has been performed. Ghosh et al. (110) used MD to calculate the melting temperature of the end members only, and found the value for PuO_2 (between 2800 K and 2825 K) to be significantly lower than the currently accepted experimental value.

Finally, Vauchy et al. (132) studied the oxygen self-diffusion in a $U_{0.55}Pu_{0.45}O_2$ sample by diffusion annealing involving gas–solid $^{18}O/^{16}O$ isotopic exchange and SIMS analysis, and concluded from the dependence of the tracer diffusion coefficients on the oxygen partial pressure at 1273 K that oxygen diffusion is a vacancy-assisted process. The oxygen self-diffusion coefficient obtained in the $U_{0.55}Pu_{0.45}O_2$ sample was found to be similar to that in nominally stoichiometric UO_2 and PuO_2, for which the values are close. Cooper et al. (74) used an MD approach to calculate the oxygen diffusivity in $(U,Pu)O_2$ as a function of uranium composition for a range of temperatures and found that below the superionic transition there is a slight but systematic enhancement of oxygen mobility compared to the linear interpolation between the UO_2 and PuO_2 end members, unlike in the $(U,Th)O_2$ system that they addressed in the same study.

6.4 Nuclear Fuel Behaviour during Irradiation

During irradiation, nuclear fuel undergoes extensive changes in its chemical composition. Each fission event creates two to three fission fragments (ions) in the mass range of about 80 at 150 amu, with peaks at 95 and 135 amu. In addition, activation of the fuel matrix, and in particular ^{238}U, through thermal neutron capture leads to the formation of transuranium elements such as plutonium and the so-called minor actinides neptunium, americium, and curium. The fuel thus evolves to a multi-element mixture.

In addition, the highly energetic fission fragments, carrying most of the ~200 MeV released in the fission process as kinetic energy, are slowed down in the fuel matrix and transfer their energy to the atoms of the crystal lattice by nuclear or electronic interactions, creating heat. Typical electronic energy losses $(dE/dx)_e$ are ~18 keV/nm for heavy fission fragments and ~22 keV/nm for light fission fragments (1, 133). During their track in the crystal lattice, the fragments create lattice defects, especially by nuclear (ballistic) collisions. Many of the fission fragments are radioactive and decay further by beta and gamma decay, and these decay modes strongly contribute to the heat production but not to lattice defects.

Another significant source of damage and heat production is the alpha decay of minor actinides produced during irradiation. In this process, the actinide nucleus $^{N}_{Z}An$ emits an α particle and transforms into a daughter nucleus $^{N-4}_{Z-2}An$, for example, $^{238}_{92}U \rightarrow ^{234}_{90}Th + ^{4}_{2}He$. The daughter nuclide has a recoil energy of about 100 keV, and shows predominantly nuclear stopping, leading to a dense collision cascade with typically about 1500 displacements within a short distance of ~20 nm. The alpha particle has an energy of about 5.5 MeV and causes a collision cascade with typically about 200 displacements within a short distance of about 15 μm (1). During irradiation, the fission products are the predominant source of radiation effects, but the alpha decay continues after the fuel is removed from the reactor core, and for used fuel the alpha decay is the main source.

The heat created in the fuel pellet must be transferred to the reactor coolant, and the heat transport through the fuel pellet is thus a critical process (2). Because UO_2 is a poor thermal conductor, the heat transport is limited and results in a relatively large temperature gradient in a fuel pellet. As shown in Figure 6.13, the temperature is about 1300 K in the centre of the pellet for a typical linear heat rate of about 250 W/cm in a LWR, and around 650 K at the pellet rim, resulting in a gradient of about 1300 K/cm.

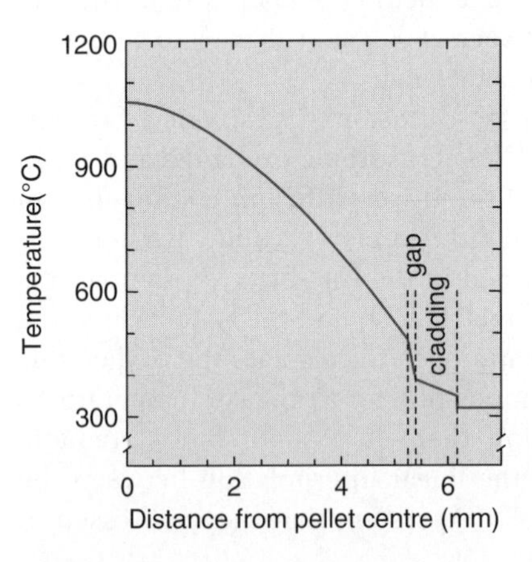

Figure 6.13 Typical temperature profile of a light water reactor (LWR) reactor fuel.

These conditions are unique to nuclear fuel and must be well understood in order to describe the evolution of the fuel microstructure, the changes in the fuel's mechanical and physical properties, and the chemical behaviour of the fission products as a function of the fuel burnup during normal operating conditions, but also during thermal transient conditions that can occur during off-normal operation and accidental conditions.

6.4.1 Radiation Effects from Fission Fragments

The path of a fission fragment in the crystal lattice – the so-called fission track – is the principal source of radiation damage in the fuel. The evolution of such a track can be divided into three phases (1):

- Primary phase – The passage of the fission fragment is very short (a few picoseconds), but it defines the initial size and cylindrical shape of the spike. Initially, the energy of the fission fragment is too high to interact with the fuel nuclei. The deposited Coulomb energy is mainly dissipated as local heating through electronic interactions, causing local heating close to or above the melting point (thermal spike). After the fission fragment has slowed down, ballistic collisions occur, and Frenkel defects are produced by secondary collision cascades.
- Second or quenching phase – Recombination of vacancies and interstitials occurs when the spike comes to thermal equilibrium. An interstitial-rich outer zone and a vacancy-rich inner zone form. The hydrostatic pressure field originally created by the quasi-molten core of the spike – contributing to the separation of interstitials from the vacancies of the Frenkel pairs formed in the primary phase – is replaced by compressive stresses in the outer zone and tensile stresses in the core.
- Third or track annealing phase – More recombination occurs, and some extended defects such as vacancy clusters are stabilised by fission gases forming embryos for bubbles that develop later.

The processes in these three phases are repeated many times throughout the volume of a homogeneous fuel, and the complete fuel is affected after a rather short irradiation time. The level of one displacement per atom (dpa) is typically reached within less than one day. The consequences are significant and lead to fission-enhanced diffusion, fission-enhanced creep, re-solution of fission gas from bubbles, etcetera (134–138), as will be discussed below.

The effect of fission tracks in the UO_2 matrix has been studied/simulated experimentally by heavy ion implantation. Wiss et al. (133) demonstrated the presence of fission tracks by TEM analysis of UO_2 irradiated with high-energy (173 MeV) Xe ions, which were not found in earlier work with lower-energy Xe ions (139, 140). These authors also found that dislocation loops form at concentrations of 7×10^{12} Xe/cm^3 at room temperature. Sonoda et al. (141) studied the effects of 100 MeV Zr and 210 MeV Xe ions on UO_2 at 573 K, and observed changes in fabrication of pores and the formation of dislocations at an ion fluence above 5×10^{14} ions/cm^2. The changes observed in the microstructure were considered to be the result of the overlapping of ion tracks, causing point defects, enhancing diffusion of point defects and dislocations, and forming sub-grains at relatively low temperature.

He et al. (142) used in-situ transmission electron microscopy combined with 300 keV Xe irradiation of single-crystal UO_2 at room temperature. They found that the dislocation microstructure evolved from nucleation and growth of dislocation loops at low irradiation doses to extended dislocations (segments and tangles) at higher doses. Xe bubbles with dimensions of 1–2 nm were observed after room-temperature irradiation. Onofri et al. (143) made a systematic TEM study of UO_2 implanted with Kr ions as a function of energy, fluence, and temperature, and found that with increasing fluence the point defects aggregate into dislocation loops and lines, up to the formation of a tangled dislocation network, similar to irradiated fuel. Dislocation loops transform progressively into lines in case of ion irradiation at elevated temperature.

Numerical simulations of collision cascades occurring in fission tracks and during ion implantation have been performed by several authors, in particular by Van Brutzel et al. (144, 145) and Martin et al. (146–148) using classical MD techniques. The thermal spikes caused by higher-energy ions, mainly through electronic excitations, were also simulated using classical MD by Huang et al. (149) and Govers et al. (150). These studies provided interesting insights into the atomic scale mechanisms and the type and numbers of defects created, as well as into the morphology of the cascades. As discussed by Jeon et al. (151), full MD simulation of fission tracks is, however, challenging for the following reasons:

- The volume of material needs to be relatively large (billions of atoms) because very high-energy ions with MeV kinetic energies can reach depths of up to several micrometres.
- Many (enough) ion track simulations need to be conducted for a statistically meaningful density profile of resting ions to be obtained, so that they can be compared to experimental ion-range data, which are a statistical composite of a very large number of ion trajectories.
- The standard empirical potentials not account for electron coupling effects in inelastic collisions needed to describe electron stopping in implantation or radiation range analysis.

In their work, Jeon et al. (151) applied a rare-event-enhanced-domain technique in combination with molecular dynamics (REED-MD) to simulate iodine implantation in polycrystalline UO_2 and compared the results to experimental range densities reported by Saidy et al. (152) for 100 keV, 440 keV, and 800 keV Iodine ions, finding very good agreement.

Finally, Martin et al. (153) compared classical MD simulation of full displacement cascades, CMD simulations of thermal spikes and the solution of a simple heat equation with a punctual thermal excitation in UO_2. They found that for displacement cascades, the conversion of half the kinetic energy into a potential energy takes place after a few tenths of picoseconds in the MD simulations, whereas energy equipartition does not occur during thermal spike simulations. These results question the relevance of thermal spike simulations for the study of radiation damage. But even if displacement cascades are far from a simple heat diffusion process, the maximum volume brought to a temperature above the melting temperature during the simulated cascade events is well reproduced by the simple thermal model applied. This volume constitutes a relevant estimate of the volume affected by a displacement cascade in UO_2.

6.4.2 Radiation Effects from Alpha Decay

Staicu et al. (107) studied the effects of α-damage in UO_2 using samples doped with ^{238}Pu (half-life of 87 years), which accelerates the alpha dose accumulation substantially. When analysing these samples with differential scanning calorimetry (DSC) after a six-month storage period, strong and broad exothermic effects were observed that could be related to the annealing/recombination of the point defects created by the alpha decay, agreeing with recovery effects in the measured thermal diffusivity.

By deconvolution, they could discriminate four peaks/effects that were assigned to the following processes:

- Oxygen vacancy/interstitial recombination, during which approximately 80% of defects created recombine.
- Uranium vacancy/interstitial cluster recombination.
- Dislocation loop growth of defects that escaped recombination.
- Void growth, as revealed by transmission electron microscopy. The void fractional volume was found to be of the order of $0.3 \pm 0.05\%$ in this specific case.

They also studied a reactor irradiated fuel sample, and the results indicate that the damage effects and the recovery phenomena during thermal annealing occur by similar mechanisms both in α- and fission-damaged UO_2.

Wiss et al. (154) studied the radiation damage of the same alpha-doped UO_2 samples using XRD and electron microscopy, revealing lattice expansion and accumulation of helium in bubbles. The lattice expansion was attributed to the ingrowth of point defects with time, saturating after a cumulative dose of about 1 dpa. The radiation-induced lattice expansion was described by the following exponential: equation:

$$\frac{\Delta a}{a} = A \times \left(1 - e^{(B \times D)}\right) \tag{6.17}$$

where D is the α-dose in dpa, $A = 0.632$ is the saturation value of (a/a_0), and $B = -2.5$ is the effective recombination volume. Hardness measurement indicated a similar saturation trend. Above a concentration of 5×10^{18} helium atoms g^{-1}, bubbles with a size of about 1 nm could be observed. Sattonnay et al. (155) observed by *in situ* He implantation and TEM characterisation that at elevated temperatures (above 873 K), bubbles of 25 nm size form.

This saturation agrees with the results by Van Brutzel and Rarivomanantsoa (156), who reported MD simulations of the primary damage state resulting from 5 keV displacement cascades due to α-self-irradiation in UO_2. Primary knock-on atom (PKA) direction and temperature were found not to influence the creation of point defects and point defects clustering. In the case of cascade overlapping, the number of created point defects saturates with increasing dose, and large clusters of vacancies were observed, which are stable within the time scale of the MD simulations.

6.4.3 Fission Product Behaviour

During irradiation, a large number of fission products is generated in the fuel. Typical burnup levels in LWR fuel are 4%–5% of the initial uranium atoms, meaning that the fission product concentration reaches 8–10 atom percent at the end of irradiation. The chemical properties of the fission products are generally different from those of uranium (valence state, oxygen affinity, and ionic size). Some fission products like Zr and the lanthanides can be accommodated in the crystal lattice sites, but most of them do not fit very well in the UO_2 crystal. This can already be observed on a microscopic scale, and examinations of irradiated fuel samples have led to the following grouping (157):

- Elements in their oxidised state that are soluble in the uranium dioxide crystal lattice, such as the rare earths zirconium and niobium.

- Inert gases (Xe and Kr) that have a very low solubility in the ceramic matrix and accumulate in gas bubbles.
- Metallic precipitates that contain noble metals (Ru, Rh, Tc, and Pd) as well as molybdenum.
- Oxide precipitates such as the so-called grey phase (Ba,Sr)(Zr,U,Pu)O_3 or caesium uranates.
- Volatiles such as caesium, iodine, and tellurium.

In view of reactor safety analysis, the fission gases and the volatiles are of prime importance as they are mobile in the fuel, and can be released from the fuel pellet. LWR fuel is designed to have a low fission gas release during normal operation, the release normally being limited to a few percent of the inventory. This low release implies that a significant amount of gas should be retained in the ceramic matrix, without causing too much swelling of the fuel pellet in order to avoid mechanical interaction with the cladding (PCMI, pellet cladding mechanical interaction).

6.4.3.1 Fission Product Dissolution in the UO_2 Matrix

The dissolution of fission products in the fuel matrix is the basic mechanism for their retention, and therefore has attracted attention for a long time. It is obvious from generic considerations that the dissolution is dependent on the atomic/ionic properties of fission products, as well as the defect state of the matrix. As discussed earlier, the UO_2 crystal lattice contains empty octahedral positions and point defects, whereas irradiation creates point defects, defect clusters, and extended defects. The question to answer is how can the fission product fit in lattice and defect sites. This atomistic information cannot be obtained easily from experiments, and computational approaches have been employed by many authors.

One of the first systematic studies was performed by Grimes et al. (45), who used an empirical pair potential approach to derive the solution energies of some key fission products (Xe, I, Cs, Rb, ...) in $UO_{2\pm x}$, considering various defect types such as oxygen vacancy, uranium vacancy, divacancy, trivacancy, and tetravacancy (see Figure 6.3). The results revealed that the neutral trivacancy is the most favourable solution site for fission products in UO_{2-x}, which can be understood from its relatively large size, while it is the uranium vacancy site for UO_{2+x}. In UO_2, both serve as solution sites depending on the fission product: the neutral trivacancy site for Xe and the uranium vacancy site for Cs and Rb. Further calculations for iodine and caesium by Grimes at al. (158) showed that their chemical state is dependent on the stoichiometry of UO_2:

- In UO_{2-x}, isolated iodine was found to be stable as I^- anions, but the bound {Cs:I} pair in a trivacancy cluster (see Figure 6.14) was found to be the most stable configuration.

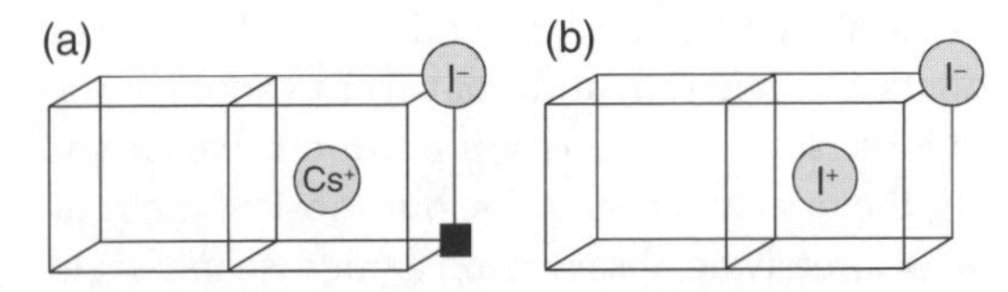

Figure 6.14 Defect clusters of Cs and I in UO_2. (a) {Cs:I} pair in a trivacancy, (b) {I:I} pair in a divacancy [after Grimes RW, Ball RGJ, Catlow CRA. Site Preference and binding of iodine and cesium in uranium-dioxide. *J Phys Chem Solids*. 1992;53(4):475–484].

- In stoichiometric UO_2, clusters of fission product species are stable, with I_2 pairs being more stable than {Cs:I} pairs.
- In UO_{2+x}, I^+ cations were found to be most stable, and as a result the association of caesium and iodine does not occur.
- At all stoichiometries, the precipitation of bulk CsI is energetically favoured over solution in the fuel.

Also, numerous DFT and DFT + U calculations have been performed to understand the incorporation energetics of fission products, especially fission gases, in UO_2 (42, 48, 64, 159–164). Petit et al. (159) were the first ones to perform electronic structure calculations to study the behaviour of Kr in point defects in uranium dioxide (DFT/LMTO-ASA). Since then, progress in the DFT technique has been substantial, and we will only discuss the DFT + U studies here in more detail, as these represent the latest state of the art.

Brillant (163) performed an extensive computational study (DFT + U/GGA) of a large number of fission products (He, Kr, Xe, I, Te, Ru, Sr, and Ce) in neutral pre-existing defects. Most FPs were found to be insoluble or poorly soluble in UO_2, except cerium, barium, and zirconium, as well as tellurium for hyperstoichiometric UO_2. They also found that iodine always forms CsI. These and the above-mentioned results by Grimes et al. should be confirmed by state-of-the-art calculations, in particular taking into account various defect charges.

Thompson (42) studied the incorporation of a series of noble gases in UO_2 and again showed the importance of Schottky defects as gas traps. The latest studies, however, take into account the effect of charge states on the incorporation. Crocombette (164) studied the incorporation of I, whereas Andersson (64) studied that of Xe, and Vathonne (76) the incorporation of Kr and Xe. They found that Kr and Xe are insoluble as interstitials in the O sublattice, but exhibit low incorporation energies in the Schottky defect and above all in U_2O_n vacancy clusters with n between 1 and 4.

A study of the behaviour of xenon and krypton implanted in UO_2 by a combination of XAS experiments and DFT + U computations was performed by Bès et al. (165) and Martin et al. (166), respectively. The XAS results for Xe (165), interpreted using the calculation results, showed that after implantation and annealing at 873 K, Xe is mainly located on a mix of uranium vacancies, interstitial sites, and Schottky defects, as well as in pressurised bubbles, but not in oxygen vacancies. After annealing at 1673 K, Xe atoms diffuse and coalesce to form highly pressurised bubbles, with the remaining dissolved Xe confined in Schottky defects. The DFT + U calculations showed that the Schottky defect oriented along the $\langle 100 \rangle$ direction is the energetically most favourable solution site for Xe. Similar results were obtained for Kr (166): before thermal annealing, the Kr atoms are incorporated in the UO_2 lattice as single atoms inside a neutral bound Schottky defect with O vacancies aligned along the $\langle 100 \rangle$ direction. After thermal treatment at 1273 K, the authors found that precipitation of dense Kr nano-aggregates occurred, which are most probably solid at room temperature, with a significant fraction of the gas atoms ($26 \pm 2\%$ for Kr) remaining in Schottky defects.

6.4.3.2 Fission Product Diffusion, Coalescence, and Precipitation

Mobility of the fission product atoms in the fuel matrix is the first step to their release, and the diffusion inside the fuel must thus be understood. Post-irradiation examinations of the fuel clearly demonstrate that the fission gases Xe and Kr, as well as the

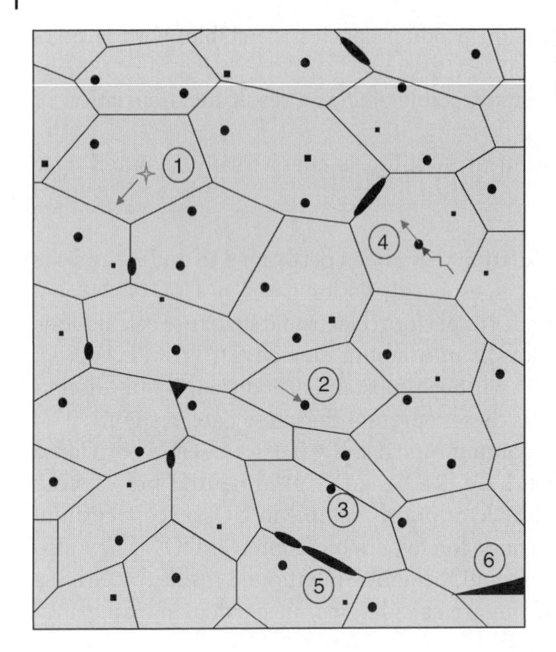

Figure 6.15 A schematic representation of the various steps in the fission gas release (After reference (1)).

volatiles I, Te, and Cs, are the most mobile elements, and as a result the investigations have focused primarily on the gas behaviour. Fission gas release is a complex process that describes the path of a fission product from its position of creation, via its mobility in the fuel, to its escape to the free volume of the fuel pin. The following stages must be distinguished (Figure 6.15):

1) Atomic diffusion of the fission product in the lattice and along grain boundaries, thermally activated and/or enhanced by radiation.
2) Capture in intragranular and intergranular cavities or bubbles that might be fabrication pores or newly formed (nanosized) cavities or bubbles that often occur along fission tracks.
3) Migration of bubbles to grain boundaries, induced by radiation and thermally assisted (at high temperatures).
4) Resolution (or re-injection) of gas from the bubbles into the fuel matrix, induced by radiation.
5) Coalescence of closed gas pores along the grain boundaries.
6) Formation of grain edge tunnels (open porosity channels) that constitute an intergranular network through which the gases can be vented to the free volume of the fuel pin.

As mentioned earlier, at the atomistic level, diffusion is a stochastic process involving jumping of atoms between vacant and/or interstitial lattice sites. However, at the temperatures in the fuel pellet, about 1300 K in the centre and 600 K at the pellet edge in a LWR, the kinetic energy of the particles is in most cases too low for chemical diffusion or thermal diffusion to occur, except in a limited part of the central zone. However, the excess of point defects created in the fission spikes will result in an effective lowering of the jump energy barrier and thus to a radiation-enhanced mobility, also called athermal

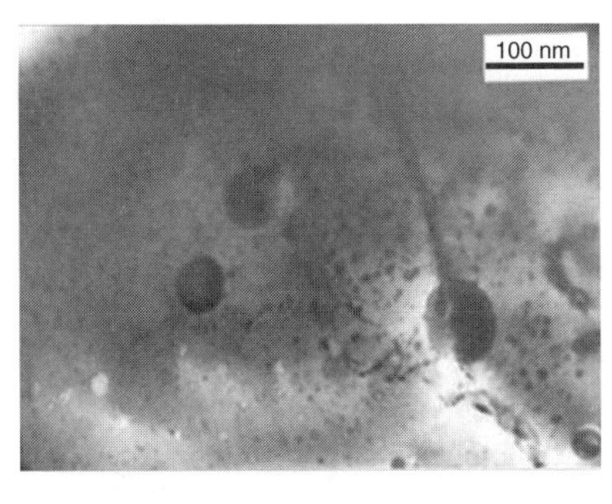

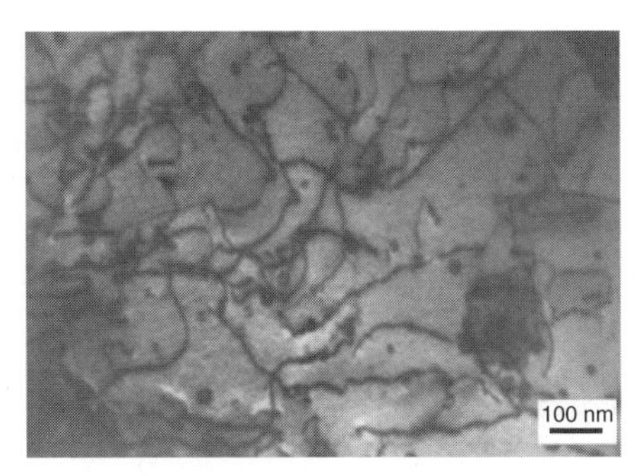

Figure 6.16 TEM micrographs of UO_2 fuel irradiated to high burnup. Left: Metallic fission product precipitates (large dark spots) and intragranular fission gas bubbles (small spots). Right: Network of dislocation lines. © European Communities, reproduced with permission.

diffusion. The elements with the highest effective diffusion coefficients will move within the grains until they are trapped at extended defects, voids, or grain boundaries, where they can aggregate to form intra- or intergranular inclusions.

As discussed by Turnbull et al. (167), the in-pile diffusion behaviour of fission gases/products can thus be divided into three regimes: (1) the atomic or lattice diffusion, which is strongly temperature dependent, (2) the athermal diffusion, which is directly related to fission spikes in the fuel, and (3) an intermediate regime in which the thermal diffusion is enhanced as a result of increased cation vacancy concentration caused by radiation (Figure 6.16).

6.4.3.2.1 Thermal and Athermal Diffusion

Post-irradiation examination of irradiated LWR fuel shows that the fission gases form intragranular bubbles in the central region (see, for example, Figure 6.16), and also at the grain boundaries in the form of lenticular bubbles that are sometimes interconnected, suggesting substantial diffusion in this part of the fuel (168). In transient-tested fuels, which have experienced an even higher temperature, bubble coalescence occurs, and bubbles surrounded by a cloud of smaller bubbles are observed by TEM. In the colder outer part of the fuel pellets, the evidence for fission gas mobility is less evident. Turnbull et al. (167) studied the gas release from irradiated single-crystal and polycrystalline UO_2 and concluded that diffusion of noble gas atoms in UO_2 during irradiation is a complex process involving more than one rate-controlling mechanism and bears a close similarity to cation self-diffusion. They also found that the low-temperature kinetics of noble gas diffusion are radically affected by irradiation damage. From their results, Turnbull et al. (167) derived values of the diffusion coefficients of Xe and Kr in UO_2 for the temperature range 535–1673 K, showing atomic diffusion in the lattice in the higher-temperature range and gas diffusion assisted by uranium vacancy mobility between 973 and 1473 K.

The thermal diffusion of Xe was investigated experimentally using ion implantation by several authors, as discussed in detail by Anderson et al. (169) in the light of the DFT + U and empirical potential calculations they performed for stoichiometric and non-stoichiometric UO_2 to simulate the experiments. The diffusion coefficients obtained as a function of non-stoichiometry are presented in Table 6.2, showing a good agreement between experiment and computations. The diffusion of Kr was studied in

Table 6.2 Measured and lowest calculated Xe activation energies (in eV) for intrinsic diffusion in $UO_{2\pm x}$. The experimental conditions (crucible, gas) are listed below each reference value; from Andersson et al. (169) and Bertolus et al. (174).

	E_a/eV		
	UO_{2-x}	UO_2	UO_{2+x}
Conditions		**Calculated**	
O_2			0.80
H_2		2.93	
W		3.94	
Mo		3.92	
Ta	6.39		
		Experimental	
Davies (175)		3.4[a]	
Cornell (176)		3.95 ± 0.61[b]	
Miekeley (177)	6.0 ± 0.1[c]	3.9 ± 0.4[d]	1.7 ± 0.4[e]
Kaimal (178)		2.87[f]	

a Mo crucible in H_2 gas flow.
b Unknown crucible in H_2 gas flow.
c Ta Knudsen cell in vacuum.
d W crucible in vacuum.
e Pt crucible in O_2 gas flow.
f Unknown crucible in He gas flow.

much less detail (see Reference (170)): only two studies give Arrhenius diffusion equations for Kr in bulk UO_2: one by Auskern from 1960 (171) and a more recent study by Michel et al. (172). The diffusion coefficients obtained are, however, very different: prefactors between 4.9×10^{-8} and 6.7×10^{-17} $m^2.s^{-1}$ and activation energies of 1.4 and 3.2 eV.

Thompson (173) investigated the pathway and energetics of xenon migration in uranium dioxide using a combination of DFT + U and empirical potential calculations. They considered the influence of Schottky defects and tetravacancies and found a path for the diffusion of xenon assisted by the tetravacancy, which has a significantly lower energy barrier that previously reported paths by nearly 1 eV. Andersson et al. (64, 169) and Vathonne (48) extended these calculations by performing systematic DFT + U/empirical potential MD studies of the vacancy-assisted diffusion of xenon and krypton, respectively, in uranium dioxide (see Figure 6.17) and considering charged defects. They determined the most favourable mechanism (trap involved) and the resulting diffusion coefficient as a function of non-stoichiometry. Depending on the extent of non-stoichiometry, the Schottky defect, the U-O bivacancy, or the U monovacancy assist Xe and Kr diffusion. For Xe, the variation of the diffusion coefficient as a function of non-stoichiometry compares very favourably with the experimental data, as shown in Table 6.2.

Iodine diffusion in UO_2 has been studied by Hocking et al. (179) using ion implantation and thermal annealing. They found that the diffusion is influenced by subtle

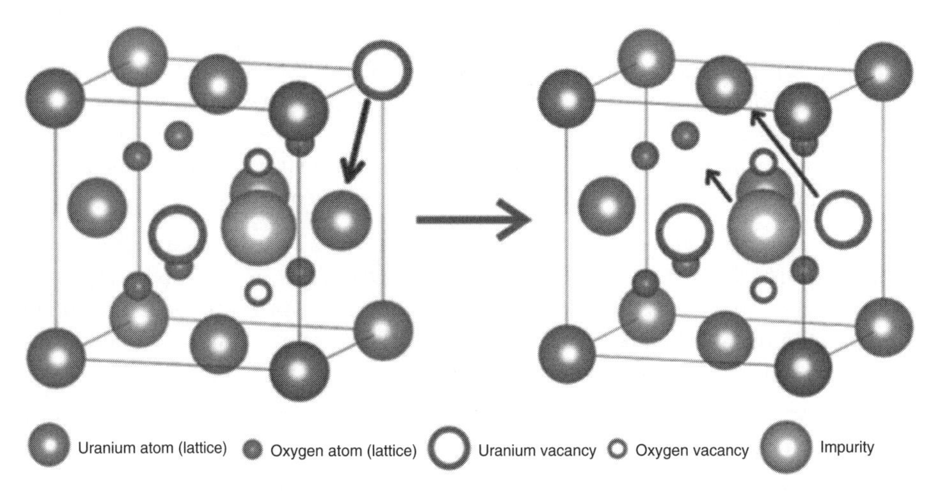

Figure 6.17 Mechanism for migration of Kr or Xe trapped in a Schottky trivacancy defect: jump of a second U vacancy (open circles) from second to first nearest neighbour of the impurity and relocation of the impurity in the middle of the vacancy cluster.

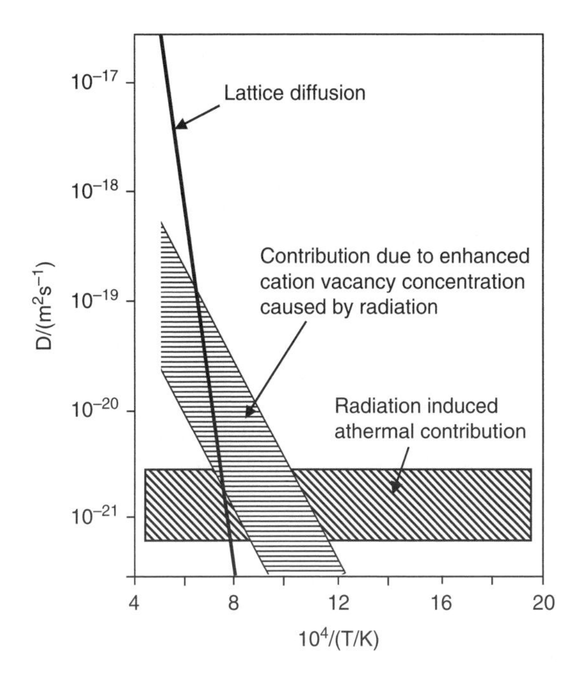

Figure 6.18 A schematic representation of the diffusion regimes for fission gases [after Turnbull JA, Friskney CA, Findlay JR, Johnson FA, Walter AJ. The diffusion-coefficients of gaseous and volatile species during the irradiation of uranium-dioxide. *J Nucl Mater*. 1982;107(2–3):168–184].

radiation damage effects and observed a pronounced concentration dependence of the thermal diffusion. At concentrations above $\sim 10^{16}$ I atoms/cm^3, a large fraction of the iodine became trapped, likely in microscopic bubbles. From the low-fluence data, the authors derived the thermal diffusion coefficients for iodine in polycrystalline UO_2 by a comparison with simulations conducted with the SRIM (Stopping and Range of Ions in Matter) software package.

6.4.3.3 Fission Gas Resolution

Gas re-solution from gas bubbles as a result of the impact of fission fragments that traverse the bubble has been suggested on the basis of post-irradiation examinations (180), but the exact re-solution mechanisms cannot be studied experimentally and can only be deduced from simulations. As discussed in the review by Olander and Wongsawaeng (181), two re-solution mechanisms must be considered:

- The heterogeneous process in which entire bubbles in the path of fission fragments are destroyed and the gas is re-injected into the matrix of the fuel as individual atoms.
- The homogeneous process in which fission gas atoms re-dissolve individually into the matrix of the fuel by singular binary scattering collisions with fission fragments and uranium recoils whose paths intersect the bubbles.

Parfitt and Grimes (182) used classical MD using a pair potential to predict the minimum energy (E_{min}) needed for krypton and xenon atoms to be resolved into uranium dioxide across a perfect (111) surface. They found that E_{min} is 53 eV for Kr and 56 eV for Xe, which is significantly less than the 300 eV value proposed by Nelson (183), who suggested a step function that assumes that the probability of a gas atom achieving re-solution is 100% above 300 eV, whereas the probability is zero below it. Parfitt and Grimes (182) defined a more representative measure, $E_{1/2}$, as the energy at which the re-solution exceeds 50% probability and found the corresponding values as 130 eV for Kr and 152 eV for Xe atoms. Nelson's value $E_{min} = 300$ eV, was found to correspond to a probability of 85%, in reasonable agreement with the 100% proposition.

Schwen and coworkers (184) simulated gas re-solution from fission gas bubbles by fission fragments also using MD combined with a pair potential, considering both homogeneous and heterogeneous re-solution mechanisms. They found that the calculated homogeneous re-solution rate is about 50 times lower than that obtained with an analytical approach (185). For ions with electronic stopping powers lower than 34 keV/nm, no bubble re-solution is observed. On the basis of this, they concluded that bubble re-solution due to the electronic stopping of fission fragments in UO_2 is insignificant compared to homogeneous re-solution. Finally, Govers et al. (150) performed atomic-scale MD simulations still using pair potentials of gas re-solution from bubbles in UO_2 and confirmed that low-energy interactions (PKA) are not effective in the re-solution process. However, the high-energy interactions destroy smaller bubbles and bring a quasi-constant number of gas atoms in re-solution when they interact with larger bubbles.

6.4.4 Helium Behaviour

The helium that is formed in the alpha-decay process[‡] is a noble gas that is inert in the fuel, similar to the fission gases Xe and Kr. It can thus contribute to the gas-induced swelling of the fuel pellets. However, He has a small atomic size, and its solubility and diffusion are substantially different from those of Kr and Xe. The behaviour of helium is particularly important for $(U,Pu)O_2$ mixed oxide fuel containing ^{241}Am as a decay

[‡] Helium is also generated in small quantities by ternary fission of the actinides. It is also the filling gas of the nuclear fuel pins, and it can diffuse into the fuel matrix under the reactor operating conditions.

Table 6.3 Helium solubility in UO_2.

		Particle size	Experimental conditions		Solubility
Authors	Material[a]	/μm	T/K	p/bar	$cm^3(STP)/(gUO_2atom)$
Belle (186)	Powder	0.088	973	1	2.28×10^{-5}
	Powder	0.16	1073	1	2.1×10^{-5}
Hasko and Szwarc (187)	Fused		1073	110	1.02×10^{-5}
Rufeh et al. (188)	Powder	4	1473	100	6.7×10^{-4}
	Powder	4	1573	100	4.2×10^{-4}
Sung (189)	SC/particle	10	1473	68	2.4×10^{-5}
Nakajima et al. (190)	SC/particle	18	1473	900	4.35×10^{-5}
Talip et al. (191)	SC/disc		1500	987	7.47×10^{-7}

a SC, single crystal.

product of ^{241}Pu as substantial quantities are generated (154), and for long-term storage of spent fuel.

The solubility of helium in UO_2 and the related solution energy have been subject of numerous studies. In these studies, helium is normally infused under high pressure into a heated sample during a sufficiently long period to attain equilibrium, after which the quantity of helium in the material is analysed, either by dissolution or thermal desorption. The results are scattered, as summarised in Table 6.3, which is not surprising in view of the very low He quantities that can be dissolved in UO_2 and the dependence of the O/U ratio and grain structure.

Olander (192, 193) studied helium solubility in uranium dioxide using a statistical-mechanical model assuming that He behaves as a harmonic oscillator in an octahedral interstitial site. His results are in the same range as the experimental results published by Rufeh (188) at the same time. Only during the last 15 years have atomistic simulations been employed to study the behaviour of helium in UO_2. Yakub (194) investigated helium incorporation in $UO_{2\pm x}$ as a function of non-stoichiometry and its temperature dependence using empirical potentials. The results reproduce most of the experimental results very well, with the exception of those by Talip et al. (191).

Three favourable sites can be envisaged for helium impurities in the crystalline lattice of UO_2: the octahedral interstitial site, the oxygen vacancy, and the uranium vacancy. The experimental results by Garrido et al. (195), who studied single-crystal UO_2 implanted at room temperature with ^3He using channelling techniques, give clear indications that the octahedral interstitial site is the preferential site for helium. Belhabib et al. (196) extended this work to the effect of temperature and found that the octahedral interstitial site remains the principal incorporation site for helium at 873 K.

Several atomistic computations have tried to shed light on this, but the results are not conclusive. DFT and DFT + U studies of the incorporation of He in UO_2 have been published by Petit et al. (197), Freyss et al. (60), Yun et al. (63, 198), Gryaznov et al. (199), and Dabrowski et al. (200). The incorporation energies obtained by the various authors

Table 6.4 Incorporation energies of He in the octahedral interstitial site ($\square_i$), the oxygen vacancy (V_O), and the uranium vacancy (V_U) in UO_2 (in eV).

	Method[a]	He-$\square_i$	He-V_O	He-V_U
Petit et al. (197)	DFT/LDA	1.3	1.8	0.2
Freyss et al. (60)	DFT/LDA	-0.1	2.4	0.4
Yun et al. (63)	DFT/GGA	2.5	3.2	1.8
Gryaznov et al. (199)	DFT + U/LDA	1.24	—	—
	DFT + U/GGA	0.81	—	—
Drabowski et al. (200)	DFT + U/GGA	4.15	0.51	0.18–032
Govers et al. (206)	EP	-0.1	-0.23	-0.2

a DFT, density functional theory; EP, empirical potentials.

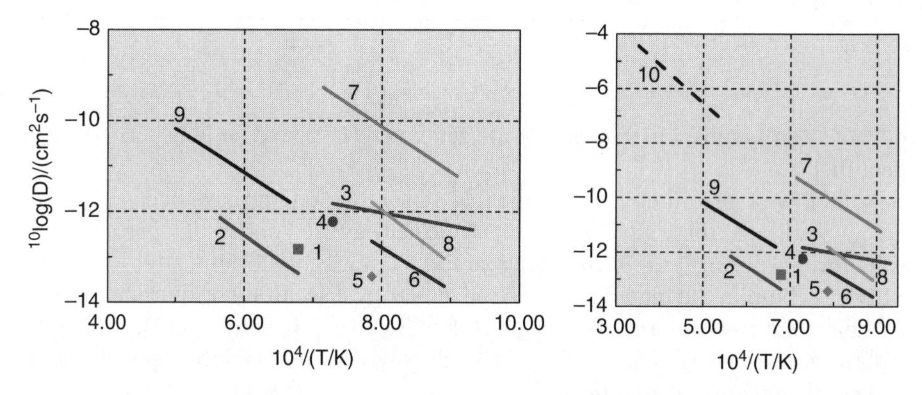

Figure 6.19 The helium diffusion coefficient in UO_2. Curves 1 to 9 show the experimental results. 1, Rufeh (188); 2, Sung (189); 3, Martin et al. (146); 4, Trocellier (209); 5, Guilbert et al. (210), 6, Roudil et al. (207); 7, Ronchi and Hiernaut (208), 8, Pipon et al. (211); 9, Nakajima et al. (190). Note that curve 7 refers to a $(U_{0.9}Pu_{0.1})O_2$ sample. Curve 10 shows the results from the MD simulations by Yakub (212). (*See insert for color representation of the figure.*)

are very scattered (see Table 6.4). This is due to the fact that (1) in several cases, metastable states were reached, as is exemplified by the metallic state observed by Gryaznov et al. (199), (2) He forms weak dispersive bonds, which are very difficult to describe using LDA and GGA functionals (201–203), and further calculations using functionals (204, 205) developed recently to treat dispersive bonds are needed. The incorporation energies obtained by Govers et al. (206) using empirical potentials are again different and show that new potentials must also be parameterised.

Helium diffusion has been studied by numerous authors, and a variety of sample types (implanted, infused, α-doped) and analytical techniques have been used. The results show appreciable scatter, as shown in Figure 6.19, though the majority of the data are grouped around the consistent curves by Nakajima et al. (190) and Roudil et al. (207) that were obtained from implanted and infused samples, respectively. The results by Ronchi and Hiernaut (208) for an alpha-doped $(U_{0.9}Pu_{0.1})O_2$ sample deviate considerably.

Martin et al. (146) and Garcia et al. (213) studied the mobility of helium in polycrystalline uranium dioxide using ion implantation with ^{3}He ions to a concentration of 0.1 at.% in the implanted region and subsequent heating at temperatures between 973 K and 1373 K. They showed that grain boundaries act as effective conduits for helium movement and release at all temperatures. At temperatures above approximately 1073 K, it was found that helium diffusion was high in areas around the grain boundaries, extending into the grain over distances of the order of microns. In areas further into the grain, diffusion proceeds much more slowly, presumably as a result of helium cluster formation. In a subsequent study, Garcia et al. (214) showed that helium behaviour is qualitatively independent of ion fluency over three orders of magnitude. Helium diffusion was found to be very slow below 1273 K within the centre of the grains, presumably as a result of helium bubble formation.

Martin et al. (215) studied the helium migration in polycrystalline uranium dioxide samples that were irradiated with 8 MeV iodine prior to the helium implantation. The results showed that for temperatures up to 1173 K, helium precipitation is enhanced in irradiated samples because of the presence of irradiation-induced defects. At temperatures above 1273 K, the precipitated helium is partly returned to the matrix and preferentially released in regions adjacent to grain boundaries. The authors suggested that grain boundaries act as defect sinks, leading to lower defect concentrations around grain boundaries and higher helium diffusivity in those regions since less defects act as traps for migrating atoms.

Govers et al. (206) performed MD simulations based on empirical pair potentials on helium diffusion in uranium dioxide as a function of stoichiometry and temperature. The results indicated two diffusion regimes with different activation energies, of which the low activation energy (0.5 eV) migration is assisted by oxygen vacancies, and provides the major contribution to diffusion when structural defects are present (extrinsic defects, imposed, e.g., by the stoichiometry). The second regime has a higher activation energy, around 2 eV, and dominates in the higher-temperature range or at perfect stoichiometry, suggesting an intrinsic migration process.

Yakub (194, 212) calculated the diffusion coefficients of helium in solid UO_{2+x} at high temperature using MD simulations based on a partly ionic model in conjunction with a polaron 'Free Hopping Approximation'. The results are shown in Figure 6.19. They are close to the diffusion constants found by Govers et al. (206), but are higher than the general trend observed in the experimental results, except for the results by Ronchi and Hiernaut (208). Yakub et al. (212) found a strong dependence of the apparent diffusion activation energy on the deviation from stoichiometry, which is a possible explanation for the large scatter observed experimentally.

6.4.5 Grain Boundary Effects

As discussed in the previous sections, grain boundaries (GBs) play an important role in the mechanisms governing defect, gas, and fission product behaviour. At high burnup this becomes particularly important for LWR fuel, as the rim zone of the pellet undergoes a substantial structural change. The original grains of 5–10 μm size transform into a structure of nanometre-sized grains with micron-sized pores, the so-called high-burnup structure (HBS) (216). Advanced LWR fuel designs employ large-grained fuels (obtained through doping, for example, with chromium) with the goal of reducing/delaying the fission gas release.

Experimental evidence for the effects of grain size and thus the grain boundary density has been presented by Turnbull (217), who observed significant lower fission gas release and mechanical deformation (swelling) in large-grain fuel (40 μm) compared to small-grain fuel (7 μm), both irradiated at 2023 K to 0.4% burnup. This was explained by the fact that the diffusion path is longer in large grains, and that the gas release to the GBs and the concomitant formation of bubbles and bubble tunnels on the grain edges is delayed. Hastings (218) studied the fission gas release from UO_2 fuel with 8, 15, and 80 μm grain size and found that this size effect on release is dependent on the linear heat rate (temperature) of the fuel. Since grain growth was not found significant at a linear power of 50 kW/m, a large initial grain size reduced the fission gas release. However, above 50 kW/m, the results were inconclusive, and a large initial grain size appeared less effective.

Sabioni et al. studied the effect of GBs on oxygen and uranium self-diffusion in UO_2 (85). They found that between 1773 and 1973 K in H_2 atmosphere, uranium diffusion in UO_2 GBs is approximately five orders of magnitude greater than uranium volume diffusion, showing that GBs constitute fast pathways for uranium diffusion in polycrystalline UO_2. On the contrary, the oxygen diffusion coefficients measured in polycrystalline and single-crystalline UO_2 between 878 and 1023 K in $H_2/N_2/H_2O$ atmosphere are similar and correspond to the volume diffusion. Arima et al. (73) studied the uranium and oxygen diffusion in GBs using MD (pair potentials) in combination with coincidence-site lattice theory for temperatures above 1600 K, exploring four grain boundary types. They found that grain boundary diffusion is much larger than bulk diffusion and that the diffusion coefficients increase and the activation energy decreased when the misorientation angle increases. For GBs with a large misorientation angle, a liquid-like disordered phase was observed.

Vincent-Aublant et al. (219) and Govers and Verwerft (220) performed empirical potential simulations of self-diffusion and Xe diffusion in UO_2 in the presence of GBs (Figure 6.20). They reached similar conclusions. An acceleration of the diffusion near GBs was observed for uranium, oxygen, and xenon atoms. The highest diffusion coefficients were observed in the grain boundary core, a zone about one nanometre thick, even if the influence of the GBs on the diffusion continues over several nanometres. This acceleration was certainly due to the presence of point defects along the GBs. MD simulations by Chiang et al. (221) revealed that GBs tend to move towards voids when they are within a few nanometres of each other, and thus act as a sink for defects, absorbing free volume. The proposed mechanism for void dissolution is through vacancy migration along the grain boundary.

Van Brutzel et al. (222) simulated displacement cascades in UO_2 bi-crystals and nano-polycrystals. They observed that a large part of the PKA energy is dissipated along the GBs and that fewer point defects are created in the grain than in monocrystalline UO_2. The grain boundary type, and more precisely, the nature of pre-existing defects along the GBs, is very important for the cascade evolution and determines if the cascade crosses the grain boundary or not.

Finally, the effect of an increased grain boundary density due to the formation of the HBS on the thermal conductivity was studied by Bai et al. (223). These authors used meso-scale simulations of the steady-state 2D heat conduction of small-grain high-burnup and large-grain un-restructured microstructures with Xe gas in the bubbles, using input from atomistic simulations (224) for the effect of Xe dissolved in the fuel

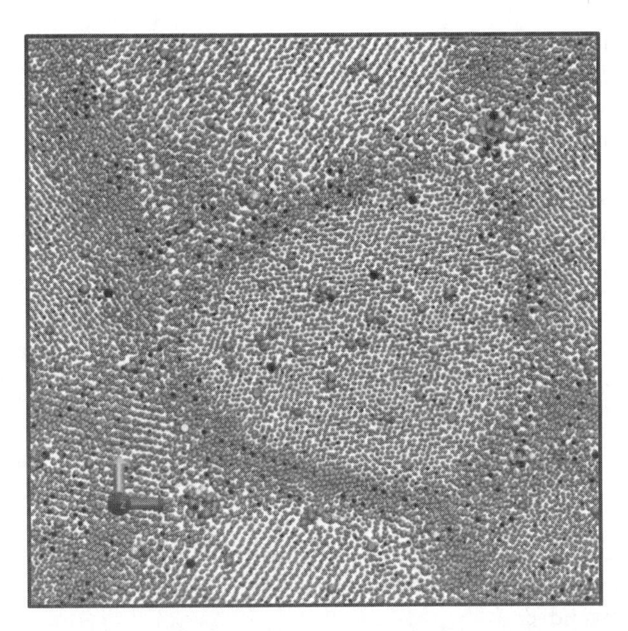

Figure 6.20 Simulation of a 2-nm-thick slab of polycrystalline UO_2 with xenon atoms inserted randomly with a concentration of 1 at.%, through substitution of uranium atoms (both in the bulk and at grain boundaries) obtained by MD optimisation (at 3000 K, after 1 ns) by Govers and Verwerft (220). Xenon clusters are formed at grain boundaries, composed mostly of grain xenon atoms initially at grain boundaries (yellow) also of bulk atoms (orange). For further information, the reader is referred to the original publication. Reproduced with permission.

matrix. The calculations showed that the interior of the HBS grains is free of dispersed Xe atoms as a result of the high density of GBs, while in un-restructured microstructures a small amount of dispersed Xe atoms remained in the grain interior. The latter effect was shown to have a larger impact on the thermal conductivity than the porosity differences. This is in line with the experimental observation for high-burnup irradiated fuel by Walker et al. (225), who found that the formation of the HBS has a very positive effect on the thermal conductivity in the outer region of the fuel, which was attributed to the removal of fission product atoms and radiation defects from the fuel lattice during recrystallisation of the fuel grains resulting from the HBS formation. Walker et al. concluded that the role of the pores of the HBS as sinks for fission gas expelled from the fuel lattice during recrystallisation is more important than their effect as barriers to heat transfer.

6.5 Concluding Remarks

During more than 50 years, a wide range of experimental techniques and modelling methods have been applied to unravel the complexity of UO_2, the principal nuclear fuel material currently in use. The result is a fairly good understanding of the thermal and equilibrium transport properties of UO_2 in the solid and liquid states, with perhaps the very high-temperature range as an exception, which can be understood from the difficulties of performing experiments and calculations for these extreme conditions.

At the lowest scale, the DFT method combined with a Hubbard-like correction (DFT + U) now achieves a good agreement with experimental observations, and key properties of UO_2 are well reproduced. The MD technique combined with empirical potentials has also achieved a maturity that allows obtaining qualitative results, in fair agreement with experiments.

Both DFT and empirical potential calculations have helped elucidate the atomistic mechanisms influencing the physico-chemical properties of UO_2, complementing the experimental work:

- The crystal and magnetic structure at low temperature, particularly the noncollinear antiferromagnetic order
- The defect chemistry, giving insight into the defect energetics
- The mechanisms of self-diffusion of oxygen and uranium
- The mechanisms of gas diffusion
- The effects of the thermally induced defect changes on heat capacity and thermal conductivity

For $(U,Pu)O_2$, relatively few calculations have been performed, but the results are encouraging, although the experimental basis to validate the atomistic calculations is much smaller, and needs to be extended.

A good qualitative knowledge exists of the behaviour of nuclear fuel under irradiation. Experimental examinations of irradiated fuels have helped identify the main effects of irradiation, such as the behaviour of fission products and helium, including their diffusion, coalescence, and release, as well as the microstructural evolution, and many of these processes are described in fuel performance codes through correlations or mechanistic models that are valid in the range of the experimental data (2).

However, to obtain a generic description with a high predictive reliability, further studies are needed to understand and describe the detailed mechanisms occurring at the atomistic and mesoscale. In particular, the following topics need further attention:

- The diffusion of fission products and its enhancement by irradiation
- The evolution of the microstructure through the accumulation of point and extended defects, eventually leading to the formation of the HBS
- The effect of GBs on the fission product behaviour
- Behaviour of the non-gaseous fission products in the UO_2 matrix
- The mechanisms governing the evolution of mechanical properties under irradiation

The main challenges for future studies are the design of experiments and (atomistic) simulations that are complementary, that is, serving the same purpose of understanding the underlying mechanisms of the nuclear fuel behaviour. This, of course, implies that the experiments must focus on the microscopic and sub-microscopic scale, and thus on techniques that allow such examinations of highly radioactive (Pu containing) and irradiated samples. In this context, it should be stressed that the very detailed knowledge that exists on UO_2 from investigations carried out during several decades must now be extended to $(U,Pu)O_2$. Microscopic and spectroscopic techniques are well suited for that purpose. However, one should not forget that the experimental studies also need to pay greater attention to very basic issues such as strict control of sample composition, through the use of well-defined conditions, in particular, to control the oxygen-to-metal ratio of the samples, which strongly influences the properties of oxide fuel as we have discussed in this chapter.

One of the main challenges for modelling is confirming the reliability of the atomic-scale methods, especially for the energetic properties that are the main input data for the higher-scale models. Concerning DFT + U, recent functionals must be used, and finite-temperature properties as well as entropy contributions must be addressed for statistically representative simulation cells. Empirical potentials are complementary methods and are highly useful for the study of the dynamic behaviour during irradiation. Potentials beyond the rigid-ion ion and 2-body approximations must be further developed and applied.

Acknowledgements

The authors would like to thank Roberto Caciuffo, Thierry Wiss (JRC Karlsruhe), and Michel Freyss (CEA Cadarache) for fruitful discussions.

References

1 Konings RJM, Wiss T, Gueneau C. Nuclear Fuels. In: Morss LR, Fuger J, Edelstein NM, editors. *The Chemistry of the Actinide and Transactinide Elements*. Volume 6. Netherlands: Springer; 2010. p. 3665–3812.
2 Van Uffelen P, Konings RJM, Vitanza C, Tulenko JS. Analysis of Reactor Fuel Rod Behavior. In: Cacuci DG, editor. *Handbook of Nuclear Engineering*. Boston, MA: Springer US; 2010. p. 1519–1627.
3 Brandt OG, Walker CT. Ultrasonic attenuation and elastic constants of uranium dioxide. *Phys Rev*. 1968;**170**(2):528.
4 Faber J, Lander GH, Cooper BR. Neutron-diffraction study of Uo-2 – Observation of an internal distortion. *Phys Rev Lett*. 1975;**35**(26):1770–1773.
5 Caciuffo R, Amoretti G, Santini P, Lander GH, Kulda J, Du Plessis PD. Magnetic excitations and dynamical Jahn-Teller distortions in UO_2. *Phys Rev B*. 1999;**59**(21):13892–13900.
6 Blackburn E, Caciuffo R, Magnani N, Santini P, Brown PJ, Enderle M, et al. Spherical neutron spin polarimetry of anisotropic magnetic fluctuations in UO_2. *Phys Rev B*. 2005;**72**(18):184411.
7 Burlet P, Rossatmignod J, Quezel S, Vogt O, Spirlet JC, Rebizant J. Neutron-diffraction on actinides. *J Less-Common Met*. 1986;**121**:121–139.
8 Santini P, Carretta S, Amoretti G, Caciuffo R, Magnani N, Lander GH. Multipolar interactions in f-electron systems: The paradigm of actinide dioxides. *Rev Mod Phys*. 2009;**81**(2):807–863.
9 Dorado B, Jomard G, Freyss M, Bertolus M. Stability of oxygen point defects in UO_2 by first-principles DFT plus U calculations: Occupation matrix control and Jahn-Teller distortion. *Phys Rev B*. 2010;**82**(3):035114.
10 Browning P. Thermodynamic properties of uranium-dioxide – A study of the experimental enthalpy and specific-heat. *J Nucl Mater*. 1981;**98**(3):345–356.
11 Clausen K, Hayes W, Macdonald JE, Osborn R, Hutchings MT. Observation of oxygen Frenkel disorder in uranium-dioxide above 2000-K by use of neutron-scattering techniques. *Phys Rev Lett*. 1984;**52**(14):1238–1241.

12 Hutchings MT, Clausen K, Dickens MH, Hayes W, Kjems JK, Schnabel PG, et al. Investigation of thermally induced anion disorder in fluorites using neutron-scattering techniques. *J Phys C Solid State*. 1984;**17**(22):3903–3940.
13 Hutchings MT. High-temperature studies of UO_2 and ThO_2 using neutron-scattering techniques. *J Chem Soc Farad T* 2. 1987;**83**:1083–103.
14 Ronchi C, Hyland GJ. Analysis of recent measurements of the heat-capacity of uranium-dioxide. *J Alloy Compd*. 1994;**213**:159–168.
15 Skinner LB, Benmore CJ, Weber JKR, Williamson MA, Tamalonis A, Hebden A, et al. Molten uranium dioxide structure and dynamics. *Science*. 2014;**346**(6212):984–987.
16 Brewer L. Energies of electronic configurations of singly, doubly, and triply ionized lanthanides and actinides. *J Opt Soc Am*. 1971;**61**(12):1666.
17 Sindzingre P, Gillan MJ. A molecular-dynamics study of solid and liquid UO_2. *J Phys C Solid State*. 1988;**21**(22):4017–4031.
18 Kurosaki K, Yamada K, Uno M, Yamanaka S, Yamamoto K, Namekawa T. Molecular dynamics study of mixed oxide fuel. *J Nucl Mater*. 2001;**294**(1–2):160–167.
19 Yakub E, Ronchi C, Staicu D. Molecular dynamics simulation of premelting and melting phase transitions in stoichiometric uranium dioxide. *J Chem Phys*. 2007;**127**(9):094508.
20 Cooper MWD, Murphy ST, Fossati PCM, Rushton MJD, Grimes RW. Thermophysical and anion diffusion properties of (U-x,Th1-x)O-2. *Proc Roy Soc A* 2014;**470**(2171):20140427.
21 Cooper MWD, Rushton MJD, Grimes RW. A many-body potential approach to modelling the thermomechanical properties of actinide oxides. *J Phys-Condens Mat*. 2014;**26**(10):105401.
22 Amoretti G, Blaise A, Caciuffo R, Fournier JM, Hutchings MT, Osborn R, et al. 5f-Electron states in uranium-dioxide investigated using high-resolution neutron spectroscopy. *Phys Rev B*. 1989;**40**(3):1856–1870.
23 Nakotte H, Rajaram R, Kern S, McQueeney RJ, Lander GH, Robinson RA. Crystal fields in UO_2 – revisited. *J Phys: Conference Series*. 2010;**251**(1):012002.
24 Liu XY, Andersson DA, Uberuaga BP. First-principles DFT modeling of nuclear fuel materials. *J Mater Sci*. 2012;**47**(21):7367–7384.
25 Dorado B, Freyss M, Amadon B, Bertolus M, Jomard G, Garcia P. Advances in first-principles modelling of point defects in UO_2: f electron correlations and the issue of local energy minima. *J Phys-Condens Mat*. 2013;**25**(33):333201.
26 Bertolus M, Krack M, Freyss M, Devanathan R. Assessment of Current Atomic Scale Modelling Methods for the Investigation of Nuclear Fuels under Irradiation: Example of Uranium Dioxide; 2015. Contract No.: NEA/NSC/R/(2015)5.
27 Yamazaki T, Kotani A. Systematic Analysis of 4f core photoemission spectra in actinide oxides. *J Phys Soc Jpn*. 1991;**60**(1):49–52.
28 Kotani A, Yamazaki T. Systematic analysis of core photoemission spectra for actinide di-oxides and rare-earth sesqui-oxides. *Prog Theor Phys Supp*. 1992(**108**):117–131.
29 Amadon B, Applencourt T, Bruneval F. Screened Coulomb interaction calculations: cRPA implementation and applications to dynamical screening and self-consistency in uranium dioxide and cerium. *Phys Rev B*. 2014;**89**(12):125110.
30 Shick AB, Pickett WE, Liechtenstein AI. Ground and metastable states in gamma-Ce from correlated band theory. *J Electron Spectrosc*. 2001;**114**:753–758.
31 Amadon B, Biermann S, Georges A, Aryasetiawan F. The alpha-gamma transition of cerium is entropy driven. *Phys Rev Lett*. 2006;**96**(6):066402.

32 Larson P, Lambrecht WRL, Chantis A, van Schilfgaarde M. Electronic structure of rare-earth nitrides using the LSDA plus U approach: Importance of allowing 4f orbitals to break the cubic crystal symmetry. *Phys Rev B.* 2007;**75**(4):045114.
33 Jomard G, Amadon B, Bottin F, Torrent M. Structural, thermodynamic, and electronic properties of plutonium oxides from first principles. *Phys Rev B.* 2008;**78**(7):075125.
34 Dorado B, Amadon B, Freyss M, Bertolus M. DFT plus U calculations of the ground state and metastable states of uranium dioxide. *Phys Rev B.* 2009;**79**(23):235125.
35 Huntzicker JJ, Westrum EF. The magnetic transition, heat capacity, and thermodynamic properties of uranium dioxide from 5 to 350 K. *J Chem Thermodynamics.* 1971; **3**(1):61–76.
36 Roy LE, Durakiewicz T, Martin RL, Peralta JE, Scuseria GE, Olson CG, et al. Dispersion in the Mott insulator UO_2: A comparison of photoemission spectroscopy and screened hybrid density functional theory. *J Comput Chem.* 2008;**29**(13):2288–2294.
37 Faber J, Lander GH. Neutron-diffraction study of UO_2 – Antiferromagnetic state. *Phys Rev B.* 1976;**14**(3):1151–1164.
38 Dzyaloshinskii IE. Magnetic-structure of UO_2. *Commun Phys.* 1977;**2**(3):69–71.
39 Ikushima K, Tsutsui S, Haga Y, Yasuoka H, Walstedt RE, Masaki NM, et al. First-order phase transition in UO_2: U-235 and O-17 NMR study. *Phys Rev B.* 2001;**63**(10): 269–278.
40 Laskowski R, Madsen GKH, Blaha P, Schwarz K. Magnetic structure and electric-field gradients of uranium dioxide: An ab initio study. *Phys Rev B.* 2004;**69**(14):140408(R).
41 Gryaznov D, Heifets E, Sedmidubsky D. Density functional theory calculations on magnetic properties of actinide compounds. *Phys Chem Chem Phys.* 2010;**12**(38):12273–12278.
42 Thompson AE, Wolverton C. First-principles study of noble gas impurities and defects in UO_2. *Phys Rev B.* 2011;**84**(13):134111.
43 Catlow CRA. Point-defect and electronic properties of uranium-dioxide. *P Roy Soc Lond a Mat.* 1977;**353**(1675):533–561.
44 Catlow CRA. Fission-gas diffusion in uranium-dioxide. *P Roy Soc Lond a Mat.* 1978;**364**(1719):473–497.
45 Grimes RW, Catlow CRA. The stability of fission-products in uranium-dioxide. *Philos T R Soc A.* 1991;**335**(1639):609–634.
46 Nerikar P, Watanabe T, Tulenko JS, Phillpot SR, Sinnott SB. Energetics of intrinsic point defects in uranium dioxide from electronic-structure calculations. *J Nucl Mater.* 2009;**384**(1):61–69.
47 Crocombette JP, Torumba D, Chartier A. Charge states of point defects in uranium oxide calculated with a local hybrid functional for correlated electrons. *Phys Rev B.* 2011;**83**(18):184107.
48 Vathonne E, Wiktor J, Freyss M, Jomard G, Bertolus M. DFT + U investigation of charged point defects and clusters in UO_2 (vol. 26, 325501, 2014). *J Phys-Condens Mat.* 2014;**26**(34):325501.
49 Willis BTM. Positions of oxygen atoms in $UO_{2.13}$. *Nature.* 1963;**197**(486):755–756.
50 Willis BTM. Crystallographic studies of anion-excess uranium-oxides. *J Chem Soc Farad T* 2. 1987;**83**:1073–1081.
51 Brincat NA, Molinari M, Parker SC, Allen GC, Storr MT. Computer simulation of defect clusters in UO_2 and their dependence on composition. *J Nucl Mater.* 2015;**456**:329–333.

52 Yakub E, Ronchi C, Staicu D. Computer simulation of defects formation and equilibrium in non-stoichiometric uranium dioxide. *J Nucl Mater.* 2009;**389**(1):119–126.
53 Ngayam-Happy R, Krack M, Pautz A. Effects of stoichiometry on the defect clustering in uranium dioxide. *J Phys-Condens Mat.* 2015;**27**(45):455401.
54 Matzke H. Diffusion-processes in nuclear-fuels. *J Less-Common Met.* 1986;**121**:537–564.
55 Matzke H. Atomic transport-properties in UO_2 and mixed oxides (U, Pu)O_2. *J Chem Soc Farad T* 2. 1987;**83**:1121–1142.
56 Dorado B, Garcia P. First-principles DFT plus U modeling of actinide-based alloys: Application to paramagnetic phases of UO_2 and (U,Pu) mixed oxides. *Phys Rev B.* 2013;**87**(19):195139.
57 Murch GE, Catlow CRA. Oxygen Diffusion in UO_2, ThO_2 and PuO_2 – A review. *J Chem Soc Farad T* 2. 1987;**83**:1157–1169.
58 Konings RJM, Beneš O. The heat capacity of NpO_2 at high temperatures: The effect of oxygen Frenkel pair formation. *J Phys Chem Solids.* 2013;**74**(5):653–655.
59 Crocombette JP, Jollet F, Nga LN, Petit T. Plane-wave pseudopotential study of point defects in uranium dioxide. *Phys Rev B.* 2001;**64**(10):104107.
60 Freyss M, Petit T, Crocombette JP. Point defects in uranium dioxide: Ab initio pseudopotential approach in the generalized gradient approximation. *J Nucl Mater.* 2005;**347**(1–2):44–51.
61 Gupta F, Brillant G, Pasturel A. Correlation effects and energetics of point defects in uranium dioxide: A first principle investigation. *Philos Mag.* 2007;**87**(16–17):2561–2569.
62 Yu JG, Devanathan R, Weber WJ. First-principles study of defects and phase transition in UO_2. *J Phys-Condens Mat.* 2009;**21**(43):435401.
63 Yun Y, Oppeneer PM, Kim H, Park K. Defect energetics and Xe diffusion in UO_2 and ThO_2. *Acta Mater.* 2009;**57**(5):1655–1659.
64 Andersson DA, Uberuaga BP, Nerikar PV, Unal C, Stanek CR. U and Xe transport in $UO_{2\pm x}$: Density functional theory calculations. *Phys Rev B.* 2011;**84**(5):054105.
65 Jonsson H, Mills G, Jacobsen KW. Nudged elastic band method for finding minimum energy paths of transitions. *Classical and Quantum Dynamics in Condensed Phase Simulations*: World Scientific; 1998. p. 385–404.
66 Henkelman G, Jónsson H. Improved tangent estimate in the nudged elastic band method for finding minimum energy paths and saddle points. *J. Chem. Phys.* 2000;**113**:9901–9904.
67 Mehrer H. *Diffusion in Solids*: Springer-Verlag Berlin Heidelberg; 2007. 654 p.
68 Auskern AB, Belle J. Oxygen ion self-diffusion in uranium dioxide. *J Nucl Mater.* 1961;**3**(3):267–276.
69 Belle J. Oxygen and uranium diffusion in uranium dioxide (a review). *J Nucl Mater.* 1969;**30**(1–2):3–15.
70 Marin JF, Contamin P. Uranium and oxygen self-diffusion in UO_2. *J Nucl Mater.* 1969;**30**(1–2):16–25.
71 Hadari Z, Kroupp M, Wolfson Y. Self-Diffusion measurement of oxygen in UO_2 by nuclear reaction O-18(p,Gamma)F-19. *J Appl Phys.* 1971;**42**(2):534–535.
72 Dorado B, Garcia P, Carlot G, Davoisne C, Fraczkiewicz M, Pasquet B, et al. First-principles calculation and experimental study of oxygen diffusion in uranium dioxide. *Phys Rev B.* 2011;**83**(3):035126.

73 Arima T, Yoshida K, Idemitsu K, Inagaki Y, Sato I. Molecular dynamics analysis of diffusion of uranium and oxygen ions in uranium dioxide. IOP Conference Series: *Mater Sci Eng.* 2010;**9**(1):012003.

74 Cooper MWD, Murphy ST, Rushton MJD, Grimes RW. Thermophysical properties and oxygen transport in the $(U_xPu_{1-x}O_{2-x})$O-2 lattice. *J Nucl Mater.* 2015;**461**:206–214.

75 Kim KC, Olander DR. Oxygen diffusion in UO_{2-x}. *J Nucl Mater.* 1981;**102**(1–2): 192–199.

76 Vathonne E. Study by Electronic Structure Calculation of Radiation Damage in UO_2 Nuclear Fuel : Behavior of the Point Defect and Fission Gases. University Aix-Marseille; 2014.

77 Bai XM, El-Azab A, Yu JG, Allen TR. Migration mechanisms of oxygen interstitial clusters in UO_2. *J Phys-Condens Mat.* 2013;**25**(1):015003.

78 Dorado B, Andersson DA, Stanek CR, Bertolus M, Uberuaga BP, Martin G, et al. First-principles calculations of uranium diffusion in uranium dioxide. *Phys Rev B.* 2012;**86**(3):035110.

79 Auskern AB, Belle J. Uranium ion self-diffusion in UO_2. *J Nucl Mater.* 1961;**3**(3):311–319.

80 Yajima S, Furuya H, Hirai T. Lattice and grain-boundary diffusion of uranium in UO_2. *J Nucl Mater.* 1966;**20**(2):162 – 170.

81 Matzke H. On uranium self-diffusion in UO_2 and UO_{2-x}. *J Nucl Mater.* 1969;**30**(1–2): 26 – 35.

82 Reimann DK, Lundy TS. Diffusion of ^{233}U in UO_2. *J Am Ceram Soc.* 1969;**52**(9): 511 – 512.

83 Matzke H. Lattice disorder and metal self-diffusion in non-stoichiometric UO_2 and (U, Pu)O_2. *J Phys Colloques.* 1973;**34**(C9):C9-317-C9-25.

84 Sabioni ACS, Ferraz WB, Millot F. First study of uranium self-diffusion in UO_2 by SIMS. *J Nucl Mater.* 1998;**257**(2):180–184.

85 Sabioni ACS, Ferraz WB, Millot F. Effect of grain-boundaries on uranium and oxygen diffusion in polycrystalline UO_2. *J Nucl Mater.* 2000;**278**(2–3):364–369.

86 Boyarchenkov AS, Potashnikov SI, Nekrasov KA, Kupryazhkin AY. Investigation of cation self-diffusion mechanisms in $UO_{2\pm x}$ using molecular dynamics. *J Nucl Mater.* 2013;**442**(1–3):148–161.

87 Dolling G, Cowley RA, Woods ADB. Crystal dynamics of uranium dioxide. *Can J Phys.* 1965;**43**(8):1397.

88 Yun Y, Legut D, Oppeneer PM. Phonon spectrum, thermal expansion and heat capacity of UO_2 from first-principles. *J Nucl Mater.* 2012;**426**(1–3):109–114.

89 Pang JWL, Buyers WJL, Chernatynskiy A, Lumsden MD, Larson BC, Phillpot SR. Phonon lifetime investigation of anharmonicity and thermal conductivity of UO_2 by neutron scattering and theory. *Phys Rev Lett.* 2013;**110**(15):157401.

90 Pang JWL, Chernatynskiy A, Larson BC, Buyers WJL, Abernathy DL, McClellan KJ, et al. Phonon density of states and anharmonicity of UO_2. *Phys Rev B.* 2014;**89**(11):115132.

91 Martin DG. The Thermal-expansion of solid UO_2 and (U, Pu) mixed oxides – A review and recommendations. *J Nucl Mater.* 1988;**152**(2–3):94–101.

92 Fink JK. Thermophysical properties of uranium dioxide. *J Nucl Mater.* 2000;**279**(1): 1–18.

93 Guthrie M, Benmore CJ, Skinner LB, Alderman OLG, Weber JKR, Parise JB, et al. Thermal expansion in UO_2 determined by high-energy X-ray diffraction. *J Nucl Mater.* 2016;**479**:19–22.

94 Willis BTM. Neutron diffraction studies of actinide oxides .2. Thermal motions of atoms in uranium dioxide and thorium dioxide between room temperature and 1100 degrees C. *Proc R Soc Lon Ser-A*. 1963;**274**(1356):134.
95 Willis BTM, Hazell RG. Re-Analysis of single-crystal neutron-diffraction data on UO_2 using 3rd cumulants. *Acta Crystallogr A*. 1980;**36**(Jul):582–584.
96 Valu SO, Beneš O, Manara D, Konings RJM, Cooper MWD, Grimes RW, et al. The high-temperature heat capacity of the $(Th,U)O_2$ and $(U,Pu)O_2$ solid solutions. *J Nucl Mater.* 2017;**484**:1–6.
97 Hiernaut JP, Hyland GJ, Ronchi C. Premelting transition in uranium-dioxide. *Int J Thermophys*. 1993;**14**(2):259–283.
98 Szwarc R. Defect contribution to excess enthalpy of uranium dioxide-calculation of Frenkel energy. *J Phys Chem Solids*. 1969;**30**(3):705.
99 Konings RJM, Beneš O, Kovacs A, Manara D, Sedmidubsky D, Gorokhov L, et al. The thermodynamic properties of the f-elements and their compounds. Part 2. The lanthanide and actinide oxides. *J Phys Chem Ref Data*. 2014;**43**(1):013101.
100 Gueneau C, Chartier A, Van Brutzel L. Thermodynamic and Thermophysical Properties of the Actinide Oxides. In: Konings RJM, editor. *Comprehensive Nuclear Materials*. 2012, Volume 2. p. 21–59.
101 International Atomic Energy Agency. Thermophysical Properties Database of Materials for Light Water Reactors and Heavy Water Reactors. Vienna: International Atomic Energy Agency; 2006.
102 Harding JH, Martin DG. A recommendation for the thermal-conductivity of UO_2. *J Nucl Mater*. 1989;**166**(3):223–226.
103 Eucken A. Die Wärmeleitfähigkeit keramischer feuerfester Stoffe; Ihre Berechnung aus der Wärmeleitfähigkeit der Betstandteile. *Forsch Gebiete Ingenieurw*. 1932;**B3**(353):16.
104 Bakker K, Konings RJM. The influence of complex pore shapes on thermal conductivity. *Nucl Technol*. 1996;**115**(1):91–99.
105 Uchida A, Sunaoshi T, Kato M, Konashi K. Thermal Properties of UO_2 by Molecular Dynamics simulation. *Prog Nucl Sci Technol.* 2011;**2**:598–602.
106 Nichenko S, Staicu D. Molecular Dynamics study of the mixed oxide fuel thermal conductivity. *J Nucl Mater*. 2013;**439**(1–3):93–98.
107 Staicu D, Wiss T, Rondinella VV, Hiernaut JP, Konings RJM, Ronchi C. Impact of auto-irradiation on the thermophysical properties of oxide nuclear reactor fuels. *J Nucl Mater*. 2010;**397**(1–3):8–18.
108 Arima T, Idemitsu K, Inagaki Y, Tsujita Y, Kinoshita M, Yakub E. Evaluation of melting point of UO_2 by molecular dynamics simulation. *J Nucl Mater*. 2009;**389**(1): 149–154.
109 Böhler R, Welland MJ, Prieur D, Cakir P, Vitova T, Pruessmann T, et al. Recent advances in the study of the UO_2-PuO_2 phase diagram at high temperatures. *J Nucl Mater*. 2014;**448**(1–3):330–339.
110 Ghosh PS, Kuganathan N, Galvin COT, Arya A, Dey GK, Dutta BK, et al. Melting behavior of $(Th,U)O_2$ and $(Th,Pu)O_2$ mixed oxides. *J Nucl Mater.* 2016;**479**:112–122.
111 Boyarchenkov AS, Potashnikov SI, Nekrasov KA, Kupryazhkin AY. Molecular dynamics simulation of UO_2 nanocrystals melting under isolated and periodic boundary conditions. *J Nucl Mater*. 2012;**427**(1–3):311–322.

112 Ronchi C, Hiernaut JP, Selfslag R, Hyland GJ. Laboratory measurement of the heat-capacity of urania up to 8000-K .1. experiment. *Nucl Sci Eng*. 1993;**113**(1):1–19.

113 Breitung W, Reil KO. The density and compressibility of liquid (U,Pu)-mixed oxide. *Nucl Sci Eng*. 1990;**105**(3):205–217.

114 Vigier JF, Martin PM, Martel L, Prieur D, Scheinost AC, Somers J. Structural Investigation of $(U_{0.7}Pu_{0.3})O_{2-x}$ Mixed Oxides. *Inorg Chem*. 2015;**54**(11):5358–5365.

115 Abe T, Asakura K. Uranium oxide and MOX production. In: Konings RJM, editor. *Comprehensive Nuclear Materials*. 2012, Volume 2. p. 393–422.

116 Ambegaokar V. Thermal resistance due to isotopes at high temperatures. *Phys Rev*. 1959;**114**(2):488–489.

117 Abeles B. Lattice thermal conductivity of disordered semiconductor alloys at high temperatures. *Phys Rev*. 1963;**131**(5):1906.

118 Fukushima S, Ohmichi T, Handa M. The effect of rare-earths on thermal-conductivity of uranium, plutonium and their mixed-oxide fuels. *J Less-Common Met*. 1986;**121**:631–636.

119 Gibby RL. Effect of plutonium content on thermal conductivity of $(U, Pu)O_2$ solid solutions. *J Nucl Mater*. 1971;**38**(2):163.

120 Schmidt HE. Progress Report No. 9. Karlsruhe: European Institute for Transuranium Elements; 1970.

121 Beauvy M. Nonideality of the solid-solution in $(U,Pu)O_2$ nuclear-fuels. *J Nucl Mater*. 1992;**188**:232–238.

122 Duriez C, Alessandri JP, Gervais T, Philipponneau Y. Thermal conductivity of hypostoichiometric low Pu content $(U,Pu)O_{2-x}$ mixed oxide. *J Nucl Mater*. 2000;**277**(2–3):143–158.

123 Yamada K, Kurosaki K, Uno M, Yamanaka S. Evaluation of thermal properties of mixed oxide fuel by molecular dynamics. *J Alloy Compd*. 2000;**307**:1–9.

124 Nichenko S, Staicu D. Molecular Dynamics study of the effects of non-stoichiometry and oxygen Frenkel pairs on the thermal conductivity of uranium dioxide. *J Nucl Mater*. 2013;**433**(1–3):297–304.

125 Philipponneau Y. Thermal-conductivity of $(U, Pu)O_{2-x}$ mixed-oxide fuel. *J Nucl Mater*. 1992;**188**:194–197.

126 Schmidt HE. Some considerations on thermal conductivity of stoichiometric uranium dioxide at high temperatures. *J Nucl Mater*. 1971;**39**(2):234–237

127 Mattys HM. Plutonium oxide as nuclear fuel. *Actin Rev*. 1968;**1**(2):165.

128 Lyon WL, Baily WE. Solid-liquid phase diagram for UO_2-PuO_2 system. *J Nucl Mater*. 1967;**22**(3):332.

129 Kato M, Morimoto K, Sugata H, Konashi K, Kashimura M, Abe T. Solidus and liquidus temperatures in the UO_2-PuO_2 system. *J Nucl Mater*. 2008;**373**(1–3): 237–245.

130 De Bruycker F, Boboridis K, Manara D, Poml P, Rini M, Konings RJM. Reassessing the melting temperature of PuO_2. *Mater Today*. 2010;**13**(11):52–55.

131 De Bruycker F, Boboridis K, Poml P, Eloirdi R, Konings RJM, Manara D. The melting behaviour of plutonium dioxide: A laser-heating study. *J Nucl Mater*. 2011;**416**(1–2):166–172.

132 Vauchy R, Robisson AC, Bienvenu P, Roure I, Hodaj F, Garcia P. Oxygen self-diffusion in polycrystalline uranium-plutonium mixed oxide $U_{0.55}Pu_{0.45}O_2$. *J Nucl Mater*. 2015;**467**:886–893.

133 Wiss T, Matzke H, Trautmann C, Toulemonde M, Klaumunzer S. Radiation damage in UO_2 by swift heavy ions. *Nucl Instrum Meth B.* 1997;**122**(3):583–588.

134 Blank H. Properties of fission spikes in UO_2 and UC due to electronic stopping power. *Physica Status Solidi A-Appl Res.* 1972;**10**(2):465.

135 Blank H, Matzke H. The effect of fission spikes on fission gas re-solution. *Radiation Eff.* 1973;**17**(1–2):57–64.

136 Brucklacher D, Dienst W. Creep behavior of ceramic nuclear fuels under neutron-irradiation. *J Nucl Mater.* 1972;**42**(3):285.

137 Ronchi C. Nature of surface fission tracks in UO_2. J Appl Phys. 1973;**44**(8):3575–3585.

138 Matzke H. Radiation enhanced diffusion in UO_2 and $(U,Pu)O_2$. *Radiat Eff Defect S.* 1983;**75**(1–4):317–325.

139 Matzke H. Radiation-damage effects in nuclear-materials. *Nucl Instrum Meth B.* 1988;**32**(1–4):455–470.

140 Matzke H. Radiation-damage in nuclear-materials. *Nucl Instrum Meth B.* 1992;**65**(1–4):30–39.

141 Sonoda T, Kinoshita M, Ishikawa N, Sataka M, Iwase A, Yasunaga K. Clarification of high density electronic excitation effects on the microstructural evolution in UO_2. *Nucl Instrum Meth B.* 2010;**268**(19):3277–3281.

142 He LF, Gupta M, Yablinsky CA, Gan J, Kirk MA, Bai XM, et al. In situ TEM observation of dislocation evolution in Kr-irradiated UO_2 single crystal. *J Nucl Mater.* 2013;**443**(1–3):71–77.

143 Onofri C, Sabathier C, Palancher H, Carlot G, Miro S, Serruys Y, et al. Evolution of extended defects in polycrystalline UO_2 under heavy ion irradiation: Combined TEM, XRD and Raman study. *Nucl Instrum Meth B.* 2016;**374**:51–57.

144 Van Brutzel L, Delaye JM, Ghaleb D, Rarivomanantsoa M. Molecular dynamics studies of displacement cascades in the uranium dioxide matrix. *Philos Mag.* 2003;**83**(36): 4083–4101.

145 Van Brutzel L, Rarivomanantsoa M, Ghaleb D. Displacement cascade initiated with the realistic energy of the recoil nucleus in UO_2 matrix by molecular dynamics simulation. *J Nucl Mater.* 2006;**354**(1–3):28–35.

146 Martin G, Garcia P, Sabathier C, Carlot G, Sauvage T, Desgardin P, et al. Helium release in uranium dioxide in relation to grain boundaries and free surfaces. *Nucl Instrum Meth B.* 2010;**268**(11–12):2133–2137.

147 Martin G, Garcia P, Sabathier C, Van Brutzel L, Dorado B, Garrido F, et al. Irradiation-induced heterogeneous nucleation in uranium dioxide. *Phys Lett A.* 2010;**374**(30):3038–3041.

148 Martin G, Garcia P, Van Brutzel L, Dorado B, Maillard S. Effect of the cascade energy on defect production in uranium dioxide. *Nucl Instrum Meth B.* 2011;**269**(14):1727–1730.

149 Huang M, Schwen D, Averback RS. Molecular dynamic simulation of fission fragment induced thermal spikes in UO_2: Sputtering and bubble re-solution. *J Nucl Mater.* 2010;**399**(2–3):175–180.

150 Govers K, Bishop CL, Parfitt DC, Lemehov SE, Verwerft M, Grimes RW. Molecular dynamics study of Xe bubble re-solution in UO_2. *J Nucl Mater.* 2012;**420**(1–3): 282–290.

151 Jeon B, Asta M, Valone SM, Gronbech-Jensen N. Simulation of ion-track ranges in uranium oxide. *Nucl Instrum Meth B.* 2010;**268**(17–18):2688–2693.

152 Saidy M, Hocking WH, Mouris JF, Garcia P, Carlot G, Pasquet B. Thermal diffusion of iodine in UO_2 and UO_{2+x}. *J Nucl Mater.* 2008;**372**(2–3):405–15.
153 Martin G, Garcia P, Sabathier C, Devynck F, Krack M, Maillard S. A thermal modelling of displacement cascades in uranium dioxide. *Nucl Instrum Meth B.* 2014;**327**:108–112.
154 Wiss T, Hiernaut JP, Roudil D, Colle JY, Maugeri E, Talip Z, et al. Evolution of spent nuclear fuel in dry storage conditions for millennia and beyond. *J Nucl Mater.* 2014;**451**(1–3):198–206.
155 Sattonnay G, Vincent L, Garrido F, Thome L. Xenon versus helium behavior in UO_2 single crystals: A TEM investigation. *J Nucl Mater.* 2006;**355**(1–3):131–135.
156 Van Brutzel L, Rarivomanantsoa M. Molecular dynamics simulation study of primary damage in UO_2 produced by cascade overlaps. *J Nucl Mater.* 2006;**358**(2–3):209–16.
157 Kleykamp H. The chemical-state of the fission-products in oxide fuels. *J Nucl Mater.* 1985;**131**(2–3):221–246.
158 Grimes RW, Ball RGJ, Catlow CRA. Site Preference and binding of iodine and cesium in uranium-dioxide. *J Phys Chem Solids.* 1992;**53**(4):475–484.
159 Petit T, Jomard G, Lemaignan C, Bigot B, Pasturel A. Location of krypton atoms in uranium dioxide. *J Nucl Mater.* 1999;**275**(1):119–23.
160 Crocombette JP. Ab initio energetics of some fission products (Kr, I, Cs, Sr and He) in uranium dioxide. *J Nucl Mater.* 2002;**305**(1):29–36.
161 Peterson KA, Figgen D, Goll E, Stoll H, Dolg M. Systematically convergent basis sets with relativistic pseudopotentials. II. Small-core pseudopotentials and correlation consistent basis sets for the post-d group 16-18 elements. *J Chem Phys.* 2003;**119**(21): 11113–11123.
162 Gupta F, Pasturel A, Brillant G. Ab initio study of solution energy and diffusion of caesium in uranium dioxide. *J Nucl Mater.* 2009;**385**(2):368–371.
163 Brillant G, Gupta F, Pasturel A. Fission products stability in uranium dioxide. *J Nucl Mater.* 2011;**412**(1):170–176.
164 Crocombette JP. First-principles study with charge effects of the incorporation of iodine in UO_2. *J Nucl Mater.* 2012;**429**(1–3):70–77.
165 Bes R, Martin P, Vathonne E, Delorme R, Sabathier C, Freyss M, et al. Experimental evidence of Xe incorporation in Schottky defects in UO_2. *Appl Phys Lett.* 2015;**106**(11):114102.
166 Martin PM, Vathonne E, Carlot G, Delorme R, Sabathier C, Freyss M, et al. Behavior of fission gases in nuclear fuel: XAS characterization of Kr in UO_2. *J Nucl Mater.* 2015;**466**:379–392.
167 Turnbull JA, Friskney CA, Findlay JR, Johnson FA, Walter AJ. The diffusion-coefficients of gaseous and volatile species during the irradiation of uranium-dioxide. *J Nucl Mater.* 1982;**107**(2–3):168–184.
168 Ray ILF, Matzke H. Fission Gas behaviour during power transients in high burn-up LWR nuclear fuels studied by electron microscopy. In: Donnelly SE, Evans JH, editors. *Fundamental Aspects of Inert Gases in Solids.* Nato Science Series B: Springer US; 1991. p. 457–466.
169 Andersson DA, Garcia P, Liu XY, Pastore G, Tonks M, Millett P, et al. Atomistic modeling of intrinsic and radiation-enhanced fission gas (Xe) diffusion in $UO_{2\pm x}$: Implications for nuclear fuel performance modeling. *J Nucl Mater.* 2014;**451**(1–3): 225–242.

170 Vathonne E, Andersson DA, Freyss M, Perriot R, Cooper MWD, Stanek CR, et al. Determination of krypton diffusion coefficients in uranium dioxide using atomic scale calculations. *Inorg Chem.* 2017;**56**(1):125–137.

171 Auskern AB. The Diffusion of Krypton-85 from Uranium Dioxide Powder. Bettis Atomic Power Laboratory; 1960. Contract No.: US Report WAPDTM-185.

172 Michel A. Etude De Comportement Des Gaz De Fission Dans Le Dioxyde D'uranium: Mecanismes De Diffusion, Nucleation Et Grossissement De Bulles. University of Caen; 2011.

173 Thompson AE, Wolverton C. Pathway and energetics of xenon migration in uranium dioxide. *Phys Rev B.* 2013;**87**(10):104105.

174 Bertolus M, Freyss M, Dorado B, Martin G, Hoang K, Maillard S, et al. Linking atomic and mesoscopic scales for the modelling of the transport properties of uranium dioxide under irradiation. *J Nucl Mater.* 2015;**462**:475–495.

175 Davies D, Long G. The Emission of Xenon-133 From Lightly Irradiated Uranium Dioxide Spheroids And Powders. United Kingdom; 1963 1963-06-15.

176 Cornell RM. Growth of fission gas bubbles in irradiated uranium dioxide. *Philos Mag.* 1969;**19**(159):539–554.

177 Miekeley W, Felix FW. Effect of stoichiometry on diffusion of xenon in UO_2. *J Nucl Mater.* 1972;**42**(3):297.

178 Kaimal KNG, Naik MC, Paul AR. Temperature-dependence of diffusivity of xenon in high-dose irradiated UO_2. *J Nucl Mater.* 1989;**168**(1–2):188–190.

179 Hocking WH, Verrall RA, Muir IJ. Migration behaviour of iodine in nuclear fuel. *J Nucl Mater.* 2001;**294**(1–2):45–52.

180 Whapham AD. Electron microscope observation of fission-gas bubble distribution in UO_2. *Nucl Appl.* 1966;**2**(2):123–130.

181 Olander DR, Wongsawaeng D. Re-solution of fission gas – A review: Part I. Intragranular bubbles. *J Nucl Mater.* 2006;**354**(1–3):94–109.

182 Parfitt DC, Grimes RW. Predicting the probability for fission gas resolution into uranium dioxide. *J Nucl Mater.* 2009;**392**(1):28–34.

183 Nelson RS. Influence of Irradiation on nucleation of gas bubbles in reactor fuels. *J Nucl Mater.* 1968;**25**(2):227.

184 Schwen D, Averback RS. Intragranular Xe bubble population evolution in UO_2: A first passage Monte Carlo simulation approach. *J Nucl Mater.* 2010;**402**(2–3): 116–23.

185 Nelson RS. Stability of Gas Bubbles in an irradiation environment. *J Nucl Mater.* 1969;**31**(2):153–161.

186 Belle J. Properties of uranium dioxide. *Proceedings of the Second United Nations International Conference on the Peaceful Uses of Atomic Energy.* 6. Geneva: United Nations; 1958. p. 569.

187 Hasko S, Szwarc R. Solubility and Diffusion of Helium in Uranium Dioxide. Washington: AEC Division of Reactor Development; 1963.

188 Rufeh F, Olander DR, Pigford TH. Solubility of helium in uranium dioxide. *Nucl Sci Eng.* 1965;**23**(4):335.

189 Sung P. Equilibrium Solubility and Diffusivity in Single-Crystal Uranium oxide. University of Washington; 1967.

190 Nakajima K, Serizawa H, Shirasu N, Haga Y, Arai Y. The solubility and diffusion coefficient of helium in uranium dioxide. *J Nucl Mater.* 2011;**419**(1–3):272–280.

191 Talip Z, Wiss T, Maugeri EA, Colle JY, Raison PE, Gilabert E, et al. Helium behaviour in stoichiometric and hyper-stoichiometric UO_2. *J Eur Ceram Soc.* 2014;**34**(5): 1265–1277.
192 Olander DR. Theory of helium dissolution in uranium dioxide. 1. Interatomic forces in uranium dioxide. *J Chem Phys.* 1965;**43**(3):779.
193 Olander DR. Theory helium dissolution in uranium dioxide. 2. Helium solubility. *J Chem Phys.* 1965;**43**(3):785.
194 Yakub E. Helium solubility in uranium dioxide from molecular dynamics simulations. *J Nucl Mater.* 2011;**414**(2):83–87.
195 Garrido F, Ibberson RM, Nowicki L, Willis BTM. Cuboctahedral oxygen clusters in U_3O_7. *J Nucl Mater.* 2003;**322**(1):87–89.
196 Belhabib T, Desgardin P, Sauvage T, Erramli H, Barthe MF, Garrido F, et al. Lattice location and annealing behaviour of helium atoms implanted in uranium dioxide single crystals. *J Nucl Mater.* 2015;**467**:1–8.
197 Petit T, Freyss M, Garcia P, Martin P, Ripert M, Crocombette JP, et al. Molecular modelling of transmutation fuels and targets. *J Nucl Mater.* 2003;**320**(1–2):133–137.
198 Yun Y, Eriksson O, Oppeneer PM. Theory of He trapping, diffusion, and clustering in UO_2. *J Nucl Mater.* 2009;**385**(3):510–516.
199 Gryaznov D, Heifets E, Kotomin E. Ab initio DFT plus U study of He atom incorporation into UO_2 crystals. *Phys Chem Chem Phys.* 2009;**11**(33):7241–7247.
200 Dabrowski L, Szuta M. Diffusion of helium in the perfect and non perfect uranium dioxide crystals and their local structures. *J Alloy Compd.* 2014;**615**:598–603.
201 Zhang YK, Pan W, Yang WT. Describing van der Waals Interaction in diatomic molecules with generalized gradient approximations: The role of the exchange functional. *J Chem Phys.* 1997;**107**(19):7921–7925.
202 Zhang R, Dinca A, Fisher KJ, Smith DR, Willett GD. Gas-phase ion-molecule reactions of metal-carbide cations MCn divided by (M = Y and La; n = 2, 4, and 6) with benzene and cyclohexane investigated by FTICR mass spectrometry and DFT calculations. *J Phys Chem A.* 2005;**109**(1):157–164.
203 van Mourik T, Gdanitz RJ. A critical note on density functional theory studies on rare-gas dimers. *J Chem Phys.* 2002;**116**(22):9620–9623.
204 Klimes J, Bowler DR, Michaelides A. A critical assessment of theoretical methods for finding reaction pathways and transition states of surface processes. *J Phys-Condens Mat.* 2010;**22**(7).
205 Klimes J, Bowler DR, Michaelides A. Van der Waals density functionals applied to solids. *Phys Rev B.* 2011;**83**(19):195131.
206 Govers K, Lemehov S, Hou M, Verwerft M. Molecular dynamics simulation of helium and oxygen diffusion in $UO_{2\pm x}$. *J Nucl Mater.* 2009;**395**(1–3):131–139.
207 Roudil D, Deschanels X, Trocellier P, Jegou C, Peuget S, Bart JM. Helium thermal diffusion in a uranium dioxide matrix. *J Nucl Mater.* 2004;**325**(2–3):148–158.
208 Ronchi C, Hiernaut JP. Helium diffusion in uranium and plutonium oxides. *J Nucl Mater.* 2004;**325**(1):1–12.
209 Trocellier P, Gosset D, Simeone D, Costantini JM, Deschanels X, Roudil D, et al. Application of nuclear reaction geometry for He-3 depth profiling in nuclear ceramics. *Nucl Instrum Meth B.* 2003;**206**:1077–1082.
210 Guilbert S, Sauvage T, Garcia P, Carlot G, Barthe MF, Desgardin P, et al. He migration in implanted UO_2 sintered disks. *J Nucl Mater.* 2004;**327**(2–3):88–96.

211 Pipon Y, Raepsaet C, Roudil D, Khodja H. The use of NRA to study thermal diffusion of helium in (U,Pu)O-2. *Nucl Instrum Meth B*. 2009;**267**(12–13):2250–2254.
212 Yakub E, Ronchi C, Staicu D. Diffusion of helium in non-stoichiometric uranium dioxide. *J Nucl Mater*. 2010;**400**(3):189–195.
213 Garcia P, Martin G, Desgardin P, Carlot G, Sauvage T, Sabathier C, et al. A study of helium mobility in polycrystalline uranium dioxide. *J Nucl Mater*. 2012;**430**(1–3): 156–165.
214 Garcia P, Gilabert E, Martin G, Carlot G, Sabathier C, Sauvage T, et al. Helium behaviour in UO_2 through low fluence ion implantation studies. *Nucl Instrum Meth B*. 2014;**327**:113–116.
215 Martin G, Sabathier C, Carlot G, Desgardin P, Raepsaet C, Sauvage T, et al. Irradiation damage effects on helium migration in sintered uranium dioxide. *Nucl Instrum Meth B*. 2012;**273**:122–126.
216 Rondinella VV, Wiss T. The high burn-up structure in nuclear fuel. *Mater Today*. 2010;**13**(12):24–32.
217 Turnbull JA. Effect of grain-size on swelling and gas release properties of UO_2 during irradiation. *J Nucl Mater*. 1974;**50**(1):62–68.
218 Hastings IJ. Effect of initial grain-size on fission-gas release from irradiated UO_2 fuel. *J Am Ceram Soc*. 1983;**66**(9):C150–151.
219 Vincent-Aublant E, Delaye JM, Van Brutzel L. Self-diffusion near symmetrical tilt grain boundaries in UO_2 matrix: A molecular dynamics simulation study. *J Nucl Mater*. 2009;**392**(1):114–120.
220 Govers K, Verwerft M. Classical molecular dynamics investigation of microstructure evolution and grain boundary diffusion in nano-polycrystalline UO_2. *J Nucl Mater*. 2013;**438**(1–3):134–143.
221 Chiang TW, Chernatynskiy A, Sinnott SB, Phillpot SR. Interaction between voids and grain boundaries in UO_2 by molecular-dynamics simulation. *J Nucl Mater*. 2014;**448**(1–3):53–61.
222 Van Brutzel L, Vincent-Aublant E, Delaye JM. Large molecular dynamics simulations of collision cascades in single-crystal, bi-crystal, and poly-crystal UO_2. *Nucl Instrum Meth B*. 2009;**267**(18):3013–3016.
223 Bai XM, Tonks MR, Zhang YF, Hales JD. Multiscale modeling of thermal conductivity of high burnup structures in UO_2 fuels. *J Nucl Mater.* 2016;**470**:208–215.
224 Tonks MR, Liu XY, Andersson D, Perez D, Chernatynskiy A, Pastore G, et al. Development of a multiscale thermal conductivity model for fission gas in UO_2. *J Nucl Mater.* 2016;**469**:89–98.
225 Walker CT, Staicu D, Sheindlin M, Papaioannou D, Goll W, Sontheimer F. On the thermal conductivity of UO_2 nuclear fuel at a high burn-up of around 100 MWd/kgHM. *J Nucl Mater*. 2006;**350**(1):19–39.

7

Ceramic Host Phases for Nuclear Waste Remediation

Gregory R. Lumpkin

Nuclear Fuel Cycle Research, Australian Nuclear Science and Technology Organisation, New South, Wales

7.1 Introduction

Large quantities of high-level waste (HLW) continue to be generated by various nuclear operations around the world, including spent fuel from commercial nuclear power stations, liquid wastes from the reprocessing of spent fuel, and wastes generated from the production of nuclear weapons. Existing weapons-grade plutonium has the potential to be recycled and used in mixed-oxide fuel (MOX), but the ultimate fate of plutonium must also be determined on the basis of the analysis of risks associated with proliferation. Recent developments, beginning formally in the year 2000, have seen a move toward more effective use of uranium-based fuels in programs that combine reprocessing, transmutation, and separations technology in advanced fuel cycles (e.g., Generation IV nuclear power systems), with the added benefit of major reductions in the "radiotoxicity" of nuclear waste. However, further rethinking of the future of nuclear energy resulted from the disaster that occurred in March of 2011 at the Fukushima Daichi Nuclear Power Plant in Japan [1]. This major accident has led to new research and development programs for accident-tolerant fuels designed to mitigate the properties of conventional UO_2 fuel and cladding in terms of the time–temperature window governing the release of potentially explosive hydrogen gas during an accident scenario. There are also major research efforts under way to solve some of the associated materials challenges in taking nuclear energy to the next level [2]. Importantly, it may also influence the decision-making process on nuclear waste form development and, more specifically, the options for direct disposal of spent fuel.

Indeed, this is an interesting period in the history of nuclear energy, and there is ongoing debate on the question of how to handle nuclear wastes in the longer term. At present, about 40% of countries with nuclear power have adopted the policy of direct disposal of spent fuel instead of reprocessing. Borosilicate glass is the currently accepted waste form of choice for many countries that reprocess their commercial spent fuel, but there exists a significant fraction of "legacy" wastes and other nuclear materials that are very complex in physical form and chemical composition (e.g., the Na, Al, and Zr rich wastes stored in tanks at sites in the United States). These complex waste materials,

Experimental and Theoretical Approaches to Actinide Chemistry, First Edition.
Edited by John K. Gibson and Wibe A. de Jong.

together with impure plutonium, and the separated fission products and actinides generated from the various partitioning strategies may be better suited for existing or new types of high-performance crystalline waste forms or glass-ceramics. Furthermore, inert matrix fuels (IMFs) represent an interesting concept for recycling of reactor-grade plutonium and minor actinides in commercial power stations followed by geological disposal [3]. This is an interesting crossover between fuels and waste forms and may be an attractive option for some countries.

Beginning in the 1970s, alternative crystalline ceramic nuclear waste forms were considered and tested, and it became clear that some of these concepts may be capable of providing a much higher level of chemical durability than borosilicate glass or spent fuel. Many of these materials have been extensively developed over the previous 20–25 years, while others are relatively new. Materials such as tailored ceramics [4], the Synroc titanate waste forms [5, 6], and related special-purpose waste forms are reasonably well developed and have been the subject of extensive leach testing and radiation damage studies. Pyrochlore is the major component of Synroc-F, a polyphase ceramic designed for partially reprocessed nuclear fuel [7] that later appeared as the principle host phase for excess weapons Pu and U in a crystalline titanate ceramic form. From the early polyphase Synroc formulations, zirconolite was designed as an ideal host phase for actinides due to a combination of crystal chemical flexibility and very high durability in aqueous fluids [8], and hollandite may provide an excellent host material for radioactive Cs for similar reasons [9]. Another Synroc host phase, perovskite, was originally included in formulations to incorporate radioactive Sr, but the aqueous durability of this phase is generally not as good as that of zirconolite and hollandite. Additional special-purpose waste forms for actinides include zircon [10], monazite [11], and zirconium-based materials having the fluorite, defect fluorite, or pyrochlore structures [12, 13]. Except for zircon, none of these materials have been studied to the same extent as the titanate waste forms, partly due to the ongoing research and development programs conducted in Australia since the 1980s. Nevertheless, monazite and Zr-based materials remain promising and are still under consideration today in view of their resistance to amorphization and excellent chemical durability. The ultimate choice of waste form materials will depend on a combination of factors including local political and industrial concerns together with a consideration of the economic and technical issues involved in their implementation.

7.2 Types of Ceramic Nuclear Waste Forms

A summary of the various types of ceramic nuclear waste forms that have been developed, to various levels, either as alternatives to borosilicate glass for HLW or as special-purpose materials for separated HLW or legacy wastes is given in Table 7.1. These materials have been designed with a number of criteria in mind [14], including compatibility with geological environments, high durability in aqueous fluids, thermodynamic stability over long periods of time, suitable physical properties (e.g., hardness, elastic modulus, and thermal conductivity), resistance to changes in properties due to radiation damage, and high waste loadings (minimum waste form volume). Together with the costs of developing the repository, the costs of the waste form engineering and production will make a major contribution to total investment in geological disposal of

Table 7.1 Summary of selected examples of ceramic nuclear waste forms proposed for the storage of a range of waste types containing actinides and other elements.

Waste form	Main phases	Application	Loading (wt.%)
Synroc-C	Zirconolite, hollandite, perovskite, rutile, metal alloys, minor oxides	Commercial HLW	15–25
Synroc-D	Zirconolite, perovskite, nepheline, ulvospinel, hercynite spinel	US defense wastes	60–70
Synroc-E	Rutile, hollandite, zirconolite, perovskite, pyrochlore, alloy	Commercial HLW	5–7
Synroc-F	Pyrochlore, hollandite, ± perovskite and UO_2	Spent fuel elements	~ 50
Supercalcine	Apatite, corundum, fluorite, spinel, monazite, pollucite	HLW from reprocessing	≤ 70
Tailored ceramics	Magnetoplumbite, zirconolite, Spinel, nepheline, UO_2	US defense wastes	~ 60
Pyrochlore	Pyrochlore, zirconolite, rutile, brannerite	Actinides, excess Pu stockpiles	~ 35
Zirconolite	Zirconolite, rutile	Actinides, excess Pu stockpiles	~ 25
Monazite	Monazite	Actinide-lanthanide wastes, Pu	~ 25
Zircon	Zircon	Excess Pu stockpiles	10
Cubic zirconia	Zirconia solid-solution	Actinide-lanthanide wastes, Pu	~10
Apatite	Apatite (phosphate and/or silicate)	Actinide-lanthanide wastes, Pu	<15
Kosnarite (NZP)	Kosnarite, minor phases	Cs, actinides, excess Pu stockpiles	<20
Garnet	Garnet (Fe-Al type, ± Si), minor phases	Actinide-lanthanide wastes, Pu	≤20
Murataite	Fluorite type (Ti, Ca, U, Ce, Zr, Mn) oxide and other phases	Actinide-lanthanidewastes, Pu	?
Crichtonite group	Crichtonite-loveringite-davidite type compositions	Actinide-lanthanide wastes, Sr	~20
AB_2O_6 oxides	Aeschynite-euxenite type phases	Actinide-lanthanide wastes	~20
Glass ceramics	Sphene (titanite), zirconolite, pyrochlore, perovskite, sodalite, nepheline, monazite, glass matrix	Legacy wastes, ILW, low actinide wastes	≤70

nuclear wastes. Increasing the waste loading, for example, of the waste form from around 20 wt% to 60 wt% can significantly reduce production costs and can potentially lower the costs of the repository itself.

Crystalline ceramic nuclear waste forms have been designed as polyphase materials for disposal of HLW from power plants [6] or as specialist materials for the encapsulation of actinides and fission products that would arise from the possible separation of HLW into various fractions or for Pu from dismantled nuclear weapons or other industrial processes. Although some waste forms are conceived as single-phase ceramics such as the pyrochlore-based materials designed circa 1997 for the US Plutonium Immobilization Project [13], the actual waste form may contain additional phases in the final product in order to meet a combination of practical requirements, including processing methods, incorporation of impurities, and criticality control [15, 16].

Ceramic materials like Supercalcine and Synroc represent the first generation of nuclear waste forms based on Th-U minerals and having the added capability of incorporating a wide range of fission products and other elements. Typical crystalline phases in Supercalcine formulations are silicate apatite, spinel, corundum, fluorite-type oxides, pollucite, and monazite. In Synroc-C, for example, zirconolite is designed to be the primary host phase for actinides, perovskite is the host phase for Sr, and hollandite is the host phase for Cs. Synroc in its various forms is perhaps the most studied ceramic waste form to date, with numerous reports and publications on the processing conditions, crystal chemistry, element partitioning, aqueous durability, and radiation damage effects [6,8,17 – 21].

In recent years, glass-ceramics have been studied extensively as possible waste forms that can be conveniently produced and encapsulated in metal containers, for example, by hot isostatic pressing (HIP) methods. The use of HIP technology provides highly densified materials within a considerable range of processing conditions. Extensive research and development has been conducted in this area, especially for legacy waste materials such as those stored in the United States (e.g., the calcined materials at Idaho National Laboratory). Sphene-glass ceramics were originally part of the Canadian program on nuclear waste forms in the 1980s to host the wastes arising from future CANDU fuel recycling [22]. In particular, the dissolution behavior of these ceramics [23] was studied with respect to the geochemical characteristics of the Canadian Shield as a possible site for the construction of the geological repository. The latest Synroc type materials produced by the Australian Nuclear Science and Technology Organisation are glass-ceramics consolidated using HIP technology. These materials have been developed with a range of applications including legacy wastes and plutonium immobilization. Notably, many of the proposed formulations include crystalline phases designed to incorporate actinides and other radionuclides. An overview of the some of the nuclear waste forms capable of incorporating actinides and fission products from nuclear waste is given in Table 7.1.

7.3 Radiation Damage Effects

In nuclear systems, the main sources of radiation damage include neutrons and alpha, beta, and gamma radiation emitted from radioactive elements. In oxide ceramics, neutrons create isolated damage in the form of Frenkel and other isolated defects as well as defect clusters, but the accumulation rate is generally insufficient to cause the material to become amorphous, especially at elevated temperatures. However, even if amorphization does not occur, there may be significant changes in the microstructure and properties of the materials [24, 25]. Alpha-decay events generally produce collision cascade damage from recoil atoms with typical energies of 80–100 keV and significant ionization and isolated defects or clusters from alpha particles with energies in the range of 4–5 MeV [26]. In the latter case, the defect clusters are found near the end of the alpha particle track, where the stopping power is predominantly nuclear. Oxide ceramics may become amorphous if they contain alpha-emitting elements, depending on the structure, composition, dose rate, and temperature [27].

There are a range of oxide crystals and ceramics with potential industrial applications whose radiation effects have been investigated. These materials include various MX, MX_2, M_2X_3, ABO_3, ABO_4, AB_2O_4, and $A_2B_2O_7$ compounds, among others. Some have also been investigated extensively for properties including electrical conductivity, ionic conductivity, and the piezoelectric effect, but a major gap exists in knowledge concerning details of how radiation damage might affect these properties. Radiation damage studies have mostly been conducted using light and heavy ion irradiation up to MeV energy levels, swift heavy ion irradiation to GeV levels, electron irradiation, and alpha particle irradiation. A few materials of interest in the nuclear fuel cycle have also been studied using actinide doping experiments with ^{238}Pu ($t_{1/2}$ = 88 years) or ^{244}Cm ($t_{1/2}$ = 18 years), but there is a large body of supporting information on radiation effects in related minerals containing long-lived uranium and thorium. It is well known that bonding and structural complexity are related to radiation tolerance in oxides, with those compounds having strong ionic bonding and simple structures tending to be more radiation tolerant than complex structures having a greater degree of covalent bonding [28–32]. Thus, the MX compounds with a rock salt structure (e.g., MgO, CaO, and SrO), MX_2 compounds with a fluorite structure (e.g., ZrO_2, CeO_2, and ThO_2), and the M_2X_3 compounds with a corundum structure (Al_2O_3, V_2O_3, Cr_2O_3, etc.) are radiation tolerant. In the following sections, we summarize some of the key findings in radiation damage of ceramics oxides using actinide doping, ion irradiation, and natural analogues, together with information obtained using atomistic simulation methods.

7.3.1 Actinide Doping Experiments

A number of important experiments using either ^{238}Pu or ^{244}Cm to induce short-term radiation damage ingrowth were conducted in the 1980s as part of nuclear waste form development programs in the United States. Researchers at Los Alamos National Laboratory [33] reported that $CaPuTi_2O_7$ ("cubic zirconolite" – in fact, this is actually pyrochlore) has a total volume expansion of 4.7%, lattice volume expansion of 2.2%, and a critical amorphization dose of 0.3×10^{16} α mg^{-1} on the basis of X-ray diffraction analysis. Further analysis of this compound showed on the basis of bulk swelling curves that $CaPuTi_2O_7$ exhibits a bulk volume expansion of 5.4% at the ambient temperature and becomes amorphous at a dose of 0.5×10^{16} α mg^{-1}. One of the most important aspects of this work, however, involved experimental runs performed at an elevated temperature (Figure 7.1). For example, samples held at 302 °C showed a bulk swelling of about 4.3% at saturation and the critical dose for amorphization was about 1×10^{16} α mg^{-1} as estimated from the swelling data. $CaPuTi_2O_7$ did not become amorphous when held at a temperature of 602 °C, but still exhibited a bulk volume swelling of 0.4% consistent with accumulation of point defects and residual, unannealed collision cascades [34]. Also during this time frame, similar studies were being conducted at Pacific Northwest National Laboratory on synthetic $Gd_2Ti_2O_7$ pyrochlore doped with 3 wt% ^{244}Cm [35]. This work reported an amorphization dose of approximately 0.4×10^{16} α mg^{-1}, a total volume expansion of ~ 5.1% at saturation, and an increase in fracture toughness together with a decrease in hardness and elastic modulus of the ceramic material. Similar pyrochlore-rich materials were also studied as part of the development program run by the

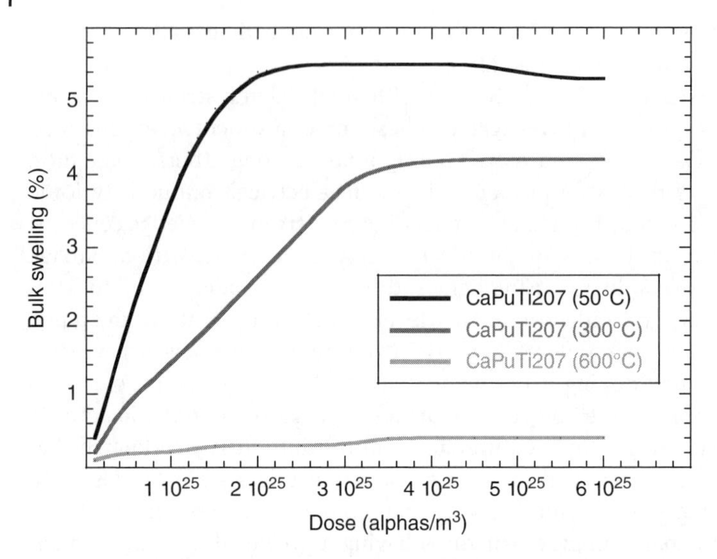

Figure 7.1 Effect of thermal annealing on radiation damage in $CaPuTi_2O_7$ produced by the decay of ^{238}Pu. Samples held at 50 °C and 300 °C become amorphous, but the saturation dose for swelling is increased, and the volume expansion is lower at the higher temperature. When held at 300 °C, the material does not become amorphous, and the bulk swelling saturates at about 0.4 vol% primarily due to the effect of retained Frenkel defects and alpha-recoil collision cascades in the crystalline lattice. Modified from Clinard, F.W., Jr., Peterson, D.E., Rohr, D.L., and Hobbs, L.W. (1984) Self-irradiation effects in 238Pu-substituted zirconolite, I: Temperature dependence of damage. *Journal of Nuclear Materials*, **126**, 245–254. (*See insert for color representation of the figure.*)

US Department of Energy for the disposal of excess weapons Pu beginning in the late 1990s. These materials have prototypical compositions close to $(Ca,Gd,Ce,Hf,U)_2Ti_2O_7$, wherein Gd and Hf were incorporated to provide criticality control. X-ray diffraction and bulk swelling measurements indicate that the critical dose for amorphization is $\sim 0.2–0.4 \times 10^{16}$ α mg^{-1}, and the material undergoes a total volume expansion of <6% with a unit cell expansion of 2.9%–4.7% [15].

The closely related synthetic $CaZrTi_2O_7$ zirconolite structure, which can be described as a pyrochlore-like structure that has been condensed on one set of (111) planes, becomes amorphous at a dose of approximately 0.5×10^{16} α mg^{-1} when doped with ~3 wt% ^{244}Cm and exhibits a total volume expansion of 6.0% at saturation. The crystalline-amorphous transformation in this material also leads to an increase in fracture toughness and a decrease in hardness and elastic modulus [35]. An earlier report on this material revealed anisotropic lattice expansion with increasing dose [36] wherein a increased by 0.3%, b increased initially by ~0.1% at low dose and very little thereafter, and c increased by 1.5%. This structural anisotropy, when weighed against the low total volume expansion (e.g., relative to zircon), is relatively inconsequential. In a similar study of synthetic $CaZrTi_2O_7$ zirconolite doped with ^{238}Pu [37], a total volume expansion of 5.5% and a critical amorphization dose of $\sim 0.5 \times 10^{16}$ α mg^{-1} was observed. Finally, in a detailed study of the effect of ^{238}Pu on the structure of three synthetic zirconolites that also contained Al, Gd, Hf, and U but no Zr, it was reported that X-ray diffraction and bulk swelling measurements of these samples indicated that the critical dose for amorphization is $\sim 0.3–0.5 \times 10^{16}$ α mg^{-1}, and the total volume expansion is ~5% [38].

Apart from the work on pyrochlore- and zirconolite-based ceramics described above, there are only a handful of other experiments conducted using short-lived actinides. In a very short-term actinide experiment, the perovskite compound $CmAlO_3$ (where 95% of the Cm is ^{244}Cm) was investigated by X-ray diffraction methods, showing that the material became amorphous after 8 days, which is equivalent to a dose of approximately $0.15–0.2 \times 10^{16}$ α mg^{-1} (~0.2–0.3 dpa) [39]. Unfortunately, this experiment was probably cut short in duration; however, the lattice parameter versus time data suggest that the lattice volume expansion, $\Delta V_c/V_{c0}$, may approach a value of around 9% at saturation. Interestingly, the volume expansion is closer to ~6% just before the material becomes amorphous, suggesting that there is continued modification of the structure of the amorphous state in this material. This is consistent with previous work reported above on actinide-doped zirconolite.

Among the ABO_4 compounds, zircon ($ZrSiO_4$) has probably received the most attention, partly due to the fact that it has natural analogues, and these are extremely important in the area of geological age dating (in which the effects of radiation damage on chemical and especially isotopic discordance need to be carefully considered). Laboratory experiments using ^{238}Pu-doped zircons revealed that the critical dose for amorphization is $\sim 1.0 \times 10^{16}$ α mg^{-1}, leading to a total volume expansion of approximately 16% [40, 41]. Apart from orthosilicates with the zircon structure, there has been considerable recent interest in ABO_4-type orthophosphate compounds with the zircon and monazite structures. For example, the zircon structure type $LuPO_4$ was doped with 1.0 wt% ^{244}Cm and stored for 18 years, reaching a cumulative dose of 5×10^{16} α mg^{-1}, but the material remained in a highly crystalline state apart from the presence of 5–10 nm defect clusters and 5–20 nm voids, which were assumed, probably correctly, to contain radiogenic He [42]. Monazite-type compounds have been studied in terms of the potential for actinide incorporation in phosphate ceramics and glass-ceramics. Experiments on synthetic $PuPO_4$ and (La,Pu)PO_4 samples doped with 7.2 and 8.1 wt% ^{238}Pu, respectively, revealed that $PuPO_4$ is heavily damaged at a dose of approximately 0.1×10^{16} α mg^{-1} and exhibits substantial volume swelling and cracking [43]. In comparison, (La,Pu)PO_4 was crystalline to a dose of $0.2–0.3 \times 10^{16}$ α mg^{-1} with a decrease in the intensity of the measured X-ray diffraction peaks, meaning that the amorphous state was not reached in this particular experiment. This may be evidence for either a dose rate effect or a compositional effect on the rate of damage recovery in these two compounds – an important issue with regard to the possible deployment of orthophospate-type waste form compounds.

Apatite, a complex phosphate-silicate compound, is also considered as a potential nuclear waste form material for actinides and fission products. For waste form applications, silicate apatites have been studied as they contain lanthanides substituting for part of the Ca, and O occupies the anion site normally occupied by monovalent species (e.g., OH, F, and Cl) in phosphate apatites. This avoids unfavorable chemical attributes of halogens and phosphorus (however, phosphate apatites may be attractive as host phases for radioactive ^{129}I). The synthetic britholite type apatite $CaNd_4(SiO_4)_3O$ doped with 1.2 wt% ^{244}Cm was found to become amorphous at a dose of $\sim 0.3 \times 10^{16}$ α mg^{-1}, and the crystalline-amorphous transformation resulted in a bulk volume expansion of 8.0%–8.5% [44, 45]. Kosnarite-type phosphates have also been studied to a limited extent, for example, $NaPu_2(PO_4)_3$ was synthesized containing ^{239}Pu or ^{238}Pu. After two years of storage, the ^{239}Pu sample accumulated a dose of 0.0091×10^{16} α mg^{-1} and remained

Table 7.2 Comparison of radiation damage data for nuclear waste form phases doped with ^{244}Cm or ^{238}Pu.

			Critical dose	Volume expansion	
Phase	Dopant	Structure	10^{16} α/mg	ΔV (total) %	ΔV (lattice) %
$Gd_2Ti_2O_7$	3 wt% ^{244}Cm	Pyrochlore	0.4	5.1	
LLNL PIP A	^{238}Pu	Pyrochlore	~0.3	<6	2.9–4.7
$CaZrTi_2O_7$	3 wt% ^{244}Cm	Zirconolite	0.5	6.0	>2.0
$CaZrTi_2O_7$	5 mol% ^{238}Pu	Zirconolite	~0.5	5.5	
LLNL PIP B	^{238}Pu	Zirconolite	~0.4	~5.0	
$CaPuTi_2O_7$	^{238}Pu	Pyrochlore	0.3	4.7	2.2
$CmAlO_3$	^{244}Cm	Perovskite	~0.3	>10?	~6
$CaNd_4(SiO_4)_3O$	1.2 wt% ^{244}Cm	Apatite	~0.3	~8	
$ZrSiO_4$	^{238}Pu	Zircon	6.7	16–18	5.1
$LuPO_4$	1.0 wt% ^{244}Cm	Zircon	>5		
$PuPO_4$	7.2 wt% ^{238}Pu	Monazite	> 0.1		
$LaPO_4$	8.1 wt% ^{238}Pu	Monazite	>0.3		
$NaPu_2(PO_4)_3$	^{238}Pu	Kosnarite	0.93		

crystalline, but the ^{238}Pu sample became amorphous at a dose of 0.93×10^{16} α mg^{-1} (conference abstract by Orlova et al 1993, cited in Reference [46]). The results of the experiments discussed above are summarized in Table 7.2 and provide a fundamental basis for comparison with ion irradiation experiments and studies of natural analogues.

7.3.2 Ion Irradiation Experiments

There have been numerous ion irradiation experiments conducted on zirconolite, pyrochlore, and defect-fluorite-type compounds since the early 1990s. Much of this output was facilitated by the HVEM and IVEM Tandem Accelerator Facility at Argonne National Laboratory, where thin transmission electron microscope (TEM) specimens are irradiated with a range of ions and energies, and the damage and recovery mechanisms are studied in situ in electron microscopes having both cryogenic and heating capabilities. Early studies of zirconolite showed that it was easily rendered amorphous at room temperature [47]. Later, heating stage runs were increasingly employed in order to study the effect of temperature on damage recovery. A study of six zirconolite compositions: $CaZrTi_2O_7$, $Ca_{0.8}Ce_{0.2}ZrTi_{1.8}Al_{0.2}O_7$, $Ca_{0.85}Ce_{0.5}Zr_{0.65}Ti_2O_7$, $Ca_{0.5}Nd_{0.5}ZrTi_{1.5}Al_{0.5}O_7$, $CaZrNb_{0.85}Fe_{0.85}Ti_{0.3}O_7$, and $CaZrNbFeO_7$ was conducted using 1.0 MeV Kr^+ in a temperature range from 25 to 973 K [48]. All the zirconolites amorphized after a similar dose at room temperature, and the amorphization dose increased with temperature, giving T_c (the "critical temperature" above which the material remains crystalline) values of 550 K for $CaZrNbFeO_7$, 590 K for $CaZrNb_{0.85}Fe_{0.85}Ti_{0.3}O_7$, 640 K for $CaZrTi_2O_7$, 900 K for $Ca_{0.8}Ce_{0.2}ZrTi_{1.8}Al_{0.2}O_7$, 1000 K for $Ca_{0.85}Ce_{0.5}Zr_{0.65}Ti_2O_7$, and 1020 K for

$Ca_{0.5}Nd_{0.5}ZrTi_{1.5}Al_{0.5}O_7$. The trend of the critical temperatures was suggested to be related to the role of Ca as a "network-modifier of the aperiodic structure formed by the polyhedra of high-valence cations," but is clearly also related to the incorporation of Nb and Fe on the Ti sites.

Within the same time period, researchers began using 1 MeV Kr ions to irradiate $A_2Ti_2O_7$ pyrochlores with A = Gd, Eu, Sm, and Y and reported critical amorphization temperatures, T_c, of 1100, 1080, 1060, and 780 K, respectively [49]. This work was followed by a study in which 0.6 MeV Bi ions were used to irradiate titanate pyrochlores with A = Gd, Sm, Y, and Lu [50]. Results of this particular study indicated that the four different compounds have similar T_c values of ~975 K. These results were thought to be an ion range problem in the case of the Bi ions [51], but are probably also related to the larger collision cascades produced by the heavier and lower-energy Bi ions. Additional work reported in Reference [51] demonstrated that the T_c values of pyrochlores irradiated with 1 MeV Kr ions increase systematically from A = Lu (T_c = 480 K) to Gd (T_c = 1120 K), followed by a decrease in T_c to 1045 K for A = Sm. A compilation and analysis of the available in situ ion irradiation data for 1.0 MeV Kr ions used in studies of the radiation response of $A_2B_2O_7$ type pyrochlore and defect fluorite compounds showed that the critical temperature for amorphization can be quantified in terms of the structural parameters, classical Pauling electronegativity difference, and disorder energies determined by atomistic simulation methods [52]. The results of this study demonstrated that radiation tolerance is correlated with a change in the structure from pyrochlore to defect fluorite, a smaller unit cell dimension, and lower cation–anion disorder energy. Radiation tolerance is promoted by an increase in the Pauling cation–anion electronegativity difference or, in other words, an increase in the ionicity of the chemical bonds. A further analysis of the data indicates that, of the two possible cation sites in ideal pyrochlore, the smaller B-site cation appears to play the major role in bonding. The results of this review and analysis provided an empirical model in which the radiation response based on T_c values can be calculated using the structural parameters and defect energies of the $A_2B_2O_7$ type pyrochlore and defect fluorite compounds and used as a predictive tool. An example of one set of calculations is shown in Figure 7.2.

Following the numerous investigations of pyrochlore and related materials, often referred to as "227" compounds, a significant body of research shifted to other stoichiometries including the "delta" phase fluorite-type compounds and the "215" group of compounds. The radiation resistance of the orthorhombic Ln_2TiO_5 (Ln = La, Nd, Sm, Gd, Dy, and Y) compounds was investigated via in situ ion irradiation of thin TEM specimens [53]. It was reported that the critical temperatures for amorphization (T_c) of La_2TiO_5, Sm_2TiO_5, and Nd_2TiO_5 were similar, whereas those for Gd_2TiO_5, Dy_2TiO_5, and Y_2TiO_5 decreased sequentially. The improved tolerance in Gd_2TiO_5, Dy_2TiO_5, and Y_2TiO_5 was explained as being due to ion-irradiation-induced phase transformation from orthorhombic to a cubic defect fluorite structure. Similar in situ ion irradiation experiments were conducted on cubic Y_2TiO_5, $YYbTiO_5$, and Yb_2TiO_5 [54]. The results of this study showed that T_c decreased from Y_2TiO_5 to $YYbTiO_5$ and then to Yb_2TiO_5. The T_c value for cubic Y_2TiO_5, 589 K, was similar to the T_c value determined above for orthorhombic Y_2TiO_5, 623 K [53]. Although this result supports the premise that the improved T_c values of the heavier, smaller Ln_2TiO_5 compounds may be due to their ability to transform into polymorphs with irradiation, the technical details involved in identifying the phase transformation using electron diffraction have been called into

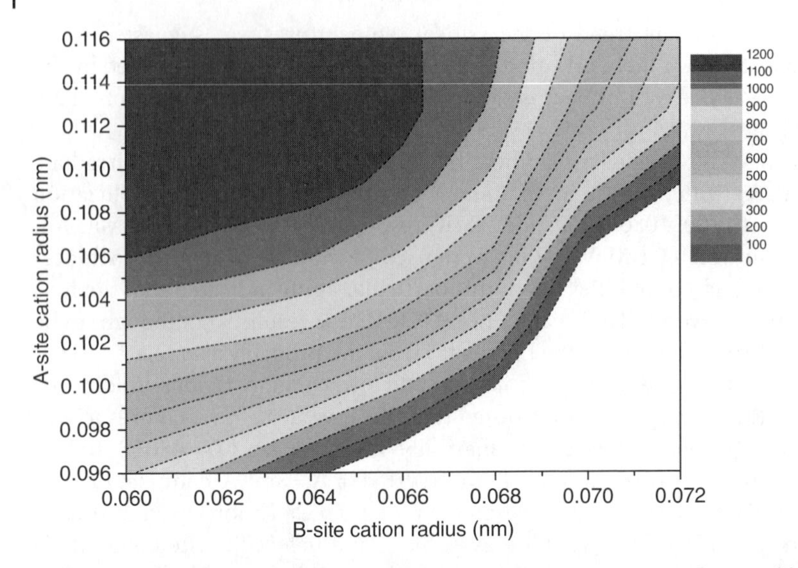

Figure 7.2 Contour map of predicted critical amorphization temperature for pyrochlore and defect fluorite compounds as a function of the radii of the A-site and B-site cations. This map was calculated from the defect energies, lattice parameters, and electronegativity data of the various compounds in this system. It can be used as a predictive tool for testing suitable compositions that are likely to remain crystalline under irradiation. The temperature scale is in degrees Celsius from most tolerant (red, 0–100) to the least tolerant (darker blue, 1100 – 1200) compositions based on $A_2B_2O_7$ stoichiometry. Unpublished data of the author, see references [Lumpkin, G.R., Pruneda, M., Rios, S., et al. (2007) Nature of the chemical bond and prediction of radiation tolerance in pyrochlore and defect fluorite compounds. *Journal of Solid State Chemistry*, **180**, 1512–1518; Lumpkin, G.R., Smith, K.L., Blackford, M.G., et al. (2009) Ion irradiation of ternary pyrochlore oxides. *Chemistry of Materials*, **21**, 2746–2754] for a general discussion and description of methods involved. *(See insert for color representation of the figure.)*

question on the basis of the physics of electron scattering in these materials when amorphous domains are present.

Recent research has been reported on the radiation resistance of a series of orthorhombic Ln_2TiO_5 compounds using in situ TEM methods [55]. In this study, samples with nominal stoichiometry Ln_2TiO_5 (Ln = La, Pr, Nd, Sm, Eu, Gd, Tb, and Dy) in the orthorhombic (*Pnma*) structure were irradiated over a range of temperatures using 1 MeV Kr ions in situ within a transmission electron microscope. In each case, the fluence was increased until a phase transition from crystalline to amorphous was observed, giving the critical dose D_c. At certain elevated temperatures, the crystallinity was maintained irrespective of fluence. This study showed that the critical temperature for maintaining crystallinity, T_c, varied nonuniformly across the series. The T_c was consistently high for La, Pr, Nd, and Sm_2TiO_5 before sequential improvement occurred on going from Eu_2TiO_5 to Dy_2TiO_5 with T_c dropping from 974 K to 712 K. In addition to the work on thin TEM specimens, bulk Dy_2TiO_5 was irradiated with 12 MeV Au ions at 300 K, 723 K, and 823 K and monitored via grazing-incidence X-ray diffraction (GIXRD). At 300 K, only amorphization is observed, with no transformation to other structure types, while at higher temperatures the specimens retained their original structure. In a second related study, the four major crystal structure types in the Ln_2TiO_5 family of compounds, namely, orthorhombic *Pnma*, hexagonal *P6₃/mmc*, cubic (pyrochlore-like) *Fd-3m*, and cubic (fluorite-like) *Fm-3m* structures were examined [56]. A series of samples, based on the stoichiometry $Sm_xYb_{2-x}TiO_5$ (x = 2, 1.4, 1, 0.6, and 0) were irradiated using 1 MeV Kr ions and characterized in situ using a transmission electron microscope. This particular study showed that the structure type plus elements of bonding are correlated to ion-irradiation tolerance. The cubic phases, Yb_2TiO_5 and $Sm_{0.6}Yb_{1.4}TiO_5$, were found to be the most radiation tolerant, with Tc values of 479 and 697 K, respectively. The improved radiation tolerance with a change in symmetry to cubic is consistent with previous studies of similar compounds. Some of the results of these studies are illustrated in Figure 7.3, showing the form of the radiation response as a function of temperature obtained by using the in situ methodology on thin TEM specimens.

The ABO_4 compounds have been studied in some detail, especially considering the importance of materials like zircon and monazite in geological sciences and industry and their reported stability in low-temperature and hydrothermal geological environments. This body of work has included Th and U compounds, for example, thorite and coffinite, as potential nuclear waste form host phases. For example, radiation damage in synthetic nanocrystalline coffinite $USiO_4$ was investigated using in situ ion irradiation with 1 MeV Kr ions [57]. They reported that the crystalline-to-amorphous transformation occurs at a relatively low dose, equivalent to about 0.27 displacements per atom (dpa) at room temperature. The critical temperature, T_c, above which coffinite cannot be rendered amorphous, was determined to be approximately 608 K. Synthetic coffinite is more stable under irradiation than zircon ($ZrSiO_4$, T_c = 1000 K) and thorite ($ThSiO_4$, T_c above 1100 K) upon ion beam irradiation at an elevated temperature, suggesting enhanced defect annealing due to the small crystallite size. This is an important observation with regard to the potential of nanoscale materials to resist amorphization caused by radiation.

Monazite- and zircon-type ABO_4 orthophosphates have also been investigated in some detail using in situ ion beam irradiation methods, including synthetic monazite samples with A = La, Pr, Nd, Sm, Eu, Gd, and the zircon structure types with A = Sc, Y, Tb, Tm, Yb, and Lu. These were irradiated using 800 keV Kr ions at temperatures of 20 to 600 K [58]. The critical amorphization temperature ranged from 350 to 485 K for orthophosphates with the monazite structure and from 480 to 580 K for those with the zircon structure. However, natural zircon ($ZrSiO_4$), on the other hand, has a much lower resistance to radiation damage and can be amorphized at over 1000 K. Within each structure type, it was shown that the critical temperature of amorphization increased with the atomic number of the lanthanide cation. Furthermore, structural topology models are consistent with the observed differences between the two structure types, but do not predict the relative amorphization doses for different compositions. The ratio of electronic-to-nuclear stopping (ENSP) correlates well with the observed sequence of susceptibility to amorphization within each structure type, consistent with previous results that electronic energy losses enhance defect recombination in the orthophosphates. In subsequent work, the zircon structured orthosilicates with A = Zr, Hf, and the monazite and zircon polymorphs with A = Th were studied in greater detail, where it was determined that all four materials became amorphous in a process that exhibited a two-stage dependence on temperature when irradiated by 800 keV Kr or Xe ions in the temperature range of 20 to 1100 K [59]. The critical amorphization temperature above which amorphization did not occur increased in the

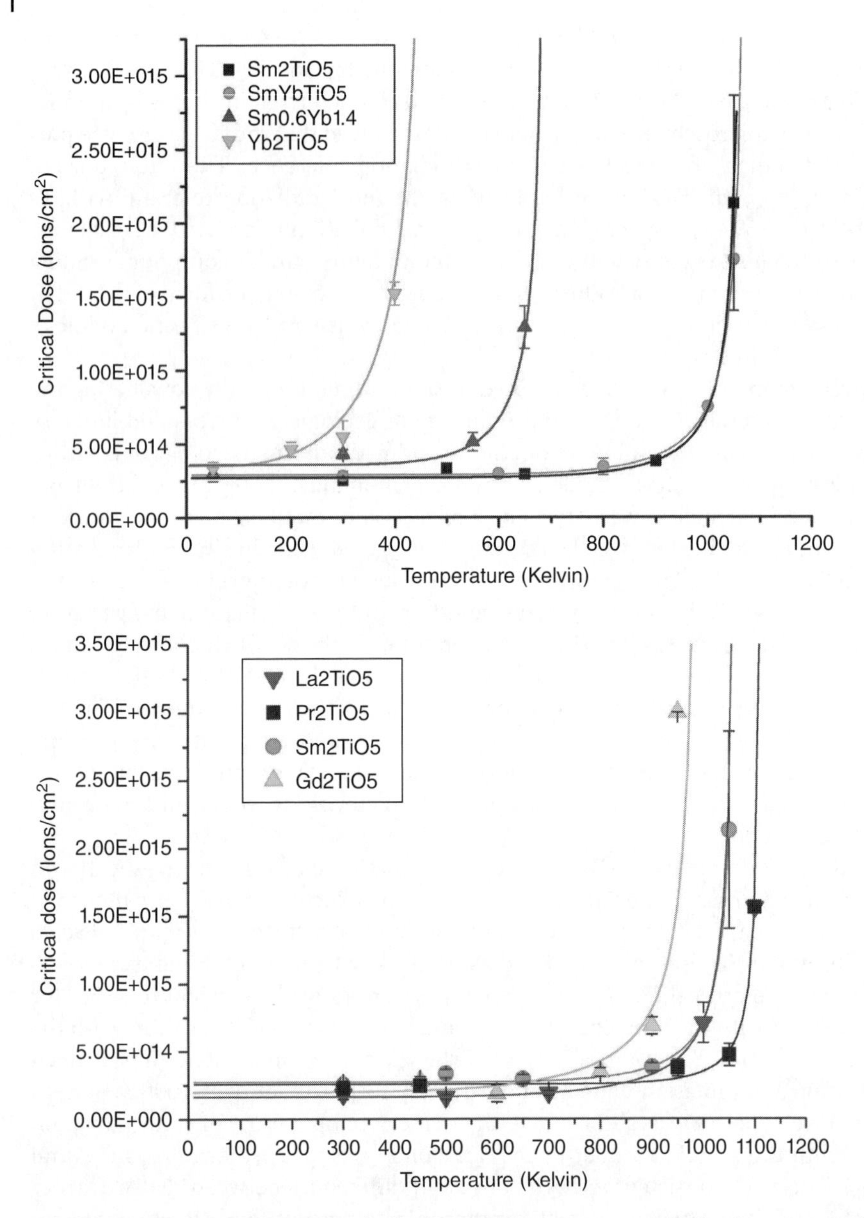

Figure 7.3 In situ ion irradiation results for the Ln_2TiO_5 type compounds. The data were obtained at the IVEM-Tandem Facility at Argonne National Laboratory using 1.0 MeV Kr ions passing through thin TEM samples. Top: Hexagonal SmYb, pyrochlore-like Sm0.6Yb1.4, and fluorite-like Yb irradiation data relative to orthorhombic Sm_2TiO_5. Bottom: Orthorhombic structures with Ln = Gd, Sm, Pr, and La. These figures are adapted from pre-publication files related to references [Aughterson, R.D., Lumpkin, G.R., Ionescu, M., et al. (2015) Ion-irradiation resistance of the orthorhombic Ln_2TiO_5 (Ln = La, Pr, Nd, Sm, Eu, Gd, Tb and Dy) series. *Journal of Nuclear Materials*, **467**, 683–691; Aughterson, R.D., Lumpkin, G.R., De los Reyes, M., et al. (2016) The influence of crystal structure on ion-irradiation tolerance in the $Sm_{(x)}Yb_{(2-x)}TiO_5$ series. *Journal of Nuclear Materials*, **471**, 17–24]. (*See insert for color representation of the figure.*)

following order: huttonite, zircon, hafnon, and thorite. At temperatures below 500 K, the tetragonal and monoclinic polymorphs of $ThSiO_4$ required approximately the same ion fluence for amorphization. Monoclinic, monazite-type $ThSiO_4$ is not more resistant to radiation damage than the tetragonal-symmetry form, but instead, its amorphous phase recrystallizes at a lower temperature. A model that accounts for the observed two-stage behavior was developed and used to calculate the recrystallization activation energies for both stages (I note here that this approach was subjected to considerable criticism due to the substantial errors on the data points – a simple smooth function can just as easily describe the data within error). When irradiated with heavy ions above a certain characteristic temperature, an unexpected decomposition was observed in which all four ABO_4 orthosilicates phase-separated into the crystalline oxides: ZrO_2, HfO_2, or ThO_2 plus amorphous SiO_2.

Garnets, a large group of cubic compounds based on the general formula $X_3Y_2Z_3O_{12}$, have also been considered as host phases for the safe immobilization of high-level nuclear waste, as they have been shown to accommodate a wide range of elements across three different cation sites, such as Ca, Y, and Mn on the X-site, Fe, Al, U, Zr, and Ti on the Y-site, and Si, Fe, and Al on the Z-site. Garnets, due to their ability to have variable composition, make ideal model materials for the examination of radiation damage and recovery in nuclear materials, including as potential waste forms. Kimzeyite, $Ca_3Zr_2FeAlSiO_{12}$, has been shown naturally to contain up to 30 wt% Zr, and has previously been examined to elucidate both the structure and ordering within the lattice. The effects of radiation damage and recovery in zirconium and hafnium garnets were studied using in situ ion beam irradiation with 1 MeV Kr ions [60]. The results of this study showed that $Ca_3Zr_2FeAlSiO_{12}$ and $Ca_3Hf_2FeAlSiO_{12}$ were relatively easily amorphized with critical temperatures for amorphization of 1129 and 1221 K, respectively, and exhibit a relatively low critical fluence for amorphization below 1000 K. A post-irradiation examination of a sample of $Ca_3Zr_2FeAlSiO_{12}$ irradiated at 1000 K was conducted using aberration-corrected STEM and found to contain discreet, nano-sized, crystalline Fe-rich particles, indicating that Fe^{3+} cations were reduced to metallic Fe during the radiation damage process. Previous work [61] on a range of garnet compositions reported critical temperatures for amorphization of 890–1130 K. At around the same time, another set of synthetic garnets with Fe on the tetrahedral site were investigated using in situ TEM methods and reported to have critical temperatures for amorphization of 820–870 K [62]. More recently, critical temperatures for amorphization of 773–873 K were determined for a set of garnet samples mainly containing Al or Fe on the tetrahedral site [63]. From these observations, it seems that the garnet structure type may be less suitable as a waste form host phase for actinides due to the susceptibility to amorphization.

7.3.3 Natural Analogues

Radiation damage effects in minerals have been studied for many years, and some have been considered as natural analogues for crystalline nuclear waste form phases. In this case, the dose rates are typically 10^5 to 10^9 times slower than the rates experienced by the materials used in the actinide doping experiments described above due to the long half-lives of ^{232}Th, ^{235}U, and ^{238}U. Therefore, the opportunity exists to examine potential dose rate effects that would be experienced by nuclear waste forms, provided that a clear understanding of the geological environment and thermal history over time is known. Pyrochlore-type oxide minerals have been the subject of numerous studies and provide a substantial knowledge basis for radiation damage effects. The analysis of the line broadening in a suite of Russian pyrochlores and showed that the strain increased from 0.0009 to 0.0035 as the crystallite dimensions decreased from 100–120 nm down to 35–40 nm in the initial stages of damage, followed by a further decrease to around 15 nm in the latter stages of damage [64]. These authors also carried out an analysis of the radial distribution function (RDF) of an amorphous sample and showed that there was no long-range order present beyond the second coordination sphere. However, peaks in the RDF representing the major M-O and M-M distances showed that the fundamental structural units (e.g., the coordination polyhedra) still existed in the amorphous state.

In a similar study [65], X-ray diffraction was also used to determine both the beginning (D_i) of the crystalline-amorphous transformation as well as the critical amorphization dose (D_c) for a large suite of pyrochlores from different localities. Here, it was shown that the transformation zone increased in dose as a function of the geological age of the samples. Analysis of the dose–age data gives an intercept value of $D_0 = 1.4 \times 10^{16}$ $\alpha\ mg^{-1}$ for the amorphization dose curve and an annealing rate of $K = 1.7 \times 10^{-9}\ yr^{-1}$ [66]. The amount of material damaged was given as $B = 2.6 \times 10^{-16}\ mg\ \alpha^{-1}$, corresponding to an average cascade radius of 2.3 nm in which a maximum of 2600 atoms are displaced (with the usual assumptions). Line broadening in these samples showed that crystallite dimensions decreased from about 500 to 15 nm with increasing dose. The strain initially increased with the dose and reached a maximum of approximately 0.003 before falling to values below 0.0005 at higher dose levels, consistent with a description of the crystalline-amorphous transformation as a type of "percolation" transition [67]. With increasing alpha-decay dose, TEM images reveal mottled image contrast due to the strain, followed by the appearance of local amorphous domains that increase in volume and begin to overlap to produce larger amorphous areas until they are connected throughout the material. This is the first percolation transition. With further increases in dose, the crystalline areas diminish in volume until they become isolated, giving way to a microstructure dominated by amorphous pyrochlore [65]. This is the second percolation transition. Local structure and bonding was also examined in selected pyrochlore samples using EXAFS-XANES methods [68–71]. The results of these studies indicate that the M-O coordination polyhedra of amorphous pyrochlores exhibit reduced bond distances, reduced coordination number, and increased distortion relative to the undamaged crystalline structure. Furthermore, there was no periodicity in evidence beyond the second coordination sphere, with some disruption of the M-M distances. From these studies, it was observed that only a slight increase in the mean M-M distance was required in order to explain the overall increase in volume caused by alpha-decay damage and that this could be facilitated by increased M-O-M angles. The potential effects of such volume swelling are illustrated in Figure 7.4.

Similar results on radiation damage zirconolite were reported on the basis of the study of amorphous and annealed zirconolite from Sri Lanka using a variety of methods, including XRD, EXAFS-XANES, TEM, and DTA [72]. Electron diffraction and high-resolution TEM studies have also suggested that amorphous zirconolite lacked periodicity beyond the second coordination sphere, consistent with a random network model of the amorphous state. EXAFS-XANES results provided more detailed

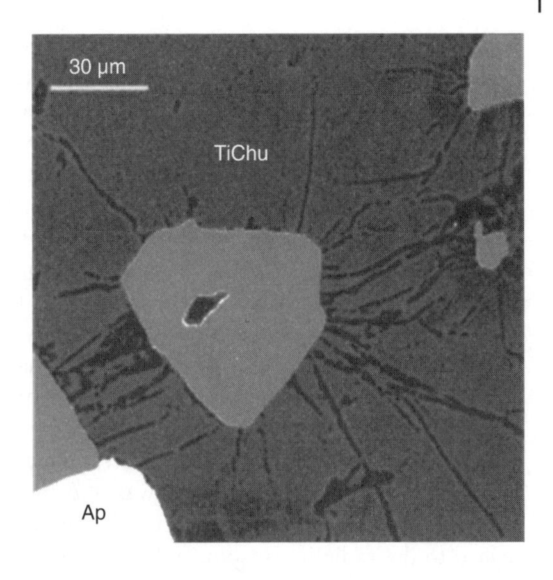

Figure 7.4 Colorized SEM map showing U-pyrochlore (red), apatite (yellow), and titanian clinohumite (blue) in hydrothermal veins from the Adamello massif, northern Italy. The minerals are approximately 40 million years old. Due to the content of ~30 wt% UO_2, the pyrochlore is amorphous due to alpha decay damage. Note the radial cracks in surrounding clinohumite due to radiation-induced volume expansion. Courtesy of Reto Gieré. (*See insert for color representation of the figure.*)

information for the Ti- and Ca-sites and indicated that amorphous zirconolite lacked periodicity beyond the first coordination sphere, with reduced M-O bond lengths, reduced coordination number, and increased distortion of the Ti-O polyhedra (determined from a prominent pre-edge feature in the XANES results). Additional results were obtained for the Zr-, Th-, and U-sites and indicate nearly identical coordination numbers and bond lengths for the amorphous and annealed samples [73]. However, a significant increase in the range of Zr-O and Th-O distances was observed, leading to the conclusion that a slight variation of the M-O-M angles can have a profound effect on long-range periodicity and medium-range order. Later work also confirmed the reduced coordination of Ti in the amorphous samples, pointing specifically to a fivefold coordination environment [74]. A detailed study of highly zoned zirconolite samples from the Bergell intrusion and the Adamello igneous complex provided the first detailed results on the full crystalline-amorphous transformation in natural zirconolites [75]. Samples from these localities contain a wide range of ThO_2 and UO_2 concentrations up to a combined maximum of over 20 wt%, thus ensuring a substantial range of alpha-decay dose. Due to the small grain size, the two suites of samples were characterized by analytical electron microscopy. TEM dark field images revealed the percolation-like behavior with increasing dose. This consists of the appearance of mottled diffraction contrast (0.08×10^{16} α mg^{-1}), followed by ingrowth of amorphous domains ($0.3–0.5 \times 10^{16}$ α mg^{-1}), followed by overlap of collision cascades to produce larger amorphous areas as crystalline domains diminish in size to less than 10 nm ($0.7–0.9 \times 10^{16}$ α mg^{-1}). A study of seven suites of zirconolite samples ranging in age from 16 Ma to 2060 Ma and with alpha-decay doses of $0.008–24.0 \times 10^{16}$ α mg^{-1} has also been reported [76]. Similar to the previous work on pyrochlore, the upward curvature of the initial and critical dose curves was interpreted as evidence for long-term annealing occurring in geological environments. Curve fits to the data gave values of $D_i = 0.11 \times 10^{16}$ α mg^{-1} and $K = 1.0 \times 10^{-9}$ yr^{-1} for the onset dose and $D_c = 0.94 \times 10^{16}$ α mg^{-1} and $K = 0.98 \times 10^{-9}$ yr^{-1} for the critical dose.

Perovskite (ideally $CaTiO_3$) is an important component of Synroc and other nuclear waste forms designed primarily for HLW from reprocessing of spent fuel and has the potential to incorporate actinides, fission products, and short-lived ^{90}Sr. Radiation damage in natural samples has not been investigated in depth, in this case due to the low Th and U contents. In a study of perovskites from the Khibina alkaline complex, Russia, it was observed that individual perovskite crystals are chemically zoned and contain 2.3–18.5 wt.% ThO_2 and very little uranium [77]. X-ray diffraction studies indicate that the cores (2.3–7.4 wt.% ThO_2) are partially crystalline and the rims (8.7–18.5 wt.% ThO_2) are completely amorphous, providing an estimate of the critical dose for amorphization of ~ 2×10^{16} α mg^{-1} based on the maximum and minimum Th contents of the cores and rims, respectively. Due to the low Th and U contents found in most natural perovskites, this result is probably the best estimate currently available for radiation-damage-induced amorphization in natural perovskite.

With regard to the minor oxide phases in titanate ceramic waste forms, brannerite (ideally UTi_2O_6) is an important constituent in some of the pyrochlore-based ceramics designed for plutonium immobilization and may account for a significant fraction of the total U and Pu inventory in these waste forms [78]. Although brannerite is present in materials studied previously by actinide doping, there are no detailed conclusions with regard to the radiation response. Most natural brannerites older than 2×10^7 years are amorphous due to the high U and Th contents and cumulative alpha-decay events in the range of 2–170 × 10^{16} α mg^{-1} [79]; however, there is a single report in the literature describing a brannerite sample from Binntal, Switzerland, which is partially crystalline [80]. This provides an estimate for the critical dose for amorphization of approximately 1×10^{16} α mg^{-1} based on the geological age of 10^6 years. Crichtonite, a complex oxide ($A_{1-x}M_{21}O_{38}$) found in early studies of Synroc, has also been suggested as a possible host phase for actinides and fission products [81]. A recent study of a small suite of U-bearing samples from several different localities (mostly *davidite* according to mineral nomenclature) reported a dose range of 0.2–44 × 10^{16} α mg^{-1} as calculated from the amounts of U and Th and the geological ages. Further analysis of the data gave an approximation for the critical dose for amorphization of ~0.8×10^{16} α mg^{-1} for the younger samples having ages of ~ 270–300 Ma [82]. However, the older samples provide evidence for annealing of alpha-decay damage, similar to the results reported above for pyrochlore and zirconolite.

Zircon ($ZrSiO_4$), thorite ($ThSiO_4$), and coffinite ($USiO_4$) are the main orthosilicates of interest for nuclear waste disposal. The early classic investigations of natural zircon samples [83, 84] elucidated a transformation from the crystalline to the amorphous state as a function of dose using X-ray diffraction patterns, calculation of lattice parameters, optical properties, and density measurements. In particular, the densities of zircons from Sri Lanka were shown to decrease by ~16% with increasing dose as the unit cell parameters increased up to a dose of ~0.6×10^{16} α mg^{-1} [84]. A more detailed analysis of the X-ray scattering in zircon samples from Sri Lanka enabled the deconvolution of peaks in the powder patterns into the Bragg and diffuse components, the latter due to the strong influence of radiation-induced defects (e.g., Frenkel pairs) in the lower dose regime [85]. This study reported anisotropic lattice expansion with increasing dose, unit cell volume expansion of 4.7%, and total volume expansion of ~18%. In a study based on a large zoned crystal of zircon from Sri Lanka, it was shown that the hardness and elastic modulus decreased by 40% and 25%, respectively, with increasing alpha-decay dose up to ~1.0×10^{16} α mg^{-1} [86]. Subsequent experimental work and atomistic modeling (see Section 7.3.4) conducted at the University of Cambridge using molecular dynamics simulations and nuclear magnetic resonance (NMR) spectroscopic methods revealed that local segregation of Si and Zr occurs within the alpha-recoil collision cascades in zircon, thereby planting the seeds for breakdown of extensively damaged zircon to ZrO_2 (monoclinic or tetragonal) and amorphous SiO_2 during natural annealing over time periods [87]. This work led to the important concept of "percolation" of domains in solids by Ekhard Salje as a physical model to explain changes in properties as a function of the radiation dose, and so on. A quantitative analysis of ^{29}Si NMR spectra of radiation-damaged, natural zircons showed that the local structure in crystalline and amorphous regions depend explicitly on the radiation dose. Nonpercolating amorphous islands of high-density "glass" within the crystalline matrix show a low interconnectivity of SiO_4 tetrahedra. This structural state is quite different from that of the high-dose, percolating regions of low-density glass with more polymerized tetrahedra. This study reported (1) a continuous nonlinear dose dependence between the high- and low-density glass states and (2) continuous evolution of the local structure of the crystalline phase up to the first percolation point in zircon. Analysis of NMR data enabled the determination of displaced atoms, ranging from approximately 3800 to 2000 atoms per decay event with increasing dose. The amorphous fractions of the metamict zircons were determined as a function of dose and are consistent with the concept of direct impact amorphization within the collision cascade. In Figure 7.5, the calculated alpha-decay doses are calculated for zircon as a function of age and two different Th and U contents representative of most natural samples. These data illustrate the importance of radiation effects in zircon in the older geological environments (e.g., $t > 2 \times 10^7$ years).

Currently there exists very little information about the radiation damage performance of natural coffinite $USiO_4$ in spite of the importance of this mineral in uranium ore deposits. The actual composition of natural coffinite samples may fall within the system defined by the end-member compositions $USiO_4$, $USiO_4{\bullet}H_2O$, and $U(SiO_4)_{1-x}(OH)_{4x}$, with the latter taking into account the substitution $Si^{4+} \Leftrightarrow 4H^+$ at the tetrahedral site of the structure. Early descriptions of coffinite suggest that the mineral shows a range of crystallinity. Samples from the U ore deposits of New Mexico range from isotropic to slightly anisotropic with refractive index (or indices) and birefringence well below that of synthetic samples [88]. In a study of thorite from Madagascar, it was noted that nanocrystalline coffinite with a composition of approximately $(U_{0.625}Th_{0.375})SiO_4$ formed via the alteration of larger, amorphous uranium-bearing thorite grains [89] during alteration and mobilization of U in the host rock. Geologically young thorite crystals with a well-defined age of 6–7 Ma have been reported to be completely amorphous [90], providing an upper limit on the critical dose of 0.8×10^{16} α mg^{-1}, consistent with data for natural zircon discussed in the previous section. However, it has also been reported that the P and V thorites from the Harding pegmatite in New Mexico retained substantial crystallinity even after sustained doses of $40–120 \times 10^{16}$ α mg^{-1} [91]. On the basis of a simple comparison of bond energies, it was proposed that thorites rich in P and V have a lower energy barrier to recrystallization than samples that are closer to $ThSiO_4$ in composition. In comparison to the zircon structure ABO_4 minerals, the orthophosphate minerals xenotime (zircon structure YPO_4) and monazite (monoclinic ABO_4 generally with A = Ca, Ln, Th, Sr and B = P, Si) are notable for retaining crystallinity over geological time scales even with estimated doses up to 10^{17} α mg^{-1}. Figure 7.6 shows the calculated doses for coffinite and thorite over geological time periods. Note that these

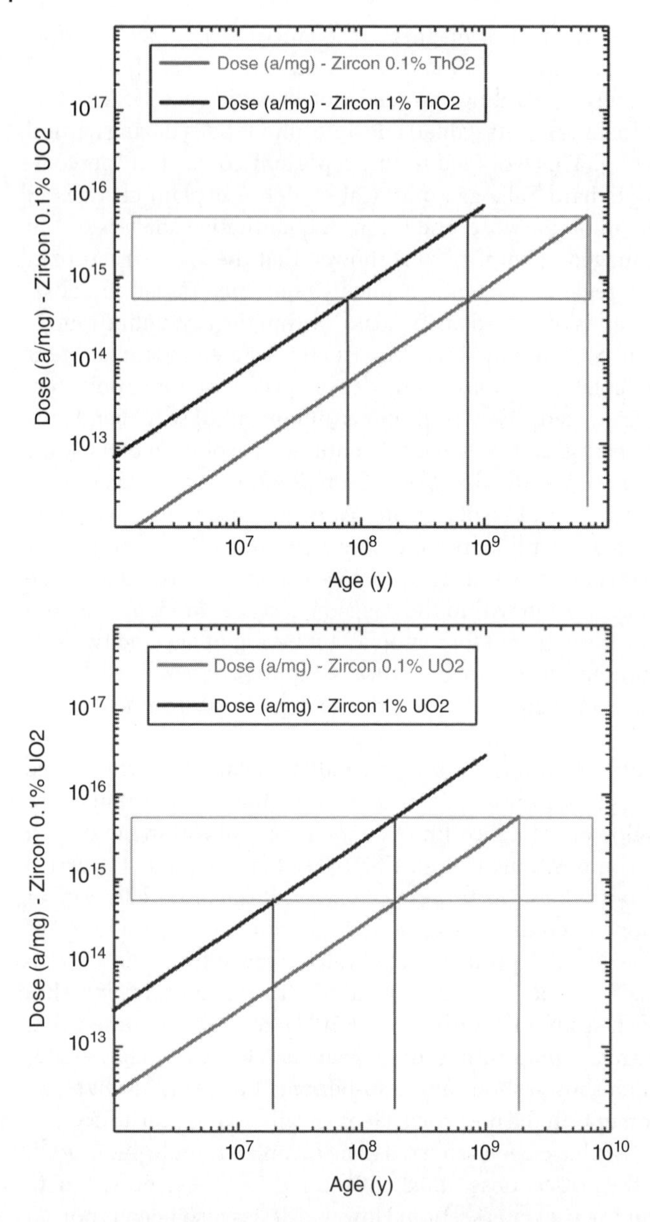

Figure 7.5 Calculated alpha-decay dose for zircon ($ZrSiO_4$) with two different concentrations of ThO_2 (top) and UO_2 (bottom). The gray lines represent the beginning and end of the crystalline-amorphous transformation zone without significant thermal annealing. Due to the low Th and U contents, zircons record radiation damage over much longer periods of time at lower dose rates relative to coffinite and thorite.

minerals appear to be susceptible to radiation effects on time scales about 1–2 orders of magnitude less than zircons.

Titanite (= sphene), ideally $CaTi(SiO_4)O$, may incorporate minor Na, lanthanides, and low levels of actinides on the Ca site, together with Fe, Al, and Nb on the Ti site.

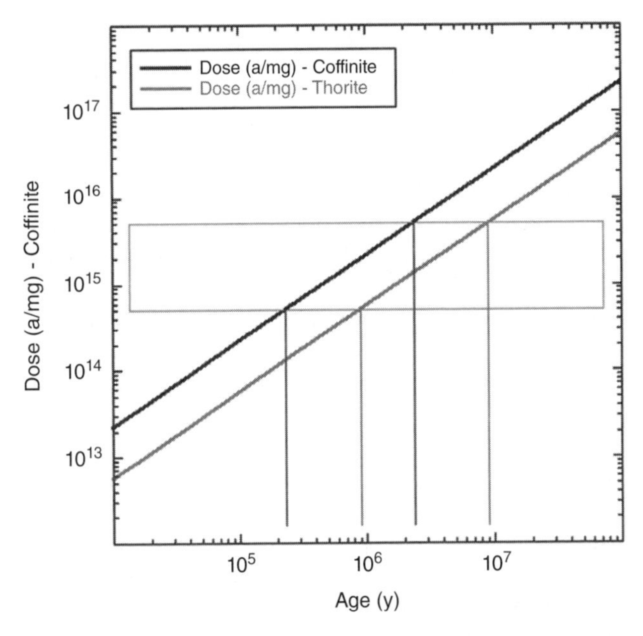

Figure 7.6 Calculated alpha-decay dose for pure coffinite ($USiO_4$) and thorite ($ThSiO_4$) as a function of geological age. The gray horizontal lines indicate the approximate beginning and end (critical dose) of the radiation damage transformation zone at low temperature (based on zircon data with no major thermal annealing). Due to the difference in the decay constants of ^{238}U and ^{232}Th, there is a major time shift in the transformation zone of thorite to longer time periods by a factor of about 5.

Additionally, significant amounts of F and OH may replace O on the anion site that is not bonded to Si. The amounts of actinides incorporated in the structure are generally below 500 ppm Th and 3000 ppm U. Due to the limited solubility of lanthanides and actinides in the structure, the titanite has seen limited use in waste forms apart from the titanite-based glass-ceramics developed for Canadian waste applications [92]. Heavily damaged samples of titanite from the Cardiff mine, Ontario, Canada, indicate that the critical dose for amorphization is greater than $0.3–0.4 \times 10^{16}\ \alpha\ mg^{-1}$ on the basis of the assumed age of 1000 Ma for this locality [93]. This estimate is in good agreement with a major study of titanite which showed that the critical dose is ~ $0.5 \times 10^{16}\ \alpha\ mg^{-1}$ and that the crystalline-amorphous transformation results in a density decrease of ~8% [94]. General features of alpha-decay damage in titanite with increasing dose include increasing thermal vibration parameters of the cations and anions, increasing the unit cell volume up to about 3% in the most heavily damaged samples, and a possible reduction of some Fe^{3+} to Fe^{2+} during electron transfer processes associated with the radioactive decay.

Radiation effects in natural apatite have not been studied extensively as the phosphates have very low Th-U contents and the silicates are quite rare in geological systems. Some information has been acquired by studying the silicate apatite known as britholite. Samples from the Eden Lake complex in Manitoba, Canada, contain about 0.9–1.3 wt% ThO_2 and 0.7–1.0 wt% UO_2 together with 52–55 wt% Ln_2O_3, 12–17 wt% CaO, 1–5 wt% P_2O_5, and 2–4 wt% F [95]. The crystals are optically isotropic but weakly crystalline according to X-ray diffraction analysis; however, in this case, the observed diffraction could be due to inclusions of associated minerals. Assuming an age of 1700 Ma for the host rocks, the estimated alpha-decay dose is ~$4–7 \times 10^{16}$ α mg^{-1} and probably represents an upper limit for the critical dose for amorphization. For comparison, the study of a series of 2100 Ma apatite-britholite samples from Ouzzal Mole, Algeria, indicated that the critical dose is ~2×10^{16} α mg^{-1} [96]. This suggestion is reasonably consistent with a recent study of six natural britholite samples from alkaline rocks in Russia, ranging in age from 320 to 2600 Ma and containing up to 12 wt% ThO_2 and UO_2. Detailed X-ray diffraction and TEM work on these samples places the critical dose for amorphization at a value close to 0.9×10^{16} α mg^{-1}. Additionally, the thermal annealing of fully amorphous samples indicated recrystallization temperatures of 500 °C–600 °C by X-ray diffraction [97].

7.3.4 Atomistic Modeling

Simulations of the structure, stability, defect behavior, and radiation damage of nuclear materials have become commonplace in materials science, including a range of existing and potential nuclear waste forms designed for the encapsulation and safe storage of actinides and fission products. This is particularly true for the titanate and related oxide waste forms, which have evolved during a time when atomistic simulation methods have become increasingly advanced. Atomistic modeling primarily involves the use of empirical interatomic pair potentials that form the basis of molecular dynamics (MD) simulations on picosecond time scales. These simulations allow for the direct observation of the formation and evolution of alpha-recoil collision cascades and defects and can be performed as a function of temperature and pressure via the use of suitable boundary conditions. Empirical potentials are also used for static calculations of structural stability and the energetics of defect formation and migration.

Density functional theory (DFT) has been increasingly applied in studies of the stability and radiation effects of nuclear materials. DFT has the ability to more accurately describe the geometry, stability, and energetics of defect formation and migration in crystalline materials on the basis of the appropriate use of a range of available density functionals and approaches to take into account the effects of exchange and correlation. All MD and DFT simulations rely on careful validation of parameters via benchmarking against the known properties of crystalline compounds. Here, we take a brief look at some of the most important background studies and key papers on atomistic modeling of nuclear waste forms and related materials in studies that have used MD and/or DFT modeling.

A convenient starting point for titanate waste forms is to look into the atomic-scale behavior of titanium dioxide (TiO_2) in several polymorphs. The threshold displacement energies (TDEs) of Ti and O were determined for rutile using empirical potentials and MD simulations [98]. For primary knock-on atoms (PKAs) introduced in various crystallographic directions at $T = 160$ K, the simulations revealed TDEs of approximately 40 eV for O and 105 eV for Ti based on a displacement probability criterion of 10%. The static and dynamic defect calculations also indicate that the radiation tolerance of rutile is related, at least in part, to low energy migration pathways for both O and Ti.

In subsequent work that was combined with in situ ion irradiation [99 – 101], it was reported that the TiO_2 polymorphs rutile, brookite, and anatase showed significant differences in radiation tolerance. At a temperature of 50 K, synthetic rutile remained crystalline up to a fluence of 5×10^{15} ions cm^{-2} and did not show convincing evidence for the onset of amorphization at this fluence level. Natural brookite and anatase, on the other hand, became amorphous at $8.1 \pm 1.8 \times 10^{14}$ and $2.3 \pm 0.2 \times 10^{14}$ ions cm^{-2}, respectively. These results correlate with the number of shared edges, the degree of octahedral distortion, and the volume properties of polymorphs, and it was shown that polyhedral distortion and volume are intimately linked through interatomic forces in the octahedral framework. MD simulations of thermal spikes and small collision cascades were performed for each polymorph. These simulations are in qualitative agreement with the experiments, with the thermal spikes showing the same relative recovery behavior between the rutile, the brookite, and the anatase structures. More recently, a quantitative confirmation of the relationship between density, volume, and defect production and recovery in rutile has been reported [101].

A series of MD studies of the fundamental properties of rutile in relation to radiation tolerance have been reported [102 – 104], and these studies are of direct relevance to procedures that are applicable to other waste form phases, provided that reliable interatomic potentials either exist or can be developed and validated. First, a general method was developed for generating the trajectories of PKAs in a uniform (non-random) distribution for any atom in the crystalline structure. This procedure allows for a proper statistical treatment of defect formation probabilities and other properties. Using this approach, TDEs were determined to be approximately 19 and 69 eV for O and Ti, respectively. It was also observed that replacement chains are a common feature of the O sublattice, playing a major role in defect formation and recovery [102]. In a second paper [103], the effect of temperature on defect formation and recovery was evaluated, showing that thermal activation of Frenkel defects at high temperature causes a reduction in the defect formation probability and an increase in the TDE for O (up to approximately 50 eV at $T = 750$ K). This work was completed using MD simulations to compare the behavior of rutile, anatase, and brookite. Results showed that all three polymorphs have comparable TDEs for O of approximately 20 eV, but reported distinctly different TDEs for Ti of 73 ± 2, 39 ± 1, and 34 ± 1 for rutile, anatase, and brookite, respectively [104]. At higher PKA energies, the defect formation probability (DFP) curves consistently show the sequence DFP anatase > brookite > rutile. Together with additional details of defect formation and migration, this work provided atomistic details for understanding differences in the radiation response of the TiO_2 polymorphs. Extending the radiation damage-focused work toward real ceramics, researchers from Los Alamos National Laboratory made the important observations, using molecular statics and temperature-accelerated dynamics methods, that grain boundaries are strong sinks for point defects in anatase and more so in rutile. Furthermore, these two polymorphs exhibit a major difference in behavior, with rutile dominated by interstitial diffusion and anatase dominated by diffusion of O vacancies and interstitials [105]. Some of the properties of rutile determined using MD simulations are illustrated in Figures 7.7 and 7.8, including TDEs and PKA displacement and DFP, and Frenkel defect production as a function of PKA energy.

Although rutile is normally used as an inert component in nuclear waste forms, it has the capacity to serve as a host phase for radionuclides under the appropriate conditions. In this regard, a study using two different DFT methods to understand the incorporation of the long-lived fission product ^{99}Tc via direct substitution for Ti in the lattice was recently reported [106]. Here, the effects of defect clustering and the transmutation of

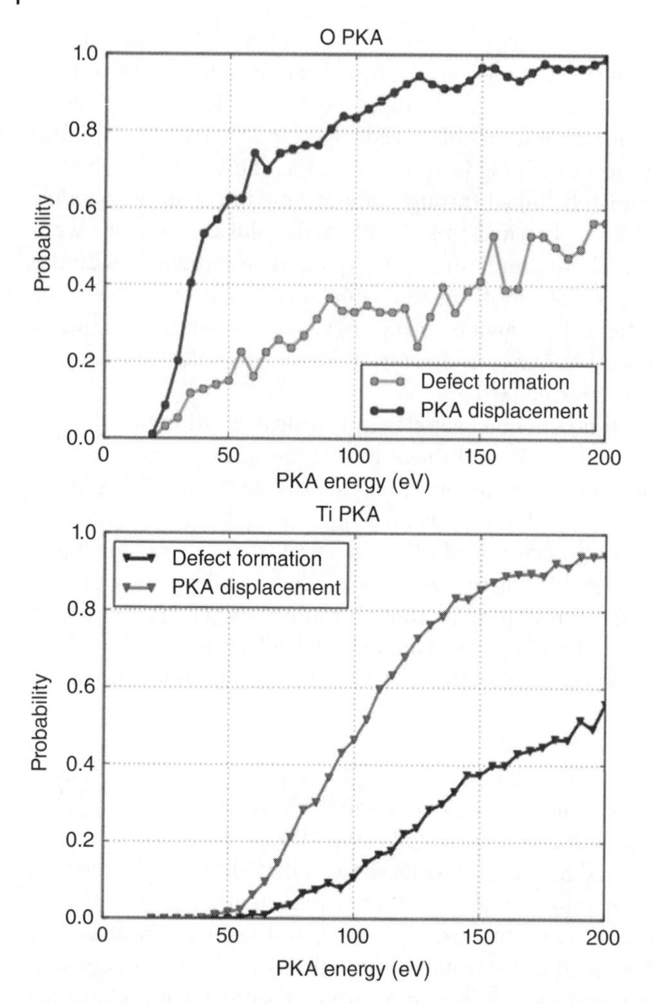

Figure 7.7 MD simulations of the probability of defect formation and PKA displacement in rutile (TiO_2). Note the major difference in the onset of displacement (threshold displacement energy, TDE) and different profiles and a function of energy for the O PKA (left) and Ti (PKA). Note the significant difference in the threshold displacement energies for O (~20 eV) and Ti (~50 eV).

Tc to Ru were investigated, demonstrating that Tc has a moderate (and temperature-dependent) solubility in rutile. However, the solubility increases when clustering is considered, with the preferred binary Tc-Tc pair having a similar distance as found in TcO_2. Importantly, the transmutation of Tc to Ru results in Ru-Ru pairs adopting a preferred second-nearest neighbor configuration. As the calculated solubility of Ru in rutile is lower than that of Tc, this study suggested the possibility of second phase formation in the host matrix if Tc is present near the solubility limit.

In stepping up to more complex materials, investigations were reported on the defect behavior of the perovskite compound $SrTiO_3$, an important titanate host phase for ^{90}Sr which is one of the crucial short-lived fission product in high-level nuclear waste [107]. In this study, the energetics of defect formation and migration were quantified using

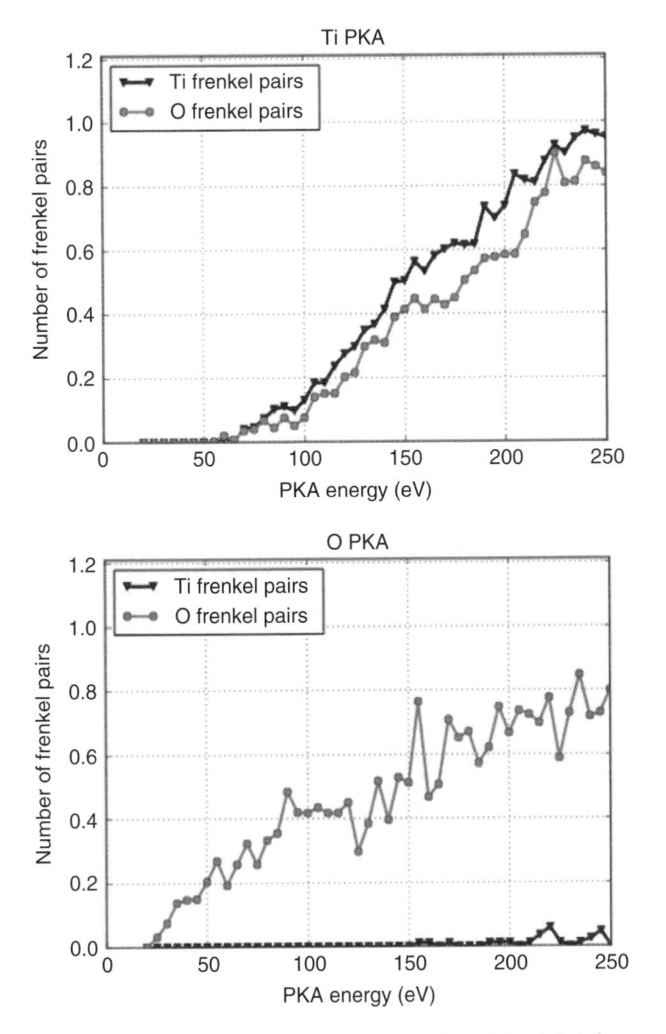

Figure 7.8 MD simulations showing the number of Frenkel defects produced in rutile (TiO_2) as a function of PKA energy for O and Ti. The Ti PKA produces slightly more Ti Frenkel pairs than O, with a similar energy profile, whereas the O PKA produces very few Ti Frenkel defect pairs.

MD and DFT methods, finding that O and Sr atoms prefer to form split interstitials and Ti atoms form interstitials in channel sites. Interstitials are more mobile than vacancies, and Sr and O have low migration barriers, indicating the potential for mobility and recombination (also refer to Reference [108] for related work). MD simulations conducted revealed that all three atom types are most easily displaced by direct replacement sequences on their own sublattices. Weighted average TDEs of 70, 140, and 50 eV were obtained for Sr, Ti, and O, respectively. However, it was also shown that there exists strong dependence of the TDEs on crystallographic direction. Similar conclusions on the energetics and types of defects and TDEs were obtained in a different study [109] via the use of ab initio MD methods. Finally, combined TEM and atomistic simulations were used to study radiation damage in $SrTiO_3$ containing Ruddlesden–Popper type faults [110]. Using empirical potentials and collision cascade simulations, the authors of this work showed that the faults are more susceptible to amorphization and that this is related to a kinetically feasible, thermodynamic driving force for defects to migrate to the faults.

Even more complex perovskite systems containing La and A-site vacancies have been investigated in considerable detail by both experimental and atomistic modeling methods [111–114]. The progression of this work was underpinned by the development of pair potentials for Sr, La, Ti, and O by fitting to experimental data and new ab initio data [111]. This work provided the atomistic basis for understanding existing experimental data on the structure [112] and radiation effects [113] in the $Sr_{1-3x/2}La_xTiO_3$ system. In particular, this solid-solution incorporates A-site vacancies according to the charge-balancing substitution $3Sr \rightarrow 2La + \square$ (where $\square$ represents a vacancy), so the role of vacancies may have important consequences for defect behavior and recovery under irradiation. In fact, the solid-solution series exhibits a minimum in the critical temperature for amorphization near $x = 0.2$ [113]. The relationships between structure, ordering, and radiation damage effects are illustrated in Figure 7.9. Using both empirical pair potentials and DFT calculations [114], it was demonstrated that the A-site cation–vacancy interactions follow electrostatic principles at low concentrations of defects, leading to the association of La ions with vacancies and dissociation of vacancy-vacancy pairs. Once long-range A-site ordering is developed at higher values of x, the defect interactions are inverted due to strain forces arising from cooperative atomic relaxations.

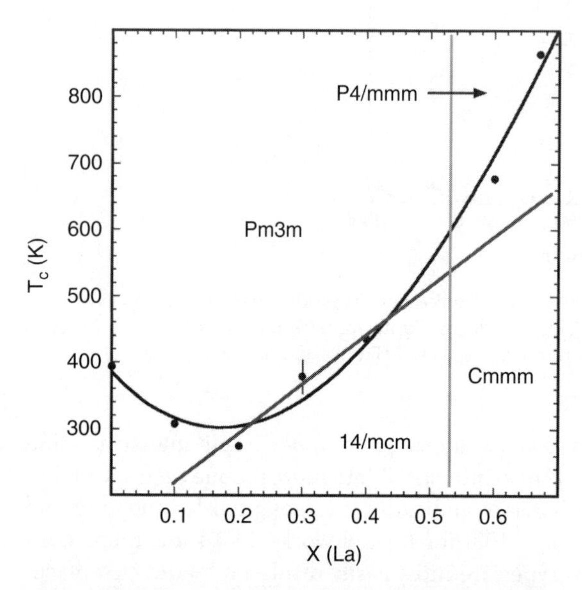

Figure 7.9 Summary of ion irradiation data for perovskites in the system $Sr_{1-1.5x}La_xTiO_3$ showing nonlinear radiation response as a function of temperature and composition. Experiments conducted in situ on thin TEM samples using 1.0 MeV Kr ions. Blue data points and curve represent the critical temperature for amorphization, the gray line is the octahedral tilting transformation, and the red line is the La-vacancy order-disorder transformation (ordered compositions to the left at $x = 0.6$ and 0.667). Modified from reference Smith, K.L., Lumpkin, G.R., Blackford, M.G., et al. (2008) In situ radiation damage studies of $La_xSr_{1-3x/2}TiO_3$ perovskites. *Journal of Applied Physics*, **103**, 083531.

Returning now to studies of zircon, using MD [115], the phases (in terms of time) of the damage production process, including the ballistic and thermal spike phases, and the effects of overlapping damage events were studied in some detail. The MD simulations of zircon indicate that the overlap of successive displacement cascades results in higher "relaxation" times and enhanced damage production in the region where cascade damage is repeated. In the core of the damaged region, it was found the SiO_4 tetrahedra may begin to polymerize to form chains. The number of connected polyhedra increases with the severity damage and is consistent with the results of NMR data discussed above. The formation of Si-O-Si chains in damaged zircon may ultimately be the cause of phase separation into discreet domains of amorphous SiO_2 and crystalline ZrO_2 as observed in many previous studies of radiation damage in zircon.

Finally, we briefly review selected contributions to the understanding of complex fluorite derivative structures zirconolite and titanate pyrochlores (with some comparison to zirconate pyrochlores). Zirconolite, ideally monoclinic $CaZrTi_2O_7$, has been the subject of very few modeling and theoretical studies. One such study [116] examined radiation damage in zirconolite using empirical, Buckingham interatomic potentials. They reported both the TDEs for individual atoms and collision cascades introduced into the crystalline lattice, finding the cascade region to consist of a disordered core and a surrounding zone of isolated point defects. Radial distribution functions of the core zone give radial distribution functions that are intermediate to those of crystalline and amorphous zirconolite. The simulations were stated to be consistent with the model of direct impact amorphization (see Reference [27]). A later DFT study [117] attempted to identify stable interstitial sites for intrinsic defects in the complicated structure of zirconolite. This study reported significant dependence on charge states for vacancies and the possible formation of O_2 in Ti and Zr vacancies. Frenkel defect energies of 1.1–3.2 eV for O (charge neutral), 1.97 eV for Ca, 8.22 eV for Zr, and 3.9–6.1 eV for Ti (all cation vacancies are in the 2+/2− charge state) were calculated. It was concluded that the low Frenkel defect energies, combined with their relative stability (high migration barriers?), provide an explanation for the ability to render zirconolite amorphous at low irradiation or alpha-decay dose. Newly developed empirical potentials have been employed in the study of zirconolite using MD to generate 70 keV single and double cascades within large simulation cells [118]. The results of this study confirmed the damage "track" after about 15 ps to consist of a highly disordered region surrounded by a zone with scattered point defects. The relative numbers of defects produced by a single cascade are 542 O, 2106 Ca, 190 Ti, and 131 Zr when normalized to stoichiometry.

The prototypical pyrochlore in nuclear waste forms is generally based on the composition $CaUTi_2O_7$; however, another generalized end-member component is $Ln_2Ti_2O_7$, which serves to accommodate lanthanide fission products from the waste stream. Most simulation studies have actually been conducted on the latter type of pyrochlore composition containing a range of lanthanides or yttrium. MD simulations of $Gd_2Ti_2O_7$ and $Gd_2Zr_2O_7$ were conducted with PKA energies up to 20 keV, showing that more defects were consistently produced in the $Gd_2Ti_2O_7$ pyrochlore composition [119]. Damage cascades in $Gd_2Zr_2O_7$ were smaller and more "ordered," whereas the larger cascades in $Gd_2Ti_2O_7$ appeared to be amorphous. Other MD simulations were used to determine the TDEs of $Gd_2Ti_2O_7$ and $Gd_2Zr_2O_7$ pyrochlores [120]. Interestingly, it was found that the TDE for Ti is >170 eV for all directions that were examined. It was hypothesized that this result is due to the higher energy required for replacement of Gd by Ti in the structure relative to replacement of Gd by Zr. The TDEs of $Y_2Ti_2O_7$ have also been investigated in considerable detail using the DFT code SIESTA, modified to incorporate the generation of a PKA in the simulation cell [121]. Average TDEs of 35.1 eV for Y, 35.4 eV for Ti, 17.0 eV for O on 48f, and 16.2 eV for O on 8b were determined for the three main cubic directions (100, 110, and 111). This work also reported a cation anti-site energy of 2.32 eV for the Y-Ti pair and a Frenkel defect energy of 8.66 eV for Y (displaced to 8b). Other Frenkel defects were found to be unstable, suggesting the possibility of correlated recombination during relaxation. Low formation energies of cation interstitials (1.6–5.3 eV for Y and 1.2–2.2 eV for Ti) may contribute to the ability to render this pyrochlore amorphous under irradiation. However, this particular study did not provide information on how the energy barriers to atomic migration might play a key role in damage recovery.

On the basis of all of these atomistic modeling studies, it is apparent that the ability of a pure crystalline solid to resist transformation to the amorphous state at the same temperature and pressure depends upon several basic properties, including the TDEs, defect formation energies, and barriers to defect migration and recombination. Even in simple oxides such as rutile, the ability to recover from cascade damage may be aided further by cooperative defect migration and recovery processes. This may also be true for the pyrochlore-defect fluorite family of compounds, in which the cation and anion defect formation and migration mechanisms may be intimately linked. A number of studies also indicate that volume-density and bonding considerations are also important in determining the ability to recover from cascade damage on short time scales. For pyrochlore-defect fluorite and related oxide systems in general, an extensive literature covers the above concepts in detail. For further information about these concepts, readers should consult References [31,122 – 125] for the detailed atomistic simulation work and also the analysis of experimental data given in References [52, 126].

7.4 Performance in Aqueous Systems

During the development cycle for use in geological repositories, nuclear waste forms are usually subjected to extensive testing to determine elemental release rates and dissolution mechanisms in aqueous solutions. Moving to higher technological readiness levels (TRLs), the prescribed tests are generally determined by the appropriate regulatory agencies. In cases where the geological repository setting is known, it is highly desirable to see tests that are directly related to the local conditions, for example, temperature, pressure, ground water flow rate, and composition. In the past, both static and flow through types of experiments have been conducted over a range of temperatures. Here we look at some of the salient laboratory experiments and a selection of experiments and observations from studies of natural analogues and their behavior in geological environments.

7.4.1 Laboratory Experiments

Some of the salient results on aqueous durability of nuclear waste forms are summarized in this section. Single-phase $Gd_2Ti_2O_7$ pyrochlores doped with ^{244}Cm were subjected to leach tests at 90 °C in pure water for 14 days using annealed, fully crystalline, and fully amorphous samples [35]. The results of these experiments indicated weight losses of approximately 0.02% and 0.05% for the crystalline and amorphous pyrochlore samples, respectively. By directly measuring the amount of Cm in the leaching fluid, the experiments also indicated that the leach rate of Cm increased by a factor of 17 as a consequence of amorphization. In a later study related to the disposition of excess weapons-grade Pu, flow through dissolution tests were used to examined the behavior of ^{238}Pu doped pyrochlore at pH = 2–12 and at temperatures of 85 °C–90 °C [15]. The results of this study showed very low release rates based on Pu and to a lesser extent U, possibly due to solubility controls on these elements. Experiments carried out on amorphous and recrystallized samples demonstrated very similar release rates for Gd at pH = 2 and 85 °C. Forward dissolution rates of $0.7–1.3 \times 10^{-3}$ g m^{-2} d^{-1} were determined for two different samples. Other experiments have been conducted in order to determine the kinetics of U release from pyrochlore, $(Ca,Gd,Ce,Hf,U)_2Ti_2O_7$, but without the complicating effects of short-lived actinides [127, 128]. These studies report that the pH dependence follows a shallow v-shaped pattern with a minimum near-neutral pH. The release rates for U, converted from the limiting rate constants given in Reference [128], range from 6×10^{-7} to 7×10^{-5} g m^{-2} d^{-1} for all experimental conditions (e.g., T = 25 °C–75 °C and pH = 2–12). With regard to zirconolite, it was shown that the forward dissolution rate of radiation-damaged material is 1.7×10^{-3} g m^{-2} d^{-1} at pH = 2 and 90 °C [38]. These dissolution tests exhibited very little dependence on pH, were not dependent on the level of radiation damage, and no cracking was observed in the zirconolite specimens. The dissolution of synthetic zirconolite without short-lived actinides has been determined as a function of pH using pure water in single pass flow through tests at temperatures of 75 °C and lower [127, 128]. The authors of these studies have independently examined a Ce-Gd-Hf zirconolite containing about 16 wt% UO_2, and the results of the two studies are similar. The release rates determined in Reference [128] for Ti and U indicate that zirconolite dissolves congruently after about 20 days following an initial period where U is released at a somewhat faster rate than Ti. The limiting rate constants are equivalent to U release rates of 6.4×10^{-7} to 1.3×10^{-5} g m^{-2} d^{-2} for zirconolite over the entire pH range of 2–12 and a temperature range of 25 °C–75 °C. The dissolution rate of zirconolite is characterized by a shallow v-shaped pattern with a minimum near pH = 8, similar to the results obtained for pyrochlore.

Hydrothermal experiments have been conducted using a natural, crystalline Ta-based pyrochlore from Lueshe near Lake Kivu of the Democratic Republic of Congo, in pure water and acidic solutions (pH = 0) at 175 °C and 200 °C [129 – 131]. The hydrothermal treatment in the acidic solutions causes the partial replacement of the pyrochlore by a new defect pyrochlore that is characterized by a larger unit cell volume, a large number of vacancies at the A site (A = Ca, Na), and anion vacancies, by molecular water, and possibly OH groups. Analyses of the experimental fluid further revealed that U was lost to the solution. TEM investigations of the interface between the new defect pyrochlore and the unreacted microlite revealed a topotactic relationship between both pyrochlore phases. Furthermore, the interface between both phases was found to be sharp on the nanoscale with a sharp, step-like decrease in the Ca and Na content at the interface toward the defect pyrochlore. TOF-SIMS and confocal micro-Raman mapping of the defect pyrochlore produced in an acidic solution that was enriched with ^{18}O (~47.5 at.%) revealed that the defect pyrochlore is strongly enriched in ^{18}O with a sharp ^{18}O gradient to unreacted areas. The authors suggested that the replacement of microlite by a defect pyrochlore occurs by a pseudomorphic reaction that involves the dissolution of the pyrochlore parent accompanied by the simultaneous reprecipitation of a defect pyrochlore at a moving dissolution-reprecipitation front, a process that has been named the interface-coupled dissolution-reprecipitation process. It is noteworthy that the treatment in pure water for 14 days at 175 °C did not produce reaction zones detectable by back-scattering electron (BSE) imaging. However, significant spectral changes in the powder IR spectra of the reaction product and the detection of Na and Ca in the experimental solution indicated that the pyrochlore also reacted in pure water. The experimental chemical and textural alteration features bear a remarkable resemblance to those seen in naturally altered samples.

In work reported in Reference [132], the hydrothermal alteration of a natural, heavily radiation-damaged pyrochlore (betafite) from a rare earth pegmatite in Southern Norway and a synthetic titanate-based pyrochlore ceramic, $(Ca_{0.76}Ce_{0.75}Gd_{0.23}Hf_{0.21})Ti_2O_7$, produced at the Lawrence Livermore National Laboratory (United States) were compared. The authors treated cuboids of both samples with edge lengths of ~3.3 mm in a 1 M HCl solution containing 43.5 atomic % ^{18}O at 250 °C for 72 h. During the experiments, both samples were transformed mainly into rutile with subordinate anatase. The degree of transformation was significantly higher for the natural radiation-damaged pyrochlore, for example, ^{18}O was highly enriched in the reaction products of both samples with a sharp gradient (on a micrometer scale) toward the unreacted pyrochlore and no apparent diffusion profile. The replacement reaction retained even fine-scale morphological features typical for pseudomorphs. On the basis of these observations, the authors suggested that the dissolution of pyrochlore is spatially and temporally coupled with the precipitation of stable (metastable) TiO_2 phases at an inwardly moving reaction front, a mechanism that is essentially the same as that proposed for the experimental replacement of crystalline material by a defect pyrochlore as discussed in the previous paragraph. The authors pointed out that their results produced under relatively extreme batch-experimental conditions show similarities with nature as well as with results derived from experiments conducted under moderate conditions rather expected in a nuclear repository. The hydrothermal alteration of crystalline ^{239}Pu-doped and X-ray amorphous ^{238}Pu-doped ($D \sim 7 \times 10^{15}$ α mg^{-1}) zirconolite ceramics with the composition $Ca_{0.87}Pu_{0.13}ZrTi_{1.74}Al_{0.26}O_7$ has also been reported [133]. A disk of each ceramic sample was treated in a Teflon© vessel with 2 mL of 1 M HCl at 200 °C for 3 days under autogeneous pressure. Analyses of the experimental fluids by ICP-OES revealed that significantly higher Ca, Al, and Pu concentrations were released into solution from the ^{238}Pu-doped than from the crystalline ^{239}Pu-doped sample. Optical and scanning electron microscope (SEM) investigations of the ^{239}Pu-doped sample after the experiment revealed no signs of alteration, whereas the X-ray amorphous ^{238}Pu-doped sample showed strong alteration features even under the optical microscope. Further examination revealed that the disk was partially covered by TiO_2 crystals. Energy-dispersive X-ray (EDX) analyses showed that the uncovered areas lost Ca, Pu, and Al and have a composition close to $ZrTiO_4$. Such an observation indicates a diffusion-controlled leaching process from the X-ray amorphous ^{238}Pu-doped zirconolite.

Experimental work on Synroc formulations and single-phase perovskite samples have shown that perovskite is very susceptible to incongruent dissolution, releasing more soluble A-site cations (especially Sr) and leaving behind TiO_2 (mainly anatase at low temperatures) as an alteration product. There is also a considerable volume reduction attending this alteration. Thermodynamic calculations and data for natural groundwaters and hydrothermal fluids (up to 300 °C) revealed that perovskite is generally unstable with respect to titanite, titanite + quartz, rutile, or rutile + calcite [134]. Measurements of the dissolution rates of two natural perovskites and synthetic $SrTiO_3$ and $BaTiO_3$ samples were obtained in pure water at 25 °C–300 °C, indicating that elemental release rates are approximately 10^{-1} to 10^{-3} g m^{-2} d^{-1} for Ca, Sr, and for Ba. Studies reported in [135] of Ce-, Nd-, and Sr-doped $CaZrO_3$ perovskite in acidic (HCl, pH = 1) and near-neutral (deionized water, pH = 5.6) solutions at 90 °C showed that the dissolution rates of the impurity elements were near 0.1 g m^{-2} d^{-1}; whereas Ca and Zr were released at rates on the order of 10^{-3} g m^{-2} d^{-1} and 10^{-3} g m^{-2} d^{-1}, respectively. Leach rates were about two orders of magnitude lower in the experiment using deionized water. An investigation of the pH dependence of the release of Ca from two perovskite samples: end-member $CaTiO_3$ and $Ca_{0.78}Sr_{0.04}Nd_{0.18}Ti_{0.82}Al_{0.18}O_3$. was reported in [136]. The results of this study, performed at 90 °C with the pH ranging from 2.1 to 12.9, demonstrated that the Ca release rates generally decrease with increasing pH. After 43 days of leaching, the Ca release rate for the end-member perovskite decreased from 8.9×10^{-2} g m^{-2} d^{-1} at pH = 2.1 to 2.2×10^{-3} g m^{-2} d^{-1} at pH = 12.9.

With regard to the ABO_4-type waste form phases, experiments were performed using natural zircon ($ZrSiO_4$) samples at 87 °C in an aqueous solution containing 5 wt% $KHCO_3$ [137]. The results of this study indicate that the dissolution rate increases by nearly two orders of magnitude from 3×10^{-8} up to 2×10^{-6} g m^{-2} d^{-1} for alpha-decay doses up to 1.0×10^{16} α mg^{-1}; for example, the zircon samples range from highly crystalline to completely amorphous. In related work [138], the forward dissolution rate of zircon has been determined at 120 °C–250 °C. On the basis of the elemental release rate of Si, it was found that the dissolution rate of zircon increases from 1.7×10^{-4} g m^{-2} d^{-1} at 120 °C to 4.1×10^{-4} g m^{-2} d^{-1} at 250 °C. With regard to the orthophosphate monazite ($LnPO_4$) as a potential host phase for actinides and (mainly) lanthanide fission products, aqueous dissolution studies in flowing solutions have shown that dissolution is generally stoichiometric for lanthanides and U over a range of pH values from 1.5 to 10 and temperatures of 50 °C–230 °C [139]. At a temperature of 70 °C, for example, the dissolution rates based on the loss of A-site cations range from 6×10^{-7} to 6×10^{-4} g m^{-2} d^{-1} with the minimum value obtained at pH = 6.

Laboratory studies of apatite dissolution at temperatures relevant to geological disposal situations are uncommon; however, dissolution experiments were performed on Cm-doped synthetic silicate apatite (britholite) $CaNd_4(SiO_4)_3O$ [44] using both fully amorphous and recrystallized samples in deionized water at 90 °C for 14 days. The results indicate that radiation-damage-induced amorphization caused an increase in the dissolution rate by approximately one order of magnitude. A similar experimental study was recently reported for synthetic apatite having the composition $Ca_{4.5}Nd_{0.5}(P_{2.5}Si_{0.5}O_4)_3F$, conducted at 25 °C and pH = 3–12 [140]. The results of this work also showed a negative correlation between the dissolution rate and pH for pH = 2–7, but a constant rate was observed for pH > 8. In these experiments, the authors discovered that the Nd release rates are slower than those of Ca, P, and F and attributed the result to the precipitation of a secondary phase, possibly rhabdophane, $NdPO_4 \cdot nH_2O$. Apatite samples doped with Nd, Th, and U were reported in Reference [141], where measured release rates of 4×10^{-4} g m^{-2} d^{-1} for Nd and 1.3×10^{-4} g m^{-2} d^{-1} were determined for Th in single-phase Nd-Th samples in experiments with 10^{-4} M HNO_3 at 90 °C. In comparison, the release rate of U was found to be 2×10^{-2} g m^{-2} d^{-1} for Nd-U doped apatite under the same conditions. This was attributed to the oxidizing conditions of the experiment and the presence of a second U phase in the Nd-U doped material.

There have been several relevant laboratory studies of the dissolution of kosnarite (NZP)-type phosphates. In Reference [46], dissolution tests were performed on undoped $NaZr_2(PO_4)_3$, several NZP samples containing Cs, Sr, Y, Nd, Gd, and Ca, and a sample containing 20 wt% simulated Purex type waste. These tests were conducted on 37–63 μm powders at 90 °C in an aqueous solution with pH = 5 and a solid surface area to solution volume ratio close to 1.0. After 28 days, the test results gave elemental release rates of 0.002–0.03 g m^{-2} d^{-1} for Na, $0.1–3 \times 10^{-7}$ g m^{-2} d^{-1} for Zr, and 0.002–0.1 g m^{-2} d^{-1} for P. The release rates of Cs and Sr were 0.002 and 0.003 g m^{-2} d^{-1}, respectively, whereas Y, Nd, and Gd were released at rates similar to that of Zr. On the basis of the measured weight losses, the authors suspected that reprecipitation had occurred during the experiments. In the sample prepared with simulated waste, release of Zr, P, Ce, Nd, and Ag were similar to those of Synroc-C, whereas other elements in the NZP showed release rates typically 1–2 orders of magnitude higher than the Synroc sample under the same conditions. Other researchers [142] have synthesized $La_{0.33}Zr_2(PO_4)_3$ and $LaPO_4$ (monazite) ceramics and determined the elemental release rates at a temperature of 96 °C. For experiments conducted with a solid surface area to solution volume ratio of ~0.1, the authors report a minimum release rate of 10^{-3} g m^{-2} d^{-1} for P, whereas the release rates of La and Zr were both $<10^{-5}$ g m^{-2} d^{-1}. In a parallel experiment, they determined that the release rates of La and P from the monazite ceramic were about one order of magnitude lower than the NZP-type compound. Readers are referred to Table 7.3 for a summary and comparison of selected aqueous durability data.

Table 7.3 Summary of dissolution data for nuclear waste forms, including spent fuel and glass for comparison.

Form or phase	Conditions	T (°C)	Mass loss (g cm^{-2}d^{-1})	Element(s)
Spent fuel	Flow pH 8–10*	19–78	$5 \times 10^{-4}–5 \times 10^{-3}$	U
SON68 glass	SPFT	90	$2 \times 10^{-2}–3 \times 10^{-1}$	Si
Synroc-C	MCC-1	95	$\sim 10^{-5}–10^{-4}$	Ti, Zr, Nd
Synroc-D	MCC-1 pH 6	90	$\sim 1 \times 10^{-4}$	U
Perovskite	SPFT pH 2–13	90	$7 \times 10^{-2}–2 \times 10^{-1}$	Ca
Apatite	Static	90	$\sim 7 \times 10^{-5}$	Cm
Brannerite	SPFT pH 2–12	70	$4 \times 10^{-4}–7 \times 10^{-2}$	U
Pyrochlore	SPFT pH 2–12	25–75	$6 \times 10^{-7}–7 \times 10^{-5}$	U
Pyrochlore	SPFT pH 2–12	85–90	$9 \times 10^{-4}–1 \times 10^{-3}$	Pu, Gd
Zircon	Soxhlet	120–250	$7 \times 10^{-5}–4 \times 10^{-4}$	Zr, Si
Zirconia	Static pH 5.6	90	$1 \times 10^{-6}–6 \times 10^{-3}$	Y, Zr, Ce, Nd, Sr
Zirconolite	SPFT pH 2–12	25–75	$6 \times 10^{-7}–1 \times 10^{-5}$	U
Monazite (natural)	Flow pH 1.5–10	70	$6 \times 10^{-7}–6 \times 10^{-4}$	REEs, Th, U

* Suite of experiments with 0.2–20 mmol/L carbonate and either 0.3 or 2% oxygen.

7.4.2 Natural Systems

Numerous investigations have demonstrated that natural pyrochlores are susceptible to alteration via reaction with aqueous fluids over a range of conditions involving pressure, temperature, and fluid composition. At higher temperatures (~300 °C–650 °C, <400 MPa) in highly evolved late stage magmatic fluids, Ca enrichment is commonly observed; whereas the main effect of alteration at moderate temperatures under hydrothermal conditions (~200 °C–350 °C, <200 MPa) is the loss of Na and F, often combined with cation exchange for Sr, Ba, REE, and Fe. Further removal of Na, F, Ca, and O may occur in low-temperature hydrothermal or weathering environments, resulting in the maximum numbers of A-site, Y-site, and X-site vacancies, maximum hydration levels, and more limited exchange large cations such as K, Sr, Cs, Ba, Ce, and Pb in certain environments [143 – 153]. Further details of the most relevant results are provided in some of these, and other studies are given below.

Ti-rich pyrochlores (betafite) from hydrothermal veins in the contact metamorphic zone adjacent to the Adamello igneous massif in northern Italy contain 29–34 wt% UO_2 and are chemically the closest natural analogues presently known for nuclear waste forms [154]. Electron microscopy and microanalytical work have revealed that these pyrochlore samples have only suffered a minor late stage hydration event, as evidenced by lower backscattered electron image contrast around the rims of the grains. The results of this study demonstrate quantitative retention of U and Th for time periods of 40 Ma, even though the crystals experienced cumulative total alpha-decay doses of $3–4 \times 10^{16} \times mg^{-1}$. In two samples of amorphous pyrochlore (betafite) from Bancroft, Ontario, Canada, it was reported that the major result of alteration was hydration, with only minor changes in elemental composition, apart from the precipitation of galena due to mobility of radiogenic Pb [147]. In contrast to these examples, the Ti- and U-rich pyrochlores from granitic pegmatites in Madagascar exhibit a range of alteration effects, including relatively high temperature, post-magmatic hydrothermal processes and lower temperature alteration [147, 155]. If the Ca content falls below 0.2–0.3 atoms *pfu*, these Ti-rich pyrochlores show various levels of recrystallization to a new phase assemblage of liandratite + rutile (or anatase). In the most severe cases documented, this may be accompanied by loss of up to 20%–30% of the original amount of U and local redistribution of the radiogenic Pb. The detailed study of a Ta-pyrochlore from Mozambique provides qualitative information on the effect of radiation damage on the alteration of pyrochlore [156]. These pyrochlore crystals exhibit a distinct growth zoning, characterized by a U-free core and a U-rich rim (up to 17 wt% UO_2). Following uplift and cooling, groundwater penetrated these fractured crystals and led to the deposition of clay minerals along both fractures and cleavage planes. This low-temperature process also led to chemical alteration of the pyrochlore, but only within the zone of the U-rich rim, resulting in the loss of Na, Ca, and F together with increased A-site vacancies (up to about 1.8 vacant A-sites *pfu*). The alteration also led to localized redistribution of radiogenic Pb and to hydration, but U remained immobile.

Previous sotopic age dating work [157] and electron microscopy studies [158] have shown that natural zirconolite exhibits closed system behavior for U, Th, and Pb for up to 650 Ma with little, if any, evidence for geochemical alteration. More recently, it was proposed [159] that zirconolite may become the principal mineral for age dating in mafic igneous rocks due to its ability to retain radiogenic Pb. The analysis of

1200 million year old dolerite intrusive rocks from Western Australia demonstrated that zirconolite returned the same $^{207}Pb/^{206}Pb$ age as zircon and baddeleyite, but the zirconolite age was much more precise (by factors of ~3.3 and 13, respectively). Work reported in References [66] and [160] discusses the alteration of amorphous zirconolite from the 2060 Ma carbonatite complex of Phalaborwa, South Africa, in somewhat greater detail. Electron microprobe analyses, element mapping, and backscattered electron images demonstrate that the alteration is localized along cracks and resulted in the incorporation of Si and loss of Ti, Ca, and Fe. However, in these samples, the Ln, Y, Th, and U contents remained relatively constant across the alteration zones. Radiogenic Pb appears to have been mobile and precipitated mainly within the altered areas as galena. In carbonatites, zirconolite may be replaced along cracks and within micron-sized domains by an unidentified Ba-Ti-Zr-Nb-ACT silicate phase, suggesting that zirconolite may not be stable in the presence of relatively low-temperature hydrothermal fluids enriched in aqueous silicate species [161, 162]. In comparison to zirconolite, perovskite commonly releases Ca in aqueous fluids even at low temperature, breaking down to one or more polymorphs of TiO_2 (generally anatase ± brookite or TiO_2-B). This is well illustrated by the alteration of perovskite to anatase, cerianite, monazite, and crandallite group minerals during severe weathering of carbonatites in Brazil [163]. On the basis of the use electron microscopy to study alteration microstructures, it was proposed [164] that the perovskite-anatase reaction mechanism involves topotactic inheritance of layers of the perovskite Ti-O framework. These observations are generally consistent with laboratory experiments on Synroc-C, which contains zirconolite and perovskite, and the more detailed investigations of the individual phases described above.

Almost all natural brannerites (ideally UTi_2O_6) described thus far are amorphous due to alpha decay of the major U component over geological time. The radiation damage effects and geochemical alteration of brannerite were covered in detail in a recent publication [79]. Brannerite is susceptible to geochemical alteration in natural environments, with substantial loss of U from altered areas and loss of Pb from unaltered areas and to a lesser extent, altered areas. An example of this is shown in Figure 7.10. The observed U loss is partly compensated by the incorporation of large amounts of Si and other elements from the attending fluid phase, including Al, P, and other metal cations. During alteration, Y is also typically removed from the altered brannerite, but the behavior of Fe, Ca, and Ln is more erratic, and these elements may be either lost or gained. Evidence was also presented for loss of U during alteration and is consistent with the observation of U-rich material located within fractures extending into the host rock matrix in two different examples. Alteration of brannerite is consistent with a dissolution-reprecipitation mechanism involving uptake of Si from the aqueous fluid phase and release of U to the local system (Figure 7.11). The major products of the alteration processes are TiO_2 and a glass-like Ti–Si–O phase. However, on the basis of the occurrence of samples with thin alteration rims in placer deposits, brannerite appears to be generally resistant to dissolution at low temperatures in relatively oxidizing environments.

As an example of radiation damage in the AB_2O_6 group of mineral compounds [165], analytical transmission electron microscopy was used to study variations in U content and alpha-recoil damage microstructures in columbite from Yinnietharra, Western Australia. The specimens investigated have inhomogeneous Ti (2.2–4.8 wt% TiO_2) and U (0.2–2.6 wt% UO_2) contents. As a result of the great age and variable U content, the

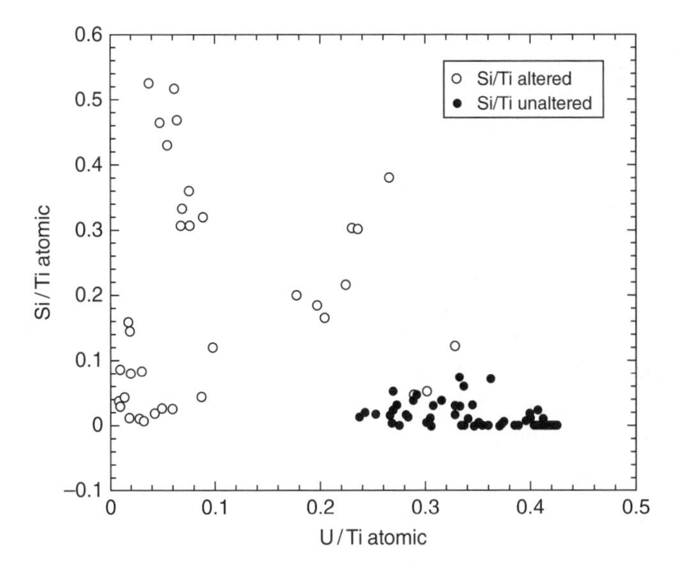

Figure 7.10 Plot of the Si/Ti and U/Ti atomic ratios showing geochemical alteration effects observed in natural brannerite samples, based on SEM-EDX analyses. Alteration is generally characterized by loss of U that may or may not be coupled with gain of Si into the radiation-damaged structure (most natural brannerites were likely to be partially damage to completely amorphous due to radiation damage prior to alteration). From Lumpkin, G.R., Leung, S.H.F., and Ferenczy, J. (2012) Chemistry, microstructure, and alpha decay damage of natural brannerite. *Chemical Geology*, **291**, 55–68.

structure ranges from highly crystalline to completely amorphous. Damage microstructures are consistent with accumulation of isolated alpha-recoil tracks, overlap of tracks producing metamict domains, and increasing metamict domain size until the structure becomes fully metamict. The structure of metamict columbite consists of a random network of edge-sharing and corner-sharing octahedra, lacking long-range periodicity beyond the second coordination sphere. On the basis of the geological age of 1800 million years, the calculated dose ranges from 10^{16} to 10^{17} α mg^{-1}, equivalent to l–12 dpa. The structure becomes fully metamict at approximately 10 dpa, suggesting that isolated alpha-recoil tracks anneal back to the crystalline structure over geological time periods.

7.5 Summary and Conclusions

Materials designed for nuclear waste disposal include a range of ceramics, glass-ceramics, and glass waste forms. The available data have been interpreted in terms of composition, radiation damage effects, and aqueous durability both in the laboratory and in nature. Where possible, atomistic simulations are being increasingly applied in this research field in order to provide both a theoretical basis and experimentally verifiable data on the stability, radiation effects, and defect formation and migration mechanisms for the different waste form phases on picosecond (plus or minus) time scales. Crystalline nuclear waste form phases have also provided the momentum for studies of minerals as a means to understand aspects of waste form crystal chemistry, behavior in aqueous systems, and radiation damage over geological periods of time. However, it is

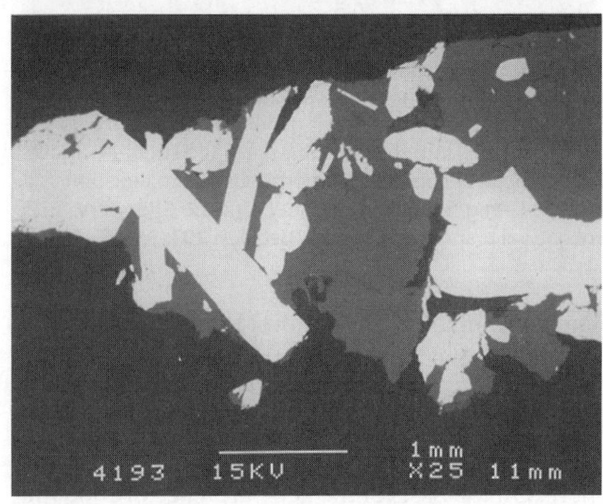

Figure 7.11 Natural brannerite samples from Crocker's Well, Australia (age = 1580 million years, top) and Ticino, Switzerland (age = 20 million years, bottom). The Australian sample exhibits the effects of geochemical alteration and possible partial recrystallization followed by a second period of amorphization due to the age of the sample. The younger Swiss sample is unaltered at the level detectable by electron microscopy and microanalysis. From Lumpkin, G.R., Leung, S.H.F., and Ferenczy, J. (2012) Chemistry, microstructure, and alpha decay damage of natural brannerite. *Chemical Geology*, **291**, 55–68.

acknowledged that there are limitations to the utility of natural analogue studies depending upon the degree of analogy to the proposed geological repository and other factors such as chemical composition.

In general, the available data suggest that host phases such as brannerite, monazite, pyrochlore, zircon, and zirconolite are generally resistant to dissolution in aqueous fluids at low temperatures. Aqueous durability may or may not extend to hydrothermal conditions, say above ~200 °C, depending upon the specifics of fluid composition, temperature, and pressure. Furthermore, radiation damage may increase the release of radionuclides from these phases by one or more orders of magnitude. At elevated temperatures, for example, the crichtonite group mineral davidite may break down to new phase assemblages including titanite, ilmenite, and rutile according to the observations made on natural samples. Perovskite is generally less resistant to dissolution at low temperatures and breaks down to TiO_2, releasing A-site cations to the aqueous fluid. Studies of radiation damage indicate that the oxide and silicate phases become amorphous as a result of the gradual accumulation of alpha-recoil collision cascades. Monazite, however, tends to remain crystalline on geological time scales, a very attractive property that may minimize changes in physical properties such as density and volume, thereby reducing the potential for cracking, which is a major concern for zircon.

In closing, it is instructive to summarize some of basic fundamental parameters that determine radiation tolerance or lack thereof in nuclear waste form phases:

- Radiation tolerance in promoted at the atomistic level and femtosecond to nanosecond time scales by relatively high TDEs, high defect formation energies, and low energy barriers to atomic migration and recovery.
- Higher levels of ionic bonding are also related to radiation tolerance in a general way. With increasing levels of covalent bonding, structures also show increasing polymerization, and connectivity/structural freedom arguments may come into play in reducing the radiation tolerance.
- Detailed studies of natural samples by experiment and atomistic simulation indicate that covalency plays another very important role in determining the response to radiation damage. During the formation and evolution of the alpha-recoil collision cascade, it is now clear that highly covalent elements like Si may polymerize, leading to a chemical segregation of elements at the atomic scale, planting the "seeds" for phase separation.
- Increasing temperature provides energy to drive recovery of damage, but this varies with structure type and composition. Radiation-assisted recovery mechanisms may play a significant role at lower temperatures. Higher temperatures also provide energy to drive phase separation.
- On geological time scales, dose rate effects may become apparent and, together with atomic diffusion, may promote increased recovery of radiation damage.

Acknowledgments

My parents were instrumental in giving me the freedom to find my own path in life, sometimes the hard way. For this I am deeply indebted to them. I am also very grateful to all of the wonderful people who have supported me over the years and the learning opportunities that were available to me through that support, together with my own love of nature, science, and music, among other things. In particular, I would like to express my gratitude to the following people: Paul Ribbe, Gerry Gibbs, Wally Lowry, Fred Read, John Reynolds, Tony Zaikowski, Clive Jones, Rod Ewing, Michael Carpenter, Simon Redfern, Lou Vance, Kath Smith, and all of my colleagues at the Australian Nuclear Science and Technology Organisation, among others. Special thanks go to Yan Gao for assistance with this project.

References

1 Amano, Y. (2015) International Atomic Energy Agency report on The Fukushima Daiichi Accident.

2 Zinkle, S.J. and Was, G.S. (2013) Materials challenges in nuclear energy. *Acta Materialia*, **61**, 735–758.

3 Degueldre, C., Kasemeyer, U., Botta, F., and Ledergerber, G. (1996) Plutonium incineration in LWR's by a once-through cycle with a rock-like fuel, in *Scientific Basis for Nuclear Waste Management XIX, Materials Research Society Symposium Proceedings*, vol. **412** (eds. W.M. Murphy and D.A. Knecht), Materials Research Society, Pittsburgh, pp. 15–23.

4 Harker, A.B. (1988) Tailored ceramics, in *Radioactive Waste Forms for the Future* (eds. W. Lutze and R.C. Ewing), North-Holland, Amsterdam, pp. 335–392.

5 Fielding, P.E. and White, T.J. (1987) Crystal chemical incorporation of high level waste species in aluminotitanate-based ceramics: Valence, location, radiation damage, and hydrothermal durability. *Journal of Materials Research*, **2**, 387–414.

6 Ringwood, A.E., Kesson, S.E., Reeve, K.D., et al. (1988) Synroc, in *Radioactive Waste Forms for the Future* (eds. W. Lutze and R.C. Ewing), North-Holland, Amsterdam, pp. 233–234.

7 Ball, C.J., Buykx, W.J., Dickson, F.J., et al. (1989) Titanate ceramics for the stabilization of partially reprocessed nuclear fuel elements. *Journal of the American Ceramic Society*, **72**, 404–414.

8 Vance, E.R. (1994) Synroc: A suitable waste form for actinides. *Materials Research Society Bulletin*, **XIX**, 28–32.

9 Carter, M.L., Vance, E.R., Mitchell, D.R.G., et al. (2002) Fabrication, characterization, and leach testing of hollandite, $(Ba,Cs)(Al,Ti)_2Ti_6O_{16}$. *Journal of Materials Research*, **17**, 2578–2589.

10 Ewing, R.C., Lutze, W., and Webber, W.J. (1995) Zircon: A host-phase for the disposal of weapons plutonium. *Journal of Materials Research*, **10**, 243–246.

11 Boatner, L.A. and Sales, B.C. (1988) Monazite, in *Radioactive Waste Forms for the Future* (eds. W. Lutze and R.C. Ewing), North-Holland, Amsterdam, pp. 495–564.

12 Kinoshita, H., Kuramoto, K., Uno, M., et al. (2000) Chemical durability of yttria-stabilized zirconia for highly concentrated TRU wastes, in *Scientific Basis for Nuclear Waste Management XXIII, Materials Research Society Symposium Proceedings*, vol. **608** (eds. R.W. Smith and D.W. Shoesmith), Materials Research Society, Warrendale, pp. 393–398.

13 Ewing, R.C., Weber, W.J., and Lian, J. (2004) Nuclear waste disposal – pyrochlore ($A_2B_2O_7$): Nuclear waste form for the immobilization of plutonium and "minor" actinides. *Journal of Applied Physics*, **95**, 5949–5971.

14 Stefanovsky, S.V., Yudintsev, S.V., Gieré, R., and Lumpkin, G.R. (2004) Nuclear waste forms, in *Energy, Waste, and the Environment: A Geochemical Perspective*, vol. **236** (eds. R. Gieré and P. Stille), Geological Society, London, pp. 36–63.

15 Strachan, D.M., Scheele, R.D., Buck, E.C., et al. (2005) Radiation damage effects in candidate titanates for Pu disposition: Pyrochlore. *Journal of Nuclear Materials*, **345**, 109–135.

16 Icenhower, J.P., Strachan, D.M., McGrail, B.P., et al. (2006) Dissolution kinetics of pyrochlore ceramics for the disposition of plutonium. *American Mineralogist*, **91**, 39–53.

17 Smith, K.L., Lumpkin, G.R., Blackford, M.G., et al. (1992) The durability of Synroc. *Journal of Materials Research*, **190**, 287–294.
18 Lumpkin, G.R., Smith, K.L., and Blackford, M.G. (1995) Partitioning of uranium and rare earth elements in Synroc: effect of impurities, metal additive, and waste loading. *Journal of Nuclear Materials*, **224**, 31–42.
19 Lumpkin, G.R. (2001) Alpha-decay damage and aqueous durability of actinide host phases in natural systems. *Journal of Nuclear Materials*, **289**, 136–166.
20 Lumpkin, G.R. (2006) Ceramic waste forms for actinides. *Elements*, **2**, 365–372.
21 Begg, B.D. (2003) Titanate ceramic matrices for nuclear waste immobilisation. *Research Advances in Ceramics*, **1**, 49–62.
22 Hayward P.J. and Cecchetto E.V. (1982) Development of sphene based glass ceramics tailored for Canadian waste disposal conditions, in *Scientific Basis for Nuclear Waste Management*, vol. **4** (ed. S.V. Topp), Elsevier, New York, 91–98.
23 Hayward, P.J., Doern, D.C., and George, I.M. (1990) Dissolution of a sphene glass-ceramic, and of its component sphene and glass phases, in Ca-Na-Cl brines. *Journal of the American Ceramic Society*, **73**, 544–551.
24 Grossbeck, M.L. (2012) Effect of radiation on strength and ductility of metals and alloys. *Comprehensive Nuclear Materials*, **1**, 99–122.
25 Zinkle, S.J. (2012) Radiation-induced effects on microstructure. *Comprehensive Nuclear Materials*, **1**, 65–98.
26 Lumpkin, G.R. and Geisler-Wierwille T. (2012) Minerals and Natural Analogues, in *Comprehensive Nuclear Materials*, vol. **5** (ed. R.J.M. Konings), Elsevier, Amsterdam, pp. 563–600.
27 Weber, W.J. (2000) Models and mechanisms of irradiation-induced amorphization in ceramics. *Nuclear Instruments and Methods in Physics Research Section B: Beam Interactions with Materials and Atoms*, **166–167**, 98–106.
28 Gupta, P.K. (1993) Rigidity, connectivity, and glass-forming ability. *Journal of the American Ceramic Society*, **76**, 1088–1095.
29 Hobbs, L.W., Sreeram, A.N., Jesurum, C.E., and Berger, B.A. (1996) Structural freedom, topological disorder, and the irradiation-induced amorphization of ceramic structures. *Nuclear Instruments and Methods in Physics Research Section B: Beam Interactions with Materials and Atoms*, **166**, 18–25.
30 Trachenko, K. (2004) Understanding resistance to amorphization by radiation damage. *Journal of Physics: Condensed Matter*, **16**, R1491–R1515.
31 Trachenko, K., Pruneda, J.M., Artacho, E., and Dove, M.T. (2005) How the nature of the chemical bond governs resistance to amorphization by radiation damage. *Physical Review B*, **71**, 184104.
32 Trachenko, K., Dove, M.T., Artacho, E., et al. (2006) Atomistic simulations of resistance to amorphization by radiation damage. *Physical Review B*, **73**, 174207.
33 Clinard, F.W., Jr., Hobbs, L.W., Land, C.C., et al. (1982) Alpha-decay self-irradiation damage in ^{238}Pu-substituted zirconolite. *Journal of Nuclear Materials*, **105**, 248–256.
34 Clinard, F.W., Jr., Peterson, D.E., Rohr, D.L., and Hobbs, L.W. (1984) Self-irradiation effects in 238Pu-substituted zirconolite, I: Temperature dependence of damage. *Journal of Nuclear Materials*, **126**, 245–254.
35 Weber, W.J., Wald, J.W., and Matzke, H.J. (1986) Effects of self-radiation damage in Cm-doped $Gd_2Ti_2O_7$ and $CaZrTi_2O_7$. *Journal of Nuclear Materials*, **138**, 196–209.

36 Wald, J.W. and Offermann, P. (1982) A study of radiation effects in curium doped $Gd_2Ti_2O_7$ (pyrochlore) and $CaZrTi_2O_7$ (zirconolite), in *Scientific Basis for Nuclear Waste Management V* (ed. W. Lutze), Elsevier Science Publishing, New York, pp. 369–378.
37 Clinard, F.W., Jr., Rohr, D.L., and Roof, R.B. (1984) Structural damage in a self-irradiated zirconolite-based ceramic. *Nuclear Instruments and Methods in Physics Research B*, **1**, 581–586.
38 Strachan, D.M., Scheele, R.D., Buck, E.C., et al. (2008) Radiation damage effects in candidate titanates for Pu disposition: Zirconolite. *Journal of Nuclear Materials*, **372**, 16–31.
39 Mosley, W.C. (1971) Self-radiation damage in curium-244 oxide and aluminate. *Journal of the American Ceramic Society*, **54**, 475–479.
40 Weber, W.J. (1990) Radiation-induced defects and amorphization in zircon. *Journal of Materials Research*, **5**, 2687–2697.
41 Weber, W.J., Ewing, R.C., and Wang, L.M. (1994) The radiation-induced crystalline-to-amorphous transition in zircon. *Journal of Materials Research*, **9**, 688–698.
42 Luo, J.S. and Liu, G.K. (2001) Microscopic effects of self-radiation damage in ^{244}Cm-doped $LuPO_4$ crystals. *Journal of Materials Research*, **16**, 366–372.
43 Burakov, B.E., Yagovkina, M.A., Garbuzov, V.M., et al. (2004) Self-irradiation of monazite ceramics: contrasting behavior of $PuPO_4$ and $(La,Pu)PO_4$ doped with Pu-238, in *Scientific Basis for Nuclear Waste Management XXVIII, Materials Research Society Symposium Proceedings*, vol. **824** (eds. J.M. Hanchar, S. Stroes-Gascoyne and L. Browning), Materials Research Society, Warrendale, pp. 219–224.
44 Weber, W.J. (1983) Radiation-induced swelling and amorphization in $Ca_2Nd_8(SiO_4)_6O_2$. *Radiation Effects*, **77**, 295–308.
45 Weber, W.J. and Matzke, H. (1986) Effects of radiation on microstructure and fracture properties in $Ca_2Nd_8(SiO_4)_6O_2$. *Materials Letters*, **5**, 9–16.
46 Zyryanov, V.N. and Vance, E.R. (1997) Comparison of sodium zirconium phosphate-structured HLW forms and synroc for high-level nuclear waste immobilization, in *Scientific Basis for Nuclear Waste Management XX, Materials Research Society Symposium Proceedings*, vol. **465** (eds. W.J. Gray and I.R. Triay), Materials Research Society, Pittsburgh, pp. 499–415.
47 Smith, K.L., Zaluzec, N.J., and Lumpkin, G.R. (1997) In situ studies of ion irradiated zirconolite, pyrochlore and perovskite. *Journal of Nuclear Materials*, **250**, 36–52.
48 Wang, S.X., Lumpkin, G.R., Wang, L.M., and Ewing, R.C. (2000) Ion irradiation-induced amorphization of six zirconolite compositions. *Nuclear Instruments and Methods in Physics Research B*, **166–167**, 293–298.
49 Wang, S.X., Wang, L.M., Ewing, R.C., and Kutty, K.V.G. (2000) Ion irradiation of rare earth and yttrium titanate pyrochlores. *Nuclear Instruments and Methods in Physics Research B*, **169**, 135–140.
50 Begg, B.D., Hess, N.J., Weber, W.J., et al. (2001) Heavy-ion irradiation effects on structures and acid dissolution of pyrochlores. *Journal of Nuclear Materials*, **288**, 208–216.
51 Lian, J., Chen, J., Wang, L.M., et al. (2003) Radiation-induced amorphization of rare-earth titanate pyrochlores. *Physical Review B*, **68**, 134107.
52 Lumpkin, G.R., Pruneda, M., Rios, S., et al. (2007) Nature of the chemical bond and prediction of radiation tolerance in pyrochlore and defect fluorite compounds. *Journal of Solid State Chemistry*, **180**, 1512–1518.

53 Zhang, J., Zhang, F., Lang, M., et al. (2013) Ion-irradiation-induced structural transitions in orthorhombic Ln_2TiO_5. *Acta Materialia*, **61**, 4191–4199.

54 Whittle, K.R., Blackford, M.G., Aughterson, R.D., et al. (2011) Ion irradiation of novel yttrium/ytterbium-based pyrochlores: the effect of disorder. *Acta Materialia*, **59**, 7530–7537.

55 Aughterson, R.D., Lumpkin, G.R., Ionescu, M., et al. (2015) Ion-irradiation resistance of the orthorhombic Ln_2TiO_5 (Ln = La, Pr, Nd, Sm, Eu, Gd, Tb and Dy) series. *Journal of Nuclear Materials*, **467**, 683–691.

56 Aughterson, R.D., Lumpkin, G.R., De los Reyes, M., et al. (2016) The influence of crystal structure on ion-irradiation tolerance in the $Sm_{(x)}Yb_{(2-x)}TiO_5$ series. *Journal of Nuclear Materials*, **471**, 17–24.

57 Lian, J., Zhang, J.M., Pointeau, V., et al. (2009) Response of synthetic coffinite to energetic ion beam irradiation. *Journal of Nuclear Materials*, **393**, 481–486.

58 Meldrum, A., Boatner, L.A., and Ewing, R.C. (1997) Displacive radiation effects in the monazite- and zircon-structure orthophosphates. *Physical Review B*, **56**, 805–814.

59 Meldrum, A., Zinkle, S.J., Boatner, L.A., and Ewing, R.C. (1999) Heavy-ion irradiation effects in the ABO_4 orthosilicates: Decomposition, amorphization, and recrystallization. *Physical Review B* **59**, 3981–3992.

60 Whittle, K.R., Blackford, M.G., Smith, K.L., et al. (2015) Radiation effects in Zr and Hf containing garnets. *Journal of Nuclear Materials*, **462**, 508–513.

61 Utsonomiya, S., Wang, L.M., Yudintsev, S., and Ewing, R.C. (2002) Ion irradiation-induced amorphization and nano-crystal formation in garnets. *Journal of Nuclear Materials*, **303**, 177–187.

62 Utsonomiya, S., Yudintsev, S., and Ewing, R.C. (2005) Radiation effects in ferrate garnet. *Journal of Nuclear Materials*, **336**, 251–260.

63 Zhang, J., Livshits, T.S., Lizin, A.A., et al. (2010) Irradiation of synthetic garnet by heavy ions and α-decay of ^{244}Cm. *Journal of Nuclear Materials*, **407**, 137–142.

64 Krivokoneva, G.K. and Sidorenko, G.A. (1971) The essence of the metamict transformation in pyrochlores. *Geochemistry International*, **8**, 113–122.

65 Lumpkin, G.R. and Ewing, R.C. (1988) Alpha-decay damage in minerals of the pyrochlore group. *Physics and Chemistry of Minerals*, **16**, 2–20.

66 Lumpkin, G.R., Hart, K.P., McGlinn, P.J., et al. (1994) Retention of actinides in natural pyrochlore and zirconolites. *Radiochimica Acta*, **66**/67, 469–474.

67 Salje, E.K.H., Chrosch, J., and Ewing, R.C. (1999) Is "metamictization" of zircon a phase transition? *American Mineralogist*, **84**, 1107–1116.

68 Greegor, R.B., Lytle, F.W., Chakoumakos, B.C., et al. (1985) An investigation of metamict and annealed natural pyrochlores by X-ray absorption spectroscopy, in *Scientific Basis for Nuclear Waste Management VIII, Materials Research Society Symposium Proceedings*, vol. **44** (eds. C.M. Jantzen, J.A. Stone and R.C. Ewing), Materials Research Society, Pittsburgh, pp. 655–662.

69 Greegor, R.B., Lytle, F.W., Chakoumakos, B.C., et al. (1985) An investigation of uranium L-edges of metamict and annealed betafite, in *Scientific Basis for Nuclear Waste Management IX, Materials Research Society Symposium Proceedings*, vol. **50** (ed. L.O. Werme), Materials Research Society, Pittsburgh, pp. 388–392.

70 Greegor, R.B., Lytle, F.W., Chakoumakos, B.C., et al. (1987) An X-ray absorption spectroscopy investigation of the Ta site in alpha-recoil damaged natural pyrochlores, in *Scientific Basis for Nuclear Waste Management X, Materials Research Society*

Symposium Proceedings, vol. **84** (eds. J.K. Bates and W.B. Seefeldt), Materials Research Society, Pittsburgh, pp. 645–658.

71 Greegor, R.B., Lytle, F.W., Chakoumakos, B.C., et al. (1989) Characterization of radiation damage at the Nb site in natural pyrochlores and samarskites by X-ray absorption spectroscopy, in *Scientific Basis for Nuclear Waste Management XII, Materials Research Society Symposium Proceedings*, vol. **127** (eds. W. Lutze and R.C. Ewing), Materials Research Society, Pittsburgh, pp. 655–662.

72 Lumpkin, G.R., Ewing, R.C., and Chakoumakos, B.C. (1986) Alpha-recoil damage in zirconolite ($CaZrTi_2O_7$). *Journal of Materials Research*, **1**, 564–576.

73 Farges, F., Ewing, R.C., and Brown, G.E. (1993) The structure of aperiodic, metamict $(Ca,Th)ZrTi_2O_7$ (zirconolite): An EXAFS study of the Zr, Th, and U sites. *Journal of Materials Research*, **8**, 1983–1995.

74 Farges, F. (1997) Five fold-coordinated Ti^{4+} in metamict zirconolite and titanite: a new occurrence shown by Ti K-edge XANES spectroscopy. *American Mineralogist*, **82**, 44–50.

75 Lumpkin, G.R., Smith, K.L., and Gieré, R. (1997) Application of analytical electron microscopy to the study of radiation damage in the complex oxide mineral zirconolite. *Micron*, **28**, 57–68.

76 Lumpkin, G.R., Smith, K.L., Blackford, M.G., et al. (1998) The crystalline-amorphous transformation in natural zirconolite: evidence for long-term annealing, in *Scientific Basis for Nuclear Waste Management XXI, Materials Research Society Proceedings*, vol. **506** (eds. I.G. McKinley and C. McCombie), Warrendale, 215–222.

77 Mitchell, R.H. and Chakhmouradian, A.R. (1998) Th-rich loparite from the Khibina alkaline complex, Kola Peninsula: isomorphism and paragenesis. *Mineralogical Magazine*, **62**, 341–353.

78 Thomas, B.S. and Zhang, Y. (2003) A kinetic model of the oxidative dissolution of brannerite, UTi_2O_6. *Radiochimica Acta*, **91**, 463–472.

79 Lumpkin, G.R., Leung, S.H.F., and Ferenczy, J. (2012) Chemistry, microstructure, and alpha decay damage of natural brannerite. *Chemical Geology*, **291**, 55–68.

80 Graeser, S. and Guggenheim, R. (1990) Brannerite from Lengenbach, Binntal (Switzerland). *Schweizerische Mineralogische und Petrographische Mitteilungen*, **70**, 325–331.

81 Gong, W.L., Ewing, R.C., Wang, L.M., and Xie, H.S. (1995) Crichtonite structure type ($AM_{21}O_{38}$ and $A_2M_{19}O_{36}$) as a host phase in crystalline waste form ceramics. *Materials Research Society Symposium Proceedings*, **353**, 807–815.

82 Lumpkin, G.R., Blackford, M.G., and Colella, M. (2013) Chemistry and radiation effects of davidite. *American Mineralogist*, **98**, 275–278.

83 Hurley, P.M. and Fairbairn, H.W. (1953) Radiation damage in zircon: a possible age method. *Geological Society of America Bulletin*, **64**, 659–674.

84 Holland, H.D. and Gottfried, D. (1955) The effect of nuclear radiation on the structure of zircon. *Acta Crystallographica*, **8**, 291–300.

85 Murakami, T., Chakoumakos, B.C., Ewing, R.C., et al. (1991) Alpha-decay event damage in zircon. *American Mineralogist*, **76**, 1510–1532.

86 Chakoumakos, B.C., Oliver, W.C., Lumpkin, G.R., and Ewing, R.C. (1991) Hardness and elastic modulus of zircon as a function of heavy-particle irradiation dose: I. In situ alpha-decay event damage. *Radiation Effects and Defects in Solids*, **118**, 393–403.

87 Farnan, I. and Salje, K.H. (2001) The degree and nature of radiation damage in zircon observed by ^{29}Si nuclear magnetic resonance. *Journal of Applied Physics*, **89**, 2084–2090.

88 Moench, R.H. (1962) Properties and paragenesis of coffinite from the Woodrow mine, New Mexico. *American Mineralogist*, **47**, 26–33.
89 Deditius, A.P., Pointeau, V., Zhang, J.M., and Ewing, R.C. (2012) Formation of nanoscale Th-coffinite. *American Mineralogist*, **97**, 681–693.
90 Foord, E.E., Cobban, R.R., and Brownfield, I.K. (1985) Uranoan thorite in lithophysal rhyolite-Topaz Mountain, Utah, USA. *Mineralogical Magazine*, **49**, 729–731.
91 Lumpkin, G.R. and Chakoumakos, B.C. (1988) Chemistry and radiation effects of thorite group minerals from the Harding pegmatite, Taos County, New Mexico. *American Mineralogist*, **73**, 1405–1419.
92 Hayward, P.J. (1988) Glass-ceramics, in *Radioactive Waste Forms for the Future* (eds. W. Lutze and R.C. Ewing), North-Holland, Amsterdam, pp. 427–493.
93 Hawthorne, F.C., Groat, L.A., Raudsepp, M., et al. (1991) Alpha-decay damage in titanite. *American Mineralogist*, **76**, 370–396.
94 Vance, E.R. and Metson, J.B. (1985) Radiation damage in natural titanites. *Physics and Chemistry of Minerals*, **12**, 255–260.
95 Arden, K.M. and Halden, N.M. (1999) Crystallization and alteration history of britholite in rare-earth-element-enriched pegmatitic segregations associated with the Eden Lake Complex, Manitoba, Canada. *Canadian Mineralogist*, **37**, 1239–1253.
96 Carpéna, J., Kienast, J.R., Ouzegane, K., and Jehanno, C. (1998) Evidence of the contrasted fission-track clock behavior of the apatites from In Ouzzal carbonatites (northwest Hoggar): the low-temperature thermal history of an Archean basement. *Geological Society of America Bulletin*, **100**, 1237–1243.
97 Yudintseva, T.S. (2006) Radiation stability of natural britholites, in *Scientific Basis for Nuclear Waste Management XXIX, Materials Research Society Symposium Proceedings*, Vol. **932** (ed. P. Van Iseghem), Materials Research Society, Warrendale, pp. 1049–1055.
98 Thomas, B.S., Marks, N.A., Corrales L.R., and Devanathan, R. (2005) Threshold displacement energies in rutile TiO_2: A molecular dynamics simulation study. *Nuclear Instruments and Methods in Physics Research B*, **239**, 191–201.
99 Marks, N.A., Thomas, B.S., Smith, K.L., and Lumpkin, G.R. (2008) Thermal spike recrystallisation: Molecular dynamics simulation of radiation damage in polymorphs of titania. *Nuclear Instruments and Methods in Physics Research B*, **266**, 2665–2670.
100 Lumpkin, G.R., Smith, K.L., Blackford, M.G., et al. (2008) Experimental and atomistic modelling study of ion irradiation damage in thin crystals of the TiO_2 polymorphs. *Physical Review B*, **77**, 214201.
101 Qin, M.J., Kuo, E.Y., Whittle, K.R., et al. (2013) Density and structural effects in the radiation tolerance of TiO_2 polymorphs. *Journal of Physics: Condensed Matter*, **25**, 355402.
102 Robinson, M., Marks, N.A., Whittle, K.R., and Lumpkin, G.R. (2012) Systematic calculation of threshold displacement energies: Case study in rutile. *Physical Review B*, **85**, 104105.
103 Robinson, M., Marks, N.A., and Lumpkin, G.R. (2012) Sensitivity of the threshold displacement energy to temperature and time. *Physical Review B*, **86**, 134105.
104 Robinson, M., Marks, N.A., and Lumpkin G.R. (2014) Structural dependence of threshold displacement energies in rutile, anatase and brookite TiO_2. *Materials Chemistry and Physics*, **147**, 311–318.
105 Uberuaga, B.P. and Bai X.M. (2011) Defects in rutile and anatase polymorphs of TiO_2: kinetics and thermodynamics near grain boundaries. *Journal of Physics: Condensed Matter*, **23**, 435004.

106 Kuo, E.Y., Qin, M.J., Thorogood, G.J., et al. (2013) Technetium and ruthenium incorporation into rutile TiO_2. *Journal of Nuclear Materials*, **441**, 380–389.
107 Thomas, B.S., Marks, N.A., and Begg, B.D. (2007) Inversion of defect interactions due to ordering in $Sr_{1-3x/2}La_xTiO_3$ perovskites: An atomistic simulation study. *Nuclear Physical Review B*, **74**, 214109.
108 Bi, Z., Uberuaga, B.P., Vernon, L.J., et al. (2013) Radiation damage in heteroepitaxial $BaTiO_3$ thin films on $SrTiO_3$ under Ne ion irradiation. *Journal of Applied Physics*, **113**, 023513.
109 Liu, B., Xiao, H.Y., Aidhy, D.S., and Weber, W.J. (2013) *Ab initio* molecular dynamics simulations of threshold displacement energies in $SrTiO_3$. *Journal of Physics: Condensed Matter*, **25**, 485003.
110 Won, J., Vernon, L.J., Karakuscu, A., et al. (2013) The role of non-stoichiometric defects in radiation damage evolution of $SrTiO_3$. *Journal of Materials Chemistry A*, **1**, 9235–9245.
111 Thomas, B.S., Marks, N.A., and Begg, B.D. (2005) Developing pair potentials for simulating radiation damage in complex oxides. *Nuclear Instruments and Methods in Physics Research B*, **228**, 288–292.
112 Howard, C.J., Lumpkin, G.R., Smith, R.I., and Zhang, Z. (2004) Crystal structures and phase transition in the system $SrTiO_3 - La_{2/3}TiO_3$. *Journal of Solid State Chemistry*, **177**, 2726–2732.
113 Smith, K.L., Lumpkin, G.R., Blackford, M.G., et al. (2008) In situ radiation damage studies of $La_xSr_{1-3x/2}TiO_3$ perovskites. *Journal of Applied Physics*, **103**, 083531.
114 Thomas, B.S., Marks, N.A., and Harrowell, P. (2006) Defects and threshold displacement energies in $SrTiO_3$ perovskite using atomistic computer simulations. *Nuclear Instruments and Methods in Physics Research B*, **254**, 211–218.
115 Trachenko, K., Dove, M.T., and Salje, E.K.H. (2001) Atomistic modelling of radiation damage in zircon. *Journal of Physics: Condensed Matter*, **13**, 1947–1959.
116 Veiller, L., Crocombette, J.P., and Ghaleb, D. (2002) Molecular dynamics simulation of the α-recoil nucleus displacement cascade in zirconolite. *Journal of Nuclear Materials*, **306**, 61–72.
117 Mulroue, J., Morris, A.J., and Duffy, D.M. (2011) Ab initio study of intrinsic defects in zirconolite. *Physical Review B*, **84**, 094118.
118 Chappell, H.F., Dove, M.T., Trachenko, K., et al. (2013) Structural changes in zirconolite under a-decay. *Journal of Physics: Condensed Matter*, **25**, 055401.
119 Purton, J.A. and Allan, N.L. (2002) Displacement cascades in $Gd_2Ti_2O_7$ and $Gd_2Zr_2O_7$: A molecular dynamics study. *Journal of Materials Chemistry*, **12**, 2923–2926.
120 Devanathan, R. and Weber, W.J. (2005) Insights into the radiation response of pyrochlores from calculations of threshold displacement events. *Journal of Applied Physics*, **98**, 086110.
121 Xiao, H.Y., Gao, F., and Weber, W.J. (2010) Threshold displacement energies and defect formation energies in $Y_2Ti_2O_7$. *Journal of Physics: Condensed Matter*, **22**, 415800.
122 Panero, W.R., Stixrude, L., and Ewing, R.C. (2004) First-principles calculation of defect-formation energies in the $Y_2(Ti, Sn, Zr)_2O_7$ pyrochlores. *Physical Review B*, **70**, 054110.
123 Pruneda, J.M. and Artacho, E. (2005) First-principles study of structural, elastic, and bonding properties of pyrochlores. *Physical Review B*, **72**, 085107.

124 Zhang, Z.L., Xiao, H.Y., Zu, X.T., et al. (2008) First-principles calculation of structural and energetic properties for $A_2Ti_2O_7$ (A = Lu, Er, Y, Gd, Sm, Nd, La). *Journal of Materials Research*, **24**, 1335–1341.
125 Jiang, C., Stanek, C.R., Sickafus, K.E., and Uberuaga, B.P. (2009) First-principles prediction of disordering tendencies inpyrochlore oxides. *Physical Review B*, **79**, 104203.
126 Lumpkin, G.R., Smith, K.L., Blackford, M.G., et al. (2009) Ion irradiation of ternary pyrochlore oxides. *Chemistry of Materials*, **21**, 2746–2754.
127 Roberts, S.K., Bourcier, W.L., and Shaw, H.F. (2000) Aqueous dissolution kinetics of pyrochlore, zirconolite and brannerite at 25, 50, and 75 °C. *Radiochimica Acta*, **88**, 539–543.
128 Zhang, Y., Hart, K.P., Bourcier, W.L., et al. (2001) Kinetics of uranium release from Synroc phases. *Journal of Nuclear Materials*, **289**, 254–262.
129 Geisler, T., Berndt, J., Meyer, H.W., et al. (2004) Low-temperature aqueous alteration of crystalline pyrochlore: Correspondence between nature and experiment. *Mineralogical Magazine*, **68**, 905–922.
130 Geisler, T., Pöml, P., Stephan, T., et al. (2005) Experimental observation of an interface-controlled pseudomorphic replacement reaction in a natural crystalline pyrochlore. *American Mineralogist*, **90**, 1683–1687.
131 Geisler, T., Seydoux-Guillaume, A.M., Pöml, P., et al. (2005) Experimental hydrothermal alteration of crystalline and radiation-damaged pyrochlore. *Journal of Nuclear Materials*, **344**, 17–23.
132 Pöml, P., Menneken, M., Stephan, T., et al. (2007) Mechanism of hydrothermal alteration of natural self-irradiated and synthetic titanate-based pyrochlore. *Geochimica et Cosmochimica Acta*, **71**, 3311–3322.
133 Pöml, P. (2008) Investigations on the suitability of pyrochlore and zirconolite compounds as nuclear waste forms. Unpublished PhD thesis, University of Münster, Germany, 87 pp.
134 Nesbitt, H.W., Bancroft, G.M., Fyfe, W.S., et al. (1981) Thermodynamic stability and kinetics of perovskite dissolution. *Nature*, **289**, 358–362.
135 Kamizono, H., Hayakawa, I., and Muraoka, S. (1991) Durability of zirconium-containing ceramic waste forms in water. *Journal of the American Ceramic Society*, **74**, 863–864.
136 McGlinn, P.J., Hart, K.P., Loi, E.H., and Vance, E.R. (1995) pH Dependence of the aqueous dissolution rates of perovskite and zirconolite at 90 °C, in *Scientific Basis for Nuclear Waste Management XVIII, Materials Research Society Symposium Proceedings*, vol. **353** (eds. T. Murakami and R.C. Ewing), Materials Research Society, Pittsburgh, pp. 847–854.
137 Ewing, R.C., Haaker, R.F., and Lutze, W. (1982) Leachability of zircon as a function of alpha dose, in *Scientific Basis for Nuclear Waste Management V* (ed. W. Lutze), Elsevier, New York, pp. 389–397.
138 Helean, K.B., Lutze, W., and Ewing, R.C. (1999) Dissolution studies of inert materials, in *Environmental Issues and Waste Management Technologies in the Ceramic and Nuclear Industries IV, Ceramic Transactions*, vol. **93** (eds. J.C. Marra and G.T. Chandler), American Ceramic Society, Westerville, pp. 297–304.
139 Oelkers, E.H. and Poitrasson, F. (2002) An experimental study of the dissolution stoichiometry and rates of a natural monazite as a function of temperature from 50 to 230 °C and pH from 1.5 to 10. *Chemical Geology*, **191**, 73–87.

140 Chaïrat, C., Oelkers, E.H., Schott, J., and Lartigue, J.E. (2006) An experimental study of the dissolution rates of Nd-britholite, an apatite-structured actinide-bearing waste storage host analogue. *Journal of Nuclear Materials*, **354**, 14–27.
141 Terra, O., Dacheux, N., Audubert, F., and Podor, R. (2006) Immobilization of tetravalent actinides in phosphate ceramics. *Journal of Nuclear Materials*, **352**, 224–232.
142 Bois, L., Guittet, M.J., Carrot, F., et al. (2001) Preliminary results on the leaching process of phosphate ceramics, potential hosts for actinide immobilization. *Journal of Nuclear Materials*, **297**, 129–137.
143 Nasraoui, M., Bilal, E., and Gibert, R. (1999) Fresh and weathered pyrochlore studies by Fourier transform infrared spectroscopy coupled with thermal analysis. *Mineralogical Magazine*, **63**, 567–578.
144 Lumpkin, G.R. and Mariano, A.N. (1996) Natural occurrence and stability of pyrochlore in carbonatites, related hydrothermal systems, and weathering environments, in *Scientific Basis for Nuclear Waste Management XIX, Materials Research Society Symposium Proceedings*, vol. **412** (eds. W.M. Murphy and D.A. Knecht), Materials Research Society, Pittsburgh, pp. 831–838.
145 Lumpkin, G.R. and Ewing, R.C. (1992) Geochemical alteration of pyrochlore group minerals: Microlite subgroup. *American Mineralogist*, **77**, 179–188.
146 Lumpkin, G.R. and Ewing, R.C. (1995) Geochemical alteration of pyrochlore group minerals: pyrochlore subgroup. *American Mineralogist*, **80**, 732–743.
147 Lumpkin, G.R. and Ewing, R.C. (1996) Geochemical alteration of pyrochlore group minerals: Betafite subgroup. *American Mineralogist*, **81**, 1237–1248.
148 Wall, F., Williams, C.T., Woolley, A.R., and Nasraoui, M. (1996) Pyrochlore from weathered carbonatite at Lueshe, Zaire. *Mineralogical Magazine*, **60**, 731–750.
149 Williams, C.T., Wall, F., Wooley, A.R., and Phillipo, S. (1997) Compositional variation in pyrochlore from the Bingo carbonatite, Zaire. *Journal of African Earth Science*, **25**, 137–145.
150 Chakhmouradian, A.R. and Mitchell, R.H. (1998) Lueshite, pyrochlore and monazite-(Ce) from apatite-dolomite carbonatite, Lesnaya Varaka complex, Kola Peninsula, Russia. *Mineralogical Magazine*, **62**, 769–782.
151 Wise, M.A. and Černý, P. (1990) Primary compositional range and alteration trends of microlite from the Yellowknife pegmatite field, Northwest Territories, Canada. *Mineralogy and Petrology*, **43**, 83–98.
152 Ohnenstetter, D. and Piantone, P. (1992) Pyrochlore-group minerals in the Beauvior peraluminous leucogranite, Massif Central, France. *Canadian Mineralogist*, **30**, 771–784.
153 Wall, F., Williams, C.T., and Woolley, A.R. (1999) Pyrochlore in niobium ore deposits in *Mineral Deposits: Processes to Processing*, vol. **1** (ed. C.J. Stanley), Balkema Publishers, Rotterdam, pp. 687–690.
154 Lumpkin, G.R., Day, R.A., McGlinn, P.J., et al. (1999) Investigation of the long-term performance of betafite and zirconolite hydrothermal veins from Adamello, Italy, in *Scientific Basis for Nuclear Waste Management XXII, Materials Research Society Symposium Proceedings*, vol. **556** (eds. D.J. Wronkiewicz and J.H. Lee), Materials Research Society, Warrendale, pp. 793–800.
155 De Vito, C., Pezzotta, F., Ferrini, V., and Aurisicchio, C. (2006) Nb-Ti-Ta oxides in the gem-mineralized and "hybrid" Anjanabonoina granitic pegmatite, central Madagascar: A record of magmatic and postmagmatic events. *Canadian Mineralogist*, **44**, 87–103.

156 Gieré, R., Buck, E.C., Guggenheim, R., et al. (2001) Alteration of Uranium-rich microlite, in, *Scientific Basis for Nuclear Waste Management XXIV, Materials Research Society Symposium Proceedings*, vol. **663** (eds. K.P. Hart and G.R. Lumpkin), Materials Research Society, Warrendale, pp. 935–944.

157 Oversby, V.M. and Ringwood, A.E. (1981) Lead isotopic studies of zirconolite and perovskite and their implications for long range synroc stability. *Radioactive Waste Management*, **1**, 289–307.

158 Ewing, R.C., Haaker, R.F., Headley, T.J., and Hlava, P.F. (1982) Zirconolites from Sri Lanka, South Africa and Brazil, in *Scientific Basis for Nuclear Waste Management V* (ed. S.V. Topp), Elsevier, New York, pp. 249–256.

159 Rasmussen, B. and Fletcher, I.R. (2004) Zirconolite: A new U-Pb chronometer for mafic igneous rocks. *Geology*, **32**, 785–788.

160 Hart, K.P., Lumpkin, G.R., Gieré, R., et al. (1996) Naturally occurring zirconolites – analogues for the long-term encapsulation of actinides in Synroc. *Radiochimica Acta*, **74**, 309–312.

161 Bulakh, A.G., Nesterov, A.R., Williams, C.T., and Anisimov, I.S. (1998) Zirkelite from the Sebl'yavr carbonatite complex, Kola peninsula, Russia: And x-ray and electron microprobe study of a partially metamict mineral. *Mineralogical Magazine*, **62**, 837–846.

162 Williams, C.T., Bulakh, A.G., Gieré, R., et al. (2001) Alteration features in natural zirconolites from carbonatites, in *Scientific Basis for Nuclear Waste Management XXIV, Materials Research Society Symposium Proceedings*, Vol. **663** (eds. K.P. Hart and G.R. Lumpkin), Materials Research Society, Warrendale, pp. 945–952.

163 Mariano, A.N. (1989) Economic geology of rare earth minerals, in *Geochemistry and Mineralogy of Rare Earth Elements, Reviews in Mineralogy*, Vol. **21** (eds. B.R. Lipin and G.A. McKay), Mineralogical Society of America, Chantilly, pp. 309–337.

164 Banfield, J.F. and Veblen, D.R. (1992) Conversion of perovskite to anatase and TiO_2 (B): A TEM study and the use of fundamental building blocks for understanding relationships among the TiO_2 minerals. *American Mineralogist*, **77**, 545–557.

165 Lumpkin, G.R. (1992) Analytical electron microscopy of columbite: A niobium-tantalum oxide mineral with zonal uranium distribution. *Journal of Nuclear Materials*, **190**, 302–311.

8

Sources and Behaviour of Actinide Elements in the Environment

M.A. Denecke[1], N. Bryan[2], S. Kalmykov[3], K. Morris[1], and F. Quinto[4]

[1] *The University of Manchester, United Kingdom*
[2] *National Nuclear Laboratory, United Kingdom*
[3] *Lomonosov Moscow State University, Department of Chemistry – Radiochemistry, Russia*
[4] *Karlsruhe Institute of Technology, Institute for Nuclear Waste Disposal, Germany*

8.1 Introduction

Actinides released to the environment are of radioecological relevance, as they are alpha-emitting radionuclides and also of varying toxicities, thereby posing a potential threat to the biosphere. Most recent perceptible public concern over actinides and other radioelements released to the environment was following the tragedy at Fukushima. Public concern seems to rise and ebb with significant events, such as Fukushima, Chernobyl, and Three Mile Island. Objective treatise of inventories and potential impact of releases is important for placing them in a factual perspective. Most environmental contamination with actinide elements is anthropogenic, due to accidental releases from nuclear reactors and from nuclear weapons, from nuclear processing plants and plutonium production activities in development of nuclear weapons, from other incidents related to other military activities (e.g. nuclear submarines and depleted uranium ammunition), and from accidents involving nuclear-powered satellites. In addition, naturally occurring actinide nuclides are distributed globally, and uranium and thorium ore milling and mining operations are another source of environmental contamination. In this chapter, we will present a short overview of inventories for these actinide sources in the environment and detail selected examples, as well as discussing potential impact of microbial activity on the mobility of actinides. A compact description of the environmental behaviour of actinides released is given, and tools for prognosis and evaluation of their migration are presented. This is intended as a primer in the sources and behaviour of actinide elements in the environment, as in this short chapter it is impossible to convey the wealth of decades of information gathered. We have tried our best to include relevant references so that the reader can find further information. We also suggest that interested readers should also refer to other recent review material, such as a book on the environmental behaviour of radionuclides associated with the nuclear fuel cycle that hence also addresses

Experimental and Theoretical Approaches to Actinide Chemistry, First Edition.
Edited by John K. Gibson and Wibe A. de Jong.

the actinides with chapters from a number of leading experts [1], as well as the reviews in [2, 3], and the review covering actinide nanoparticles in the environment [4].

8.2 Naturally Occurring Actinides

There are around 340 nuclides comprising the 90 naturally occurring elements. Of these, 254 nuclides form the so-called primordial nuclides, the word *primordial* stemming from the Latin *primordialis*, meaning 'of the beginning'. The primordial nuclides or isotopes formed before our solar system itself was formed ($\sim 4.6 \times 10^9$ years ago) as a result of the cosmogenic processes at the beginning of the known universe and nucleosynthesis reactions occurring during stellar evolution. Of the 254 primordial nuclides, about 34 are radioactive, including isotopes of the actinide elements ^{232}Th, ^{235}U, ^{238}U, and ^{244}Pu. These actinide isotopes likely formed from neutron capture reactions without ß- decay in high neutron flux densities of supernova explosions. The half-lives ($t_{1/2}$) of these nuclides are necessarily long for them to still be present on Earth in quantifiable amounts. For example, ^{238}U has a $t_{1/2}$ of 4.47×10^9 years, which is only 3% less than the age of our solar system. ^{235}U has a shorter lifetime ($t_{1/2} = 0.7 \times 10^9$ years), which is 85% less than the age of our solar system, much less than that for ^{238}U. The shorter-lived ^{235}U is found in a much lower relative abundance (modern natural abundance of $^{235}U = 0.72\%$ and $^{238}U = 99.27\%$).

These naturally occurring primordial actinide nuclides are the source of what are called radiogenic nuclides, which are nuclides that form from radioactive decay (products are 'daughters' or 'progeny'). Important decay chains of primordial isotopes ^{232}Th, ^{235}U, and ^{238}U are shown in Figure 8.1. You can see that some daughters in these decay chains are also actinide nuclides. All of the daughters have half-lives that are much shorter than their primordial mother isotope. Well known in geochronology is uranium-lead dating, which relies on determining the ratio of two radiogenic lead isotopes from the radium and actinium series and their mother uranium nuclides, $^{206}Pb/^{238}U$ and $^{207}Pb/^{235}U$ and known decay rates (see, e.g. Ref. [5]). This was the method used for the first radiometric dating of rocks to help seal the debate on Earth's age at the beginning of the last century [6].

There are naturally occurring actinides that are neither primordial nor a product of simple radioactive decay, but result from naturally occurring nuclear reactions. These reactions involve neutrons and alpha particles stemming from, for example, spontaneous fission and alpha-emitting nuclides. The products of such reactions are the so-called nucleogenic nuclides. For example, nucleogenic ^{236}U and ^{239}Pu are created by neutron capture on natural ^{235}U and ^{238}U, respectively, in uranium ores.[7] The ^{236}U isotope is produced largely by thermal neutron capture, and ^{239}Pu is produced via epithermal neutron capture in the resonance region [8], so that the relative amounts of ^{236}U (on the order of 10's to 100's ppt, based on total U content [9]) and ^{239}Pu (around hundreds of ppq [10]) found in ores are dependent on the presence of water (and other factors affecting thermalisation) and neutron absorbers such as Gd [11]. Note that the main sources of ^{239}Pu in the environment are not natural but anthropogenic in origin, stemming from atmospheric weapons testing and spent nuclear fuel reprocessing operations (see Section 8.3.3). The most famous source of nucleogenic ^{239}Pu is the Oklo and Bangombé uranium deposits in Gabon, Africa, where nuclear fission spontaneously

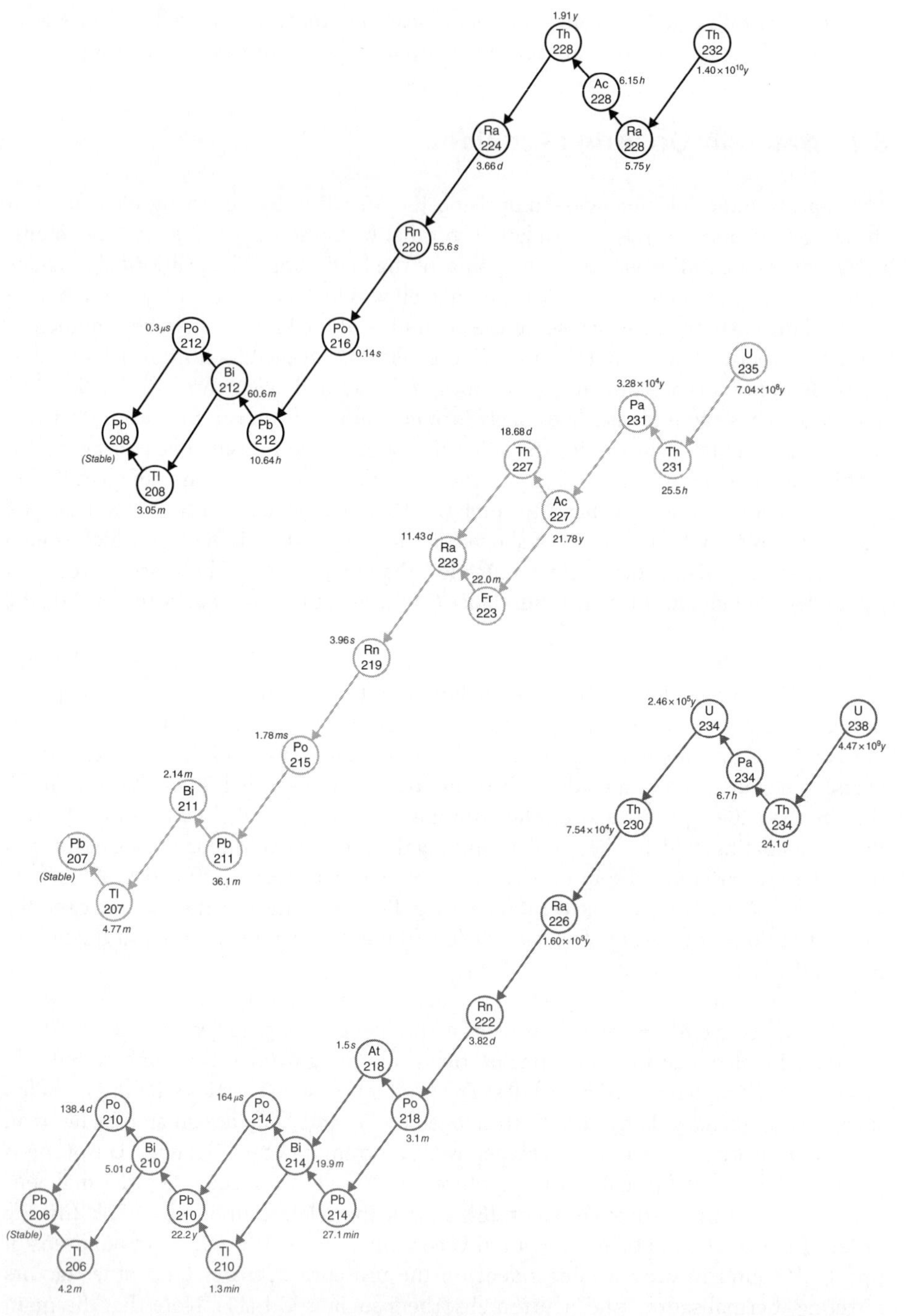

Figure 8.1 Decay series for ^{232}Th, ^{235}U, and ^{238}U, often respectively referred to as the thorium series, the actinium series, and the radium or uranium series, or as the 4n, 4n + 3, and 4n + 2 series (where n is an integer, 4 is the mass change associated with alpha decay and 0, and 3 and 2 are the remaining number of protons and neutrons left over following removal of all possible alpha particles). (*See insert for color representation of the figure.*)

began around two billion years ago, and activity lasted for around one million years [12]. An estimated 2–3 tonnes ^{239}Pu was generated [13]. Fission was possible because two billion years ago, the uranium still contained a significant amount of fissionable ^{235}U present in uranium ore (between 3% and 4%). This was a subsurface 'natural nuclear' reactor, and the fission reaction products formed have been largely contained for two billion years (plutonium was observed to have migrated less than 3 m [2]), which provides indirect evidence that migration of radionuclides was limited at this site and gives an indication that long-term deep geologic disposal of nuclear waste is feasible [12].

8.2.1 Commercial Uses of Naturally Occurring Actinides

Uranium dominates the present commercial importance of natural actinides, as it is the natural resource mined for ultimate fabrication of fuels for the current fleets of commercial nuclear-power-generating reactors. Thorium has potential application in an alternative fuel cycle economy, which finds enthusiastic support in Norway and India [14, 15]. The Indian interest in a thorium-based fuel cycle lies in the large thorium ore deposits within their national borders. The next sections are meant to be brief introductions to uranium resources mining and its environmental impact, as well as a short description of thorium resources.

In addition to its use for the production of electricity, the main use of uranium and depleted uranium (which is left following ^{238}U enrichment processing) is for ship and submarine propulsion, research, as ship ballast, and counterweights for aircraft and military applications. Thorium finds use in a number of industrial catalysts and high-refractive-index specialty glasses [16]. Due to the high melting point of its oxide, it glows intensely white at the temperature of the burning gas, and so thorium was once used in incandescent gas lamp (such as camping gas) mantles, but its use has historically decreased over time. Likewise, uranium was used as a colourant in glass, glazes, and enamels but then disappeared after it was discovered to be a radioelement [17]. More modern uses of uranium were for dental porcelains of dentures and surfaces of crowns up to the 1980s, but it was then replaced by rare earth elements, and for counterweight applications in aircraft, but here it was replaced by tungsten starting in the 1980s [18]. The commercial use of other naturally occurring actinides is limited, due to low concentrations and radio- and chemo-toxicities.

8.2.2 Uranium Resources and Mining

Uranium was first discovered by Klaproth in the mineral pitchblende as a by-product of silver mining in St. Joachimsthal (Jachymov, Czech Republic) in 1789. Uranium was the first element discovered to be radioactive, by Henri Becquerel in 1896. It is a dense (19.05 g/m^3; more dense than lead) metal with an average abundance in Earth's crust of ~2.5 ppm. It is, in fact, more abundant than gold, silver, mercury, antimony, or cadmium. It is often found in higher relative naturally amounts in watersheds containing felsic igneous rocks such as granite. Uranium has been reported in oxidation states (0) to (VI); U(III) to U(VI) are found naturally, and U(IV) and U(VI) are the most common in the environment. The environmental chemistry of uranium is similar to that of other metals that come in varying oxidations states and involves a combination of complexation with organic and inorganic ligands, sorption-desorption equilibria, and redox reactions (see Section 8.4). U(VI) is quite soluble, and its solubility is the underlying reason for the relative high amount of uranium found in the oceans (which contain around 4.5 billion tonnes of uranium [19]). Uranium under reducing conditions (in natural environments, the Eh of the U(IV)/U(VI) generally lies in the range of −0.042 to 0.086 V [20]) is present as U(IV), which is very insoluble. The extreme insolubility of the U(IV) mineral uraninite at neutral pH and low Eh makes it practically immobile in such environments [21]. The solubilisation of U(VI) under oxidising conditions and the precipitation of U(IV) under reducing conditions are essentially the dominant processes that concentrate uranium into deposits that are of high enough grade or quality for economic mining and extraction. They are also key mechanisms by which redox control is used for remediation and immobilisation strategies for mitigation of contaminants. Investigations of the evolution of uranium deposits are of great interest not only for exploration of resources but also for development of process understanding and associated modelling tools for long-term assessment of the radiological safety of geological disposal sites for spent nuclear fuel and nuclear waste, as investigation of natural systems can underpin advances made in understanding the long-term behaviour of U in complex environments [22–24].

Uranium occurs in over 200 different minerals, but only a handful are presently economically feasible for mining and processing (for an in-depth and current report of global uranium resources and their classification, see the newest edition of the 'Redbook' [25]; a less recent overview of uranium mineralogy is found in [26]). The profitability of a uranium resource is primarily associated with two key factors, namely, the cost of mining and processing [18]. The top uranium mining countries in 2013 were Kazakhstan, Canada, and Australia, accounting for 65% global production (38%, 16%, and 11%, respectively). The majority of economically important uranium minerals are the tetravalent minerals – uraninite and pitchblende (UO_2; [27]), coffinite ($U(SiO_4)_{1-x}(OH)_{4x}$), and brannerite (U, Ca, Y, Ce, La)(Ti, Fe^{2+})$_2$$O_6$; [28]) – but minerals with uranium in its hexavalent form such as carnotite ($K_2(UO_2)_2(VO_4)_2 \cdot 3H_2O$), autinite ($Ca(UO_2)_2(PO4)_2 \cdot 10–12H_2O$), and uranophane $CaH_2(SiO_4)_2(UO_2) \cdot 5H_2O$ are also important.

The Redbook [25] uses the 15 major geological categories of uranium deposits defined by the International Atomic Energy Agency (IAEA) in 1996 [29]. These are (with prominent examples):

1) Sandstone (presently the most economically important deposit type conducive to in situ leaching, including numerous roll-front deposits in Kazakhstan).
2) Proterozoic unconformity (the two major districts are the Athabasca Basin in Canada and the Pine Creek Orogen in Australia).
3) Polymetallic Fe-oxide breccia complex (includes as the main example the world's largest uranium resource, Olympic Dam, Australia).
4) Paleo-quartz-pebble conglomerate (an example is the largest know gold reserves in the Witwatersrand basin, South Africa, where uranium is mined as a by-product).
5) Granite-related (as true ore veins or disseminated mineralisation in hydrothermally altered granite, episyenite).
6) Metamorphite.
7) Intrusive (includes one of the oldest and deepest open-pit uranium mines, Rössing Mine in the Namib Desert, Namibia).

8) Volcanic-related (the most significant example being Streltsovska caldera in the Russian Federation).
9) Metasomatic (includes areas of tectonomagmatic activity of the Precambrian shields in Central Ukraine, such as the first large deposit discovered there, Michurinske);
10) Surficial (being near surface are open-pit mines, such as the Yeelirrie deposit owned by Cameco);
11) Carbonate (an example being the Tummalapalle uranium reserve in the Proterozoic Cuddapah basin, state of Andhra Pradesh, India).
12) Collapse breccia-type (e.g. breccia pipes in Mississippian Redwall Limestone near the Grand Canyon, United States).
13) Phosphate (typically low-grade ores such as in the Land Pebble District in the Polk and Hillborough Counties, Florida).
14) Lignite and coal (e.g. Freital near the Ore Mountains, Germany).
15) Black shale (including the Gera-Ronneburg deposit, abandoned in 1990, which was mined in former East Germany by the WISMUT Company, the largest single producer of uranium ore for the former USSR). Presently, sandstone and Proterozoic unconformities account for more than 75% of the world's uranium production [30].

Global uranium resources are classified according to geological certainty and costs of production. There are two broad classes: (1) identified conventional resources and (2) undiscovered resources. The former includes both uranium deposits known with a high degree of confidence in terms of estimates of grade and tonnage, the so-called reasonably assured resources (RARs), and inferred sources, whose grade and tonnage and are less well known. Undiscovered resources are not known and also similarly divided into two groups based on levels of confidence in their existence: prognosticated and speculative sources. In 2013, the identified conventional resources recoverable at costs less than US\$100/lbU_3O_8 yellow cake (or less than US\$260/kg U) totalled 7,635,200 tonnes of uranium. Approximately the same tonnage is estimated to be the undiscovered resources. The total uranium used to fuel the 437 commercial nuclear reactors worldwide (net generating capacity of 372 GWe in 2012) was about 62,000 tonnes, and the total tonnage uranium mined and produced in 21 countries amounted to 59,500 tU in 2013. Assuming a constant uranium demand (constant generating capacity) in the future, the global uranium reserves, both identified and undiscovered, would be sufficient for nearly 250 years. It is often pointed out that the world's oceans offer a near inexhaustible uranium reserve, containing over 4 billion tonnes of uranium. However, because of the low uranium concentration in seawater (3–4 ppb), cost-effective extraction is as yet not technologically feasible.

Another classification of uranium reserves one finds is according to the mining or production method, 'conventional' excavation underground and open-pit mining, and production using (in situ and in-place) leaching methodologies [31]. Around 50% of uranium produced stemmed from in situ leaching in 2014, and 40% was mined using hydrometallurgical treatment; the remaining 10% was produced as a co-product from other activities or stemmed from unconventional or unknown resources. The year 2014 marked the first year that leaching was the major production method. Because leaching is relatively less expensive, this has led to a corresponding decrease in recoverable RAR costs. Uranium is mined in open pits (or open cuts) for shallow ore bodies [32], typically less than 200 m below the surface, such as the Rössing Mine in Namibia. Deeper high-grade, dense ore bodies are mined underground. One of the largest underground uranium mines in the world is in the Canadian McArthur River. Underground mining varies in specific details for each site but generally involves excavating a deep shaft, from which horizontal levels to access the ore body are drilled and the ore removed. Often, the first ore processing steps of grinding and milling are done inside the mine. Modern mining techniques integrate use of remote control equipment to minimise mine worker exposure to radiation.

Leaching is an inexpensive mining method applicable for porous geological material such as ore bodies in sandstone deposits. So-called in situ leaching is different from conventional open-pit and underground mining methods, in that the ore and host rock are not excavated for subsequent uranium recovered; rather, the uranium is recovered in situ or dissolved from the ore/host rock itself. Generally, an acid solution[1] with added oxidant is injected into the deposit from a grid of injection wells called a wellfield. The injected solution is often referred to as a lixivant, and its function is to dissolve the uranium ore so that it can be simply pumped to the surface, and the uranium is then removed using solvent extraction or ionic exchange. After uranium removal from the lixivant, the solution is re-used or recycled. In situ leaching avoids waste tailings, which arise from processing of ores from open-pit and underground mines. In addition, it results in a smaller footprint of surface disruption from mining activities compared to conventional mining.

The process used to isolate the uranium from the solid ore and rocks in conventional mining begins with grinding and milling to increase the surface area to facilitate subsequent hydrometallurgical treatment. The uranium is then separated from the milled ore using chemical means that are in essence the same as for leaching mining technologies: dissolution with a lixivant and isolation of the uranium using solvent extraction or ion exchange. The uranium in the ion exchange or solvent is stripped, and liquor is then isolated by precipitating it out by adding either ammonia, hydrogen peroxide, caustic soda, or caustic magnesia. The resulting solid is dried and then fired at as uranium oxide 'yellow cake'. For low-grade ores such as the Sheep Mountain Project in central Wyoming, where the uranium concentration is ≥0.01 % [33], or the black shale mined at Ronneburg in the former German Democratic Republic in the years 1946–1990 [34], the ore is/was 'heap-leached', meaning that the lixivant is trickled or percolated through the ore that has been placed in a heap. The resulting uranium loaded solution is collected and then further processed [35].

8.2.3 Environmental Impacts of Uranium Mining and Milling

Among the by-products of uranium mining and milling that are of environmental concern are waste rock, mill tailings, and residues and solids generated on-site by waste treatment processes. One refers to the transformation of naturally occurring radioactive materials (NORMs) into technologically enhanced naturally occurring radioactive materials (TENORMs). Through concentration and exposure, TENORMs pose an increased potential environmental risk compared to NORMs. Most of the mined ore remains undissolved in the leaching process as rock or other minerals. The remaining

1 Sometimes alkaline carbonate-bicarbonate solutions with an added oxidising agent are used.

solids left over after separating out uranium-rich solution are called 'tailings'. Further by-products are also generated, for example, co-precipitated solids resulting in neutralisation of acidic mine/mill effluents. It depends on the ore grade, but on average about 2% of the mined material volume is product; the rest is tailings and contains most of the radioactivity (5% of the uranium and 85% of the original radioactivity) [36]. For example, the >50 million tonnes uranium mill tailings at Helmsdorf tailings near Zwickau, former East Germany, contains 80 tonnes of uranium [37]. Even for higher-grade ores, the volume of by-products produced is larger than one might expect, as high-grade ore is diluted with normal rock before milling for radiological reasons. It is difficult to find an estimate of the global tailings burden,[2] but a IAEA report suggests that approximately 2.2 million tonnes of uranium was produced worldwide up to the year 2004 [39]. Using an average tonnage of tailings per tonne uranium produced (~556 t tailing/t U[3]; [36]) and assuming that the global production is around 55,000 t/year, we find that 366,960,000 tonnes tailings have accumulated over the past 12 years. Adding this to the 1.2×10^9 tonnes of tailing accumulated up to 2004, an estimate of a current global tailing inventory of around 1.6 billion tonnes is made. These large inventories of material with high surface area and complicated chemistry from processing (oxidising agents, extractant and lixivant solutions, neutralising agents) must be managed to minimise harmful environmental impacts. Note that often the element that is the central constituent of concern for tailing management is neither uranium nor its daughters, so that appropriate measures must be taken to prevent the negative environmental impact due to other tailing constituents, including toxic heavy metals and compounds. For example, at the AREVA-run JEB Tailings Management Facility (JEB TMF) at McClean Lake, Saskatchewan, an important problematic constituent of concern is arsenic, and a number of measures are implemented to reduce its concentration in tailing pore water [40].

A number of uranium sites such as historic legacy mining sites and mines/mills associated with the cold war and the early years of nuclear power generation, which are now abandoned, did not properly dispose of or manage mining by-products and tailings due to lack of regulatory control. This has led not only to health issues of mine workers but also to negative environmental impact, for example, mobilisation of U(VI) entering groundwater and water supplies [41]. In particular, the poor effectiveness of the uranium extraction technology applied during the late 1940s, up to the early 1960s, tended to make their tailing rather high in uranium content [42]. Notorious examples are found in the Ferghana Valley of Central Asia. The one example with likely the most significant health impacts is the Taboshar legacy uranium facility in northern Tajikistan, which is an abandoned mining and processing facility that delivered uranium for USSR military purposes during the cold war. Poor mining and milling practices, combined with lack of regulation, led to significant contamination extending over 400 ha. The milling and processing of uranium ore from different sources with different compositions, hence requiring varying processing treatments and chemicals, began in 1942. From 1942 to 1945, 10,000 tonnes ore was processed, which increased rapidly; by 1950, 600,000 tonnes

2 In 2004, the IAEA estimated the volume of tailing to be nearly 1000 million cubic meters [38].
3 In 1992, 36,000 tonnes uranium was produced, along with 20 million tonnes of tailings. Other interesting estimate values are 17,000 tonnes of 1 wt% uranium ore is needed to produce 20 tonnes of metallic U or ~ 22.7 tonnes of UO_2. 100,000 tonnes of uranium ore is needed to produce 1 GWe.

of ore per year was being mined and processed. The tailings and other wastes generated were not contained, and no remediation plan was implemented when operations stopped and discontinued. Use of the site and materials on it, for example, contaminated water for agricultural irrigation and has led to spread of contaminants. In June of 2016, the European Commission provided €8 million of funds to finance the remediation of Tajikistan and neighbouring Kyrgyzstan and Uzbekistan sites [43].

Modern rock waste and tailing management practice include a range of engineering and chemical treatment efforts to prohibit leaching of heavy metals (including the actinides uranium and thorium) by natural precipitation percolating through the tailings and prevent their migration to surface and groundwaters over time. In addition, mining and milling facilities are regulated and monitored, to ensure the effectiveness of these practices. Best practice is safe, long-term (1000 years after decommissioning [38]) isolation; tailings are isolated by containing them as heaps or piles in impounded, lined, and covered areas. Tailings are contained either above ground, which is most common, below ground level, or subaqueous, that is, at the bottom of a deep lake [38]. Underground solutions either involve depositing tailings in former underground or open-pit mines, where measures are taken to prevent contact with groundwater (impermeable geologic or man-made lining). If an acid tailings repository remains unlined, migration of dissolved uranium can be significant. Groundwater uranium concentrations up to 2.08 mg/L, moving at rates of several tens of metres per year, have been reported [44]. In alkaline mill tailings, migration of U into groundwater is facilitated by the formation of stable uranyl carbonate complexes ($UO_2(CO_3)_2^{2-}$ and $UO_2(CO_3)_3^{4-}$) [44] and can create a contaminated plume, which in one case was observed to move 15 m per year. Modern tailing management measures also include compacting tailings and using rubble or French drains to direct water away. The static and dynamic stability of the tailings impoundment is critical. The embankment or tailings dam of an aboveground impoundment must withstand increasing fill levels and possible natural catastrophic events (heavy rain, earthquake, or flood). A well-known example of embankment failure occurred in 1979 at the Churchrock uranium mine/mill in New Mexico, when 370,000 m^3 of tailings was discharged. Contamination entered the nearby river, Rio Puerco, and was carried downstream for ~100 km [45]. Note that uranium mobility in a tailing repository can be impacted by microbial activity through reduction of soluble U(VI) to insoluble U(IV). In 1991, Lovely et al. [46] reported that microorganisms capable of transferring electrons to [Fe(III)] and reducing it to [Fe(II)] can also metabolise U(VI), thereby reducing it to U(IV), which since then has led to numerous investigations into the microbial redox processes of uranium. Currently, this ability of bacteria to reduce U(VI) to U(IV) is being capitalised in bioremediation technologies [47], including below the ground in situ remediation that is being developed, for example, at the Old Rifle site in Colorado with contamination stemming from mill tailings piles and ponds legacies. More on this topic is presented in Section 8.4.

Water leached through acid mill tailings and mine rock piles often has low pH due to acid mine drainage. Acid mine drainage is not limited to uranium mines; it is caused by oxidation of sulphide minerals in the ore that become exposed when the ore body extracted, generating acidic waters. The acidic waters can dissolve radionuclides and heavy metals from the tailings and rocks, including thorium and uranium. To avoid this, during mine and milling operations the amount of oxygen reaching tailings (that could oxidise sulphide minerals) is limited by a cover of water over the tailings within an impoundment area. In addition, acidic tailings and waters are collected and treated for pH neutralisation using lime [48]. In the decommissioning phase of tailings ponds, the pond water and seepage are removed (dewatering) and treated to meet water quality standards prior to any off-site release, the dams re-contoured, a final covering installed, and then landscaped with vegetation [49].

8.2.4 Thorium Resources and Potential Use as Fuel

Thorium was discovered by Berzelius in 1828 from a huttonite, $ThSiO_4$, sample and is named after the Norse God of thunder, Thor [50]. Natural thorium is mainly ^{232}Th with near negligible amounts of its short-lived isotopes ^{234}Th, ^{231}Th, ^{230Th}, and ^{228}Th as radioactive decay daughters (Figure 8.1). It is generally stable in the tetravalent oxidation state (Th^{4+}) and because of its high charge readily forms complexes, is strongly hydrated, and readily hydrolysed. Its high charge and associated tendency to sorb to surfaces also determines its behaviour during weathering of rocks and minerals; it is less soluble and mobile than uranium and hence has a lower concentration in surface water (sea water <1 ppb).

Thorium-232 itself is not a fissile nuclide. However, it is 'fertile', meaning you can use it to breed a fissile nuclide, in this case ^{233}U. Neutron absorption by ^{232}Th produces ^{233}Th ($t_{1/2} = 27$ d), which undergoes two successive beta decays to form ^{233}U (via ^{233}Pa; $t_{1/2} \sim 22$ min). There are advantages to a thorium-based fuel cycle; one of the primary reasons is its greater abundance in nature than uranium. Thorium is three to four times more abundant than uranium in Earth's crust. The worldwide thorium inventory is estimated to be up to 4.5 million tonnes [2], with the largest reserved in India, Australia, Brazil, and the United States [25]. Should an increase in thorium demand be stimulated by establishment of appropriate reactor designs, it can be produced from a variety of deposit types. The most abundant of these are thorium-bearing monazite, apatite, zircon and titanite, and heavy-mineral sands, such as the thorium-rich sands in the southern and southwest coastal areas of Kerala and Tamil Nadu, India. The global inventory for monazite (a phosphate mineral with the composition (RE,Th)(PO_4,SiO_4), RE = rare earth, containing 1–10 wt% Th) is around 12 million metric tonnes [2]. Granite contains thorium in variable amounts typically ranging between 8 and 94 ppm [51]. Huttonite and thorianite, ThO_2, are minerals with high thorium content but are not common.

8.3 Anthropogenic Actinides Release

All the nuclides of transuranium elements (TRUs) are considered as anthropogenic, despite the fact that ultratrace concentrations of ^{236}U and ^{239}Pu are found in uranium ores as a result of neutron capture by ^{235}U and ^{238}U (see Section 8.2). The first mention of the existence of TRUs is a reference in the early works of Enrico Fermi in the early 1930s [52]. However, the first TRU was synthesised in 1940 by McMillan and Abelson through (n,γ) reaction on ^{238}U nuclei: $^{238}U(n,\gamma)^{239}U(\beta^-)^{239}Np$. The first plutonium nuclide, ^{238}Pu, was synthesised the same year in a cyclotron through (d,2n) reaction and subsequent β-decay of ^{238}Np: $^{238}U(d,2n)^{238}Np(\beta^-)^{238}Pu$ [53]. In 1941, the same team headed by Glenn Seaborg synthesised and separated ^{239}Pu [54, 55]. Since that time, about 500 tonnes of chemically separated plutonium has accumulated worldwide as a result of weapon-related activities and civil nuclear power production [55]. Anthropogenic actinides releases due to accidents and intentional release have occurred from nuclear weapon testing, accidental releases from reactors, enrichment[4] and reprocessing plants, accidents with nuclear-powered satellites, and release from storage containment. A comprehensive review of the worldwide release of actinides to the environment is given by Runde and Neu [2]. Dispersion following release can be via either an aqueous or atmospheric route. In the following, we discuss releases associated with accidents and intentional discharges from reprocessing that resulted in environmental contamination with actinides, in which plutonium plays a central role due to accidents and lack of radiological best practice during the era of the nuclear weapons race. Major accidents of both power generating and plutonium breeder nuclear reactors with reactor core damage occurred in Windscale (1957), Three Mile Island (1979), Chernobyl (1986), and Fukushima Daiichi (2011), which also resulted in global release of man-made radionuclides and in the large-scale environmental dispersion of actinide elements. These releases and other accidental releases such as satellites are discussed in Section 8.3.2. Since 1945, 2056 nuclear tests have been conducted worldwide that have resulted in the release of 0.1 tonnes of plutonium in underground tests and ~3.5 tonnes in atmospheric tests, with global fallout asymmetrically distributed in the Northern and Southern hemispheres [57]. This global-fallout inventory of actinide elements resulting from nuclear weapons is discussed in detail in Section 8.3.3.

8.3.1 Releases from Nuclear Reprocessing Facilities

Releases from the radiochemical facilities for plutonium production and spent nuclear fuel (SNF) reprocessing are large-scale sources of man-made actinides in the environment. Some of the most prominent examples of such nuclear facilities are the Hanford, Rocky Flats, and Oak Ridge in the United States; Sellafield in the United Kingdom; La Hague in France; and the Mayak, Tomsk, and Krasnoyarsk sites in Russia. The releases from these sites included both airborne and liquid releases [2, 58]. Such releases were either intentional or unintentional, and continual release from plutonium production legacies still occurs today, for example, radioactive liquid and sludge waste remaining in single shell underground storage tanks at the Hanford site [59].

One of the largest Russian (former Soviet) radiochemical facilities is the Mayak Production Association (PA), which was established in 1947 for the production of weapons-grade plutonium. Several accidental and routine discharges took place, as summarised by Degteva et al. [60] and Napier et al. [61]. The absence of reliable waste management technology during the initial period of Mayak operation resulted in significant radioactive contamination of the Techa River. One year before the plutonium production plant was launched, Complex C with an assembly of tanks for high-level waste (HLW) for the first stages of radiochemical technology was constructed. Complex C included three separate underground canyons with stainless steel tanks that were cooled by a continuous by-pass water supply from the Techa River. Leakage of HLWs into the canyons occurred as a result of damage to pipes used for solution transfer.

4 Inventory from enrichment and conversion activities is by comparison minimal and will not be discussed further. Even the criticality accident in 1999 at the Tokai Plan in Japan only resulted in small and spatially limited ratio anomalies concentrated in the easily extractable portions of soils and botanical samples [56].

Techa River contamination was not monitored, since the design did not consider potential radioactive contamination of the cooling waters.

Radiochemical processing resulted in a large amount of liquid radioactive wastes with an accumulation rate of about 200 m^3 per day. Low-level wastes formed in the final stages of plutonium production were planned to be released into the Techa River. Other types of low-level and medium-activity waste were planned to be released into closed water reservoirs designated as B-9 (known as Lake Karachay) and B-17 (known as Old Swamp). However, the amount of radioactive wastes generated was underestimated, and evolution of large amounts of several types of medium-activity wastes was not anticipated. A method of alkaline precipitation of iron and chromium hydroxides and co-precipitation of uranium, plutonium, and some fission products was used, which allowed volume reduction of HLW by a factor of 30–40 but created problems with medium-activity waste volumes. Starting in autumn 1950, the radioactive solution was then released into the Techa River after the precipitates settled. The total routine release was 52 PBq; the total release from accidental overflows from Complex C was 63 PBq. The release was made up of a large number of radionuclides, dominated by relatively short-lived fission products.

The most contaminated reservoirs at Mayak are Lake Karachay and Old Swamp. According to the results of an inspection in 2002, all components of the open part of Lake Karachay (water, silts, and loams from the reservoir bed) contain approximately 30 million Ci (1110 PBq) of β-emitters and 1 million Ci (37 PBq) of α-emitters, including plutonium [62]; the majority of radionuclides are found in the silts and loams. Water in the Karachay Lake has high salt content (16–145 g/L) and a pH of 7.9 to 9. 3, favouring coagulation of colloids and actinide cation hydrolysis. During operations, a layer of anthropogenic silts composed of Al, Fe, and Mn hydroxides and carbonates with a fraction of organic substances was formed. The concentration of $^{239,240}Pu$ in these silts is about 10^7–10^8 Bq/kg [62]. In the mid-1960s, due to the dry summer season, the level of the pond surface lowered, exposing the contaminated bottom sediments. During a high-wind event in 1967, 22 TBq of radionuclides were distributed in the regions of the Mayak Facility [63].

Lake Karachay is connected to a 55- to 100-m-thick groundwater zone, in which fluids flow through fractured Silurian and Devonian metavolcanic rocks with andesitic and basaltic composition [64]. This favours subsurface migration of contaminated fluids. A study by Novikov et al. [65] reported that plutonium migrated over significant distances from the Karachay Lake with groundwaters. At the source, the plutonium activity is ~1000 Bq/L and is still 0.16 Bq/L at a distance of 3 km. Seventy to 90 mol% of this mobilised plutonium was sorbed onto colloids, confirming that colloids are responsible for the long-distance transport of plutonium. Using a combination of transmission electron microscopy (TEM) and secondary ion mass spectrometry (SIMS), plutonium was observed to be primarily bound to colloidal particles of amorphous ferrihydrite. Results of the sequential extraction of colloidal particles from groundwaters of the Karachay contamination aureole also indicate that plutonium occurs in the poorly soluble form predominantly bound to the fraction of amorphous iron oxides.

The most important atmospheric release at Kyshtym, in the Mayak Complex, was the explosion of an overheated tank containing HLW in September 1957. About 740 PBq was released from the tank, with about 90% of the ejected material falling out in close proximity (Table 8.1) [61], covering an area of 20,000 km^2, which is now referred to as

Table 8.1 Radionuclides released as a result of HLW tank explosion in September 1957 [61].

Radionuclide	Content, %
$^{90}Sr/^{90}Y$	5.4
$^{95}Zr/^{95}Nb$	24.8
$^{106}Ru/^{106}Rh$	3.7
$^{137}Cs/^{137m}Ba$	0.35
$^{144}Ce/^{144}Pr$	65.8
$^{239,240}Pu$	0.002

the East Urals Radioactive Trace (EURT). The material released included about 1 kg of weapons-grade plutonium [66].

Another radiochemical facility is the Krasnoyarsk Mining and Chemical Industrial Complex (MCIC), where so-called direct-flow reactors were operated over decades, resulting in radioactive contamination of the Yenisei River. Radionuclides entered the river ecosystem and mostly accumulated in bottom sediments and bottomland soils [67]. The maximum concentration of $^{239,240}Pu$ in samples studied by Kuznetsov et al. [67] was 17.8 Bq/kg. Highly stable 'hot' particles were found in the bottom sediments and riverside soils of the Yenisei River near the Krasnoyarsk facility. Using sequential extraction, it was shown that plutonium has low mobility in most cases, existing in residual, barely soluble fractions as plutonium-containing hot particles [68, 69]. This was confirmed by a study by Skipperud et al. [70], who determined the forms in which plutonium exists in Yenisei and Ob[5] rivers. It was found that the major fraction of plutonium in the bottom sediments of these rivers exists in fractions soluble in H_2O_2 (a mild oxidant) and HNO_3 and is incorporated in natural organic compounds, but a significant fraction of plutonium exists as hot particles in the insoluble residue.

8.3.2 Inventories of Releases from Accidents and Incidents

Actinides have been released to the environment due to accidents and incidents at civilian nuclear power industries, the most recent examples being Chernobyl and Fukushima nuclear power plants (NPPs). Releases from NPP accidents mirror the atomic and isotopic composition of the nuclear fuel at the time of the accident. During reactor operation, neutron flux conditions for power generation result in formation of $^{240}Pu/^{239}Pu$, $^{241}Pu/^{239}Pu$, and $^{242}Pu/^{239}Pu$ isotopic ratios, which are generally higher than those derived from nuclear testing (cf. Section 8.3.3.2) and increase as fuel burnup increases. $^{240}Pu/^{239}Pu$ isotopic ratios of 0.38–0.42 [71–73] and 0.303–0.330 [74] from the Chernobyl accident and Fukushima incidents, respectively, reflect the irradiation history of the nuclear fuel. At the same time, decay of the short-lived plutonium isotopes, ^{241}Pu and especially ^{243}Pu, is favoured over the neutron capture reactions, so that high proportions of ^{241}Am and ^{243}Am are formed. Following successive neutron capture reactions and ß-decays, ^{242}Cm (162.94 d) and ^{244}Cm (18.10 y) are also formed.

5 Most of the nuclear fuel reprocessing and weapons testing facilities in the former Soviet Union (Mayak, Tomsk-7, and Semipalitinsk) are located within the Ob drainage basin.

In particular, ^{244}Cm rather than ^{244}Pu is preferentially produced in a nuclear reactor [75]. The proportion of transuranium elements with mass number exceeding 240 can be very substantial, and the total activity of ^{241}Am, ^{242}Cm, and ^{244}Cm can be a 100- to a 1000-fold greater than that of ^{239}Pu [76].

The Chernobyl (Ukraine, 26 April 1986) accident is to date the most serious accident involving an NPP. A sudden power surge during a test caused the thermal destruction of the reactor of type RBMK-1000 (*'reaktor bolshoy moshchnosty kanalny'* of Soviet design). The subsequent steam explosion and the ignition of the graphite moderator, with the graphite fire lasting 10 days, led to a significant and widely dispersed radionuclide contamination to the Northern hemisphere. The temperature in part of the core is thought to have been higher than 2773 K. Estimated inventories of the actinides released in the accident are based on a 1.5% fuel particle release of the Unit 4 reactor core at the time of the accident [77, 78] (50% lower than previous estimates [79]) and under the assumption that no actinide fractionation occurred during debris and the fallout release. The atomic ratios of $^{236}U/^{239}Pu$, $^{237}Np/^{239}Pu$, $^{241}Pu/^{239}Pu$, and $^{241}Am/^{239}Pu$ can be obtained from the inventory of the actinides in the reactor core at the time of the accident and are 8.7, 0.024, 0.113, and 0.0032, respectively [79]. Upon comparing these atomic ratios with the ^{239}Pu release inventory (5.6 kg, [78]), the inventories for ^{236}U and ^{237}Np can be obtained as 48.5 kg and 0.136 kg, respectively. The main contribution to the ^{241}Am inventory originates from the decay of ^{241}Pu, which presently amounts to about 0.52 kg. Such inventories and atomic and isotopic composition are listed in Table 8.2. The composition of the Chernobyl fallout was observed to vary with distance from the accident.

Table 8.2 Inventories of ^{236}U, ^{237}Np, ^{239}Pu, and ^{241}Am in kilograms from global fallout and from the nuclear accidents at Chernobyl and Fukushima, together with the related $^{236}U/^{238}U$, $^{236}U/^{239}Pu$, $^{237}Np/^{239}Pu$, and $^{240}Pu/^{239}Pu$ ratios. These nuclides represent the isotope of the corresponding actinide dominantly produced in nuclear testing and accidents. The inventories of the actinides are in this work expressed as kilograms rather than as activities, thereby allowing for a direct comparison of the amounts of actinides; see the text for discussion.

Nuclide	Global fallout		Chernobyl nuclear accident		Fukushima nuclear incident	
^{236}U (kg)	900–1698	[110], [123], [119], [124], [125]	48.5	[79], [78]	5×10^{-4}	[85]
^{237}Np (kg)	1500	[107]	0.136	[79], [78]	/	
^{239}Pu (kg)	2800–3108	[114], [57], [88]	5.6	[77], [78]	2×10^{-4}; 5×10^{-4}	[74]
^{241}Am (kg)	41–43	[116], [126], [this study]	0.52	[79], [78]	/	
Atomic ratio	**Global fallout**		**Chernobyl nuclear accident**		**Fukushima nuclear incident**	
$^{236}U/^{238}U$	Not measured in stratospheric debris		0.019	[79]	/	
$^{236}U/^{239}Pu$	0.22–0.30	[110]	8.7	[79]	1.1	[85]
$^{237}Np/^{239}Pu$	0.44–0.59	[104], [107], [109]	0.024	[79]	/	
$^{240}Pu/^{239}Pu$	0.18 (integrated value)	[117]	0.38–0.42	[71], [72], [73]	0.303–0.330	[74]

Fractionation of debris composition with distance from the Chernobyl NPP was observed. In the near zone within 100 km from the reactor, hot particles deposited reflected mainly the radionuclide composition of the fuel and cladding material, with large particles containing the refractory elements (Zr, Mo, Ce, ...), as well as uranium, plutonium, and the heavier actinides produced by neutron activation. In soils sampled 15 km west of Chernobyl and in lichen samples collected near the city in 1990, $^{243,244}Cm$ activities of 11.4 to 86 mBq and activity ratios of $^{243}Cm/^{244}Cm$ and $^{243}Am/^{241}Am$ in the ranges 0.066–0.076 and $2.6–2.7 \times 10^{-3}$, respectively, were observed [80]. In the far zone, namely, from 100 km to ~2000 km, mainly the volatile fission products (I, Te, and Cs) were dispersed in the form of condensation-generated particles [79]. However, fallout of actinides was also recorded significantly distant from the near zone limit. At more than 1100 km north-west from the Chernobyl NPP in Sweden, about 1 μm diameter hot spherical particles composed of variable proportions of refractory elements were found in air filters shortly after the accident. One type of particle contained about 70 Bq of ^{239}Np [81], which has, in the meantime, decayed to about 2×10^7 atoms of ^{239}Pu. From 29 April to 10 May 1986, more than 1300 km west of the Chernobyl NPP, in Bavaria state of Germany, the ground deposition of actinides from the accident was clearly observed. In particular, cumulative depositions of ^{237}U, ^{239}Np, ^{238}Pu, $^{239,240}Pu$, and ^{242}Cm (27, 1000–1300, 0.021, 0.0051, and 0.60 Bq/m^2, respectively) were reported [82].

The incident at the Fukushima Daiichi NPP (FDNPP, 11 March 2011) released radionuclides as a consequence of controlled venting operations, three massive oxy-hydrogen explosions that damaged the buildings of Unit 1, 3, and 4, and a hydrogen explosion in the condensation chamber of Unit 2. Melting of the core material occurred, but it was estimated that the core temperature remained below 2670 K, so that refractory elements were mobilised only to a minor extent [83]. Plutonium derived from the accident of Fukushima shows isotopic ratios of $^{240}Pu/^{239}Pu$ equal to 0.303–0.330 and of $^{241}Pu/^{239}Pu$ equal to 0.103–0.135, as listed in Table 8.2. This isotopic composition likely reflects plutonium releases from the Unit 3 reactor, which had a mixed core containing both uranium fuel and uranium and plutonium mixed-oxide (MOX) fuel, which would be expected to result in a higher proportion of ^{241}Pu produced from the ^{239}Pu fuel [74]. Airborne releases of ^{239}Pu, ^{240}Pu, ^{241}Pu, and ^{236}U from FDNPP were identified in air filters placed 120 km from FDNPP [84]. The corresponding maximum activity of ^{239}Pu and ^{236}U in the air was 130 ± 21 nBq/m^3 and 0.43 ± 0.16 nBq/m^3, respectively. The amount of $^{239,240}Pu$ released by the Fukushima accident has been estimated to lie between 1×10^9 Bq and 2.4×10^9 Bq [74] (Table 8.2), corresponding to a ^{239}Pu mass release of 0.1 g and 0.5 g. This estimate is based on the total amount of ^{137}Cs released and the average of $^{137}Cs/^{239,240}Pu$ activity ratios observed in litter samples collected about 30 km from the FDNPP. The inventory of ^{236}U released to the atmosphere has been estimated from analyses of 'black substances' collected from the roadside in the Fukushima prefecture. Using the $^{236}U/^{239,240}Pu$ and $^{236}U/^{137}Cs$ activity ratios observed in these samples, a ^{236}U release of 1.2×10^6 Bq, corresponding to 0.5 g, is estimated [85] (Table 8.2).

Other accidents leading to significant plutonium release to the environment include release of weapons-grade ^{239}Pu during the fire at Windscale and crash of military bombers, as well as ^{238}Pu release from a nuclear-powered satellite, 'SNAP-9A.' At the Windscale nuclear complex in Cumbria, United Kingdom, on October 1957 about 8.6 g of ^{239}Pu and 60 kg uranium were released to the environment following a fire involving graphite

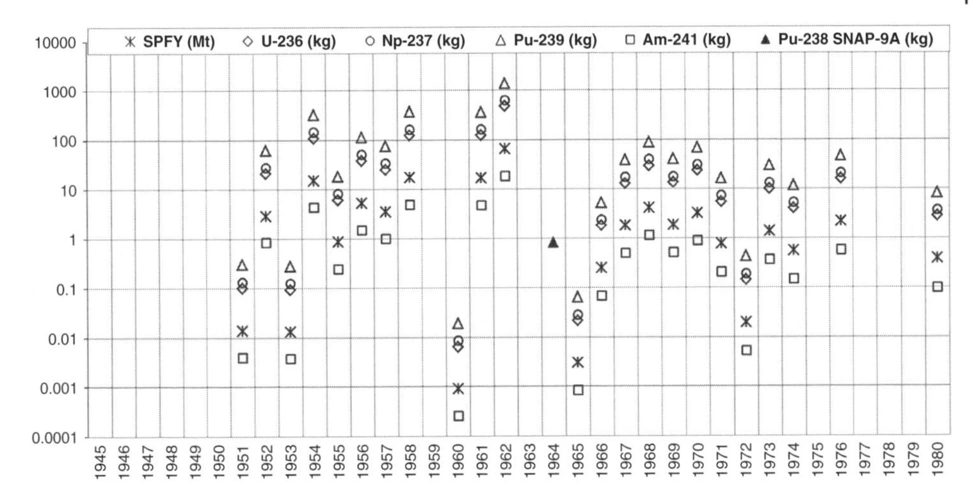

Figure 8.2 Stratospheric-partitioned fission yield in Mt (SPFY values from [114]) per year from atmospheric testing of thermonuclear devices and corresponding estimates for produced yearly inventories of ^{236}U, ^{237}Np, ^{239}Pu, and ^{241}Am in kilograms for 1 January 2016, from this work (see text for discussion). No global fallout occurred from atmospheric testing before 1951, as between 1945 (the first nuclear bomb 'Trinity' test) and 1951 only kt-range A-bombs, and no thermonuclear devices were detonated. The ~0.811 kg ^{238}Pu inventory from the 1964 SNAP-9A accident has been added and does not include the total of about 0.449 kg ^{238}Pu produced by nuclear testing.

and uranium fuel elements and flooding of the core with water [86]. The beaches on the Cumbrian coast are monitored for radioactive particulates, and although the number of alpha-emitting particles found has decreased in recent years, persistent ^{241}Am signatures remain [87]. In addition, plutonium characterised by an isotopic ratio $^{240}Pu/^{239}Pu$ equal to 0.05 was released as consequence of two aircraft accidents, one at Palomares in January 1966, when about 240 g of ^{239}Pu was dispersed over an area of 200 ha, and a second accident at Thule in January 1968, when about 2.7 kg of ^{239}Pu was released to the environment [88]. An accidental stratospheric injection of ^{238}Pu occurred in April 1964, when a System for Nuclear Auxiliary Power generator, SNAP-9A, completely burned up upon re-entry to the atmosphere in the Southern hemisphere, turning into small particulate fragments at an altitude of about 50 km. SNAP-9A contained about 1 kg of ^{238}Pu, and it was estimated that as a consequence of the accident about 13.9 kCi (514 TBq) of ^{238}Pu was deposited over Earth's surface by the end of 1970. This ^{238}Pu corresponds to 0.811 ± 0.128 kg inventory, and this addition nearly tripled the inventory when added to the 0.449 ± 0.053 kg ^{238}Pu fallout stemming from atmospheric nuclear testing [57] (Figure 8.2).

8.3.2.1 Source-Dependent Speciation and Behaviour of Released Actinides

Migration and bioavailability in the environmental behaviour of radionuclides, including actinide nuclides, determine their potential hazard for the biosphere and in accident scenarios is largely defined by their source-dependent speciation and less by local geochemical conditions [89, 90]. A major fraction of refractory radionuclides can be present as radioactive or hot particles, whose element, radionuclide, and phase composition, as well as size and properties like the dissolution rate, depend on the source and the release scenarios; therefore, such information should be included in the source term. A feature of actinides and plutonium in particular – unlike other radionuclides – is that they may form extremely sparingly soluble crystalline micron- and submicron-sized radioactive particles –hot particles – at extreme conditions such as high temperatures. One such example of radioactive particles formed in the Chernobyl accident was mentioned in the previous section. Nuclear explosions [90], accidents with melting of the active reactor zone [91, 92], fires [93], use of ammunitions containing depleted uranium [94, 95], and other high-temperature activities result in uranium or plutonium environmental entry in the form of hot particles. The behaviour of hot particles in the environment is to a large extent determined by the conditions they are formed in, that is, temperature, presence of oxygen and other elements, and so on, and their dissolution rate is largely determined by the size and morphology of the particles themselves. Thus, actinide nuclides in hot particles are present in kinetically stable forms that are not in thermodynamic equilibrium with the surrounding environment.

In a detailed study by Batuk et al. [89], the speciation of actinides in samples from various contaminated sites was reported. These included soils, grounds, and bottom sediments from near the Chernobyl NPP and nuclear complexes at Hanford, Rocky Flats, PA Mayak, and in the territory of McGuire Air Force Base (United States).[6] Actinides speciation and structural characterisation was studied by a combination of X-ray absorption fine structure (XAFS) spectroscopy and X-ray fluorescence (XRF) element mapping. Actinide speciation in the samples studied varied with the source terms and sample histories, revealing the complex chemistry and reactivity of U and Pu on the angstrom to micrometre scales in their initial release and after environmental exposure: relatively well-ordered PuO_{2+x} and UO_{2+x} that had equilibrated with O_2 and H_2O under both ambient conditions and in fires or explosions; instances of small, isolated UO_{2+x}, U_3O_8, and U(VI) particles, coexisting in close proximity after decades in the environment; alteration phases of uranyl with other elements, including ones that would not have come from soils; and mononuclear Pu – O species and novel PuO_{2+x}-type compounds incorporating additional elements that may have been produced because the Pu was exposed to extreme chemical conditions, such as acidic solutions released directly into soil or concrete.

Actinide-containing hot particles in soils near the damaged reactor at Chernobyl NPP contain various phases with various elemental and radionuclide compositions [91]. During the explosion, various particles formed at high temperatures >2600 °C [97] with uranium in its reduced form as UO_2 and UO_2-ZrO_2 solid solutions with varying U/Zr ratios (from $(U_{0.985}Zr_{0.015})O_2$ to $(Zr_{0.995}U_{0.005})O_2$). The hot particles that were released over a few days following the explosion, and during the fire contained uranium in higher oxidation states. These particles manifested considerably smaller kinetic stability and underwent relatively fast oxidation to yield U(VI) compounds [92].

The behaviour of hot particles in the vadose zone depends on weathering and the rate of their dissolution, associated with the transition of radionuclides to a soluble form

6 The site of what is referred to as the BOMARC Missile Accident Site in New Jersey, where an explosion and ensuing fire in 1960 partially consumed a long-range nuclear-equipped warhead of the BOMARC type and released weapons-grade plutonium over an area limited to the base. Plutonium was released into water used to extinguish the fire, and the average Pu-239 contamination level of the area is 32 pCi/g (1.2 Bq/g). The ratio of $^{239}Pu/^{241}Am$ (^{241}Am resulting from ^{241}Pu decay) activity in 1992 was 5.9 [96].

[4, 98]. For example, dissolution of hot particles in radioactive waste, contaminated soil, and vegetation buried in shallow trenches in the 'Red Forest' in the vicinity of the fourth power unit of the Chernobyl NPP resulted in migration of plutonium by 5 to 15 m during 1987–2005, apparently in association with low-molecular organic compounds. Higher ratios between americium and plutonium isotopes in groundwater monitoring wells in the area compared to the average value in the fuel component of the Chernobyl fallout indicate that americium was more mobile than plutonium in the aquifer.

8.3.3 Burden from Nuclear Testing

In addition to reprocessing releases and environmental actinide burden from major nuclear accidents, atmospheric nuclear bomb testing between 1945 and 1980 and the Hiroshima and Nagasaki nuclear bombs dropped in 1945 are sources for a significant actinide contamination that was spread worldwide. Actinide releases from underground testing (1945 to the present) and from accidents related to plutonium weapons and weapons production, according to the available literature, have resulted in a relatively modest and mainly confined contamination. Among the actinides, the major actinide radioisotopes produced and dispersed into the environment were plutonium nuclides. ^{239}Pu is the most abundant actinide nuclide present in global fallout, followed by ^{237}Np, ^{236}U, ^{240}Pu, ^{241}Am, ^{242}Pu, ^{241}Pu, ^{238}Pu, and ^{244}Pu. The actinide inventory from the accident of Fukushima (2011) is negligible compared to these sources (cf. Table 8.2).

Early, extensive studies of fallout inventories and dynamics were possible for radionuclides with specific activities high enough to be detected with radiometric techniques, for example, the fission products ^{90}Sr and ^{137}Cs. Since the main actinide nuclides produced in nuclear testing and accidents are characterised by long half-lives and trace concentrations in the environment, investigation of their deposition occurred later but has progressed with innovation in mass spectrometric techniques, like inductively coupled plasma mass spectrometry (ICP-MS), thermal ionisation mass spectrometry (TIMS), and accelerator mass spectrometry (AMS). With the development of more sensitive analytical techniques, the environmental contamination from the long-lived and extremely rare actinides (^{236}U, ^{237}Np, and $^{242-244}$Pu) has been possible, and new data are continuously produced. In the following, the generation of actinide nuclides from nuclear bombs and accidents are explained, and insights from analytical characterisation concerning their dispersion and assessment of their inventories are presented.

8.3.3.1 Nuclear Testing

According to the Preparatory Commission for the Comprehensive Test Ban Treaty Organization,[7] a total of 2056 nuclear tests have been performed between 1945 and January 2016, including the last Democratic People's Republic of Korea (DPRK; North Korea) test. The majority of tests were in the Northern hemisphere, with a significant number of tests in the polar region; about 15% were in the Southern hemisphere and exclusively in the equatorial region [99]. Nuclear weapons tests were performed in the atmosphere, at high altitudes, underwater, and underground; the type of test

7 Established in 1996 with headquarters in Vienna, Austria, aimed at ensuring verification of the Comprehensive Nuclear-Test-Ban Treaty (CTBT).

determined the magnitude of dispersion of the fallout debris. While underground testing resulted in releases to a mostly confined environment, atmospheric testing was responsible for a long range, as well as worldwide distribution of the fallout debris and is the major anthropogenic source of actinides in the environment. Nearly 500 tests were performed in the atmosphere, including about 20 high-altitude tests [100]. The highest frequency of atmospheric nuclear testing occurred from 1951 to 1957 (49 tests), in 1958 (99 tests), 1961 (58 tests), and 1962 (116 tests). Tests were interrupted between November 1958 and September 1961, when the United States and the former USSR observed a moratorium on testing. Atmospheric testing declined following 1963, with the United States, former USSR, and the United Kingdom becoming Parties of the Partial Test Ban Treaty (PTBT[8]). Sixty-one atmospheric tests were carried out by non-PTBP signatory nuclear nations between 1963 and 1974, the last of which was conducted by France. An additional eight tests were performed up to 1980, after which China also interrupted atmospheric testing. The inventories of the actinides released by atmospheric tests each year is best correlated to the yearly yield of testing, rather than to testing frequency, as will be discussed in Section 8.3.3.4.

8.3.3.2 Actinides Released in Nuclear Testing

In nuclear testing, the excess free neutrons produced both by fission and fusion reactions are captured within a fraction of a second either in materials inside the weapon or in the environment of the explosion, producing a variety of induced radioactive nuclides [101]. In particular, nuclei of the employed fissile or fissionable material, namely, ^{235}U, ^{238}U, and ^{239}Pu, undergo not only fission, but also neutron capture reactions, by which the long-lived ^{236}U, ^{237}Np, heavier plutonium isotopes, and other actinide nuclides are produced. Weapons involving only a fission stage, the so-called A-bombs or atomic bombs, are based on the fission of weapons-grade uranium (with an enrichment of ^{235}U $\geq$ 90%), ^{239}Pu, or a composite of the two. The majority of the fissile material remains unchanged upon detonation; small portions are involved in fission and neutron capture reactions. For instance, it is estimated [102] that about 1.8% of the ^{235}U used in the explosion of the Hiroshima A-bomb underwent fission and about 0.13% underwent fast neutron capture, producing ^{236}U ($t_{1/2} = 2.342 \times 10^7$ y):

$$^{235}U(n,\gamma)^{236}U \tag{8.1}$$

Similarly, in the explosion of the Nagasaki A-bomb, between 8% and 12% of the ^{239}Pu core underwent fission, while the rest was released as the original weapon ^{239}Pu [103] and as the heavier plutonium isotopes built up by successive neutron capture on ^{239}Pu ($t_{1/2} = 2.411 \times 10^4$ y):

$$^{239}Pu(n,\gamma)^{240}Pu(n,\gamma)^{241}Pu(n,\gamma)^{242}Pu(n,\gamma)^{243}Pu(n,\gamma)^{244}Pu \tag{8.2}$$

Similar to ^{239}Pu and with the exception of ^{241}Pu ($t_{1/2} = 14.35$ y) and ^{243}Pu ($t_{1/2} = 4.956$ h), most of these isotopes are characterised by a long half-life: ^{240}Pu ($t_{1/2} = 6563$ y), ^{242}Pu ($t_{1/2} = 3.733 \times 10^5$ y), and ^{244}Pu ($t_{1/2} = 8.08 \times 10^7$ y). Following ß-decay of ^{241}Pu and ^{243}Pu, ^{241}Am ($t_{1/2} = 432.2$ y) and ^{243}Am ($t_{1/2} = 7370$ y), respectively, are formed.

---

8 Signed by the United States, the former USSR, and the United Kingdom on 5 August 1963, the PTBT banned nuclear testing in the atmosphere, in outer space, and under water.

Thermonuclear weapons, the so-called H-bombs or hydrogen bombs, were composed of three stages: fission, fusion, and fission. The x-rays released by the primary fission device (also called the trigger) compress and ignite the secondary fusion device, where the fusion of deuterium and tritium produces helium and high-energy (14 MeV) neutrons. The tamper (or blanket) surrounding the fusion device was, in many cases, made of natural or depleted uranium. For instance, a uranium tamper was employed in the second large-scale thermonuclear test, 'Bravo', conducted at the Bikini Atoll (1 March 1954) [104]. The high-energy neutrons produced by the fusion cause the ^{238}U blanket to undergo both fission and neutron capture, generating ^{239}Pu:

$$^{238}U(n,\gamma)^{239}U(\beta-)\xrightarrow{23.4m}{}^{239}Np(\beta-)\xrightarrow{2.4d}{}^{239}Pu \tag{8.3}$$

The heavier plutonium isotopes are then produced as described in Section 9.3 above, but mainly via successive neutron capture reactions on uranium isotopes and subsequent ß-decay to the corresponding plutonium isobars[9] [105]. The amounts and the isotopic compositions of uranium, plutonium, and trans-plutonium elements left after the explosion of a nuclear bomb vary according to the completeness of the fission chain reaction and the generated neutron flux. A $^{240}Pu/^{239}Pu$ isotopic ratio of 0.32 ± 0.03 and 0.363 ± 0.004 was measured in the debris from the Bravo and Mike tests [105], indicating neutron-rich events typical for the US thermonuclear tests conducted between 1951 and 1958. Mostly lower $^{240}Pu/^{239}Pu$ isotopic ratios (0.18) were characteristic of the Soviet thermonuclear tests in 1961–1962 and of the French ones (0.09) in 1966–1974 [106].

If weapons-grade uranium fission is used as the trigger of a thermonuclear device, ^{237}Np ($t_{1/2} = 2.144 \times 10^6$ y) can form via two successive (n, γ) reactions on ^{235}U and ^{236}U and subsequent β-decay of ^{237}U [107]. However, generation of ^{237}Np is most likely to occur in thermonuclear explosions from the ^{238}U in the tamper, via the (n, 2n) reaction:

$$^{238}U(n,2n)^{237}U(\beta-)\xrightarrow{6.7d}{}^{237}Np \tag{8.4}$$

While the ratio of the (n, 2n)/(n, γ) reactions on ^{238}U (given by the ratio of Reactions (8.4) to (8.3)) upon thermonuclear device detonation ranges between the values of 0.5 and 1, the measured $^{237}Np\backslash^{239}Pu$ in the debris of Bravo was 0.42 ± 0.04, indicating a contribution from unburned ^{239}Pu [104].

The isotope ^{238}Pu (87.7 y) is produced by the ß-decay of ^{238}Np (2.117 d) that can originate by neutron capture on ^{237}Np.

Similarly to ^{237}Np, ^{236}U is formed in thermonuclear explosions by the fast neutron capture reaction on ^{238}U:

$$^{238}U(n,3n)^{236}U \tag{8.5}$$

Comparing both the cross section for the Reactions (8.4) and (8.5) (0.8 and 0.4 barn, respectively [108]) and the measured atomic ratio of $^{237}Np\backslash^{239}Pu$ in global fallout of 0.44–0.59 [104, 107, 109], an atomic production ratio of $^{236}U/^{239}Pu$ of about 0.22–0.30 has been estimated [110]. Compared to ^{239}Pu, the heavier plutonium isotopes, as well as

9 Isobars are nuclides with the same number of nucleons or mass number and run along the nuclide chart at a diagonal from lower right to upper left.

the heavier actinides, were produced in several-orders-of-magnitude smaller amounts. In the thermonuclear test Mike [105] (Pacific Ocean; 1 November 1952) and in the low-yield underground test Barbel [111] conducted at the Nevada test site, the isotopic composition of the actinides from plutonium to fermium was analysed. The nuclide yield declines with increasing mass according to a general exponential trend, and a slightly higher abundance of the nuclides with even mass number until mass 250 is observed.

8.3.3.3 Debris and Fallout of Actinides from Atmospheric Nuclear Testing

Actinides associated with debris produced by the explosion of nuclear weapons were released. At the temperatures reached by such explosions, for example, 5×10^7 K [105], the nuclear fuel and structural components of the device, as well as the surrounding air and environmental materials, were vaporized during the first few thousandths of a second. With the rapid expansion and consequent cooling down of the fireball, the gaseous and melted materials solidified, producing radioactive debris whose characteristics were determined by the height at which the tests were performed. Debris of atmospheric testing in which the fireball did not touch the ground originated from the vapour of fissionable elements, their products, and structural materials that oxidised and condensed. The condensed debris tended to be in the form of spherical aerosol particles prominently constituted by magnetite (Fe_3O_4) and other oxides, with a size distribution between 10 μm and 0.01 μm [88]. In high-altitude explosions (between 40 and 540 km), no appreciable nucleation or particle formation is expected to occur, and the debris observed was either gaseous or made up of extremely small particles, such as clusters of some hundreds of atoms. In underground testing and when the fireball intersected the ground or water surface, the same kinds of particles present in an air explosion were found incorporated into larger particles of variously altered soil, marine sand, corals, or water slurry drops of up to 250 μm [101]. For example, some of the debris produced by the series of barge and surface thermonuclear tests at the Bikini Atoll in 1954 was composed of coral reef, the so-called Bikini ash. Bikini ash was found in samples of hemp-palm leaves of Bontenchiku, the fishing gear of the Fifth-Fukuryu-Maru, exposed to one of the thermonuclear tests. Samples of the hemp-palm leaves collected in 1995 exhibited $^{239,240}Pu$, ^{241}Am, and ^{237}Np specific activities of 5.3, 3.13, and 0.0115 Bq/g dry weight, respectively [104]. Debris collected after atmospheric tests above the tropopause[10] usually show lower activities and smaller particle sizes than samples collected in the troposphere [112].

Tests of thermonuclear devices, characterised by yields in the range of megatonnes (Mt), released enough energy to inject debris into the stratosphere. In contrast, debris from A-bomb explosions with kilotonnes (kt)-range yields remained in the troposphere [113], except for tests performed at altitudes high enough to allow the cloud to reach the tropopause [114]. The apportionment of debris between the troposphere and stratosphere determined the dynamics of debris deposition on Earth's surface, namely, the fallout. Such partitioning of debris in the atmospheric compartments can be

10 The tropopause is the boundary between the troposphere and the stratosphere or the lowest major atmospheric layer and the overlying stratosphere.

considered proportional to the fission yield of atmospheric testing, since the fission yield is proportional both to the generation of fission products and to that of nuclides by neutron capture, including the actinides. The total fission yield of global atmospheric testing was equal to 189 Mt, of which 29 Mt was partitioned to the local or regional environment, 16 Mt to the troposphere, and 144 Mt (equivalent to 76%) to the stratosphere [114].

Since the 144 Mt fission yield partitioned to the stratosphere stems from fission of the uranium tamper by neutrons released by the fusion stage of the thermonuclear devices, tests of thermonuclear devices were responsible for the production of 76% of the actinides inventory from nuclear testing and for global fallout of actinides. In fact, once injected into the stratosphere, due to the atmospheric transport processes, the actinides were distributed worldwide, with mean residence times in the atmosphere dependent on altitude, latitude, and season. Half times between 30 and 33 months are required for aerosols to descend by gravity from the high atmosphere and the upper stratosphere. The exchange of air masses between the equatorial and polar regions occurs with a half-time of 12 months, while the exchange between the hemispheres at the equator occurs in spring, fall, and winter with a half-time of 24 months. In this way, the dispersion of radioactive debris occurred predominantly over the hemisphere where the tests were performed. Since most of the atmospheric tests were conducted in the Northern hemisphere, the fallout was found to be about 79% greater there than in the Southern hemisphere [57, 88, 114, 115]. The mean residence time of aerosols in the lower stratosphere[11] ranges from 3 to 12 months in the polar regions and from 8 to 24 months in the equatorial regions [114, 116]. Debris remains in the troposphere for a period of a few months before being deposited mainly with precipitation [113]. In fact, fallout deposition has been shown to be strongly correlated with annual average rainfall, so that inventories in arid regions are expected to be lower than those recorded in wetter climates at the same latitude [115]. Because of the preferential exchange of air between the stratosphere and the troposphere in the mid-latitudes (40°–70°), there was enhanced deposition in the temperate regions and decreased deposition in the equatorial and polar regions. Such a pattern has been observed both for fission products, for example, ^{137}Cs [115] and plutonium. In fact, the highest deposition density for $^{239,240}Pu$ occurred between 30° and 70° N latitudes, with a maximum of about 80 Bq/m^2 at the latitude band of 40°–50° N, and a minimum of about 3 Bq/m^2 at 0°–10° N. In the Southern hemisphere, a more intense fallout was observed between 0° and 10° S and between 20° and 50° S, with a maximum of about 15 Bq/m^2 at 30–40° S and the minimum of ca. 0.4 Bq/m^2 extrapolated for the latitude band of 80°–90° S [57]. Similar to $^{239,240}Pu$, the deposition of global-fallout ^{237}Np is observed to be up to one order of magnitude higher in soils sampled in several locations of the Northern hemisphere compared to locations in the Southern hemisphere [117]. Comparing the global-fallout depositions of the main actinide nuclides in number of atoms per square metre, the levels of global fallout of

11 The height of the tropopause varies with latitude and season. For modelling purposes in EML-348 (1978) and in UNSCEAR (2000), it is assumed that the troposphere has an average altitude of 9 km, the lower stratosphere is confined between 9 and 17 km, and the upper stratosphere between 17 and 50 km in the polar regions (0° to 30° latitude), while in the equatorial regions (30° to 90° latitude) the same regions of the atmosphere are found at heights of 17 km, 17–24 km, and 24–50 km, respectively.

^{239}Pu and ^{237}Np range from 10^{12} to several 10^{13} atoms/m^2 and those for ^{240}Pu range from 10^{11} to 10^{12} atoms/m^2. These values stem from 52 soil samples collected in 1970/1971 to a depth of 30 cm from locations representing the various latitude bands of the hemispheres [57, 117]. Similarly, the global-fallout deposition of ^{236}U at latitudes of 47° and 48° N captured in soil [118] and a peat core [119] has a cumulative areal inventory of 10^{12}–10^{13} atoms/m^2. At the same latitudes in peat profiles, the global-fallout-derived ^{241}Am and ^{238}Pu exhibit a deposition density of 10^{11} and 10^9 atoms/m^2 [120]. ^{241}Pu/^{239}Pu, ^{242}Pu/^{239}Pu, and ^{244}Pu/^{239}Pu isotopic ratios measured in soil samples at a latitude of 47° N were equal to $(1.43 \pm 0.06) \times 10^{-3}$, $(3.22 \pm 0.05) \times 10^{-3}$, and $(5.7 \pm 1.0) \times 10^{-5}$, respectively [75]. From these data and considering the aforementioned deposition densities of ^{239}Pu, areal inventories for ^{241}Pu and ^{242}Pu in the range of 10^9–10^{10} atoms/m^2 are obtained and for ^{244}Pu 10^7–10^8 atoms/m^2. These results have been confirmed for ^{241}Pu and ^{242}Pu from the plutonium isotopic composition found in a peat core at a similar latitude [119]. Note that the partitioning of the SNAP-9A ^{238}Pu fallout (Section 8.3.2) exhibited a pattern opposite to that from nuclear bomb tests, with 78% of the total in the Southern hemisphere (0.630 kg) and the remaining ~22% (0.180 kg) in the Northern hemisphere. ^{238}Pu/^{239,240}Pu activity ratios up to 0.2 are currently observed in the stratosphere, which are higher than those expected from global fallout (0.02–0.03), and indicate that the ^{238}Pu injected by the burnup of SNAP-9A in the upper stratosphere is still circulating [121].

8.3.3.4 Inventories of Actinides from Atmospheric Nuclear Testing

The inventories of ^{236}U, ^{237}Np, ^{239}Pu, and ^{241}Am from global fallout are discussed in the following paragraphs. These inventories are summarised in Table 8.2 and compared to the nuclear accidents at Chernobyl and Fukushima, together with the corresponding ^{236}U/^{238}U, ^{236}U/^{239}Pu, ^{237}Np/^{239}Pu, and ^{240}Pu/^{239}Pu atomic ratios. As mentioned in Section 8.3.2, the ^{240}Pu/^{239}Pu, ^{241}Pu/^{239}Pu, and ^{242}Pu/^{239}Pu isotopic ratios from NPP accidents are generally higher than those related to nuclear tests, offering a means of differentiating actinide sources. The inventories of the actinides in this work are expressed in kilograms rather than as activities, thereby allowing for a direct comparison of the amounts of actinides.

The extensive study of Hardy et al. [57] provided the first estimate of the global inventory of fallout plutonium from measurements of the levels of ^{239,240}Pu and ^{238}Pu in soils from various latitude bands of the globe. Those values, converted to kilograms and taking into account that the ^{240}Pu/^{239}Pu isotopic ratio of the integrated global fallout is 0.18, correspond to 3108 ± 344 kg of ^{239}Pu and 0.449 ± 0.053 kg of ^{238}Pu. Given that the activity ratio ^{239,240}Pu/^{90}Sr in stratospheric samples was essentially constant, Harley estimated the inventory of ^{239,240}Pu by referencing the measured ^{90}Sr inventory, obtaining a production of about 400 kCi, one fifth of which is considered to be deposited near the test sites [88]. Such inventories correspond to 3060 kg of ^{239}Pu distributed worldwide and 765 kg of near detonation (or 'close-in') site fallout. In UNSCEAR (1993) [122] and UNSCEAR (2000) [114], the production of transuranic radionuclides was inferred from their ratios to ^{90}Sr measured at their deposition. In this way, a lower value was obtained for the global-fallout inventory of ^{239}Pu, about 2800 kg. We can expect that the various actinides behave similarly to each other during the processes of atmospheric debris formation and deposition. We can furthermore assume that the actinides in global fallout have, to a certain extent, the same atomic composition independent of the amount of precipitation and latitude. Under these assumptions, the inventory of an actinide, for example, ^{237}Np, can be estimated from the known inventory of another actinide, for example, ^{239}Pu, and their atomic ratio $^{237}Np/^{239}Pu$ measured in an environmental sample accounting for the entire history of global fallout in a given region. In a terrestrial environment, such global-fallout inventories can be extrapolated from analysis of undisturbed soils or peat profiles at locations where remobilisation of the deposited radionuclides from human activities, overland runoff, and wind-forced re-suspension is minimal. Using such an approach, an inventory of ^{237}Np from stratospheric fallout equal to 1500 kg was obtained in 1998 from measurements in several soil samples [107]. Similarly, the first estimate of the inventory of ^{236}U of 900 kg was reported in 2009 from a study on a soil core and considering a $^{236}U/^{239}Pu$ atomic ratio of about 0.23 [110]. An investigation of ^{236}U in a coral core collected in the Caribbean Sea reported a production of 7.4 ± 1.6 kg ^{236}U per Mt thermonuclear device, with a related global ^{236}U fallout inventory of 1060 ± 210 kg [123]. A later study in an ombrotrophic peat profile believed to be representative of the period of stratospheric fallout [119] reported a global inventory of 1698 ± 850 kg ^{236}U. Within the reported uncertainties, the global-fallout inventory of ^{236}U estimates inferred from terrestrial environmental reservoirs with respect to plutonium [110, 119] are in agreement with that obtained in the coral core using a different approach of considering the production of ^{236}U per Mt [123]. A higher estimate of about 2100 kg ^{236}U was obtained by the analysis of two ocean water profiles in the equatorial Atlantic [124]. This is interpreted as being probably due to the transport of additional ^{236}U with the North Atlantic Deep Water into the sampling location. Higher levels of ^{236}U than those expected solely from global fallout were also found in a study involving a transect of nine seawater depth profiles spanning 64° to 2.54° N along the Northwest Atlantic Ocean [125]. The authors concluded that 115–250 kg ^{236}U in excess of their estimate of global-fallout inventory (1000 and 1400 kg) was found in the North Atlantic Ocean due to contributions from the nuclear reprocessing plants of Sellafield and La Hague.

We can estimate the yearly production of ^{239}Pu from thermonuclear devices by comparing the values of the yearly partitioned fission yields to the stratosphere provided in Table 4 of the UNSCEAR report (2000) [114] with the global-fallout ^{239}Pu inventory of 3110 kg [57]. Similarly, the yearly production and dispersion to the stratosphere of ^{237}Np and ^{236}U can be estimated by reference to that for ^{239}Pu and $^{237}Np/^{239}Pu$ and $^{236}U/^{239}Pu$ atomic ratios in the integrated global fallout (Table 8.2 of the UNSCEAR report [114]). We can assume that the dominant path for production of ^{241}Am in nuclear testing is the ß-decay of ^{241}Pu [116]. Therefore, by calculating the yearly production of ^{241}Pu, we can estimate the yearly inventory of global-fallout-derived ^{241}Am. The isotopic ratio $^{241}Pu/^{239}Pu$ at formation is approximately 0.014 [88]. By multiplying this value by the yearly inventory of ^{239}Pu, the yearly production of ^{241}Pu is obtained, and from these values the corresponding yearly ingrowths of the daughter ^{241}Am is calculated. In this way, a global-fallout inventory of ^{241}Am from thermonuclear testing of about 41 kg is estimated. This inventory should be reduced to about 38 kg by 1 January 2016, due to the decay of ^{241}Am. Our estimate of ^{241}Am is in agreement with the value of about 43 kg ^{241}Am obtained by considering a production rate of ^{239}Pu equal to 1.1 kCi/Mt (corresponding to ~17.4 kg/Mt) and a $^{241}Pu/^{239}Pu$ isotopic ratio of 0.0138 for the 1961–1962 testing period and otherwise 0.0118 [116]. These values are similar to the global-fallout ^{241}Am inventory of 42.3 kg given in UNSCEAR (1982) [126].

Figure 8.2 depicts the yearly SPFY in Mt [114] from atmospheric testing of thermonuclear devices and the corresponding estimates for produced yearly inventories of ^{236}U, ^{237}Np, ^{239}Pu, and ^{241}Am in kilograms. This figure pinpoints the inventories of the main actinide nuclides that were distributed and deposited worldwide; four main periods of atmospheric nuclear testing can be identified. Between 1945 and 1951, only fission weapons were tested that, due to their low yield in the range of kt, produced mainly tropospheric fallout deposited at the latitude band of the test sites. Starting in 1951, high-yield thermonuclear devices were also tested, releasing debris into the stratosphere, and the fallout was spread worldwide. Tests conducted before the moratorium in 1958 to 1961 were responsible for the release of the 31% (45.1 Mt) of the total global fallout of actinides. Most actinides were released immediately after the moratorium when, during the years between 1961 and 1962, 57% (83.1 Mt) of the global-fallout inventory was produced. The period following the PTBT in 1963 until 1980 contributed only 12% (16.8 Mt) to the total stratospheric inventory of the actinides.

Production of ^{239}Pu, ^{237}Np, ^{236}U, and ^{241}Am appears in Figure 8.2 in the same relative proportions each year. Especially in the case of ^{241}Am, this is a conservative assumption resulting from the use of the integrated ^{241}Pu/^{239}Pu isotopic ratio, although the plutonium isotopic composition changed significantly. The period 1951–1958 was dominated by neutron-rich thermonuclear tests performed by the USA, releasing plutonium with a higher proportion of heavier isotopes compared to the later tests. In fact, in ice strata corresponding to the pre-moratorium testing period (1951–1958) of two glaciers from the Arctic and two from the Antarctic, ^{240}Pu/^{239}Pu isotopic ratios in the range 0.21-0.34 and ^{241}Pu/^{239,240}Pu activity rations in the range 16–29 were found. Lower ^{240}Pu/^{239}Pu isotopic ratios of 0.09-0.22 and ^{241}Pu/^{239,240}Pu activity ratios of 10-13 were observed in ice layers associated with the post-moratorium testing period (1961–1980) [106, 127]. From these values, ^{241}Pu/^{239}Pu isotopic ratios in the range 0.017–0.038 in the pre-moratorium and in the range of 0.008–0.014 in the post-moratorium period are calculated. In particular, a ^{241}Pu/^{239}Pu isotopic ratio equal to 0.013 for the period 1961–1962 is obtained. Considering the significantly higher ^{241}Pu/^{239}Pu isotopic ratios in the period 1951–1958 compared to the value of 0.014 used in our calculation, we expect a higher ingrowth of ^{241}Am from this period than that calculated in our conservative estimate. Our global-fallout inventory of ^{241}Am should therefore be considered a lower limit.

An estimate of the inventory of ^{243,244}Cm was given by Beasley and Ball [128] from their analysis of Columbia river sediments. The authors determined that the curium deposition from the fallout lay between 1% and 3% that of ^{239,240}Pu. The ^{239,240}Pu inventory for the global fallout of 325 kCi [57] translates to between 3 and 10 kCi (or $1.1–3.7 \times 10^{11}$ Bq) ^{243,244}Cm activity resulted from weapons testing.

8.3.3.5 Environmental Behaviour of Fallout Actinides

A recent study on plutonium and ^{137}Cs in stratospheric aerosol samples collected over Switzerland from 1970 until 2011 [121] has pinpointed two of the determinant processes in the behaviour of radioactive debris in the atmosphere: thermal stratification and interaction with natural aerosols. A large variability in the levels of ^{239,240}Pu in the stratosphere, spanning 0.1 to 100 μBq/m^3, has been observed over the past decades, with values up to five orders of magnitude higher than those in ground-level aerosols and a local maximum corresponding to the eruption of the Eyjafjallajökull volcano in

2010. These observations can be explained by a low mixing rate of stratospheric air masses and/or a low input to the troposphere. Furthermore, a significant fraction of the radioactive debris injected above the tropopause is possibly associated with particles <0.1 μm, which remains in the stratosphere for several decades with a mean residence time of 2.5–5 years. The injection of fine-grained ash and gases (e.g. SO_2) up to the lower stratosphere during the Eyjafjallajökull eruption can be responsible for an increased scavenging of plutonium and ^{137}Cs aerosols in the stratosphere and a related increased transport into the troposphere by sedimentation [121]. We assume that the same atmospheric behaviour of plutonium can be attributed to ^{236}U, ^{237}Np, ^{241}Am, and the other actinide nuclides derived from global fallout.

Global-fallout-derived actinides have been reported both in aquatic and terrestrial environments. A study of ^{236}U in a depth profile of the western equatorial Atlantic Ocean [124], in a transect in the North Atlantic Ocean [125], and in a yearly resolved coral core from the Caribbean Sea [123] have elucidated the conservative behaviour of this nuclide in the ocean and demonstrated its use as tracer for oceanic water. Unlike uranium, plutonium does not exhibit conservative behaviour in ocean water. Pu(IV) and Pu(V) are the most common oceanic species. While the Pu(IV) form is highly particle-reactive and therefore scavenged easily by suspended matter and colloids, the oxidised form, Pu(V), is more soluble and can be transported in the dissolved phase over long distances. Since the isotopic composition of plutonium, especially the $^{240}Pu/^{239}Pu$ isotopic ratio (Table 8.2), is strongly related to its source term, estimation of the contribution for a particular source term can be used in understanding oceanographic processes, as discussed in Ref. [129].

Colloid-mediated transport of plutonium from an underground nuclear test conducted at the Nevada test site was discovered 30 years after the test 1.3 km from the test location [130], and the mobility of ^{237}Np in the same arid region was recently demonstrated [131]. These studies have proved that even though underground testing did not cause wide dispersion of radioactive debris inherent to atmospheric testing, the nuclear contamination was not limited to the immediate area of the test. Global-fallout $^{239,240}Pu$ and ^{241}Am of a pine forest soil from the Bavaria region of Germany presented similar rates of vertical migration, generally <1 cm/year, and migration behaviour in the various soil horizons was observed to be dominated by the formation of complexes between plutonium, americium, and humic substances [132]. In the acidic, organic-rich environment of an ombrotrophic peat core from the Black Forest, the post-depositional downward migration of global-fallout ^{236}U and plutonium isotopes has been observed. In particular, the mobility of ^{236}U was lower compared to that of plutonium, and the position of the ^{236}U bomb influx was preserved in the chronology of the peat formation. In contrast, the position of the plutonium maximum concentration was translated ~10 cm vertically down the peat profile relative to uranium [119, 133]. Plant uptake of global-fallout ^{236}U versus plutonium isotopes in species representative of a forested environment in the Australian Capital Territory has been investigated [134]. ^{236}U was found to be preferentially taken up by plants with enrichment factors $(^{236}U/^{239}Pu)_{veg}/(^{236}U/^{239}Pu)_{soil}$ ranging between 7 and 52. In a previous study [135], the uptake and incorporation of global ^{236}U and plutonium fallout by European deer antlers was demonstrated. Since the measured $^{239}Pu/^{236}U$ atomic ratios in the antlers were generally lower than in global fallout, the authors concluded that the higher antler ^{236}U content was derived from the preferential uptake of ^{236}U in plant tissue [134].

We have seen that contamination sources are characterised by specific isotopic and atomic ratios, the analysis of which in environmental samples allows information to be gained of the nuclear contamination origin. Likewise, if the contamination source is known, any deviation of the atomic ratios measured in an environmental sample is an indicator of different relative geochemical behaviour of the actinides in question. When investigating the behaviour of global-fallout actinides in the environment, it is important to identify any stratospheric fallout contribution from a local or regional fallout, which would change the atomic and isotopic composition of the contamination. Tropospheric fallout can be characterised by lower $^{240}Pu/^{239}Pu$ isotopic ratios compared to stratospheric fallout; therefore, in regions affected by both tropospheric and stratospheric contributions, such as nuclear test sites and the equatorial regions, a lower $^{240}Pu/^{239}Pu$ ratio is generally determined [117, 136].

The isotopic vector of plutonium is a more sensitive indicator of mixed contamination sources than the levels of ^{236}U. For example, in a recent study on soil profiles collected in the black-rain area around Hiroshima, no contribution of the close-in ^{236}U fallout produced by the Hiroshima A-bomb was found. The authors conclude that global fallout dominated the current levels of ^{236}U and $^{239,240}Pu$ in the investigated area [102].

8.4 Radionuclide Biogeochemistry – Contaminated Land and Radioactive Waste Disposal

As we have seen, there is a significant global legacy of radioactively contaminated land as a consequence of nuclear operations with releases to the environment from both authorised and accidental emissions. In addition, decommissioning and clean-up of defunct nuclear facilities and processing and management of SNF has led to a range of higher-activity radioactive wastes, the majority of which are in storage prior to final disposal. The environmental mobility of actinides (An) potentially released from contaminated environments or radioactive waste is determined by a number of parameters (cf. Section 8.5), many of which impact their oxidation state as a dominating factor. In groundwater at ambient pH, as a contaminant transport medium, the lower actinide oxidation states, $An^{3+}$and An^{4+}ions, are present as hydrated ions, whereas the higher V and VI oxidation states form linear trans-dioxo(actinyl) cations, $AnO_2^{+}$and AnO_2^{2+}, which generally exhibit higher solubilities and hence mobilities. Therefore, one of the major concerns of actinides in an environmental context is control of the redox conditions. In Figure 8.3, the expected oxidation states of uranium, neptunium, and plutonium as a function of the Eh at neutral pH is shown [137]. The Eh will vary depending on concentration and the presence/proportion of other components in the system (e.g. the Pu(IV)/Pu(III) couple for the Pu(IV)-EDTA complex is slightly positive and that for hydrous plutonium oxide is a few tenths of a volt negative [138]). Also shown are the approximate potentials for some microbial respiratory processes. It is clear that Fe(III)-reducing bacteria show the potential to reduce these actinides from higher oxidation to lower oxidation states enzymatically. Alternatively, Fe(II) being more electronegative than the potentials for Pu(V)/Pu(IV), Np(V)/Np(IV) and U(VI)/U(IV) formed from the metabolic activity of Fe(III)-reducing bacteria may also act as a chemical reductant in an indirectly, although the speciation of the Fe(II) itself (aqueous, sorbed, or associated within mineral phases) impacts its reactivity.

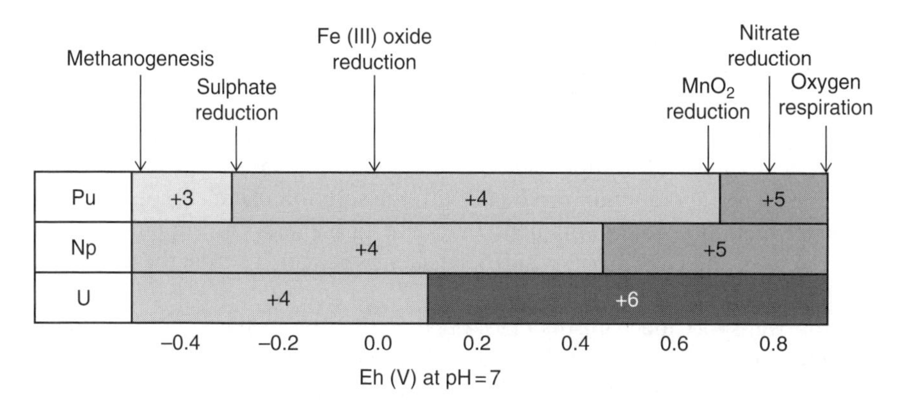

Figure 8.3 Expected oxidation states of uranium, neptunium, and plutonium as a function of the Eh at pH = 7. Adapted from Lloyd JR, Chesnes J, Glasauer S, Bunker DJ, Livens FR, et al. (2002) Reduction of actinides and fission products by Fe(III)-reducing bacteria. *Geomicrobiology Journal*, **19**(1), 103–120.

In this section, we discuss the speciation of actinides resulting from microbial interactions in the context of contaminated land and waste disposal. Specifically, we discuss the redox-active and long-lived uranium, neptunium, plutonium, and technetium radionuclide interactions with microbes and minerals in conditions relevant to both natural and engineered environments, with a focus on their speciation and mobility. This highlights the biogeochemical processes which will have a profound effect on the mobility of radionuclides in environmental and radioactive waste disposal settings.

8.4.1 Bioreduction Processes

Microbial reactions are key controls on the biogeochemical evolution of natural and engineered environments [139]. These reactions involve the breakdown, and hydrolysis of organic matter, the destruction/formation of inorganic compounds and minerals and the chemical modification of sediment pore waters. In pure culture experiments, bacteria capable of oxidising organic matter coupled to Fe(III)- or Mn(IV)-reduction were first described in the late 1980s [140–142]. The microbial isolate from freshwater sediments, *Geobacter metallireducens*, was under anaerobic conditions able to couple oxidation of acetate to CO_2 to reduction of ferrihydrite to magnetite/vivianite and obtain energy for growth [141]. The significance of these processes in the context of the biogeochemistry of iron and manganese, where the reduction of Fe(III) and Mn(IV) was demonstrated as predominantly an enzymatic process rather than a chemical one, was highlighted in this early work and has been an enduring theme ever since. The further impact of anaerobic microbial metabolism on contaminant and specifically radionuclide behaviour was quickly recognised [46].

8.4.2 Uranium Biogeochemistry

Soon after the early work with *Geobacter metallireducens* demonstrating coupled acetate oxidation with Fe(III) and Mn(IV) reduction mentioned above, pure culture experiments with *Geobacter* and another freshwater sediment isolate, *Shewanella oneidensis,* U(IV) as soluble uranyl carbonate species and acetate as the electron donor showed enzymatic reduction to insoluble U(IV) as uraninite, UO_2 ([46], cf. Section 8.2.3). In the intervening years, work on pure culture systems has highlighted that a relatively diverse range of prokaryotes can reduce U(VI) to U(IV), with UO_2 identified as the reduction product in many systems [143, 144]. Interestingly, recent work has identified poorly ordered "non-crystalline" U(IV) polymers coordinated to phosphate and/or carboxylate groups as enzymatic reduction products, with the solution chemistry observed to influence the reduction product obtained [145]. Not all bacteria can support growth via U(VI) reduction, and *Geobacter sp.* and *S. oneidensis* remain, to date, the most well studied of these microorganisms [143, 146].

The electron transport mechanisms of U(VI) bioreduction in pure culture are not fully understood. There is debate on the role of electron carriers such as flavins and of conductive cell surface appendages or "pili" in U(VI) reduction, with several studies presenting evidence for only a marginal role for these in U(VI) reduction [143, 147]. A marked complexity of U speciation in enzymatic reduction systems has also been observed. Work on enzymatic reduction with *Geobacter sp.* showed that the reduction was likely mediated via a one electron transfer with U(VI) being reduced to U(V) [148, 149]. This U(V) was then unstable with respect to disproportionation and formed U(IV) and U(VI), with the U(IV) undergoing hydrolysis to the insoluble U(IV)product. This one electron transfer hypothesis was tested using Np(V), which is stable with respect to disproportionation. As expected, the Np(V) was not reduced enzymatically, supporting the one electron transfer mechanism identified with U, and at the same time demonstrating surprising actinide specificity to enzymatic reduction [148]. Although in enzymatic systems U(V) is considered to be transient, recent work suggests that U(V) may be stabilised in certain environmental media, notably when incorporated into the Fe(II)- and Fe(III)-bearing mineral magnetite [150, 151]. Indeed, recent appreciation of the complexity of U speciation in biogeochemical systems is noteworthy, and it is clear that the early model of precipitation to uraninite is evolving into a more complex understanding of uranium speciation and fate.

In more complex environmental media, where uranium contamination is present, U(VI) reduction facilitated via stimulation of indigenous sediment microorganisms by addition of an electron donor such as acetate to sediments has been widely reported (e.g. [143, 146]). Much of this work has focussed on remediation of soluble U(VI) in uranium-contaminated groundwaters via stimulation of the subsurface with an electron donor such as acetate. The indigenous microbial population develops reducing conditions with concomitant U(VI) reduction to poorly soluble U(IV) [146, 152]. Early studies highlighted the potential of this non-invasive, in situ process to treat U(VI) groundwater contamination via reduction to poorly soluble U(IV), identified as uraninite [153]. Interestingly, for U(VI) the mechanisms of reduction seem to include both direct, enzymatic reduction, and indirect, abiotic reduction via reaction with microbial reduction products such as Fe(II) and sulphide [146, 154], but with enzymatic reduction certainly significant in many situations. Non-crystalline U(IV) species, observed as reduction products in pure culture experiments [145], have also been recognised as being significant in environmental systems [154–158]. In these cases, U(IV) is thought to be complexed to biomass via carbon and/or phosphate, and it is becoming clear that an intricate interplay between enzymatic and abiotic Fe(II) and sulphide reduction pathways are controlling factors for the fate of U(IV) upon reduction [154, 159]. It is worth noting that although the vast majority of work on uranium bioreduction has been performed in the context of contaminated land, recent work has highlighted the potential role that these processes may play in deep geological disposal of radioactive wastes. Microbial activity at elevated pH representative of conditions expected in disposal of cement containing intermediate level wastes has been studied. When sediments from a historical, industrially impacted lime working site were stimulated with electron donor, a range of bioreduction processes occurred up to pH ~11–12 [160, 161]. This is relevant to cementitious intermediate level waste disposal, as such waste will contain electron donors that can support biodegradation processes such as cellulose and hydrogen gas, which evolves during corrosion of metals present in the wastes [162, 163]. The potential for U(VI) reduction was tested in lactate-stimulated sediment microcosms at pH 10–11 [155]. U(VI) was observed to be only partially reduced in experiments where no additional Fe(III) was present in the sediments. In contrast, when Fe(III) was added to the starting sediment, U(VI) was fully reduced to U(IV) with both uraninite and non-crystalline U(IV) species present as reduction products. Interestingly, the microbial populations that were stimulated during bioreduction included Gram-positive species, in contrast to circumneutral pH conditions where Gram-negative species dominate. The Gram-positive species lack an outer membrane with c-type cytochromes that are implicated in Fe(III)- and U(VI)- reduction at neutral pH. This suggests a potentially fundamental difference in bioreduction pathways for Fe(III) and U(VI) at elevated pH conditions [155].

An important aspect of radionuclide behaviour in systems exploiting bioreduction for contaminant immobilisation concerns the stability of the reduced products. The bioreduction products need to be durable if in situ remediation is to be effective. In many sediment microcosm and column experiments, U(IV) re-oxidation via aerated waters occurs over a period of days to weeks, and under these conditions, oxidative remobilisation to solution is often near complete for uranium [158, 164–166]. Recent studies have reported differences in re-oxidation rates for non-crystalline versus uraninite U(IV), with non-crystalline U(IV) undergoing more facile re-oxidation in air [167]. In contrast, work examining U(IV) re-oxidation rates by aeration of progressively aged sediments showed a change in U(IV) speciation with time, with evidence for uraninite formation over 15 months. Despite this change in U(IV) speciation, there was little difference in the re-oxidation rates reported for the differently aged samples [158]. In addition to air re-oxidation, workers have also examined microbially mediated nitrate re-oxidation of radionuclides, as nitrate is often a common co-contaminant in groundwaters at nuclear facilities and will be present in significant quantities in some radioactive wastes [158, 160, 168, 169]. In microcosms, nitrate typically causes significant, but incomplete re-oxidation of Fe(II) [158, 170]. U(IV) nitrate-mediated oxidative remobilisation is also significant; typically ~50% uranium is remobilised to solution under nitrate re-oxidising conditions [158, 166, 168]. Recently, a dual approach of bioreduction coupled with biomineralisation to treat uranium contamination has been described with amendment using glycerol phosphate [171, 172]; the glycerol moiety functions as the available electron donor to stimulate microbial reduction, and after cleavage of the glycerol phosphate molecule, the soluble phosphate is available for biomineralisation to form insoluble uranium-phosphate phases. Interestingly, in systems where U(VI) was significantly sorbed to sediments prior to bioreduction, a poorly soluble U(VI)-phosphate formed [172], whereas in sediments where U(VI) was present in solution, U(IV)-phosphate (as a 'ningyoite-like' phase) formed [171]. The U(IV)-phosphate reduction product was considerably more recalcitrant to oxidative remobilisation with both air and nitrate than the U(IV) products of microbial U(VI) reduction formed on biostimulation with acetate [171]. Generally, for the U(IV), bioreduction products formed when stimulation is via simple electron donor, oxidative remobilisation is significant, suggesting that long-term amendments with electron donor will be necessary to retain reducing conditions and thus U(IV) in sediments. New approaches targeting bioreduction and biomineralisation of uranium (and by implication other actinides) show the potential for tailored in situ remediation approaches yielding durable biominerals within which the contaminant species is immobilised. A promising approach where analysis of natural analogues indicates immobilisation of U(IV) in carbonate phases for at least 1 million years has recently been described [173].

Microbial mediated U(VI)-reduction processes offers a potentially positive outlook for conditions relevant to deep geological disposal of radioactive wastes in that indigenous subsurface microorganisms can be stimulated to create a zone of bioreduction around the geological disposal facility that will reductively scavenge key radionuclides from groundwater, including uranium. In geological disposal, both direct enzymatic reduction processes and abiotic processes may be relevant. For example, microbial reduction may cause Fe(III) reduction in key minerals in the subsurface, thus priming them for reactivity to oxidised mobile species including radioactive contaminants [150]. This concept was explored for the phyllosilicates biotite and chlorite, whereby the minerals were reacted with *Geobacter sulfurreducens* to promote Fe(III)-reduction. The bioreduced minerals were then washed and abiotically reacted with U(VI), Tc(VII), and Np(V) with reductive precipitation of the reduced form of the bioreduced minerals enhancing radionuclide reduction in all cases [174].

8.4.3 Technetium Biogeochemistry

Although not an actinide element, ^{99}Tc is a long-lived (2.11×10^5 y), high-yield fission product, which in common with actinide elements is significant in remediation of contaminated land and disposal of nuclear waste. In terms of its environmental chemistry, technetium displays a highly soluble oxic species, pertechnetate (TcO_4^-), and a common reduced form, Tc(IV), which forms relatively poorly soluble precipitates including hydrous TcO_2 phases. When Tc(IV) is undersaturated with respect to hydrous TcO_2 formation, it shows strong sorption to solids [175–177]. Under sulphate-reducing conditions, TcS_2 phases can potentially form [178, 179]. Tc(VII) is widely reported to be scavenged to sediments under Fe(III)-reducing conditions and across a wide range of Tc(VII) concentrations [175, 180, 181]. In systems amenable to spectroscopic analysis, the reduction products in environmental systems where Fe(II) is present have been identified as short-chain polymeric/hydrous TcO_2 phases, potentially with Fe interactions [174, 182–184]. Recent work has also highlighted that in abiotic systems, structural incorporation of Tc(IV) in the Fe(II)-bearing mineral magnetite can occur, which is relevant because magnetite is a biological reduction product in some environments [161, 185]. Development of sulphate-reducing conditions with sulphidogenic conditions in pure culture systems can lead to production of TcS_2 [186, 187]. In sediment systems, development of sulphate-reducing conditions does not seem to significantly affect the speciation of hydrous TcO_2 over short timescales [180], with recent work highlighting the potential for modest TcS_2 ingrowth to sediments over a timescale of months, potentially with colloid formation also occurring [179]. Re-oxidation of Tc(IV) bioreduction products in sediment aeration experiments leads typically to ~50% of Tc being oxidatively remobilised over a few months [169, 188, 189], with the sediment lithology a significant determinant in the re-oxidation rate [182]. Generally, a significant fraction of Tc(IV) is retained in solids even upon air re-oxidation, suggesting incorporation of Tc(IV) into the mineralogical Fe oxidation products and/or occlusion of Tc(IV) by newly formed Fe(III) minerals [178, 185, 190].

8.4.4 Neptunium Biogeochemistry

Neptunium is redox active, and its speciation and associated environmental mobility are affected by biogeochemistry and redox conditions in the environment [191]. Under environmental conditions, neptunium's oxidised neptunyl form, $Np(V)O_2^+$, is much more mobile than the reduced species, Np(IV). In pure bacteria culture, Np(V) shows a surprising specificity for reduction with different microorganisms; *Geobacter sp.* is unable to facilitate Np(V) reduction, whereas *Shewanella sp.* exhibits some propensity for reduction, but forms potentially soluble reduction products [148, 192, 193].

In sediment systems, microbial mediated redox transformations of Np(V) to Np(IV) occurs under bioreducing conditions and with possible abiotic reduction, coupling to both manganese and iron reduction [191, 194]. Interestingly, abiotic reduction of soluble Np(V) is also possible by reaction with Fe(II) in previously bioreduced phyllosilicates, leading to a nanoparticulate NpO_2 reaction product [174]. Finally, Np(V) reduction was also facilitated by an alkaline-tolerant microbe population stemming from a high pH lime workings site [195]. This again suggests a potential positive role for bioreduction in reducing Np(V) solubility in alkaline radioactive waste disposal scenarios.

8.4.5 Plutonium Biogeochemistry

Plutonium as an element is toxic, and its nuclides have long half-lives, making it an element of environmental concern. There are, however, fewer studies of the microbial redox reaction of plutonium because of the high radiotoxicity of Pu. The redox chemistry of Pu is relatively rich; it can coexist in several oxidation states simultaneously, but Pu(IV) is generally the dominant oxidation state in environmentally contaminated sites. Due to the high charge and low solubility of Pu(IV), it is expected to sorb onto subsurface and soil mineral and rock surfaces and/or precipitate out. However, microbial activity leading to bioreduction of Pu(IV) to much more soluble Pu(III) can enhance migration of plutonium in contaminated environments [196]. The earliest study of bacterially mediated Pu(IV) reduction using iron-reducing *Bacillus polymyxa* and *B. circulans*, demonstrating solubilisation of solid Pu(IV) hydrous oxides, was reported in Ref. [197]. Another study reports that *G. metallireducens* and *S. oneidensis* activity in the presence of complexing ligands leads to enhanced reduction of solid Pu(IV) hydrous oxides to soluble Pu(III) species. Without addition of complexant molecules, only 1%–8% of Pu(IV) was reduced, whereas in the presence of complexant the majority of Pu(IV) was reduced quickly [138]. A more recent investigation of this process revealed that addition of quinone compound increased the amount of Pu(III) formed and that the outer membrane c-type cytochrome MtrC of these bacteria is responsible for the electron transfer [198]. Flavins can be secreted by *Shewanella oneidensis* MR-1 and have been shown to act as electron shuttles promoting anoxic growth coupled to accelerated reduction of poorly crystalline Fe(III)-oxides, and their role in facilitating U(VI), Tc(VII), Np(V), and Pu(IV) reduction has also recently been investigated [199]. In experimental systems, addition of environmentally relevant concentrations of riboflavin in bioreduction experiments with *Shewanella oneidensis* MR-1 enhanced reduction rates of Tc(VII), Pu(IV), and to a lesser extent Np(V), but had little impact on U(VI) reduction, suggesting that flavins may be a controlling factor in reduction of key anthropogenic radionuclides in environmental systems. These studies indicate that targeted strategies are needed when considering microbial bioremediation of sites with the potential for co-treatment of several species but with reductive solubilisation of Pu(IV) to Pu(III) also requiring careful attention.

In a study of the impact of indigenous bacterial activity on amended plutonium contaminated soil samples from the Nevada test site, the type of carbon amendment was reported to affect plutonium solubilisation; citrate amendment led to formation of a Pu polymer, whereas glucose amendment led to increased solution plutonium concentration due to plutonium release from biotically generated reduced magnesium and iron species, along with an increase in overall acidity promoting dissolution [200]. This is also an example of microbial activity indirectly affecting plutonium solubility in the environment by means other than enzymatic redox reactions, that is, by changing the pH of the medium. Microbial production of excretion/metabolic products, which are strong complexing agents such as siderophores, is another indirect route to mobilisation of plutonium in contaminated sites without concomitant reduction. Also, plutonium complexation onto functional groups of bacterial cells can determine migration and transport behaviour [58].

8.5 Transport and Surface Complexation Modelling

8.5.1 Key Processes in Actinide Transport

A complex interacting series of effects determines the transport of the actinides in the environment; those other than microbiological discussed in the previous section are shown schematically in Figure 8.4. Although there are many processes, actinide transport is essentially determined by the partition between the solid and the solution phase. The key to successful transport modelling is to predict that distribution, taking into account the various processes in Figure 8.4. Some of these processes are better understood than others (in particular, complex formation with simple ligands and precipitation), and thermodynamic databases are available to enable them to be calculated. The others are less well understood.

8.5.2 Interactions of Actinides with Inorganic Phases

Sorption is the general term for the removal of a solute from solution by interaction with a surface, and a number of processes contribute to actinide sorption [201]:

1) **Surface complexation**, where an ion interacts with a surface. In outer sphere complexes, the ion is physically attracted by the surface charge, but there is no direct chemical bond. For inner sphere complexes, there can be an electrostatic interaction,

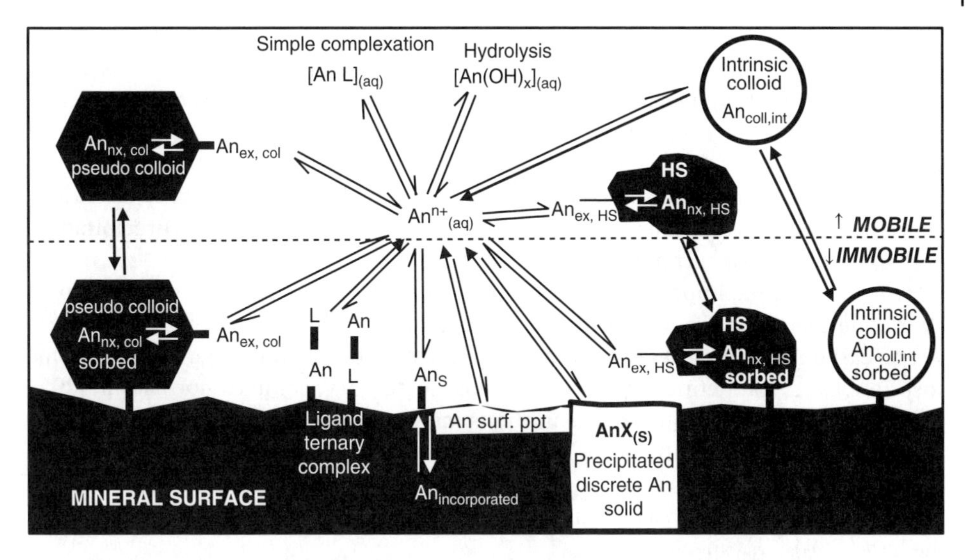

Figure 8.4 Processes that control actinide mobility in the environment; half arrows (⇀) represent processes that are generally fast; full arrows represent reactions that are or may be slow (→).

but a direct chemical bond also forms. Inner sphere surface complexation tends to be stronger than outer sphere complexation.

2) **Surface precipitation**, where a precipitate nucleates on a surface.
3) **Incorporation**, where an ion is incorporated within the structure of a solid. This process is often slow, as is the subsequent release of the incorporated ion.
4) **Ternary Complex formation**, where a ligand and a bound metal ion attach to a surface. The ligand or the metal can act as the bridge to the surface.

In most cases, sorption will tend to reduce transport by fixing the actinide in the immobile phase. However, if the surface is colloidal, then transport may be enhanced. Available data on the environmental speciation of actinides demonstrate that sorption onto various solid matrixes (e.g. soils, rocks, engineered-barrier of a geological disposal facility (GDF), and corrosion products) is one of the determinant processes that control their environmental behaviour.

The trivalent and tetravalent actinides form such strong chemical interactions that inner sphere surface complexation dominates. Trivalent actinide ions sorb strongly to surfaces with oxygen donor atoms, including clays, iron oxides, and other mineral surfaces, and the highly charged tetravalent ions have very strong interactions with most surfaces, almost regardless of their surface chemistry [201].

For the hexavalent 'yl' ions, although the inner sphere mechanism still dominates, outer sphere complexation can take place, particularly at low pH [201]. For pentavalent species, outer sphere sorption has been observed, for example, Np(V) interactions with smectite [202]. However, inner sphere complexation is expected to be important for most surfaces, including iron and aluminium oxides. For inner sphere sorption to surfaces, the order of affinity is expected to be [203, 204]

$$AnO_2^{+} < AnO_2^{2+} / An(III) < An(IV) \tag{8.6}$$

The tetravalent actinides have very low intrinsic solubilities under neutral and alkaline conditions due to hydrolysis, and at neutral to alkaline pH the $An(OH)_4$ species dominates the true solution speciation. Oxyhydroxide cluster and colloid formation are highly favourable in the environment [202, 205]; at the same time, An(IV) cations interact strongly with mineral surfaces [201, 206]. Therefore, in the absence of strongly complexing ligands, their fate will be determined by competition between oxyhydroxide cluster/colloid formation and inner sphere surface complexation or surface precipitation, perhaps with some incorporation.

Hydrolysis can be important for hexavalent species, although the tendency for oxyhydroxide cluster and colloid formation is less than for the tetravalent ions. They form strong carbonate complexes, which can affect sorption behaviour, either by suppressing sorption as a competing ligand or by surface ternary complex formation [201]. Generally, as pH and An(VI) concentrations increase, there is an increasing tendency for hydrolysis, ternary surface complex formation, and finally (surface) precipitation. For example, van Veelen et al. [207] found evidence for carbonate ternary complex formation on brucite, with a coordination environment that is essentially the same as in the $UO_2(CO_3)_3{}^{4-}{}_{(aq)}$ species.

8.5.2.1 Examples of Actinide Interfacial Redox Behaviour

As pointed out in Section 8.4.1.5, plutonium can be present in multiple oxidation states; this lends plutonium remarkable chemistry among other elements and its complex physical and chemical properties. Therefore, the interfacial behaviour of plutonium can be accompanied by associated redox reactions. Changes in the oxidation state of plutonium in the course of sorption on both redox-active and redox-inactive minerals have been demonstrated repeatedly in the literature.

Among the first works demonstrating a change in the oxidation state of plutonium at the water–mineral interface are the works by Sanchez et al. [208] and Keeney-Kennicutt et al. [209], who demonstrated that the sorption of Pu(V) onto the mineral goethite (α-FeOOH) is a fairly rapid process, which results in the stabilisation of Pu(IV) at the surface under steady-state conditions. Reduction of Pu(V) at the magnetite (Fe_3O_4)–water interface was demonstrated by Powell et al. [210]. At pH 5–8, reduction occurred in the near-surface layer of magnetite, with sorption being a rate-limiting step, while the plutonium fraction in solution was Pu(V). This was interpreted as being due to the interaction of Pu(V) on the surface with ferrous ions of magnetite. Another mineral-surface-mediated redox reaction was demonstrated for plutonium sorption onto Mn-containing minerals including manganese dioxide that, unlike magnetite, exhibits oxidising properties. Morgenstern and Choppin [211] studied Pu(IV) oxidation kinetics in pyrolusite (β-MnO_2) suspensions. They found that Pu(IV) oxidises to Pu(V,VI) at pH 2–3.5, whereas at pH 8.1, a significantly lower Pu(V) concentration was detected over time. The authors relate this to the stabilisation of Pu(IV) on the surface at higher pH.

The sorption of Pu(V) on volcanic tuff containing iron oxides and impurities of manganese oxide was studied by μ-XRF, and it was revealed that Pu(V) is predominately associated with manganese oxides (rancieite, $(Ca,Mn)OMn^{IV}O_2 \cdot 3H_2O$) and smectites rather than with iron oxides and zeolites [212]. Plutonium L_3-edge X-ray absorption near the edge structure (XANES) spectra of the surface-sorbed plutonium showed that plutonium retains its higher oxidation state after one day of equilibration; however following long-term equilibration, Pu(IV) was detected as the dominant redox species [213]. The same effect of fast oxidation to Pu(V/VI) at the interface followed by slow reduction to Pu(IV) was confirmed in several studies onto well-characterised phases of pyrolusite [213], manganite (MnOOH), and hausmannite (Mn_3O_4) [214]. In these studies, the small fraction of plutonium found in solution in equilibrium with the solids was Pu(V). Analogous results were obtained by Kersting et al. [215], who applied XAFS techniques for redox speciation (XANES and extended X-ray absorption fine structure, EXAFS) and observed that sorption of Pu(V) on colloidal particles of bernessite and pyrolusite is accompanied by its gradual reduction to Pu(IV) on the surface. Powell et al. [216] and Romanchuk et al. [217] demonstrated that for Pu(V)–hematite interaction under steady-state conditions Pu(IV) is stable on the surface, while plutonium remaining in solution is Pu(V). Various mechanisms of interfacial redox reactions were proposed including disproportionation of Pu(V) in a double electric layer at the surface because of its concentration on sorption, autoreduction of Pu(V,VI) due to radiolysis products, reduction of Pu(V,VI) by trace amounts of Fe(II) occurring in hematite, and reduction of Pu(V,VI) due to the photocatalytic properties of hematite. However, the same redox behaviour was established for plutonium at a concentration of about 10^{-14} M using ^{237}Pu, a short-lived EC[12] nuclide, which enables elimination of both disproportionation and radiolysis effects mechanisms. Possibilities of determining the trace concentrations of Fe(II) are rather limited by the sensitivity of Moessbauer spectrometry, XRF analysis, and EPR techniques. Identification of the mechanism of such redox reactions on the interface is yet to be resolved, so that predictive transport modelling tools for processes including these reactions are underdeveloped.

Surprisingly, reduction of Pu(V) to Pu(IV) was also found upon sorption on the surface of other minerals, which are not semiconductors and do not exhibit any redox properties. Kersting et al. [215], Zavarin et al. [218], and Hixon et al. [219] studied the sorption and speciation of plutonium at the surface of bernessite, calcite, montmorillonite, clinoptilolite, goethite, and quartz. For quartz and clinoptilolite, extremely slow reduction to Pu(IV) and sorption occur.

The formation of hydrated plutonium oxide precipitates at the mineral interface is possible due to the low solubility of oxygen compounds of Pu(IV), which are stabilised at the surface of minerals as a result of the redox reaction accompanying sorption. Romanchuk et al. [217] assumed the formation of Pu(IV) polynuclear species at the surface of hematite upon sorption of Pu(IV,V,VI) even at a relatively low total concentration of about 10^{-10} M. Kumar et al. [220] explained their observation of a decrease in Pu(IV) sorption rate at the surface of smectite by the slow formation of polynuclear plutonium complexes. Using HRTEM, Powell et al. [221] demonstrated formation of 2- to 5-nm-sized plutonium-containing particles upon addition of a Pu(IV) solution to suspensions of goethite and quartz. The nanoparticles formed at the water–mineral interface were Pu_4O_7 in the case of goethite. The similarity of lattice parameters is the underlying reason for epitaxial growth of this plutonium oxide at the goethite surface. PuO_2 nanoparticles were observed to form at the surface of quartz. Slow formation of

12 Electron capture.

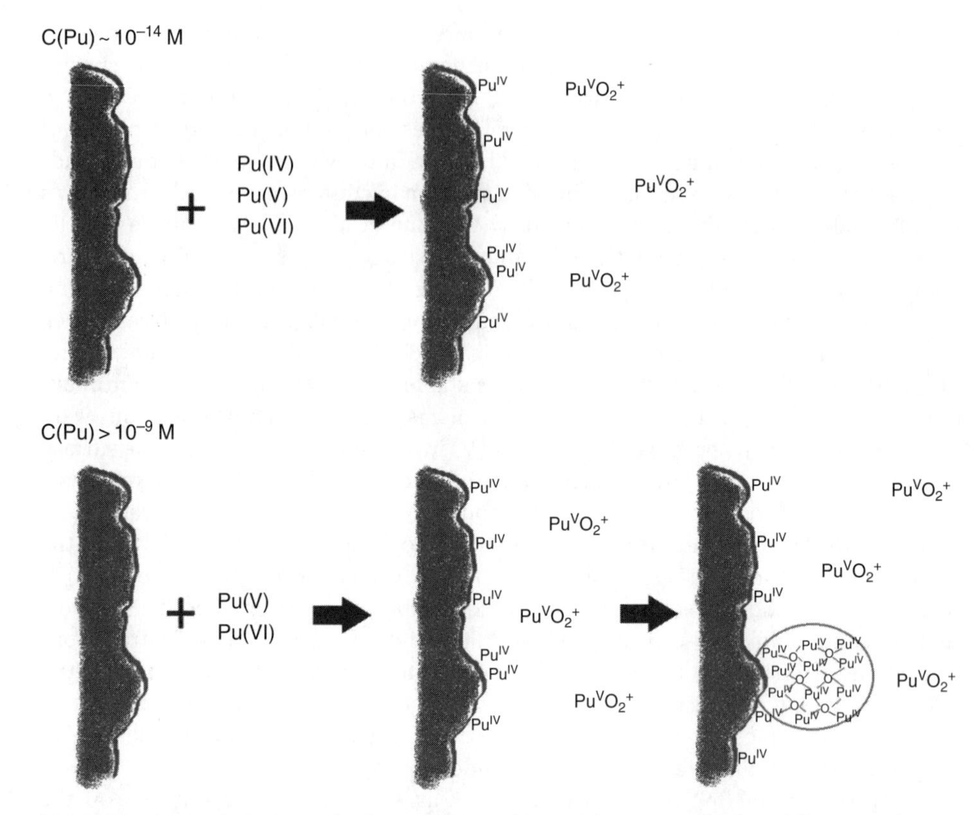

Figure 8.5 The general scheme for the interaction of Pu with hematite colloids at different total plutonium concentrations and the distribution of different Pu redox states [222].

aggregates of $PuO_{2+x} \cdot nH_2O$ crystalline nanoparticles upon the interaction of Pu(V,VI) with hematite at neutral pH was reported from XAFS and HRTEM results [222]. In the absence of a solid phase under analogous conditions, Pu(V) remains stable in solution for a long time (up to 3 years of observation). Pathways explaining the variation of physicochemical speciation of plutonium at the water–hematite interface at different total plutonium concentrations were proposed (Figure 8.5).

8.5.3 Surface Complexation Modelling

Surface complexation controls actinide environmental transport in many cases, and so there is a need to predict its extent in support of transport modelling. A generalised modelling approach may be used, and the same general reactions take place whether the surface is part of a bulk phase or a colloid. When inorganic oxides are placed in an aqueous solution, there is a layer of amphoteric hydroxyl functional groups (≡S-OH) at the surface:

$$\equiv S-OH \rightleftarrows \equiv S-O^- + H^+ \tag{8.7}$$

$$\equiv S-OH + H^+ \rightleftarrows \equiv S-OH_2^+ \tag{8.8}$$

where $\equiv S$ indicates that the species is attached to a bulk or colloidal surface [223]. A pair of pK_a values describes these interactions:

$$pK_{a1} = -\log_{10}\left(\frac{[\equiv S-OH][H^+]}{[\equiv S-OH_2^+]}\right) \quad pK_{a2} = -\log_{10}\left(\frac{[\equiv S-O^-][H^+]}{[\equiv S-OH]}\right) \tag{8.9}$$

Depending upon the amounts of $\equiv S\text{-}O^-$ and $\equiv S\text{-}OH_2^+$, the surface can have a net negative or positive charge. As a result of its charge, the surface generates an electrical potential, which attracts ions of opposite charge, affecting surface–ion interactions [224, 225]. For a surface with more than one type of functional group, there is a pair of pK_a values for each type: for goethite, there are thought to be three different types of surface hydroxyl groups, depending upon whether it is bound to 1, 2, or 3 iron atoms [226]. Metal cations (M^{n+}) such as actinide cations bind to surface groups with a number of possible coordination environments that will depend upon the surface and ion. Complexes have been suggested with denticities up to tetradentate [227].

The key to modelling surface complexation reactions is their dual nature. The chemical component depends upon the intrinsic chemical affinity of the surface for the cation, and the physical component is due to the electrical potential. The total free energy change for cation binding, $\Delta G^\theta{}_T$, is the sum of the chemical and physical components, $\Delta G^\theta{}_{Chem}$ and $\Delta G^\theta{}_{Phys}$, respectively:

$$\Delta G^\theta{}_T = \Delta G^\theta{}_{Chem} + \Delta G^\theta{}_{Phys} \tag{8.10}$$

Similarly, the overall equilibrium constant, K_T, is the product of the two components:

$$K_T = K_{Chem} \cdot K_{Phys} \tag{8.11}$$

The electrostatic effect may be understood in terms of the ion concentration at the binding sites. If ψ_0 is the surface electrical potential, the free cation concentrations at a binding site($[H^+]_S$; $[M^{n+}]_S$) will be different to those in the bulk ($[H^+]_{BULK}$; $[M^{n+}]_{BULK}$):

$$[H^+]_S = [H^+]_{BULK} \exp\left(\frac{-e\psi_0}{kT}\right) \tag{8.12}$$

$$[M^{n+}]_S = [M^{n+}]_{BULK} \exp\left(\frac{-ne\psi_0}{kT}\right) \tag{8.13}$$

Therefore:

$$K_T = K_{Chem} \exp\left(\frac{-ne\psi_0}{kT}\right), \tag{8.14}$$

where K_{Chem} is the intrinsic equilibrium constant, which would be observed if the surface charge were zero. Because of the dual nature, special surface complexation models (SCMs) must be used. For aqueous speciation and solubility, there are universal conventions. Unfortunately, this is not the case for SCM, and a variety of models may be applied [201]. A selection of SCM examples for modelling actinide sorption behaviour is discussed below.

The CD-MUSIC model (Charge Distribution MUlti-SIte Complexation) is a unique approach to surface complexation modelling [226]. It considers the crystal structure and the crystallographic plane exposed at the surface. For example, for a goethite 110 surface, there are three different types of surface oxygen atom, each with different pK_a

values [226]. To model uranyl binding to ferrihydrite, Hiemstra et al. [228] defined two surface functional groups, corresponding to hydroxyls bonded to one or three iron atoms. It is possible to rationalise goethite and ferrihydrite surface charge from the relative contributions of the two types of groups [229]. For uranyl binding to ferrihydrite, Hiemstra et al. [228] then defined two coordination modes, a bidentate 'normal' site, where the uranyl coordination was completed with water molecules, and another where the uranyl was bound as a tris-carbonato ternary complex.

The most widely applied SCM is that of Dzombak and Morel [223] for ferrihydrite. They collated a number of experimental datasets and fitted them with a single model having two types of functional groups: weak and strong. Waite et al. [230] used XAFS to experimentally define the uranyl-ferrihydrite surface complex coordination geometry, and using the Dzombak and Morel model, they found that the bidentate surface complex ($[(\equiv SO)_2UO_2]$) coordination geometry determined spectroscopically agreed well in modelling binding to ferrihydrite. They also defined ternary complexes ($[(\equiv SO)_2UO_2CO_3{}^{2-}]$) that allowed prediction of the effect of carbonate. The Dzombak and Morel approach was adapted by Sverjensky [227] to simulate the interaction of Group II metal ions (M^{2+}) with iron oxides. Two general types of surface complex were used, the first based on a tetradentate site: $(\equiv SOH)_2(\equiv SO)_2M$; $(\equiv SOH)_2(\equiv SO)_2M(OH)^-$, and the second monodentate: $\equiv SO\text{-}M^+$; $\equiv SO\text{-}M(OH)$.

A number of models have been developed to explain and predict radionuclide surface complexation on bentonite (montmorillonite), for example, [231], but the two site protolysis non-electrostatic surface complexation and cation exchange (2SPNE SC/CE) model of Bradbury and Baeyens has proved particularly successful (e.g. [232–234]). In this model, there are three inner sphere surface complexes for the trivalent actinides, $\equiv SO\text{-}An^{2+}$, $\equiv SO\text{-}An(OH)^+$, and $\equiv SO\text{-}An(OH)_2$, in addition to ion exchange sites. It is a quasi-mechanistic model, with spectroscopic evidence for the predicted surface speciation (e.g. [235]).

Different SCM approaches can give equivalent fits to sorption data, despite the fact that they use different surface complexes, acid base reactions, and electrostatic models to describe the surface complexation reactions taking place. Hence, at least some surface complexation modelling is semi-empirical, particularly where spectroscopic evidence in unavailable to confirm the nature of the surface complexes.

There has been much success in the application of SCMs to single phases. However, surfaces in the environment are heterogeneous, with multiple phases present. Currently, averaged empirical distribution coefficients (K_d) are generally used in calculations of actinide mobility in the environment. In order to enable reliable predictions, rigorous physicochemical models for heterogeneous systems are required.

There are two general approaches to predicting surface complexation on heterogeneous materials [236]:

1) ***Generalised Composite*** – Empirical sorption constants are used. This approach can provide a good fit to experimental data, but because it is empirical, it cannot be applied outside of the parameters used for its calibration.
2) ***Component Additivity*** –An attempt is made to describe the sorption in terms of a weighted average of the individual components. It can be applied with more confidence outside of its calibration; however, relating sorption to a mineral composition defined in terms of mass can be challenging, because it is the surface area that determines reactivity.

Natural sand is a complex material that can include significant inclusions as strong actinide binding phases, for example, iron phases. In some circumstances, simple generalised composite approaches can work successfully for modelling such systems. For example, Warwick et al. [237] used this approach to simulate europium transport in a system where there was strong competition between sand and humic substances. Gamerdinger et al. [238] studied U(VI) transport through columns of sand. Although simple linear adsorption isotherms were observed in batch experiments, U(VI) transport behaviour could not be predicted using the batch experiment K_d; different constants were required. Conversely, Shang et al. [239] reported success in modelling uranyl transport through sediment, even though the surface chemistry of the sediment was not treated explicitly in the model.

Dong and Wan [236] modelled uranium sorption to sediment, which was approximately 96% quartz with minor contributions from kaolinite and goethite. Humic substance adsorption was used to derive separate specific surface areas, which were used with surface complexation equations to predict the sorption behaviour of the whole sediment. At low pH (<6), all three (quartz, kaolinite, and goethite) phases were predicted to make a contribution, but at neutral and alkaline pH the goethite dominated with a single bidentate carbonate ternary complex, despite the fact that the iron content of the sand was low (~1%). Efstathiou and Pashalidis [240] also found a correlation between uptake of U(VI) to sea sand samples and iron content. Hence, for natural materials, minor mineralogical components may determine sorption and transport behaviour.

Slow reaction kinetics can complicate the modelling of natural sediments. For example, Handley-Sidhu et al. [241] found that the transport of uranyl could only be simulated if slow dissociation ($\approx 10^{-8}$ s^{-1}) was included in the model. There was rapid uptake to the surface, followed by slow transfer to two different forms that were slow to dissociate. This slow transfer could be transfer of uranium from one phase to another or incorporation.

8.5.4 Incorporation

Incorporation processes could have a significant impact on transport, since incorporated ions will be slow to dissociate. For bulk solids, this will retard actinide mobility, and any calculation that excludes incorporation could overestimate transport. Conversely, incorporation within the structure of a colloid will tend to promote transport.

Farr et al. [242] studied the binding of Pu(IV) by brucite. The apparent interaction strength was observed to increase with time, which was associated with incorporation of Pu(IV) by the diffusion of $Pu(OH)_4$ through a surface gel layer to form a mixed region of Mg/Pu hydroxide. Using EXAFS, Macé et al. [243] found that calcium silicate hydrates (CSHs) dominated the interactions of U(VI) in cement. Harfouche et al. [244] proposed that U(VI) could be incorporated into the CSH in a uranophane-like structure, while Tits et al. [245] found four different environments for U(VI), including some indicative of incorporation. Np(V) also interacts strongly with cementious minerals, although the mechanism is uncertain. Some evidence suggests a subsequent reduction to Np(IV) [246], but other studies have suggested that the neptunium remains in a higher oxidation state [247]. Incorporation of Np(V) into calcite has been reported and characterised using surface diffraction techniques [248]. There is also evidence for the incorporation of tetravalent actinides in cement phases [249, 250] and also calcite [251]. Trivalent actinide incorporation has been observed for aluminosilicates, gypsum, calcite, and CSH [201, 252].

8.5.5 Humic Substances

Humic substances (HS) are natural organic species derived from the decomposition of plant matter [253]. They are polydisperse and heterogeneous polyelectrolytes that are capable of binding virtually all metal ions, including the actinides. They are able to act as a vector for the transport of radionuclides [254], because of the unique way in which they interact with cations (e.g. [255]). On first contact, they bind metal ions exchangeably ($An_{ex,\,HS}$), but with time the cation is transferred to a state where it is hidden from solution and surface interactions ($An_{nx,\,HS}$). In the exchangeable fraction, the cation is strongly bound, but if it encounters a stronger sink, for example, on a competing surface, then it can dissociate instantaneously [255]. However, due to its internalisation in the humics, dissociation from the 'non-exchangeable' fraction is slow, regardless of the strength of the competing sink [255]. Distinct approaches are used to simulate the two binding modes.

For the exchangeable interaction, a number of models have been developed. In some, the HS is treated as a mixture of ligands with a common equilibrium constant, for example, the charge neutralisation model, which has been used to simulate the binding of actinides [256]. More complex models take into account the large negative charge of the HS, due to the dissociation of its acidic functional groups. As for surface complexation, there are chemical and physical (electrostatic) components, although SCMs are not used to simulate uptake by HSs. Two approaches have been used most widely. MODEL VII by Tipping et al. [257] simulates the chemical heterogeneity of the HS using a finite number of discrete binding sites, whereas the NICA[13]–Donnan model uses a continuous distribution [258]. Although HSs are heterogeneous mixtures, their metal ion binding behaviour is largely the same for HSs from different sources, and generic parameters have been developed for both models.

For the non-exchangeable metal ions bound to HSs, there is a continuum of dissociation rates, but also a distinct, slowly dissociating fraction that accounts for 10%–80% of the non-exchangeably bound radionuclide, and which may be described with a single first-order rate constant [255]. The observed rate constants vary little from one cation to another; they are typically within an order of magnitude, 10^{-8}–$10^{-7}\ s^{-1}$ [259].

The role of the non-exchangeable interaction and its slow dissociation was revealed in americium column experiments reported by Artinger et al. [260], where the amount of actinide eluting from the column was observed to decrease with increasing residence time. Americium elution also increased with increasing pre-equilibration time, which was due to transfer of the exchangeable to the non-exchangeable bond ions [255]. Similar behaviour was observed for: uranium column experiments [261]; Eu(III)/sand column experiments [237]; transport of europium through intact sandstone cores [262]; and radionuclide transport through Boom clay (Belgium) [263, 264].

Although models of the exchangeable interaction have been capable of simulating the effects of HS on radionuclides in static systems, attempts to model HS-mediated radionuclide transport using equilibrium-only models were unsuccessful [237]. The inclusion of non-exchangeable binding is essential if transport is to be simulated [265]. In some cases, non-exchangeably bound radionuclides behave as a conservative tracer

13 Consistent NICA; NICA = non-ideal competitive adsorption.

[237, 255], probably due to pre-equilibration of the column packing with relatively large concentrations of HSs.

The sorption of HSs themselves onto mineral surfaces is significant in modelling, because this retards HS-bound radionuclide transport. HS/mineral sorption reactions show significant hysteresis and slow desorption [266–268]. Some of these effects are the result of chemical fractionation [269]. van de Weerd et al. [267, 268] developed a model of competitive HS sorption/desorption, which simulates an HS as a small number of discrete fractions (i), each with its own kinetic Langmuir isotherm. The fractional surface coverage, θ_i, is given by

$$\theta_i = \frac{Q_i}{Q_{i,\max}} \quad (8.15)$$

where Q_i is the amount sorbed, and $Q_{i.max}$ is the amount that could be bound if the entire surface were saturated. The rate of change is given by

$$\frac{dQ_i}{dt} = k_{a,HA,i}\left[HA_i\right]\left(Q_{i,\max}\left(1-\theta_T\right)\right) - k_{d,HA,i}Q_i \quad (8.16)$$

where $[HA_i]$ is the concentration of the fraction in solution, $k_{a,HA,i}$ is the adsorption rate constant, $k_{d,HA,i}$ is the desorption rate constant, and θ_T is the total fractional surface coverage over all fractions. This approach has been used to model HS uptake by iron phases [267], as well as laboratory column and larger-scale transport experiments [268].

In most column experiments that were pre-equilibrated, the residence time of the solution is too low for displacement of sorbed HS by HS in the eluent solution [265]. However, examination of the elution peaks in americium [270] and uranium [261] column experiments did indicate some HS sorption [255]. Note that much longer residence and equilibration times in the environment than in a laboratory column experiment may mean that the sorption of HS/radionuclide complexes can be more important. When Bryan et al. [262] modelled the transport of europium and radiolabelled HS through intact sandstone cores, it was necessary to include HS sorption using an approach closely related to that of van de Weerd et al. [267, 268]. McCarthy et al. [271, 272] observed rapid migration of americium and curium at the Oak Ridge site in the United States, which was attributed to interactions with HS. Subsequent analysis [255] has shown that the behaviour is consistent with the transport of non-exchangeably bound ions.

8.5.6 Colloids

Colloids are small particles that are too large to be considered part of the true solution phase, but are too small to sediment under gravity. Because they are suspended in solution, they and any material associated with them are potentially mobile [130]. There are two types:

- **Intrinsic colloids**: Formed from the element of interest; for example, UO_2 colloids are uranium intrinsic colloids.
- **Pseudo-colloids**: Formed from a phase that does not include the element of interest, but which have sequestered the element, either incorporated or on the surface.

8.5.6.1 Intrinsic Colloids

Historically, tetravalent actinides were considered immobile due to their low solubility. However, although the ions themselves have very low mobility, their intrinsic colloids can be highly mobile [273]. An(IV) colloids form by hydrolysis. Initially, $An(OH)_x(H_2O)_y^{(4-x)+}$ monomeric species are formed, followed by small hydroxide clusters, $An_a(OH)_x(H_2O)_y^{(4a-x)+}$, and finally oxyhydroxide colloids, $An_a(OH)_x(O)_z(H_2O)_y^{(4a-x-2z)+}$ [4]. Significantly for transport, stable nanometre sized (and hence mobile) An(IV) oxyhydroxide colloids are known [274, 275]. For An(IV), the contribution of the colloids to the total concentration can be orders of magnitude higher than that of the actinide in the true solution [4]; for example, Th(IV) and Pu(IV) colloid concentrations of $10^{-6.3}$ and $10^{-7.9}$ M have been observed. For tetravalent oxyhydroxide colloids, the colloid population seems to be in equilibrium with the true solution phase [273], which suggests that they may be unavoidable in saturated systems, but could disperse in undersaturated conditions, for example, dilution or upon reduction in pH [273]. However, the dissolution process can be slow [4].

The actinides of greatest environmental concern, for example, Pu, are unlikely to reach sufficient concentrations to produce intrinsic colloids outside of a repository near-field. However, mixed Th(IV)/U(IV) phases, $Th_{1-x}U_xO_{2+y\ (s)}$, can form [276], and natural uraninite (UO_2) samples incorporate thorium [229]. If mixed Th(IV)/U(IV) phases form in the environment, then mixed An(IV) colloid (Np(IV)/Pu(IV)/U(IV) or Np(IV)/Pu(IV)/Th(IV)) production and transport are possible. Kaminski et al. [277] did observe colloids containing U(IV) and Pu(IV) ($U_{1-x}Pu_xO_2$).

An(IV) form stable intrinsic colloids with silicate [273, 278]. For example, U(IV) and Th(IV) silicate colloids can form at neutral pH, resulting in apparent actinide solubilities of 10^{-3} M. These particles can be stable for many years, and they can have sizes <20 nm [273, 278]. Mixed oxyhydroxide/silicate species form, with varying ratios of actinide:silicon. These colloids are stable, even in conditions where the surface charge of the pure actinide oxyhydroxide phases would be low, due to a surface coating of ionised silanol groups that keeps the colloids in solution [273, 278].

Due to their lower tendency to hydrolysis, An(VI) ions were thought to be much less susceptible to colloid formation. However, An(VI) ions can form colloids under alkaline conditions, even at low concentrations. Bots et al. [279] observed U(VI) colloids that formed within hours in synthetic cement leachate at pH > 13, which remained stable for periods of years. The primary colloids were approximately 1.5–1.8 nm in size, but associated 20–60 nm loose aggregates were also observed. XAFS and TEM suggested that they were related to the clarkeite (sodium–uranate type) structure, although the fit to the X-ray data was improved by the inclusion of K and Ca, and so it seems that the structure of the colloids was $(Na,K,Ca)UO_2O(OH)\cdot(H_2O)_{0-1}$. Smith et al. [280] observed similar U(VI) colloids in calcite systems that delayed surface sorption by their slow rate of dissociation.

Actinide colloids may be retarded by surface interactions [201]. van der Lee et al. [281] predicted that in reduced systems the adsorption of UO_2 colloids was likely to be the chief retention mechanism for uranium. However, little is known of the interaction of tetravalent or hexavalent actinide colloids with surfaces. PuO_2 colloids (2–5 nm) have been shown to sorb to goethite (FeOOH) and quartz [221]. Schmidt et al. [282] used surface X-ray diffraction and reflectivity analysis to show that Pu(IV) nanoparticles ($[Pu_{38}O_{56}Cl_x(H_2O)_y]^{(40-x)+}$) interact with a muscovite surface, probably due to the electrostatic interaction between the positively charged colloids and the negatively charged surface. Some particles were attached to the surface, but others were held out in the double layer. Modelling colloid sorption is complex, because it can be affected by surface roughness, and samples of the same material can show different colloid sorption [283].

8.5.6.2 Pseudo-colloids

Most solid phases that can bind actinide elements are also capable of forming potentially problematic pseudo-colloids (e.g. iron oxyhydroxides and clays), particularly if they can incorporate actinide ions. Like HS, metal ions can interact exchangeably ($An_{ex, col}$) or non-exchangeably ($An_{nx, col}$) with colloids. Theoretical calculations [284] have shown that slow dissociation from either colloid interaction may significantly influence radionuclide mobility, but experimental evidence from studies is lacking for most systems. One exception is the bentonite colloid system, because bentonite has been identified as a potential backfill for radioactive waste repositories. Although equilibrium models have been proposed for radionuclide movement through bulk bentonite (e.g. [285]), kinetic processes are significant for bentonite-colloid-mediated transport. U(VI) and Np(V) do not associate strongly with bentonite colloids, whereas Th(IV), Pu(IV), and Am(III) do [286]. In the latter case, slow dissociation was observed for Am(III) and Pu(IV), with dissociation rate constants of $1.0–2.5 \times 10^{-6}$ s^{-1} and 3.9×10^{-7} to 2.4×10^{-6} s^{-1}, respectively [286]. Bouby et al. [287] found that Th(IV) dissociation was incomplete, even after three years. It was suggested that this might be due to the formation of a surface precipitate.

Transport experiments through a Grimsel rock core revealed that Th(IV) was more permanently associated with the clay colloids during transport compared to Am(III) and Tb(III) [288]; the Th(IV) was found to elute with the colloids. Missana et al. [289] found that Eu(III) transport occurred in association with bentonite colloids, with limited dissociation taking place; for Pu(IV), the dissociation was slower. Geckeis et al. [290] found that Am(III) and Pu(IV) transport could only be explained by slow dissociation from colloids.

8.5.7 Damkohler Analysis of HS/Colloid-Mediated Transport

The importance of colloids and HSs in the transport of radionuclides depends on their ability to bind radionuclides 'non-exchangeably' and critically on the magnitude of the associated rate constant of cation disassociation from the colloid. An approach is required to assess the likely probability of transport, and the same procedure may be used for colloids and HSs. For this, Möri et al. [291] developed the 'Colloid Ladder'. An adapted version is shown in Figure 8.6.

Assessing the potential mobility of actinide bourne colloids or colloid-mediated actinide transport requires the prediction of actinide cation distribution in the ternary system (mineral surface–colloid–metal cation). Any radionuclides that are bound to the colloid/HS exchangeably, and hence may dissociate instantaneously, are expected to be quickly removed by the available mineral surface binding sites, which will be present in excess [255]. For non-exchangeably bound radionuclides, colloid disassociation is slow, so that surface sorption processes potentially preventing transport are less important. For inorganic colloids and HSs, the mechanisms responsible for slow dissociation are different, but the net effect is the same. Radionuclides will be transported provided that

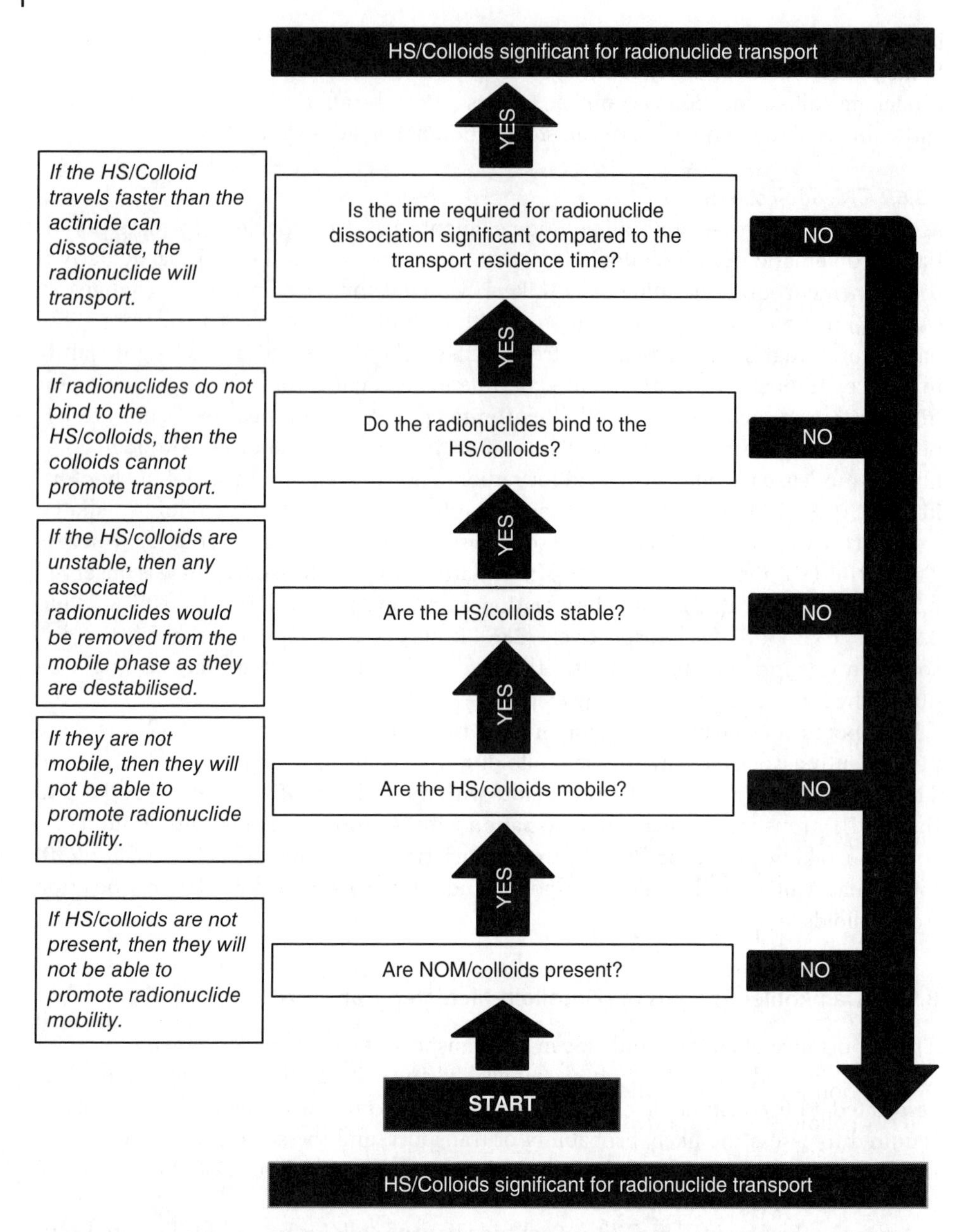

Figure 8.6 The amended 'Colloid Ladder'. Adapted from Möri A, Alexander WR, Geckeis H, Hauser W, Schäfer T, et al. (2003) The colloid and radionuclide retardation experiment at the Grimsel Test Site: Influence of bentonite colloids on radionuclide migration in a fractured rock. *Colloids and Surfaces A: Physicochemical and Engineering Aspects*, **217**(1–3), 33–47.

the time required for them to dissociate from the colloid or HS mediating the transport is of the same order or greater than the transport residence time, t_{res}, which is given by

$$t_{res} = \frac{L}{V} \tag{8.17}$$

where L is the distance, and V is the linear velocity of the mobile phase. The transport behaviour may be rationalised using Damkohler numbers [255].

For a system with a first-order rate constant that represents the rate determining step for dissociation (k_b), the amount of radionuclide that will remain bound to the colloid/HS depends only on k_b and the time available for dissociation (t_{res}). The dimensionless Damkohler number for a radionuclide in the non-exchangeable fraction, D_M, is defined by

$$D_M = \frac{L}{V} k_b = t_{res} \cdot k_b \tag{8.18}$$

Systems with the same D_M show analogous behaviour.

For any system, there are three classes of transport calculation:

1) **Large D_M**: Dissociation is sufficiently fast and/or transport time sufficiently long that dissociation may be simulated with an equilibrium constant (high k_b and/or high t_{res}).
2) **Low D_M**: Dissociation is sufficiently slow or residence time sufficiently short that effectively dissociation does not take place (low k_b and/or low t_{res}).
3) **Intermediate D_M**: Dissociation can only be described accurately by the use of rate equations (intermediate k_b and/or t_{res}).

D_M may be used to judge the impact of dissociation kinetics and assign a system to classes 1–3 and hence judge when it is necessary to include kinetics explicitly in transport calculations. As D_M decreases, non-exchangeably bound radionuclide tends towards the behaviour of a conservative tracer, provided that the colloid/HS itself is not retarded. In this case, virtually no radionuclide leaves the non-exchangeable fraction in the time of the calculation. Therefore, the reaction that connects the exchangeable and non-exchangeable fractions may be removed from the calculation, and the two fractions can be treated as independent species. The advantage of this approach is that it is inherently conservative. At high D_M, an equilibrium approximation is possible, although this will tend to underestimate migration [255], the error decreasing with increasing D_M.

Processes that retard the HS or colloids can have a disproportionate effect on radionuclide transport, since they have the effect of increasing the effective residence time, allowing more radionuclide to dissociate [255, 265]. This can complicate transport modelling, because HS and colloid desorption can be slow. In this case, the radionuclide dissociation and the desorption kinetics are coupled. However, when the sorption of the HS/colloid may be treated as an equilibrium, an adapted Damkohler number may be used to assess the kinetics [255]:

$$D_M^{eff} = \frac{k_b}{V} L(1 + K_C) \tag{8.19}$$

where K_C is the equilibrium distribution coefficient for HS/colloid sorption.

8.6 Conclusions and Outlook

In this consolidated overview of sources of actinide elements in the environment and their fundamental geochemical and biogeochemical behaviour, we have addressed sources of naturally occurring actinide elements, those naturally occurring sources that have been technologically enhanced and have anthropogenic sources. The contamination of Earth's environment by intentional and accidental anthropogenic actinide release from early nuclear weapons production and detonations during the cold war is by far the most significant. Releases from accidents and incidents at civil nuclear power stations are a number of orders of magnitude smaller.

The potential impact that environmental actinide releases has is dependent on a number of parameters, and knowledge of the conditions upon release and the actinide's speciation, particularly redox conditions and state for multi-valent actinide nuclides, is essential in predicting their environmental transport behaviour and, hence, in assessing their impact and in managing, containing, and remediating released contamination. The present status of underpinning science gathered over the past 70 years, and particularly in the past two decades, is advanced enough to successfully model competing processes responsible for actinide partitioning between solid and solution phases at a range of spatial and time scales in relatively simple laboratory systems, in order to predict transport. This indeed demonstrates great progress; however, work remains to be done in order to develop tools sophisticated enough to reliably model actinide transport in real natural systems. Sophisticated models are needed that are capable of simulating actinide behaviour in increasingly complex natural systems, that is, of describing integrated biogeochemical system behaviour. Such theoretical modelling tools should link individual processes temporally and spatially, for example, by including predictive descriptions of actinide interaction with colloids coupled to colloid-mediated transport, these in turn being linked to other thermal–mechanical–chemical processes under possibly varying (cycling, extreme events) conditions, and so on. Presently, in order to circumvent such challenges in modelling transport in natural systems, conservative estimates are used, with the understanding that the results may not be accurate, presumably yielding an overestimation of actinide mobility and environmental impact and thus leading to overdesign in remediation of radionuclide contaminated systems. In order to enable more sophisticated modelling tools, fundamental research on determinant processes must continue, in order to feed into modelling development. Furthermore, existing contaminated sites offer potential as case studies to test developed tools.

To further complicate challenges in modelling actinide transport, the influence of living organisms and their metabolic products clearly has significant influence on radionuclide transport. Understanding how actinide speciation is influenced by biogeochemical processes involving actinides is crucial. The assumption that bacteria will not survive toxic environments and radiation fields in the vicinity of actinide contaminants does not hold generally, and we have highlighted in our discourse the profound impact biogeochemical processes have on actinide behaviour in natural and engineered environments. Our discussion focussed on the influence of biogeochemical processes on the behaviour of the risk driving radionuclides, uranium, technetium, neptunium, and plutonium, but it is clear these processes apply to other radionuclides. The present outlook of harnessing microbial processes to treat mobile radionuclides in radioactively contaminated groundwaters is positive. In particular, dual bioreduction and biomineralisation systems offer promise for select treatment of problematic radionuclides to optimise formation of durable bioreduction products. It is becoming clear that microbial processes are likely to have a profound effect on radionuclide mobility in many geological disposal scenarios [160, 292]; however, the outlook in this context can be viewed in a positive light, as microbial activity can lead to a range of benefits, including breakdown of organic complexant species in wastes which hitherto have been considered potential vectors for radionuclide transport [162]; reduction to reactive mineral phases such as magnetite and reduced phyllosilicates; and reduction of radionuclides to form sparingly soluble products. Overall, understanding biogeochemical processes as a controlling factor in the behaviour of actinides and other radionuclides in both natural and engineered environments will ultimately enable the utilisation of these processes to our benefit, which undoubtedly will be critical for the clean-up and management of our global radioactive waste legacy.

We close with a brief comment on increasing experimental sophistication as the driver for improving our understanding of actinide environmental behaviour. A number of examples in this chapter were presented, where speciation investigations using synchrotron-based techniques have allowed successfully modelling of the behaviour of surface complexation, characterisation of actinide nanoparticle and colloid structure, and identification of actinide immobilisation processes in heterogeneous systems. Presently, advances in application of high-resolution X-ray emission spectroscopy techniques such as RIXS (resonant inelastic X-ray scattering) and PFY-XANES (partial fluorescence yield XANES, also called HR-XANES and HERFD-XANES, for high-resolution and high-energy resolution fluorescence detection) [293, 294] will help refine our knowledge of, for example, redox coupled reactions involving actinide elements. Additionally, developments of laboratory source instruments may soon find a niche in actinide research to complement synchrotron source experimental efforts. Developments in highly sensitive analytical techniques, such as ICP-MS, TIMS and AMS, made in the last two decades now allow the determination of ultratrace levels of actinides originating from global fallout in a variety of environmental compartments, both in the atmosphere and in aquatic and terrestrial systems. These are powerful tools for the direct investigation of the behaviour and speciation of uranium, neptunium, plutonium, and americium nuclides in environmental systems, complementing and validating the findings of laboratory-scale experiments and modelling studies. With such sensitive analytical capability, the long-term impact of colloid-mediated transport of actinides within the frame of large-scale in situ radionuclide tracer tests has also become possible [295].

List of Acronyms

2SPNE SC/CE	Two site protolysis non-electrostatic surface complexation and cation exchange
AMS	Accelerator mass spectrometry
CD-MUSIC	Charge Distribution, MUlti-SIte Complexation
CTBT	Comprehensive Nuclear-Test-Ban Treaty
DPRK	Democratic People's Republic of Korea; North Korea
EC	Electron Capture
EURT	East Urals Radioactive Trace
EXAFS	Extended X-ray absorption fine structure
FDNPP	Fukushima Daiichi Nuclear Power Plant
HERFD-XANES	High-energy resolution fluorescence detection X-ray absorption near edge structure

HLW	High-level waste
HR-XANES	High-resolution X-ray absorption near edge structure
ICP-MS	Inductively coupled plasma mass spectrometry
K_d	Distribution coefficient
MOX	Uranium and plutonium mixed-oxide
NORM	Naturally occurring radioactive materials
NPP	Nuclear power plant
PA	Production Association
PFY-XANES	Partial fluorescence yield X-ray absorption near edge structure
PTBT	Partial Test Ban Treaty
RAR	Reasonably assured resources
SCM	Surface complexation models
SIMS	Secondary ion mass spectrometry
SPFY	Stratospheric-partitioned fission yields
$t_{1/2}$	Half-life
TEM	Transmission electron microscopy
TENORM	Technologically Enhanced Naturally Occurring Radioactive Materials
TIMS	Thermal ionisation mass spectrometry
TMF	Tailings Management Facility
XAFS	X-ray absorption fine structure
XANES	X-ray absorption near edge structure
XRF	X-ray fluorescence

References

1 Poinssot C, Geckeis H. (2012) Overview of radionuclide behaviour in the natural environment, in *Radionuclide Behaviour in the Natural Environment* (eds. C Poinssot, H Geckeis), Woodhead Publishing, Oxford (UK), pp. 1–10.

2 Runde W, Neu MP. (2010) Actinides in the Geosphere, in *The Chemistry of the Actinide and Transactinide Elements*, 4th edn. (eds. LR Morss, NM Edelstein, J Fuger), Springer, Dordrecht (Netherlands), pp. 3475–593.

3 Hu Q-H, Weng J-Q, Wang J-S. (2010) Sources of anthropogenic radionuclides in the environment: A review. *Journal of Environmental Radioactivity*, **101**(6), 426–437.

4 Walther C, Denecke MA. (2013) Actinide colloids and particles of environmental concern. *Chemical Reviews*, **113**(2), 995–1015.

5 Yuan J. (2015) Slope year for the U-Pb dating method and its applications. *Open Journal of Geology*, **5**, 351–366.

6 Boltwood BB. (1907) On the ultimate disintegration products of the radioactive elements. Part II. The disintegration products of uranium. *American Journal of Science*, **23**(134), 77–88.

7 Richter S, Alonso A, Boll WD, Wellum R, Taylor PDP. (1999) Isotopic "fingerprints" for natural uranium ore samples. *International Journal of Mass Spectrometry*, **193**, 9–14.

8 European Commission Joint Research Centre; Institute for Reference Materials and Measurements (2014) Schillebeeckx P, Becker B, Harada H, Kopecky S. Neutron Resonance Spectroscopy for the Characterisation of Materials and Objects. EUR 26848 EN.

9 Berkovits D, Feldstein H, Ghelberg S, Hershkowitz A, Navon E, et al. (2000) 236U in uranium minerals and standards. *Nuclear Instruments and Methods in Physics Research, Section B: Beam Interactions with Materials and Atoms,* **172**(1–4), 372–376.
10 Wilcken KM, Fifield LK, Barrows TT, Tims SG, Gladkis LG. (2008) Nucleogenic 36Cl, 236U and 239Pu in uranium ores. *Nuclear Instruments and Methods in Physics Research Section B: Beam Interactions with Materials and Atoms,* **266**(16), 3614–3624.
11 Wilcken KM, Barrows TT, Fifield LK, Tims SG, Steier P. (2007) AMS of natural 236U and 239Pu produced in uranium ores. *Nuclear Instruments and Methods in Physics Research Section B: Beam Interactions with Materials and Atoms,* **259**(1), 727–732.
12 Mervine E. Nature's Nuclear Reactors: The 2-Billion-Year-Old Natural Fission Reactors in Gabon, Western Africa 2011 [cited 2016]. http://blogs.scientificamerican.com/guest-blog/natures-nuclear-reactors-the-2-billion-year-old-natural-fission-reactors-in-gabon-western-africa/.
13 Maher K, Bargar JR, Brown GE. (2013) Environmental speciation of actinides. *Inorganic Chemistry,* **52**(7), 3510–3532.
14 http://thorenergy.no/[accessed 2016].
15 Vijayan PK. 2013 [cited 2016]. http://thoriumenergyconference.org//sites/default/files/pdf/Overview%20of%20the%20Thorium%20Programme%20in%20India%20-%20PK%20Vijayan.pdf.
16 Dikshith TSS. (2013) *Hazardous Chemicals: Safety Management and Global Regulations.* Boca Raton, FL: CRC Press.
17 Strahan D. Uranium in glass, glazes, and enamels: History, identification, and handling. *27th Annual Meeting Objects Specialty Group Session,* Saint Louis, Missouri, American Institute for Conservation of Historic and Artistic Works.
18 Betti M. (2003) Civil use of depleted uranium. *Journal of Environmental Radioactivity,* **64**(2–3), 113–119.
19 Gabriel S, Baschwitz A, Mathonnière G, Fizaine F, Eleouet T. (2013) Building future nuclear power fleets: The available uranium resources constraint. *Resources Policy,* **38**(4), 458–469.
20 Wall JD, Krumholz LR. (2006) Uranium reduction. *Annual Review of Microbiology,* **60**, 149–166.
21 Langmuir D. (1978) Uranium solution-mineral equilibria at low temperatures with applications to sedimentary ore deposits. *Geochimica et Cosmochimica Acta,* **42**(6, Part A), 547–569.
22 Noseck U, Tullborg E-L, Suksi J, Laaksoharju M, Havlová V, et al. (2012) Real system analyses/natural analogues. *Applied Geochemistry,* **27**(2), 490–500.
23 Goldstein SJ, Abdel-Fattah AI, Murrell MT, Dobson PF, Norman DE, et al. (2010) Uranium-series constraints on radionuclide transport and groundwater flow at the Nopal I uranium deposit, Sierra Peña Blanca, Mexico. *Environmental Science & Technology,* **44**(5), 1579–1586.
24 Duro L, Bruno J. (2012) 11 – Natural analogues of nuclear waste repositories: Studies and their implications for the development of radionuclide migration models A2 – Poinssot, Christophe, in *Radionuclide Behaviour in the Natural Environment,* (ed. H Geckeis), Woodhead Publishing, Cambridge (UK), pp. 411–45.
25 OECD Nuclear Energy Agency and the International Atomic Energy Agency (2014) Uranium 2014: Resources, Production and Demand. **7209**.
26 (1999) *Uranium: Mineralogy, Geochemistry, and the Environment,* (ed. P Ribbe), Virginia, Mineralogy Society of America. 679 p.

27 Bhargava SK, Ram R, Pownceby M, Grocott S, Ring B, et al. (2015) A review of acid leaching of uraninite. *Hydrometallurgy*, **151**, 10–24.
28 Gilligan R, Nikoloski AN. (2015) The extraction of uranium from brannerite – A literature review. *Minerals Engineering*, **71**, 34–48.
29 IAEA, Blaise J, Boitsov S, Bruneton P, Ceyhan M, Dahlkamp F, et al. (2009) World Distribution of Uranium Deposits (UDEPO) with Uranium Deposit Classification. IAEA-TECDOC-1629.
30 Bruneton P, Cuney M. (2016) Geology of uranium deposits, in *Uranium for Nuclear Power*, (ed. I Hore-Lacy), Woodhead Publishing, Oxford (UK), pp. 11–52.
31 (2016), in *Uranium for Nuclear Power: Resources, Mining and Transformation to Fuel*, (ed. I Hore-Lacy), Woodhead Publishing, Oxford (UK).
32 Woods PH. (2016) Uranium mining (open cut and underground) and milling A2 – Hore-Lacy, Ian, in *Uranium for Nuclear Power*, Oxford, Woodhead Publishing, pp. 125–156.
33 United States Department of the Interior; Geological Survey, Houston RS, Patten LL, Gersic J. (1983) Mineral Resource Potential of the Sheep Mountain Wilderness, Albany County, Wyoming. Open-File Report 83–468.
34 Carlsson E, Büchel G. (2005) Screening of residual contamination at a former uranium heap leaching site, Thuringia, Germany. *Chemie der Erde – Geochemistry*, **65**, Supplement 1, 75–95.
35 Petersen J. Heap (2016) Leaching as a key technology for recovery of values from low-grade ores – A brief overview. *Hydrometallurgy*, **165**(1), 206–212.
36 (1994) Waggitt P. A Review of Worldwide Practices for Disposal of Uranium Mill Tailings. Technical Memorandum 48.
37 http://www.wise-uranium.org/
38 IAEA (2004) The Long Term Stabilization of Uranium Mill Tailings: Final Report of a Coordinated Research Project 2000–2004. TECDOC-1403.
39 International Atomic Energy Agency (2008) Estimation of Global Inventories of Radioactive Waste and Other Radioactive Materials. IAEA-TECDOC-1591.
40 AREVA Resources Canada Inc., Rowson J, Schmid B, Langmuir D, Mahoney J, Rinas C, et al. (2011) McClean Lake Operation, Tailings Optimisation and Validation Program. Validation of Long Term Tailings Performance Report (2009).
41 Bopp IV CJ, Lundstrom CC, Johnson TM, Sanford R, Long PE, et al. (2010) Uranium 238U/235U Isotope Ratios as Indicators of Reduction: Results from an in situ Biostimulation Experiment at Rifle, Colorado, United States. *Environmental Science & Technology*, **44**, 5927–5933.
42 Merkel BJ, Hoyer M. (2012) 16 – Remediation of sites contaminated by radionuclides A2 – Poinssot, Christophe, in *Radionuclide Behaviour in the Natural Environment*, (ed H Geckeis), Woodhead Publishing, pp. 601–45.
43 World Nuclear News. 2015. [cited 2016]. Available from: http://www.world-nuclear-news.org/UF-EBRD-launches-uranium-mining-legacy-fund-1806157.html.
44 Abdelouas A. (2006) Uranium mill tailings: Geochemistry, mineralogy, and environmental impact. *Elements*, **2**(6), 335–41.
45 ICOLD (International Commission on Large Dams) (2001) Lagoons CoTDaW. Tailing Dams Risks of Dangerous Occurrences: Lessons learnt from practical experiences. Bulliten 121.

46 Lovley DR, Phillips EJP, Gorby YA, Landa ER. (1991) Microbial reduction of uranium. *Nature*, **350**(6317), 413–416.

47 Campbell KM, Gallegos TJ, Landa ER. (2015) Biogeochemical aspects of uranium mineralization, mining, milling, and remediation. *Applied Geochemistry*, **57**, 206–235.

48 Virginia CoUMi, Resources CoE, Resources BoESa, Studies DoEaL, Council NR. (2012) *Scientific, Technical, Environmental, Human Health and Safety, and Regulatory Aspects of Uranium Mining and Processing in Virginia*. Washington DC: The National Academies Press.

49 Neudert A, Barnekow U. (2006) Decommissioning of Uranium mill tailings ponds at WISMUT (Germany), in *Uranium in the Environment: Mining Impact and Consequences*, (eds. BJ Merkel, A Hasche-Berger), Berlin, Heidelberg, Springer Berlin Heidelberg, pp. 415–23.

50 Krishnaswami S. (1998) Thorium: Element and geochemistry, in *Geochemistry*, Dordrecht, Springer Netherlands, pp. 630–635.

51 Oak Ridge National Laboratory, Brown KB, Hurst FJ, Crouse DJ, Arnold WD. (1963) Review of Thorium Reserves in Granitic Rock and Processing of Thorium Ores. Technology – Raw Materials. ORNL-3495.

52 Holden NE. (2001) A Short History of Nuclear Data and Its Evaluation, Upton, NY, National Nuclear Data Center, Brookhaven National Laboratory.

53 Clark DL, Hecker SS, Jarvinen GD, Nue MP. (2011) *The Chemistry of the Actinide and Transactinide Elements*, 4th edition, (eds. LR Morss, NM Edelstein, J Fuger, JJ Katz), Dordrecht, the Netherlands, Springer, pp. 856.

54 Seaborg GN, McMillan EM, Kennedy JW, Wahl AC. (1946) Radioactive element 94 from Deuterons on Uranium. *Physical Review*, **69**, 366.

55 Seaborg GT, Wahl AC, Kennedy JW. (1946) Radioactive element 94 from Deuterons on Uranium. *Physical Review*, **69**, 367.

56 Tanaka A, Doi T, Uehiro T. (2000) Uranium isotope ratios in the environmental samples collected after a criticality accident in the uranium conversion facilities of JCO. *Journal of Environmental Radioactivity*, **50**(1–2), 151–160.

57 Hardy EP, Krey PW, Volchok HL. (1973) Global inventory and distribution of fallout plutonium. *Nature*, **241**(5390), 444–445.

58 Kersting AB. (2013) Plutonium transport in the environment. *Inorganic Chemistry*, **52**(7), 3533–3546.

59 Pacific Northwest Laboratories, Triplett MB, Watson DJ, Wellman DM. (2013) Risks from Past, Current, and Potential Hanford Single Shell Tank Leaks DE-AC05-76RL01830.

60 Degteva MO, Shagina NB, Vorobiova MI, Anspaugh LR, Napier BA. (2012) Re-evaluation of waterborne releases of radioactive materials from the "Mayak" Production Association into the Techa River in 1949–1951. *Health Physics*, **102**, 25–30.

61 Napier BA. (2014) Joint US/Russian studies of population exposures resulting from nuclear production activities in the Southern Urals. *Health Physics*, **106**, 294–304.

62 Myasoedov BF, Drozhko EG. (1998) Up-to-date radioecological situation around the 'Mayak' nuclear facility. *Journal of Alloys and Compounds*, **271–273**, 216–220.

63 Degteva MO, Kozheurov, VP, Burmistrov DS, Vorobiova MI, Valchuk VV, Bougrov NG, Shishkina NA. (1996) An approach to dose reconstruction for the Urals population. *Health Physics*, **71**, 71–76.

64 Solodov IN, Zotov, AV, Khoteev AD, Mukhamet-Galeev AP, Tagirov BR, Apps JA. (1998) Geochemistry of natural and contaminated subsurface waters in fissured bed rocks of the Lake Karachai area, Southern Urals, Russia. *Applied Geochemistry*, **13**, 921–939.

65 Novikov AP, Kalmykov SN, Utsunomiya S, Ewing RC, Horreard F, Merkulov A, Clark SB, Tkachev VV, Myasoedov BF. (2006) Colloid transport of plutonium in the far-field of the Mayak Production Association, Russia. *Science*, **314**, 638–641.

66 Aarkrog A, Dahlgaard H, Nielsen SP, Trapeznikov AV, Molchanova IV, et al. (1997) Radioactive inventories from the Kyshtym and Karachay accidents: Estimates based on soil samples collected in the South Urals (1990–1995). *Science of the Total Environment*, **201**(2), 137–154.

67 Kuznetsov YV, Legin VK, Strukov VN, Novikov AP, Goryachenkova TA, Shilov AE, Savitsky Yu V. (2000) Transuranium elements in flood-land deposits of the Yenisei river. *Radiochemistry*, **42**, 519–528.

68 Bolsunovsky AY, Cherkizyan VO, Barsukova KV, Myasoedov BF. (2000) Characterization of high-level soil samples and hot particles collected from the flood plain of the Yenisei river. *Radiochemistry*, **42**, 620–624.

69 Vlasova IE, Kalmykov SN, Konevnik Yu, V, Simakin SG, Simakin IS, Anokhin A Yu, Sapozhnikov Yu A. (2008) Alpha track analysis and fission track analysis for localizing actinide-bearing micro-particles in the Yenisey River bottom sediments. *Radiation Measurements*, **43**, 303–8.

70 Skipperud L, Brown J, Fifield LK, Oughton DH, Salbu B. (2009) Association of plutonium with sediments from the Ob and Yenisey Rivers and Estuaries. *Journal of Environmental Radioactivity*, **100**, 290–300.

71 Muramatsu Y, Rühm W, Yoshida S, Tagami K, Uchida S, et al. (2000) Concentrations of 239Pu and 240Pu and their isotopic ratios determined by ICP-MS in soils collected from the Chernobyl 30-km zone. *Environmental Science & Technology*, **34**(14), 2913–2917.

72 Boulyga SF, Erdmann N, Funk H, Kievets MK, Lomonosova EM, et al. (1997) Determination of isotopic composition of plutonium in hot particles of the Chernobyl area. *Radiation Measurements*, **28**(1–6), 349–352.

73 Boulyga SF, Becker JS. (2002) Isotopic analysis of uranium and plutonium using ICP-MS and estimation of burn-up of spent uranium in contaminated environmental samples. *Journal of Analytical Atomic Spectrometry*, **17**(9), 1143–1147.

74 Zheng J, Tagami K, Watanabe Y, Uchida S, Aono T, et al. (2012) Isotopic evidence of plutonium release into the environment from the Fukushima DNPP accident. *Scientific Reports*, **2**, 304.

75 Steier P, Hrnecek E, Priller A, Quinto F, Srncik M, et al. (2013) AMS of the minor plutonium isotopes. *Nuclear Instruments and Methods in Physics Research Section B: Beam Interactions with Materials and Atoms*, **294**, 160–164.

76 Holm E, Persson BRR. (1978) Global fallout of curium. *Nature*, **273**(5660), 289–290.

77 Kashparov VA, Lundin SM, Zvarych SI, Yoshchenko VI, Levchuk SE, et al. (2003) Territory contamination with the radionuclides representing the fuel component of Chernobyl fallout. *Science of the Total Environment*, **317**(1–3), 105–19.

78 (2008) UNSCEAR (United Nations Scientific Committee on the Effects of Atomic Radiation), New York. Sources and effects of ionizing radiation. UNSCEAR 2008 report to the General Assembly, Scientific Annex D.

79 (2000) UNSCEAR (United Nations Scientific Committee on the Effects of Atomic Radiation), New York. Sources and effects of ionizing radiation. UNSCEAR 2000 report to the General Assembly, Volume II: Effects.

80 Mitchell PI, Holm E, León Vintró L, Condren OM, Roos P. (1998) Determination of the 243Cm/244Cm ratio alpha spectrometry and spectral deconvolution in environmental samples exposed to discharges from the nuclear fuel cycle. *Applied Radiation and Isotopes*, **49**(9–11), 1283–1288.

81 Devell L, Tovedal H, Bergstrom U, Appelgren A, Chyssler J, et al. (1986) Initial observations of fallout from the reactor accident at Chernobyl. *Nature*, **321**(6067), 192–193.

82 Rosner G, Hötzl H, Winkler R. (1988) Actinide nuclides in environmental air and precipitation samples after the Chernobyl accident. *Environment International*, **14**(4), 331–333.

83 Steinhauser G, Brandl A, Johnson TE. (2014) Comparison of the Chernobyl and Fukushima nuclear accidents: A review of the environmental impacts. *Science of the Total Environment*, **470–471**, 800–17.

84 Shinonaga T, Steier P, Lagos M, Ohkura T. (2014) Airborne plutonium and non-natural uranium from the Fukushima DNPP found at 120 km distance a few days after reactor hydrogen explosions. *Environmental Science & Technology*, **48**(7), 3808–3814.

85 Sakaguchi A, Steier P, Takahashi Y, Yamamoto M. (2014) Isotopic compositions of U-236 and Pu isotopes in "Black Substances" collected from roadsides in Fukushima prefecture: Fallout from the Fukushima Dai-ichi Nuclear Power Plant Accident. *Environmental Science & Technology*, **48**(7), 3691–3697.

86 Garland JA, Wakeford R. (2007) Atmospheric emissions from the Windscale accident of October 1957. *Atmospheric Environment*, **41**(18), 3904–3920.

87 UK Nuclear Decommissioning Authority (2013) Particles in the Environment Annual Report 2012/12.

88 Harley JH. (1980) Plutonium in the Environment: A Review. *Journal of Radiation Research*, **21**(1), 83–104.

89 Batuk ON, Conradson SD, Aleksandrova ON, Boukhalfa H, Burakov BE, et al. (2015) Multiscale speciation of U and Pu at Chernobyl, Hanford, Los Alamos, McGuire AFB, Mayak, and Rocky flats. *Environmental Science & Technology*, **49**(11), 6474–6484.

90 Wendel CC, Skipperud, L., Lind, O. C., Steinnes, E., Lierhagen, S., Salbu, B. (2015) Source attribution of Pu deposited on natural surface soils. *Journal of Radioanalytical and Nuclear Chemistry*, **304**, 1243–1252.

91 Shabalev SI, Burakov BE, Anderson EB. (1997) General classification of "hot" particles from the nearest Chernobyl contaminated areas. Materials Research Society Symposia Proceedings, **465**, 1343–1350.

92 Schubert-Bischoff P, Lutze W., Burakov BE.(1997) Properties and genesis of hot-particles from the Chernobyl reactor accident. *Materials Research Society Symposia Proceedings*, **465**, 1319–1325.

93 LoPresti V, Conradson SD, Clark DL. (2007) XANES identification of plutonium speciation in RFETS samples. *Journal of Alloys and Compounds*, **444**, 540–543.

94 Lind OC, Salbu B., Skipperud L, Janssens K, Jaroszewicz J, De Nolf W. (2009) Solid state speciation and potential bioavailability of depleted uranium particles from Kosovo and Kuwait. *Journal of Environmental Radioactivity*, **100**, 301–307.

95 Danesi PR, Markowicz A, Chinea-Cano E, Burkart W, Salbu B, Donohue D, Ruedenauer F, Hedberg M, Vogt S, Zahradnik P, Ciurapinski A. (2003) Depleted uranium particles in selected Kosovo samples. *Journal of Environmental Radioactivity*, **64**, 143–154.
96 Record of Decision Bomarc Missile Accident Site McGuire Air Force Base, NEW JERSEY. In: US Airforce, editor. 1992.
97 Burakov BE, Anderson EB, Shabalev SI, Strykanova EE, Ushakov SV, Trotabas M, Blanc J-Y, Winter P, Duco J. (1997) The behaviour of nuclear fuel in first days of the Chernobyl accident. Materials Research Society Symposium Proceedings, **465**, 1297–1308.
98 Levchuk S, Kashparov V, Maloshtan I, Yoschenko V, Van Meir N. (2012) Migration of transuranic elements in groundwater from the near-surface radioactive waste site. *Applied Geochemistry*, **27**, 1339–1347.
99 http://www.ctbto.org/nuclear-testing/.
100 (2008) UNSCEAR (United Nations Scientific Committee on the Effects of Atomic Radiation), New York. Sources and effects of ionizing radiation. UNSCEAR 2008 report to the General Assembly, Scientific Annex B.
101 (1962) UNSCEAR (United Nations Scientific Committee on the Effects of Atomic Radiation), New York. Report of the United Nations Scientific Committee on the Effects of Atomic Radiation, Scientific Annex F.
102 Sakaguchi A, Kawai K, Steier P, Imanaka T, Hoshi M, et al. (2010) Feasibility of using 236U to reconstruct close-in fallout deposition from the Hiroshima atomic bomb. *Science of the Total Environment*, **408**(22), 5392–5398.
103 Kudo A, Mahara Y, Santry DC, Suzuki T, Miyahara S, et al. (1995) Plutonium mass balance released from the Nagasaki A-Bomb and the applicability for future environmental research. *Applied Radiation and Isotopes*, **46**(11), 1089–1098.
104 Yamamoto M, Ishiguro T, Tazaki K, Komura K, Kaoru U. (1996) 237Np in Hemp-Palm leaves of Bontenchiku for fishing gear used by the fifth Fukuryu-Maru: 40 years after "Bravo". *Health Physics*, **70**(5), 744–748.
105 Diamond H, Fields PR, Stevens CS, Studier MH, Fried SM, et al. (1960) Heavy isotope Abundances in Mike thermonuclear device. *Physical Review*, **119**(6), 2000–2004.
106 Koide M, Bertine KK, Chow TJ, Goldberg ED. (1985) The 240Pu 239Pu ratio, a potential geochronometer. *Earth and Planetary Science Letters*, **72**(1), 1–8.
107 Beasley TM, Kelley JM, Maiti TC, Bond LA. (1998) 237Np239Pu atom ratios in integrated global fallout: a reassessment of the production of 237Np. *Journal of Environmental Radioactivity*, **38**(2), 133–146.
108 Shibata K, Kawano T, Nakagawa T, Iwamoto O, Katakura J-I, et al. (2002) Japanese evaluated nuclear data library version 3 revision-3: JENDL-3.3. *Journal of Nuclear Science and Technology*, **39**(11), 1125–1136.
109 Yamamoto M, Kofuji H, Tsumura A, Yamasaki S, Yuita K, et al. (1994) Temporal feature of global fallout 237Np deposition in paddy field through the measurement of low-level 237Np by high resolution ICP-MS. *Ract*, **64**(3–4), 217.
110 Sakaguchi A, Kawai K, Steier P, Quinto F, Mino K, et al. (2009) First results on (236)U levels in global fallout. *Science of the Total Environment*, **407**(14), 4238–4242.
111 Balagna JP, Barnes JW, Barr DW, Bayhurst BP, Browne CI et al. (1965) Production of very heavy elements in thermonuclear explosions – Test Barbel. *Phys Rev Lett*, **14**(23), 962–965.

112 Sisefsky J. (1961) Debris from tests of nuclear weapons: Activities roughly proportional to volume are found in particles examined by autoradiography and microscopy. *Science*, **133**(3455), 735–40.
113 Allkofer OC, Fox JM. (1966) New conceptions on the meteorology of stratospheric fallout from nuclear weapon tests. Archives for *Meteorology, Geophysics,* and *Bioclimatology A*, **15**(3–4), 299–317.
114 (2000) UNSCEAR (United Nations Scientific Committee on the Effects of Atomic Radiation), New York. Sources and effects of ionizing radiation. UNSCEAR 2000 report to the General Assembly, Volume I, Scientific Annex C.
115 Aoyama M, Hirose K, Igarashi Y. (2006) Re-construction and updating our understanding on the global weapons tests 137Cs fallout. *Journal of Environmental Monitoring*, **8**(4), 431–438.
116 Bennet BG. (1978) Environmental Aspects of Americium. EML-348.
117 Kelley JM, Bond LA, Beasley TM. (1999) Global distribution of Pu isotopes and 237Np. *Science of the Total Environment*, **237–238**, 483–500.
118 Ketterer ME, Groves AD, Strick BJ, Asplund CS, Jones VJ. (2013) Deposition of 236U from atmospheric nuclear testing in Washington state (USA) and the Pechora region (Russian Arctic). *Journal of Environmental Radioactivity*, **118**, 143–149.
119 Quinto F, Hrnecek E, Krachler M, Shotyk W, Steier P, et al. (2013) Measurements of (236)U in ancient and modern peat samples and implications for postdepositional migration of fallout radionuclides. *Environmental Science & Technology*, **47**(10), 5243–5250.
120 Łokas E, Mietelski JW, Ketterer ME, Kleszcz K, Wachniew P, et al. (2013) Sources and vertical distribution of 137Cs, 238Pu, 239 + 240Pu and 241Am in peat profiles from southwest Spitsbergen. *Applied Geochemistry*, **28**, 100–108.
121 Alvarado JAC, Steinmann P, Estier S, Bochud F, Haldimann M, et al. (2014) Anthropogenic radionuclides in atmospheric air over Switzerland during the last few decades. *Nature Communications*, **5**.
122 UNSCEAR. (1993) Sources and Effects of Ionizing Radiation. United Nations Scientific Committee on the Effects of Atomic Radiation. Annex B. Exposure from man-made sources of radiation.
123 Winkler SR, Steier P, Carilli J. (2012) Bomb fall-out 236U as a global oceanic tracer using an annually resolved coral core. *Earth and Planetary Science Letters*, **359–360**, 124–130.
124 Christl M, Lachner J, Vockenhuber C, Lechtenfeld O, Stimac I, et al. (2012) A depth profile of uranium-236 in the Atlantic Ocean. *Geochimica et Cosmochimica Acta*, **77**, 98–107.
125 Casacuberta N, Christl M, Lachner J, van der Loeff MR, Masqué P, et al. (2014) A first transect of 236U in the North Atlantic Ocean. *Geochimica et Cosmochimica Acta*, **133**, 34–46.
126 UNSCEAR. (1982) Ionizing Radiation: Sources and Biological Effects. Annex E: Exposures resulting from nuclear explosions.
127 Koide M, Goldberg ED. (1981) 241Pu/239 + 240Pu ratios in polar glaciers. *Earth and Planetary Science Letters*, **54**(2), 239–247.
128 Beasley TM, Ball LA. (1980) 243,244Cm in Columbia River sediments. *Nature*, **287**(5783), 624–625.

129 Lindahl P, Lee S-H, Worsfold P, Keith-Roach M. (2010) Plutonium isotopes as tracers for ocean processes: A review. *Marine Environmental Research*, **69**(2), 73–84.

130 Kersting AB, Efurd DW, Finnegan DL, Rokop DJ, Smith DK, et al. (1999) Migration of plutonium in ground water at the Nevada Test Site. *Nature*, **397**(6714), 56–59.

131 Zhao P, Tinnacher RM, Zavarin M, Kersting AB. (2014) Analysis of trace neptunium in the vicinity of underground nuclear tests at the Nevada National Security Site. *Journal of Environmental Radioactivity*, **137**, 163–172.

132 Bunzl K, Kracke W, Schimmack W, Auerswald K. (1995) Migration of fallout 239 + 240Pu, 241Am and 137Cs in the various horizons of a forest soil under pine. *Journal of Environmental Radioactivity*, **28**(1), 17–34.

133 Quinto F, Hrnecek E, Krachler M, Shotyk W, Steier P, et al. (2013) Determination of (239)Pu, (240)Pu, (241)Pu and (242)Pu at femtogram and attogram levels – evidence for the migration of fallout plutonium in an ombrotrophic peat bog profile. *Environmental Science Processes & Impacts*, **15**(4), 839–847.

134 Froehlich MB, Dietze MMA, Tims SG, Fifield LK. (2016) A comparison of fallout 236U and 239Pu uptake by Australian vegetation. *Journal of Environmental Radioactivity*, **151**(3), 558–562.

135 Froehlich MB, Steier P, Wallner G, Fifield LK. (2016) European roe deer antlers as an environmental archive for fallout 236U and 239Pu. *Journal of Environmental Radioactivity*,**151**(3), 587–592.

136 Bu W, Ni Y, Guo Q, Zheng J, Uchida S. (2015) Pu isotopes in soils collected downwind from Lop Nor: regional fallout vs. global fallout. *Scientific Reports*, **5**, 12262.

137 Lloyd JR, Chesnes J, Glasauer S, Bunker DJ, Livens FR, et al. (2002) Reduction of actinides and fission products by Fe(III)-reducing bacteria. *Geomicrobiology Journal*, **19**(1), 103–120.

138 Boukhalfa H, Icopini GA, Reilly SD, Neu MP. (2007) Plutonium(IV) reduction by the metal-reducing bacteria Geobacter metallireducens GS15 and Shewanella oneidensis MR1. *Applied & Environmental Microbiology*, **73**(18), 5897–5903.

139 Konhauser KO. (2009) *Introduction to Geomicrobiology*, John Wiley & Sons, Malden (MA).

140 Lovley DR, Stolz JF, Nord Jr GL, Phillips EJP. (1987) Anaerobic production of magnetite by a dissimilatory iron-reducing microorganism. *Nature*, **330**(6145), 252–254.

141 Lovley DR, Phillips EJP. (1988) Novel mode of microbial energy metabolism: organic carbon oxidation coupled to dissimilatory reduction of iron or manganese. *Applied & Environmental Microbiology*, **54**(6), 1472–1480.

142 Myers CR, Nealson KH. (1988) Bacterial manganese reduction and growth with manganese oxide as the sole electron acceptor. *Science Science*, **240**(4857), 1319–1321.

143 Williams KH, Bargar JR, Lloyd JR, Lovley DR. (2013) Bioremediation of uranium-contaminated groundwater: A systems approach to subsurface biogeochemistry. *Current Opinion in Biotechnology*, **24**(3), 489–497.

144 Suzuki Y, Kelly SD, Kemner KM, Banfield JF. (2002) Nanometre-size products of uranium bioreduction. *Nature*, **419**(6903), 134.

145 Bernier-Latmani R, Veeramani H, Vecchia ED, Junier P, Lezama-Pacheco JS, et al. (2010) Non-uraninite products of microbial U(VI) reduction. *Environmental Science and Technology*, **44**(24), 9456–9462.

146 Newsome L, Morris K, Lloyd JR. (2014) The biogeochemistry and bioremediation of uranium and other priority radionuclides. *Chemical Geology*, **363**, 164–184.

147 Cherkouk A, Law GT, Rizoulis T, Law K, Renshaw J, et al. (2016) Influence of riboflavin on the reduction of radionuclides by Shewanella oneidenis MR-1. *Dalton Transactions*, **45**, 5030–5037.

148 Renshaw JC, Butchins LJC, Livens FR, May I, Charnock JM, et al. (2005) Bioreduction of uranium: Environmental implications of a pentavalent intermediate. *Environmental Science and Technology*, **39**(15), 5657–5660.

149 Jones DL, Andrews MB, Swinburne AN, Botchway SW, Ward AD, et al. (2015) Fluorescence spectroscopy and microscopy as tools for monitoring redox transformations of uranium in biological systems. *Chemical Science*, **6**(9), 5133–5138.

150 Marshall TA, Morris K, Law GT, Mosselmans JFW, Bots P, et al. (2015) Uranium fate during crystallization of magnetite from ferrihydrite in conditions relevant to the disposal of radioactive waste. *Mineralogical Magazine*, **79**(6), 1265–1274.

151 Huber F, Schild D, Vitova T, Rothe J, Kirsch R, et al. (2012) U(VI) removal kinetics in presence of synthetic magnetite nanoparticles. *Geochim Cosmochim Acta* **96**, 154–173.

152 Anderson RT, Vrionis HA, Ortiz-Bernad I, Resch CT, Long PE, et al. (2003) Stimulating the in situ activity of Geobacter species to remove uranium from the groundwater of a uranium-contaminated aquifer. *Applied and Environmental Microbiology*, **69**(10), 5884–5891.

153 Lovley DR, Phillips EJP. (1992) Bioremediation of uranium contamination with enzymatic uranium reduction. *Environmental Science and Technology*, **26**(11), 2228–2234.

154 Bargar JR, Williams KH, Campbell KM, Long PE, Stubbs JE, et al. (2013) Uranium redox transition pathways in acetate-amended sediments. *Proceedings of the National Academy of Sciences of the United States of America*, **110**(12), 4506–4511.

155 Williamson AJ, Morris K, Law GTW, Rizoulis A, Charnock JM, et al. (2014) Microbial reduction of U(VI) under alkaline conditions: Implications for radioactive waste geodisposal. *Environmental Science and Technology*, **48**(22), 13549–13556.

156 Kelly SD, Kemner KM, Carley J, Criddle C, Jardine PM, et al. (2008) Speciation of uranium in sediments before and after in situ biostimulation. *Environmental Science and Technology*, **42**(5), 1558–1564.

157 Alessi DS, Lezama-Pacheco JS, Janot N, Suvorova EI, Cerrato JM, et al. (2014) Speciation and reactivity of uranium products formed during in situ bioremediation in a shallow alluvial aquifer. *Environmental Science and Technology*, **48**(21), 12842–12850.

158 Newsome L, Morris K, Shaw S, Trivedi D, Lloyd JR. (2015) The stability of microbially reduced U(IV); impact of residual electron donor and sediment ageing. *Chemical Geology*, **409**, 125–135.

159 Stylo M, Neubert N, Roebbert Y, Weyer S, Bernier-Latmani R. (2015) Mechanism of uranium reduction and immobilization in Desulfovibrio vulgaris biofilms. *Environmental Science and Technology*, **49**(17), 10553–61.

160 Rizoulis A, Steele HM, Morris K, Lloyd JR. (2012) The potential impact of anaerobic microbial metabolism during the geological disposal of intermediate-level waste. *Mineralogical Magazine*, **76**(8), 3261–3270.

161 Williamson AJ, Morris K, Shaw S, Byrne JM, Boothman C, et al. (2013) Microbial reduction of Fe(III) under alkaline conditions relevant to geological disposal. *Applied and Environmental Microbiology*, **79**(11), 3320–3326.

162 Bassil NM, Bryan N, Lloyd JR. (2015) Microbial degradation of isosaccharinic acid at high pH. *ISME Journal*, **9**(2), 310–320.

163 Rizoulis A, Milodowski AE, Morris K, Lloyd JR. (2016) Bacterial diversity in the Hyperalkaline Allas Springs (Cyprus), a natural analogue for cementitious radioactive waste repository. *Geomicrobiology Journal*, **33**(2), 73–84.

164 Hee SM, Komlos J, Jaffé PR. (2007) Uranium reoxidation in previously bioreduced sediment by dissolved oxygen and nitrate. *Environmental Science and Technology*, **41**(13), 4587–4592.

165 Begg JDC, Burke IT, Lloyd JR, Boothman C, Shaw S, et al. (2011) Bioreduction behavior of U(VI) sorbed to sediments. *Geomicrobiology Journal*, **28**(2), 160–171.

166 Law GTW, Geissler A, Burke IT, Livens FR, Lloyd JR, et al. (2011) Uranium redox cycling in sediment and biomineral systems. *Geomicrobiology Journal*, **28**(5–6), 497–506.

167 Alessi DS, Uster B, Veeramani H, Suvorova EI, Lezama-Pacheco JS, et al. (2012) Quantitative separation of monomeric U(IV) from UO_2 in products of U(VI) reduction. *Environmental Science and Technology*, **46**(11), 6150–6157.

168 Wilkins MJ, Livens FR, Vaughan DJ, Beadle I, Lloyd JR. (2007) The influence of microbial redox cycling on radionuclide mobility in the subsurface at a low-level radioactive waste storage site. *Geobiology*, **5**(3), 293–301.

169 McBeth JM, Lear G, Lloyd JR, Livens FR, Morris K, et al. (2007) Technetium reduction and reoxidation in aquifer sediments. *Geomicrobiology Journal*, **24**(3–4), 189–197.

170 Burke IT, Boothman C, Lloyd JR, Livens FR, Charnock JM, et al. (2006) Reoxidation behavior of technetium, iron, and sulfur in estuarine sediments. *Environmental Science and Technology*, **40**(11), 3529–3535.

171 Newsome L, Morris K, Trivedi D, Bewsher A, Lloyd JR. (2015) Biostimulation by glycerol phosphate to precipitate recalcitrant uranium(IV) phosphate. *Environmental Science and Technology*, **49**(18), 11070–11078.

172 Salome KR, Green SJ, Beazley MJ, Webb SM, Kostka JE, et al. (2013) The role of anaerobic respiration in the immobilization of uranium through biomineralization of phosphate minerals. *Geochimica et Cosmochimica Acta*, **106**, 344–363.

173 Suzuki Y, Mukai H, Ishimura T, Yokoyama TD, Sakata S, et al. (2016) Formation and geological sequestration of uranium nanoparticles in deep granitic aquifer. *Scientific Reports*, **6**.

174 Brookshaw DR, Pattrick RA, Bots P, Law GT, Lloyd JR, et al. (2015) Redox Interactions of Tc (VII), U (VI), and Np (V) with microbially reduced biotite and chlorite. *Environmental Science & Technology*, **49**(22), 13139–13148.

175 Lear G, McBeth JM, Boothman C, Gunning DJ, Ellis BL, et al. (2010) Probing the biogeochemical behavior of technetium using a novel nuclear imaging approach. *Environmental Science & Technology*, **44**(1), 156–162.

176 Burke IT, Livens FR, Lloyd JR, Brown AP, Law GTW, et al. (2010) The fate of technetium in reduced estuarine sediments: Combining direct and indirect analyses. *Applied Geochemistry*, **25**(2), 233–241.

177 Thorpe CL, Lloyd JR, Law GTW, Williams HA, Atherton N, et al. (2016) Retention of 99mtc at ultra-trace levels in flowing column experiments – Insights into bioreduction and biomineralization for remediation at nuclear facilities. *Geomicrobiology Journal*, **33**(3–4), 199–205.

178 Wharton MJ, Atkins B, Charnock JM, Livens FR, Pattrick RAD, et al. (2000) An X-ray absorption spectroscopy study of the coprecipitation of Tc and Re with mackinawite (FeS). *Applied Geochemistry*, **15**(3), 347–354.

179 Lee J-H, Zachara JM, Fredrickson JK, Heald SM, McKinley JP, et al. (2014) Fe(II)- and sulfide-facilitated reduction of Tc-99(VII) O-4(-) in microbially reduced hyporheic zone sediments. *Geochimica et Cosmochimica Acta*, **136**, 247–264.

180 Burke IT, Boothman C, Lloyd JR, Mortimer RJG, Livens FR, et al. (2005) Effects of progressive anoxia on the solubility of technetium in sediments. *Environmental Science and Technology*, **39**(11), 4109–4116.

181 Fredrickson JK, Zachara JM, Kennedy DW, Kukkadapu RK, McKinley JP, et al. (2004) Reduction of TcO4 – by sediment-associated biogenic Fe(II). *Geochimica et Cosmochimica Acta*, **68**(15), 3171–3187.

182 Fredrickson JK, Zachara JM, Plymale AE, Heald SM, McKinley JP, et al. (2009) Oxidative dissolution potential of biogenic and abiogenic TcO_2 in subsurface sediments. *Geochimica et Cosmochimica Acta*, **73**(8), 2299–2313.

183 Morris K, Livens F, Charnock J, Burke I, McBeth J, et al. (2008) An x-ray absorption study of the fate of technetium in reduced and reoxidised sediments and mineral phases. *Applied Geochemistry*, **23**, 603–617.

184 Liu J, Pearce CI, Qafoku O, Arenholz E, Heald SM, et al. (2012) Tc(VII) reduction kinetics by titanomagnetite (Fe3 – xTixO4) nanoparticles. *Geochimica et Cosmochimica Acta*, **92**(0), 67–81.

185 Marshall TA, Morris K, Law GTW, Mosselmans JFW, Bots P, et al. (2014) Incorporation and retention of 99-Tc(IV) in magnetite under high pH conditions. *Environmental Science and Technology*, **48**(20), 11853–11862.

186 Henrot J. (1989) Bioaccumulation and chemical modification of Tc by soil bacteria. *Health Physics*, **57**(2), 239–245.

187 Lloyd JR, Nolting HF, Solé VA, Bosecker K, Macaskie LE. (1998) Technetium reduction and precipitation by sulfate-reducing bacteria. *Geomicrobiology Journal*, **15**(1), 45–58.

188 Geissler A, Law GTW, Boothman C, Morris K, Burke IT, et al. (2011) Microbial communities associated with the oxidation of iron and technetium in bioreduced sediments. *Geomicrobiology Journal*, **28**(5–6), 507–518.

189 Begg JDC, Burke IT, Morris K. (2007) The behaviour of technetium during microbial reduction in amended soils from Dounreay, UK. *Science of the Total Environment*, **373**(1), 297–304.

190 Zachara JM, Heald SM, Jeon B-H, Kukkadapu RK, Liu C, et al. (2007) Reduction of pertechnetate [Tc(VII)] by aqueous Fe(II) and the nature of solid phase redox products. *Geochimica et Cosmochimica Acta*, **71**(9), 2137–2157.

191 Law GTW, Geissler A, Lloyd JR, Livens FR, Boothman C, et al. (2010) Geomicrobiological redox cycling of the transuranic element neptunium. *Environmental Science and Technology*, **44**(23), 8924–8929.

192 Lloyd JR, Yong P, Macaskie LE. (2000) Biological reduction and removal of Np(V) by two microorganisms. *Environmental Science & Technology*, **34**(7), 1297–1301.

193 Icopini GA, Boukhalfa H, Neu MP. (2007) Biological reduction of Np(V) and Np(V) citrate by metal-reducing bacteria. *Environmental Science and Technology*, **41**(8), 2764–2769.

194 Thorpe CL, Morris K, Lloyd JR, Denecke MA, Law KA, et al. (2015) Neptunium and manganese biocycling in nuclear legacy sediment systems. *Applied Geochemistry*, **63**, 303–309.

195 Williamson AJ, Morris K, Boothman C, Dardenne K, Law GT, et al. (2015) Microbially mediated reduction of Np (V) by a consortium of alkaline tolerant Fe (III)-reducing bacteria. *Mineralogical Magazine*, **79**(6), 1287–1295.

196 Renshaw JC, Law N, Geissler A, Livens FR, Lloyd JR. (2009) Impact of the Fe(III)-reducing bacteria Geobacter sulfurreducens and Shewanella oneidensis on the speciation of plutonium. *Biogeochemistry*, **94**(2), 191–196.

197 Rusin PA, Quintana L, Brainard JR, Strietelmeier BA, Tait CD, et al. (1994) Solubilization of Plutonium Hydrous Oxide by Iron-Reducing Bacteria. *Environmental Science & Technology*, **28**(9), 1686–1690.

198 Plymale AE, Bailey VL, Fredrickson JK, Heald SM, Buck EC, et al. (2012) Biotic and Abiotic Reduction and Solubilization of $Pu(IV)O_2 \cdot xH_2O(am)$ as Affected by Anthraquinone-2,6-disulfonate (AQDS) and Ethylenediaminetetraacetate (EDTA). *Environmental Science & Technology*, **46**(4), 2132–2140.

199 Cherkouk A, Law GTW, Rizoulis A, Law K, Renshaw JC, et al. (2016) Influence of riboflavin on the reduction of radionuclides by Shewanella oneidenis MR-1. *Dalton Transactions*, **45**(12), 5030–5037.

200 Francis AJ, Dodge CJ. (2015) Microbial mobilization of plutonium and other actinides from contaminated soil. *Journal of Environmental Radioactivity*, **150**, 277–285.

201 Geckeis H, Lützenkirchen J, Polly R, Rabung T, Schmidt M. (2013) Mineral–water interface reactions of actinides. *Chemical Reviews*, **113**(2), 1016–1062.

202 Zavarin M, Powell BA, Bourbin M, Zhao P, Kersting AB. (2012) Np(V) and Pu(V) ion exchange and surface-mediated reduction mechanisms on montmorillonite. *Environmental Science & Technology*, **46**(5), 2692–2698.

203 Geckeis H, Rabung T. (2002) *Solid–Water Interface Reactions of Polyvalent Metal Ions at Iron Oxide–Hydroxide Surfaces*, New York, Marcel Dekker.

204 Righetto L, Bidoglio G, Azimonti G, Bellobono IR. (1991) Competitive actinide interactions in colloidal humic acid-mineral oxide systems. *Environmental Science & Technology*, **25**(11), 1913–1919.

205 Choppin GR. Environmental behavior of actinides. *Czechoslovak Journal of Physics*, **56**(1), D13–D21.

206 Clark MW, Harrison JJ, Payne TE. (2011) The pH-dependence and reversibility of uranium and thorium binding on a modified bauxite refinery residue using isotopic exchange techniques. *Journal of Colloid and Interface Science*, **356**(2), 699–705.

207 van Veelen A, Copping R, Law GTW, Smith AJ, Bargar JR, et al. (2012) Uranium uptake onto Magnox sludge minerals studied using EXAFS. *Mineralogical Magazine*, **76**(8), 3095–3104.

208 Sanchez AL, Murray, J. W., Sibley, T. H. (1985) The adsorption of plutonium IV and V on goethite. *Geochim Cosmochim Acta* **49**, 2297–2307.

209 Keeney-kennicutt WL, Morse, J. W. (1985) The redox chemistry of $Pu(V)O_2^+$ interaction with common mineral surfaces in dilute solutions and seawater. *Geochim Cosmochim Acta* **49**, 2577–2588.

210 Powell BA, Fjeld RA, Kaplan DI, Coates JT, Serkiz SM. (2004) $Pu(V)O_2^+$ adsorption and reduction by synthetic magnetite (Fe_3O_4). *Environmental Science & Technology*, **38**, 6016–6024.

211 Morgenstern A, Choppin GR. (2002) Kinetics of the oxidation of Pu(IV) by manganese dioxide. *Radiochim Acta*, **90**, 69–74.

212 Duff MC, Hunter DB, Triay IR, Bertsch PM, Reed DT, Sutton SR, Shea-McCarthy G, Kitten J, Eng P, Chipera SJ, Vaniman DT. (1999) Mineral associations and average oxidation states of sorbed Pu on Tuff. *Environmental Science & Technology*, **33**, 2163–9.

213 Powell BA, Duff MC, Kaplan DI, Fjeld RA, Newville M, Hunter DB, Bertsch PM, Coates JT, Eng P, Rivers ML, Serkiz SM, Sutton SR, Triay IR, Vaniman DT. (2006) Plutonium oxidation and subsequent reduction by Mn(IV) minerals in Yucca Mt. Tuff. *Environmental Science & Technology*, **40**, 3508–3514.

214 Shaughnessy DA, Nitsche H, Booth CH, Shuh DK, Waychunas GA, Wilson RE, Gill H, Cantrell KJ, Serne RJ. (2003) Molecular interfacial reactions between Pu(VI) and manganese oxide minerals manganite and hausmannite. *Environmental Science & Technology*, **37**, 3367–3374.

215 Kersting AB, Reirmus, P.W. (2003) Colloid-Facilitated Transport of Low-Solubility Radionuclides: A Field, Experimental, and Modeling Investigation. Lawrence Livermore National Laboratory.

216 Powell BA, Fjeld RA, Kaplan DI, Coates JT, Serkiz SM. (2005) $Pu(V)O_2^+$ adsorption and reduction by synthetic hematite and goethite. *Environmental Science & Technology*, **39**, 2107–2114.

217 Romanchuk AY, Kalmykov SN, Aliev RA. (2011) Plutonium sorption onto hematite colloids at femto- and nanomolar concentrations. *Radiochim Acta*, **99**, 137–144.

218 Zavarin M, Roberts SK, Hakem N, Sawvel AM, Kersting AB. (2005) Eu(III), Sm(III), Np(V), Pu(V), and Pu(IV) sorption to calcite. *Radiochim Acta*, **93**, 93–102.

219 Hixon AE, Arai Y, Powell BA. (2013) Examination of the effect of alpha radiolysis on plutonium(V) sorption to quartz using multiple plutonium isotopes. *Journal of Colloid and Interface Science*, **403**, 105–112.

220 Kumar S, Kasar SU, Bajpai RK, Kaushik CP, Guin R, Das SK, Tomar BS. (2014) Kinetics of Pu(IV) sorption by smectite-rich natural clay. *Journal of Radioanalytical and Nuclear Chemistry*, **300**, 45–59.

221 Powell BA, Dai Z, Zavarin M, Zhao P, Kersting AB. (2011) Stabilization of plutonium nano-colloids by epitaxial distortion on mineral surfaces. *Environmental Science & Technology*, **45**(7), 2698–2703.

222 Romanchuk A. Yu, Kalmykov SN, Egorov AV, Zubavichus YV, Shiryaev AA, Batuk ON, Conradson SD, Pankratov DA, Presnyakov IA. (2013) Formation of crystalline $PuO_{2+x} \cdot nH_2O$ nanoparticles upon sorption of Pu(V,VI) onto hematite. *Geochim Cosmochim Acta*, **121**, 29–40.

223 Dzombak DA, Morel FMM. (1990) *Surface Complexation Modeling: Hydrous Ferric Oxide*, John Wiley & Sons, New York.

224 Hiemenz PC, Rajagopalan R. (1997) *Principles of Colloid and Surface Chemistry.* Third Edition. Marcel Dekker, New York.

225 Tanford C. (1961) *Physical Chemistry of Macromolecules*, Wiley, New York.
226 Hiemstra T, Van Riemsdijk WH. (1996) A surface structural approach to ion adsorption: The charge distribution (CD) model. *Journal of Colloid and Interface Science*, **179**(2), 488–508.
227 Sverjensky DA. (2006) Prediction of the speciation of alkaline earths adsorbed on mineral surfaces in salt solutions. *Geochimica et Cosmochimica Acta*, **70**(10), 2427–2453.
228 Hiemstra T, Riemsdijk WHV, Rossberg A, Ulrich K-U. (2009) A surface structural model for ferrihydrite II: Adsorption of uranyl and carbonate. *Geochimica et Cosmochimica Acta*, **73**(15), 4437–4451.
229 Hiemstra T, Van Riemsdijk WH. (2009) A surface structural model for ferrihydrite I: Sites related to primary charge, molar mass, and mass density. *Geochimica et Cosmochimica Acta*, **73**(15), 4423–4436.
230 Waite TD, Davis JA, Payne TE, Waychunas GA, Xu N. (1994) Uranium(VI) adsorption to ferrihydrite: Application of a surface complexation model. *Geochimica et Cosmochimica Acta*, **58**(24), 5465–5478.
231 Kowal-Fouchard A, Drot R, Simoni E, Ehrhardt JJ. (2004) Use of spectroscopic techniques for uranium(VI)/montmorillonite interaction modeling. *Environmental Science & Technology*, **38**(5), 1399–1407.
232 Bradbury MH, Baeyens B. (1999) Modelling the sorption of Zn and Ni on Ca-montmorillonite. *Geochimica et Cosmochimica Acta*, **63**(3–4), 325–36.
233 Bradbury MH, Baeyens B, Geckeis H, Rabung T. (2005) Sorption of Eu(III)/Cm(III) on Ca-montmorillonite and Na-illite. Part 2: Surface complexation modelling. *Geochimica et Cosmochimica Acta*, **69**(23), 5403–5412.
234 Bradbury MH, Baeyens B. (2006) Modelling sorption data for the actinides Am(III), Np(V) and Pa(V) on montmorillonite *Radiochim Acta*, **94**, 619–625.
235 Rabung T, Pierret MC, Bauer A, Geckeis H, Bradbury MH, et al. (2005) Sorption of Eu(III)/Cm(III) on Ca-montmorillonite and Na-illite. Part 1: Batch sorption and time-resolved laser fluorescence spectroscopy experiments. *Geochimica et Cosmochimica Acta*, **69**(23), 5393–5402.
236 Dong W, Wan J. (2014) Additive Surface Complexation modeling of uranium(VI) adsorption onto quartz-sand dominated sediments. *Environmental Science & Technology*, **48**(12), 6569–6577.
237 Warwick PW, Hall A, Pashley V, Bryan ND, Griffin D. (2000) Modelling the effect of humic substances on the transport of europium through porous media: A comparison of equilibrium and equilibrium/kinetic models. *Journal of Contaminant Hydrology*, **42**(1), 19–34.
238 Gamerdinger AP, Kaplan DI, Wellman DM, Serne RJ. (2001) Two-region flow and decreased sorption of uranium (VI) during transport in Hanford groundwater and unsaturated sands. *Water Resources Research*, **37**(12), 3155–3162.
239 Shang J, Liu C, Wang Z, Zachara JM. (2011) Effect of grain size on uranium(VI) surface complexation kinetics and adsorption additivity. *Environmental Science & Technology*, **45**(14), 6025–6031.
240 Efstathiou M, Pashalidis I. (2013) A comparative study of the adsorption of uranium on commercial and natural (Cypriot) sea sand samples. *Journal of Radioanalytical and Nuclear Chemistry*, **298**(2), 1111–1116.

241 Handley-Sidhu S, Bryan ND, Worsfold PJ, Vaughan DJ, Livens FR, et al. (2009) Corrosion and transport of depleted uranium in sand-rich environments. *Chemosphere*, **77**(10), 1434–1439.

242 Farr JD, Neu MP, Schulze RK, Honeyman BD. (2007) Plutonium uptake by brucite and hydroxylated periclase. *Journal of Alloys and Compounds*, **444–445**, 533–539.

243 Macé N, Wieland E, Dähn R, Tits J, Scheinost Andreas C. (2013) EXAFS investigation on U(VI) immobilization in hardened cement paste: Influence of experimental conditions on speciation. *Ract*, **101**(6), 379.

244 Harfouche M, Wieland E, Dähn R, Fujita T, Tits J, et al. (2006) EXAFS study of U(VI) uptake by calcium silicate hydrates. *Journal of Colloid and Interface Science*, **303**(1), 195–204.

245 Tits J, Geipel G, Macé N, Eilzer M, Wieland E. (2011) Determination of uranium(VI) sorbed species in calcium silicate hydrate phases: A laser-induced luminescence spectroscopy and batch sorption study. *Journal of Colloid and Interface Science*, **359**(1), 248–256.

246 Sylwester E, Allen P.G., Zhao P, Viani BE. (2000) Interactions of uranium and neptunium with cementitious materials studied by XAFS. *Materials Research Society Symposia Proceedings*, **608**, 307–312.

247 Gaona X, Wieland E, Tits J, Scheinost AC, Dähn R. (2013) Np(V/VI) redox chemistry in cementitious systems: XAFS investigations on the speciation under anoxic and oxidizing conditions. *Applied Geochemistry*, **28**, 109–118.

248 Heberling F, Denecke MA, Bosbach D. (2008) Neptunium(V) coprecipitation with calcite. *Environmental Science & Technology*, **42**(2), 471–476.

249 Gaona X, Dähn R, Tits J, Scheinost AC, Wieland E. (2011) Uptake of Np(IV) by C–S–H phases and cement paste: An EXAFS study. *Environmental Science & Technology*, **45**(20), 8765–8771.

250 Tits J, Gaona X, Laube A, Wieland E. (2014) Influence of the redox state on the neptunium sorption under alkaline conditions: Batch sorption studies on titanium dioxide and calcium silicate hydrates. *Ract*, **102**(5), 385.

251 Sturchio NC, Antonio MR, Soderholm L, Sutton SR, Brannon JC. (1998) Tetravalent uranium in calcite. *Science*, **281**(5379), 971–973.

252 Schlegel ML, Pointeau I, Coreau N, Reiller P. (2004) Mechanism of Europium retention by calcium silicate hydrates: An EXAFS Study. *Environmental Science & Technology*, **38**(16), 4423–4431.

253 Jones MN, Bryan ND. (1998) Colloidal properties of humic substances. *Advances in Colloid and Interface Science*, **78**(1), 1–48.

254 Choppin Gregory R. (1988) Humics and radionuclide migration. *Ract*, **44–45**(1), 23.

255 Bryan ND, Jones DLM, Keepax RE, Farrelly DH, Abrahamsen LG, et al. (2007) The role of humic non-exchangeable binding in the promotion of metal ion transport in groundwaters in the environment. *Journal of Environmental Monitoring*, **9**(4), 329–347.

256 Kim JI, Czerwinski KR. (1996) Complexation of metal ions with humic acid: Metal ion charge neutralization model. *Ract*, **73**(1), 5.

257 Tipping E, Lofts S, Sonke JE. (2011) Humic Ion-Binding Model VII: A revised parameterisation of cation-binding by humic substances. *Environmental Chemistry*, **8**(3), 225–235.

258 Kinniburgh DG, van Riemsdijk WH, Koopal LK, Borkovec M, Benedetti MF, et al. (1999) Ion binding to natural organic matter: Competition, heterogeneity, stoichiometry and thermodynamic consistency. *Colloids and Surfaces A: Physicochemical and Engineering Aspects*, **151**(1–2), 147–166.

259 Bryan N, Jones D, Keepax R, Farrelly D, Abrahamsen L, et al. (2015) Factors affecting the dissociation of metal ions from humic substances. *Mineralogical Magazine*, **79**(6), 1397–1405.

260 Artinger R, Kienzler B, Schussler W, Kim JI. (1998) Effects of humic substances on the Am-241 migration in a sandy aquifer: Column experiments with Gorleben groundwater/sediment systems. *Journal of Contaminant Hydrology*, **35**, 261–275.

261 Artinger R, Rabung T, Kim JI, Sachs S, Schmeide K, et al. (2002) Humic colloid-borne migration of uranium in sand columns. *Journal of Contaminant Hydrology*, **58**(1–2), 1–12.

262 Bryan ND, Barlow J, Warwick P, Stephens S, Higgo JJW, et al. (2005) The simultaneous modelling of metal ion and humic substance transport in column experiments. *Journal of Environmental Monitoring*, **7**(3), 196–202.

263 Maes N, Wang L, Hicks T, Bennett D, Warwick P, et al. (2006) The role of natural organic matter in the migration behaviour of americium in the Boom Clay – Part I: Migration experiments. *Physics and Chemistry of the Earth, Parts A/B/C*, **31**(10–14), 541–547.

264 Maes N, Bruggeman C, Govaerts J, Martens E, Salah S, et al. (2011) A consistent phenomenological model for natural organic matter linked migration of Tc(IV), Cm(III), Np(IV), Pu(III/IV) and Pa(V) in the Boom Clay. *Physics and Chemistry of the Earth, Parts A/B/C*, **36**(17–18), 1590–1599.

265 Bryan ND, Abrahamsen L, Evans N, Warwick P, Buckau G, et al. (2012) The effects of humic substances on the transport of radionuclides: Recent improvements in the prediction of behaviour and the understanding of mechanisms. *Applied Geochemistry*, **27**(2), 378–389.

266 Gu B, Schmitt J, Chen Z, Liang L, McCarthy JF. (1994) Adsorption and desorption of natural organic matter on iron oxide: Mechanisms and models. *Environmental Science & Technology*, **28**(1), 38–46.

267 van de Weerd H, van Riemsdijk WH, Leijnse A. (1999) Modeling the dynamic adsorption/desorption of a NOM mixture: Effects of physical and chemical heterogeneity. *Environmental Science & Technology*, **33**(10), 1675–1681.

268 van de Weerd H, van Riemsdijk WH, Leijnse A. (2002) Modeling transport of a mixture of natural organic molecules: Effects of dynamic competitive sorption from particle to aquifer scale. *Water Resources Research*, **38**(8), 33-1--19.

269 Pitois A, Abrahamsen LG, Ivanov PI, Bryan ND. (2008) Humic acid sorption onto a quartz sand surface: A kinetic study and insight into fractionation. *Journal of Colloid and Interface Science*, **325**(1), 93–100.

270 Kim JI, Delakowitz B, Zeh P, Klotz D, Lazik D. (1994) A column experiment for the study of colloidal radionuclide migration in Gorleben aquifer systems. *Ract*, **66–67**(s1), 165.

271 McCarthy JF, Czerwinski KR, Sanford WE, Jardine PM, Marsh JD. (1998) Mobilization of transuranic radionuclides from disposal trenches by natural organic matter. *Journal of Contaminant Hydrology*, **30**(1–2), 49–77.

272 McCarthy JF, Sanford WE, Stafford PL. (1998) Lanthanide field tracers demonstrate enhanced transport of transuranic radionuclides by natural organic matter. *Environmental Science & Technology*, **32**(24), 3901–3906.
273 Zänker H, Hennig C. (2014) Colloid-borne forms of tetravalent actinides: A brief review. *Journal of Contaminant Hydrology*, **157**, 87–105.
274 Soderholm L, Almond PM, Skanthakumar S, Wilson RE, Burns PC. (2008) The structure of the plutonium oxide nanocluster $[Pu_{38}O_{56}Cl_{54}(H_2O)_8]_{14}$. *Angewandte Chemie (International ed in English)*, **47**(2), 298–302.
275 Rothe J, Walther C, Denecke MA, Fanghänel T. (2004) XAFS and LIBD Investigation of the formation and structure of colloidal Pu(IV) hydrolysis products. *Inorganic Chemistry*, **43**(15), 4708–4718.
276 Rousseau G, Fattahi M, Grambow B, Boucher F, Ouvrard G. (2002) Coprecipitation of thorium with UO_2. *Ract*, **90**(9–11/2002), 523.
277 Kaminski MD, Dimitrijevic NM, Mertz CJ, Goldberg MM. (2005) Colloids from the aqueous corrosion of uranium nuclear fuel. *Journal of Nuclear Materials*, **347**(1–2), 77–87.
278 Dreissig I, Weiss S, Hennig C, Bernhard G, Zänker H. (2011) Formation of uranium(IV)-silica colloids at near-neutral pH. *Geochimica et Cosmochimica Acta*, **75**(2), 352–367.
279 Bots P, Morris K, Hibberd R, Law GTW, Mosselmans JFW, et al. (2014) Formation of stable uranium(VI) colloidal nanoparticles in conditions relevant to radioactive waste disposal. *Langmuir*, **30**(48), 14396–14405.
280 Smith KF, Bryan ND, Swinburne AN, Bots P, Shaw S, et al. (2015) U(VI) behaviour in hyperalkaline calcite systems. *Geochimica et Cosmochimica Acta*, **148**, 343–359.
281 van der Lee J, Ledoux E, de Marsily G. (1992) Modeling of colloidal uranium transport in a fractured medium. *Journal of Hydrology*, **139**(1), 135–158.
282 Schmidt M, Wilson RE, Lee SS, Soderholm L, Fenter P. (2012) Adsorption of plutonium oxide nanoparticles. *Langmuir*, **28**(5), 2620–2627.
283 Darbha GK, Fischer C, Luetzenkirchen J, Schäfer T. (2012) Site-Specific retention of colloids at rough rock surfaces. *Environmental Science & Technology*, **46**(17), 9378–9387.
284 NAGRA (1994) Report on the Long-Term Safety of the L/ILW Repository at the Wellenberg Site. NAGRA Technical Report. NTB 94-06.
285 Ochs M, Lothenbach B, Shibata M, Sato H, Yui M. (2003) Sensitivity analysis of radionuclide migration in compacted bentonite: A mechanistic model approach. *Journal of Contaminant Hydrology*, **61**(1–4), 313–328.
286 Huber F, Kunze P, Geckeis H, Schäfer T. (2011) Sorption reversibility kinetics in the ternary system radionuclide–Bentonite colloids/nanoparticles–granite fracture filling material. *Applied Geochemistry*, **26**(12), 2226–2237.
287 Bouby M, Geckeis H, Lützenkirchen J, Mihai S, Schäfer T. (2011) Interaction of bentonite colloids with Cs, Eu, Th and U in presence of humic acid: A flow field-flow fractionation study. *Geochimica et Cosmochimica Acta*, **75**(13), 3866–3880.
288 Schäfer T, Geckeis H, Bouby M, Fanghänel T. (2004) U, Th, Eu and colloid mobility in a granite fracture under near-natural flow conditions. *Ract*, **92**(9-11/2004), 731.
289 Missana T, Alonso Ú, García-Gutiérrez M, Mingarro M. (2008) Role of bentonite colloids on europium and plutonium migration in a granite fracture. *Applied Geochemistry*, **23**(6), 1484–1497.

290 Geckeis H, Schäfer T, Hauser W, Rabung T, Missana T, et al. (2004) Results of the colloid and radionuclide retention experiment (CRR) at the Grimsel Test Site (GTS), Switzerland – Impact of reaction kinetics and speciation on radionuclide migration. *Ract,* **92**(9-11/2004), 765.

291 Möri A, Alexander WR, Geckeis H, Hauser W, Schäfer T, et al. (2003) The colloid and radionuclide retardation experiment at the Grimsel Test Site: Influence of bentonite colloids on radionuclide migration in a fractured rock. *Colloids and Surfaces A: Physicochemical and Engineering Aspects,* **217**(1–3), 33–47.

292 Behrends T, Krawczyk-Bärsch E, Arnold T. (2012) Implementation of microbial processes in the performance assessment of spent nuclear fuel repositories. *Applied Geochemistry,* **27**(2), 453–462.

293 Bès R, Rivenet M, Solari P-L, Kvashnina KO, Scheinost AC, et al. (2016) Use of HERFD–XANES at the U L3- and M4-edges to determine the uranium valence state on $[Ni(H_2O)_4]_3[U(OH,H_2O)(UO_2)_8O_{12}(OH)_3]$. *Inorganic Chemistry,* **55**(9), 4260–4270.

294 Vitova T, Kvashnina KO, Nocton G, Sukharina G, Denecke MA, et al. (2010) High energy resolution x-ray absorption spectroscopy study of uranium in varying valence states. *Physical Review B,* **82**(23), 235118.

295 Quinto F, Golser R, Lagos M, Plaschke M, Schafer T, et al. (2015) Accelerator mass spectrometry of actinides in ground- and seawater: An innovative method allowing for the simultaneous analysis of U, Np, Pu, Am, and Cm isotopes below ppq levels. *Analytical Chemistry,* **87**(11), 5766–573.

9

Actinide Biological Inorganic Chemistry: The Overlap of 5f Orbitals with Biology

Peter Agbo, Julian A. Rees, and Rebecca J. Abergel

Chemical Sciences Division, Lawrence Berkeley National Laboratory, United States

9.1 Introduction

Even among chemists, a dearth in familiarity with the elements of the f-block persists, particularly the actinides. Indeed, the status of the f group as rather alien elements of the periodic table is reflected in the field of biological inorganic chemistry, where the vast majority of studied interactions between metals and biological systems derive from the d-block transition metals. It may be tempting to assign this asymmetry to the relative differences in natural abundance of these groups of metals, with the implication that the actinide's low impacts on biological processes stem from their relatively low representation in natural environments. Such arguments, however, fail to explain why a greater body of literature exists on the biological inorganic chemistry of iridium (Ir), one of the rarest elements in the terrestrial continental crust, than on uranium (U) or thorium (Th), actinides known to persist in living organisms, and that exist at concentrations in Earth's crust 100,000 times greater than Ir (Figure 9.1) [1].

Among the 15 elements of the actinide series (atomic numbers 89 through 103, from actinium – Ac – to lawrencium – Lr), only Th and U occur naturally in significant amounts. Both are widely dispersed in Earth's crust as long-lived isotopes ^{232}Th ($t_{1/2} = 1.405 \times 10^{10}$ yr; essentially 100% of natural Th), ^{238}U ($t_{1/2} = 4.468 \times 10^{9}$ yr; about 9.3% of natural U), and ^{235}U ($t_{1/2} = 7.04 \times 10^{8}$ yr; about 0.7% of natural U). Monazite, a phosphate mineral that also contains large amounts of lanthanides, is the main Th ore and can be composed of up to about 20% Th. Other rarer Th sources include the ThO_2-containing thorianite and the $ThSiO_4$-based thorite. U is found in hundreds of minerals, with pitchblende (U_3O_8) being the most common form. Other early actinides Ac and protactinium (Pa) are also present in trace amounts, if ever, as decay products of ^{235}U and ^{238}U, and minute amounts of neptunium (Np) and plutonium (Pu) isotopes may occur in uranium minerals, as neutron capture products. However, for practical reasons, actinides other than Th and U are regarded as man-made with no biological function.

Invoking the fact that even naturally occurring actinides tend to play far less, if any, significant roles in the biochemistry of organisms provides a more convincing explanation for the disparity in the body of relevant work dedicated to 5f elements versus other

Experimental and Theoretical Approaches to Actinide Chemistry, First Edition.
Edited by John K. Gibson and Wibe A. de Jong.

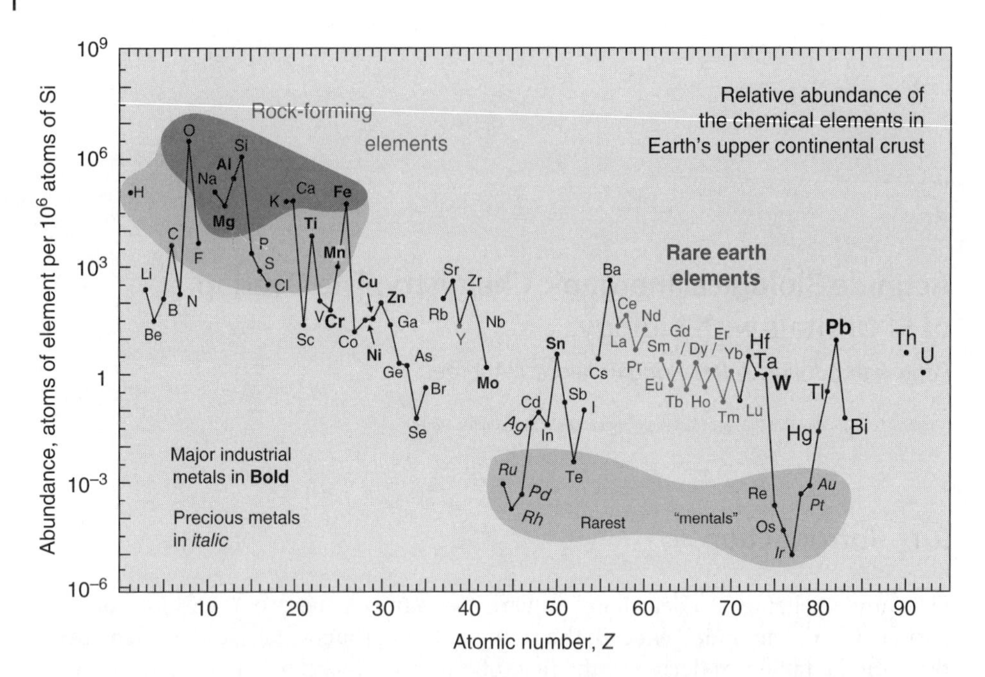

Figure 9.1 Relative abundance (expressed as atoms of element per 10^6 atoms of Si) of the elements in the terrestrial upper continental crust as a function of atomic number. Many of the elements are classified into the following (partially overlapping) categories: (1) rock-forming elements (major elements in green field and minor elements in light green field); (2) rare earth elements (lanthanides, La–Lu, and Y; labeled in blue); (3) major industrial metals (global production > ~3×10^7 kg per year; labeled in bold); (4) precious metals (italic); and (5) the nine rarest "metals"—the six platinum group elements plus Au, Re, and Te (a metalloid). Actinides Th and U are also represented. Reproduced with permission from US Geological Survey, Department of the Interior. Accessed from website April 18, 2017. http://pubs.usgs.gov/fs/2002/fs087-02/

metals. Nevertheless, considering that much of biological inorganic chemistry now focuses on the development of biosynthetic systems, including the incorporation of non-native metals into macromolecular ligand scaffolds, the current representation of the 5f elements in this arena is remarkably low. It is also necessary to remark that the ever-increasing use of actinides in the civilian industry and defense sectors over the past seven decades has resulted in persistent occurrence and future possibilities of widespread release of these elements in aquatic and terrestrial environments, with subsequent contamination of living organisms. Although controlled processing and disposal of wastes from the nuclear fuel cycles are the main source of actinide dissemination, significant quantities of these elements have also been dispersed as a consequence of nuclear weapons testing, nuclear power plant accidents, and compromised storage of nuclear materials [2]. In addition, events of the last 15 years have heightened public concern that large amounts of actinides may be released as the result of potential terrorist use of radiological dispersal devices or after a natural disaster affecting nuclear power plants or nuclear material storage sites [3, 4]. It is therefore essential to investigate the interactions between naturally occurring species and the actinides in environmentally and biologically relevant systems, to develop new

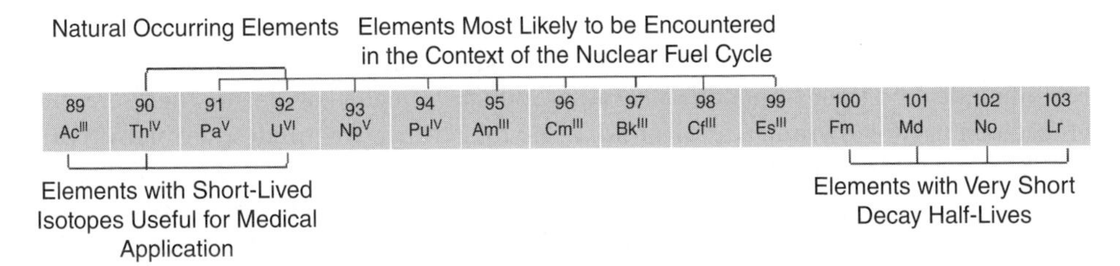

Figure 9.2 The actinide series encompasses the 15 chemical elements with atomic numbers from 89 to 103, actinium (Ac) to lawrencium (Lr). Among those elements, only uranium (U) and thorium (Th) occur naturally. Most elements from Th to einsteinium (Es) may be produced in large enough quantities to warrant investigating their biochemical properties. Elements beyond Es may not have any relevance in biological and environmental processes due to the very rapid decay of their respective isotopes. A few short-lived isotopes of Ac, Th, and U are currently under investigation for potential medical use. The metal oxidation states most relevant for biological and environmental considerations are indicated.

prevention and remediation strategies, and also implement biomimetic technologies that would exhibit actinide selectivity and specificity properties only found in biology.

The chasm between the attention paid to main group elements and the actinides in academia is starkly reflected in the current makeup of the RCSB Protein Data Bank (PDB), which is the single worldwide repository of information about the 3D structures of large biological molecules, including proteins and nucleic acids. As of this writing, the PDB boasts almost 130,000 structures. Of these, over one fifth represent metalloprotein structures of d-block metals. By contrast, a mere 16 entries can be found for metalloprotein binding of actinides or actinide complexes, providing an apt illustration of the knowledge gap dividing our understanding of d- versus f-element behavior in biochemical systems. A large proportion of these reports include the protein transferrin (Tf), a eukaryotic iron-transport protein that has been found to display binding to a number of free actinides and actinide–small ligand complexes. In addition to this, a number of heme-type cytochromes and various metal reductases have been implicated in the biomineralization of some actinides, particularly uranium, through the conversion of U(VI) to U(IV), through either direct, inner-sphere electron transfer mechanisms or, more commonly, through outer-sphere electron transfer mechanisms that include one or several biochemical reductants [5–14].

One aim of this chapter is to provide a general overview of known interactions between actinides and biological molecules in the context of environmental dissemination and migration, as well as mammalian contamination. The first section is therefore dedicated to broad surveys of those elements (Fig. 9.2) that would be most likely encountered in natural systems, either because they occur naturally (U, Th) or because they are produced in significant quantities in the nuclear fuel cycle and need to be disposed of (Np, Pu, americium (Am), and curium (Cm)). The following section highlights specific molecular systems that are known to transport essential alkaline-earth and transition metal ions such as calcium (Ca) or iron (Fe). These metallo-transporters have been proposed as key players in actinide biological transport mechanisms and may provide platforms for the development of new synthetic models for separation and sequestration of actinides. Finally, the last large section takes a different approach in probing the relevance of actinides in biology, as the last two decades have witnessed a surge in the use of short-lived actinide isotopes such as ^{225}Ac, ^{227Th}, or ^{230}U for medical applications. From that perspective, a thorough understanding of the *in vivo* behavior of these radionuclides and their respective daughter products is needed, to ensure their safe and efficacious use for therapy and diagnosis. The four heaviest actinides, fermium (Fm), mendelevium (Md), nobelium (No), and Lr, may not have any relevance in biological and environmental processes due to the very rapid decay of their respective isotopes and will not be discussed in this context. Finally, the end of this chapter will focus on the need for extended computational methods to support experimental work in the field of biological actinide chemistry. Experimental studies are not only limited by the radioactivity and sometimes scarcity of the actinide isotopes of interest, but restrictions also arise from the inherent complexity of biological matrices, and the lack of methodologies to thoroughly study metal speciation *in vivo* or *in situ* at trace levels.

9.2 Interactions between Actinides and Living Systems

The biological and environmental behavior of the f-elements has been reviewed several times in the last several decades [15–20]. From those comprehensive reviews, it quickly appears that the speciation of these elements in living organisms must be determined in order to understand what drives their *in vivo* transport and storage processes. In addition, while we often refer to the 5f-elements or the actinides throughout this chapter, each of the investigated elements clearly behaves differently, as expected from their respective solution chemistry properties. Even though the range of solution parameters (pH, ionic strength, concentration, etc.) seems to be limited in biological systems, the oxidation (Fig. 9.2) and aggregation states of the metal ions will still determine their speciation, which can vary considerably. Under most environmental and physiological conditions, oxidation state III is observed for Ac, as well as for Am and heavier actinides. Oxidation state IV is encountered with Th and Pu, while U and Np are mostly observed as the actinyl species $U(VI)O_2^{2+}$ and $Np(V)O_2^{+}$.

A first approach to determining the speciation of these elements is through modeling, using available thermodynamic databases that contain data for common components of biological fluids. In such models, macromolecules such as proteins are usually rendered as simple ligands competing with inorganic ions (mostly carbonate and phosphate anions) and small organic molecules. Such computational simulations using macroscopic thermodynamic models were performed recently for U(VI) [21–23], Pu(IV) [24], and Am(III) [25] with parallel experimental results sometimes enabling the verification of these models. However, while *in silico* techniques based on thermodynamics and quantum mechanics will allow predictive speciation under different environmental conditions, determining actinide speciation in biological media must be taken beyond the consideration of simple endogenous ligands known to bind essential metals and assessing their respective affinity for actinide ions. Analytical and spectroscopic tools will provide structural, molecular, elemental, and isotopic characterization. In addition, new techniques currently grouped under the relatively young field of functional genomics will allow elucidating the biomolecular pathways associated with actinide dissemination. It is only through a combined used of all these methods that we will be able to explore the intricacy of actinide chemistry in biological media, where composition and equilibria are driven by a variety of dynamic processes. Herein, selected examples are given from those different approaches to provide a cursory overview of the biological behavior of actinide ions, with most of the available data stemming from studies of U and Pu. These examples were mostly chosen to highlight macromolecular species and small ligands with some specificity for metal ions that were identified as potential actinide binders, beyond the generic inorganic and small organic ions.

9.2.1 Uranium in a Geochemical Context

Application of proteomics to the field of biogeochemistry has revealed the involvement of c-type cytochromes in microbe-mediated mineralization of UO_2. In particular, members of the *Geobacter* and *Shewanella* genera have been shown to constitute bacterial species (dissimilatory metal-reducing bacteria) capable of reducing a wide range of inorganic compounds and elemental species under anaerobic conditions [6, 7, 26]. Furthermore, statistical analysis of 16S rRNA isolated from microbial communities at a US Department of Energy (DOE) field testing center in Tennessee unveiled a strong association between biological reduction of Fe(III), inorganic sulfur, and U in *Desulfovibrio*, *Anaeromyxobacter*, and *Desulfosporosinus genera* [13]. In fact, the biochemical activity of some of these organisms for enzyme-mediated uranyl reduction is so prolific that they have become one of the foci in the DOE efforts to facilitate bioremediation of contaminated watersheds [5, 12]. Here, the operating principle for bioremediation of sites contaminated with inorganic, rather than organic, pollutants relies on the idea of using microbial populations to promote the biochemical conversion of soluble forms of metals into their insoluble counterparts, which reduces their bioavailability [5]. In aerobic water environments, U exists primarily as UO_2^{2+} and UO_2OH in complexation with carbonate and phosphate. To date, most research on microbial dissimilatory uranyl reduction has shown a sharp dependence on the presence of carbonate, and decreased activity in the presence of nitrate [5]. It is worth noting, however, that the microbe *Thiobacillius denitrificans* has been demonstrated to actually couple nitrate-dependent cell respiration with the oxidation of insoluble uraninite [5]. Krumboltz and coworkers reported speciation modeling indicating that in aerobic marine environments, where carbonate concentrations are generally lower, the $UO_2(OH)_3CO_3^-$ and $UO_2(OH)_3^-$ should dominate uranium speciation equilibria between pH 5.0 and 9.5 [5]. Groundwater, however, would be subject to a different ion equilibrium, with $UO_2(CO_3)^{2-}$ and $UO_2(CO_3)_3^{4-}$. Formal potentials for aqueous forms of uranyl generally lie in the range of −42 to 86 mV versus NHE [5].

Work in the past decade on dissimilatory uranyl reduction has revealed the role of outer-membrane-bound, c-type cytochromes in the formation of UO_2 (uraninite) nanoparticles by the *MR-1* strain of the bacterium *Shewanella oneidensis* [7]. Bioinformatic inspection of this bacterium's genome has already revealed the existence of as many as 42 c-type cytochrome genes, among which can be found a grouping of three genes (*MtrA, B,* and *C*), an operon encoding dodecaheme proteins whose specific evolutionary purpose has been assigned to metal reduction [7, 26, 27]. Transcription of this operon ultimately results in the assembly of the Mtr–protein complex, comprising MtrB, a transmembrane protein that is incorporated in the outer cell membrane, MtrA, which docks with MtrB's exposed periplasmic side, and MtrC, which binds to the hydrophilic, extracellular region of MtrB. Assembly of this enzyme supercomplex constitutes an electron transfer relay, allowing for charge exchange between MtrA and

MtrC, with the membrane-bound MtrB functioning as a conduit [26]. Work by Frederickson and coworkers has provided further insight into the role that outer-membrane cytochromes play in dissimilatory reduction of U(VI) to U(IV). Their construction of a mutant featuring a global interruption in the heme-incorporation machinery for c-type cytochromes was found to yield the anticipated apoforms of the expressed cytochromes, with cells expressing these heme-depleted enzymes displaying no competence for U(VI) (uranyl carbonate) reduction [7]. Targeted deletion of individual genes, coupled with *in vitro* assays, have revealed that the 10-subunit enzyme MtrC is responsible for a large share of the uranyl reductase activity of *S. oneidensis*. Suppressed expression of MtrC and various other outer-membrane cytochromes *in vivo* was found to influence the distribution of 1–5 nm, polycrystalline UO_2 nanoparticles produced by cellular reductase activity in the presence of uranyl, with uraninite precipitation occurring in either the periplasmic space, in association with a polymeric substance excreted by the cells, or as solution aggregates [7].

Furthermore, these redox processes seem to have some reversibility in nature. Organisms, including *Shewanella* species, that are capable of excreting soluble redox mediators such as flavin mononuclotide (FMN) have also been shown to reduce/oxidize insoluble forms of metal oxides. These small molecules act as a charge mediator between particulate inorganic oxides and conserved CX_nC disulfide bond motifs in surface-displayed, outer-membrane, redox-active cytochromes such as omcA and MtrC [26]. While small-molecule redox mediators are generally assumed to act as freely diffusing compounds, work has also shown that these cytochromes may act on inorganic substrates by first forming FMN–enzyme complexes, which, in turn, reduce the metal oxides [26]. Such biochemical activity carries important implications for the successful geochemical remediation of U-contaminated areas: these charge transport mechanisms suggest that while biogeochemical uranyl reduction to uraninite has been well studied, the resulting U(VI) species may still be bioaccessible, particularly in environments where the local microbiota are largely defined by the ubiquitous *Shewanella* species, allowing for the possibility that oxidation of U(IV) to U(VI) may compete with dissimilatory U reduction [5, 7].

Chromate reductase, an FMN- and NAD(P)H-dependent terminal electron donor found in the electron transport chain of *Gluconacetobacter hansenii*, has also been shown to be capable of uranyl reduction. Reductive catalysis by this protein proceeds through the self-assembly of an enzyme homotetramer, which binds U(VI) as $UO_2(CO_3)_3^{4-}$ in a pocket formed at the junction of the multiple protein subunits (Fig. 9.3) [28]. Crystal structures reveal an FMN binding site in each monomer, with the anionic FMN ribityl group housed in a positively charged binding pocket, while the cofactor's alloxazine ring lies close to a collection of hydrophobic residues. Co-crystallization of the protein with NAD(P)H cofactors and uranyl was unsuccessful. However, a putative metal-binding site in the vicinity of a critical R101 residue was deduced through site-directed mutagenesis studies [28]. These investigations suggested that uranyl (as the uranyl carbonate anion) may bind near the R101 residue, which the authors noted would likely position the uranyl complex within favorable distance of the FMN aromatic ring and a bound NAD(P)H to facilitate hydride transfer [28].

The ability of *G. hansenii* chromate reductase to engage in uranyl reduction is generally considered to be the result of promiscuous chemical reactivity, which allows the enzyme to bind and reduce chemically similar substrates, rather than the result of

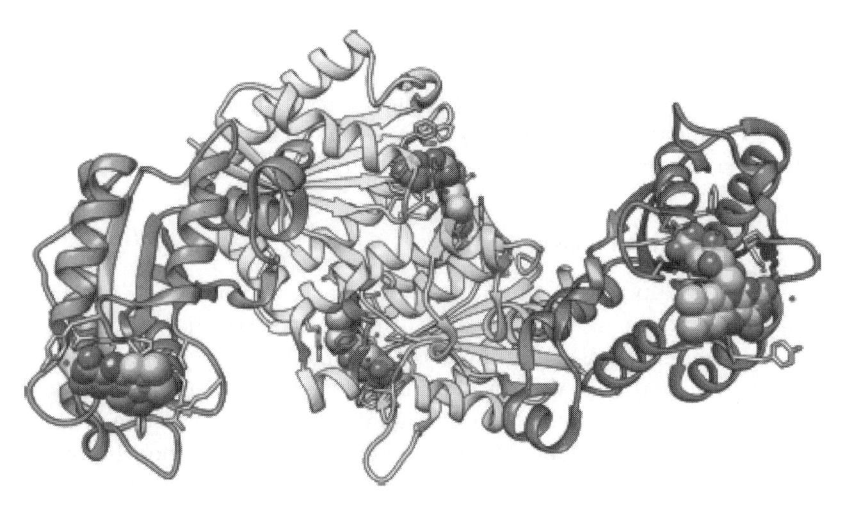

Figure 9.3 Crystal structure of the *G. hansenii* chromate reductase homotetramer (PDB: 3S2Y). Individual monomer chains are differentiated by color, with bound flavin mononucleotide illustrated as spheres. (*See insert for color representation of the figure.*)

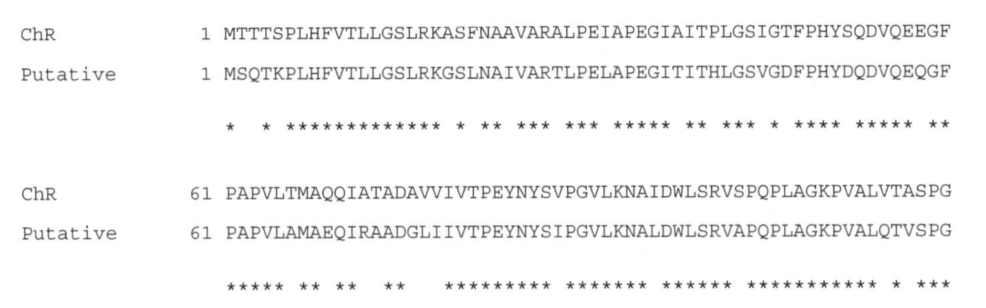

```
ChR          1 MTTTSPLHFVTLLGSLRKASFNAAVARALPEIAPEGIAITPLGSIGTFPHYSQDVQEEGF
Putative     1 MSQTKPLHFVTLLGSLRKGSLNAIVARTLPELAPEGITITHLGSVGDFPHYDQDVQEQGF

               *  * ************* * ** *** *** ***** ** *** * **** ***** **

ChR         61 PAPVLTMAQQIATADAVVIVTPEYNYSVPGVLKNAIDWLSRVSPQPLAGKPVALVTASPG
Putative    61 PAPVLAMAEQIRAADGLIIVTPEYNYSIPGVLKNALDWLSRVAPQPLAGKPVALQTVSPG

               ***** ** **  **   ********* ******* ****** *********** * ***
```

Figure 9.4 A protein alignment (ExPASy) of the *G. hansenii* chromate reductase monomer (ChR; PDB: 3S2Y) against a putative chromate reductase (accession number XP_017046536) found in *Drosophila ficusphilia*. Asterisks are used to highlight conserved residues. Alignment yields a 74% sequence identity between the two protein sequences, with an E-value of 2×10^{-98}. The high degree of sequence homology between the two proteins suggests that *D. ficusphilia* chromate reductase may to display some degree of uranyl reductase activity as well. Alignment was conducted using the BLOSUM62 comparison matrix with an open gap penalty of 12 and a gap extension penalty of 4.

evolution [28]. Given the lack of high U concentrations in the environments endemic to *G. hansenii*, it is unlikely that its chromate reductase would have been under any type of evolutionary selective pressure to utilize U(VI) as an electron acceptor to support biological respiration. This fact is bolstered by the results of protein alignments that indicate that the *G. hansenii* chromate reductase has significant sequence homology with putative chromate reductase enzymes in organisms as distant as fruit flies. Specifically, the *G. hansenii* chromate reductase displays a 74% sequence identity with that of the putative chromate reductase from *Drosophila ficusphilia* (Fig. 9.4). As of this writing, no rigorous biochemical characterization of this enzyme's activity has been conducted. However, the likelihood that it displays some uranyl reductase activity is supported by the demonstration of uranyl reductase activity in *Pseudomonas putida* chromate reductase, a protein sharing only 57% sequence similarity with the *G. hansenii* isoform, indicating an even more disparate evolutionary relationship than the fly homologue [28]. Such evidence suggests that observed U reduction in many biological systems is likely the result of comparatively weak enzymatic discrimination between inorganic substrates, rather than a product of evolutionary pressure.

9.2.2 Uranium in Larger Mammalian Systems

Though U is a common trace element and some intake and accumulation in living organisms is natural, both α-emitters ^{235}U and ^{238}U can induce radiation damage and carcinogenesis, as well as chemical damage in the kidneys and liver [15]. Among the four oxidation states of uranium (III, IV, V, and VI), the uranyl ion (UO_2^{2+}, U(VI)) is the most stable form in aqueous solutions and *in vivo*, and uranyl compounds have therefore be the focus of most pharmacology studies [15, 16]. In humans, biokinetic models predict the following distribution of U, 24 h after an intake: skeleton 15%, kidneys 17%, other tissues 5%, and urine 63% [29]. The U initially deposited in the kidneys has been shown to pass into the urine with a half-life of about 7 days, until nearly all of the original U deposit has been eliminated. Renal injury is the primary chemical damage caused by U-poisoning. Nearly all of the long-retained U in the body remains in the bones [15].

Many U compounds with broad ranges of composition and solubility have been tested in animals either through oral administration, injection, or wound implantation [15]. The fractions of U metal fragments that dissolve into the blood are distributed in tissues (skeleton and kidneys) and excreted through the renal pathway similarly to injected $UO_2(NO_3)_2$ [30–32]. The kidney is the only organ displaying histopathological changes that are characteristic of U-poisoning. Plasma clearance of intravenously injected UO_2^{2+} has been studied in three animal species and in humans [33–36]. In all cases, 90% of UO_2^{2+} is cleared from the plasma in 10 to 80 min, and quantitatively cleared by 700 min. The large fraction of circulating UO_2^{2+} that is rapidly excreted in urine is indicative of the formation of ultra-filterable complexes with low-molecular-weight ligands, with bicarbonate presumed to be the strongest ligand available in body fluids. Bicarbonate ions are abundant in plasma and tissue fluids, and uranyl carbonate complexes are more stable (log K_f values of 16.2, 21.6, and 28.1 for $[UO_2(CO_3)_2]^{2-}$, $[UO_2(CO_3)_3]^{4-}$, and $[Ca_2UO_2(CO_3)_3]_{aq}$, respectively) [37] than those formed with small phosphate ions, citric acid, as well as with non-filterable proteins such as the ferric ion transporter Tf (log $K_f = 16$) [16]. As plasma UO_2^{2+} is depleted by glomerular filtration of the low-molecular-weight complexes, the thermodynamic equilibrium between the filterable and non-filterable complexes shifts to continuously clear filterable UO_2^{2+} by renal filtration and binding to the skeleton. The dominant species in the tubular fluids is $[Ca_2UO_2(CO_3)_3]_{aq}$. However, when present in high concentrations, UO_2^{2+} is lost from the transiting fluid phase and deposits in the renal tubular system, thereby promoting cellular damage.

9.2.3 Pentavalent Actinides Neptunium and Protactinium

The body of research centered on the biological inorganic chemistry of Np remains small and that of Pa is nonexistent. Of the little work that has been produced in this arena, most has involved studies of Np decorporation in mammalian systems [38–40], yielding results that suggest this actinide behaves distinctly from Pu and Am *in vivo*

[39]. In particular, investigations *in vivo* are generally believed to involve Np(V), the most stable neptunium oxidation state, though some evidence for the persistence of Np(IV) in biological systems has been reported [41].

Studies have shown that this metal is rapidly expelled from the bloodstream in mammals, with one study reporting the decay of ^{237}Np concentrations in the bloodstream of rats to 0.2% of their initial levels after only 24 h [42]. Work conducted with human males injected with ^{239}Np citrate showed that initial removal of the metal from the blood was fast, with an average 20% of the Np being excreted in urine within the first day. The rate of this process decays rapidly however, with a cumulative fraction (~33%) being removed by the eighth day. A model of Np biodistribution and kinetics developed by the International Commission on Radiation Protection (ICRP) suggests 45% of the Np exiting the bloodstream deposits into the skeleton of adults, while another 10% is eventually concentrated in the liver [42]. The reverse-equilibration between bone and intravenous Np is believed to occur slowly, with the ICRP model suggesting a bone half-life as long as 22 years [42]. In accordance with the earlier mentioned findings for Np excretion in human trials, the ICRP models 32% of neptunium leaving the human bloodstream as passing through the kidneys, with a minor fraction (~2%) depositing on various renal tissues [42].

Neptunium residence in the liver is estimated to have a half-life of roughly two years [42]. Studies of ^{237}Np nitrate injection into liver cells suggest that the liver's retention mechanisms for this element are highly time dependent, with Np predominating in the cellular cytosol during the first hour following administration, and concentrating in nuclei and lysosomes at longer times [39]. Inspection of cytoplasmic fractions reveals that the metal is sequestered in the iron-storage protein, ferritin, and an unassigned 200 kDa protein during the first day of injection [39]. Evaluation of Np in cytosolic fractions at longer timescales (7–40 days) resulted in an altered distribution, with Np bound primarily to ferritin, and low-molecular-weight molecules (under 1,500 Da) [39, 41]. The finding marked a notable difference from the biodistributions of Pu and Am, which displayed no time-dependent behavior [39]. Furthermore, the total mass of injected Np was found to influence its biodistribution in liver cell, whereas metal valency was found to be a non-factor, with both the neptunium (IV) and (V) oxidation states assuming similar distributions *in vivo* [39, 42].

The stable oxidation state of Pa, one of the most poorly studied actinides, is (V) *in vivo*, with an eight-coordinate ionic radius smaller than those of its tetravalent neighbors and a stronger tendency to hydrolyze at lower pH. Stability constants with biologically relevant ligands are not available for Pa^{5+}, but it is expected to form protein complexes as stable, if not more stable, than those formed with Pu^{4+}. Complete biodistribution studies of Pa^{5+} have been conducted in rats [15, 43], resulting in predominant deposition in the skeleton and low uptake in the liver (about 46% and 8% of the injected dose, respectively) and very little excretion (mostly fecal), which closely resembles the distribution patterns observed with Th^{4+}.

9.2.4 Tetravalent Actinides Plutonium and Thorium

While not as extended as that seen for U, the body of data available for Pu behavior in biological systems is greater than for most f-elements due to the heavy use of this metal in nuclear materials. Similar to the other transuranic elements, early research with Pu has focused on understanding contamination patterns, with contaminated wounds and inhalation serving as the principal routes of Pu entry into the body; contaminated intact skin and ingestion are considered secondary, as they do not result in significant uptake. A number of postmortem studies have provided information on Pu distribution in humans, including both workers occupationally exposed[44, 45] and the general public [46, 47]. In those studies, about 95% of systemically absorbed Pu is localized in the liver, skeleton, and spleen, with the remaining fraction in skeletal muscle. However, in the cases of contamination by inhalation, substantial fractions of the total Pu body burden were found in the lungs, lymph nodes, and respiratory tract, exceeding those of other tissues [45]. Unfortunately, such measurements only reflect body distribution profiles at autopsy, and do not represent initial distributions of absorbed Pu or redistribution of Pu over time. More accurate biokinetic profiles can only be obtained from animal studies, and have been explored thoroughly in a wide variety of models, including rodents, dogs, and nonhuman primates [15]. Laboratory animal studies have demonstrated the impact of contaminant chemical form on Pu biokinetics and biodistribution, with more insoluble species resulting in greater tissue accumulation in comparison to more soluble contaminants such as Pu–ligand complexes [15, 48, 49]. Overall, the intake, transport, and excretion patterns in mammalian organisms following Pu exposure are highly dependent on the complex physico-chemical properties of Pu. Plutonium biokinetics have been reviewed in depth by the ICRP [50] and by the US-based National Council on Radiation Protection and Measurements (NCRP) [29].

Once absorbed from the different sites of intake, Pu circulates in the blood predominantly as Pu(IV) tightly complexed with plasma proteins or small polycarboxylate ligands such as citrate or lactate [51]. Pu then gets distributed to sites rich in blood sinusoids, such as the liver and the red bone marrow, and deposits in the tissues as an ion complexed by protein metal-binding sites or as insoluble particulates. It has been found associated with hemosiderin (in bone marrow macrophages), ferritin (in liver, spleen, and bone marrow), or as insoluble aggregates encapsulated in fibrotic material in the lungs [51]. Changes in the distribution patterns then occur upon regeneration of injured tissue or slow relocation via systemic circulation. Little excretion occurs, predominantly through the fecal pathway, and mechanisms of excretion are still poorly understood.

Among the actinides, Pu marks the first synthetic element, with virtually no natural representation in the terrestrial crust or oceans. As a result, it is reasonable to assume that this element's interactions with biological systems are adventitious in nature, arising from its chemical similarities to naturally abundant metals. In particular, Pu's observed chemical behavior has been found to mirror much of Fe biochemistry [52]. This fact accounts for reports tabulating the chelation of Pu^{4+} by multiple proteins in mammalian cells, including ferritin and Tf [38, 52]. Recent findings, however, have indicated that understanding the biochemistry of other metals may be pertinent to fully understanding Pu interactions in biological systems. In one study, analysis of 2D gel electrophoresis coupled with liquid chromatography/mass spectrometry revealed the adventitious binding of Pu to a Ca-dependent chaperone protein (p32 trimer), nucleoside diphosphate kinase B (a native divalent magnesium (Mg^{2+}) binder), an isoform of M2 pyruvate kinase, and the B chain of alpha crystallin, a heat-shock protein known to natively bind Ca^{2+}, zinc (Zn^{2+}), and copper (Cu^{2+}) cations [52]. In the case of the p32 trimer, crystal structures of the apo-protein have shown the protein complex has a highly asymmetric charge distribution, with a high density of negatively charged residues being grouped on one side of the trimer ring [52]. The anionic

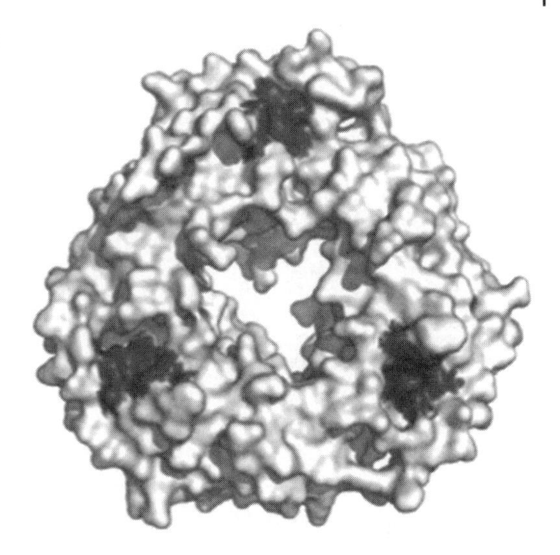

Figure 9.5 Putative Pu^{4+} binding sites on the human p32 trimer (PDB: 1P32). The sites (in red) are formed on the negatively charged side of the trimer from residues E89, E93, L231, D231, and Y268 in each chain. (*See insert for color representation of the figure.*)

face of the trimer includes a grouping of aspartic and glutamic acid residues that form a shallow binding pocket that has been speculated to form a binding site for Ca^{2+} and possibly Pu^{4+} (Fig. 9.5).

The potential for Th, in particular its 232 isotope ($t_{1/2} \sim 14$ billion yr), to serve as a nuclear fuel in advanced breeder reactors has brought along with it an increased likelihood of chemical accidents and contaminations involving this element [53]. Notably, the advent of breeder reactors has corresponded with an increased number of cases reported for ^{232}Th overexposure for workers in nuclear facilities in countries including the United States, Russia, India, Spain, and Brazil [54]. As a result, many efforts in Th research now center on preempting the effects of its biological contamination, prompting a need to understand its fundamental biochemical behavior, especially in humans [38, 53, 54]. This urgency is compounded by the various routes available for Th contamination, which include oral, skin, open wound, and respiratory modes of intake [53, 54]. The very low specific activity of ^{232}Th points toward its chemical, rather than radiological properties, as being the cause for its observed toxicity. Work in the area of Th bio-contamination has shown that bone and liver tissue are preferential sites for accumulation of this element, with one study showing over 50% of total injected Th concentrating in the liver, and another demonstrating the ability of accumulated ^{232}Th to alter liver function in mice [38, 53, 54]. Experiments with ^{232}Th have also revealed that the element crosses the blood-brain barrier in mice and can lead to downstream neurological defects [54]. Examinations of Th interactions at the molecular level unveiled chemical interplay between ^{232}Th and glycophorin, a protein expressed on the surface of human erythrocytes. The study also found that Th interaction with cell surface-bound sialic acid was related to past observations of cell aggregation and lysis occurring upon Th exposure [53, 54]. Related studies probing the molecular mechanisms for cellular aggregation and hemolysis proposed a scheme where at low Th/cell ratios, Th binds to surface sialic acid, neutralizing the surface charge of erythrocytes. Higher Th/cell ratios promote conformational changes in glycophorin that result in small breaks in the cell lipid bilayer, allowing for potassium ion (K^+) leakage. The result is a high cation flux which destabilizes cellular osmotic pressure, leading to lysis (Fig. 9.6) [54].

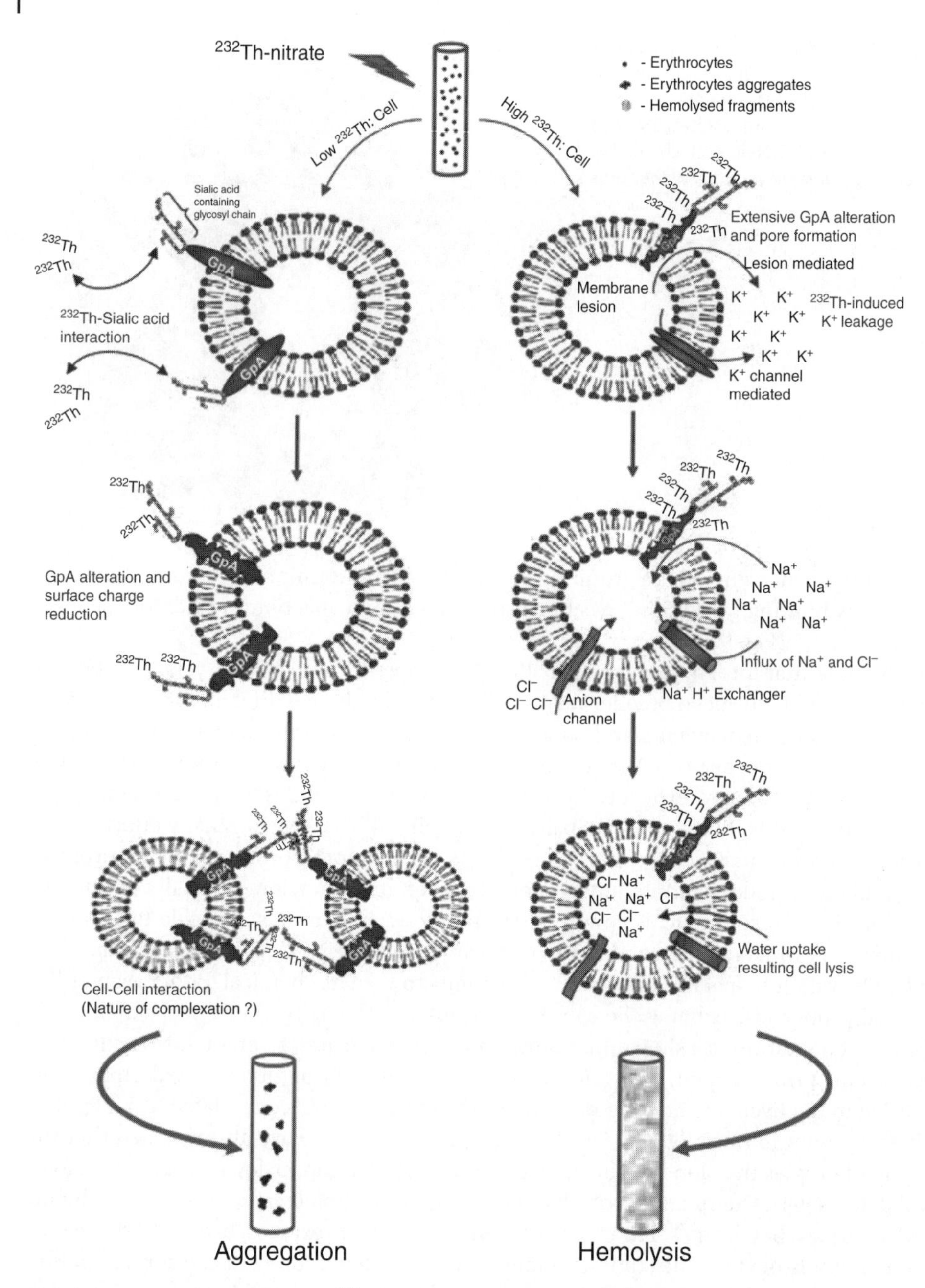

Figure 9.6 A proposed scheme for ^{232}Th-driven aggregation and lysis of erythrocytes. The ^{232}Th:erythrocyte ratio determines aggregation or hemolytic effect of ^{232}Th. At low ^{232}Th:cell ratio, ^{232}Th binds to negatively charged membrane sialic acid of GpA resulting in surface charge reduction and GpA alterations, which consequently induces erythrocytes aggregation. At high ^{232}Th:cell ratio, ^{232}Th binding to GpA induces GpA alterations and membrane pore formation, leading to significant K^+ leakage. Increased membrane permeability of K^+ (TEA and PEG1000 sensitive) generates an electrolyte imbalance by excessive entry of Na^+ and Cl^- ions through NHE and anion channels respectively, which results in an influx of water and subsequent hemolysis. Reproduced with permission from Ref [54].

Research conducted into the effects of ^{232}Th exposure to modified liver cell lines (HepG2 cells) showed that Th affected liver cell proliferation through modification of cell signaling pathways involving insulin-like growth factor 1 receptor (IGF-1R), and two kinases, phosphotidyl-inositol-3 (PI3K/Akt) and mitogen-activated protein (MAPK) [53]. A positive correlation was found between Th concentration and cell density in serum-free media for concentrations up to 10 μM ^{232}Th, presumably through an increase in the steady-state value of liver cells undergoing the DNA replication (S and G2-M) phases of mitosis [53].

9.2.5 Trivalent Metals from Americium to Einsteinium

Because transplutonium trivalent actinides do not occur naturally, biochemical studies of these elements have mainly focused on biokinetic aspects in the context of potential human contamination events, and less attention has been paid to understanding specific chemical mechanisms and speciation [15]. Biokinetic studies in animals have been conducted in several species (mice, rats, dogs, and nonhuman primates) on the actinide sequence from Am^{3+} to Es^{3+} [15, 55, 56]. For the whole sequence, overall initial distribution patterns are similar, with bone and liver being the major deposition sites. The larger animals (dogs and baboons) retained more Am^{3+}, Cm^{3+}, Cf^{3+}, and Es^{3+} in the bulk soft tissues, mainly muscle and pelt, than did the rats and mice.

Renal excretion of the trivalent elements was found to be limited, indicating that metal uptake into the target tissues is rapid and that the major fraction of circulating metal is bound to non-filterable protein. In addition, skeletal fraction and renal excretion were found to be somewhat greater for the heavier elements. The partitioning between deposition in bone and liver shifts from predominant liver uptake to bone deposition with the decrease in ionic radius seen along the series [15, 55, 57]. This pattern, however, was not correlated to the gradual affinity of these metal ions to bone surface ligands, but rather to the ability of the cations to occupy both metal-binding sites of Tf, the iron-transport protein of mammalian blood plasma [58].

The plasma clearance of intravenously injected Am^{3+} (as citrate or nitrate solutions of ^{241}Am) has been studied in five laboratory animals, while that of Cm^{3+} (as citrate solutions of ^{243}Cm or ^{244}Cm) has been investigated in three species [15, 55]. Clearances of Am^{3+} and Cm^{3+} are fast and almost identical in all animals: they were 90% complete in 60 min and 99% complete in less than 600 min [59]. Tissue uptake is correspondingly fast with only small amounts excreted within the first 24 h. In comparison, clearance of $^{249,252}Cf^{3+}$ and $^{253}Es^{3+}$ from the plasma of beagle dogs was 90% complete in about 100 min (almost twice the time observed for the two lighter and larger Am and Cm ions), and the time required to achieve 99% completion was about 800 min [60]. The slower plasma clearance rates of the heavier Cf and Es ions are consistent with the hypothesis that smaller metal centers would form more stable protein complexes. Provisional formation constants ($\log K_1$) values of 5.3 and 6.5 have been calculated for the Tf complexes of Am^{3+} and Cm^{3+}, respectively [38]. Such species are therefore weak non-filterable complexes that prevent renal filtration into urine but cannot compete with stronger binding molecules found in hepatic cells or on bone surfaces.

A number of soft tissues other than liver were found to accumulate significant concentrations of trivalent actinides, with the initial deposits located extracellularly in the connective tissue components, which contain potential ligands such as glycogen or glycoproteins [61]. Additional evidence for strong binding of trivalent actinides by glycoproteins was provided through *in vitro* affinity measurements between purified bone glycoprotein and Am^{3+} [62, 63].

9.3 Molecular Interactions of Actinides with Biological Metal Transporters

The multivalent actinides, particularly Pu(IV), share some metabolic properties with Fe(III). When present in the blood, both Pu(IV) and Np(IV) are quantitatively bound (>90%) to Tf, whereas U(VI), Am(III), and Np(V) are only weakly bound to Tf [15, 38, 64]. While U(VI) is thought to bind to hemoglobin and red blood cells, Am(III) can associate with albumin, and Np(V) is believed to complex carbonate ions [15, 38]. Some structural investigations have recently been initiated to characterize Tf–actinide complexes; however, no crystallographic data are available yet [22, 64–67]. The processes of actinide transfer from the blood to the liver and skeleton tissue have not been elucidated yet. Surprisingly, it has been argued that exogenous metal species may not be taken up into cells through the Tf-mediated receptor mechanism identified for Fe(III) [68–70], and a few studies describe possible binding of Pu(IV), Np(IV), and Am(III) to high-molecular-weight proteins such as calmodulin, ferritin, and lipofuscin [71, 72]. Using mineralized bones or uncalcified matrix models for bones, Duffield and Taylor have described other proteins such as sialoproteins, chondroitin sulfate–protein complexes, and glycoproteins interacting with actinides [73]. Studying the actinide coordination chemistry of these biomolecules is a critical step to unravel the detailed mechanisms of actinide uptake, transport, and intracellular storage. The following sections provide further details on selected recent molecular investigations of actinide biological coordination in mammalian and bacterial systems.

9.3.1 Transferrin-Mediated Metal Uptake Pathways

Among the several biochemical targets for actinides, the mammalian iron transporter Tf has generated the most characterization [38], and studies have reported the binding of Tf to many metals other than iron, including a number of therapeutic metals (titanium (Ti^{4+}), vanadium (VO^{2+}), chromium (Cr^{3+}), ruthenium (Ru^{3+}), and bismuth (Bi^{3+})), radio-therapeutic agents (gallium (Ga^{3+}), indium (In^{3+})), and toxic metals (aluminum (Al^{3+}), lanthanides (Ln^{3+}), thallium (Tl^{3+}), Th^{4+}, UO_2^{2+}, Np^{4+}, Pu^{4+}, Am^{3+}, and Cm^{3+}) [74]. Under normal conditions, this glycoprotein binds iron reversibly and shuttles the essential metal as Fe^{3+} between sites of uptake, utilization, and storage. Human serum Tf is an ~80 kDa protein folded into a single polypeptide chain with two homologous lobes (*N*-lobe and *C*-lobe) [75]. Each lobe can tightly coordinate a single ferric ion in a binding pocket, through two tyrosine residues, one histidine and one aspartate, and in the presence of a bidentate synergistic carbonate anion (Fig. 9.7) [76]. The two iron-binding pockets of Tf are similar, with 40% sequence identity, each creating an octahedral environment suitable to stabilize not only Fe^{3+} but also other metal ions, including trivalent and tetravalent actinides [77]. Metal-free apo-transferrin (apoTf) has an open cleft conformation, and Fe^{3+}-binding triggers cleft closure to yield either of two monoferric species Fe_CTf and Fe_NTf (in which a single Fe^{3+} is bound to the *C*-lobe or the

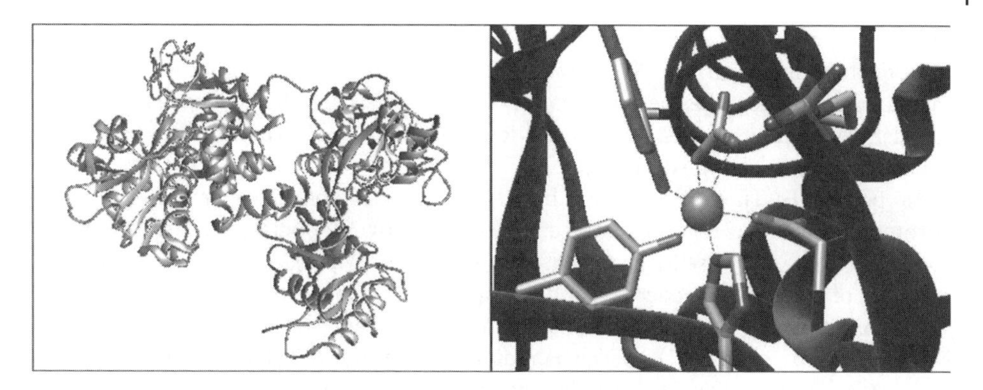

Figure 9.7 Human serum transferrin is a ~80 kDa protein folded into a single polypeptide chain with two homologous lobes (left). Each lobe can tightly coordinate a single ferric ion in a binding pocket, through two tyrosine residues, one histidine, and one aspartate, in the presence of a bidentate synergistic carbonate anion (right). (*See insert for color representation of the figure.*)

Table 9.1 Calculated conditional actinide-Tf stability constants derived from metal hydrolysis constants, with carbonate as a synergistic anion (where log K_C and log K_N represent the C- and N-terminal site binding constants, respectively) [16]. Comparative experimental values are available and provided in parentheses for Cm^{3+} [67], Pu^{4+} [83], and UO_2^{2+} [78].

Ion	Am^{3+}	Cm^{3+}		Pu^{4+}		U^{4+}	Th^{4+}	Np^{4+}	NpO_2^{+}	UO_2^{2+}	
Log K_C	10.1	10.3	*(8.8)*	23.2	*(22.5)*	23.1	19.4	23.4	2.5	14.1	*(16.0)*
Log K_N	8.5	8.7	*(7.0)*	21.3		21.9	18.2	22.3	0.6	12.6	

N-lobe, respectively), or the diferric species Fe_2Tf. When this mechanism is applied to the binding of Pu, the reported conditional stability constant log $K = 22.5$ for the binding of a first Pu^{4+} ion is within the same order of magnitude as those constants known for Fe^{3+}, albeit slightly higher. Furthermore, assuming coordination of Pu^{4+} in the well-described Fe^{3+}-binding sites, extended X-ray absorption fine structure (EXAFS) and X-ray absorption near edge spectroscopy (XANES) studies have suggested that mixed hydroxo–Tf complexes are formed with Pu^{4+} to stabilize the cation and complete the larger coordination of the actinide polyhedron [64] in comparison to Fe. Ansoborlo and coworkers used solution thermodynamic analogies with lanthanide ions and hydrolysis constant correlations [16] to estimate the conditional stability constants of the actinide–Tf complexes (Table 9.1), with scarce experimental values (for UO_2^{2+} and Cm^{3+}) [67, 78] available for confirmation. The approximate concentration of Tf in a normal individual is 30 μM, and it is estimated that only 30% of the protein is saturated with Fe^{3+}, with the following calculated distribution: 27% Fe_2Tf, 23% Fe_NTf, 10% Fe_CTf, and 40% apoTf [79]. The protein can therefore potentially transport up to 34 μM of exogenous actinides acquired as the result of internal contamination [74].

Physiologically, Fe^{3+} is transported inside cells through recognition of Fe_2Tf by the cognate Tf receptor TfR, a type II transmembrane homodimeric receptor [80] that is located on the extracellular surface of actively dividing cells. TfR–Fe_2Tf complexes ($K_d \sim 1$ nM, 1:2 stoichiometry) [81] undergo endocytosis and subsequently release iron inside the lower-pH cell compartments. The TfR–apoTf complexes are then shuttled back to the cellular surface [81]. Though tightly controlled for iron homeostasis, TfR-mediated endocytosis has been extensively discussed as a possible mechanism for effective cellular uptake of exogenous metal ions [67–69, 82]. However, extensive *in vitro* experiments have demonstrated that Pu_2Tf complexes do not follow the same mechanistic path as Fe_2Tf. A new method based on size-exclusion high-performance liquid chromatography was recently derived to determine two dissociation constants, corresponding to the stepwise formation of 1:1 and 1:2 TfR:$(M_xTf)_y$ adducts [70]. The small amounts of metal ions needed to conduct these assays allowed the authors to probe the TfR recognition of Tf loaded with transition metals (Fe^{3+} and Ga^{3+}) and lanthanides (La^{3+}, Nd^{3+}, Gd^{3+}, Yb^{3+}, and Lu^{3+}), as well as radioactive actinides ($^{232}Th^{4+}$, $^{238}UO_2^{2+}$, $^{242}Pu^{4+}$, and $^{248}Cm^{3+}$). The strongest affinity was recorded for Fe_2Tf, ($K_{d1,\ 2}=5$ and 20 nM), whereas the lowest affinity was found for Tf loaded with two Pu^{4+} ions ($K_{d1,\ 2}=282$ and 1766 nM). While the binding of a metal ion to human Tf mostly depends on its oxidation state, charge, and size, the recognition of M_2Tf by TfR is driven by other factors including shape and conformation of the metal-bound Tf. The source of intracellular Pu had remained elusive until a recent study by Jensen and coworkers used small-angle X-ray scattering, receptor binding assays, and synchrotron X-ray fluorescence microscopy of rat adrenal gland (PC12) cells to characterize the conformation of Pu-bound Tf [69]. The scattering profiles of different metallo-transferrin species revealed that TfR only binds the metallated Fe_2Tf and Pu_CFe_NTf, in which both Tf lobes are locked in a closed conformation, whereas Pu_2Tf and Fe_CPu_NTf retained one open lobe. While the Pu^{4+} only Tf adduct does not induce the closure of the Tf lobes (resulting in the di-substituted Pu_2Tf not being recognized by TfR), mammalian cells were found to acquire Pu through TfR-mediated endocytosis of the mixed Pu_CFe_NTf species. Even though a considerable portion of human Tf is unsaturated by Fe^{3+} and is thus available to bind a large number of toxic metals, the receptor TfR plays a role in the translocation mechanism of exogenous metals by acting as an additional selective barrier for cellular entry. It is therefore now understood that mixed-metal-Tf species may be the unsuspected key to actinide cellular uptake. However, the intracellular pathways involved in the putative release of actinides from Tf in the endosomes and subsequent retention in the cytoplasm have yet to be identified.

9.3.2 Ferric Ion Binding Proteins

The liver and skeleton are the principal deposition sites for trivalent and tetravalent actinides in mammalian species. Within hepatocytes, the lysosomes form the major intracellular deposition sites, with some association with the cell nuclei and cytosol. Ferritin, a ubiquitous 24 kDa globular protein, is the primary intracellular iron-storage protein and has therefore been suggested as the main storage molecule for ions such as Pu^{4+}. This insight has come through chromatographic analyses of soluble subcellular fractions of rat and canine liver organelles, as well as through the co-localization of plutonium with hemosiderin, which appears to be a complex of ferritin, denatured ferritin, and other material [38, 71, 84]. Cytosolic ferritin is able to enzymatically oxidize reactive Fe(II) and sequester the Fe(III) product within a nanocrystal of inorganic iron embedded within the central pore of the ferritin holoprotein. The intrinsic intricacy of the ferritin structure, made of twelve four-helical subunits, complicates theoretical models for the binding of Pu, but suggests different potential fixation mechanisms: through sorption onto the iron oxyhydroxy core, or direct binding to phosphate or carboxylic groups. EXAFS analysis of Np(IV) complexation by horse ferritin suggested that Np(IV) does not interact with the oxyhydroxy iron core of the protein but rather with carboxylate functions [85]. Structural analyses of Pu–ferritin complexes have not been performed; however Pu-labeled ferritin has been prepared successfully for *in vitro* metal-release studies [86] and described preparation methods could therefore be used for future molecular studies aimed at characterizing the binding of Pu by ferritin in more detail.

Ferritin and Tf, the iron-binding proteins typically proposed for the mammalian transport and deposition of actinides, use conserved motifs to directly bind metal ions. However, the mammalian immune system incorporates proteins that sequester ferric complexes formed with microbial siderophores; small molecules produced by invading bacteria to scavenge iron [87]. All of the known or hypothesized members of this functional group of proteins belong to the lipocalin family of binding proteins, and so are known as "siderocalins" ("siderophore-binding lipocalins"). The conserved lipocalin structure is an eight- stranded β-barrel fold, encompassing a highly sculpted, cup-shaped binding site known as a calyx. The archetype of the family, siderocalin (Scn, also known as neutrophil gelatinase-associated lipocalin (NGAL), Lipocalin 2 (Lcn2) or 24p312), was revealed recently as a new essential component of Fe trafficking and is unique in its extremely high affinity for ferric iron complexed by siderophores or siderophore-derived cofactors [87]. Scn was also shown to bind a Pu complex formed with the microbial siderophore enterobactin, providing the first crystallographic characterization of a Pu-protein species (Fig. 9.8) [88]. In addition, controlled *in vitro* assays with kidney proximal tubule LLCPK cells demonstrated that intracellular Pu^{4+} uptake could be through Scn-mediated endocytosis. These results identified siderocalins as potential new players in the biological trafficking of Pu and potentially other actinides, but with the particularity of a secondary ligand-based mechanism for metal sequestration. Endogenous siderophore-like molecules that are recognized by Scn, such as catechol, could also be associated with the low-molecular-weight small fraction of circulating Pu^{4+} in the plasma. New contemporary hyphenated LC-MS/MS techniques would be the most adequate to test this hypothesis.

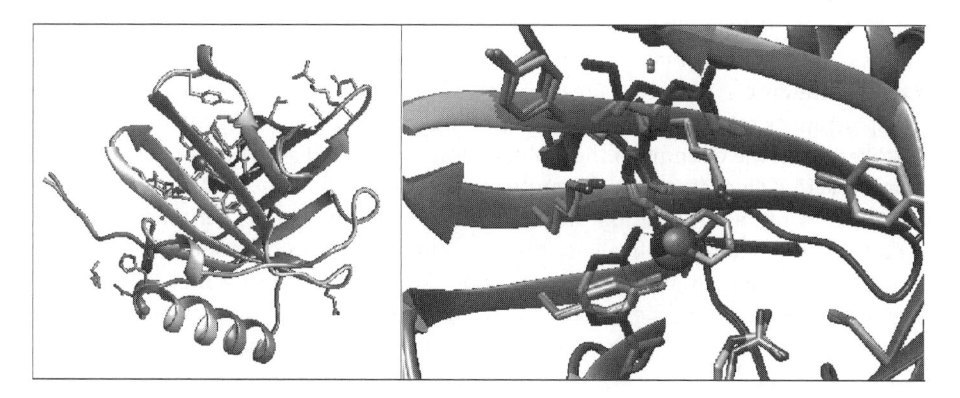

Figure 9.8 Structural superposition of the Scn adducts formed with Pu(IV)-enterobactin and Fe(III)-enterobactin, with a zoom into the Scn calyx (right panel) showing the high rigidity of the protein calyx and the exact same position for the metal ions. (*See insert for color representation of the figure.*)

Beyond probing the binding of Pu by the natural Scn–enterobactin pair, it was also found that Scn can recognize synthetic analogues of enterobactin–metal complexes with high affinity, providing that those complexes are negatively charged and contain catecholate-like aromatic rings, which then enable those key binding cyclic cation-π and electrostatic interactions [88]. The synthetic octadentate 3,4,3-LI(1,2-HOPO) ligand is an analogue of enterobactin better suited for actinide coordination, as described further below in Section 9.3.6. Upon deprotonation and subsequent metalation with trivalent ions such as Am^{3+}, Cm^{3+}, and Cf^{3+}, it forms negatively charged complexes that were shown to tightly bind the trilobal calyx of Scn [88]. X-ray diffraction data of the resulting Scn-ligand-metal adducts provided unparalleled information on the coordination center of these metals and revealed the use of protein crystallography as a new tool for the characterization of actinide coordination, since protein crystals usually require much less material than small-molecule crystals and may be used in a systematic metal-substitution manner as in this particular case. Moreover, the lack of interactions between Scn and the 3,4,3-LI(1,2-HOPO) complex of berkelium (Bk) evidenced the stabilization of the tetravalent ion Bk^{4+} by the siderophore analogue in solution [89]. The nonrecognition of the neutral [Bk^{IV}3,4,3-LI(1,2-HOPO)] by Scn suggested innovative procedures to separate Bk from M(III) by passing a 3,4,3-LI(1,2-HOPO) solution of the irradiated mixture through a Scn-containing medium, followed by discrimination using either size, mass, affinity, polarity, or solubility differences [89]. While these studies seem to have a sole fundamental character, they actually are the basis for a new protein-/ligand-based separation concept, where separation occurs on the basis of the macromolecular recognition properties of metal complexes instead of using the subtle differences in metal–ligand binding only. This is also an example of how applied processes may be inspired from the intricate biological chemistry of metal ions.

9.3.3 Divalent Metal Ion Transport Pathways

Divalent metal ions such as Ca^{2+} may also present physiological coordination properties similar to those of actinide cations. In mammalian systems, Ca^{2+} acts as a signal transmitter by reversibly binding to specific proteins and triggering conformational changes that in turn result in the activation of biochemical pathways. Calmodulin, a relatively small protein (~17 kDa), is the major intracellular receptor for Ca^{2+} [90]. Its structure contains two globular domains connected by a solvent-exposed α-helix, each incorporating two similar Ca^{2+}-binding sites through a helix-loop-helix structural motif. Here, the loop arranges seven oxygen atoms in a pentagonal-bipyramidal geometry. Binding to Ca^{2+} results in the opening of the globular domains as well as exposure of hydrophobic amino acid side chains. Initially motivated by the need for new structural tools to characterize proteins in solution, neutron resonance scattering experiments demonstrated that Pu^{3+} binds specifically to the Ca^{2+} binding sites of calmodulin, most likely due to the similar ionic radii of the two cations [91]. The average distance between the closest Pu^{3+} ions nested in the two Ca^{2+}-binding sites on the same globular domain was determined as 11.8 ± 0.4 Å, which agrees perfectly with the average distance between Ca^{2+} ions in calmodulin (11.7 Å) determined crystallographically. The $(Pu^{3+})_4$–calmodulin complex is stable in solution for several days, indicating tight binding but also suggesting enhanced stability with the more physiologically relevant Pu^{4+} cation, as the charge-to-radius ratio increases from Ca^{2+} to Pu^{3+} to Pu^{4+}. Tight Pu binding at low internal contamination concentrations could then permanently displace Ca^{2+} from calmodulin and disrupt normal kinetic processes associated with cellular regulation, resulting in toxicity. Binding to calmodulin could also facilitate Pu delivery to intracellular DNA-containing sites, leading to radioactive damage to DNA. Such mechanisms of toxicity have been speculated but never evidenced. In addition, many regulatory proteins contain Ca^{2+}-binding sites similar to those of calmodulin, providing a number of targets for actinides.

While there is no characterization of Pu complexes of Ca^{2+}-binding proteins other than those neutron resonance scattering data, synthetic aspartyl-rich pentapeptides have been prepared to mimic the Ca^{2+}- and Fe^{3+}- binding sites of calmodulin and Tf, respectively [92]. Such peptides were found by nuclear magnetic resonance (NMR) and EXAFS spectroscopies to fill the Pu(IV) coordination sphere, forming molecular species in which the metal is linked by hydroxo or oxo bridges. Recent efforts by Vidaud and coworkers have also focused on the identification of serum and bone proteins that may be involved in the transport of the uranyl ion specifically [93]. A rapid screening method based on surface plasmon resonance (SPR) was developed to determine the apparent affinities of major serum proteins for U(VI) [94]. After ranking the proteins according to both their serum concentrations and affinities for the actinyl, fetuin-A, a bovine protein that shares more than 70% homology with the human α-2-HS glycoprotein with a 30-nM affinity for U(VI), was estimated to carry more than 80% of the metal. Chromatographic and spectroscopic investigations demonstrated that the protein can bind 3 U(VI) at different binding sites. The non-collagenous protein osteopontin, which also plays an important role in bone homeostasis, was investigated as well for its uranyl-binding properties [95]. It contains a polyaspartic acid sequence and numerous patterns of alternating acidic and phosphorylated residues, and is mainly found in the extracellular matrix of mineralized tissues but also in body fluids such as milk, blood, and urine. Native phosphorylated osteopontin proteins were shown to bind UO_2^{2+} with a nanomolar affinity, with conformational changes enabling the formation of stable complexes with six and nine equivalents of uranium per mol of protein for bovine and human osteopontin, respectively. Phosphorylated residues were speculated to influence the affinity for the metal ion and the stability of the resulting complexes. Because fetuin-A and osteopontin have been shown to participate in bone mineralization and be included in the bone matrix during calcification processes, they may play a major role in uranyl deposition in the skeleton. Considering the relative lack of structural data available for biochemical complexes of actinides, these studies add significant information to our understanding of the biological coordination of these heavy metal ions.

9.3.4 Skeleton Deposition: The Role of the Bone Matrix

Bone is one of the main target organs for actinides, with the resulting intraskeletal distribution being a defining characteristic of actinide biological behavior [17]. The deposition patterns in the skeleton and the evolution of UO_2^{2+}, Pu^{4+}, Am^{3+}, and Np, at the macro- and microscopic levels have been studied and reviewed [15, 17, 96–98]; however, little effort has been directed toward the chemical and structural characterization of actinide coordination in the bone [99].

Similar to other tetravalent cations, Pu^{4+} was shown through autoradiography to deposit at endosteal surfaces, with particular selectivity and affinity for the highly vascularized trabecular bone. In addition, endosteal deposition is not uniform, with higher concentrations on fully mineralized resorbing surfaces, which are composed of a low crystalline and nonstoichiometric form of carbonated hydroxyapatite [$Ca_{10}(PO_4)_6(OH)_2$], with some substitution of Ca^{2+} cations by Mg^{2+}, and of hydroxide groups by carbonates. Incorporation of exogenous Pu^{4+} cations into such a structure is feasible through substitution of Ca^{2+}, as shown through various structures obtained by ceramic synthesis. Two different sites are available for calcium atoms in the hexagonal unit cell of crystalline hydroxyapatite ($P6_3/m$ space group). The first one coordinates Ca^{2+} through nine oxygen atoms, with a polyhedron volume of 30 Å^3, while the second one uses seven oxygens resulting in a smaller volume (22 Å^3). A 3/1 preference was shown for the binding of Pu^{4+} in the second site over the first site, in mixed calcium/lanthanide britholite models generated through density functional theory (DFT) calculations [99]. While Pu is known to bind stably *in vitro* to Ca phosphate and orthophosphate structures, there is currently no available structural data on the incorporation of Pu^{4+} to native or synthetic hydroxyapatite after synthesis under biologically relevant conditions. In contrast to Pu^{4+}, which tends to initially deposit entirely on endosteal surfaces, Am^{3+}, UO_2^{2+}, and NpO_2^{+} have been shown through autoradiography to deposit on all types of bone surfaces – endosteal, periosteal, and the vascular canal surfaces of compact bone. In addition, while deposition of Am^{3+} is somewhat more intense on resorbing and resting surfaces, uranyl and neptunyl are found preferentially on actively growing surfaces.

In addition to its mineral component, the organic components of the bone matrix can also participate in Pu complexation. Chipperfield and Taylor used gel filtration studies with fractions extracted from bovine cortical bone to demonstrate the binding of actinides by collagen and five distinct glycoproteins [62, 100, 101]. While not quantitative, the results of these studies indicated that bone sialoprotein, a highly acidic 60–80 kDa flexible protein containing a large amount of glutamic acid residues, and a bone protein complex formed with chondroitin sulfate [102], a sulfated glycosaminoglycan composed of a chain of alternating sugars, both bind Pu^{4+} with significantly higher affinities than transferrin. Three other less well-characterized glycoprotein fractions displayed Pu^{4+} binding properties similar to those of transferrin. A pH-dependent behavior was observed for the binding of Pu^{4+} to these bone glycoproteins, suggesting that the metal may be coordinated to carboxyl groups on the protein's amino acid side chains, and furthermore the bone sialoprotein's binding affinity decreased in the absence of terminal sialic acid moieties, which is indicative of additional binding through sialic acid groups.

9.3.5 Small-Molecule Metallophores

As in larger living systems encountered in the biosphere, accurate predictions of actinide migration in the geosphere are critically dependent on identification of the biological, chemical, and physical processes which affect actinide mobility, but in this case in soil and water. As an example, Pu most likely enters into soils as Pu(IV) hydroxides and oxides. These forms of Pu are thought to pose little risk for contaminating ground water and/or becoming bioavailable because they have very low solubility, and uncomplexed Pu(IV) strongly sorbs to mineral surfaces. However, compounds that solubilize the Pu may significantly increase its environmental migration and thereby also increase its bioavailability. As briefly mentioned above, siderophores are low-molecular-weight organic ligands produced by microbes to chelate Fe(III) in response to low availability of soluble Fe. Siderophores contain anionic hydroxamate or catecholate functional groups containing hard oxygen donors that strongly bind to Lewis acids, resulting in complexes with remarkably high-stability constants [103]. Because of the charge/ionic radius ratio similarities between Fe(III) and tetravalent actinides, it was suggested early that siderophores could bind actinides with high complex formation constants ($K_a > 10^{16}$) [104].

Concentrations of microbial siderophores ranging up to 240 µg per kg of soil have been measured in bioassays [105]. The prevalence of siderophore-producing microbes in soil makes siderophores likely to influence actinide mobility in the environment, as suggested by the determination of rate constants for the solubilization of hydrous Pu oxide by the siderophores enterobactin and desferrioxamine B (DFO, Fig. 9.9) and selected carboxylate, amino polycarboxylate, and catecholate ligands [106]. On a per molecule basis, enterobactin was found to be ~10^3 times more effective than the other chelators tested in increasing the rate of solubilization of hydrous Pu oxide. Notably, ferric–siderophore complexes were more effective in solubilizing actinide oxides than the siderophores in the absence of Fe [106]. In addition, studies of UO_2^{2+} with DFO show the presence of the U-DFO species through a large pH range [107]. No reduction of UO_2^{2+} by DFO was observed, and three species were detected spectrophotometrically: UO_2DFOH_2, UO_2DFOH, and $UO_2OHDFOH$, where $DFOH_4$ is fully protonated DFO. For the trivalent actinides, complexation experiments with Am^{3+} have been performed and the complexation constant estimated to be ~ log 16 for the Am–DFO complex [107]. Desferrioxamine siderophores display high affinity for Pu(IV); for Pu(IV)-DFO, $\log\beta_{110} = 30.8$, and the macrocyclic analog Pu(IV)-DFOE is sufficiently stable that its crystal structure was determined using X-ray diffraction methods [108]. By utilizing NMR techniques typically used for protein and large molecule chemistry, Boggs and coworkers were able to characterize the structure of two distinct Pu(IV) organic complexes formed with DFO, with XAS analysis supporting the identification of a Pu dimer at high pH [109]. Studies with the DFO-producing bacterium *Microbacterium flavescens* JG-9 demonstrated that the Pu(IV)–DFO complex was taken up by living, metabolically active bacteria [110]. As reviewed recently [111], a number of subsequent studies have shown that, through the use of siderophores, microorganisms can dissolve actinides and significantly affect their subsurface and environmental distribution [112].

The high affinity of siderophores for actinide ions has critical implications in the fields of environmental contamination and nuclear waste management. Many countries plan to deposit spent nuclear fuel in deep geological repositories, and microorganisms are present in these subterranean environments that could potentially affect the repository. Understanding the effect of microbial siderophores on the dissolution behavior of spent nuclear fuel pellet fragments is crucial. From this perspective, DFO and pyoverdin (Fig. 9.9) siderophores, isolated from cultures of *Pseudomonas fluorescens*, were used to demonstrate the feasibility of Np and Pu capture from fuel pellets [113]. Differences in Pu(IV) sorption to montmorillonite, a group of clay minerals investigated for geological repositories, have also been observed in the presence of organic matter such as DFO, dependence on pH, order of ligand addition, and stability of the metal–ligand complex. Sorption of DFO was found to be controlled by interactions with the

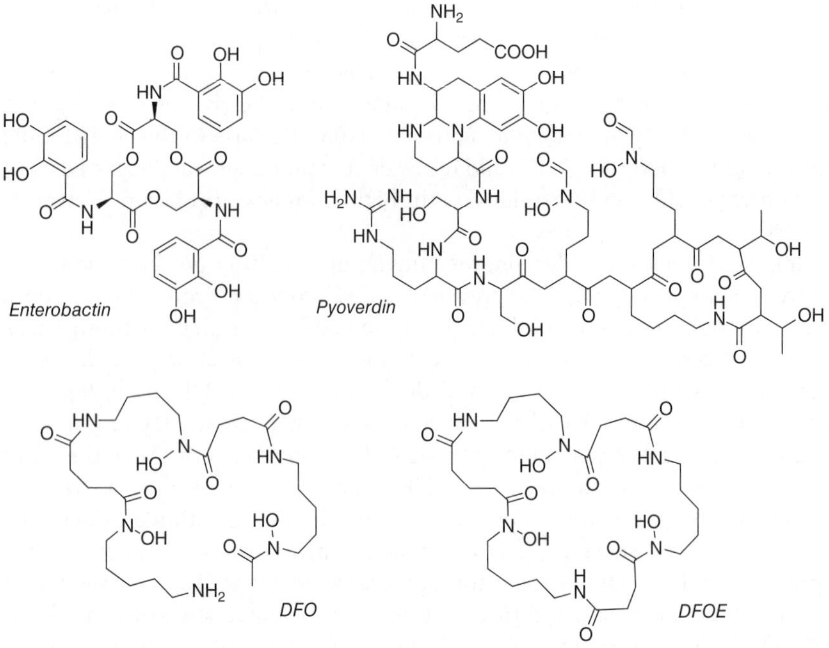

Figure 9.9 Molecular structures of the common natural siderophores discussed in the chapter.

interlayer of montmorillonite, with the Pu(IV)–DFO complex remaining irreversibly bound in the interlayer, which would make it inaccessible to desorption processes and enhance colloid-facilitated Pu transport [114]. Interactions between pyoverdine released by *P. fluorescens* isolated from the granitic rock aquifers and Cm(III) have also been described [115]. The complexation of Cm(III) with this siderophore is stronger than the complexation with ethylenediaminetetraacetic acid (EDTA, Fig. 9.10), hydroxide, or carbonate. Along those lines, a recent study investigated Pu concentrations in wetland surface sediments collected downstream of a former nuclear processing facility in F-Area of the Savannah River Site (SRS), United States [116]. Results suggested that while hydroxamate siderophores are a very minor component in sediment particulate/colloidal fractions, their concentrations greatly exceed those of ambient Pu, and those siderophores may play an especially important role in Pu immobilization/remobilization in wetland sediments [116]. All these data show that the influence of siderophores secreted by microorganisms on the migration processes of any toxic metal in our environment must be taken into account in strategies for the risk assessment of potential waste disposal sites [117].

Inspired by the high affinity of siderophore molecules for actinide ions, several research efforts have focused on using natural siderophores and/or synthetic analogues for applications ranging from separation to decontamination [118]. A separation example would be for the Purex process, which is used commercially to reprocess irradiated nuclear fuel by solvent extraction techniques and separate U and Pu for reuse from fission products [119]. Throughout the process, U and Pu flow into the solvent and become contaminated with Np. Since siderophores have been shown to allow for the

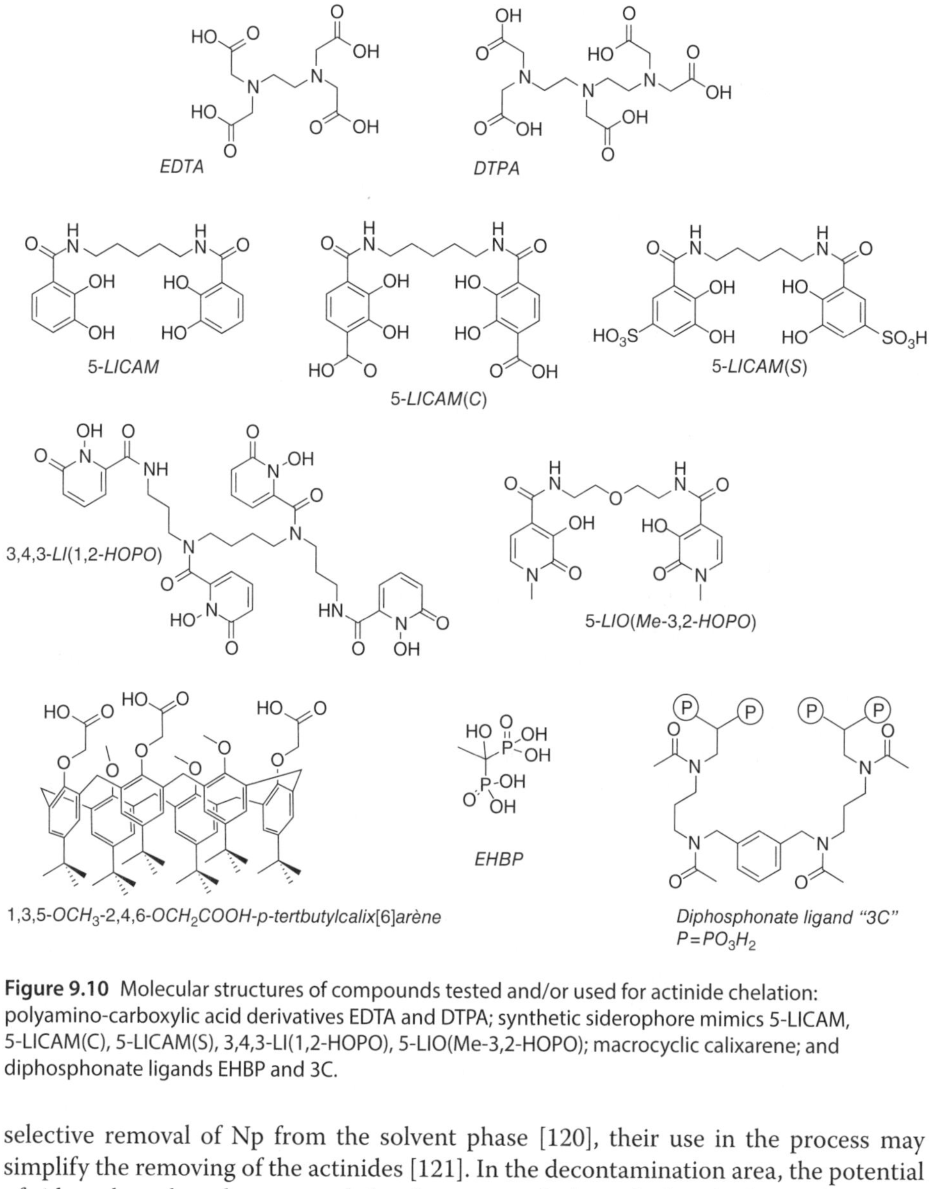

Figure 9.10 Molecular structures of compounds tested and/or used for actinide chelation: polyamino-carboxylic acid derivatives EDTA and DTPA; synthetic siderophore mimics 5-LICAM, 5-LICAM(C), 5-LICAM(S), 3,4,3-LI(1,2-HOPO), 5-LIO(Me-3,2-HOPO); macrocyclic calixarene; and diphosphonate ligands EHBP and 3C.

selective removal of Np from the solvent phase [120], their use in the process may simplify the removing of the actinides [121]. In the decontamination area, the potential of siderophore-based compounds has been extensively probed and advanced over the past three decades since Raymond and coworkers made the parallel between Fe(III) and Pu(IV) sequestration [104]. The strategic development of siderophore analogues for therapeutic actinide sequestration in humans is discussed in greater detail below.

9.3.6 Siderophore Analogues for Chelation Therapy

The hazard one typically associates with actinides is acute radiation sickness caused by internal exposure to large quantities of radiation, but the chemical and radioactive properties of these elements are such that they can also cause long-term damage and induce cancer in the tissues in which they deposit. The only practical therapy to reduce the health consequences of internal actinide contamination is aggressive, and often protracted, treatment with chemical agents that can form excretable low-molecular-weight chelates with the contaminants [122][123]. The hexadentate polyaminocarboxylic acid EDTA (Fig. 9.10), widely used for a variety of industrial chemical applications by the early 1950s and known for the high stabilities of its metal chelates [124], was among the first chelators to be tested for actinide chelation [125]. The calcium di-sodium salt of EDTA ($CaNa_2$-EDTA), which does not deplete Ca and is therefore not as toxic as the protonated version of EDTA, was shown to enhance the excretion of Pu(IV) and Am(III) in rats. However, *in vivo* actinide chelation was only achieved after high doses and repeated injections, reaching toxic levels [126]. The octadentate analogue of EDTA, diethylenediaminepentaacetic acid (DTPA, Fig. 9.10), was purposely designed to overcome these limitations [127]. Because of the higher denticity of this ligand, the stabilities of the actinide complexes formed with DTPA are higher than those of the corresponding EDTA compounds, with a better selectivity over divalent metal ions such as Ca^{2+} [128]. The $CaNa_3$-DTPA salt was quickly recognized for its better ability to decorporate Pu(IV) from rats and has been studied extensively *in vitro* and *in vivo*, and used in humans since then [29, 129]. Chelation therapy with intravenous or nebulized pentetate calcium trisodium (Ca-DTPA) and pentetate zinc trisodium (Zn-DTPA) was recently approved by the US Food and Drug Administration (FDA) for the treatment of individuals with known or suspected contamination with Pu, Am, or Cm [130], and has been the treatment of reference for decades. However, DTPA salts still exhibit significant limitations for use as therapeutics in contamination scenarios, such as their lack of efficacy for U and Np contaminants. While the results of several studies in laboratory animals have shown that Ca-DTPA was ineffective at promoting Np elimination, independent of the isotope used, the ligand dosage, or the mode of administration, Ca-DTPA and Zn-DTPA are still the only recommended substances for Np decorporation, despite some added risk of nephrotoxicity [29]. The substance recommended by the NCRP for U decorporation is sodium bicarbonate, although there may be undesirable side effects such as hypokalemia and alkalosis [29]. The use of DTPA salts is contraindicated for treatment of U contamination because of the added risk of renal damage. Finally, the approved chelation treatments Ca-DTPA and Zn-DTPA can only be administered intravenously or through a nebulizer, which would make chelation therapy in mass casualty situations cumbersome and challenging and is a considerable limitation for the treatment of actinide contamination of a very large population of contaminated individuals in a crisis setting [4, 131].

The previously mentioned hexadentate siderophore DFO (Desferal, Fig. 9.9), has been used for several decades as a therapeutic Fe chelator for the treatment of Fe-overload diseases [132]. However, DFO displayed lower efficacy than DTPA at removing Pu(IV) *in vivo*, presumably due to its lower denticity, and the weaker acidity of its hydroxamate Fe-binding units [133]. With the hypothesis that synthetic ligands adapted from siderophores but with increased denticity would form extremely stable actinide complexes and structures suitable for *in vivo* metal scavenging [123], Raymond and coworkers screened over 60 multidentate synthetic analogs for their *in vivo* Pu(IV) decorporation properties and potential toxicity, as described in detail in the literature [122, 123]. These chelating structures used siderophore-inspired bidentate chelating units such as functionalized catecholamides (CAM, carboxylated CAM(C), and sulfonated CAM(S)) attached to a variety of molecular polyamine backbones, enabling the development of new and improved chelating agents to evolve by providing an understanding of some of the relationships underlying the efficacy of a ligand for the decorporation of actinides (e.g., denticity, binding group acidity, backbone flexibility, and solubility) [122]. Representative examples of these ligands are displayed in Figure 9.10. The isomeric hydroxypyridinone metal-binding groups 1,2-HOPO and Me-3,2-HOPO are ionized at lower pH than catecholamide moieties, making them better ligands for the actinides, which are more Lewis-acidic than Fe. Ligands incorporating these groups were among those most selective and efficacious at removing actinides *in vivo*, with little to no observed toxicity in animals [122, 123, 134]. After extensive toxicity and efficacy studies in mice and a limited number of tests in dogs and baboons, two particular molecules, 3,4,3-LI(1,2-HOPO) and 5-LIO(Me-3,2-HOPO) (Fig. 9.10), were selected as promising candidates for orally available actinide decorporation agents [133, 135–144]. Both compounds were found to be up to 30 times more potent than DTPA for the decorporation of Pu(IV), and to sequester a wider spectrum of radionuclides, including U and Np, as well as particulate contaminants from mixed oxide fuel [134, 145]. In addition, unlike DTPA, both molecules have the advantage of being efficacious in the oral delivery format [134]. Over the last 10 years, remarkable progress has been made in the preclinical development of both agents. With 5-LIO(Me-3,2-HOPO) remaining in the pipeline, the most efficacious octadentate structure 3,4,3-LI(1,2-HOPO) was taken forward through a series of nonclinical efficacy, safety pharmacology, and toxicology assessments, necessary to demonstrate its viability as a therapeutic product [146–152].

Because actinide ions are attracted by phosphate groups on the bone mineral surface [17], a series of cyclic and linear polyphosphates and polyphosphonic or phosphinic acids have also been evaluated, resulting in enhanced urinary excretion of U but little reduction of U-induced nephrotoxicity in rats [123, 153]. Ethane-1-hydroxy-1,1-bisphosphonate (EHBP, also known as etidronic acid, Fig. 9.10), a treatment for bone remodeling disorders, stood out as it inhibits bone resorption. The efficacy of EHBP was demonstrated in rats contaminated with 50% lethal amounts of uranyl nitrate: EHBP prevented mortality with a 100% rate, significantly increased urinary U excretion, and reduced U content in kidneys when injected intramuscularly at a wound site or intraperitoneally promptly after contamination [154]. Another screening study identified a series of uranyl-binding diphosphonate ligands through state-of-the-art high-throughput methods [155, 156]. Some of these molecules significantly reduced uranium burden in the kidney (up to 50% for the structure named "C3" as shown in Fig. 9.10), liver, and skeleton of rats contaminated with uranyl.

Finally, another take on actinide chelation has been the use of macrocyclic calixarene structures, which were previously designed for the selective extraction of actinides [157]. Uranyl retention in rats was not affected after injection of the 1-hydroxy-4-sulfonatobenzene hexamer or octamer [158]. Nevertheless, these cage-like molecules provide conformational flexibility and can be functionalized with chelating groups. Substitution of the hexamer *p-tert*-butylcalix[6]arene structure with three carboxylic groups arranged in C3 symmetry results in 1,3,5-OCH_3-2,4,6-OCH_2COOH-*p-tert*-butylcalix[6]arene, a compound with a much higher affinity for uranyl that shows promise for U decontamination [159, 160].

9.4 Actinide Coordination for Radiopharmaceutical Applications

Over the past two decades, the design of specific chelators for f-block metal ions has found new applications beyond radionuclide separation and decontamination, since the production of short-lived radioisotopes from the 4f and 5f element series has gained prominence in medical physics and radiation oncology, as detailed below. Treating cancer with radiation is extremely effective, but can have devastating side effects. In order to improve the effectiveness and minimize the consequences of radiotherapy, many strategies have been developed to target radiation specifically to tumors. One of these strategies, targeted alpha therapy (TAT) holds tremendous potential as a cancer treatment, as it offers the potential of delivering a highly cytotoxic dose to targeted cells while minimizing damage to the surrounding healthy tissue, due to the short range and high linear energy transfer (LET) of α particles [161]. Moreover, TAT overcomes several barriers to traditional radiation therapy such as adaptive resistance and cell cycle progression. Currently, a number of radionuclides that emit single α particles are under investigation, including the halogen ^{211}At and the metallic isotopes ^{213}Bi, and ^{212}Pb. However, a growing subset of medical isotopes in this field includes the *in vivo* α-generator radionuclides ^{225}Ac, ^{223}Ra, ^{227}Th, and ^{230}U; isotopes that emit multiple α particles in their decay chains, dramatically increasing the potential delivered dose [162]. This principle was recently exploited in the development of Alpharadin, $^{223}RaCl_2$, a drug for bone metastases [163]. While Alpharadin relies on the natural propensity of ^{223}Ra for bone, other specifically targeted α-radiation delivery strategies based on metallic isotopes use constructs formed with a chelating agent to complex the α-emitter and a cancer site targeting vector such as tumor antigen binding monoclonal antibodies (mAbs) and cell surface receptor binding peptides. In this case, the chelating agent is termed "bifunctional" as it has the capacity to link the α-emitting radionuclide to the targeting molecule. However, even when bifunctional, the chelator must rapidly form kinetically inert and thermodynamically stable complexes with the radioisotopes, in order to ensure delivery of the radiation dose at the targeted site. Beyond stable and rapid chelation, ideal properties of the resultant targeting metallo-construct include facile internalization into the target cell and retention of the decay products within the target cell to achieve full cytotoxic potential and reduce the loss of the daughters to non-target tissues, thereby mitigating systemic radiotoxic events.

Given their respective half-lives, large alpha particle emission energies, and favorable rapid decay chains, the two actinide isotopes ^{225}Ac and ^{227}Th have been the subject of extensive research efforts over the past decade, with promising advances made from production to preclinical and clinical demonstrations of efficacy. Currently, the only method of generating ^{225}Ac is through the decay of long-lived ^{229}Th ($t_{1/2}$ = 7880 y), a natural decay product of ^{233}U [164]. Because of the limited worldwide availability of ^{229}Th, the DOE has invested significant resources in developing new methods to directly generate ^{225}Ac through the high-energy proton (78–192 MeV) irradiation of ^{232}Th at high-current, high-energy linear accelerator facilities [165]. A single decay of ^{225}Ac ($t_{1/2}$ = 10.0 d; 6 MeV α particle) yields four alpha and three beta disintegrations, producing a cascade of six key radionuclide daughters to stable ^{209}Bi (Fig. 9.11). ^{227}Th ($t_{1/2}$ = 18.7 d, 5.9 MeV average) is a member of the ^{235}U series, obtained from the decay of ^{227}Ac ($t_{1/2}$ = 21.8 y), which is itself produced by thermal neutron irradiation of ^{226}Ra followed

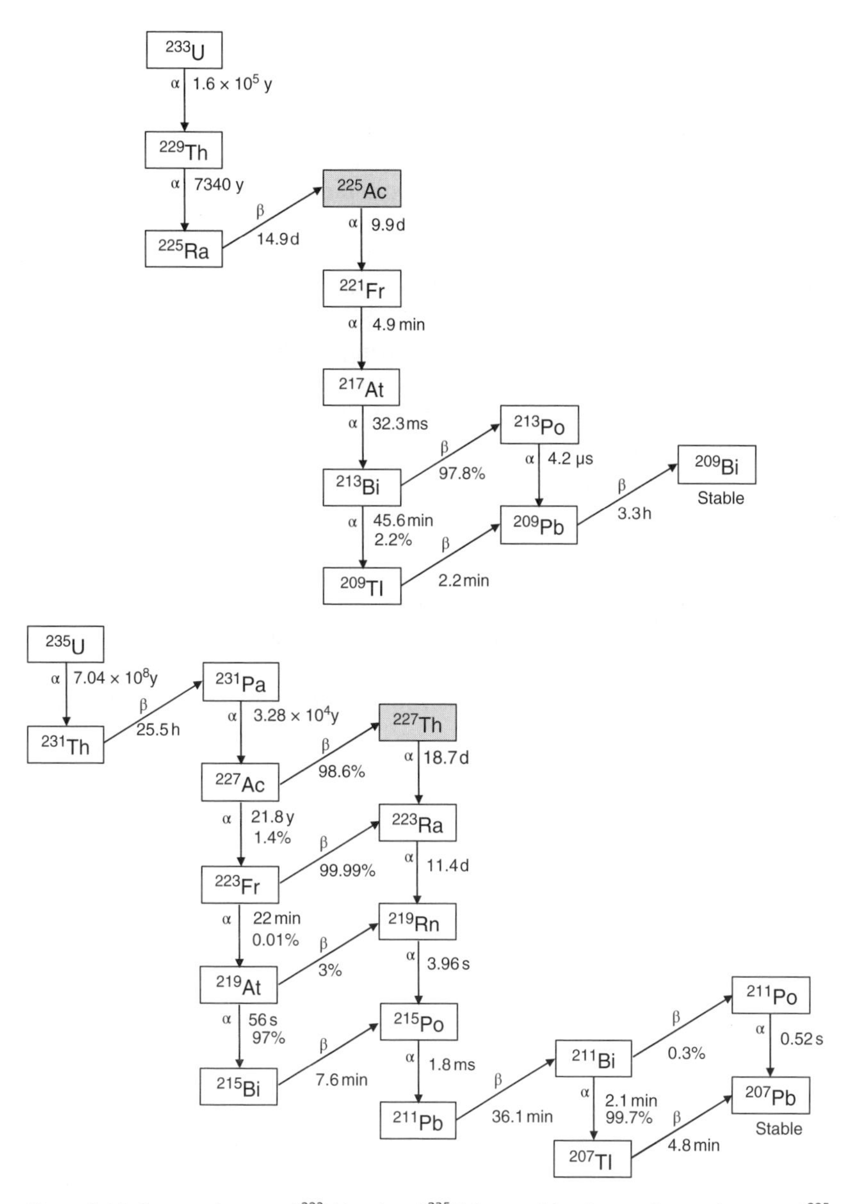

Figure 9.11 Decay schemes of ^{223}U (top) and ^{235}U (bottom) leading to the production of ^{225}Ac and ^{227}Th, respectively.

by beta decay of ^{227}Ra. ^{227}Th decays to stable ^{207}Pb through an average of five alpha particles and two beta particles, with ^{223}Ra being its first daughter (Fig. 9.11) [166]. Some examples of promising α-generator immunoconjugates are the lintuzumab conjugate ^{225}Ac-DOTA-HuM195 for myeloid leukemia treatment[167, 168] and ^{227}Th-DOTA-trastuzumab for treating HER-2 positive breast and ovarian cancer [169, 170]. The following sections highlight ^{225}Ac and ^{227}Th chelating strategies to achieve the full cytotoxic potential of these radiometals.

9.4.1 Common and Most Promising New Bifunctional Chelators for ^{225}Ac and ^{227}Th

For a number of years, the potential of ^{225}Ac as a therapeutic radionuclide was not fully exploited due to a limited knowledge of its coordination chemistry and to the lack of a suitable chelating agent [171]. Despite its large ionic radius (1.120 Å vs. 0.975 Å for Am) [172] and early studies revealing differences between the solution behavior of Ac(III) and that of other trivalent lanthanide and actinide ions [173], it was suggested that Ac(III) is another hard, oxophilic cation that would bind hard donating ligands with similar affinities than other elements from the 5f-series. However, the past few years have seen a resurgence of interest in characterizing Ac coordination chemistry in order to design better ligands. Ferrier and coworkers reported Ac complex coordination features distinct from those of corresponding trivalent actinide species, including unexpectedly long Ac-O bond distances and a higher coordination number than predicted in the aquo ion, thereby highlighting the need to tailor ligand features to this bigger ion [174, 175].

For the most part, bifunctional chelating agents based on polyaminocarboxylate ligands (Fig. 9.12) and commonly used for other radionuclides such as ^{90}Y, ^{111}In, ^{166}Ho, ^{177}Lu, or ^{212}Bi and ^{213}Bi have been complexed with ^{225}Ac [176, 177]. These ligands can be divided into two classic types: linear backbone analogues such as those based on EDTA and DTPA structures, and macrocyclic backbone analogues such as those based on structures of 1,4,7,10-tetraazacyclododecane-1,4,7,10-tetraacetic

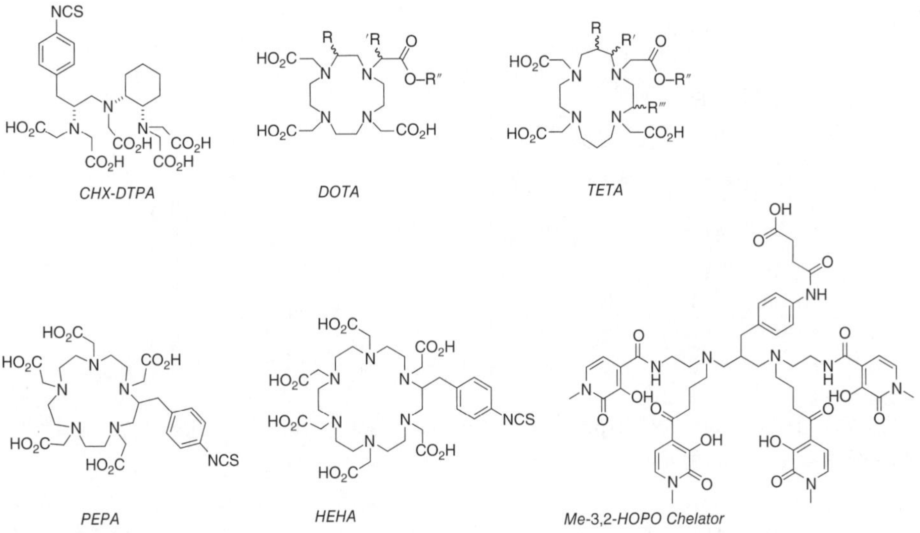

Figure 9.12 Structural scaffolds of common chelators used for linking alpha-emitting radionuclides to targeting biological molecules: CHX-DTPA contains three carbon stereocenters, resulting in several possible stereoisomers; DOTA can be easily functionalized in three common positions (R, R′, and R″); TETA can be easily functionalized in four common positions (R, R′, R″, and R‴); PEPA and HEHA have been commonly used as isothiocyanate bifunctional chelators; the Me-3,2-HOPO bifunctional chelator was developed specifically for the delivery of ^{227}Th.

acid (DOTA), 1,4,8,11-tetraazacyclotetradecane-1,4,8,11-tetraacetic acid (TETA), 1,4,7,10,13-pentaazacyclopentadecane-1,4,7,10,13-pentaacetic acid (PEPA), or 1,4,7,10,13,16-hexaazacyclooctadecane-1,4,7,10,13,16-hexaacetic acid (HEHA), with an early assumption that ^{225}Ac complex stability is higher but metalation kinetics are slower with macrocyclic chelates than with linear agents (Fig. 9.12) [176]. Because of its well-known ability to chelate metal ions, it is not surprising that DTPA was considered a candidate of choice for the targeted delivery of ^{225}Ac, with early in vivo stability studies investigating the cyclohexyl-substituted, pre-organized analogue CHX-DTPA [178]. An advantage of the macrocyclic structures is the potential for easy tuning of ligand denticity, macrocycle size, overall charge, and complexation kinetics, while preserving a common binding ability through carboxylic acids and amine functionalities. Despite its intrinsically smaller binding cavity size and relatively slower metal-binding rate, DOTA was found to form the most stable ^{225}Ac complex *in vivo*, leading to the selection of isothiocyanate-functionalized derivatives of DOTA as the most promising to pursue for coupling to antibody molecules. The bifunctional molecules MeO-DOTA-NCS (α-(5-isothiocyanato- 2-methoxyphenyl)-1,4,7,10-tetraazacyclododecane-1,4,7,10-tetraacetic acid) and 2B-DOTA-NCS (2-(p-isothiocyanatobenzyl)-1,4,7,10-tetraazacyclododecane-1,4,7,10-tetraacetic acid) were both evaluated and compared as their respective ^{225}Ac-DOTA-antibody constructs [177]. A two-step labeling method had initially been developed using several different antibodies, with chelation followed by bioconjugation to meet the requirement for high-temperature labeling of the DOTA chelate without sacrificing the biological activity of the antibody [176]. More recently, a single-step procedure at 37 °C was reported, with higher radiolabeling yields than those obtained with the two-step method [179]. While subsequent *in vitro* and *in vivo* studies have culminated in the clinical evaluation of ^{225}Ac-DOTA-HuM195 as a potential leukemia treatment [167, 168], the architecture of this construct is not optimal. A better chelator would address the thermodynamic and kinetic deficiencies of DOTA, in that it would allow for room temperature radiolabeling of an already-formed chelating bioconjugate. It would also increase the *in vitro* and *in vivo* stability of the final construct and allow for better radiolabeling yields.

As in the case of ^{225}Ac, ^{227}Th is known to form complexes with the common chelators listed above. Even though it exists as Th(IV) under physiologically relevant conditions and a much larger body of work is available on its chemistry, DOTA had also been considered the best chelating agents for ^{227}Th until recently [170, 180]. Functionalized ^{227}Th-DOTA constructs have been studied extensively, with promising preclinical properties when linked to trastuzumab for the treatment of HER-2 positive breast and ovarian cancer [169, 170]. However, similarly to ^{225}Ac, the complexation step must either be performed as a two-step process or directly at elevated temperatures in order to achieve sufficient labeling of DOTA-coupled antibodies. Those harsh conditions used for direct labeling are still incompatible with biological macromolecule stability. Thus, extensive efforts have been focused in the past decade on the development of new chelators that can complex ^{227}Th in near quantitative yield at ambient temperatures [181, 182]. In particular, a bifunctional octadentate chelator was prepared and characterized by Bayer AS, utilizing the siderophore-inspired 3-hydroxy-N-methyl-2-pyridinone (Me-3,2-HOPO) metal-binding units (Fig. 9.12) [182]. This chelator displays remarkably high yields and fast kinetics of radiolabeling, as well as enhanced *in vivo* stability. Preclinical safety and efficacy studies were reported on this chelator conjugated to (1) lintuzumab, which binds to the sialic acid receptor CD33, for the treatment of acute myeloid leukemia, and to (2) a CD70-targeting antibody, for the treatment of B-cell lymphomas and renal cell carcinoma, demonstrating the formidable promise of the ^{227}Th constructs [183][184]. Moreover, ongoing clinical studies are now evaluating the corresponding epratuzumab ^{227}Th conjugate for the treatment of relapsed or refractory, CD-22 positive, non-Hodgkin's lymphoma.

9.4.2 Maximizing Radiometal Delivery and Minimizing Damage Through Chemistry

Although initial preclinical and clinical studies demonstrated the therapeutic potential of these targeting constructs, kinetics of formation and stability of the Ac or Th metal complexes, as well as retention at the target site of α-emitting daughter products, have arisen as major challenges of targeted α-generator therapy: the radionuclides are not necessarily bound to the parent agent due to coordination chemistry specificities and high recoil energies, resulting in possible migration and toxic dose to non-target tissue. Two main strategies can be considered to address this recoil problem. The first one relies on the fast radionuclide cellular uptake and incorporation using internalizing targeting antibodies or peptides, with the assumption that tumor cells would retain the recoils [185]. Not yet established clinically, this strategy is highly dependent on the characteristics of the targeting vectors as well as on the fate of the parent nuclide once apoptosis is achieved. The second approach involves recoil retention within macromolecular architectures such as nanoparticles [186], liposomes [187, 188], polymer vesicles (polymersomes) [189, 190], or carbon nanotubes [191]. Significant progress [186, 192] was made by using multilayered gold-coated lanthanide phosphate nanoparticles doped with ^{225}Ac to contain the recoiling daughters. While this platform did enhance the site-retention of the α particles and offers versatility for multi-functionalization with targeting moieties and simultaneous imaging, the non-specificity of the described doping process did not fully prevent redistribution of the radiation to other organs. In soft materials, Sofou and coworkers [187] observed only a slight correlation between the retention of ^{213}Bi and the ^{225}Ac-carrying liposome size, with an overall recoil escape of more than 90%. Recently, Thijssen and coworkers computed the best polymersome design for the retention of the recoil daughters ^{221}Fr and ^{213}Bi [189]. Follow-up experimental work conducted by Wang and coworkers indicated that larger polymersomes are needed to attain satisfactory retention of recoiling radionuclides [190]. In addition to potentially retaining radionuclide daughter products, the other advantage of nano-carrier structures is the potential for an increased load of radionuclide, equivalent to an increase in cytotoxic dose. In a more recent study, Matson and coworkers employed ultrashort, single-walled carbon nanotubes to successfully sequester $^{225}Ac^{3+}$ ions in the presence of Gd^{3+} ions and retain them after a human serum challenge, making those "^{225}Ac@gadonanotubes" candidates for radioimmunotherapeutic delivery of $^{225}Ac^{3+}$ ions at higher concentrations than is currently possible for traditional ligand carriers [191]. While encouraging, these elegant studies have not yet been pursued in thorough preclinical studies that would validate their feasibility for clinical applications.

9.5 Approaching Actinide Biochemistry from a Theoretical Perspective

In much of bioinorganic chemistry, a theoretical treatment of metal sites in biomolecules presents some distinct challenges. Proteins generally contain a sufficient number of atoms such that their treatment at the quantum mechanical (QM) level is somewhat intractable – thus, computational methods such as DFT, coupled cluster, and other higher-level ab initio methods are not typically the desired approach when addressing questions such as protein structure. Instead, classical methods such as molecular mechanics (MM) can often provide good structural solutions by accounting for electrostatic interactions within biomolecules. However, classical methods can succeed only insofar as the electrostatic potential of the component atoms and bonds are "ideal"; for example, every backbone amide bond has the same properties.

In the case of actinide metals, as well as many first-row transition metals, the electrostatic model is woefully insufficient. In almost all cases, classical calculations of metal geometries will fail badly, as important metal–ligand bonding interactions are overlooked. Instead, QM methods are needed to correctly determine the energy associated with metal bonding; that is, the nephelauxetic effect, and ultimately arrive at correct geometric and electronic structures. To achieve a QM treatment of the metal site in the presence of the protein environment, there are presently two methods that are commonly utilized.

First, one can simply merge the QM and MM approaches, where a region to be treated at the QM level is selected, and the remainder of the molecule is treated classically [193–197]. To use the previously detailed Scn protein as an example, one might opt to describe the bound siderophore or siderophore-derived metal complex at the DFT level of theory, and use an MM approach for the protein (Fig. 9.13, left panel). However, such a computational method neglects all non-Coulombic interactions between the ligand and the protein. While this may be acceptable in certain instances, it is by no means a sufficient

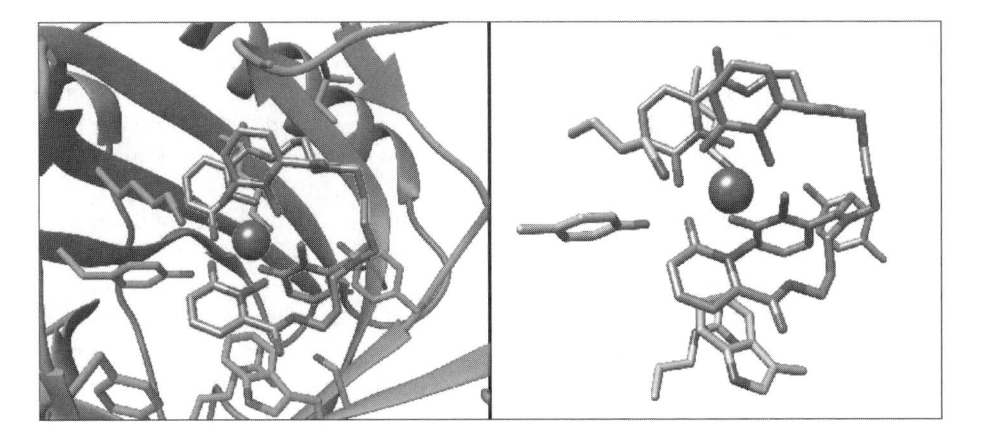

Figure 9.13 Structure of $[Am(3,4,3\text{-}LI(1,2\text{-}HOPO))]^-$ bound in the Scn calyx (PDB 4ZHG). Left: In a QM/MM calculation, the metal and ligand (highlighted with a gray backbone) would be treated using a QM approach, while the protein (aqua) would be treated using MM. Right: The protein structure is removed except for the residues that ligate the complex (same coloring scheme), and the entire model is treated at the QM level. (*See insert for color representation of the figure.*)

treatment for all cases. Care should be taken to evaluate the validity of the selected QM/MM approach, by incremental expansion of the QM region and/or comparison of calculated and experimental properties. An alternative approach is to simply truncate the model by including only the amino acid residues that interact with the actinide complex. The remainder of the protein structure is simply removed from the model, and the residues are passivated, usually by methylation or removal of the amine and carboxyl termini (Fig. 9.13, right panel) [198–201]. It should be noted, however, that if the geometry of the metal complex is of interest, it is necessary to fix the positions of at least the side-chain alpha carbons, and frequently all non-hydrogenic atoms, to obtain meaningful results. With this approach, the size of the problem, in number of atoms, is decreased, potentially allowing the entire model to be treated quantum mechanically. Truncating the protein structure requires that one bring a chemist's intuition to bear on the problem regarding the selection of relevant residues to include, and thus some degree of user bias is inherent in the results. Furthermore, this method requires prior knowledge of the protein structure, or at least the immediate vicinity of the metal-binding site, and as a final noteworthy drawback, insight into global protein structure is clearly not attainable.

For metal-centric problems however, the truncated model approach can provide an accurate description of geometric and electronic structure [198, 201–203], and at fairly modest computational expense depending on the size of the model. It is well suited for investigating trends in actinide bonding, assessing protein influence on actinide complexes, and providing insight into spectroscopic results. In contrast, the QM/MM approach allows exploration of changes to protein structure in the presence of an actinide ion or complex. As an example, investigation of possible metal-binding sites on a protein with a known apo structure could be accomplished by comparing the energies of QM/MM results with each putative binding site. Ideally this could be supplemented by calculation of spectroscopic results for comparison to experiment. In some cases, the de novo prediction or design of protein structure may even be possible, in particular if one has obtained prior knowledge of a well-defined metal-binding site by spectroscopic means [204–207].

As detailed above, an emerging area of biological actinide research is the use of various radionuclides for therapeutic and diagnostic applications [208–210]. Radioimmunotherapy frequently involves tethering actinide complexes to biomolecules such as monoclonal antibodies (mAbs), affibodies, and assorted fusion proteins; however atomically precise structures of these bioconjugates are largely unknown [211, 212]. While this may seem an ideal void to be filled using theoretical methods, these molecules present substantial challenges for which new approaches may be needed. The substantial size of typical mAbs, approximately 150 kDa, presents a challenge in and of itself, but one that may be addressed using classical methods. Typical conjugation strategies, however, are designed to minimally interfere with the protein structure, and are likely somewhat solvent-exposed as evidenced by the existence of protease-cleavable linkers [212, 213]. Thus, a comprehensive model would need to account for not only the protein but also the surrounding solvent shell, linker, and actinide complex. The flexible nature of most linker molecules all but eliminates a well-defined energetic minimum for the bioconjugate, and at present determining the structures of these complexes via theoretical means is largely intractable and remains an important and outstanding problem for theoretical bioactinide chemistry.

A final noteworthy topic is the development of a uranyl-binding protein via computational screening of more than 10,000 structures in the PDB [214]. This engineered protein is able to scavenge uranyl from (synthetic) seawater with femtomolar affinity, and can be immobilized on resin for facile metal sequestration. In contrast to the methods discussed thus far, a QM description of the actinide is not a prerequisite for a successful study; rather, structural data from existing complexes of the uranyl ion were used to define motifs likely to bind uranyl. Such an approach sidesteps the complex problem of calculating the energy of a large system via classical or quantum methods, and instead utilizes available structural data to search for a suitable scaffold protein. Subsequent mutation of the appropriate residues in the proposed binding pocket to anionic side-chain amino acids positioned to coordinate to uranyl effected a dramatic increase in binding affinity. In sum, computational methods can make important and diverse contributions to biological actinide chemistry, and their continued development may help solve future problems in areas such as actinide sequestration and bioconjugate structure and design.

References

1 Haxel GB, Hedrick JB, Orris GJ, Stauffer PH, Hendley II JW (2002) Rare earth elements: Critical resources for high technology. USGS Fact Sheet, 087–02.

2 Albright D, Berkhout F, Walker W (1997) *Plutonium and Highly Enriched Uranium, 1996: World Inventories, Capabilities and Policies*. Oxford University Press, New York.

3 Pellmar TC, Rockwell S (2005) Priority list of research areas for radiological nuclear threat countermeasures. *Radiat Res* **163**:115–123.

4 Cassatt DR, Kaminski JM, Hatchett RJ, DiCarlo AL, Benjamin JM, Maidment BW (2008) Medical countermeasures against nuclear threats: Radionuclide decorporation agents. *Radiat Res* **170**:540–548.

5 Wall JD, Krumholz LR (2006) Uranium reduction. *Annu Rev Microbiol* **60**:149–166.

6 Lovley DR, Phillips EJ (1992) Bioremediation of uranium contamination with enzymatic uranium reduction. *Environ Sci Technol* **26**:2228–2234.

7 Marshall MJ, Beliaev AS, Dohnalkova AC, Kennedy DW, Shi L, Wang Z, Boyanov MI, Lai B, Kemner KM, McLean JS (2006) c-Type cytochrome-dependent formation of U (IV) nanoparticles by Shewanella oneidensis. *PLoS Biol* **4**:e268.

8 VanEngelen MR, Field EK, Gerlach R, Lee BD, Apel WA, Peyton BM (2010) UO 22+ speciation determines uranium toxicity and bioaccumulation in an environmental Pseudomonas sp. isolate. *Environ Toxicol Chem* **29**:763–769.

9 Abdelouas A, Lutze W, Gong W, Nuttall EH, Strietelmeier BA, Travis BJ (2000) Biological reduction of uranium in groundwater and subsurface soil. *Sci Total Environ* **250**:21–35.

10 Kubicki JD, Halada GP, Jha P, Phillips BL (2009) Quantum mechanical calculation of aqueuous uranium complexes: Carbonate, phosphate, organic and biomolecular species. *Chem Cent J* **3**:10.

11 Sivaswamy V, Boyanov MI, Peyton BM, Viamajala S, Gerlach R, Apel WA, Sani RK, Dohnalkova A, Kemner KM, Borch T (2011) Multiple mechanisms of uranium immobilization by Cellulomonas sp. strain ES6. *Biotechnol Bioeng* **108**:264–276.

12 Luo W, Wu W-M, Yan T, Criddle CS, Jardine PM, Zhou J, Gu B (2007) Influence of bicarbonate, sulfate, and electron donors on biological reduction of uranium and microbial community composition. *Appl Microbiol Biotechnol* 77:713–721.
13 Cardenas E, Wu W-M, Leigh MB, Carley J, Carroll S, Gentry T, Luo J, Watson D, Gu B, Ginder-Vogel M (2010) Significant association between sulfate-reducing bacteria and uranium-reducing microbial communities as revealed by a combined massively parallel sequencing-indicator species approach. *Appl Environ Microbiol* **76**: 6778–6786.
14 Newsome L, Morris K, Lloyd JR (2014) The biogeochemistry and bioremediation of uranium and other priority radionuclides. *Chem Geol* **363**:164–184.
15 Durbin PW (2008) Actinides in animals and man. In: *Chemistry of the Actinide and Transactinide Elements.* Springer, pp. 3339–3440.
16 Ansoborlo E, Prat O, Moisy P, Den Auwer C, Guilbaud P, Carriere M, Gouget B, Duffield J, Doizi D, Vercouter T (2006) Actinide speciation in relation to biological processes. *Biochimie* **88**:1605–1618.
17 Vidaud C, Bourgeois D, Meyer D (2012) Bone as target organ for metals: The case of f-elements. *Chem Res Toxicol* **25**:1161–1175.
18 Maher K, Bargar JR, Brown Jr GE (2012) Environmental speciation of actinides. *Inorg Chem* **52**:3510–3532.
19 Banaszak J, Rittmann B, Reed D (1999) Subsurface interactions of actinide species and microorganisms: Implications for the bioremediation of actinide-organic mixtures. *J Radioanal Nucl Chem* **241**:385–435.
20 Duffield JR, Taylor DM, Williams DR (1994) The biochemistry of the f-elements. *Handb Phys Chem Rare Earths* **18**:591–621.
21 Berto S, Crea F, Daniele PG, Gianguzza A, Pettignano A, Sammartano S (2012) Advances in the investigation of dioxouranium (VI) complexes of interest for natural fluids. *Coord Chem Rev* **256**:63–81.
22 Montavon G, Apostolidis C, Bruchertseifer F, Repinc U, Morgenstern A (2009) Spectroscopic study of the interaction of U (VI) with transferrin and albumin for speciation of U (VI) under blood serum conditions. *J Inorg Biochem* **103**:1609–1616.
23 Sutton M, Burastero SR (2004) Uranium (VI) solubility and speciation in simulated elemental human biological fluids. *Chem Res Toxicol* **17**:1468–1480.
24 Duffield JR, May PM, Williams DR (1984) Computer simulation of metal ion equilibria in biofluids. IV. Plutonium speciation in human blood plasma and chelation therapy using polyaminopolycarboxylic acids. *J Inorg Biochem* **20**:199–214.
25 Bion L, Ansoborlo E, Moulin V, Reiller P, Collins R, Gilbin R, Février L, Perrier T, Denison F, Cote G (2005) Influence of thermodynamic database on the modelisation of americium (III) speciation in a simulated biological medium. *Radiochim Acta* **93**:715–718.
26 Edwards MJ, White GF, Norman M, Tome-Fernandez A, Ainsworth E, Shi L, Fredrickson JK, Zachara JM, Butt JN, Richardson DJ (2015) Redox linked flavin sites in extracellular decaheme proteins involved in microbe-mineral electron transfer. *Sci Rep* **5**:11677.
27 Hartshorne RS, Jepson BN, Clarke TA, Field SJ, Fredrickson J, Zachara J, Shi L, Butt JN, Richardson DJ (2007) Characterization of Shewanella oneidensis MtrC: A cell-surface decaheme cytochrome involved in respiratory electron transport to extracellular electron acceptors. *JBIC J Biol Inorg Chem* **12**:1083–1094.

28 Jin H, Zhang Y, Buchko GW, Varnum SM, Robinson H, Squier TC, Long PE (2012) Structure determination and functional analysis of a chromate reductase from Gluconacetobacter hansenii. *PLoS One* 7:e42432.

29 Bair W, Bloch W, Dickerson W, Eckerman K, Goans R, Karem A, Leggett R, Lipsztein J, Stabin M, Wiley AL (2008) *Management of Persons Contaminated with Radionuclides: Handbook*, NCRP (National Council on Radiation Protection and Measurements), Bethesda, MD (USA).

30 Hahn FF, Guilmette RA, Hoover MD (2002) Implanted depleted uranium fragments cause soft tissue sarcomas in the muscles of rats. *Environ Health Perspect* **110**:51.

31 Li WB, Roth P, Wahl W, Oeh U, Höllriegl V, Paretzke HG (2005) Biokinetic modeling of uranium in man after injection and ingestion. *Radiat Environ Biophys* **44**:29–40.

32 Hartmann HM, Monette FA, Avci HI (2000) Overview of toxicity data and risk assessment methods for evaluating the chemical effects of depleted uranium compounds. *Hum Ecol Risk Assess* **6**:851–874.

33 Durbin PW, Kullgren B, Xu J, Raymond KN (1997) New agents for in vivo chelation of uranium (VI): Efficacy and toxicity in mice of multidentate catecholate and hydroxypyridinonate ligands. *Health Phys* **72**:865–879.

34 Lipsztein JL (1981) *Improved Model for Uranium Metabolism in the Primate.* New York Univ., NY (United States).

35 Bernard S, Struxness E (1957) A Study of the Distribution and Excretion of Uranium in Man. An interim report. Oak Ridge National Lab., Tenn.

36 Stevens W, Bruenger F, Atherton D, Smith J, Taylor G (1980) The distribution and retention of hexavalent 233U in the beagle. *Radiat Res* **83**:109–126.

37 Bernhard G, Geipel G, Reich T, Brendler V, Amayri S, Nitsche H (2001) Uranyl (VI) carbonate complex formation: Validation of the Ca_2UO_2 $(CO_3)_3$ (aq.) species. *Radiochim Acta* **89**:511–518.

38 Taylor DM (1998) The bioinorganic chemistry of actinides in blood. *J Alloys Compd* **271**:6–10.

39 Paquet F, Ramounet B, Métivier H, Taylor D (1998) The bioinorganic chemistry of Np, Pu and Am in mammalian liver. *J Alloys Compd* **271**:85–88.

40 Wirth R, Taylor D, Duffield J (1985) Identification of transferrin as the principal neptunium-binding protein in the blood serum of rats. *Int J Nucl Med Biol* **12**:327–330.

41 Paquet F, Verry M, Grillon G, Landesman C, Masse R, Taylor D (1995) Subcellular and intranuclear localization of neptunium-237 (V) in rat liver. *Radiat Res* **143**:214–218.

42 Taylor D, Stradling G, Henge-Napoli M (2000) The scientific background to decorporation. *Radiat Prot Dosimetry* **87**:11–18.

43 Lanz H, Scott K, Crowley J, Hamilton J (1946) The Metabolism of Thorium, Protoactinium and Neptunium in the Rat. MDDC-648.

44 Filipy RE, Kathren RL (1996) Changes in soft tissue concentrations of plutonium and americium with time after human occupational exposure. *Health Phys* **70**:153–159.

45 McInroy J, Kathren R, Swint M (1989) Distribution of plutonium and americium in whole bodies donated to the United States Transuranium Registry. *Radiat Prot Dosimetry* **26**:151–158.

46 Ibrahim S, Warren G, Whicker F, Efurd D (2002) Plutonium in Colorado residents: Results of autopsy bone samples collected during 1975–1979. *Health Phys* **83**:165–177.

47 Kawamura H, Tanaka G-I (1983) Actinides concentrations in human tissues. *Health Phys* **44**:451–456.

48 Park J, Bair W, Busch R (1972) Progress in beagle dog studies with transuranium elements at Battelle-Northwest. *Health Phys* **22**:803–810.
49 Morin M, Nenot J, Lafuma J (1972) Metabolic and therapeutic study following administration to rats of 238Pu nitrate-a comparison with 239Pu. *Health Phys* **23**: 475–480.
50 International Commission on Radiological Protection (1994) Age-Dependent Doses to Members of the Public from Intake of Radionuclides: Part 2 Ingestion Dose Coefficients: A Report of a Task Group of Committee 2 the International Commission on Radiological Protection. Elsevier Health Sciences.
51 Taylor DM (1973) Chemical and physical properties of plutonium. In: *Uranium, Plutonium, Transplutonic Elements*. Springer, pp. 323–347.
52 Aryal BP, Paunesku T, Woloschak GE, He C, Jensen MP (2012) A proteomic approach to identification of plutonium-binding proteins in mammalian cells. *J Proteomics* **75**:1505–1514.
53 Ali M, Kumar A, Pandey BN (2014) Thorium induced cytoproliferative effect in human liver cell HepG2: Role of insulin-like growth factor 1 receptor and downstream signaling. *Chem Biol Interact* **211**:29–35.
54 Kumar A, Ali M, Pandey BN, Hassan PA, Mishra KP (2010) Role of membrane sialic acid and glycophorin protein in thorium induced aggregation and hemolysis of human erythrocytes. *Biochimie* **92**:869–879.
55 Durbin PW (1973) Metabolism and biological effects of the transplutonium elements. In: *Uranium, Plutonium, Transplutonic Elements*. Springer, pp. 739–896.
56 Scott KG, Axelrod DJ, Hamilton JG (1949) The metabolism of curium in the rat. *J Biol Chem* **177**:325–335.
57 Durbin PW (1962) Distribution of the transuranic elements in mammals. *Health Phys* **8**:665–671.
58 Pecoraro VL, Harris WR, Carrano CJ, Raymond KN (1981) Siderophilin metal coordination. Difference ultraviolet spectroscopy of di-, tri-, and tetravalent metal ions with ethylenebis [(o-hydroxyphenyl) glycine]. *Biochemistry (Mosc)* **20**:7033–7039.
59 Taylor D (1962) Some aspects of the comparative metabolism of plutonium and americium in rats. *Health Phys* **8**:673–677.
60 Stevens W, Bruenger F (1972) Interaction of 249Cf and 252Cf with constituents of dog and human blood. *Health Phys* **22**:679–683.
61 Taylor D (1972) Interactions between transuranium elements and the components of cells and tissues. *Health Phys* **22**:575–581.
62 Chipperfield A, Taylor D (1968) Binding of plutonium and americium to bone glycoproteins. *Nature* **219**:609–610.
63 Chipperfield AR, Taylor DM (1972) The binding of thorium (IV), plutonium (IV), americium (III) and curium (III) to the constituents of bovine cortical bone in vitro. *Radiat Res* **51**:15–30.
64 Jeanson A, Ferrand M, Funke H, Hennig C, Moisy P, Solari PL, Vidaud C, Den Auwer C (2010) The role of transferrin in actinide (IV) uptake: Comparison with iron (III). *Chem- Eur J* **16**:1378–1387.
65 Michon J, Frelon S, Garnier C, Coppin F (2010) Determinations of uranium (VI) binding properties with some metalloproteins (transferrin, albumin, metallothionein and ferritin) by fluorescence quenching. *J Fluoresc* **20**:581–590.

66 Bauer N, Fröhlich DR, Panak PJ (2014) Interaction of Cm (III) and Am (III) with human serum transferrin studied by time-resolved laser fluorescence and EXAFS spectroscopy. *Dalton Trans* **43**:6689–6700.

67 Sturzbecher-Hoehne M, Goujon C, Deblonde GJ-P, Mason AB, Abergel RJ (2013) Sensitizing curium luminescence through an antenna protein to investigate biological actinide transport mechanisms. *J Am Chem Soc* **135**:2676–2683.

68 Vidaud C, Gourion-Arsiquaud S, Rollin-Genetet F, Torne-Celer C, Plantevin S, Pible O, Berthomieu C, Quéméneur E (2007) Structural consequences of binding of UO22+ to apotransferrin: Can this protein account for entry of uranium into human cells? *Biochemistry (Mosc)* **46**:2215–2226.

69 Jensen MP, Gorman-Lewis D, Aryal B, Paunesku T, Vogt S, Rickert PG, Seifert S, Lai B, Woloschak GE, Soderholm L (2011) An iron-dependent and transferrin-mediated cellular uptake pathway for plutonium. *Nat Chem Biol* **7**:560–565.

70 Deblonde GJ-P, Sturzbecher-Hoehne M, Mason AB, Abergel RJ (2013) Receptor recognition of transferrin bound to lanthanides and actinides: A discriminating step in cellular acquisition of f-block metals. *Metallomics* **5**:619–626.

71 Taylor DM, Seidel A, Planas-Bohne F, Schuppler U, Neu-Mlüler M, Wirth RE (1987) Biochemical studies of the interactions of plutonium, neptunium and protactinium with blood and liver cell proteins. *Inorganica Chim Acta* **140**:361–363.

72 Paquet F, Frelon S, Cote G, Madic C (2003) The contribution of chemical speciation to internal dosimetry. *Radiat Prot Dosimetry* **105**:179–184.

73 Duffield J, Taylor D (1991) The biochemistry of the actinides. In: Freeman AJ, Keller C (Eds.), *Handbook on the Physics and Chemistry of the Actinides*, Elsevier Science, Amsterdam.

74 Vincent JB, Love S (2012) The binding and transport of alternative metals by transferrin. *Biochim Biophys Acta BBA-Gen Subj* **1820**:362–378.

75 Wally J, Halbrooks PJ, Vonrhein C, Rould MA, Everse SJ, Mason AB, Buchanan SK (2006) The crystal structure of iron-free human serum transferrin provides insight into inter-lobe communication and receptor binding. *J Biol Chem* **281**:24934–24944.

76 Baker E (1994) Structure and reactivity of transferrins. *Adv Inorg Chem* **41**:389–463.

77 Aisen P, Wessling-Resnick M, Leibold EA (1999) Iron metabolism. *Curr Opin Chem Biol* **3**:200–206.

78 Scapolan S, Ansorborlo E, Moulin C, Madic C (1998) Uranium (VI)-transferrin system studied by time-resolved laser-induced fluorescence. *Radiat Prot Dosimetry* **79**: 505–508.

79 Williams J, Moreton K (1980) The distribution of iron between the metal-binding sites of transferrin human serum. *Biochem J* **185**:483–488.

80 Cheng Y, Zak O, Aisen P, Harrison SC, Walz T (2004) Structure of the human transferrin receptor-transferrin complex. *Cell* **116**:565–576.

81 Li H, Sun H, Qian ZM (2002) The role of the transferrin–transferrin-receptor system in drug delivery and targeting. *Trends Pharmacol Sci* **23**:206–209.

82 Du X, Zhang T, Yuan L, Zhao Y, Li R, Wang K, Yan SC, Zhang L, Sun H, Qian Z (2002) Complexation of ytterbium to human transferrin and its uptake by K562 cells. *Eur J Biochem* **269**:6082–6090.

83 Duffield JR, Taylor DM (1987) A spectroscopic study on the binding of plutonium (IV) and its chemical analogues to transferrin. *Inorganica Chim Acta* **140**:365–367.

84 Schuler F, Taylor DM (1987) The subcellular distribution of and239 Pu in primary cultures of rat hepatocytes. *Radiat Res* **110**:362–371.

85 Den Auwer C, Llorens I, Moisy P, Vidaud C, Goudard F, Barbot C, Solari P, Funke H (2005) Actinide uptake by transferrin and ferritin metalloproteins. *Radiochim Acta* **93**:699–703.

86 Taylor DM, Kontoghiorghes GJ (1986) Mobilisation of plutonium and iron from transferrin and ferritin by hydroxypyridone chelators. *Inorganica Chim Acta* **125**:L35–L38.

87 Clifton MC, Corrent C, Strong RK (2009) Siderocalins: Siderophore-binding proteins of the innate immune system. *Biometals* **22**:557–564.

88 Allred BE, Rupert PB, Gauny SS, An DD, Ralston CY, Sturzbecher-Hoehne M, Strong RK, Abergel RJ (2015) Siderocalin-mediated recognition, sensitization, and cellular uptake of actinides. *Proc Natl Acad Sci* **112**:10342–10347.

89 Deblonde GJ-P, Sturzbecher-Hoehne M, Rupert PB, An DD, Illy M-C, Ralston CY, Brabec J, de Jong WA, Strong RK, Abergel RJ (2017) Chelation and stabilization of berkelium in oxidation state+ IV. *Nat. Chem.*

90 Finn BE, Forsén S (1995) The evolving model of calmodulin structure, function and activation. *Structure* **3**:7–11.

91 Seeger P, Rokop S, Palmer P, Henderson S, Hobart D, Trewhella J (1997) Neutron resonance scattering shows specific binding of plutonium to the calcium-binding sites of the protein calmodulin and yields precise distance information. *J Am Chem Soc* **119**:5118–5125.

92 Jeanson A, Berthon C, Coantic S, Den Auwer C, Floquet N, Funke H, Guillaneux D, Hennig C, Martinez J, Moisy P (2009) The role of aspartyl-rich pentapeptides in comparative complexation of actinide (IV) and iron (III). Part 1. *New J Chem* **33**:976–985.

93 Huynh TS, Bourgeois D, Basset C, Vidaud C, Hagège A (2015) Assessment of CE-ICP/MS hyphenation for the study of uranyl/protein interactions. *Electrophoresis* **36**: 1374–1382.

94 Basset C, Averseng O, Ferron P-J, Richaud N, Hagège A, Pible O, Vidaud C (2013) Revision of the biodistribution of uranyl in serum: Is fetuin-A the major protein target? *Chem Res Toxicol* **26**:645–653.

95 Qi L, Basset C, Averseng O, Quéméneur E, Hagège A, Vidaud C (2014) Characterization of UO 2 2+ binding to osteopontin, a highly phosphorylated protein: Insights into potential mechanisms of uranyl accumulation in bones. *Metallomics* **6**:166–176.

96 Sontag W (1993) Microdistribution of 237Np in the skeleton of female rats. *Int J Radiat Biol* **63**:383–393.

97 Priest N, Howells G, Green D, Haines J (1982) Uranium in bone: Metabolic and autoradiographic studies in the rat. *Hum Exp Toxicol* **1**:97–114.

98 Priest N, Howells G, Green D, Haines J (1983) Pattern of uptake of americium-241 by the rat skeleton and its subsequent redistribution and retention: mplications for human dosimetry and toxicology. *Hum Toxicol* **2**:101–120.

99 Brandel V, Dacheux N (2004) Chemistry of tetravalent actinide phosphates—Part II. *J Solid State Chem* **177**:4755–4767.

100 Chipperfield A, Taylor D (1970) Binding of plutonium to glycoproteins in vitro. *Radiat Res* **43**:393–402.

101 Chipperfield AR, Taylor DM (1972) The binding of thorium (IV), plutonium (IV), americium (III) and curium (III) to the constituents of bovine cortical bone in vitro. *Radiat Res* **51**:15–30.

102 Gilbert I, Myers N (1960) Metal binding properties of chondroitin sulphate. *Biochim Biophys Acta* **42**:469–475.

103 Harris WR, Carrano CJ, Cooper SR, Sofen SR, Avdeef AE, McArdle JV, Raymond KN (1979) Coordination chemistry of microbial iron transport compounds. 19. Stability constants and electrochemical behavior of ferric enterobactin and model complexes. *J Am Chem Soc* **101**:6097–6104.

104 Raymond KN (1990) Biomimetic metal encapsulation. *Coord Chem Rev* **105**:135–153.

105 Bossier P, Hofte M, Verstraete W (1988) Ecological significance of siderophores in soil. In: *Advances in Microbial Ecology*. Springer, pp. 385–414.

106 Brainard JR, Strietelmeier BA, Smith PH, Langston-Unkefer PJ, Barr ME, Ryan RR (1992) Actinide binding and solubilization by microbial siderophores. *Radiochim Acta* **58**:357–364.

107 Mullen L, Gong C, Czerwinski K (2007) Complexation of uranium (VI) with the siderophore desferrioxamine B. *J Radioanal Nucl Chem* **273**:683–688.

108 Neu MP, Matonic JH, Ruggiero CE, Scott BL (2000) Structural characterization of a plutonium (IV) siderophore complex: Single-crystal structure of Pu-Desferrioxamine E. *Angew Chem Int Ed* **39**:1442–1444.

109 Boggs MA, Mason H, Arai Y, Powell BA, Kersting AB, Zavarin M (2014) Nuclear magnetic resonance spectroscopy of aqueous plutonium (IV) desferrioxamine B complexes. *Eur J Inorg Chem* **2014**:3312–3321.

110 John SG, Ruggiero CE, Hersman LE, Tung C-S, Neu MP (2001) Siderophore mediated plutonium accumulation by Microbacterium flavescens (JG-9). *Environ Sci Technol* **35**:2942–2948.

111 Reed DT, Deo RP, Rittmann BE (2010) Subsurface interactions of actinide species with microorganisms. In: *Chemistry of the Actinide and Transactinide Elements*. Springer, pp. 3595–3663.

112 Johnstone TC, Nolan EM (2015) Beyond iron: Non-classical biological functions of bacterial siderophores. *Dalton Trans* **44**:6320–6339.

113 Johnsson A, Ödegaard-Jensen A, Skarnemark G, Pedersen K (2008) Leaching of spent nuclear fuel in the presence of siderophores. *J Radioanal Nucl Chem* **279**: 619–626.

114 Boggs MA, Dai Z, Kersting AB, Zavarin M (2015) Plutonium (IV) sorption to montmorillonite in the presence of organic matter. *J Environ Radioact* **141**:90–96.

115 Moll H, Johnsson A, Schäfer M, Pedersen K, Budzikiewicz H, Bernhard G (2008) Curium (III) complexation with pyoverdins secreted by a groundwater strain of Pseudomonas fluorescens. *Biometals* **21**:219–228.

116 Xu C, Zhang S, Kaplan DI, Ho Y-F, Schwehr KA, Roberts KA, Chen H, DiDonato N, Athon M, Hatcher PG (2015) Evidence for hydroxamate siderophores and other N-containing organic compounds controlling 239,240 Pu immobilization and remobilization in a wetland sediment. *Environ Sci Technol* **49**:11458–11467.

117 Schalk IJ, Hannauer M, Braud A (2011) New roles for bacterial siderophores in metal transport and tolerance. *Environ Microbiol* **13**:2844–2854.

118 Ahmed E, Holmström SJ (2014) Siderophores in environmental research: Roles and applications. *Microb Biotechnol* 7:196–208.

119 Taylor R, May I (1999) The reduction of actinide ions by hydroxamic acids. *Czechoslov J Phys* **49**:617–621.

120 Taylor R, May I, Wallwork A, Denniss I, Hill N, Galkin BY, Zilberman BY, Fedorov YS (1998) The applications of formo-and aceto-hydroxamic acids in nuclear fuel reprocessing. *J Alloys Compd* **271**:534–537.

121 Renshaw JC, Robson GD, Trinci AP, Wiebe MG, Livens FR, Collison D, Taylor RJ (2002) Fungal siderophores: Structures, functions and applications. *Mycol Res* **106**:1123–1142.

122 Gorden AE, Xu J, Raymond KN, Durbin P (2003) Rational design of sequestering agents for plutonium and other actinides. *Chem Rev* **103**:4207–4282.

123 Durbin PW (2008) Lauriston S. Taylor Lecture: The quest for therapeutic actinide chelators. *Health Phys* **95**:465–492.

124 Schwarzenbach G, Ackermann HCV (1947) Ethylenediaminetetraacetic acid. *Helv Chim Acta* **30**:1798.

125 Hamilton JG, Scott KG (1953) Effect of calcium salt of versene upon metabolism of plutonium in the rat. *Proc Soc Exp Biol Med* **83**:301–305.

126 Schubert J (1955) Removal of radioelements from the mammalian body. *Annu Rev Nucl Sci* **5**:369–412.

127 Kroll H (1955) Development of chelating agents potentially more effective than EDTA in radioelement removal. pp. 150–151.

128 National Institute of Standards and Technology. Gaithersburg M., Smith RM (2004) NIST Critically Selected Stability Constants of Metal Complexes: Version 8.0. NIST.

129 Volf V (1978) Treatment of incorporated transuranium elements, IAEA Technical Report 184.

130 US Food and Drug Administration (2004) Guidance for industry calcium DTPA and zinc DTPA drug products—submitting a new drug application. Silver Spring MD US Food Drug Adm.

131 Maher C, Hu-Primmer J, MacGill T, Courtney B, Borio L (2012) Meeting the challenges of medical countermeasure development. *Microb Biotechnol* **5**:588–593.

132 Porter JB (2001) Practical management of iron overload. *Br J Haematol* **115**: 239–252.

133 Durbin PW, Jeung N, Rodgers SJ, Turowski PN, Weitl FL, White DL, Raymond KN (1989) Removal of 238Pu (IV) from mice by poly-catechoylate,-hydroxamate or-hydroxypyridinonate ligands. *Radiat Prot Dosimetry* **26**:351–358.

134 Abergel RJ, Durbin PW, Kullgren B, Ebbe SN, Xu J, Chang PY, Bunin DI, Blakely EA, Bjornstad KA, Rosen CJ (2010) Biomimetic actinide chelators: An update on the preclinical development of the orally active hydroxypyridonate decorporation agents 3, 4, 3-LI (1, 2-HOPO) and 5-LIO (Me-3, 2-HOPO). *Health Phys* **99**:401.

135 White DL, Durbin PW, Jeung N, Raymond KN (1988) Specific sequestering agents for the actinides. 16. Synthesis and initial biological testing of polydentate oxohydroxypyridinecarboxylate ligands. *J Med Chem* **31**:11–18.

136 Stradling G (1994) Recent progress in decorporation of plutonium, americium and thorium. *Radiat Prot Dosimetry* **53**:297–304.

137 Stradling G, Gray S, Ellender M, Moody J, Hodgson A, Pearce M, Wilson I, Burgada R, Bailly T, Leroux Y (1992) The efficacies of 3, 4, 3-LIHOPO and DTPA for enhancing the excretion of plutonium and americium from the rat: Comparison with other siderophore analogues. *Int J Radiat Biol* **62**:487–497.

138 Stradling G, Gray S, Pearce M, Wilson I, Moody J, Burgada R, Durbin P, Raymond K (1995) Decorporation of thorium-228 from the rat by 3, 4, 3-LIHOPO and DTPA after simulated wound contamination. *Hum Exp Toxicol* **14**:165–169.

139 Poncy J, Rateau G, Burgada R, Bailly T, Leroux Y, Raymond K, Durbin P, Masse R (1993) Efficacy of 3, 4, 3-LIHOPO for reducing the retention of 238Pu in rat after inhalation of the tributyl phosphate complex. *Int J Radiat Biol* **64**:431–436.

140 Paquet F, Poncy J, Rateau G, Burgada R, Bailly T, Leroux Y, Raymond K, Durbin P, Masse R (1994) Reduction of the retention of 238Pu inhaled as the tributylphosphate complex in rats treated by 3, 4, 3-LIHOPO. *Radiat Prot Dosimetry* **53**:323–326.

141 Paquet F, Poncy J, Métivier H, Grillon G, Fritsch P, Burgada R, Bailly T, Raymond K, Durbin P (1995) Efficacy of 3, 4, 3-LIHOPO for enhancing the excretion of plutonium from rat after simulated wound contamination as a tributyl-n-phosphate complex. *Int J Radiat Biol* **68**:663–668.

142 Volf V (1996) Treatment with 3, 4, 3-LIHOPO of simulated wounds contaminated with plutonium and americium in rat. *Int J Radiat Biol* **70**:109–114.

143 Volf V, Burgada R, Raymond K, Durbin P (1997) Chelation therapy by DFO-HOPO and 3, 4, 3-LIHOPO for injected Pu-238 and Am-241 in the rat: Effect of dosage, time and mode of chelate administration. *Occup Health Ind Med* **3**:105.

144 W. Durbin BK J Xu, KN Raymond, P (2000) Multidentate hydroxypyridinonate ligands for Pu (IV) chelation in vivo: Comparative efficacy and toxicity in mouse of ligands containing 1, 2-HOPO or Me-3, 2-HOPO. *Int J Radiat Biol* **76**:199–214.

145 Paquet F, Chazel V, Houpert P, Guilmette R, Muggenburg B (2003) Efficacy of 3, 4, 3-LI (1, 2-HOPO) for decorporation of Pu, Am and U from rats injected intramuscularly with high-fired particles of MOX. *Radiat Prot Dosimetry* **105**: 521–525.

146 Chang P, Bunin D, Gow J, Swezey R, Shinn W, Shuh D, Abergel R (2012) Analytical methods for the bioavailability evaluation of hydroxypyridinonate actinide decorporation agents in pre-clinical pharma-cokinetic studies. *J Chromatogr Sep Tech S* **4**:196.

147 Jarvis EE, An DD, Kullgren B, Abergel RJ (2012) Significance of single variables in defining adequate animal models to assess the efficacy of new radionuclide decorporation agents: Using the contamination dose as an example. *Drug Dev Res* **73**:281–289.

148 Bunin DI, Chang PY, Doppalapudi RS, Riccio ES, An D, Jarvis EE, Kullgren B, Abergel RJ (2013) Dose-dependent efficacy and safety toxicology of hydroxypyridinonate actinide decorporation agents in rodents: Towards a safe and effective human dosing regimen. *Radiat Res* **179**:171–182.

149 Kullgren B, Jarvis EE, An DD, Abergel RJ (2013) Actinide chelation: Biodistribution and in vivo complex stability of the targeted metal ions. *Toxicol Mech Methods* **23**:18–26.

150 An DD, Villalobos JA, Morales-Rivera JA, Rosen CJ, Bjornstad KA, Gauny SS, Choi TA, Sturzbecher-Hoehne M, Abergel RJ (2014) 238Pu elimination profiles after delayed treatment with 3, 4, 3LI (1, 2HOPO) in female and male Swiss-Webster mice. *Int J Radiat Biol* **90**:1055–1061.

151 Choi TA, Endsley AN, Bunin DI, Colas C, An DD, Morales-Rivera JA, Villalobos JA, Shinn WM, Dabbs JE, Chang PY (2015) Biodistribution of the multidentate hydroxypyridinonate ligand [14C]-3, 4, 3-LI (1, 2-HOPO), a potent actinide decorporation agent. *Drug Dev Res* **76**:107–122.

152 Choi TA, Furimsky AM, Swezey R, Bunin DI, Byrge P, Iyer LV, Chang PY, Abergel RJ (2015) In vitro metabolism and stability of the actinide chelating agent 3, 4, 3-LI (1, 2-HOPO). *J Pharm Sci* **104**:1832–1838.

153 Henge-napoli M, Ansoborlo E, Houpert P, Mirto H, Paquet F, Burgada R, Hodgson S, Stradling G (1998) Progress and trends in in vivo chelation of uranium. *Radiat Prot Dosimetry* **79**:449–452.

154 H. Henge-Napoli EA V Chazel, P Houpert, F Paquet, P Gourmelon, M (1999) Efficacy of ethane-1-hydroxy-1, 1-bisphosphonate (EHBP) for the decorporation of uranium after intramuscular contamination in rats. *Int J Radiat Biol* **75**:1473–1477.

155 Sawicki M, Siaugue J, Jacopin C, Moulin C, Bailly T, Burgada R, Meunier S, Baret P, Pierre J, Taran F (2005) Discovery of powerful uranyl ligands from efficient synthesis and screening. *Chem Eur J* **11**:3689–3697.

156 Sawicki M, Lecerclé D, Grillon G, Le Gall B, Sérandour A-L, Poncy J-L, Bailly T, Burgada R, Lecouvey M, Challeix V (2008) Bisphosphonate sequestering agents. Synthesis and preliminary evaluation for in vitro and in vivo uranium (VI) chelation. *Eur J Med Chem* **43**:2768–2777.

157 Dinse C, Baglan N, Cossonnet C, Bouvier C (2000) New purification protocol for actinide measurement in excreta based on calixarene chemistry. *Appl Radiat Isot* **53**:381–386.

158 Archimbaud M, Henge-Napoli M, Lilienbaum D, Desloges M, Montagne C (1994) Application of calixarenes for the decorporation of uranium: Present limitations and further trends. *Radiat Prot Dosimetry* **53**:327–330.

159 Araki K, Hashimoto N, Otsuka H, Nagasaki T, Shinkai S (1993) Molecular design of a calix [6] arene-based super-uranophile with C3 symmetry. High UO22+ selectivity in solvent extraction. *Chem Lett* **22**:829–832.

160 Spagnul A, Bouvier-Capely C, Phan G, Rebière F, Fattal E (2010) A new formulation containing calixarene molecules as an emergency treatment of uranium skin contamination. *Health Phys* **99**:430–434.

161 Kim Y-S, Brechbiel MW (2012) An overview of targeted alpha therapy. *Tumor Biol* **33**:573–590.

162 Knapp FR, Dash A (2016) Alpha radionuclide therapy. In: *Radiopharmaceuticals forTherapy*. Springer, New Delhi (India), pp. 37–55.

163 Kluetz PG, Pierce W, Maher VE, Zhang H, Tang S, Song P, Liu Q, Haber MT, Leutzinger EE, Al-Hakim A (2014) Radium Ra 223 dichloride injection: US Food and Drug Administration drug approval summary. *Clin Cancer Res* **20**:9–14.

164 Boll RA, Garland MA, Mirzadeh S (2008) Production of Thorium-229 at the ORNL High Flux Isotope Reactor. Oak Ridge National Laboratory (ORNL); High Flux Isotope Reactor.

165 Griswold JR, Medvedev DG, Engle JW, Copping R, Fitzsimmons JM, Radchenko V, Cooley J, Fassbender M, Denton DL, Murphy KE (2016) Large scale accelerator production of 225 Ac: Effective cross sections for 78–192MeV protons incident on 232 Th targets. *Appl Radiat Isot* **118**:366–374.

166 Morgenstern A, Abbas K, Bruchertseifer F, Apostolidis C (2008) Production of alpha emitters for targeted alpha therapy. *Curr Radiopharm* **1**:135–143.

167 Schwartz J, Jaggi J, O'donoghue J, Ruan S, McDevitt M, Larson S, Scheinberg D, Humm J (2011) Renal uptake of bismuth-213 and its contribution to kidney radiation dose following administration of actinium-225-labeled antibody. *Phys Med Biol* **56**:721.

168 Miederer M, McDevitt MR, Sgouros G, Kramer K, Cheung N-KV, Scheinberg DA (2004) Pharmacokinetics, dosimetry, and toxicity of the targetable atomic generator, 225Ac-HuM195, in nonhuman primates. *J Nucl Med* **45**:129–137.

169 Heyerdahl H, Krogh C, Borrebæk J, Larsen Å, Dahle J (2011) Treatment of HER2-expressing breast cancer and ovarian cancer cells with alpha particle-emitting 227 Th-trastuzumab. *Int J Radiat Oncol Biol Phys* **79**:563–570.

170 Heyerdahl H, Abbas N, Brevik EM, Mollatt C, Dahle J (2012) Fractionated therapy of HER2-expressing breast and ovarian cancer xenografts in mice with targeted alpha emitting 227 Th-DOTA-p-benzyl-trastuzumab. *PLoS One* **7**:e42345.

171 Lambrecht RM (1983) Radionuclide generators. *Radiochim Acta* **34**:9–24.

172 Shannon R t (1976) Revised effective ionic radii and systematic studies of interatomic distances in halides and chalcogenides. *Acta Crystallogr A* **32**:751–767.

173 Diamond R, Street Jr K, Seaborg GT (1954) An ion-exchange study of possible hybridized 5f bonding in the actinides1. *J Am Chem Soc* **76**:1461–1469.

174 Ferrier MG, Stein BW, Batista ER, Berg JM, Birnbaum ER, Engle JW, John KD, Kozimor SA, Pacheco JSL, Redman LN (2017) Synthesis and characterization of the actinium aquo ion. *ACS Cent Sci* **3**:176.

175 Ferrier MG, Batista ER, Berg JM, Birnbaum ER, Cross JN, Engle JW, La Pierre HS, Kozimor SA, Pacheco JSL, Stein BW (2016) Spectroscopic and computational investigation of actinium coordination chemistry. *Nat. Commun.* **7**: 12312.

176 McDevitt MR, Ma D, Simon J, Frank RK, Scheinberg DA (2002) Design and synthesis of 225 Ac radioimmunopharmaceuticals. *Appl Radiat Isot* **57**:841–847.

177 Miederer M, Scheinberg DA, McDevitt MR (2008) Realizing the potential of the Actinium-225 radionuclide generator in targeted alpha particle therapy applications. *Adv Drug Deliv Rev* **60**:1371–1382.

178 Davis I, Glowienka K, Boll R, Deal K, Brechbiel M, Stabin M, Bochsler P, Mirzadeh S, Kennel S (1999) Comparison of 225 actinium chelates: Tissue distribution and radiotoxicity. *Nucl Med Biol* **26**:581–589.

179 Maguire WF, McDevitt MR, Smith-Jones PM, Scheinberg DA (2014) Efficient 1-step radiolabeling of monoclonal antibodies to high specific activity with 225Ac for α-particle radioimmunotherapy of cancer. *J Nucl Med* **55**:1492–1498.

180 Price EW, Orvig C (2014) Matching chelators to radiometals for radiopharmaceuticals. *Chem Soc Rev* **43**:260–290.

181 Washiyama K, Amano R, Sasaki J, Kinuya S, Tonami N, Shiokawa Y, Mitsugashira T (2004) 227 Th-EDTMP: A potential therapeutic agent for bone metastasis. *Nucl Med Biol* **31**:901–908.

182 Ramdahl T, Bonge-Hansen HT, Ryan OB, Larsen Å, Herstad G, Sandberg M, Bjerke RM, Grant D, Brevik EM, Cuthbertson AS (2016) An efficient chelator for complexation of thorium-227. *Bioorg Med Chem Lett* **26**:4318–4321.

183 Hagemann UB, Wickstroem K, Wang E, Shea AO, Sponheim K, Karlsson J, Bjerke RM, Ryan OB, Cuthbertson AS (2016) In vitro and in vivo efficacy of a novel CD33-targeted thorium-227 conjugate for the treatment of acute myeloid leukemia. *Mol Cancer Ther* **15**:2422–2431.

184 Hagemann UB, Mihaylova D, Uran SR, Borrebaek J, Grant D, Bjerke RM, Karlsson J, Cuthbertson AS (2017) Targeted alpha therapy using a novel CD70 targeted thorium-227 conjugate in in vitro and in vivo models of renal cell carcinoma. *Oncotarget* **8**:56311–56326.

185 McDevitt MR, Ma D, Lai LT, Simon J, Borchardt P, Frank RK, Wu K, Pellegrini V, Curcio MJ, Miederer M (2001) Tumor therapy with targeted atomic nanogenerators. *Science* **294**:1537–1540.

186 McLaughlin MF, Woodward J, Boll RA, Wall JS, Rondinone AJ, Kennel SJ, Mirzadeh S, Robertson JD (2013) Gold coated lanthanide phosphate nanoparticles for targeted alpha generator radiotherapy. *PLoS One* **8**:e54531.

187 Sofou S, Thomas JL, Lin H, McDevitt MR, Scheinberg DA, Sgouros G (2004) Engineered liposomes for potential α-particle therapy of metastatic cancer. *J Nucl Med* **45**:253–260.

188 Chang M-Y, Seideman J, Sofou S (2008) Enhanced loading efficiency and retention of 225Ac in rigid liposomes for potential targeted therapy of micrometastases. *Bioconjug Chem* **19**:1274–1282.

189 Thijssen L, Schaart D, De Vries D, Morgenstern A, Bruchertseifer F, Denkova A (2012) Polymersomes as nano-carriers to retain harmful recoil nuclides in alpha radionuclide therapy: A feasibility study. *Radiochim Acta Int J Chem Asp Nucl Sci Technol* **100**:473–482.

190 Wang G, de Kruijff R, Rol A, Thijssen L, Mendes E, Morgenstern A, Bruchertseifer F, Stuart M, Wolterbeek H, Denkova A (2014) Retention studies of recoiling daughter nuclides of 225 Ac in polymer vesicles. *Appl Radiat Isot* **85**:45–53.

191 Matson ML, Villa CH, Ananta JS, Law JJ, Scheinberg DA, Wilson LJ (2015) Encapsulation of α-particle–emitting 225Ac3+ ions within carbon nanotubes. *J Nucl Med* **56**:897–900.

192 Symonds P, Jones D (2013) Advances in Clinical Radiobiology. *Clin Oncol* **10**:567–568.

193 Kirchner B, Wennmohs F, Ye S, Neese F (2007) Theoretical bioinorganic chemistry: The electronic structure makes a difference. *Curr Opin Chem Biol* **11**:134–141.

194 Blomberg MRA, Borowski T, Himo F, Liao R, Siegbahn PEM (2014) Quantum chemical studies of mechanisms for metalloenzymes. *Chem Rev* **114**:3601–3658.

195 Proos Vedin N, Lundberg M (2016) Protein effects in non-heme iron enzyme catalysis: Insights from multiscale models. *JBIC J Biol Inorg Chem* **21**:645–657.

196 Gascon JA, Leung SSF, Batista ER, Batista VS (2006) A Self-consistent space-domain decomposition method for QM/MM computations of protein electrostatic potentials. *J Chem Theory Comput* **2**:175–186.

197 Shaik S, Kumar D, de Visser SP, Altun A, Thiel W (2005) Theoretical perspective on the structure and mechanism of cytochrome P450 enzymes. *Chem Rev* **105**: 2279–2328.

198 Lancaster KM, Roemelt M, Ettenhuber P, Hu Y, Ribbe MW, Neese F, Bergmann U, DeBeer S (2011) X-ray emission spectroscopy evidences a central carbon in the nitrogenase iron-molybdenum cofactor. *Science* **334**:974–977.

199 Bjornsson R, Lima FA, Spatzal T, Weyhermüller T, Glatzel P, Bill E, Einsle O, Neese F, DeBeer S (2014) Identification of a spin-coupled Mo(<scp>iii</scp>) in the nitrogenase iron–molybdenum cofactor. *Chem Sci* **5**:3096–3103.

200 Siegbahn PEM (2016) Model calculations suggest that the central carbon in the FeMo-cofactor of nitrogenase becomes protonated in the process of nitrogen fixation. *J Am Chem Soc* **138**:10485–10495.

201 Roos K, Siegbahn PEM (2013) Activation of dimanganese class Ib ribonucleotide reductase by hydrogen peroxide: Mechanistic insights from density functional theory. *Inorg Chem* **52**:4173–4184.

202 Rees JA, Bjornsson R, Kowalska JK, Lima FA, Schlesier J, Sippel D, Weyhermüller T, Einsle O, Kovacs JA, DeBeer S (2017) Comparative electronic structures of nitrogenase FeMoco and FeVco. *Dalton Trans* **46**:2445–2455.

203 Bjornsson R, Neese F, DeBeer S (2017) Revisiting the Mössbauer isomer shifts of the FeMoco cluster of nitrogenase and the cofactor charge. *Inorg Chem* acs.inorgchem.6b02540.

204 Mills JH, Khare SD, Bolduc JM, Forouhar F, Mulligan VK, Lew S, Seetharaman J, Tong L, Stoddard BL, Baker D (2013) Computational Design of an Unnatural Amino Acid Dependent Metalloprotein with Atomic Level Accuracy Computational Design of an Unnatural Amino Acid Dependent Metalloprotein with Atomic Level Accuracy. doi: 10.1021/ja403503m.

205 Qian B, Raman S, Das R, Bradley P, McCoy AJ, Read RJ, Baker D (2007) High-resolution structure prediction and the crystallographic phase problem. *Nature* **450**:259–264.

206 Mocny CS, Pecoraro VL (2015) *De Novo* protein design as a methodology for synthetic bioinorganic chemistry. *Acc Chem Res* 150803114526003.

207 Yu F, Cangelosi VM, Zastrow ML, Tegoni M, Plegaria JS, Tebo AG, Mocny CS, Ruckthong L, Qayyum H, Pecoraro VL (2014) Protein design: Toward functional metalloenzymes. *Chem Rev* **114**(7):3495–578.

208 Speer TW (2011) *Targeted Radionuclide Therapy*, 1st ed. Lippincott Williams & Wilkins, Philadelphia, PA.

209 Walter RB, Press OW, Pagel JM (2010) Pretargeted radioimmunotherapy for hematologic and other malignancies. *Cancer Biother Radiopharm* **25**:125–142.

210 Larson SM, Carrasquillo JA, Cheung N-K V, Press OW (2015) Radioimmunotherapy of human tumours. *Nat Rev Cancer* **15**:347–360.

211 Tolmachev V, Orlova A, Pehrson R, et al (2007) Radionuclide therapy of HER2-positive microxenografts using a 177Lu-labeled HER2-specific affibody molecule. *Cancer Res* **67**:2773–2782.

212 Sievers EL, Senter PD (2013) Antibody-drug conjugates in cancer therapy. *Annu Rev Med* **64**:15–29.

213 Carter PJ, Senter PD (2008) Antibody-drug conjugates for cancer therapy. *Cancer J* **14**:154–169.

214 Zhou L, Bosscher M, Zhang C, et al (2014) A protein engineered to bind uranyl selectively and with femtomolar affinity. *Nat Chem* **6**:236–241.

Index

Note: Page numbers in *italics* indicate figures.

a

Experimental and Theoretical Approaches to Actinide Chemistry, First Edition.
Edited by John K. Gibson and Wibe A. de Jong.

b

d

g

h

j

k

l

m

o

p

q

r

u